SURFACE ENGINEERING

This volume contains papers presented at the International Conference on Surface Engineering: Current Trends and Future Prospects, held at the University of Toronto, Toronto, Ontario, Canada, 25–27 June 1990.

The event was sponsored by the following organisations:

Dofasco, Canada
Ontario Hydro, Canada
Pan Abrasive Inc., Welland, Canada
Surface Engineering Society, UK
British Surface Treatment Association, UK
SAE Fatigue Design and Evaluation Committee, USA
National Aeronautical Establishment, National Research Council, Canada
Natural Sciences and Engineering Research Council (NSERC), Canada
Manufacturing Research Corporation of Ontario (MRCO), Canada
Centre for Hydrogen and Electrochemical Studies (CHES), Canada

SURFACE ENGINEERING

Edited by

S. A. MEGUID

Department of Mechanical Engineering,
University of Toronto, Toronto, Ontario, Canada

ELSEVIER APPLIED SCIENCE
LONDON and NEW YORK

ELSEVIER SCIENCE PUBLISHERS LTD
Crown House, Linton Road, Barking, Essex IG11 8JU, England

Sole Distributor in the USA and Canada
ELSEVIER SCIENCE PUBLISHING CO., INC.
655 Avenue of the Americas, New York, NY 10010, USA

WITH 96 TABLES AND 458 ILLUSTRATIONS

Softcover reprint of the hardcover 1st edition 1990

British Library Cataloguing in Publication Data

Surface engineering.
1. Materials. Surfaces
I. Meguid, S. A.
620.11

ISBN-13: 978-94-010-6834-5 e-ISBN-13: 978-94-009-0773-7
DOI: 10.1007/978-94-009-0773-7

Library of Congress CIP data applied for

Preface

Of those disciplines associated with product design/development, surface engineering has probably made the most significant advances in recent years. It will undoubtedly have a major influence on manufactured products and will influence the economics of the manufacturing nations in the next century. Important and dramatic improvements in surface and component integrity are possible by the use of the latest advances in surface engineering.

The purpose of this international meeting is to bring together scientists and engineers from all over the world to present their latest experimental and theoretical results over a broad range of topics dealing with the engineering of surfaces of metallic and non-metallic materials.

The papers presented at the Conference are included in these proceedings. Both the quantity and the quality of the submitted articles as well as the vigour of the topics covered are indicative of the heightened interest in this important field.

The proceedings of this Conference should help design and manufacturing engineers, metallurgists and materials scientists, engineers involved in surface sciences, coatings, welding, research and development, and graduate students to become aware of recent developments in surface engineering. It addresses six important topics: surface characterisation, surface preparation and protection, adhesive bonding and bondface characterisation, surface mechanical integrity, surface treatment, and developments in surface engineering.

I should like to express my thanks to the authors for their contributions and to the many—necessarily anonymous—reviewers for their time, efforts and constructive criticism of the manuscripts. The level of quality of the proceedings is a direct function of their respective efforts and opinions. Furthermore, I wish to express my sincere thanks to the sponsoring organisations for their commitment and financial support to this international gathering; and to Brenda Fung, Julie Verduci, Nigel Easby and Hélène Vaillancourt for their help at the different stages of the meeting.

Finally, I wish to acknowledge the support, patience and understanding of my wife, Eli, daughter, Jenna Danielle and our newly born son, Michael Jonathan, during the different stages of this project.

S. A. Meguid

Préface

De toutes les disciplines associées au développement et au design de produits, l'étude des surfaces est probablement celle qui a progressé de la façon la plus significative ces dernières années. Cela aura certainement une influence importante sur les produits qui seront manufacturés dans les années 90 ainsi que sur l'économie des pays industrialisés au cours du prochain siècle. Les récents développements dans le domaine de l'étude des surfaces rendent possible d'importantes améliorations en ce qui concerne la qualité et l'intégrité des surfaces.

Le but de ce congrès international est de rassembler des chercheurs et des ingénieurs du monde entier afin qu'ils présentent leurs plus récents résultats théoriques et expérimentaux touchant l'étude des surfaces des matériaux métalliques et non-métalliques.

Les compte rendus de ce congrès sont rassemblés dans ce document. La qualité et la quantité des articles soumis ainsi que la grande étendue de sujets couverts montrent bien l'intérêt marqué pour cet important domaine.

Les compte rendus devraient aider les ingénieurs en design et production, les métallurgistes, les spécialistes en matériaux, les ingénieurs impliqués dans l'étude des surfaces, des revêtements et de la soudure, ceux en recherche et développement ainsi que les étudiants gradués à se tenir au courant des récents développements dans l'étude des surfaces. Six sujets importants sont traités: caractérisation des surfaces, préparation et protection des surfaces, adhésion et caractérisation des surfaces, intégrité des surfaces, traitement des surfaces et développements dans l'étude des surfaces.

J'aimerais remercier tous les auteurs pour leur contribution ainsi que les nombreuses personnes qui ont révisés les articles pour leur temps, leur travail et leurs critiques constructives. C'est grâce à leurs efforts que les compte rendus sont de si haute qualité. Je souhaite sincèrement remercier les organismes promoteurs pour leur engagement et leur support financier sans lesquels cette rencontre internationale n'aurait été possible, ainsi que Brenda Fung, Julie Verduci, Nigel Easby et Hélène Vaillancourt pour leur précieuse aide au cours de la préparation de cette rencontre.

Finalement, j'aimerais exprimer ma reconnaissance à ma femme Eli, à ma fille Jenna Danielle ainsi qu'à mon fils nouveau-né Michael Jonathan pour leur support, leur patience et leur compréhension tout au long de ce projet.

S. A. MEGUID

Contents

Adhesive Bonding and Bondface Characterisation

Surface Effects on Implantation

Surface Integrity of Aerospace Alloys

Laser Surface Treatment

Surface Integrity: Fatigue, Fretting, Corrosion and Wear

Surface Residual Stresses

Techniques in Surface Engineering

Developments in Surface Engineering

Trends in Surface Engineering

Advances in Surface Engineering

List of Contributors

H. A. Ahmed, Department of Mineral, Petroleum and Metallurgical Engineering, Faculty of Engineering, Cairo University, Giza, Egypt

M. M. I. Ahmed, Department of Mineral, Petroleum and Metallurgical Engineering, Faculty of Engineering, Cairo University, Giza, Egypt

K. F. Amouzouvi, Fuel Channel Components Branch, Whiteshell Nuclear Research Establishment, Pinawa, Manitoba R0E 1L0, Canada

A. Ankara, Department of Metallurgical Engineering, Middle East Technical University, Ankara, Turkey

P. Au, National Aeronautical Establishment, National Research Council Canada, Ottawa, Ontario K1A 0R6, Canada

P. G. Bailey, GE Aircraft Engines, General Electric Company, One Neumann Way, Evendale, Ohio 45215, USA

G. Banaś, Department of Civil Engineering, University of Illinois at Urbana-Champaign, 205 North Mathews Avenue, Urbana, Illinois 61801, USA

J. Barralis, Laboratoire MécaSurf, ENSAM, 13617 Aix-en-Provence Cedex, France

A. Basu, Department of Mechanical Engineering, University of Wollongong, PO Box 1144, Wollongong, New South Wales 2500, Australia

H. Berns, Institute for Materials, Faculty of Mechanical Engineering, Ruhr-University Bochum, Universitätsstrasse 150, D-4630 Bochum, Federal Republic of Germany

W. J. Bratina, Department of Metallurgy and Materials Science, University of Toronto, 184 College Street, Toronto, Ontario M5S 1A4, Canada

P. M. Bronsveld, Department of Applied Physics, University of Groningen, Materials Science Centre, Nijenborgh 18, 9747 AG Groningen, The Netherlands

C. A. Brown, Mechanical Engineering Department, Worcester Polytechnic Institute, Worcester, Massachusetts 01609, USA

I. A. Bucklow, The Welding Institute, Abington, Cambridge CB1 6AA, UK

C. J. Budziak, Department of Mechanical Engineering, University of Toronto, 5 King's College Road, Toronto, Ontario M5S 1A4, Canada

A. F. Budzyński, Akademia Techniczno-Rolnicza, Katedra Obrabiarek i Robotów, ul. Olszewskiego 20, 85-225 Bydgoszcz, Poland

H. J. Busscher, Laboratory for Materia Technica, University of Groningen, Antonius Deusinglaan 1, 9713 AV Groningen, The Netherlands

L. Castex, Laboratoire MécaSurf, ENSAM, 13617 Aix-en-Provence Cedex, France

T. Chandra, Department of Materials Engineering, University of Wollongong, PO Box 1144, Wollongong, New South Wales 2500, Australia

J. Charrier, Laboratoire de Mécanique et de Physique des Máteriaux, URA au CNRS 863—Groupe MECAMAT, ENSMA, rue Guillaume VII, 86034 Poitiers, France

C. Christofides, Photoacoustic and Photothermal Sciences Laboratory, Department of Mechanical Engineering, University of Toronto, 5 King's College Road, Toronto, Ontario M5S 1A4, Canada

L. J. Clegg, Fuel Channel Components Branch, Whiteshell Nuclear Research Establishment, Pinawa, Manitoba R0E 1L0, Canada

J. R. Conrad, Department of Nuclear Engineering and Engineering Physics, University of Wisconsin—Madison, Madison, Wisconsin 53706, USA

J. J. Daly, Metal Improvement Company, 10 Forest Avenue, Paramus, New Jersey 07656, USA

D. F. Dauw, Applied Research Department, Charmilles Technologies SA, CH-1217 Meyrin, Switzerland

D. P. Davies, Westland Helicopters Ltd, Yeovil, Somerset BA20 2YB, UK

J. Th. M. de Hosson, Department of Applied Physics, University of Groningen, Materials Science Centre, Nijenborgh 18, 9747 AG Groningen, The Netherlands

P. Dickstein, PO Box 11871, Tel Aviv 61166, Israel

R. Ding, Department of Mechanical Engineering, Southeast University, Nanjing 210018, People's Republic of China

R. A. Dodd, Department of Materials Science and Engineering, University of Wisconsin—Madison, Madison, Wisconsin 53706, USA

W. Doswell, Carleton University, Ottawa, Ontario K1S 5B6, Canada

J. R. Easterday, Application Engineering, Kolene Corporation, 12890 Westwood Avenue, Detroit, Michigan 48223, USA

J. S. Eckersley, Metal Improvement Company, 41200 Coca Cola Drive, Belleville, Michigan 48111, USA

J. E. Elder, Hawker Siddeley Canada Inc., Orenda Division, Mississauga, Ontario, Canada

M. R. El-Koussy, Department of Petroleum, Mining and Metallurgy, Faculty of Engineering, Cairo University, Giza, Egypt

M. B. Elmoselhi, Research Division, Ontario Hydro, 800 Kipling Avenue, Toronto, Ontario M8Z 5S4, Canada

F. A. Elrefaie, Department of Mineral, Petroleum and Metallurgical Engineering, Faculty of Engineering, Cairo University, Giza, Egypt

H. E. Elsayed-Ali, Laboratory for Laser Energetics, University of Rochester, 250 East River Road, Rochester, New York 14623, USA

J. Fan, Yunnan Institute of Technology, Kunming, People's Republic of China

A. Fischer, Institute for Materials, Faculty of Mechanical Engineering, Ruhr-University Bochum, Universitätsstrasse 150, D-4630 Bochum, Federal Republic of Germany

J. F. Flavenot, Fatigue–Fracture Group, CETIM, 52 Avenue Félix-Louat, 60300 Senlis, France

A. J. Fletcher, Sheffield City Polytechnic, Pond Street, Sheffield S1 1WB, UK

C. Gasc, Laboratoire de Mécanique et de Physique des Máteriaux, URA au CNRS 863—Groupe MECAMAT, ENSMA, rue Guillaume VII, 86034 Poitiers, France

P. E. Gaultier, Department of Mechanical Engineering, University of Toronto, 5 King's College Road, Toronto, Ontario M5S 1A4, Canada

J. Gaydos, Department of Mechanical Engineering, University of Toronto, 5 King's College Road, Toronto, Ontario M5S 1A4, Canada

S. X. Gong, Department of Mechanical Engineering, University of Toronto, 5 King's College Road, Toronto, Ontario M5S 1A4, Canada

D. Guan, Institute of Materials and Technology, Dalian Maritime University, Dalian, People's Republic of China

D. W. Hammond, British Aerospace plc, Sowerby Research Centre, FPC 266, PO Box 5, Filton, Bristol, UK

Th. Happle, Lehrstuhl für Metallurgie und Metallkunde, Technische Universität München, Arcisstrasse 21, D-8000 München 2, Federal Republic of Germany

T. Hatsushika, Department of Mechanical Engineering, Faculty of Engineering, Yamanashi University, Kofu 400, Japan

M. Hirbod, Research and Development, ABB Stal AB, S-612 20 Finspong, Sweden

D. W. Hoeppner, Mechanical Engineering Department, Quality and Integrity Design Engineering Center, University of Utah, Salt Lake City, Utah 84112, USA

B. Hong, Analysis and Measurement Centre, Harbin Institute of Technology, Harbin 150006, People's Republic of China

J. Horyna, Department of Material and Metallurgy, Technical University in Liberec, Halkova 6, 461 17 Liberec, Czechoslovakia

R. Huang, Department of Mechanical Engineering, Southeast University, Nanjing 210018, People's Republic of China

J. B. Hull, Sheffield City Polytechnic, Pond Street, Sheffield S1 1WB, UK

I. M. Hutchings, Department of Materials Science and Metallurgy, University of Cambridge, Trumpington Street, Cambridge CB2 3QZ, UK

I. S. Jawahir, Carboloy Inc., 11177 East Eight Mile Road, Warren, Michigan 48089-3071, USA

B. L. Jiang, Department of Metals and Technology, Harbin Institute of Technology, Harbin 150006, People's Republic of China

C. K. Jun, Carboloy Inc., 11177 East Eight Mile Road, Warren, Michigan 48089-3071, USA

F. Jungblut, Lehrstuhl B und Institut für Werkstoffkunde, RWTH Aachen, Templergraben 55, D-5100 Aachen, Federal Republic of Germany

R. Kejžar, ZRMK, Institute for Materials Ljubljana, Dimičeva 12, 61109 Ljubljana, Yugoslavia

A. Khelawan, Sinclair & Valentine Co., 4590 Dufferin Street, Downsview, Ontario M3H 5S5, Canada

O. Knotek, Lehrstuhl B und Institut für Werkstoffkunde, RWTH Aachen, Templergraben 55, D-5100 Aachen, Federal Republic of Germany

P. J. Koblischek, INTERACRYL GmbH, Schultheissenweg 105c, D-6000 Frankfurt/Main 90, Federal Republic of Germany

S. Korichi, Materials Science Centre, University of Manchester Institute of Science and Technology, Grosvenor Street, Manchester M1 7HS, UK

A. K. Koul, National Aeronautical Establishment, National Research Council Canada, Ottawa, Ontario K1A 0R6, Canada

J. Kozak, Politechnika Warszawska, Wydzial Mechaniczny Technologiczny, ul. Andersena 3/98, 01-917 Warszawa, Poland

P. Kuba, Technical University, Křemencova 10, 110 00 Prague 10, Czechoslovakia

D. Kung, Powertech Laboratories Inc., Surrey, British Columbia V3W 7R7, Canada

F. V. Lawrence Jr, Department of Civil Engineering, University of Illinois at Urbana–Champaign, 205 North Mathews Avenue, Urbana, Illinois 61801, USA

T. C. Lei, Department of Metals and Technology, Harbin Institute of Technology, Harbin 150006, People's Republic of China

D. Li, Department of Mechanical Engineering, University of Toronto, Toronto, Ontario M5S 1A4, Canada

Z. Li, Department of Mechanical Engineering, Southeast University, Nanjing 210018, People's Republic of China

G. J. Lian, Department of Mechanical Engineering, Daqing Petroleum Institute, Anda, Heilongjiang 151400, People's Republic of China

J. A. Little, Department of Materials Science and Metallurgy, University of Cambridge, Trumpington Street, Cambridge CB2 3QZ, UK

W. Liu, Department of Metals and Technology, Harbin Institute of Technology, Harbin 150006, People's Republic of China

F. Löffler, Lehrstuhl B und Institut für Werkstoffkunde, RWTH Aachen, Templergraben 55, D-5100 Aachen, Federal Republic of Germany

J. Lu, Fatigue–Fracture Group, CETIM, 52 Avenue Félix-Louat, 60300 Senlis, France

E. Lunarska, Institute of Physical Chemistry, Polish Academy of Sciences, Kasprzaka 44, 01-224 Warsaw, Poland

J. Mackie, Extrude Hone Ltd, 42 Alston Drive, Bradwell Abbey, Milton Keynes MK13 9HB, UK

M. Madapura, Department of Materials Science and Engineering, University of Wisconsin—Madison, Madison, Wisconsin 53706, USA

A. Mandelis, Photoacoustic and Photothermal Sciences Laboratory, Department of Mechanical Engineering, University of Toronto, 5 King's College Road, Toronto, Ontario M5S 1A4, Canada

Y. Mao, Department of Mechanical Engineering, Southeast University, Nanjing 210018, People's Republic of China

L. J. Martínez-Miranda, Moore School of Electrical Engineering, University of Pennsylvania, Philadelphia, Pennsylvania 19104-6390, USA

I. Masuda, Research and Development Laboratories—III, Nippon Steel Corporation, 1-1-1 Edamitsu, Yawata-Higashi-ku, Kitakyushu 805, Japan

K. Masugata, Laboratory of Beam Technology, Nagaoka University of Technology, Nagaoka, Niigata 940-21, Japan

S. A. Meguid, Department of Mechanical Engineering, University of Toronto, 5 King's College Road, Toronto, Ontario M5S 1A4, Canada

K. E. Mohamed, Atomic Energy Authority, Inchas, Egypt

M. O. A. Mokhtar, Mechanical Engineering Department, Cairo University, Giza, Egypt

P. Morrell, Manufacturing Technology, Rolls Royce plc, Elton Road, Derby, UK

D. J. D. Moyer, Center for Adhesive and Sealant Science, Department of Chemistry, Virginia Polytechnic Institute and State University, Blacksburg, Virginia 24061-0212, USA

P. C. Nautiyal, Indian Institute of Petroleum, Dehardrun, India

A. W. Neumann, Department of Mechanical Engineering, University of Toronto, 5 King's College Road, Toronto, Ontario M5S 1A4, Canada

Ch. v. Nguyen, Institute for Materials, Faculty of Mechanical Engineering, Ruhr-University Bochum, Universitätsstrasse 150, D-4630 Bochum, Federal Republic of Germany

J. Noordhuis, Department of Applied Physics, University of Groningen, Materials Science Centre, Nijenborgh 18, 9747 AG Groningen, The Netherlands

Y. Oshida, Department of Mechanical and Aerospace Engineering, Syracuse University, Syracuse, New York 13244-1240, USA

T. A. Osman, Mechanical Engineering Department, Cairo University, Giza, Egypt

W. Otaño, Moore School of Electrical Engineering, University of Pennsylvania, Philadelphia, Pennsylvania 19104-6390, USA

L. Pataki, Thermodynamics and Kinetics Laboratory, Department of Mechanical Engineering, University of Toronto, 5 King's College Road, Toronto, Ontario M5S 1A4, Canada

M. R. Piggott, Department of Chemical Engineering and Applied Chemistry, University of Toronto, Toronto, Ontario M5S 1A4, Canada

R. Pishva, Hawker Siddeley Canada Inc., Orenda Division, Mississauga, Ontario, Canada

R. Priestner, Materials Science Centre, University of Manchester Institute of Science and Technology, Grosvenor Street, Manchester M1 7HS, UK

V. R. Puttagunta, Department of Chemical Engineering, Lakehead University, Thunder Bay, Ontario P7B 5E1, Canada

S. Raebel, Department of Materials Science and Engineering, University of Wisconsin—Madison, Madison, Wisconsin 53706, USA

N. A. A. Rahim Aly, Department of Petroleum, Mining and Metallurgy, Faculty of Engineering, Cairo University, Giza, Egypt

M. D. Raizenne, Structures and Materials Laboratory, National Aeronautical Establishment, National Research Council Canada, Ottawa, Ontario K1A 0R6, Canada

A. S. Rao, Powertech Laboratories Inc., Surrey, British Columbia V3W 7R7, Canada

T. Rasul, Dr A. Q. Khan Research Laboratories, Government of Pakistan, PO Box 502, Rawalpindi, Pakistan

J. Reby, Département Matériaux, CETIM, 52 Avenue Félix-Louat, 60300 Senlis, France

L. J. Rhoades, Extrude Hone Corporation, 8075 Pennsylvania Avenue, Irwin, Pennsylvania 15642, USA

J. M. Rigsbee, Department of Materials Science and Engineering, University of Illinois at Urbana–Champaign, 1304 West Green, Urbana, Illinois 61801, USA

P. M. Rogers, Department of Materials Science and Metallurgy, University of Cambridge, Trumpington Street, Cambridge CB2 3QZ, UK

A. R. Rosenthal, Laser Zentrum Hannover eV, Vahrenwalder Strasse 7, D-3000 Hannover 1, Federal Republic of Germany

R. Roux, Laboratoire de Mécanique et de Physique des Máteriaux, URA au CNRS 863—Groupe MECAMAT, ENSMA, rue Guillaume VII, 86034 Poitiers, France

R. Rudziejewski, Centre for Advanced Manufacturing and Industrial Automation, University of Wollongong, PO Box 1144, Wollongong, New South Wales 2500, Australia

Z. S. Safar, Mechanical Engineering Department, Cairo University, Giza, Egypt

E. Sánchez-Carambot, Moore School of Electrical Engineering, University of Pennsylvania, Philadelphia, Pennsylvania 19104-6390, USA

J. J. Santiago-Avilés, Moore School of Electrical Engineering, University of Pennsylvania, Philadelphia, Pennsylvania 19104-6390, USA

G. Savary, Department of Materials, Swiss Federal Institute of Technology, CH-1007 Lausanne, Switzerland

J. M. Schakenraad, Laboratory for Materia Technica, University of Groningen, Antonius Deusinglaan 1, 9713 AV Groningen, The Netherlands

J. A. Schey, Department of Mechanical Engineering, University of Waterloo, Waterloo, Ontario N2L 3G1, Canada

Kh. G. Schmitt-Thomas, Lehrstuhl für Metallurgie und Metallkunde, Technische Universität München, Arcisstrasse 21, D-8000 München 2, Federal Republic of Germany

V. Sedláček, Department of Material, Technical University in Prague, Křemencova 10, 110 00 Prague, Czechoslovakia

X. Shi, Department of Mechanical Enginering, Southeast University, Nanjing 210018, People's Republic of China

Y. Shi, Moore School of Electrical Engineering, University of Pennsylvania, Philadelphia, Pennsylvania 19104-6390, USA

S. H. Shim, Ohio State University, 190 North Oval Mall, Columbus, Ohio 43210, USA

M. M. Sinnott, Reliability Design and Testing Branch, Institute for Biomedical Engineering: Artificial Heart Research Laboratory, University of Utah, 803 North 300 West, Salt Lake City, Utah 84103, USA

B. S. Sitaramam, Department of Chemical Engineering, Lakehead University, Thunder Bay, Ontario P7B 5E1, Canada

K. Sridharan, Department of Materials Science and Engineering, University of Wisconsin—Madison, Madison, Wisconsin 53706, USA

K. Stevens, Surface Technology, National Centre of Tribology, AEA Technology, Risley, Warrington, Cheshire WA3 6AT, UK

I. Stokroos, Laboratory for Materia Technica, University of Groningen, Antonius Deusinglaan 1, 9713 AV Groningen, The Netherlands

R. C. Styles, Fuel Channel Components Branch, Whiteshell Nuclear Research Establishment, Pinawa, Manitoba R0E 1L0, Canada

B. Sutter, Département Matériaux, CETIM, 52 Avenue Félix-Louat, 60300 Senlis, France

T. Suzuki, Department of Mechanical Engineering, Faculty of Engineering, Yamanashi University, Kofu 400, Japan

Z. Szklarska-Smialowska, Ohio State University, 190 North Oval Mall, Columbus, Ohio 43210, USA

T. Takaai, Department of Mechanical Engineering, Faculty of Engineering, Yamanashi University, Kofu 400, Japan

H. M. Tensi, Technische Universität München, Institut für Werkstoff- und Verarbeitungswissenschaften, Munich, Federal Republic of Germany

H. K. Tönshoff, Laser Zentrum Hannover eV, Vahrenwalder Strasse 7, D-3000 Hannover 1, Federal Republic of Germany

S. A. Trengove, Sheffield City Polytechnic, Pond Street, Sheffield S1 1WB, UK

C.-S. Tsai, US National Center for Earthquake Engineering Research, State University of New York at Buffalo, Buffalo, New York 14260, USA

S. Uchida, Research and Development Laboratories—III, Nippon Steel Corporation, 1-1-1 Edamitsu, Yawata-Higashi-ku, Kitakyushu 805, Japan

N. Uddin, Carboloy Inc., 11177 East Eight Mile Road, Warren, Michigan 48089-3071, USA

B. J. van Brussel, Department of Applied Physics, University of Groningen, Materials Science Centre, Nijenborgh 18, 9747 AG Groningen, The Netherlands

H. C. van der Mei, Laboratory for Materia Technica, University of Groningen, Antonius Deusinglaan 1, 9713 AV Groningen, The Netherlands

G. Vasseur, Département Matériaux, CETIM, 52 Avenue Félix-Louat, 60300 Senlis, France

I. A. Vitkin, Photoacoustic and Photothermal Sciences Laboratory, Department of Mechanical Engineering, University of Toronto, Toronto, Ontario M5S 1A4, Canada

W. Wallace, Structures and Materials Laboratory, National Aeronautical Establishment, National Research Council Canada, Ottawa, Ontario K1A 0R6, Canada

C. A. Ward, Thermodynamics and Kinetics Laboratory, Department of Mechanical Engineering, University of Toronto, 5 King's College Road, Toronto, Ontario M5S 1A4, Canada

B. D. Warr, Research Division, Ontario Hydro, 800 Kipling Avenue, Toronto, Ontario M8Z 5S4, Canada

K. Watanabe, Research and Development Laboratories—III, Nippon Steel Corporation, 1-1-1 Edamitsu, Yawata-Higashi-ku, Kitakyushu 805, Japan

R. B. Waterhouse, Department of Materials Engineering and Materials Design, University of Nottingham, Nottingham NG7 2RD, UK

G. Weatherly, Department of Metallurgy and Materials Science, University of Toronto, 184 College Street, Toronto, Ontario M5S 1A4, Canada

J. M. Whalen, GE Aircraft Engines, General Electric Company, One Neumann Way, Evendale, Ohio 45215, USA

J. P. Wightman, Center for Adhesive and Sealant Science, Department of Chemistry, Virginia Polytechnic Institute and State University, Blacksburg, Virginia 24061-0212, USA

M. Wittmann, Technische Universität München, Institut für Werkstoff- und Verarbeitungswissenschaften, Munich, Federal Republic of Germany

F. J. Worzala, Department of Materials Science and Engineering, University of Wisconsin—Madison, Madison, Wisconsin 53706, USA

J. Y. Xu, Department of Applied Physics, University of Groningen, Materials Science Centre, Nijenborgh 18, 9747 AG Groningen, The Netherlands

L. Yang, Institute of Materials and Technology, Dalian Maritime University, Dalian, People's Republic of China

K. Yatsui, Laboratory of Beam Technology, Nagaoka University of Technology, Nagaoka, Niigata 940-21, Japan

X. Yi, Analysis and Measurement Centre, Harbin Institute of Technology, Harbin 150006, People's Republic of China

S. B. Young, Department of Metallurgy and Materials Science, University of Toronto, 184 College Street, Toronto, Ontario M5S 1A4, Canada

Z. Yu, Institute of Materials and Technology, Dalian Maritime University, Dalian, People's Republic of China
W. S. Yun, Development Company of Energy Source Applications, Daqing Petroleum Administration Bureau, Daqing, Heilongjiang 163000, People's Republic of China
J. Zamecnik, Department of Material and Metallurgy, Technical University in Liberec, Halkova 6, 461 17 Liberec, Czechoslovakia
H. Zhang, Institute of Metal Research, Academia Sinica, Shenyang, People's Republic of China

INTERNATIONAL SCIENTIFIC COMMITTEE

C. Barrett, Chairman of Surface Enhancement Division, SAE Fatigue Design and Evaluation Committee, USA
H. W. Bergamann, Lehrstuhl Werkstoffwissenschaft, Universität Erlangen-Nürnberg, Federal Republic of Germany
H. Berns, Institut für Werkstoffe, Lehrstuhl für Werkstofftechnik, Ruhr-Universität Bochum, Federal Republic of Germany
G. L. Boerman, Director, Materials Engineering, FMC Corporation, USA
I. A. Bucklow, Head of Surface Technology, The Welding Institute, Cambridge, UK
L. Castex, Laboratoire Microstructure et Mécanique des Matériaux, ENSAM, Paris, France
J. J. Daly, Senior Vice President, Research and Development, Metal Improvement Co., USA
J. F. Flavenot, CETIM, Senlis, France
H. Fukuoka, Department of Mechanical Engineering, Osaka University, Japan
E. D. Goddard, Speciality Chemicals Division, Research and Development, Union Carbide Corporation, USA
D. W. Hoeppner, Chairman, Department of Mechanical and Industrial Engineering, University of Utah, USA
B. H. Kear, Chairman, Department of Mechanics and Materials Science, Rutgers University, USA
S. A. Meguid, Department of Mechanical Engineering, University of Toronto, Canada
V. Milot, Senior Vice-President, Research and Development, Pan Abrasive Inc., Welland, Canada
B. K. Mukherjee, Unit Head, Non-Destructive and Fracture Evaluation, Ontario Hydro, Canada
A. W. Neumann, Department of Mechanical Engineering, University of Toronto, Canada
P. Norton, Department of Chemistry, University of Western Ontario, Canada
W. Wallace, Head of Materials, National Aeronautical Establishment, National Research Council, Canada
C. A. Ward, Department of Mechanical Engineering, University of Toronto, Canada
J. M. White, Center for Materials Chemistry, University of Texas at Austin, USA
J. P. Wightman, Department of Chemistry, Virginia Polytechnic Institute and State University, USA

Characterisation and Quality Control of Surface Treatments and Coatings

K. Stevens

AEA Technology
Risley
UK

ABSTRACT

This paper presents an overview of surface treatment and coating processes for ferrous materials aimed at controlling friction, wear and corrosion. They are described in terms of the key physical and metallurgical properties produced at the surface as well as the requirements for quality assurance and quality control. Finally, the paper highlights recent progress with some particular characterisation techniques being explored at the National Centre of Tribology; X-ray fluorescence, scanning acoustic microscopy, ultra low-load hardness testing and metallographic sectioning; discussing their use as quality control tools.

THE SURFACE ENGINEERING CONCEPT

Modern industrial processing presents engineers with difficult tribological problems. Often it is the structural or environmental constraints which dictate the choice of substrate material and subsequent problems with wear or friction are tackled as an afterthought. In the chemical and petrochemical industries there is corrosion, abrasion and, particularly when stainless steel is used, adhesive wear and galling. For instance, the processing of filled plastics produces significant abrasive wear. In the extraction of oil and gas, the pump and valve manufacturers have to cope with sour conditions combined with abrasive, erosive silica slurries. Even in food and confectionery processing, the extruding and cutting machinery is subjected to severe abrasive wear. In mining and processing of fossil fuels the problems are mainly of aqueous corrosion and abrasion, whereas the nuclear and power generating industries are faced with a full range of tribological problems, including abrasion, corrosion, fretting, adhesive wear and galling.

In many cases it is the combined effect of wear and corrosion that is so damaging. Whilst the individual solution to either a problem of abrasion or corrosion might be quite straightforward, when the two occur together a completely different approach may be required. The passivating layer that provides corrosion protection is removed in a wearing situation and a surface selected for its resistance to abrasion can be quickly undermined by corrosion, no matter how hard it is.

A Surface Engineering approach to problem solving is often the only step that can be taken. If the bulk material is decided on the grounds of strength or environmental considerations, the component can then be coated or surface treated to overcome the tribological problems. This is the case whether the substrate be a wrought or cast material or a powder metallurgy material. Even so, the designers and engineers are faced with a bewildering choice of surface treatments, covering a wide range of thickness or case depth, and selection must be made on a rational basis. The subsequent surface treatment must not impair the properties of the substrate for which it was originally chosen; for instance, it should not produce unacceptable distortion or dimensional changes, reduce the hardness and load carrying capability or impair the corrosion resistance. Equally, the choice of treatment must relate to the problem; a shallow thermochemical diffusion treatment will not give protection in a high-stress abrasive situation and, in the other extreme, it is inappropriate to resort to a massive weld overlay coating when the wear process involves only light loading. Additional problems are presented by the treatment of powder metallurgy, sintered products. Residual porosity can cause poor adhesion of coatings or can absorb undesirable chemicals from the treatment process.

The user of surface treatments and coatings has, therefore, many requirements to satisfy. He has to identify the particular surface properties needed to solve his specific problem, to select an appropriate surface treatment and, above all, be reassured that the process can produce consistent and predictable results. In many cases he will demand certification from the supplier, backed by a series of quality control evaluation techniques. He may prefer these techniques to be non-destructive and be applicable to actual components rather than to tab samples.

SURFACE ENGINEERING PROCESSES

Surface engineering covers a wide spectrum of processes, involving treatment depths ranging from as little as 0.1 micron up to several centimetres (Figure 1). To rationalise the processes, they can be sub-divided into three categories:

i Those involving a physical or metallurgical surface modification.

ii Those involving a chemical and metallurgical change.

iii Those involving a new material deposited on to the surface (coatings).

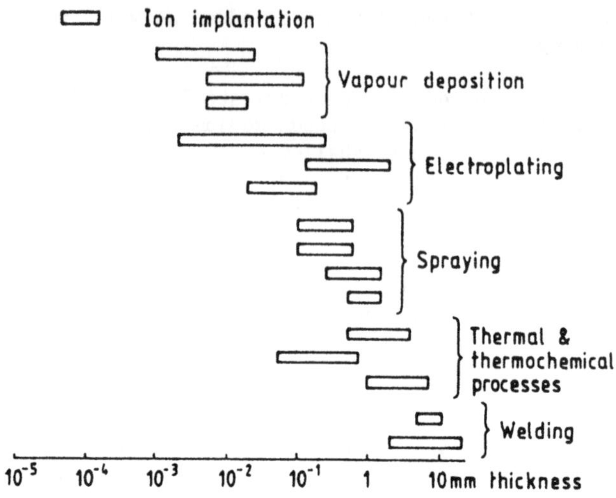

Figure 1. Coating Thickness Produced by Different Processes

i. Modifying the Surface without changing the chemistry.

The most common method is to transformation harden the outermost region of a steel component by localised heating. This can be achieved by **induction, direct flame impingement** or by using a **laser** or **electron beam.** In each case, the key properties of interest are surface hardness, case-depth and the occurrence of cracking. Such treatments are chosen because they produce minimum distortion and preserve the properties of the substrate (eg. high toughness). They can also be applied to discrete areas of a component; for instance, just those which require wear resistance.

Shot peening, grit blasting and **specialised machining techniques** can be used to produce physical changes in the surface. In this case, the object is to create work hardening, the key property being an increase in the surface residual stress; with benefits to both fatigue resistance and surface stress corrosion.

ii. Changing the Surface Chemistry

The surface chemistry of ferrous materials can be altered by the introduction of interstitial elements like carbon, nitrogen or boron, or metallic elements like chromium, aluminium or silicon.

Treatments fall into two main categories, those at high temperatures with the steel in an austenitic condition (~900°C), and those at lower temperature with the steel in a ferritic condition (below 600°C). Of the high temperature processes, **carburising (also called case-hardening)** is by far the most common. The carbon is introduced into the surface by pack, gas or plasma processes and the outer regions (1 or 2 mm) can then be

transformation hardened. (When nitrogen is added as well, the process is called **carbonitriding**).

Boronising can be applied to almost any steel and relies on the diffusion of boron, from a pack or gaseous medium, to form a surface layer of iron borides. The case is more shallow (~0.05mm) that that produced by carburising but can generate hardness values as high as 1500Hv. Because of the high temperature, the desired core properties must be reclaimed by post-heat treatment.

Chromising, Aluminising and **Siliconising** are high temperature pack processes aimed at improving wear and corrosion at elevated temperatures in hostile environments. Usually, the key surface properties are case depth and integrity (lack of cracking or other defects which could permit corrosion) with hardness of secondary importance.

The final high temperature process under this heading can be termed the **Carbide Diffusion Treatment** (commonly known under the trade name of the Toyota Diffusion Process). It is a salt bath treatment at around 1030°C designed for high carbon steels. It produces a layer of vanadium carbide on the surface which appears as a coating rather than a diffusion treatment (Figure 2). It is relatively thin (approximately 15 microns) but with hardness values above 1500Hv. Thickness and layer distribution are the critical properties with hardness of secondary concern. Adhesion is not usually a problem.

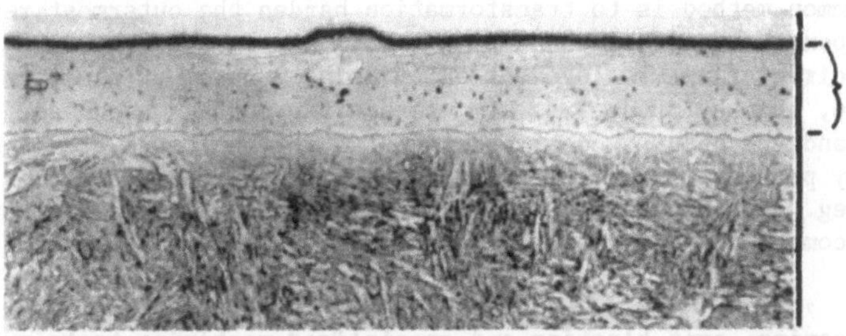

Figure 2. Carbide Diffusion Treatment Applied to a Tool Steel (X1000)

At lower temperatures, in the range 450 to 570°C, the most important treatments are **nitriding** and nitrocarburising. **Nitriding** involves the formation of alloy nitrides at the surface, either by gas, salt-bath or plasma processes. It requires the presence of alloying elements like chromium and aluminium in the steel and produces case-depths of around 0.05mm. With high alloy steels (eg. austenitic stainless), the case is very hard, with a sharp diffusion gradient into the substrate. In this case, treatment distribution (around corners and down holes) is an issue, as well as the potential for surface cracking and case-spalling.

Nitrocarburising involves the diffusion of carbon as well as nitrogen and, on alloy steels, produces similar effects to nitriding. On low alloy steels it is used to produce highly specialised surface 'compound layers' of iron nitride. They are thin (around 30 microns) and are often produced with surface porosity which can be waxed, oiled or oxidised and polished (Figure 3). Hardness, thickness distribution and layer structure are all key properties.

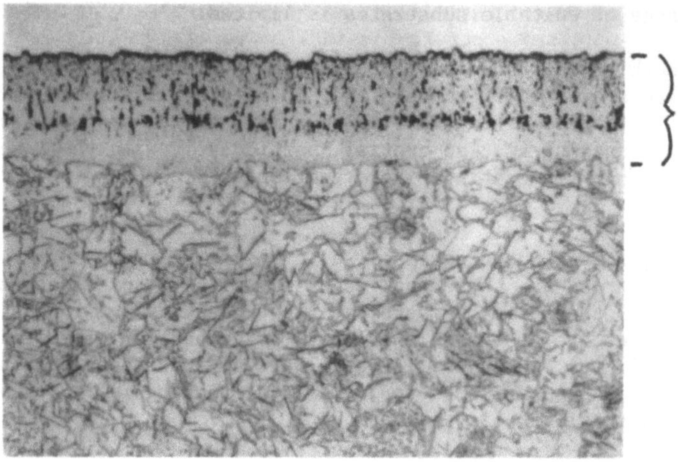

Figure 3. Gas Nitrocarburised 0.4%C Steel (X1000)

Finally, at even lower temperature (around 150°C), **ion implantation** can be said to be a diffusion process. It is based on bombardment by a high energy beam of ions (usually nitrogen) which are forced into the top-most regions of the component. Case-depths are shallow (0.1 micron) but volume fractions of the implanted species are high (up to 30%) with a theoretical increase in surface hardness and compressive stress. Obviously, proving that such effects have actually occurred in such a shallow depth is a major problem.

iii. Adding a layer of new material

There is a long list of processes which involve coating a component with an extra layer of material. They range in thickness from just one or two microns to several centimetres. They are applied for a variety of reasons; to reduce wear or friction, to give corrosion protection or to reclaim a component.

The thinnest coatings are applied by **physical vapour deposition; (PVD)** and are usually between 2 and 4 microns thick (Figure 4). The process uses a glow discharge in a vacuum chamber. A range of materials can be deposited by evaporative PVD or by sputtering but, for tribological purposes, the most important coating is titanium nitride (TiN). The deposition temperature is around 400°C so many components used in the process industries can be coated without distortion or tempering. The coating gives excellent resistance to adhesive wear and, because it is so hard (3000 Hv), to low stress abrasion. However, the coating requires good support and should not be applied to soft substrates in high load application. TiN cannot be relied on to give corrosion protection.

Chemical vapour deposition (CVD) is used to deposit metals and ceramics by the decomposition of a reactive gas at the surface of components placed within the reaction chamber. It is used to deposit materials like TiN and TiC, as well as CrN (usually to a depth of about

5-10 microns). However, the deposition temperature is high (above 800°C) and the range of suitable substrates is limited.

For both these processes (PVD and CVD), thickness and coating distribution are of major concern and, in many applications, performance is critically dependent on there being a minimum value of coating thickness. Adhesion is also crucial, particularly with the lower temperature thin PVD coatings where poor quality can lead to very rapid failure.

The majority of coatings applied to resist wear and corrosion in the process industries are those applied by **electrolytic** or **electrochemical** techniques. Both are low temperature (less than 100°C) processes. Hard **chromium plates** are also used when there is a need for oil retention, as in cylinder liners. Hardness is not always an advantage; soft plates of tin, silver, or lead are used to facilitate 'running-in', prevent fretting and confer corrosion resistance.

In corrosive wear situations, **electroless nickel** can prove effective. As deposited, it has a hardness of 550 Hv but this can be increased to 1000 Hv by heat treatment for 1hr at 400°C. Electroless nickel plate is also available with additions of materials like SiC or Al_2O_3 for high temperature applications, and with PTFE for reduced friction. (Figure 5).

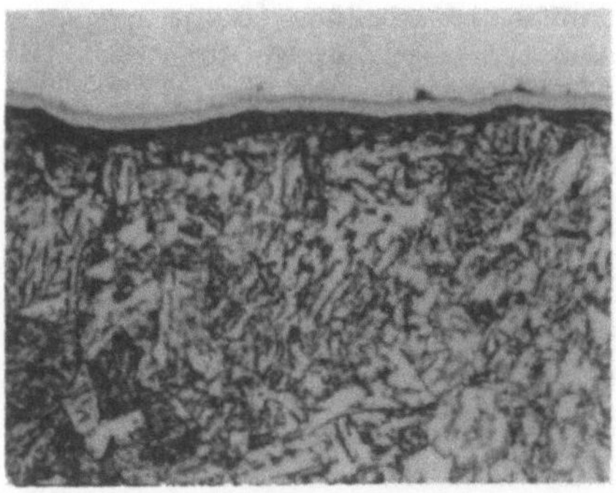

Figure 4. PVD TiN (X1000)

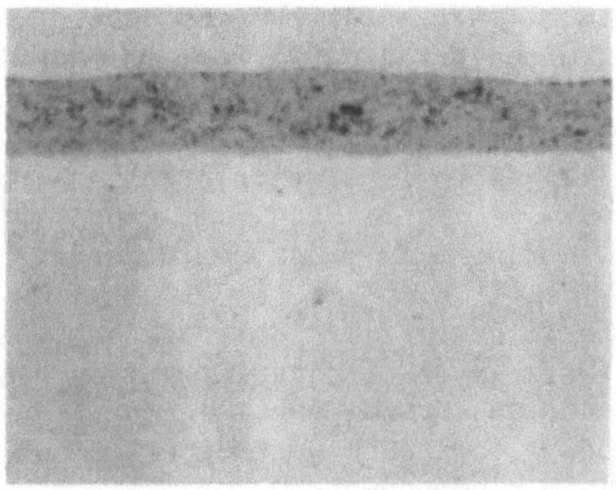

Figure 5. Electroless Nickel/PTFE (X1000)

Coatings are also applied by **spray** techniques. Powders are heated to a semi-molten state and deposited at high velocities onto the component surface. Coating thicknesses vary from about 0.05mm to 1.00 millimetre. The techniques can be divided broadly into **flame gun, electric arc, plasma-arc** and **high velocity** processes.

Electric arc spraying is used to deposit metals for either wear resistance, corrosion resistance or reclamation. The coating metal is fed as two wires and an electric arc is struck between them to cause melting. The molten metal is then propelled onto the substrate by compressed air.

In **flame spraying,** the source of heat is a burning gas, such as acetylene, and the coating material is fed into the gun either as a wire or a powder. It is a relatively cheap process and gives a high deposition rate but, in general, the bond strength is lower and the porosity is

higher than that achieved by the electric arc process. The **spray-fuse** process takes the technique a step further by first spraying and then fusing with a second heat source, such as flame, torch or by induction. This is the basic technique used for nickel or cobalt-based alloys and gives excellent resistance to corrosion, erosion, abrasion and fretting. Alloys can have hardness values up to 900 Hv with near zero porosity. There will, of course, be substrate distortion and the parts will require finish-grinding. Coating thickness are typically up to 0.5mm.

The **plasma spray** process makes use of an ionised gas (usually argon or nitrogen) to produce much higher temperatures than those created during flame spraying. This allows deposition of higher melting point materials such as metal oxides or metal/ceramic mixtures. Such coatings have relatively good substrate adhesion and porosity levels are usually in the range 2 to 4% (Figure 6). A typical coating thickness would be 0.1mm and the process finds a wide range of applications in resisting abrasive wear. Substrate heating is minimal.

There are several variations on the **plasma arc** technique. The process can be carried out in a partially evacuated chamber **(Low Pressure Plasma Spraying - LPPS)**, giving reduced porosity, reduced oxidation of the coating material and, because the substrate reaches a higher temperature, better adhesion. The same advantages can be gained by shrouding the arc in a non-oxidising gas **(Inert Atmosphere Plasma Spraying)** and both these processes are now used increasingly to deposit the nickel-cobalt-chromium-aluminium-ytrium (MCrAlY) type alloys which are used extensively in the aircraft industry to resist high temperature oxidation, erosion and fretting. A third variation is the **Transferred Plasma-Arc** process in which a secondary electric current is established between the arc and the workpiece. This promotes substrate heating and surface melting and gives more dense and more adherent coatings. Deposition rates can be very high and the technique is typically used to deposit thick (up to 10mm) abrasion and erosion resistant coatings for use in applications like mining and agriculture. The substrate must be electrically conducting and be able to withstand some thermal distortion.

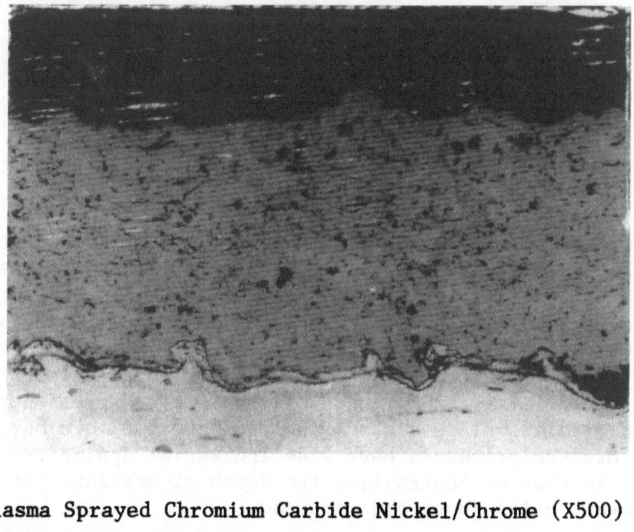

Figure 6. Plasma Sprayed Chromium Carbide Nickel/Chrome (X500)

The ultimate spray techniques are those producing the highest particle temperatures and velocities **(High Velocity and Detonation Processes)**. They are based on combustion of high octane fuels. Porosity is very low (less than half of one percent) and substrate adhesion is excellent. They are used to deposit tough abrasion and erosion resistant coatings such as chromium carbide or tungsten carbide/cobalt cermets. (Figure 7.) Thickness is typically 0.1mm and substrate heating is minimal.

With all sprayed coatings, the most disputes about quality occur over substrate adhesion and coating porosity. Both have a major impact on wear and corrosion performance and considerable effort is directed at demonstrating their values, usually on tab samples. Coating hardness and surface finish are also important.

Figure 7. Detonation Gun WC/Co Coating (X200)

Laser processes are now used increasingly to deposit coatings. For **cladding**, the powdered material is blown directly into the laser-generated melt pool. **Laser alloying** is a similar process, except that the energy is increased to produce more substrate melting and complete surface alloying with the powdered material. The reaction area is shielded with an inert gas. A particular advantage of such techniques is that specific areas can be treated, thus minimising component distortion. Corrosion resistant surfaces and metallic glasses are being produced but more development is required before the processes have wide industrial application. One of the problems is that of controlling the depth of heating, particularly when thin layer fusion zones are required.

The **thickest coatings** are applied by **welding** and **roll-cladding**; typically up to 2 or 3cm. **Welding** can be used to good effect in tribological situations where high stress abrasive wear is the problem, such as coating digger teeth, tank tracks and on ore-handling equipment.

Cladding is usually associated with corrosion or mild wear problems that are encountered in the chemical, food processing or printing industries. The two processes, roll-cladding and weld cladding, are complementary; hard abrasion resistant materials are difficult to fabricate and are best deposited by welding, whilst the more corrosion resistant materials are based on ductile austenites and are amenable to roll-cladding and forming.

SURFACE CHARACTERISATION

Market research by NCT has indicated the following order of priority given by treatment users to the various surface properties.

Top priority - Coating/substrate adhesion
- Thickness/case depth
- Hardness
- Porosity
- Surface texture

Lower priority - Structure and composition

i Coating Adhesion

There are many traditional methods of assessing coating adhesion which are used for quality control. Most are qualitative, involving bending, scratching, impacting or thermal shocking, followed by a visual inspection for flaking and cracking. Some coating companies employ tensile tests, using coated pegs glued together with high-strength adhesive, to determine a real bond strength. It is often more a test of the adhesive than of the coating and has little relevance to actual industrial components.

For non-destructive characterisation of treated components, the **Scanning Acoustic Microscope (SAM)** is now showing considerable potential. It functions by transmitting acoustic signals into the surface and analysing the reflections from interfaces and imperfections within the material. Water is used as the couplant and the parts are scanned by an acoustic probe. Provided that a sufficiently large water bath can be constructed, and the probe can be correctly manipulated, there is no limit on the size of component that can be examined.

Work at NCT has shown that SAM can easily detect reduced adhesion of sprayed coatings which has been deliberately produced by compromising the deposition procedure. For instance, it can detect areas which were contaminated by grease or that are given inadequate preparatory grit blasting. The SAM image is presented as a colour-coded or shaded contour map to indicate the degree of reflection, a complete dis-bond giving maximum reflection. Figure 8 shows an image for a plasma sprayed coating.

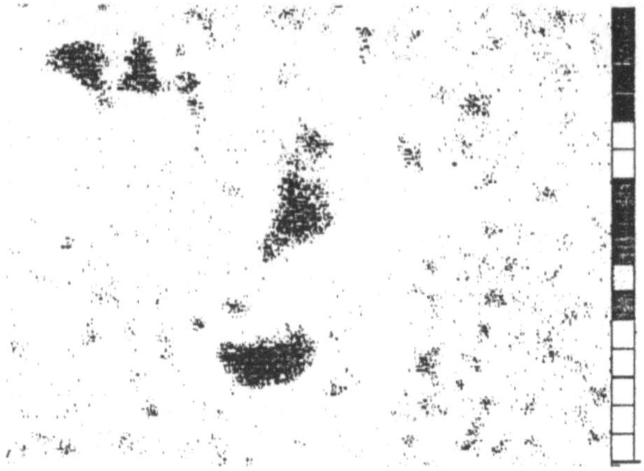

Figure 8. SAM Image Showing Discrete Areas of Poor Adhesion (X10)

ii **Thickness**

There are well established traditional methods of measuring thickness using electrical techniques based on eddy currents or capacitance. However, with the advent of much thinner coatings such as those produced by PVD and CVD, techniques need to be more discriminating. NCT has found that **x-ray fluorescence (XRF)** is an ideal tool and Figure 9 shows the results of a large sampling exercise on TiN-coated twist drills, indicating the spread in thickness produced by three different PVD processes.

XRF employs the principle of bombarding the sample with an x-ray source and monitoring the characteristic x-rays which are excited from the substrate and coating atoms. Provided there is atomic number contrast and calibration standards are available, the relative intensity of coating and substrate x-rays gives a measure of thickness. It is an accurate and quick measuring technique, relatively insensitive to geometry or surface finish and an ideal quality control tool, increasingly used by coating specialists. It can determine thicknesses as low as 0.1 microns but is limited at the higher end to about 30 microns. As yet, therefore, it is not applicable to sprayed ceramic or cermet coatings.

iii **Hardness**

For thick coatings and deep diffusion treatments, hardness measuring is a routine exercise, with either macro indentations at several Kg load (Vickers diamond impressions) or micro hardness tests (particularly hardness profiles on polished sections through the treatment) at loads down to about 25gms.

For thin coatings and treatments like ion implantation, **ultra-low hardness** tests are required. The technique uses a conventional diamond indenter with loads between 0.5 and 100 gms. It relies on a measurement of the displacement of the indenter into the surface rather than a visual assessment of the impression, as in more traditional techniques.

Figure 10 illustrates some results obtained by NCT on thin PVD coatings. It shows the variation in hardness for different coating thicknesses, the values at low load being those of the coating itself and those at higher load being increasingly influenced by the substrate.

Work is now continuing with ion-implanted layers, where the low load hardness technique promises to be one of the very few methods of characterising such thin treatments.

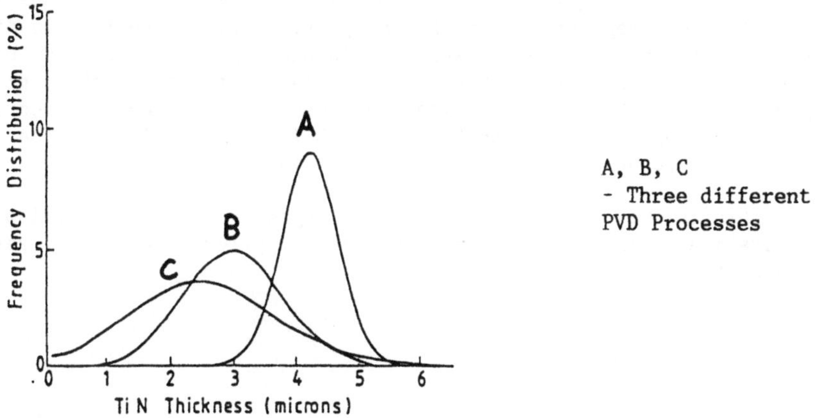

Figure 9. Thickness of TiN on Twist Drills.

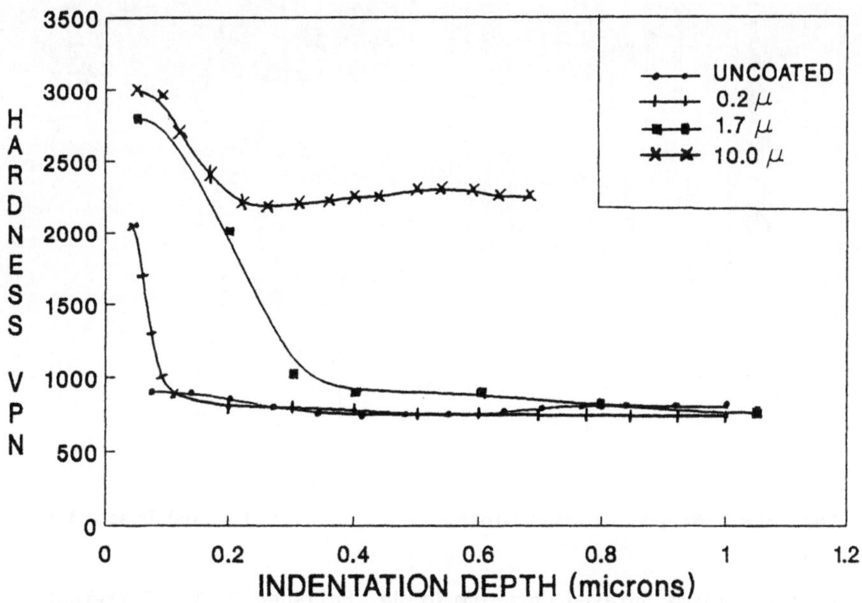

Figure 10. Hardness as a Function of Coating Thickness for PVD TiN

iv Porosity, Structure and Composition

Most companies continue to rely on metallographic sectioning, either of sacrificial components or tab samples, to reveal coating porosity and structure. To date, NCT has failed to identify any non-destructive methods for determining the porosity, specifically of sprayed coatings where the issue promotes most contention. XRF failed to discriminate between coatings of identical composition (Chromium oxide, 100 microns thick) but with widely different porosity (10% and 2%). Likewise, vacuum impregnation and weighing techniques fail because some porosity is isolated.

Even with metallographic sectioning there is heated debate about polishing techniques, particularly the creation of 'false porosity' by pull-out or the disguising of true porosity by smearing. NCT has made a wide study of preparation techniques, particularly those for the difficult sprayed ceramic and cermet coatings. To highlight the problems that can be encountered, Figure 11 shows two samples of the same coating prepared by different routes. It takes careful scanning electron microscopy and elemental mapping to show that Figure 11b is a false picture, the preparation technique having smeared material into the pores.

This is one area where there is an urgent need to standardise preparation techniques, so that users and suppliers can compare results with confidence.

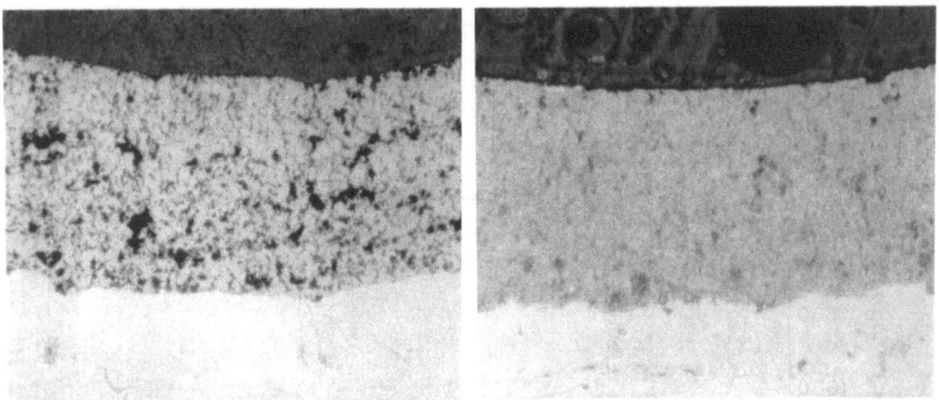

(a) Short-Napped Diamond Cloth (b) Hard Diamond Pad

Figure 11. Plasma Sprayed Chromium Oxide, Polished by Two Different Routes (X200)

v Surface Texture

The stylus profilometer manufacturers have turned the power of the computer towards ever more sophisticated analysis of the surface profile. There is a wealth of new parameters to add to the traditional Ra and Rq values.

The experience at NCT is that there are three parameters most relevant to the tribological behaviour of surfaces, namely Ra, Δq (the RMS slope) and RSK (the skew or shape of the profile). Between them, they provide a comprehensive picture of the profile, in terms of profile height, wavelength and shape. For instance, Δq values can be used to determine the friction of surfaces used in textile applications, and surfaces with a -ve RSK can be used to retain lubricants.

For quality control, there is now a need for profilometer manufacturers to offer standards which can be used to calibrate these new parameters as well as the traditional Ra value.

ENSURING QUALITY

Although considerable effort is devoted to researching and developing quality control tools for evaluating the key properties of the final product, this must not be seen as the desired objective. Ideally, the need for costly and wasteful quality control inspection of large numbers of components should be eliminated by rigorous quality assurance of the process itself. Once the treatment technique is fully understood, all the variables controlled, the component material correctly authenticated and the procedure defined and documented, then the foundations of a 'zero inspection' policy are established. To achieve it, the quality assurance principle must be applied to all links in the chain; delivery and documentation of components, pre-treatment procedures, handling and jigging, treatment parameters, system operation and monitoring, post-treatment procedures, packaging and certification and, finally, dispatch.

In many of the traditional surface engineering industries, this need for minimal quality control inspection is being achieved. For instance, the principles of electroplating are so well understood that knowledge of the substrate material and geometry, and the bath composition, current density and temperature are sufficient to guarantee a precise coating thickness and hardness.

In the emerging new technologies, however, deposition and treatment processes are less well understood. Until all the aspects of a particular technique have been researched it is impossible to define precise operational parameters and the required quality assurance principles. It is here, therefore, that reliable quality control techniques are required, both to reassure potential users and help a new technology into the market place, but to assist in continued process development towards 'quality perfection'.

Microstructural Preparation and the Achievement of a Representative Surface

I. A. Bucklow

The Welding Institute
Cambridge UK

and P. Morrell

Rolls Royce plc
Derby UK

ABSTRACT

The complete non-destructive-assessment of engineering surfaces tailored to exhibit engineering properties is, in general, not yet possible. In consequence, sectioning and microstructural examination is widely used as a method of specifying and checking such surfaces, particularly where the thicker coatings have been applied. Microstructural preparation of a sectioned coating is always difficult; although preparation procedures have been devised for many materials, they are usually specific to one manufacturer's equipment and consumables, and therefore it is difficult to transpose those procedures between equipment. An attempt is being made in the UK to formulate a universal microstructural preparation language which is independent of commercial nomenclature and which will allow the specification of procedures designed to achieve surfaces of the highest integrity.

INTRODUCTION

It has long been acknowledged that microstructural preparation is all too easily performed as a largely uncontrolled process; from the earliest strictures outlined by Sorby (1) to the careful procedures detailed by Vilella (2), the danger of introducing artefacts during microstructural preparation has been recognised. Individual manufacturers of preparation equipment and materials have made heroic strides (reflected in heroic prices) in automatic and reproducible methods for revealing the true surface of sections of a wide range of materials, but it is virtually impossible to translate the preparation route detailed by one equipment supplier into the route (for the same material) as described by the supplier of different equipment. This is not to say that both recipes will not produce the same result, but when seen from the point of view of a preparation laboratory, or the purchaser of the material being examined, the situation is far from satisfactory. In these days of increasingly strict quality control and assurance, the purchaser of material must be able to specify the acceptance procedures for that material, and the preparation laboratory must be free to use whatever equipment it possesses (and can afford). Furthermore, the material supplier must be confident that he can meet - and be shown to meet - the acceptance standards of more than one customer. The foregoing highlights problems that have always been with us to some degree in the field of metals, but these problems are now assuming a major importance in the field of new materials in general, and of tailored surfaces in particular.

Surfaces tailored to exhibit specific engineering properties are produced by eg, heat treatment, or cladding, or by coating. The complete non-destructive assessment of these surfaces is a problem that has yet to be solved and so quality assurance of many of the surface properties must depend upon sampling and destructive examination and testing. Microstructural examination - usually of a section through the surface and its substrate - is widely used as one such method of both specifying and assessing a treatment or coating, but microstructural preparation of any edge is difficult - and when that edge has properties very different from those of the substrate, the problem is compounded. Similar problems exist in the preparation of composites, of honeycombs, of porous media, and indeed of many multi-phase materials. The difficulties faced by the QA engineer when examining a section can be summed up in two questions, thus: "How can we specify a preparation route that can be followed by any laboratory and be referenced back to an acknowledged standard?" and "Is what we see a true representation of the structure, without artefacts?"

The integrity of any prepared section must always be in some doubt but it should be possible to arrive at a concensus that allots a minimum of significance to that doubt provided that the preparation route has been based on sound engineering principles. Many of those principles have been enunciated in the pioneering work of L E Samuels (3) in Australia, and it is upon those principles that an attempt is being made in the UK to formulate a universal scheme for highest integrity. Briefly, the preparation scheme assumes that the depth of significant damage below a surface is proportional to the rate at which that surface is machined, further, it assumes that any combination of equipment plus consumables for grinding/lapping/etc can be characterised for any substance in terms of material removal rate. It follows that material removal rate is the major criterion on which to base a preparation procedure, and that such a procedure can therefore be specified in terms that are independent of any given combination of equipment and consumables.

It will be necessary to undertake removal rate characterisation for each material treated by each combination of equipment and consumable, but once done, the route is defined, it is independent of manufacturers' apparently arbitrary recipes, and is universal. The approach acknowledges that some materials will require more gentle treatment than others, and that ultimate judgement of a "truly representative surface" will always be subjective.

The proposed system broadly follows three arguments, viz: that the depth of sub-surface damage inflicted during cutting and grinding is proportional to the rate of material removal, that a surface which is representative of the bulk is produced by a cutting action, and that polishing is largely a cosmetic operation and can inflict further damage. It is assumed that the relationship between grinding rate and depth of damage is linear (probably not so for many materials, but it is as good a starting point as any), and that therefore systematic preparation routes to an irreducible minimum of damage can be devised. Given constant pressure, sliding speed, etc., each combination of equipment and consumables can, in theory, be characterised for each material in terms of grinding rate, and therefore any route to the true surface (ie, of 100% integrity) can be described in terms that are "traceable" and independent of commercial reference. Such a surface will always be scratched to a degree, but polishing should be kept to the absolute minimum.

One difficulty is to identify the "true surface"; the methods involved will depend to some extent on the material, but porosity determination of sprayed coatings is a useful example to illustrate the concepts described here. Porosity determination often forms part of an acceptance specification for sprayed coatings and it also can be measured by image analysis (preparation damage usually involves pull-out and so results in an increase in the apparent porosity). Work done by Rolls Royce on sprayed ceramic coatings using the "surface integrity" concept on applied to two types of commercial equipment has proved to be very encouraging.

THE SYSTEM

Definitions

Platen - the object that supports or carries the abrasive or polishing medium. The name is a generic one because the platen may be naked plate, or covered by a cloth, etc. It is important to note that the nature of the platen influences the way that any given abrasive behaves, all other factors being constant, Fig. 1.

Surface - refers solely to the surface being examined.

Surface Integrity - the approach to the "true" surface, however that may be measured or defined.

Damage Level - the degree of departure from the true surface. It may be defined in terms of the stock removal rate, eg Damage Level 10 is that given by a sample stock removal rate of 10µm/min in the given preparation system, Damage Level 0 = 100% surface integrity. This term is little more that the inverse of surface integrity, but it is useful as a means of characterizing the nature of a platen, eg, a "Damage Level 5 Platen", or simply a "Level 5 Platen".

Preparation systems - the combination of lubricant, abrasive medium, platen, conditions, etc.

Grinding, Lapping and Polishing - see Fig 2. Grinding with a fixed abrasive (eg, grinding wheel or abrasive paper) gives rise to high shear forces in the surface of the specimen. Grinding with a charged abrasive results in the abrasive embedding in the softest phase present and gives a lower impact energy. Lapping employs rolling abrasive grains that do not embed but create high compressive, and low shear, forces in the surface; lapping has the lowest removal rate. Polishing is accomplished by fine particles that are free to rise and fall in the material of the platon surface.

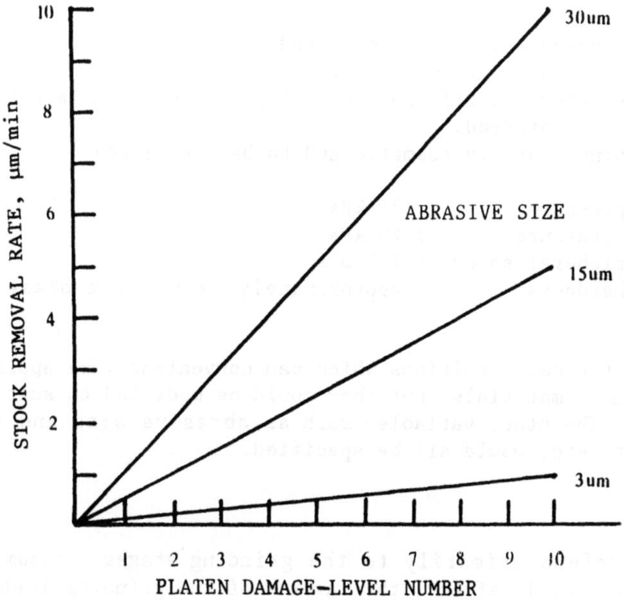

Fig. 1 Influence of the platen on material removal rate.

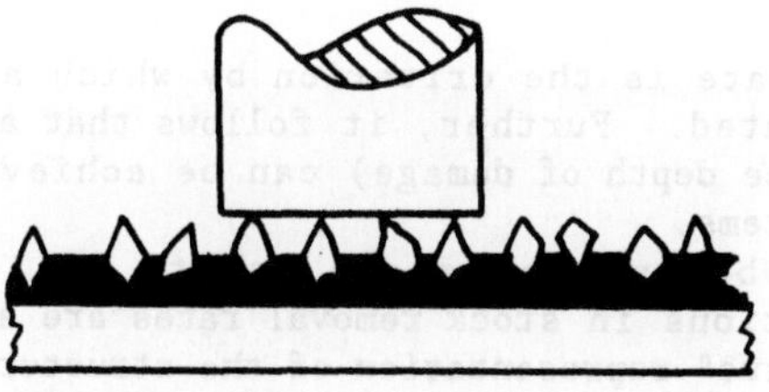

Fig. 2a Fixed abrasive grinding

Fig. 2c Lapping

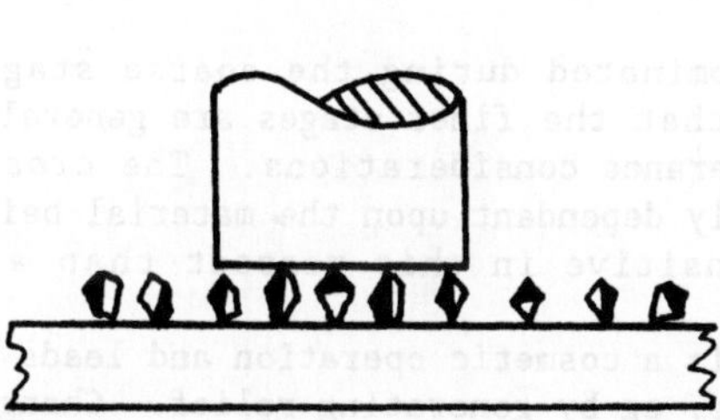

Fig. 2b Charged abrasive grinding

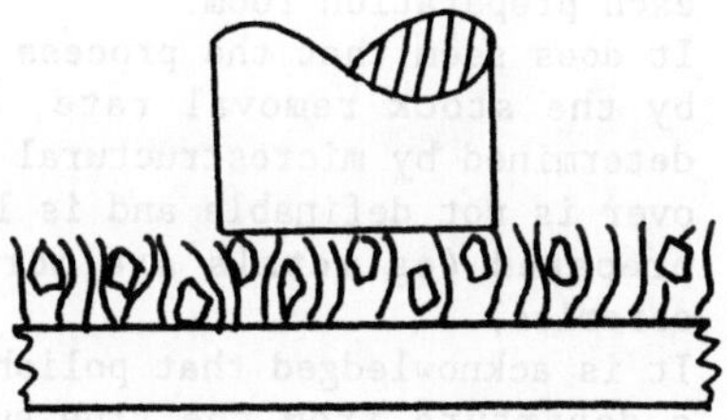

Fig. 2d Polishing

Conventions

Three stages of preparation are recognised:

a) planar grinding (fast stock removal)
b) sample integrity grinding (grinding, rather than lapping, is preferred)
c) polishing (largely cosmetic and to be minimised)

Grinding pressure	:	35 kPa
Polishing pressure	:	20 kPa
Platen peripheral speed	:	1.5 m/s
Abrasive hardness	:	approximately 2x hardest phase in sample

(3)

These are typical conditions which can conveniently be applied in most equipment to most materials, but they could be modified to suit particular circumstances. The other variables such as abrasive size and type, feed rate, lubricant, etc, would all be specified.

The Argument

The argument refers primarily to the grinding stages - assumptions made about polishing are, in effect, the inverse of the grinding argument but in principle only.

1. A given combination of abrasive, abrasive support system, lubricant, pressure, etc, gives a certain stock removal rate measured in μm/min.

2. Significant damage is created to a depth that is proportional to the stock removal rate. As a starter, that proportionality is assumed to be linear, and a stock removal rate of 10μm/min is taken as the reference value.
3. It follows that stock removal rate is the criterion by which all preparation systems are calibrated. Further, it follows that any given stock removal rate (and hence depth of damage) can be achieved by many different preparation systems.
4. Providing that each stage cuts below the damage create by its predecessor, progressive reductions in stock removal rates are all that is necessary to achieve a "true" representation of the structure, ie equivalent to a zero stock removal (depths of damage are not necessarily additive, but it would be better to assume that they are).
5. The rate at which one progresses towards the zero rate depends upon the material being prepared and upon the particular circumstances of each preparation room.
6. It does seem that the process is dominated during the coarse stages by the stock removal rate, but that the finer stages are generally determined by microstructural appearance considerations. The cross-over is not definable and is largely dependant upon the material being prepared (eg metals are more sensitive in this respect than are ceramics).
7. It is acknowledged that polishing is a cosmetic operation and leads to a departure from the true surface, eg by generating relief. Changes in the integrity of the polished surface cannot be equated to some simple parameter in the same way that is used to assess grinding damage.
8. To define a preparation route for a given material (various parameters agreed and held constant), trial-and-error is used to find a grinding

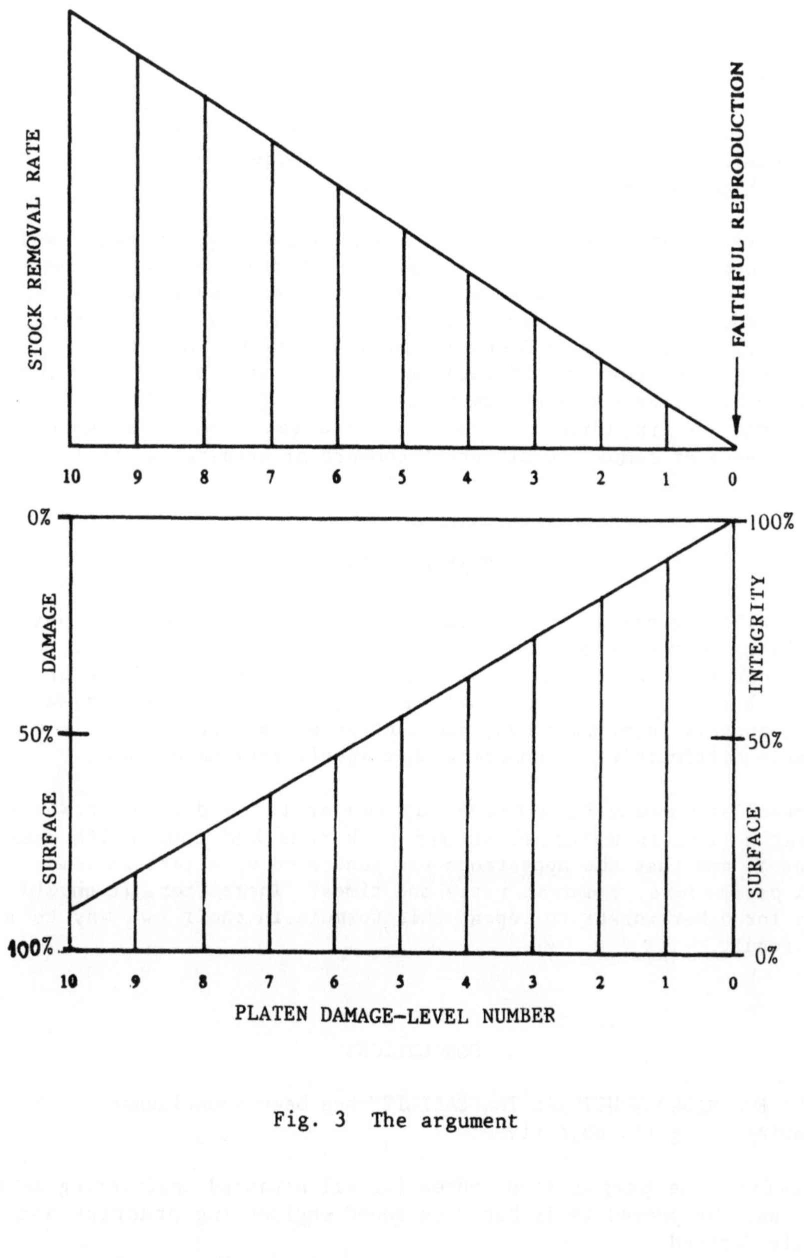

Fig. 3 The argument

combination that gives a removal rate of 10μm/min; this is the reference level. Less coarse stages are used - each defined by its removal rate - to approach to a very low removal rate and/or the stage at which no microstructural changes are seen. This is taken as the "true" surface of 100% integrity, and the same result should be obtainable by different combinations of equipment and consumables - each characterised by its removal rate.

The system hinges upon devising a non-commercial language which will specify, among other things, the nature of the platen onto which the abrasives are fixed or charged. The terms employed must also allow the cross-referencing of platens made by different manufacturers, but this cross-referencing can only be done by trial and error. There is nothing new in any of these concepts (except, possibly, the use of material removal rate as the determinant) and, indeed, they are promoted in a commercial context, (4), but their formulation in the manner outlined in this paper does allow them to be used as a non-commercial means of comparing different commercial preparation systems.

TRACEABILITY

As well as allowing comparisons to be made, the system offers the possibility of specifying a universal preparation procedure, independent of trade names etc, that can be referenced to agreed visual standards whose preparation history is known. This is of particular importance to organisations who receive components, etc, from many sub-contractors and who currently experience considerable difficulties in ensuring that specifications are met.

Traceability would be achieved by two or three different workers who agree that a certain material should look thus and thus (with examples to demonstrate), and that the appearance was generated by a certain combination of standard parameters, removal rates and times. Thereafter, it should only be necessary for other worker to repeat this formula in their own way to achieve the same result.

CONCLUSIONS

A SOCIETY FOR MICROSTRUCTURAL TRACEABILITY has been established in the UK with the following among its objectives:

To define the preparation routes for all advanced engineering materials, ensuring that the procedure is based on sound engineering practice and is not empirically derived.

To create a universal language enabling a non-commercialised preparation system to be operated.

To reference all materials by a traceable standard which would include, as necessary, samples of actual material, photomicrographs, image analysis results and other appropriate documentation.

To clarify the products of the commercial manufacturers of metallographic equipment and consumables in terms of the universal language.

Certainly, there are many technical criticisms of the assumptions made in the scheme. Furthermore, it is appreciated that the proposed scheme is neither easy nor cheap, but the alternatives will ultimately be more expensive to those organisations concerned with either advanced materials or engineering surfaces. The objectives of the society are ambitious and the present ideas may ultimately prove to be unrealistic, but some form of universal qualitative control must eventually be formulated.

REFERENCES

1 Sorby, H.C., J Iron and Steel Institute., 1887, 31, 255.

2 Vilella, J. R., Metallographic Techniques for Steels, American Society for Metals, Cleveland, USA, 1938.

3 Samuels, L.E., Metallographic Polishing by Mechanical Methods, (3rd ed.), American Society for Metals, Cleveland, USA, 1982.

4 Bousfield, B., The New Concept: a systematic approach to sample preparation, Buehler (UK) Ltd, Coventry, UK.

Photothermal Reflectance Investigation of Implanted Silicon

I. A. Vitkin, C. Christofides and A. Mandelis

University of Toronto
Toronto
Ontario
Canada

ABSTRACT

Silicon films damaged by ion implantation have been examined using the Photothermal Reflectance (*PTR*) technique. Data are presented on the sensitivity of this method to the implant parameters such as dose and species, and on the dependence on thermal annealing. It has been shown that the technique provides much valuable information about the state of the implanted layer, and is a sensitive probe for monitoring the annihilation of the induced damage as a function of the annealing temperature. The activation energy of the *local* annealing recovery mechanism has been found to be 0.15*eV*. The presense of negative annealing has been detected at about 500 °C.

INTRODUCTION

In recent years, thermal wave techniques have been introduced as a method of characterization of ion implanted materials [1,2]. In this paper we will show and analyze some results concerning implanted silicon films, obtained with the laser-induced photothermal reflectance (*PTR*) method at ambient temperature.

A useful alternative to high temperature diffusion doping is the direct implantation of energetic dopant ions into the semiconductor. Ion implantation, with its capability for great uniformity, impurity profile control, and shallow penetration, is a doping method particularly attractive for the production of shallow p-n junctions, which are crucial in modern integrated circuits and solar cell technology. One disadvantage of ion implantation is its generation of defects in the lattice which strongly influence the electronic, optical, and thermal properties of semiconductors. The importance of implanted semiconductor analysis has led to the development of many characterization techniques.

The aim of this paper is to study the dependence of modulated photothermal reflectance (*PTR*) method on the implanted dose, Φ (ion/cm^2), on the type of the implanted impurities (As^+, P^+, and B^+), and on thermal annealing. The obtained information can be related to the physical state of the implanted layer, and its evolution under different implantation regimes and different annealing conditions.

Theory

The mechanism of the generation of the photothermal signal can be understood in terms of the induced modulation of the refractive index of the illuminated material. In electronically inactive (or low active) materials, the modulated reflectance of the probe beam is primarily based on the temperature dependence of the sample's optical reflectivity. In general, in such solids a small local temperature excursion (ΔT) about average sample temperature, T, brought about by a light pulse from the heating laser, will cause a corresponding fractional change in the reflectivity of the probe beam:

$$\frac{\Delta R}{R_o} = \frac{1}{R_o} \left(\frac{\partial R}{\partial T}\right) \Delta T \tag{1}$$

where $1/R_o\ (\partial R/\partial T)$ is the temperature coefficient of reflectivity. It is generally a function of the material properties of the sample, its temperature, and the wavelength of the probe beam. In silicon, the introduction of impurities and disorder by the ion implantation process will influence the local temperature rise in several ways: the optical absorption coefficient will increase, increasing the amount of light energy absorbed at the surface; the local thermal and electronic transport coefficients will decrease, allowing less heat to be dissipated from the point of excitation [3]; and the presence of near midgap defect and impurity states will cause an increase in the local nonradiative recombination rate of the photoexcited carriers, thus providing additional heat to the surrounding lattice [4]. Additional heat generating processes, like non-linear two-photon absorption, may also be dependent on the degree of local disorder [5]. Thus, the presence of impurities and defects in an ion-implanted layer will cause an increase in the local ac sample temperature, and, therefore, in the local reflectivity, over that of the crystalline material. In effect, the "thermal wave" produced in the sample by periodic heating will be strengthened in local amplitude but diminished in spatial extend by the ion implantation process.

In semiconductors, moreover, there is an additional effect brought about by the heating laser whose light energy exceeds the sample's band-gap. As pointed out by Opsal *et al.* [6], the electron-hole plasma can also contribute to the photothermal reflectance signal in several ways. There is the dependence of the complex index of refraction on the charge carrier density; there is also the photoelectroreflectance effect, whereby the photogenerated plasma may influence the wafer reflectivity by surface band bending; in addition, the thermalization of the plasma with the lattice, and the subsequent inter-band recombination will also contribute indirectly to the *PTR* signal through the above mentioned ΔT term. Regardless of the particular mechanism of the plasma effect on reflectivity, we can write for a semiconductor:

$$\frac{\Delta R}{R_o} = \frac{1}{R_o} \left[\left(\frac{\partial R}{\partial N}\right) \Delta N + \left(\frac{\partial R}{\partial T}\right) \Delta T\right] \tag{2}$$

where $1/R_o\ (\partial R/\partial N)$ is the plasma coefficient of reflectivity, and ΔN is the local variation in the plasma density brought about by the heating laser. As pointed out by several investigators, the thermal and the plasma wave coefficients of reflectance are often comparable in magnitude and opposite in sign, resulting in a partial net cancellation of the observed signal [7]. In order to separate the plasma and the thermal contributions to modulated reflectivity, one must vary the product $\omega\tau$, where $\omega=2\pi f$, with f the modulation frequency, and τ is the lifetime of the photoexcited carriers. Despite a certain ambiguity in assigning a precise value to this relaxation time, it is generally agreed that when $\omega\tau \ll 1$, the *PTR* signal has predominantly thermal origins [8]. On the other hand, for $\omega\tau \gg 1$, the plasma effect may become comparable to the thermal effect, or even dominate it. In our experiment, with f=100 kHz, we expect to detect contributions from both thermal and plasma terms for the non-implanted crystalline wafer ($\tau \approx 10^{-6}$ *sec*) [9]; impurities and defects introduced via ion implantation should decrease the plasma contribution by decreasing the Drude and the surface band bending effects, since the $\omega\tau$ term in the damaged surface layer is now much less than unity ($\tau < 10^{-9}$ *sec* for inhomogeneous or amorphous silicon). However, the photogenerated plasma may still contribute indirectly through the thermal term via lattice thermalization and nonradiative recombination routes.

Another possible way to separate the thermal and the plasma wave effects is to examine the temperature response of the photothermal reflectance signal. As both effects have their own dependences (the thermal term is predicted to be an increasing function of T, [10] and the excess plasma term is temperature independent (at least to first order [11]), the temperature evolution of the experimental *PTR* signal should reveal the dominant contribution, as well as elucidate the details of the underlying physical mechanisms.

Experimental

A. Silicon samples

The samples which have been used for this study were obtained from two different sources. (100)-oriented n-type Si wafers were obtained from Northern Telecom Electronics in Ottawa in 1984. The experimental batch consisted of 13 pairs of wafers implanted through an oxide layer with B^+, P^+, or As^+ ions matrixed with doses between 1×10^{12} and 1×10^{16} *ions/cm*^{2}, and energies between 50 and 150 *keV*. The oxide was then etched away, and one member of each pair was thermally annealed at 1100 °C for 15 minutes in N_2.

The second experimental batch consisted of (100), p-type (6 Ω*cm*) silicon wafers implanted through a thick oxide layer, at room temperature, with arsenic (dose $\Phi = 5\times10^{14}$ *ions/cm*$^{-2}$; energy E = 150 *keV*). They were procured from the Microelectronic Center of Grenoble (CIME), France, in 1987. Upon removal of the SiO_2 overlayer, some of the samples were thermally annealed isochronally (1 hour) under different temperatures (400, 500, 600 and 800 °C). Using spreading resistance measurements, and the SUPREM simulation program, the junction depths were determined to be approximately 0.4 μ*m*.

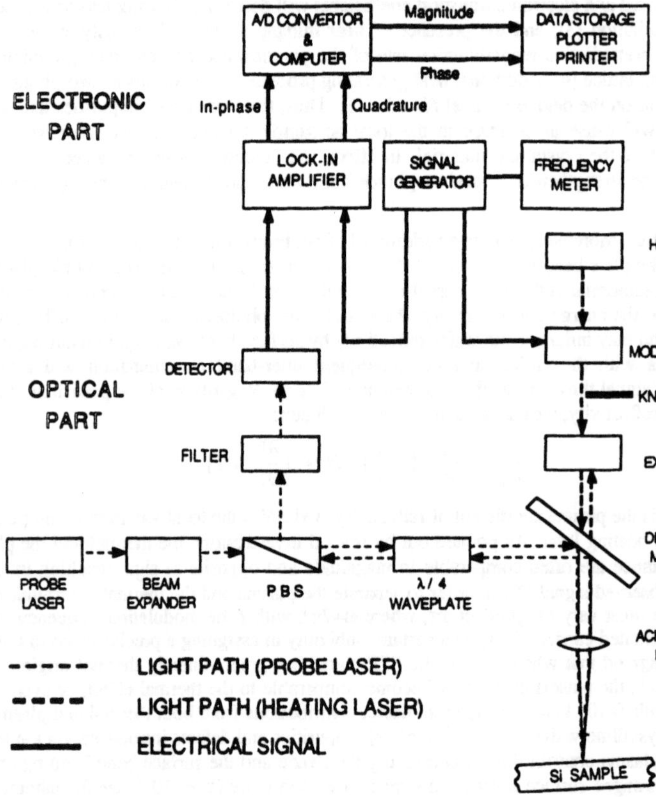

Figure 1 Experimental set-up for the laser-induced photothermal reflectance (*PTR*) measurements.

B. Experimental set-up

The apparatus used for the *PTR* studies is shown in Fig. 1; it is similar to the Therma-Probe 150 system, first introduced by Opsal *et al* [12]. The periodic sample heating is obtained with the *Ar*-ion laser beam (488 nm), acousto-optically modulated at 100 kHz. The beam, of incident power approximately 25 mW, is focussed normally onto the silicon surface to a spot size of about 30 μm. The changes in the reflectivity of a collinear *HeNe* laser beam (632.8 nm), resulting from the thermal and the plasma waves generated by the heating laser, are measured with a silicon photodiode detector (UDT), filtered by a 632.8 nm band-pass interference filter. In order to reject any signal contributions arising from the thermoelastic deformation of the sample surface, the detector is deliberately underfilled, and is operated in the sum mode. The output of the detector is monitored with a fast lock-in amplifier (EG&G 5202), synchronous with Ar^+ laser modulation of 100 kHz. The in-phase and quadrature components of the signal are stored in the computer for subsequent analysis and display. For a given wafer, about 15 different readings on 3 different spots of the surface are recorded, and then normalized to the signal from the reference crystalline Si sample under identical focusing conditions.

Results and Discussion

Fig. 2 shows the measured photothermal signal as a function of phosphorus dose, over the range 10^{12} to 10^{16} $ions/cm^2$. The solid curve corresponds to the as-implanted samples, and the dashed one shows the

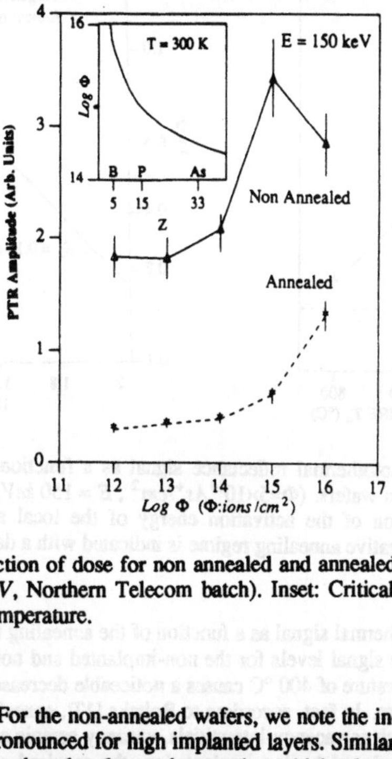

Figure 2 *PTR* signal as a function of dose for non annealed and annealed (1100 °C, 15 min) P^+ implanted samples ($E = 150\ keV$, Northern Telecom batch). Inset: Critical dose Φ_c as a function of atomic number Z at room temperature.

response of the annealed ones. For the non-annealed wafers, we note the increase of *PTR* signal as a function of dose which becomes more pronounced for high implanted layers. Similar behavior of the photothermal signal has been reported [11,14], as has the observed saturation which takes place for dose over 10^{15} $ions/cm^2$ [14]. The observed sensitivity to doping levels is lower than that obtained by the above authors; this has to do with our experimental choice of modulation frequency and laser beam spot sizes [11]. The increase of the *PTR* signal with P^+ dose is consistent with the higher degree of lattice damage caused by the implantation. In order to better understand the meaning of the observed results, we present the inset of Fig. 2, which shows the variation of the critical dose of amorphization, Φ_c, as a function of the atomic number, Z, of the ions implanted into silicon at room temperature. We note that Φ_c increases with the decreasing atomic number. According to Christel *et al.* [15], if the implanted ions displace 10% of the target atoms, the sample can be considered to be amorphous. When the layer is not completely amorphous, it can be considered to be an inhomogeneous layer composed of amorphous clusters within a crystalline matrix. Thus, the implanted layer's amorphization increases with the dose up to the critical dose Φ_c; additional implanted impurities will likely not cause any additional structural modification. This fact can explain the saturation of the *PTR* signal of the non-annealed wafers at high fluences of Fig.2, implying that for phosphorus doses of 10^{15} *ions/cm*^{2} or greater, the crystalline-to-inhomogeneous-to-amorphous transition has been completed. Also visible is a slight decrease of the *PTR* amplitude at very high dose of $10^{16}P^+/cm^2$. This implies two possibilities: i) the decrease is real, indicating some physical alteration from the dose 10^{15} to 10^{16} *ions/cm*^{2} ii) the decrease is an experimental artifact falling within the range of the experimental error, as indicated by the error bars on Fig.2; in this case we simply observe the signal saturation regime in the amorphous state, as described above.

The photothermal response of samples annealed at 1100 °C for 15 minutes is somewhat reminiscent of their non-annealed counterparts in that the signal magnitude increases with the ion dose, especially in the high-fluence regime. It is clear that this set of annealing conditions is quite effective in returning the layer to a relatively undamaged state for doses less than 10^{14} *ions/cm*^{2} but not for the higher ones. Ishikawa *et al.* [14], using a similar thermal wave technique, also report the inability of such short time, high temperature annealing cycle to annihilate the As^+ induced damage for doses higher than 10^{15} *ions/cm*^{2}. The large post-annealing *PTR* signal of the sample purported to undergo self-annealing during implantation is consistent with the finding that crystallinity restoration is more difficult to achieve in a heavily damaged but not amorphous material, than in a disordered and amorphous one [16].

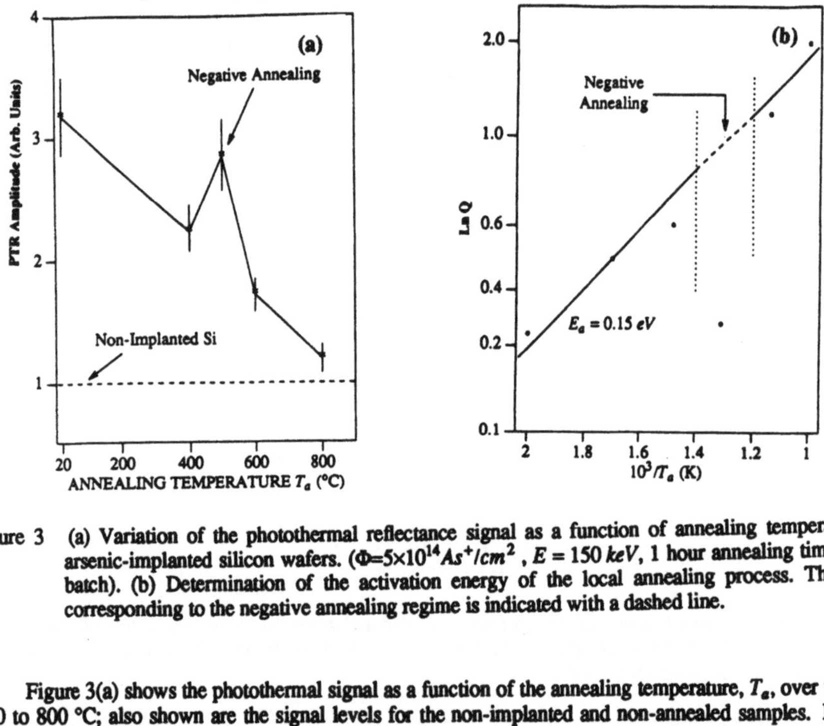

Figure 3 (a) Variation of the photothermal reflectance signal as a function of annealing temperature for arsenic-implanted silicon wafers. (Φ=$5\times10^{14}As^+/cm^2$, $E = 150$ *keV*, 1 hour annealing time, CIME batch). (b) Determination of the activation energy of the local annealing process. The region corresponding to the negative annealing regime is indicated with a dashed line.

Figure 3(a) shows the photothermal signal as a function of the annealing temperature, T_a, over the range 400 to 800 °C; also shown are the signal levels for the non-implanted and non-annealed samples. Note that annealing at a relatively low temperature of 400 °C causes a noticeable decrease of the *PTR* signal, implying a significant decrease in local disorder. In fact, according to Boltaks [17], annealing at 400 °C is able to annihilate many kinds of point defects such as vacancy-interstitials, vacancy-arsenic and divacancies. In Table 1 [18] we present the different defects induced by ion implantation, the required annealing temperature for their annihilation, and the corresponding activation energies. Also noticeable in the figure is the increase of the *PTR* signal vs T_a between 400 and 500 °C, which may be associated with the formation of arsenic complexes. This phenomenon is known as negative or retrograde annealing. Higher annealing temperatures (>550 °C) are required to dissociate these arsenic multivacancies. As evident from the figure, for high annealing temperatures (>600 °C), the signal approaches that of the non-implanted silicon, indicating a high degree of crystallinity restoration. In fact, according to Table 1, only line and loop dislocations could survive after an annealing at 800 °C. It seems from the similarity of the photothermal signal of the sample annealed at 800 °C and the non implanted sample that the dislocation-type defects do not significantly influence the photothermal signal. That is, either the dislocations are present predominantly in the bulk of the wafer, where the surface-sensitive photothermal technique is not affected by them, or, if they are present at the surface, their effect on the local optical, thermal and electronic parameters is not sufficient to influence the observed signal. This may be because the dislocations are too few or too small for the beam size averaged signal to "see" them. A *PTR* study as a function of modulation frequency could give a more definite answer as to whether the effects of the crystalline bulk are significant.

The kinetics of the annihilation of the damaged layer, as monitored by the *PTR* method, are consistent with a *local* annealing process in which the observed photothermal reflectance signal could be described by the relaxation-type relationship [18-20]:

$$\Delta R(T_a) = (\Delta R_i - \Delta R_\infty)\ exp(-t/\tau_a) + \Delta R_\infty \qquad (3)$$

where $\Delta R(T_a)$ is the photothermal signal from the annealed wafer at a given temperature, ΔR_i is the signal for the as implanted sample, ΔR_∞ is the signal corresponding to non implanted wafer (i.e. equivalent to a very high annealing temperature), t is the annealing duration, and τ_a is the relaxation time of the local annealing process. τ_a is energy activated:

$$\tau_a = \tau_{a_0}\ exp(E_a/k_B\,T_a) \qquad (4)$$

where E_a is the activation energy of the defect annealing process, and k_B is Boltzmann's constant. Rearranging Eq. (3), the reduced photothermal signal, Q, could be defined as:

$$Q = \frac{\Delta R_i - \Delta R_\infty}{\Delta R(T_a) - \Delta R_\infty} = exp(t/\tau_a) \qquad (5)$$

From Eqs. (4) and (5):

$$Ln\,(Q) = (t/\tau_{a_0})\exp(-E_a/k_B\,T_a) \qquad (6)$$

TABLE 1

Defects induced by ion implantation, required annealing temperatures for their annihilation, and the corresponding activation energies.

Type of damage	T_a (°C) t = 30 to 60 min	$E_a(eV)$
Interstitial vacancy	180	0.2-0.5
Arsenic vacancy	200	0.8
Divacancy	300	1
Layer completely "amorphous"	550-650	2-3
Layer incompletely "amorphous"	700	3-4
Line dislocations	800-1000	5-8
Loop dislocations	800-1000	5-8

Thus, plotting the double logarithm of Q versus $1/T_a$ should yield a straight line of slope $-E_a/k_B$. To construct this plot, we must ignore the experimental point in the negative annealing regime, because this "global amorphization" process is not governed by Eq. (3). Performing the appropriate calculations, we obtain the plot shown in Fig.3(b), where the region in the 450 - 550 °C range is indicated with a dashed line. The measurement of the slope yields $E_a \approx 0.15\ eV$. The small value of E_a is most likely attributable to a local migration of point defects (see Table 1), and differs from the usual value (2-3 eV) which is characteristic of a mechanism involving propagation of the recrystallization front from the substrate towards the surface. Thus, it appears that our technique is indeed sensitive to the local rearrangement of the defects such as the migration of the interstitials atoms.

Conclusions

In this paper, we have used photothermal reflectance experiments to study the influence of implantation of silicon by arsenic, phosphorus, and boron impurities (at different doses and energies). The crystallinity restoration effectiveness of short time, high temperature anneal (1100 °C, 15 min) has been examined; in addition, we have reported the influence of thermal annealing under different temperatures (400, 500, 600, and 800 °C for 1 hour) in order to study the annihilation of induced defects. Qualitative discussions relating the magnitude of the *PTR* signal to the local conditions of the layer (presence and types of defects, inhomogeneous / amorphous transition, etc) have also been performed.

The main results of our study can be summarized as follows: (1) The strong influence of implant dose on the photothermal signal which points out the degree of the induced disorder in the damaged layer. (2) The strong increase of the *PTR* signal around the critical dose and its subsequent saturation, which indicates the crystalline-to-inhomogeneous-to-amorphous transformation of the implanted layer. (3) The *PTR* sensitivity to residual damage after annealing and its ability to detect anomalous ("negative") annealing temperature regimes. (4) The large effect of the annealing temperature on the photothermal signal of the implanted layers. Annealing times longer than 15 minutes appear to be necessary to recrystallize heavily damaged layers, even at T_a = 1100 °C. The activation energy of the local annealing recovery mechanism in arsenic-damaged silicon has been found to be $E \approx 0.15\,eV$. It seems that the 100 kHz photothermal signal (induced by a laser beam of 30 μm spot size) is not sensitive to the presence of dislocation-type defects,

References

1 W.L. Smith, A. Rosencwaig, and D. Willenborg, Appl. Phys. Lett. **47**, 584 (1985).

2 D. Guidotti and H.M. van Driel, Appl. Phys. Lett. **47**, 1336 (1985).

3 U. Zammit, M. Marinelli, F. Scudierry, and S. Martellucci, Appl. Phys. Lett. **50**, 830 (1987).

4 P. Kireev, *Physics of Semiconductors* (Mir, Moscow, 1975), ch. 6.

5 I.W. Wylie and E.P.C. Lai, Chemtronics **Vol. 1**, 69 (1986).

6 J. Opsal, M.W. Taylor, W.L. Smith, and A. Rosencwaig, J. Appl. Phys. **61**, 240 (1987).

7 J. Opsal and A. Rosencwaig, Appl. Phys. Lett. **47**, 498 (1985).

8 D. Guidotti and H.M. van Driel, Appl. Phys. Lett. **49**, 301 (1986).

9 B.G. Streetman, *Solid State Electronic Devices, 2nd edition* (Prentice-Hall, New York, 1980), ch.4.

10 O. Doka, A. Miklos, and A. Lorintz, J. Appl. Phys. **63**, 2156 (1988).

11 I.A. Vitkin, C. Christofides, and A. Mandelis, J. Appl. Phys. (in the press).

12 J. Opsal, A. Rosencwaig, and D. Willenborg, Appl. Opt. **22**, 3169 (1983).

13 W.L. Smith, A. Rosencwaig, D.L. Willenborg, J. Opsal, and M.W. Taylor, Solid State Technol. **29**, 85 (1986).

14 K. Ishikawa, M. Yoshida, and M. Inoue, Jap. J. Appl. Phys. **26**, L1089 (1987).

15 L.A. Christel, J.F. Gibbons, and T.W. Sigmon, J. Appl. Phys. **52**, 7143 (1981).

16 J.F. Gibbons, Proc. IEEE **60**, 1062 (1972).

17 B. Boltaks, *Diffusion in Semiconductors* (Mir, Moscow, 1977).

18 C. Christofides, G. Ghibaudo, and H. Jaouen, J. Appl. Phys. **65**, 4840 (1989).

19 C. Christofides, G. Ghibaudo, and H. Jaouen, Rev. Phys. Appl. **22**, 407 (1987).

20 C.S. Fuller, *Semiconductors*, edited by N.B. Hannay (Reindhold, New York, 1959), ch. 5.

The Erosion-Corrosion of a Nitrocarburized Low Alloy Steel in a Simulated FBC Environment

P. M. Rogers, I. M. Hutchings and J. A. Little
University of Cambridge
Cambridge
UK

ABSTRACT

There is interest in the use of surface treatments to reduce the wastage caused by erosion-corrosion of in-bed heat exchanger tubing in fluidized bed combustors. The erosion-corrosion of a nitrocarburized low alloy steel was examined in a laboratory scale test rig based on a fluidized bed. It was found that the value and temperature of maximum wastage was reduced on nitrocarburizing the steel. Wastage of the compound layer of the nitrocarburized steel was found to be comparable or greater than that of the untreated steel, suggesting the improvements in wastage resistance observed were due to the diffusion layer. At low temperatures, the wastage of the nitrocarburized steel was greater than the untreated steel. It was proposed that this was due to the nitrocarburized steel having a greater initial oxidation rate.

INTRODUCTION

In fluidized bed combustion (FBC), the fuel is burned within a fluidized bed of inert solid particles which are supported by air blown into the bed from below. The bed becomes fluidized at a gas velocity when the drag force on the particles supports their weight. There is rapid heat and mass transfer within the fluidized bed and the bed behaviour is comparable to that of a fluid. Advantages of fluidized bed boilers in comparison to conventional pulverized coal boilers include lower emissions of oxides of nitrogen and sulphur, higher combustion efficiencies and the ability to burn lower grade fuels, such as coals with a high ash content, peat, or waste materials.

A major problem with FBC technology has been that the erosive action of the bed, combined with high temperature oxidation causes high wastage rates of the in-bed heat exchanger tubing. For industrial boilers, low alloy steels are preferred for the heat exchanger tubing because of economic considerations. Surface treatments or coatings may be cheaper than using a more expensive alloy for the whole tube, and may be worthwhile if an increase of lifetime of a few thousand hours can be obtained.

The surface treatment investigated in this work is nitrocarburizing, in which both nitrogen and carbon are diffused into the specimen surface to give improved wear, fatigue and corrosion resistance. Nitrocarburizing has the advantage over nitriding of shorter treatment times. Three commercial treatments are used : gaseous, salt-bath and plasma nitrocarburizing. [1] In the present work, plasma nitrocarburizing was investigated as a method of reducing the erosion-corrosion rate of a low alloy steel ; the wastage rates were determined in a laboratory-scale FB test rig.

MATERIALS AND METHODS

Specimen preparation

Specimens were prepared from a low-alloy nitriding steel (B.S.970: 722M24T, for composition see table 1), as machined cylinders 25mm in diameter and 80mm long. They were internally threaded at one end for attachment to the central shaft of the erosion-corrosion rig (see below) and there was a drilled centre at the other end. The specimens were ground cylindrically between dead centres.

Plasma nitrocarburizing was performed at 480°C for 10 hours in a nitrogen and methane enriched plasma (Klöckner Ionen, Leverkusen, West Germany), and specimens were cooled in the plasma furnace.

TABLE 1

Composition of BS970: 722M24T steel

Element	C	Mn	Cr	Mo
Nominal composition (wt%)	0.24	0.58	3.2	0.55

Microscopy and metallography

Scanning electron microscopy was performed with a Camscan S2 electron microscope. Optical microscopy was performed with an Olympus Model BHM microscope. Microhardness measurements were made with a Leitz model 60-366.002 microhardness tester. Nitrocarburized specimens were etched with 2% nital solution.

Erosion-corrosion test rig

A test rig has been designed and built in Cambridge and is based upon a commercial fluidized bed for heat treatment. (figure 1) Detailed descriptions of the rig and of the experiments carried out to evaluate its performance have been published [2],[3]. The cylindrical specimens are mounted on coaxial vertical shafts, so that they lie within the fluidized bed of abrasive. One pair of specimens rotates clockwise in the bed, and the other pair anticlockwise. This prevents gross rotation of the bed material and ensures that the impact velocity of the particles on the specimens is almost solely due to the motion of the rotors and not the motion of the particles in the very lightly fluidized bed.

The cylindrical geometry of the specimens was chosen so that they would experience a range of impact angles (from normal to glancing incidence) around the forward half circumference. The particle impact velocity on the cylindrical specimen also increases along its length in proportion to the distance from the axis of rotation. Thus the design allows a range of impact velocities and angles to be studied in each experiment.

The specimens are rotated at 250 rpm to give a maximum specimen rotor tip speed of about 2.5 ms^{-1} which is comparable to the particle impact velocities found in fluidized bed combustors. The abrasive was alumina grit, of 170 µm mean size, with a measured indentation hardness of approximately 2100 H_V.

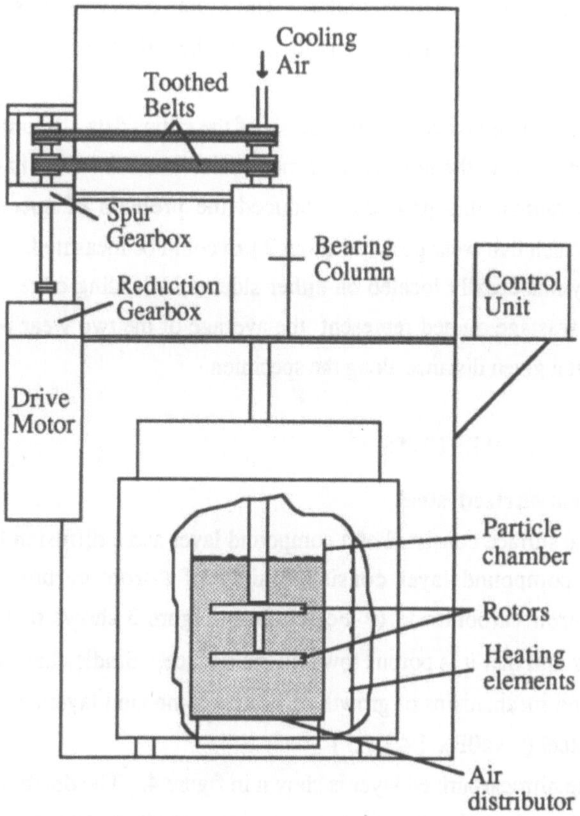

Figure 1. Erosion-corrosion test rig

Air is passed through the distributor plate at the bottom of the bed and the fluidization conditions are controlled using an air flow meter with a pressure gauge. The air flow rate giving the minimum fluidization velocity (U_{mf}) for the bed with a given abrasive was determined by measuring pressure drops using a simple liquid manometer, and the bed was then run at a fluidizing velocity of 1.2 U_{mf}

Experimental procedure

The tests each lasted for 72 hours. In order to counteract the effect of degration of the bed abrasive, conditioned alumina was used and approximately 0.75 kg of abrasive was replaced by new material every 24 hours, while slightly less than 0.75 kg was removed (the difference replacing material eluitriated from the bed) [3]. Test specimens were used only for the top two rotors on the test rig, while dummy specimens were used for the bottom two.

Specimen radii were measured before and after testing. They were rotated about their axes between the dead centres while the radius was measured with a linear variable displacement transducer system. This was interfaced to a microcomputer, which also controlled the specimen rotation through a stepper motor. The radius measurements were taken at 6mm intervals; 400 data points were taken per rotation, and these data were all stored on magnetic disc.

The specimen wastage was obtained by computer analysis of the radius data. In order to obtain an accurate value of wastage peak, the unworn specimen radius was subtracted in each case from the worn specimen radius; this procedure reduced the problem of specimen eccentricity. The accuracy was such that wear peaks of about 2 μm could be measured. Each specimen showed wear scars symmetrically located on either side of its leading edge when rotating in the bed. Values of wastage quoted represent the average of the two wear peaks (maximum reduction of radius) at a given distance along the specimen.

RESULTS

Microstructure of the nitrocarburized steel

After nitrocarburizing the surface consisted of a compound layer and a diffusion layer below this (figure 2). The compound layer consists mainly of ε-iron carbonitride (ε-$Fe_2(N,C)_{1-x}$) with some γ'-iron carbonitride (γ'-$Fe_4(N,C)$). Figure 3 shows that it is between 15 μm and 20 μm thick and that it is porous towards the surface. Studies have been reported of the microstructure and mechanisms of growth of similar compound layers on salt-bath nitrocarburized 722M24T steel (EN40B). [4], [5]

A hardness profile for the nitrocarburized layer is shown in figure 4. The depth of the diffusion layer below the surface is 320±10 μm measured by optical microscopy and

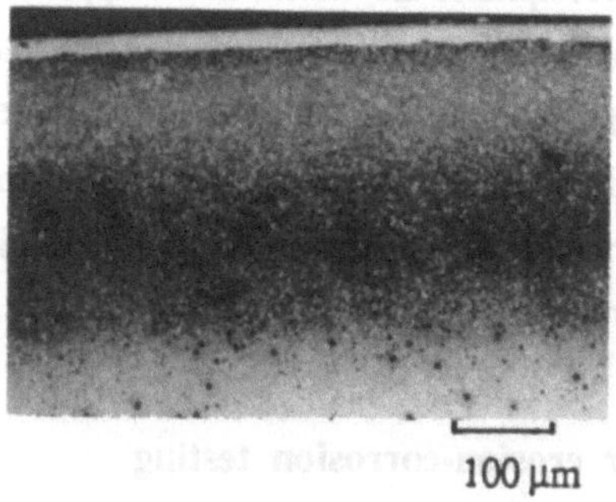

Figure 2. Optical micrograph of nitrocarburized 722M24T steel.

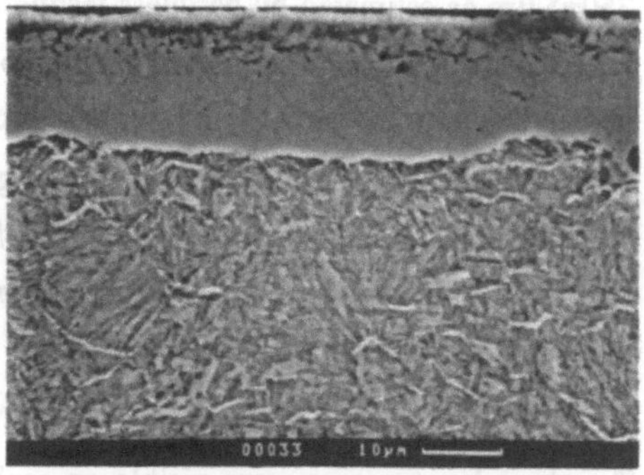

Figure 3. SEM backscattered image of the compound layer and the diffusion layer immediately below it on nitrocarburized 722M24T steel.

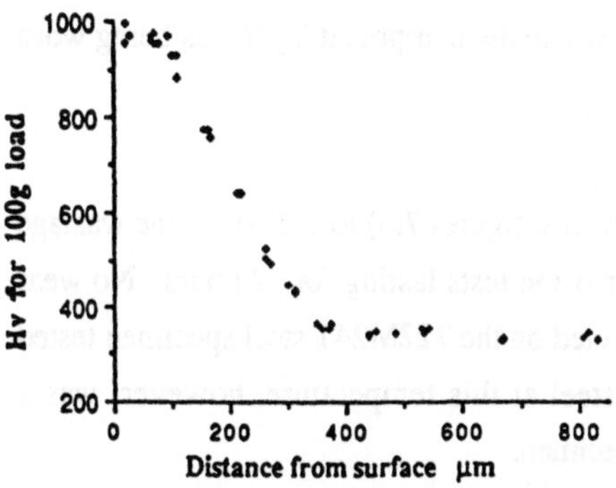

Figure 4. Hardness profile of nitrocarburized 722M24T steel.

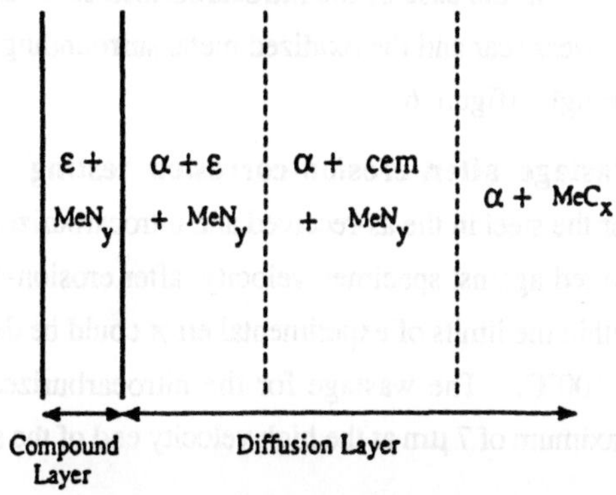

Figure 5. Schematic diagram of the sequence of phases formed on an alloy steel (after Slycke and Sproge, [1]).

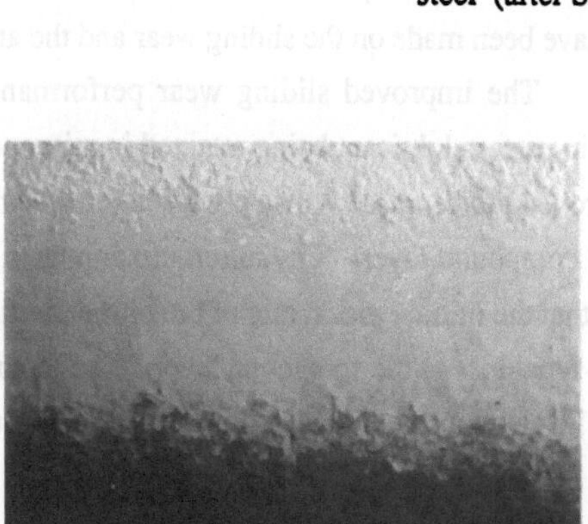

Figure 6. SEM backscattered image of the surface of nitrocarburized 722M24T steel after erosion-corrosion at 300°C.

350±5 μm as estimated by hardness testing. Precipitates are visible in the upper diffusion zone (figure 3) which have grown parallel to the surface (since the material is free to expand perpendicular to the surface [1]). These precipitates consist of ε-phase and cementite, while precipitates of cementite form around prior grain boundaries lower in the diffusion zone. A schematic diagram of the sequence of phases formed is shown in figure 5. [1] The diffusion zone matrix probably consists of bainite and martensite, while the bulk of both the steel and the nitrocarburized steel consists of bainite.

Microstructure of specimen surface after erosion-corrosion testing

Two broad wear scars were visible on each of the specimens which were approximately symmetrical on either side of the specimen front. Maximum wear for each scar usually occured at an angle between 60° and 70° away from the specimen front (however for some of the lowest temperature specimens it was between 80° and 90° away).

In the case of the nitrocarburized specimen there was a thin boundary region between the wear scar and the oxidized metal surrounding it where the compound layer was being worn through. (figure 6)

Wastage after erosion-corrosion testing

For the steel in the as-received and nitrocarburized forms, figures 7(a) to 7(d) show the wastage plotted against specimen velocity after erosion-corrosion tests lasting for 72 hours. No wear within the limits of experimental error could be detected on the 722M24T steel specimen tested at 100°C. The wastage for the nitrocarburized steel at this temperature, however, was a maximum of 7 μm at the high velocity end of the specimen.

DISCUSSION

Little work has been reported on the erosion performance or the oxidation of nitrocarburized steels. A number of studies have been made on the sliding wear and the atmospheric corrosion performance. [6] - [10] The improved sliding wear performance under lubricated conditions is likely to be due in part to lubricant being retained in microporosity at the surface of the compound layer. Improved performance is also observed for dry sliding wear and again it can be ascribed to the hard compound layer. Cavitation-erosion tests on a nitrocarburized low alloy steel [11] showed that the initial erosion rate of the nitrocarburized steel was greater than the untreated steel due to rapid loss of the compound layer. The removal of the compound layer revealed the diffusion layer which then showed a lower erosion rate than that of the untreated steel.

For the tests at 500°C and 400°C, figures 7(a) and 7(b) show that at low velocities the rate of wastage of the compound layer is approximately the same as the that of the steel. At higher velocities the wastage of the untreated steel is greater than that of the surface treated

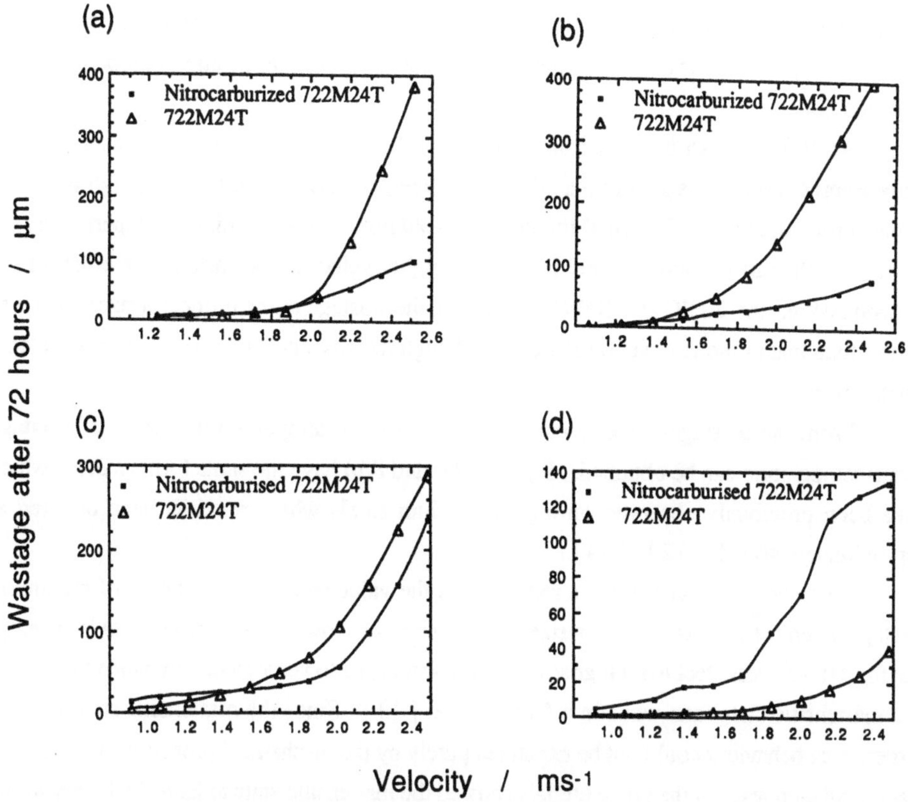

Figure 7. (a) Wastage-velocity curves after erosion-corrosion at 500°C
(b) Wastage-velocity curves after erosion-corrosion at 400°C
(c) Wastage-velocity curves after erosion-corrosion at 300°C
(d) Wastage-velocity curves after erosion-corrosion at 200°C

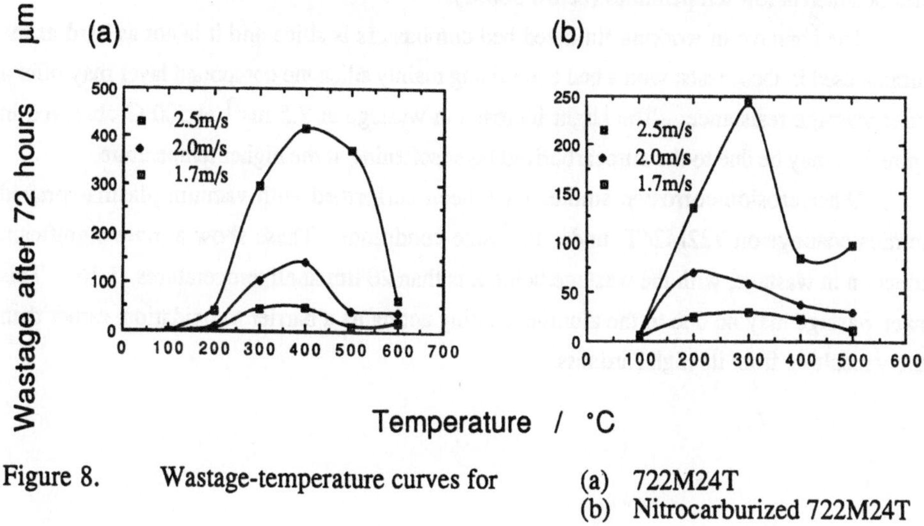

Figure 8. Wastage-temperature curves for (a) 722M24T
(b) Nitrocarburized 722M24T

steel. Since the change in the wastage between the steel and the nitrocarburized steel occurs after the compound layer is worn through, the reduction is likely to be due to properties of the diffusion zone.

At 300°C, it can be seen from figure 7(c) that at low velocities the wastage of the nitrocarburized surface is greater than that of the steel, and can be ascribed to a high wear rate of the compound layer, of original thickness 15 - 20 μm. As at the higher temperatures, the wastage of the nitrocarburized steel is lower at higher velocities because of the effect of the diffusion zone. At 200°C and 100°C, in contrast, the wastage of the nitrocarburized steel was greater than that of the steel at all velocities, although the absolute values of wastage were now much lower.

From the wastage-velocity curves at the different temperatures, curves of wastage versus temperature can be obtained. Figures 8(a) and 8(b) have the same form as those which have been previously observed on high chromium steels where maxima have occured at approximately 400°C. [12] - [14]

It can be seen that there is a reduction in the value and the temperature of maximum wastage when the steel is nitrocarburized. Studies have shown that the maximum in wastage rate for 410 stainless steel has a higher value and occurs at a lower temperature when the steel is in a martensitic state than when in a ferritic state. [13] Since the martensitic state was the harder, such behaviour could not be explained purely by the mechanical properties of the steels. Static oxidation tests on the two materials showed the martensitic state to have the higher initial oxidation rate [15] and it was suggested that the initial oxide growth rate is the more important factor affecting the wastage [13]. It is possible that the initial oxidation rate of the nitrocarburized steel is greater than that of the untreated steel, and that this has caused a similar reduction in the temperature of maximum wastage, and is responsible for the higher wastage rates obtained at low temperatures (below 300°C).

The abrasive in working fluidized bed combustors is silica and it is not as hard as the alumina used in these tests; with a bed comprising mainly silica the compound layer may offer a better wastage resistance. The slight increase in wastage at 2.5 ms^{-1} at 500°C observed in figure 8(b) may be due to the nitrocarburized layer softening at the higher temperature.

Other erosion-corrosion studies have been performed with vacuum plasma-sprayed alumina coatings on 722M24T under the same conditions. These show a more significant reduction in wastage, with the wastage being less than 20 μm at all temperatures. [16] This lower wastage may be due to the alumina coating acting as a barrier to oxidation, rather than purely resulting from its high hardness.

CONCLUSIONS

1. Nitrocarburizing of 722M24T steel causes a reduction in the maximum wastage under simulated FBC conditions.
2. The temperature of maximum wastage is reduced when 722M24T is nitrocarburized which may result from the nitrocarburized layer showing a greater initial oxidation rate than the untreated steel.
3. At high temperatures the compound layer of the nitrocarburized steel wore away at approximately the same rate as the untreated steel. The improved erosion-corrosion performance of the nitrocarburized steel is therefore likely to be due to the properties of the diffusion layer rather than the compound layer.
4. Although a reduction in wastage occurs when nitrocarburizing 722M24T other studies suggest that there may be better surface treatments for this application. Suitable surface characteristics for erosion-corrosion resistance under these conditions may involve a low initial oxidation rate as well as good mechanical properties.

ACKNOWLEDGEMENTS

P.M. Rogers wishes to thank British Coal and the Science and Engineering Research Council for their support via a CASE studentship. The authors are also grateful to Dr. P. A. Dearnley and the management of Klöckner Ionon, Leverkusen (in particular to Dr. Remges) for the provision of specimens.

REFERENCES

1. Slyke, J. and Sproge, L., Kinetics of gaseous nitrocarburising process. Surf. Eng., 1989, **5**, 125-40

2. Little, J.A., Hutchings, I.M. and Entwisle, M.J., Erosion-corrosion testing of fluidised bed combustor materials : A new experimental method, Proc. Int. Conf. Corrosion-Erosion-Wear of Materials at elevated temperatures, Berkeley, California, 1986, ed A.V. Levy, NACE

3. Ninham, A.J., Entwisle, M.J., Hutchings, I.M. and Little, J.A., A laboratory-scale fluidized bed rig for high-temperature tube wastage studies, Proc. 10th International FBC Conference, San Francisco, 1989, ASME

4. Mittemeijer, E.J., Rozendaal, H.C.F., Colijn, P.F., van der Schaaf, P.J., and Furnée, R.Th., Microstructure of nitrocarburized steels. In Heat Treatment '81, The Metals Society, London, 1983, pp.107-15

5. Rozendaal, H.C.F., Colijn, P.F and Mittemeijer, E.J, Morphology, composition and residual stresses of compound layers of nitrocarburized iron and steels. Surf. Eng., 1985, **1**, 30-42

6. Bell, T., Ferritic Nitrocarburising. Heat. Treat. Met., 1975.2, 39-49

7. Bell, T., Ferritic thermochemical treatments. In Heat Treatment Shanghai '83, ed. T. Bell, The Metals Society, London, 1984, pp.1.1 - 1.10

8. Dawes, C., Tranter, D.F., and Smith, C.G., Reappraisal of nitrocarburizing and nitriding when applied to design and manufacture of non-alloy steel automobile components. Met. Tech., 1979, **6**, 345-53

9. Wells, A., and Shaw, M.P., An investigation into dry sliding wear behaviour of the ferritic nitrocarburized surface of a plain low carbon steel. Wear, 1985, **103**, 29-42

10. Gleave, C., and Farrow, M., Tribological investigation of compound layers formed by ferritic thermochemical diffusion treatments. In Heat Treatment '81, The Metals Society, London, 1983, pp.123-29

11. Heathcock, C.J., Ball, A., Yamey, D., and Protheroe, B.E., Cavitation erosion of cobalt based stellite alloys, cemented carbides and surface treated low alloy steels. Proc. Int. Conf. Wear of Materials, San Francisco, 1981, ed. K.C. Ludema, ASME, pp. 597-606

12. Ninham, A.J., Hutchings, I.M., and Little, J.A., Erosion-corrosion of austenitic and ferritic alloys in a fluidised bed environment. Proc. Int. Conf. NACE Corrosion '89, New Orleans, 1989

13. Hutchings, I.M., Little, J.A., and Ninham, A.J., Low velocity erosion-corrosion of steels in a fluidized bed. Proc. Int. Conf. Corrosion-Erosion-Wear of Materials at Elevated Temperatures, Berkeley, California, 1990, ed. A.V. Levy, NACE (in press)

14. Stott, F.H., Stack, M.M., and Wood, G.C., The role of oxides in the erosion-corrosion of alloys under low velocity conditions. Proc. Int. Conf. Corrosion-Erosion-Wear of Materials at Elevated Temperatures, Berkeley, California, 1990, ed. A.V. Levy, NACE (in press)

15. Entwisle, M.J., Erosion-corrosion of iron-chromium alloys in a fluidized bed environment, PhD Thesis, University of Cambridge, 1990

16 Rogers, P.M., Shipway, P.S., Hutchings, I.M., and Little, J.A., The erosion-corrosion of sprayed alumina coatings on a low chromium steel. Proc. Int. Conf. Corrosion-Erosion-Wear of Materials at Elevated Temperatures, Berkeley, California, 1990, ed. A.V. Levy, NACE (in press)

Comparison of Fractal and Conventional Topographic Analysis of Electric Discharge Machined Ceramic Surfaces

C. A. Brown

Worcester Polytechnic Institute
Worcester Massachusetts USA

D. F. Dauw

Charmilles Technologies SA
Meyrin Switzerland

and G. Savary

Swiss Federal Institute of Technology
Lausanne Switzerland

ABSTRACT

Eleven materials, including eight ceramics, were electric discharge machined using two different conditions. The surface topographies were examined qualitatively, and profiles were recorded. Four conventional roughness parameters were calculated, in addition to two parameters derived from fractal plots: slope and threshold. These parameters are analyzed for covariance, coefficient of variation and their abilities to differentiate materials and machining conditions. The conventional roughness parameters all correlate reasonably well with each other, as do Rz and Rpm with the slope of the fractal plot, indicating that they contain superfluous information. None of the parameters examined are able to differentiate the two machining conditions. All the parameters, with the exception of the Rpm:Rz ratio, are able to differentiate the materials, significant at greater than 97.5% confidence. The slope of the fractal plot appears to be an indication of complexity of the topographies, and of the amount of TiC or TiN phase in some ceramic-ceramic composites.

INTRODUCTION

A central problem in surface engineering is to correlate surface formation phenomena with the resulting topography, and then to correlate the topography with surface behavior. Because residual stresses, and superficial gradients in composition and microstructure are factors which also influence surface behavior, and because it is experimentally difficult to adjust the surface topography independently of these other factors, the resolution of the second part of this problem is difficult. Nevertheless, the resolution of both parts depends to a large degree on finding relevant parameters that characterize the topography. One objective of this work is the comparison of parameters produced by conventional and fractal analysis. This comparison is based on their sensitivity to topographies resulting from the interaction of electric discharge machining with, primarily, ceramic workpieces.

Electric discharge machining (EDM), also called spark erosion, is a thermo-electrical material removal process often applied in the tool and die making business. Sparks are generated between a workpiece and a tool electrode, through a liquid dielectric. This produces strong local heating, and material removal is achieved principally by melting, and evaporation in metals, although thermal spalling [1] and falling out of solid particles [2] have also been proposed as material removal mechanisms in ceramics.

In cases where conventional machining processes are ineffective because the workpieces are too hard or brittle, EDM is often a promising candidate. Two techniques are currently available: die sinking and wire cutting. EDM has been shown to be effective in machining ceramics provided their conductivity is greater than about 0.01 S/cm [2]. Currently, engineering ceramics are commercially available which feature sufficient electrical conductivity as well as good mechanical properties. Machining speeds, when EDM is applied to these ceramics, can be better than those obtained with conventional machining. The extension of EDM to ceramics provides designers with interesting possibilities, although the relationships between machining parameters and surface roughness, that exist for metals, need to be developed for ceramics.

Fractal analysis may provide some insight into the relation between surface creation processes, workpiece material, and surface topography. It has been proposed [3,4] that fractal geometry provides a means for analyzing the irregularity of engineering surface topographies over a wide range of scale. An object is said to be fractal, or to exhibit fractal behavior over a range of scales of observation, if, over that range, the irregularity increases as an exponential function of the decrease in scale. In that range such objects would demonstrate self-similarity or self-affinity with scale; that is, after magnification (different amounts in the horizontal and vertical directions in the case of self-affine fractals) the object looks essentially the same as it did before.

The insensitivity of conventional surface roughness analysis to fine scale flaws arising from machining has been recognized for some time [5,6]. One manifestation of this problem is that, despite the existence and wide spread use of surface roughness measuring methods for over forty years [7], there is a paucity of studies providing correlations with behavior or surface creation mechanisms [8]. The sensitivity of parameters derived from fractal analysis to fine scale surface features [4] can be particularly important when analyzing surfaces of parts with small dimensions or where fine scale surface roughness may have a first order effect, e.g., surfaces of data storage media, and surface preparation for adhesion.

A second objective of the present work is to advance our understanding of electric discharge (ED) machined surfaces on ceramic materials, in particular, in regard to the fine scale surface features resulting from the interaction of the ED process with the various ceramic constituents. Profiles of surface topographies, created using two different EDM conditions, are analyzed by conventional and by fractal methods. The surfaces are also studied using scanning electron microscopy (SEM) in an attempt to provide a more complete understanding of the surface features.

METHODS

The machining was performed by die sinking on a Charmilles Technologies FORM 20. The machining conditions can be characterized by features of the time histories of the pulse electric potential and current which power the spark. Typical histories are shown in Fig. 1.

Two sets of conditions were selected, and are listed in Table 1. Both are typical of roughing conditions for metals. Regime A uses less discharge current, which results in a lower discharge power. Regime A also has a longer discharge duration, which results in greater total discharge energy. Conversely, regime B has, in addition to a greater intensity and shorter duration, a shorter pulse interval, or off, time.

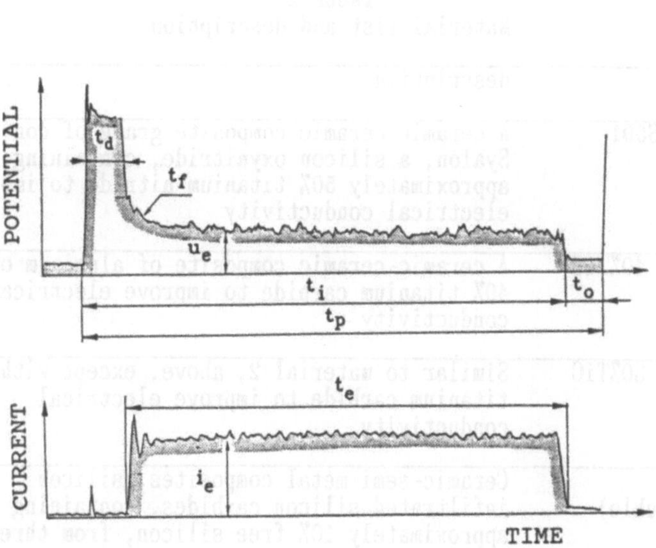

Figure 1. Example of a typical pulse occurring during EDM: electric potential (u) versus time above, and current (i) versus time below (td: ignition delay time, tf: voltage fall time, ti: pulse duration, tp: pulse cycle time).

TABLE 1
EDM Working Conditions

regime	ie(A)	ue(V)	te(μs)	to(μs)	Pe(W)	We(mJ)
A	8	28	100	400	224	22
B	12	28	50	200	336	17

ie: discharge current; ue: discharge voltage; te: discharge duration; to: pulse interval time; Pe: discharge power; We: discharge energy

Eight electric discharge machinable ceramics were selected (Table 2) along with three better known materials for EDM: steel (K107), copper, and graphite. The ceramics can be divided into three categories: 1) ceramic-ceramic composites with a large difference in electric conductivity in the components; 2) ceramic-semi-metal composites and 3) monolith ceramics.

Profiles from the ED machined surfaces were analyzed using conventional as well as fractal analysis. Six traces 0.56 mm long were made at random orientations and locations. A Perthometer S6P profilometer (Feinpruf GmbH, Gottingen, FRG) with an RFHTB-50 stylus (5 um radius spherical tip) was used. Tracing lengths shorter than normal

TABLE 2
Material list and description

material	description
1. Syalon S501	A ceramic-ceramic composite grade of conventional Syalon, a silicon oxynitride, containing approximately 50% titanium nitride to improve electrical conductivity
2. Al203 + 40%TiC	A ceramic-ceramic composite of aluminum oxide with 40% titanium carbide to improve electrical conductivity
3. Al203 + 30%TiC	Similar to material 2, above, except with 30% titanium carbide to improve electrical conductivity
4. SiSiC (Feldmuhle) 5. SiSiC (Charmilles FRG) 6. SiSiC (DIDIER)	Ceramic-semi-metal composites, silicon infiltrated silicon carbides, containing approximately 10% free silicon, from three different suppliers, as indicated
7. TiB2	Titanium diboride, a monolith ceramic
8. B4C	Boron carbide, a monolith ceramic

were used in order to improve the sensitivity to fine scale detail by increasing the horizontal resolution, to the maximum available with the equipment: 88 nm. Horizontal resolution has been found to be an important parameter in the fractal analysis [9]. These short tracing lengths should have an effect on the conventional roughness parameters similar to that of a shorter critical wave length, i.e., slight reductions in the calculated roughness values.

The following conventional parameters [10] were chosen for analysis: the arithmetic average roughness, Ra, because of its wide spread use; average peak height or leveling depth, Rpm; average peak-to-valley roughness, Rz; and the Rpm:Rz ratio, because it can be used to characterize the fullness of the profile [11]. Mean values from six traces were calculated.

The fractal analysis used here is a type of compass method [4,9,12,13], applied to approximately 34,000 profile points in total from the six traces. This analysis is based on a log-log plot of the apparent profile lengths versus the scales of observation, i.e., the lengths of the rulers, or separation of the compass, used for measuring the apparent profile lengths. Shorter ruler lengths tend to make the profile appear longer, as they conform to more of the irregularities. These plots have three regions: 1) at relatively long ruler lengths the profile length is approximately equal to its projected length, and the plot tends to be parallel to the ruler-length axis; 2) at intermediate ruler lengths the log of the apparent profile length increases approximately linearly with the log of the ruler length; 3) at shorter ruler lengths, as the horizontal resolution is approached, the slope decreases. Two parameters, based on the composite plots from the six measured profiles, are reported: the slope of the log-log plot in region 2, which may be related to the fractal dimension [14]; and the ruler length that corresponds to the transition between regions 1 and 2. This transition ruler length will be referred to as the roughness threshold, i.e., the scale of observation, or ruler length, below which the surface appears rough, above which it is smooth.

In this study, the slope of region 2, where the surface might be said to demonstrate fractal behavior, is the slope of a best fit line to the plotted points, over the decade of ruler lengths in which a linear regression analysis gives the greatest r value. In this study the r value was always 0.98 or greater. The roughness threshold is taken to be the ruler length at which the relative length the first exceeds 10% of its maximum calculated value.

RESULTS

A profile based on data taken from one of the six traces on the surface on the Al_2O_3-30%TiC composite, ED machined according to the conditions for regime A, is shown in Fig. 2. Note that in the figure the magnification is greater in the vertical direction than in the horizontal. This difference in magnifications improves aesthetics in the figure, but is not used in the analysis.

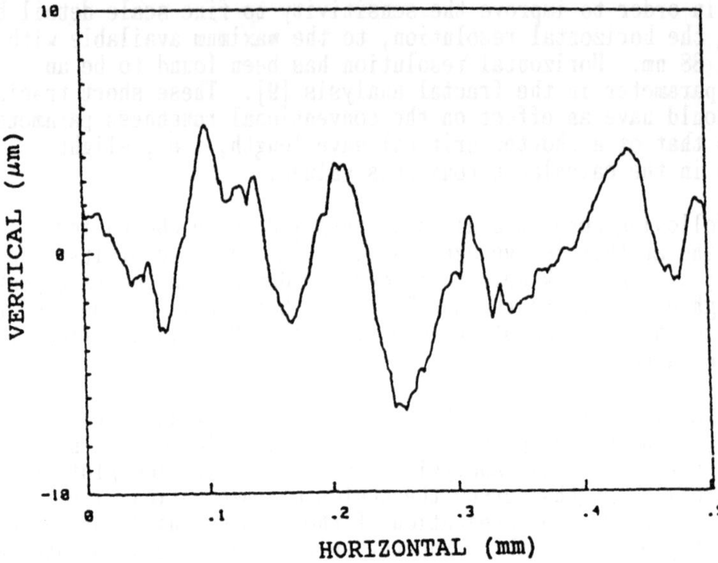

Figure 2. Profil from the surface of Al203+30%TiC, ED machined under conditions in regime A.

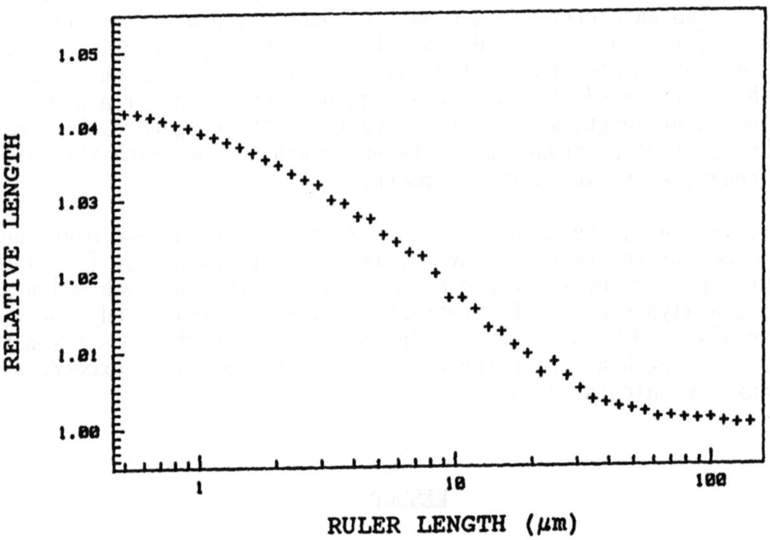

Fig. 3 Compass method fractal plot based on data taken from profils from the surface of Al203+30%TiC, ED machined under the conditions in regime A.

The compass method fractal analyses were based on profile data with identical horizontal and vertical magnifications. A compass method fractal plot from the Al2O3-30%TiC composite, ED machined according to the conditions for regime A, is shown in Fig. 3. The vertical axis is the relative length, i.e., the apparent profile length divided by its projected length. Therefore, at long ruler lengths the relative length tends towards one. The slope of this plot is -0.0116, based on the portion of the plot between ruler lengths of 2.31 and 24.38μm. The roughness threshold is 30.85μm. A SEM micrograph of this surface is shown in Fig. 4a.

TABLE 3
Material, EDM regime, and roughness parameters

Material	Regime	Ra	Rz	Rpm	Rpm/Rz	threshold	slope
		mean*	mean*	mean*		10%*	-%
Syalon 501	A	1.67	7.62	4.40	0.577	34.71	2.62
	B	1.62	7.23	3.18	0.440	39.05	2.25
Al203+40% TiC	A	1.71	7.81	3.48	0.446	43.93	2.11
	B	1.26	5.42	2.48	0.458	43.93	1.18
Al203+30% TiC	A	1.11	5.05	2.62	0.519	30.85	1.16
	B	1.34	5.33	2.51	0.471	49.42	1.09
SiSiC Feld.	A	1.48	6.44	3.17	0.492	49.42	1.52
	B	1.58	6.76	3.15	0.466	49.42	1.40
SiSiC Charm.	A	1.46	6.09	2.78	0.456	55.60	1.44
	B	1.54	6.04	3.01	0.498	43.49	1.46
SiSiC Didier	A	1.47	6.60	3.04	0.461	39.05	1.69
	B	1.43	5.91	2.83	0.479	39.05	1.73
TiB2	A	2.07	8.59	4.00	0.466	79.16	2.35
	B	1.79	8.45	3.57	0.422	34.71	3.05
B4C	A	2.11	6.85	2.99	0.436	142.66	1.36
	B	1.72	6.36	3.17	0.498	89.06	1.18
Steel	A	2.33	8.39	3.94	0.470	89.06	2.09
	B	2.39	9.00	3.88	0.431	79.16	2.03
Copper	A	0.84	4.06	1.88	0.463	30.85	0.74
	B	1.05	4.68	2.38	0.509	34.71	0.96
Graphite	A	1.95	8.13	3.60	0.443	62.55	2.21
	B	1.47	5.99	2.94	0.491	49.42	1.58

* units: μm

Conventional and fractal parameters describing the topography of the surfaces for the 11 materials and the 2 machining conditions are listed in Table 3. The correlation coefficients between the several roughness parameters are presented in a fashion similar to that of Nowicki [15] in Table 4. Parameters which are highly correlated, e.g., Ra and Rz, or Rz and Rpm, give essentially the same, or redundant, information about the surface, whereas, parameters that do not correlate, e.g., the roughness threshold (Th) and the slope (Sl), give complementary information. From the linear regression analysis the relationship between the two couples of parameters that have correlation coefficients above 0.9 can be expressed as (units are μm):

$$Rpm = 2.06\ Rz + 0.19 \qquad [1]$$
$$Ra = 3.13\ Rz + 1.63 \qquad [2]$$

The results of F tests for significance [16] of the ability of each of the several parameters to differentiate the two machining conditions, regardless of the material, and to differentiate the eleven materials, regardless of the machining conditions, are shown in Table 5. None of the parameters can differentiate the two machining conditions. But, with the exception of the Rpm:Rz ratio, all the parameters can differentiate the materials with better than 97.5% confidence.

TABLE 4
Covariance, correlation coefficients for all eleven materials and both machining conditions

	Rz	Rpm	Rpm/Rz	Th	Sl
Ra	0.906	0.813	-0.358	0.699	0.625
Rz		0.918	-0.343	0.381	0.861
Rpm			0.054	0.287	0.841
Rpm/Rz				-0.302	-0.176
Th					-0.0002

The inability of the Rpm:Rz ratio to differentiate the materials chosen in this study, and Rpm:Rz ratio's low correlation with the other parameters (Table 4) appears to consistent with a relatively low coefficient of variation (Table 6). This indicates that, in comparison to the other parameters, the Rpm:Rz ratio varies little in this investigation.

It is not attempted to quantitatively evaluate based the SEM micrographs used in this study. Qualitatively, it appears that, for both machining conditions, the increased concentration of the conductive phases in the ceramic-ceramic composites, i.e., 30% and 40% TiC, in the Al2O3-TiC, and 50% TiN in the Syalon, corresponds to a finer topographic structure observed on the micrographs shown in Fig. 4.

The increase in the fineness of the topographic structure with TiN or TiC content also corresponds to a steeper negative slope on the fractal plots and to a higher average peak heights (Table 3). Furthermore, for these three ceramic-ceramic composites, the machining conditions in regime A, with the longer discharge duration, also correspond to steeper negative slopes and to higher peak heights.

TABLE 5
F tests, mean square ratios

machining conditions		materials	
Ra:	0.291	Ra:	7.960
Rz:	0.480	Rz:	5.909
Rpm:	0.982	Rpm:	3.680
Rpm/Rz:	0.157	Rpm/Rz:	0.474
Th:	0.716	Th:	4.707
Sl:	0.246	Sl:	6.769
(F 0.01 = 2.97)		(F 0.10 = 2.25) (F 0.025 = 3.53)	

TABLE 6
Coefficients of variation, ratios of standard deviations to means

	all materials	ceramics only
Ra	0.238	0.162
Rz	0.199	0.154
Rpm	0.188	0.159
Rpm/Rz	0.072	0.077
Th	0.473	0.509
Sl	0.337	0.328

DISCUSSION

The failure of the examined roughness parameters to differentiate the machining conditions (Table 5) stems, either from their inability to characterize the essential topographic features, or from a restriction of range, i.e., because there is no significant difference between the topographies resulting from the two machining conditions.

The high degree of correlation between Ra and Rz, agrees with the result of Nowicki [15] and suggests that it is not necessary to report both values. Nowicki did not report Rpm values, where we also find a good correlation with Rz. Similar strong correlations between Rpm and Rz should not, we expect, hold for a wide variety of engineering surfaces. Certainly, the relationships noted between the parameters

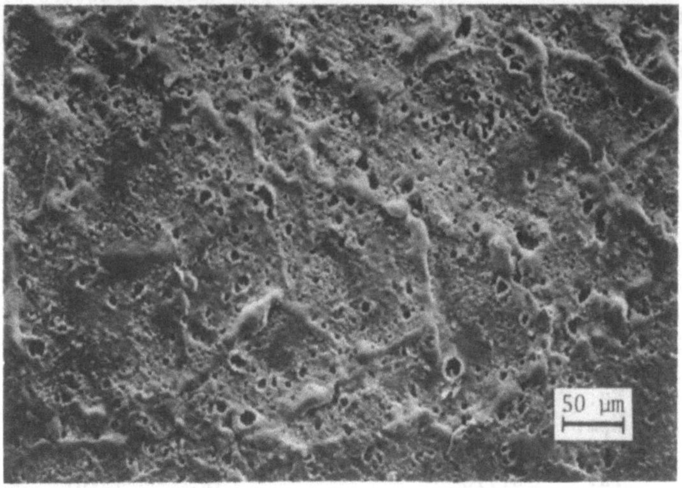

Figure 4a. SEM micrograph of Al203+30%TiC machined under the conditions in regime A.

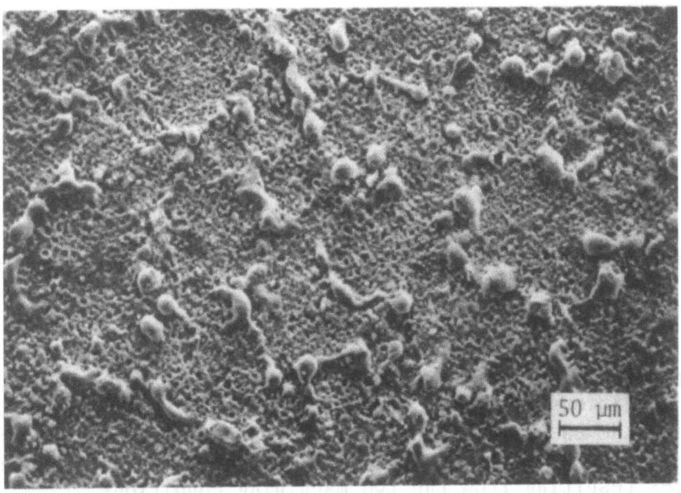

Figure 4b. SEM micrograph of Al203+40%TiC machined under the conditions in regime A.

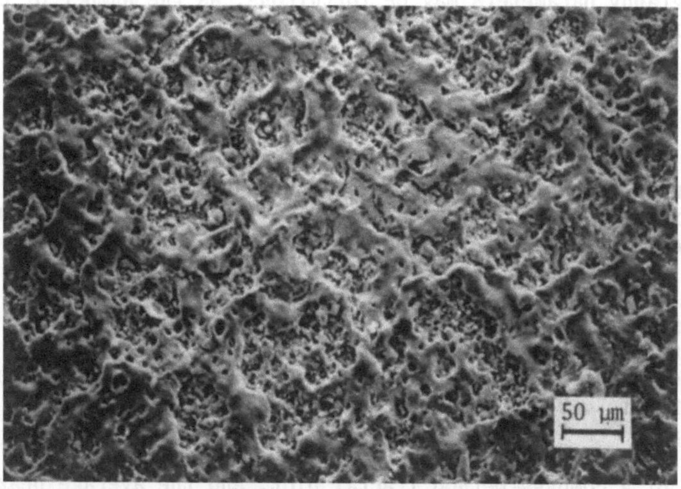

Figure 4c. SEM micrograph of Syalon S501 (approximately 50% TiN) machined under the conditions in regime A.

examined here, e.g., equations 1 and 2, and the relatively low coefficient of variance of the Rpm:Rz ratio, are, to some extent, dependent on the special sorts of surfaces examined in this study. Peters et al. [11], in a study involving several machining processes, noted larger ranges of values of similar, conventional parameters, particularly the Rpm:Rz ratio. This ratio was found to be indicative of the fullness of the surface or skew of the profile. It is supposed that the ED machined surfaces examined here could all have approximately the same degree of fullness.

The apparent ability of the slope of the fractal plots to characterize the relative fineness of surface topography in the two Al2O3-TiC composites and the Syalon S501 is consistent with the claims that the fractal analysis is sensitive to fine scale topographic detail. It is, however, not unique in this ability, as the average peak height, Rpm, also consistently ranks the three composites based on percentage of TiC or TiN. The response of these composites to EDM, that causes this apparent correlation between the amount of one phase and the surface topography, could be a preferential erosion of that phase, similar to that observed in EDM of cermets. ED machined surface roughness on WC-Co has been observed to increase with increasing cobalt content [17]. In the same study, increasing WC grain size with constant cobalt content was also noted to increase roughness. Grain sizes were not monitored in the present study, and there are probably other factors contributing to the development of the surface topography of ED machined ceramics, such as those indicated by features on the SEM micrographs that suggest the formation of glassy phases during EDM. Clearly, the relations found here between the concentrations of TiC or TiN, surface appearances, and surface topography parameters may be fortuitous, as the number of cases examined is rather small.

CONCLUDING REMARKS

This preliminary comparison of fractal and conventional surface roughness parameters, and investigation in to the nature of ED machined surfaces on ceramics, suggests three insights that could be the attention of further research:

1. The roughness threshold and slope, derived from compass-method fractal analysis, appear to provide potentially interesting information about surface topography, and this information is largely independent of that provided by conventional surface roughness parameters.

2. The slope of a compass-method fractal plot, and the average peak height, Rpm, appear to be indicative of the extent of preferential erosion of one phase during EDM of some types of ceramic-ceramic composites.

3. Preferential erosion of one phase could provide part of the support for developing a model for predicting ED machined surface topography based on the interaction of a ceramic-ceramic composite with the spark erosion process.

ACKNOWLEDGEMENTS

The authors would like to thank Prof. B. Ilschner, head of the Mechanical Metallurgy Laboratory at the Swiss Federal Institute of Technology in Lausanne, and Dr. J. Gauger former head of the Materials Engineering Department at CERAC in Ecublens, Switzerland for their support of this project, Prof. J. Shealy, of the Rochester Institute of Technology, for his valuable suggestions on statistical analysis, Mr. W. Barozzi and Dr. G. Heath, both formerly of CERAC for work with the SEM micrographs, and Mrs. P. Scavone and Mr. G. Schmidt, Jr. both of WPI for assistance in preparing the manuscript.

REFERENCES

1. N.F. Petrofes, A.M. Gadalla, Electrical discharge machining of advanced ceramics, Ceramic Bulletin, 1988, 67/6

2. W. Konig, D.F. Dauw, G. Levy, U. Penten, EDM-Future steps towards the machining of ceramics, Annals of the CIRP, 1988, 37/2, 1-9

3. Dubuc, B., Quiniou, J.F., Rocques-Carmes, C., Tricot, C., Zrucker, S.W., Evaluating the fractal dimension of profiles, Physical Review A, 1989, 39/3, 1500-1512

4. C.A. Brown, G. Savary, Fractal aspects of machined surface topography determined by stylus profilometry, in Fractal Aspects of Materials: Disordered Systems, D.A. Weitz, L.M. Sander, eds., B.B. Mandelbrot, Materials Research Society, Philadelphia, 1988, pp. 275-278

5. Bailey, J.A., On surface damage during machining of AISI 4340 steel, Wear, 1974, 27, 161-173

6. Bailey, J.A., Surface damage during machining of annealed 18% nickel maraging steel Part I- Unlubricated conditions, Wear, 1977, 42, 277-296

7. T.R. Thomas, Surface roughness: the next ten years, Surface Topography 1, 1988, 3-9

8. C.A. Brown, F. Dupont, B. Senior, Machined surface flaws on the high strength aluminum alloys 2024 and 7075, Materials Science and Engineering, 1989, A118, 53-59

9. C.A. Brown, G. Savary, Describing ground surface roughness using contact profilometry and fractal analysis, out for review, 1990

10. ANSI B46.1-1978, Surface Texture: Surface Roughness, Waviness and Lay, ASME, New York, 1978

11. J. Peters, P. Vanherck, M Sastrodinoto, Assessment of surface typology analysis techniques, Annals of the CIRP, 1979, 28/1, 539-554

12. B.B. Mandelbrot, Self-affine fractals and fractal dimension, PHYS SCI 32, 1985, 257-260

13. E.E. Underwood, K. Banerji, Fractals in fractography, MATERIALS SCIENCE AND ENGINEERING 80, 1986, 1-14

14. B.B. Mandelbrot, Fractals Form, Chance and Dimension, W.H. Freeman and Company, San Francisco, 1977, p. 32

15. B. Nowicki, Multiparameter representation of surface roughness, Wear, 1985, 102, 161-176

16. C. Lipson, N.J. Sheth, Statistical Design and Analysis of Engineering Experiments, McGraw Hill Book Company, New York, 1973

17. A.M. Gadalla and W. Tsai, Electrical discharge machining of tungsten carbide-cobalt composites, Journal American Ceramic Society, 1989, 72/8, 1396-1401

Corrosion Protection through Salt Bath Nitriding

J. R. Easterday

Kolene Corporation
Detroit
Michigan
USA

ABSTRACT

An unique salt bath nitriding/oxidizing process has been developed that provides excellent corrosion resistance and cosmetic appeal for ferrous based components. The treatment, identified as QPQ™, also enhances the other engineering properties normally associated with salt bath nitriding, i.e. wear resistance, lubricity, and fatigue strength. QPQ is currently specified for many automotive applications where corrosion resistance is pre-requisite to successful performance.

INTRODUCTION

A new metallurgical process has been developed that can substantially replace chromium and other critical materials in many applications where plating has traditionally been used to improve cosmetic appearance, wear properties and, most important, corrosion resistance. The new technique requires no critical materials, and also avoids serious waste disposal problems associated with some plating operations.

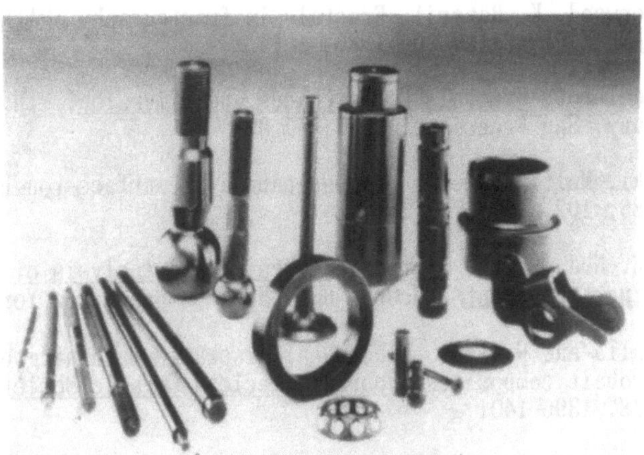

Fig. 1

The foundation of the new treatment termed "QPQ™" (quench-polish-quench) is based on the well-established salt-bath nitriding process known commercially as "MELONITE®" (AMS 2753). While MELONITE alone assures a degree of added corrosion protection, its primary engineering objective has been to improve wear and fatigue properties. The new QPQ adjunct to the process not only retains these highly valued engineering properties, but also develops an attractive cosmetic appearance, Fig. 1, and enhances the resistance to corrosion. The overall beneficial influence of the QPQ treatment on mechanical and chemical properties is shown in Fig. 2.

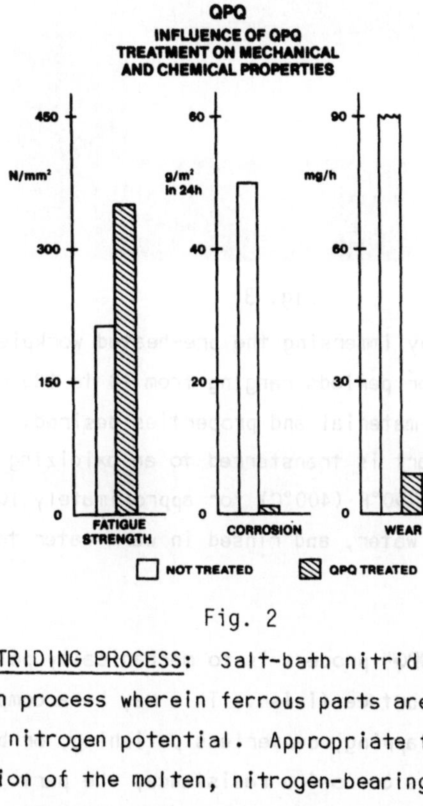

Fig. 2

SALT-BATH NITRIDING PROCESS: Salt-bath nitriding is a thermal-chemical diffusion process wherein ferrous parts are immersed in salts with a particular nitrogen potential. Appropriate formulation and continuous agitation of the molten, nitrogen-bearing cyanate salts assure a controlled and uniform release of nitrogen at the interface of the materials being processed. The nitrogen, and some carbon, resulting from the disassociation of the cyanate (CNO^-) diffuses into the metal chemically combining with iron at the surface to form a tough, ductile compound zone of epsilon iron nitride, Fe_3N. The subjacent zone, termed the diffusion zone, contains a much lower concentration of nitrogen in solid solution with the base material.

Being a diffusion process, the depth of nitrogen penetration is influenced by dwell time at the operating temperature and the chemistry of the material being treated. As an example, compound zones of 0.0004" - 0.0008" (10 - 20 microns) and diffusion zones (total penetration) of .020" - .040" (0.5 - 1.0 mm) are typical for unalloyed steels with processing time of 90 minutes. The resulting microstructure, Fig. 3, shows both of the distinct zones, i.e. compound zone (Fe_3N) and a portion of the diffusion zone. Each zone imparts particular engineering properties to the workpiece.

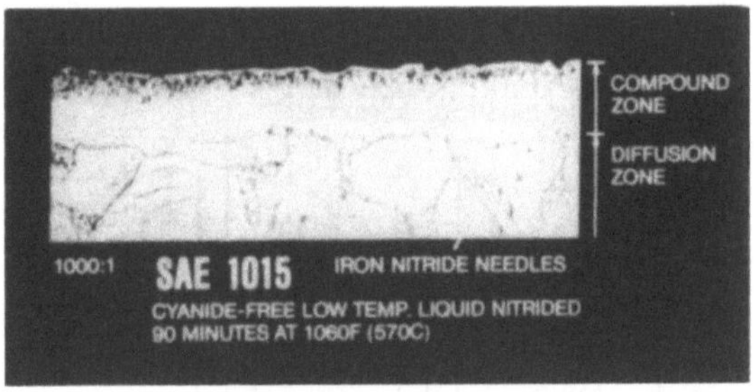

Fig. 3

The treatment begins by immersing the pre-heated workpiece in an aerated nitriding salt bath for periods ranging from 10 to 180 minutes at 1075°F (580°C), depending on material and properties desired. Following the nitriding bath, the part is transferred to an oxidizing salt-bath quench, KQ500®, maintained at 750°F (400°C) for approximately 10 minutes. It is then cooled in air or water, and rinsed in warm water to remove salt residues.

The next step of the QPQ™ process is to mechanically polish the part to restore the original surface finish. This may be accomplished by vibratory polishing, lapping, centerless polishing, or by other similar means. To optimize the corrosion resistance, the part is then reimmersed in the quench bath for 20 - 30 minutes, rinsed and oil dipped. Thus, the term QPQ for quench-polish-quench.

The change in surface chemistry of the workpiece resulting from this sequence of nitriding/oxidizing salt bath treatments may be determined by tracking the nitrogen, carbon, and oxygen concentrations through use of Auger analyses. Results are shown in Fig. 4.

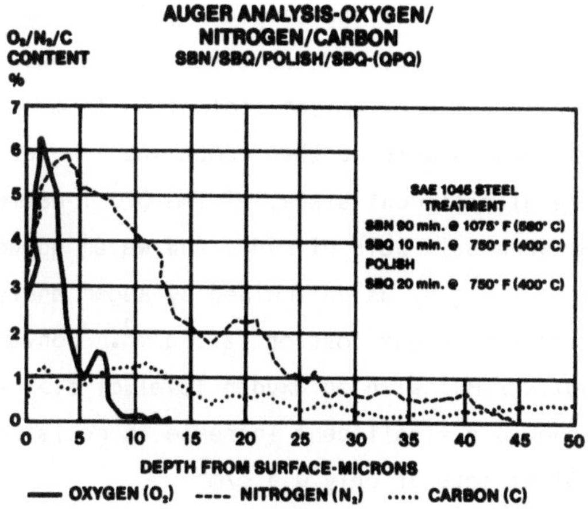

Fig. 4

ENGINEERING PROPERTIES AND APPLICATIONS

Many engineering properties of ferrous base components are developed or enhanced as a result of salt bath nitriding and its QPQ™ adjunct. With one exception these properties are attributable to the presence of the nitride and/or oxide formation at the exposed surfaces. In addition to the cosmetic appeal of the QPQ treated surface, benefits include increased corrosion resistance, wear resistance, and lubricity.

The other engineering property significantly enhanced through salt bath nitriding is fatigue strength. This strengthening effect is the result of the solid solution of nitrogen in the base material, previously defined as the diffusion zone.

Each engineering property alone may provide for a particular application function, e.g. corrosion resistance, wear resistance, etc. However, experience indicates that it is more likely that two or more properties will be utilized in combination to insure successful performance of a part. Consequently, most of the applications discussed will relate to corrosion resistance in conjunction with another engineering property.

CORROSION RESISTANCE

The epsilon iron nitride formed during the nitriding process along provides some degree of corrosion resistance. However, corrosion protection is significantly enhanced by continuing the polishing and re-oxidation, i.e. the complete QPQ™ treatment. It is the presence of oxygen within the nitride zone and at the outer most surfaces of the part that provides the excellent protection from corrosion. This influence of oxygen becomes evident when comparing the levels of corrosion resistance at different stages of the QPQ treatment Fig. 5. The test specimens that were salt bath nitrided and water quenched (SBN™/WQ) had a maximum level of oxygen, as determined by Auger analysis, of 1% and a corresponding corrosion weight loss of 12.3 g/m^2. Conversely, the QPQ samples (SBN/SBQ/polish/SBQ) with an oxygen level of 6.3% and the presence of Fe_3O_4 showed a significant increase in resistance to corrosion with a weight loss of only 0.3 g/m^2.

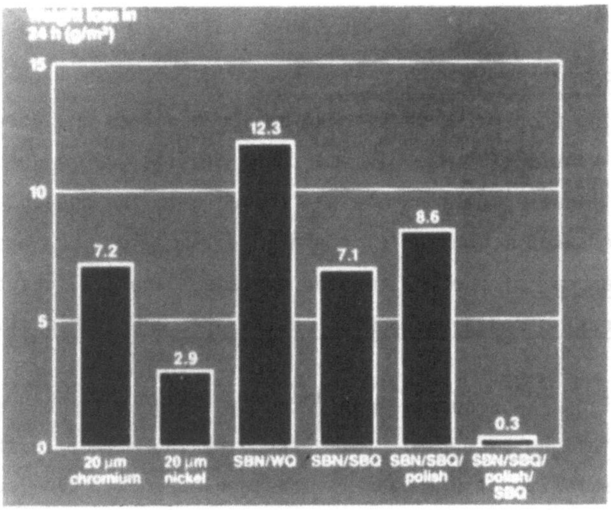

Fig. 5

Evaluation of this series of surface treatments made use of the German test procedure DIN 50905 part 4 in which the test specimens are totally immersed in a 3% solution of NaCl plus an additive of 3 g/liter of H_2O_2. As a measure of corrosion, weight loss in g/m^2 in 24 hours is used.

One investigation as to the relative level of corrosion protection developed by the QPQ process made use of spool shafts used in an automotive power steering system. Three surface treatments were selected as candidates for corrosion testing-chromium plating, electroless nickel plating, and QPQ. Two test specimens of each of the three surface treatments were subjected to the salt spray environment, ASTM B-117. The test results shown in Fig. 6 indicate the significantly greater corrosion protection provided by the QPQ™. Even after extending exposure time in the salt spray environment to 336 hours, the two QPQ specimens demonstrated more protection than the other surface treatments in less exposure time.

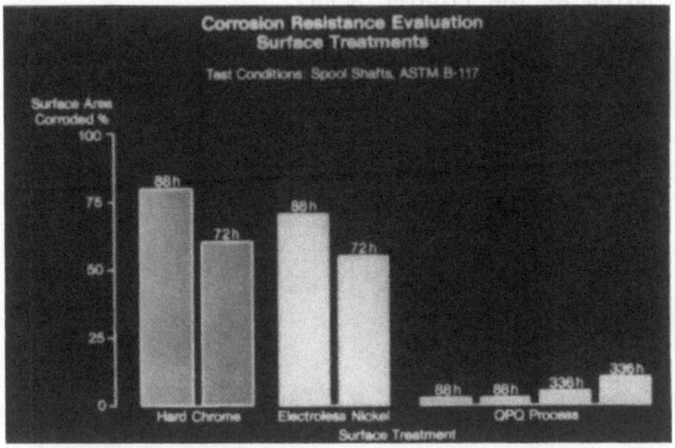

Fig. 6

<u>CORROSION RESISTANCE AND WEAR RESISTANCE</u>

Corrosion and wear are probably the most common combination of operational variables in normal applications. This may occur where ever relative motion takes place in an unprotected environment, e.g. under atmospheric conditions, in aggressive gases, in water or other corrosive mediums.

One type of application that deals particularly with corrosion and wear relates to hydraulic and pneumatic systems, specifically piston rods. The wear resistance (and lubricity) of the surface of the rod is important to maintain the designed characteristics of the sealing system, whether it be a gas or liquid filled cylinder. In addition, in many applications, the piston rod is exposed to various degrees of environmental corrosion especially when in the extended position.

One series of tests dealt with comparing the performance of chrome plated shock absorber piston rods with rods treated with the QPQ process. Following the appropriate surface treatments of the rods, the assembled shocks were subjected to 300,000 cycles in an axial endurance test mode that included side loading to introduce intimate contact of rod surface to seal. After the performance test, the shocks were disassembled and the rods exposed to 96 hours of salt spray, ASTM B-117. Evaluation consisted of a measure of the density of the resulting corrosion sites. As evident in Fig. 7 showing segments of the tested rods, the QPQ™ surface demonstrated complete resistance to corrosion, even at the seal contact region noted as the lighter area.

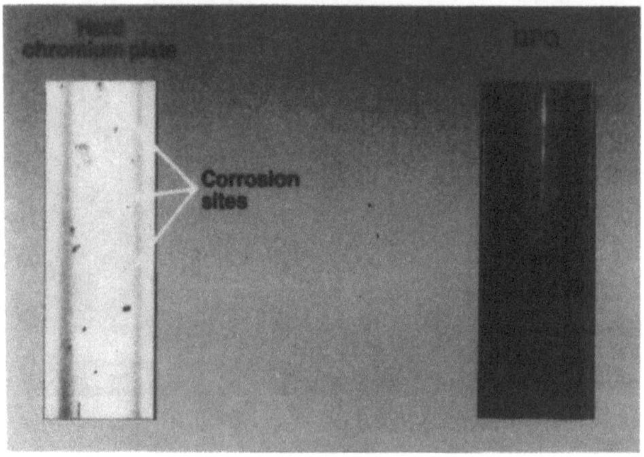

Fig. 7

A prominent internationally accepted application of the QPQ treatment is for piston rods used in pneumatic springs as shown in Fig. 8. These power-assisting devices require very particular engineering characteristics to achieve the designed performance; QPQ surface provides the appropriate corrosion and wear resistance along with low frictional properties.

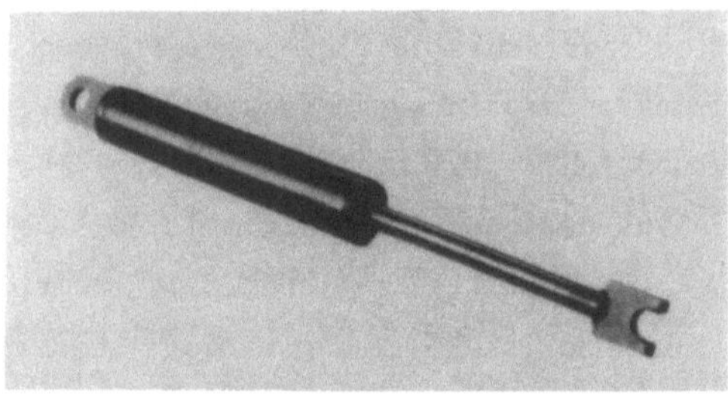

Fig. 8

CORROSION RESISTANCE AND LUBRICITY

One of the intrinsic properties of epsilon iron nitride formed during salt bath nitriding is its relatively low coefficient of friction. Subsequent polishing and re-oxidation, i.e. the QPQ adjunct, develops the degree of lubricity as shown in Fig. 9. In both the "dry" state as well as under lubrication, the frictional characteristics as determined using

Amsler test disks, indicate some significant increase in lubricity of the QPQ™ surface as compared to chrome plating.

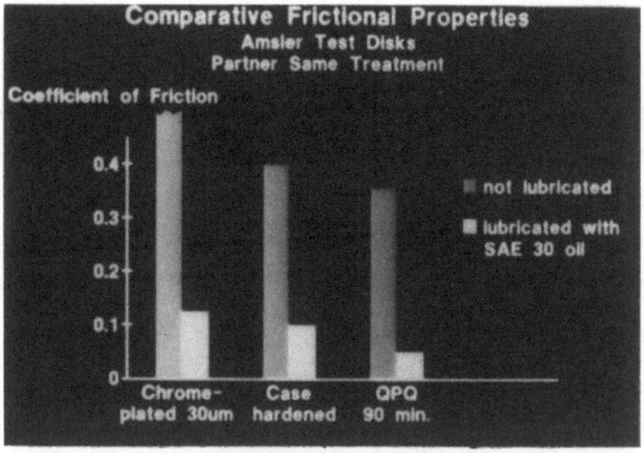

Fig. 9

An automotive application that makes use of lubricity and corrosion resistance of the QPQ surface is the link ball shown in Fig. 10

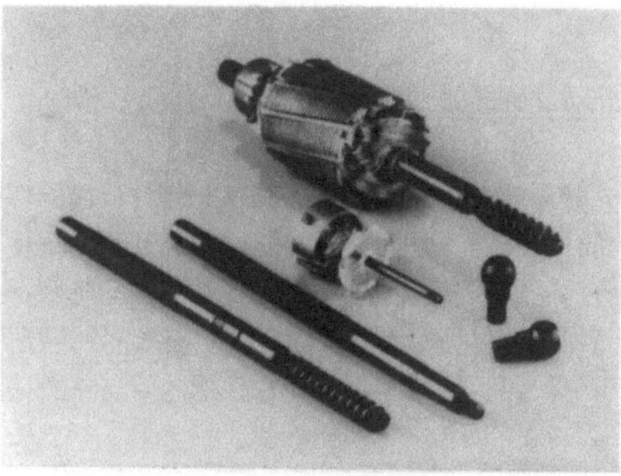

Fig. 10

These components are part of the transmission linkage for windshield wiper systems. It was determined through appropriate evaluation that the original material and method of manufacturing, i.e. machined from stainless steel, could be changed to a cold formed, low carbon steel component and treated by the QPQ process, with a significant saving in manufacturing costs.

The armature shafts also shown in Fig. 10, are salt bath nitrided to introduce surface lubricity at the splined area and thus facilitate assembly into the armature. Result was an appreciable increase in productivity, and lower manufacturing costs.

Cast irons may also be QPQ™ treated to produce engineering properties. The ductile iron housing shown in Fig. 11 is part of a gear pump used to dehydrate gas as it comes from gas wells. The pump system handles glycol through a wide range of temperatures and viscosities. Since glycol has virtually no lubricity and can be highly corrosive especially at elevated temperatures, the QPQ surface provides the necessary frictional and corrosion resistant properties to enable successful pump performance.

Fig. 11

CORROSION RESISTANCE AND FATIGUE STRENGTH

The final engineering property to be considered is fatigue strength, particularly under bending and torsional loading. It is known that the increase in fatigue strength is dependent on the state of precipitation of the nitrogen in the diffusion zone. Fatigue strengthening resulting from the QPQ treatment is shown in Fig. 12 as compared to the untreated condition.

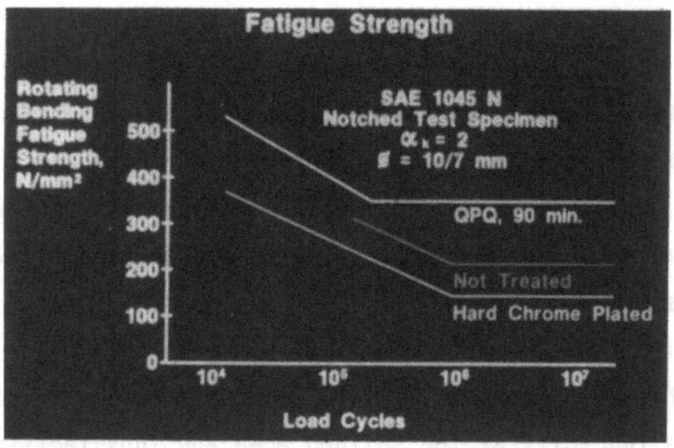

Fig. 12

The importance of the QPQ treatment becomes more evident when components are subjected to oscillating loads (alternating cyclic conditions) in a corrosive environment. As the test results demonstrate in Fig. 13, the improvement in fatigue strength under normal atmospheric conditions is increased by 135% through using QPQ compared to the more significant increase of 5700% under corrosive conditions, although at a lower base line.

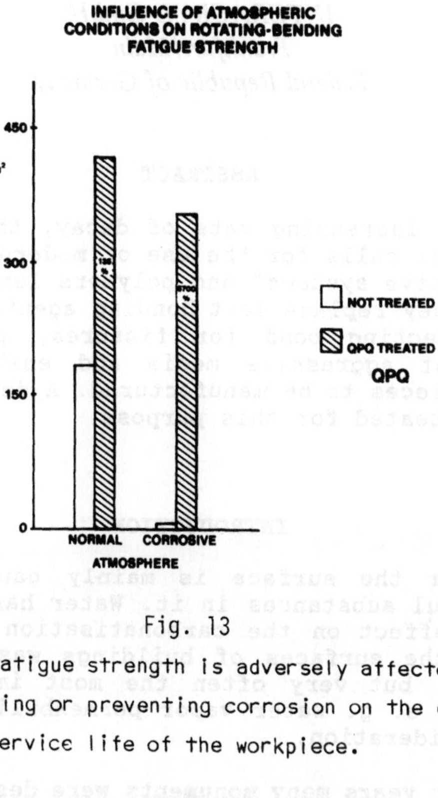

Fig. 13

With knowledge that fatigue strength is adversely affected by corrosion, it follows that reducing or preventing corrosion on the component surface is important to the service life of the workpiece.

SUMMARY

The QPQ process introduces substantial increase in corrosion resistance to ferrous base materials while still providing wear resistance, fatigue strengthening, lubricity, and cosmetic appeal.

Furthermore, plain carbon and low alloy steels, QPQ treated, may be substituted for high alloy grades with attendant cost savings. QPQ is an environmentally compatible surface treatment that provides corrosion resistance without the use of critical materials.

Protection of Surfaces of Natural Stone and Concrete through Polymers

P. J. Koblischek

INTERACRYL GmbH
Frankfurt/Main
Federal Republic of Germany

ABSTRACT

Due to the ever increasing rate of decay, the conservation of building material calls for the use of modern processes. Above all "reactive systems" and polymers lend themselves to this purpose. They replace lost bonding agents in the material, provide a connecting bond for fissures, protect the upper surfaces against aggressive media and enable more weather-resistant workpieces to be manufactured. A description is given of the systems tested for this purpose.

INTRODUCTION

The attack over the surface is mainly caused by water and dissolved harmful substances in it. Water has also on concrete a precipitant effect on the carbonatisation through CO_2. The protection of the surfaces of buildings was mainly attempted through paints, but very often the most important physical-building rules, e. g. water vapor permeability, have not been taken into consideration.

In the past years many monuments were destroyed more due to the wrong methods of treatment and insufficient materials than through environmental damage.

But the chemistry offers possibilities for a very effective and long-lasting protection of the materials and therefore also of the buildings from water and it`s harmful substances.

Through proper treatment of the surfaces of natural stones, concrete and a selection of products, one can achieve to the greatest possible extent, e.g. to water vapor permeability, needed for the requirements. This can be either an impregnation with monomer silanes, reacting with the inner surface of the material to polysiloxanes, or a coating with a mortar with a reacting resin as a binder.

By an impregnation only the water vapor permeability is hardly influenced; compared with coating, where it becomes watertight.

In addition to the physical consequences of a treatment of the surfaces, the chemical property of the used products must also been taken in consideration. In contrast to concrete, there occurs with the weathering of natural stone a loss of binder, which causes a reduction of the stability.

The modern restoration grants every building the right of being old. Therefore, it is necessary to restore the lost binder in the natural stone and so the stability. The achieved profile of consolidation must be the same as for the not yet effloresced stone. Is this not the case, tensions arise and crackings and cracking-offs are the consequences. The loss of old material would be larger than before the treatment. In order to avoid this, the silicic acid ester must penetrate deeply and must not be catalyzed.

HISTORY

The protection of monuments and buildings made of natural stones is not a development of modern times, it has been used since millenniums. Probably the oldest description of a chemical provision for the protection of a monument can be find in the Bible:

"Jacob got up early next morning, took the stone that was under his head, and set it up as a memorial. Then he poured olive oil on it".

As an example for mechanical protection of a building we can look at the pyramids in Gizeh. The Cheop's pyramid, raised from 2551 till 2528 B. C. is covered with plates made of Tura limestone, which remained till today, except the damages done in the Middle Ages, where these plates were used as building material.

We know from Plinius the Eldest (24 -79 A. C.):

"Marble needs great care, although it has a good durability".

In the antiquity the stones were treated with linseed oil or hot wax to protect the surface from the access of water.
We come to the following conclusion:

"RAIN PROTECTION IS PRESERVATION OF MONUMENTS"

This is possibly due to chemical methods, structural precautions or combination of both.

As another example for structural precautions, we can look at gargoyles on churches, still in use today; their purpose is to keep rainwater away from the wall.

We will only talk here about chemical protective measures. Up to the 19th Century it was common to give buildings, made of natural stones, a protective coat. This could be either of plaster, distemper or paint. The paints were based on oil-colour, tempera, casein or lime casein. Casein, the most important protein component of milk, was used since the antiquity as a bonding agent for inorganic pigments. Lime casein is formed out of casein and lime milk and produces a tough waterproof painting for outside walls. A combination with alkaline oil colour was possible. So each component was protected on the surface. The paintings used to be, according to the surroundings, coloured or monochrome. During the 19th Century, the Romantic period discovered the natural beauty of the colours and structures of natural stone. Two circumstances have come together, speeding up the decay: the renunciation of the protective coat and the beginning of the industrialization.

TABLE 1
Diagram by Winkler [1] with attached time table of the Romanticism.

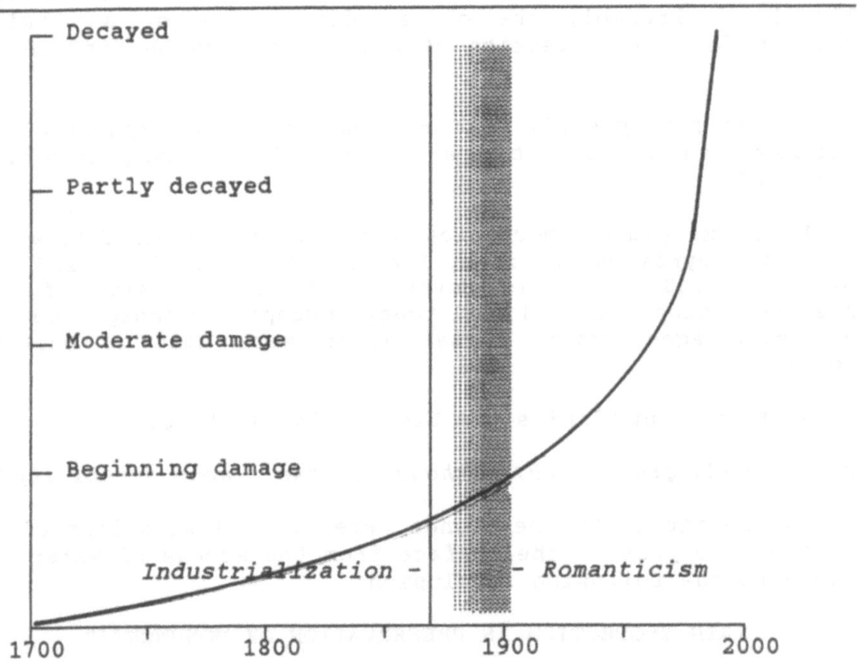

The knowledge of Winkler, that the destruction of surfaces of natural stones since the beginning of the industrialization has speeded up rapidly, has been corroborated everywhere.

PRESERVATION

Until recently, only two alternatives were known to the building specialist as regards the conservation of old buildings and monuments: Rebuilding the damaged areas of the natural stone, and replacing defective natural stonework with new natural stone. A building renovated via this customary manual method looked like new at the end of the work, and there was no more worry for the next 30 - 40 years. Each generation carried out one renovation. This situation has altered in recent years due to environmental damage, and the interval before the next renovation becomes necessary has shrunk to 5 - 10 years. Not only is this linked with an enormous rise in costs, but - as experienced in Berne, for example - there are not enough masons available for this work. Damage therefore occurs more rapidly than renovations can be carried out.

New, simple methods therefore had to be developed for renovation purposes; in other words, an overall concept permitting an historic property not only to be renovated but also to be restored. This is being helped by a rethinking in the field of future monument maintenance, inasmuch as the aim now is to retain as much of the old building material as possible.
No fresh damage should be caused by the renovation measures, either immediately or in future. Without going into the details of damage analysis, there should be discussion of the products which are envisaged for renovating historic buildings. [2]

A pre-requisite for the use of the systems described above is that overall conditions and climatic surroundings of the object to be renovated be taken into account; basic damage analysis and preliminary tests on the original materials must also be carried out.

APPLICATION

TABLE 2
Survey of the various systems and their applications.

	Impreg-nation	Paint-ting	Coating	Inject-ion	Polymer-concrete
PMMA-solutions	+	+			
PMMA + Silanes	+	+			
Silanes	+	+			
Silicic acid est.	+				
PUR-systems	+	+	+	+	+
EP-resins mod.			+	+	
PMMA-resins			+	+	+
EP-resins pure			+		+

PMMA-solutions
The solutions of polymethylmethacrylate, or of the copolymers are employed as raw materials for paints, are mass-produced by the chemical industry, by a hot-setting process: They thus contain no aromatic residue of the starter system and are therefore absolutely non-yellowing, at the same time showing excellent UV-resistance. Special types of polymers are used dissolved in appropriate solutions, and become effective after evaporation of the solvent by forming films of varying strength. This is regulatable according to the choice of concentration and the number of applications. As these PMMA-films are permeable to water vapor but resistant to air-borne pollutants. We are using a special modified polymethyl-butyl-copolymer with reactive groups.

TABLE 3

Diffusion Resistance Number [3] of MOTEMA-FINISH
$\mu\ CO_2 = 6\ 100\ 000$
$\mu\ H_2O = 11\ 000$

This product reduces the carbonatization of concrete, without reducing the water vapor permeability. It can also be pigmented. It is used with great success also as a protective paint for paintings, frescoes, marble, gold-platings and for preserving stonework. [4] After all these good results and experiences, it was obvious to combine this product with silanes.

PMMA-solution + silane = MOTEMA-FINISH-SIL
A combination is also possible with the silanes, and so PMMA-solutions also match the reactive system. Its test results are shown together with those of silanes.

Silanes
This silicon compound contains ester groups which form silanols during hydrolysis:

$$R\text{-}Si(OCH_3)_3 + 3\ H_2O \longrightarrow R\text{-}Si(OH)_3 + 3\ CH_3OH$$

This alkyl-silanol is not stable and reacts immediately with another molecule, but also with OH-groups in the building material - which every silica building material contains - with a dehydration - process taking place. In this way, the silanol molecule becomes cross-linked or linked with a silicon atom in the building material. Through the intermediate stage of siloxanes, polysiloxanes (silicon resin) are then formed. These now show an excellent water-repellent finish, and protect the building material from penetration by water and the pollutants dissolved therein. At the same time, however, there is only an insignificant reduction in water vapor permeability.

Through the use of monomeric silanes with a molecular length of approx. 10 Å, deep penetration into the material is made possible. Gel pores in concrete have an average size of 25 Å, the percentage in cement stone runs up to 20 - 30% per vol. Polysiloxanes (silicon resins) have a molecular length between 150 and 1000 Å, so they are too large for gel pores. Also silicon resins do not fit in the capillary pores, connecting the gel pores, because the capillary pores measure 10 - 100 Å.

10 Å = 1/1,000,000 mm = 1 mμ.

TABLE 4
General View of the Proportions

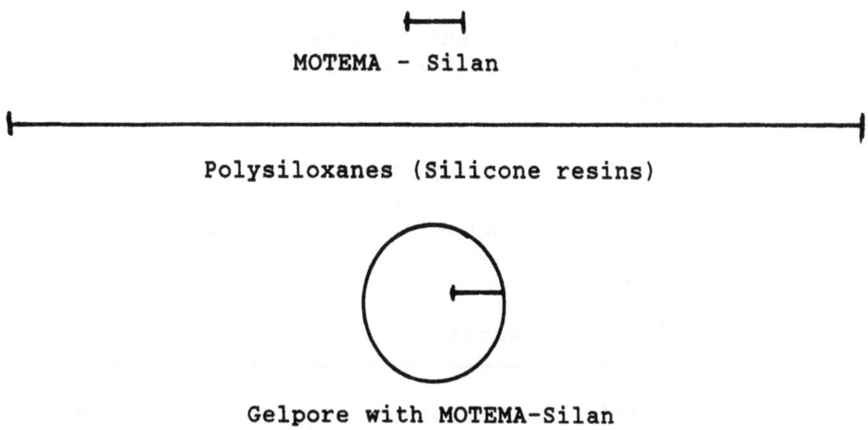

Scale: 10 Å = 1 mμ = 10 mm: ⊢—⊣

The capillary absorptive capacity is reduced drastically through hydrophobizing. Thereby the frost-resistance has been fundamentally improved and the penetration of harmful substances e.g. chloride-ions is stopped effectively. This applies to concrete as well as to natural stones. Due to the hydrophobizing, the steel corrosion in concrete is stopped. It can not be excluded, that the carbonatization in fresh concrete is speeded up slightly. Therefore also here a combination with MOTEMA-FINISH. The depth of penetration and the effectiveness was specified on DIN-standard-mortar 0-6 mm. They verified:

- the water absorption coefficient A,
- the depth of penetration through grinding off and repeated examination,
- the alkaline resistance through an exposure (14 days) at +60 C in saturated calcium-hydroxide-solution.

TABLE 5

Water Absorption Coefficient A: kg / $m^2 \cdot h^{0,5}$	Silane	Finish-Sil
Mortar untreated	0,200	0,200
Mortar surface treated	0,038	0,039
grind-off: - 1 mm	0,054	0,057
: - 2 mm	0,055	0,067
: - 3 mm	0,094	0,649
after 14 days $Ca(OH)_2$ + 60°C	0,057	0,053

The water absorption coefficient A is reduced by MOTEMA-Silane by 5,3 times and by MOTEMA-Finish-Sil by 5,1 times. The alkaline resistance exists.

TABLE 6

Water Vapor Diffusion Resistance Factor:	kg / $m^2 \cdot h^{0,5}$
Mortar untreated	0,150
Silane treated	0,029
Finish-Sil treated	0,029

The water vapor diffusion resistance factor is 81% under the value of the untreated sample.

TABLE 7

Equivalent of Diffusion Resistance Factor:	Sd in m
Mortar untreated	0,75
Silane treated	0,75
Finish-Sil treated	0,74

TABLE 8

Product Sd • A:	µ 50 - 100
Mortar untreated	60
Silane treated	60
Finish-Sil treated	59

The fundamential demand for a hydrophobizing: no change of the water vapor permeability is met by both products.

Silicic acid esters

In large parts of central Europe, the most frequently used natural stone is sandstone. The bonding agents holding the individual sand granules together are washed out over the course of years, with the result that the sandstone loses its firmness. In order to retain as much as possible of the building substance, it is now necessary to replace the lost bonding agent, i.e. to compact the sandstone. Silicic acid ester reacts as follows with water:

$$Si(OC_2H_5)_4 + 4\ H_2O \longrightarrow SiO_2 + 4\ C_2H_5OH$$

Through hydrolysis, an anorganic "polymer" is produced here from an anorganic ester. Silicium dioxide is the purest quartz; the ethanol (ethyl- alcohol) which comes about as a by-product is not damaging either the building material or the building structure. Of importance is the fact that the hydrolysis of the silicic acid ester only takes place slowly, so that the silicic acid ester can penetrate deeply into the stone to be strengthened; only in this way can a uniform degree of strength be achieved. In no event may the silicic acid ester be so reactive that it already reacts in the upper level, causing a shell to be formed. The penetration should therefore be evident for up to a depth of 50 mm. Serious differences of the stability profil, e.g. a rapid decrease of the effectiveness in already a depth of 40 mm [6] as it occurs by a catalyzed silicic acid ethyl ester, have not been found in any of the numerous tests. Because the large pores remain unfilled, the porosity and so the important water vapor permeability is preserved.

Reactive binder system

Restorers wishes for a non alkali binder for colour pigments insted of water-glass, led to the development of a reactive methacrylic-silan-silicic acid-system: it is a binder and at the same time a water proofing agent. [7]
The same tests have been carried out here, too. Sandstone plates with a thickness of 12 mm and a colour coat of 0.5 mm.

TABLE 9

Water Absorption Coefficient A:	$kg / m^2 \cdot h^{0,5}$
Sandstone untreated	1,020
treated w. Mineral colour	0,111
treated w. MOTEMA-Binder	0,017

In comparision with an ordinary mineral-colour on a potassium water-glass basis, the water absorption is reduced through this new bonding agent by 6.5 times.

TABLE 10

Equivalent of Diffusion of Air Film Thickness:	Sd in m
Sandstone untreated	1,02
treated w. Mineral colour	1,08
treated w. MOTEMA-Binder	1,15

TABLE 11

Product	Sd • A:	μ 50 - 100
Sandstone	untreated	81
	treated w. Mineral colour	86
	treated w. MOTEMA-Binder	92

The restorer has access to a complete, coordinated renovation system for natural stones which is free of harmful substances.

Polymer concrete (PC)
Many different problems arise in connection with renovation, and these also require different products. Only EP and PUR resins permit the manufacture of porous, i.e. water vapour permeable systems, synthetic substitute stone materials. MMA resins are not suitable to this purpose, because these do not polymerize due to oxygen inhibition. [8]

TABLE 12

Reactive Resins for Polymer Concrete (PC) and Coatings	
Epoxi resins	EP
Methacrylic resins	MMA
Polyester resins (unsaturated)	UP
Polyurethane resins	PUR

Epoxi resins harden by polyaddition; i.e. by separation of the epoxide groups and by addition = addition of accelerators with active hydrogen atoms. Resins and accelerators must be dosed in stoichiometric amounts. The hardening reaction is temperature-dependent. Due regard must be paid to the fact that there are no non-yellowing epoxi resins and, if they are for example used in weather-exposed areas, that the stone replacement materials made from them must be weatherproofed with silanes

Methacrylic resins are distinguished from all the other reaction resins by the fact that in its case monomeric methylmethacrylate is already polymerisable with a viscosity of 1 mPa·s (identical to water). If required, however, the viscosity can be set at any desired level by means of dissolving polymethacrylates. A further advantage of MMA resins is their high reactivity, even at temperatures of 0°C and below. For special applications like synthetic white marble, we have now a redoxsystem without any yellowing.

Polyurethane resins harden through polyaddition, and, as a rule, form elastic polymers. The hardening is strongly catalyzed through moistness. Using aliphatic isocyanates in conjunction with aliphatic poly-alcohols, light- and weatherproof polyurethane systems are produced. The resistance to UV has to be improved by adding UV- inhibitors.

CONCLUSION

The destruction of surfaces of natural stones and concrete has speeded up rapidly and can be corroborated everywhere. The attack on the surface is primarily caused by water and in it dissolved harmful substances.

In past years many monuments were destroyed more due to the wrong methods of treatment and insufficient materials than through environmental damage. But the responsible chemistry offers possibilities for a very effective and long-lasting protection of the materials and therefore also for the buildings and monuments.

The alternative is to do nothing and wait the further dilapidation of the monuments.

REFERENCES

1. Winkler, E.M., Stone Properties, Durability in Man's Environment. Springer Verlag Wien - New York, 1979.

2. Koblischek, P.J., Reaktionsharze in der Denkmalsanierung. Proc. Internationales Kolloquium Werkstoffwissenschaften und Bausanierung, Esslingen, 6.- 8. Sept. 1983, pp. 341-3.

3. Engelfried, R., Carbonatisation von Beton, ihre Bedeutung und ihre Beeinflussung durch Beschichtunge, defacet No. 9 1977, pp. 353-9.

4. Furlan, V., Pancella, R., Properties d'un grès tendre traité avec des silicates d'èthyle et un polymere acrylique. Proc. International Symposium, Bologna, 27.- 30. Oct. 1981.

5. Koblischek, P.J., Polymere in der Denkmalsanierung. Proc. IV. Internationaler Kongreß "Polymere und Beton" - Technische Hochschule Darmstadt, 19.- 21. Sept. 1984, pp. 441-6.

6. Ettl,H., Schuh, H., Konservierende Festigung von Sandsteinen mit Kieselsäureester. Bautenschutz + Bausanierung, 12,1989, pp. 35-8.

7. Koblischek, P.J., Organisches, hydrophobierendes Bindemittel für Mineralfarben. Proc. 2. Internationales Kolloquium Werkstoffwissenschaften und Bausanierung, Esslingen, 2.-4. Sept. 1986, pp. 533-8.

8. Koblischek, P.J., Polymere bei der Sanierung von Bauwerken aus Natursteinen. Kunststoffe im Bau, 20, Vo. 4, 1985, pp. 185-9.

The Influence of Directivity of Machining Marks on Grindability and Wear of Honed Surfaces

A. F. Budzyński

Akademia Techniczno-Rolnicza
Bydgoszcz
Poland

ABSTRACT

This paper deals with the development of an experimental setup which is used to validate mathematical models in the form of a regression equation of the influence of an intersection angle of machining marks (α=0-180°). Typical of the case investigated are hydraulic cylinders and their seals using diameter increment δd. This factor gives evidence of wear of the tested surface, while a mass decrement δm of seal rings characterizes their grindability. The changes in mating surfaces roughness were tested and leaks of a medium through the seal are also described. These leaks reflect the grinding-in of tested kinematic pairs. Finally, boundary values for correct performance of kinematic pairs are given.

INTRODUCTION

Particular surfaces of machine elements can be shaped by means of point, envelope curved and shaping tools. The review of fundamental (necessary) movements during shaping of rotary cylinder surface during honing process is presented in Figure 1 and Figure 2. If we assume that:

$$\begin{aligned} &\bar{v} \neq 0, \\ &\bar{v}_f = 0 \quad \text{or} \quad \bar{v}_f = \bar{v}_{fC}, \\ \text{or} \quad &\bar{v}_f \neq 0, \\ &\bar{v} = 0 \quad \text{or} \quad \bar{v} = \bar{v}_{0C} \end{aligned} \tag{1}$$

where: $\bar{v}$ - peripheral (transverse) speed,

$\bar{v}_f$ - feed (longitudinal, axial) speed,

$\bar{v}_{0C}$- discrete value $\bar{v}$,

$\bar{v}_{fC}$- discrete value $\bar{v}_f$,

then the directivity of unevenness (irregularity of surface) is parallel or perpendicular in relation to the cylinder axis. Such kind of honing is

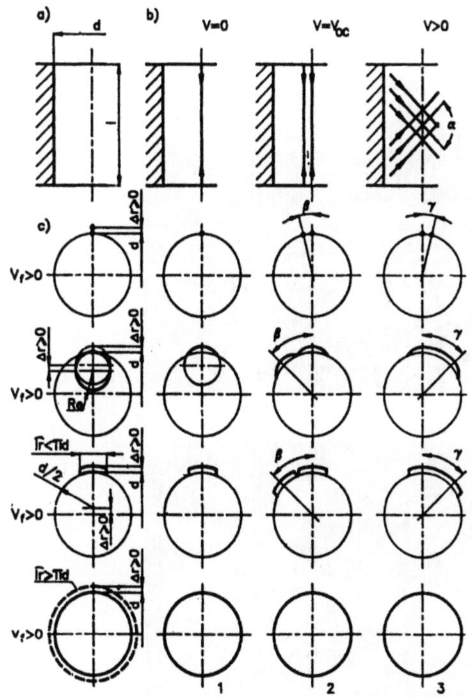

Figure 1. Basic methods of surface shaping of rotational cylinder with point, envelope and shaped tools: a - shaped hole, b - machining marks, c - cross-section of tools.

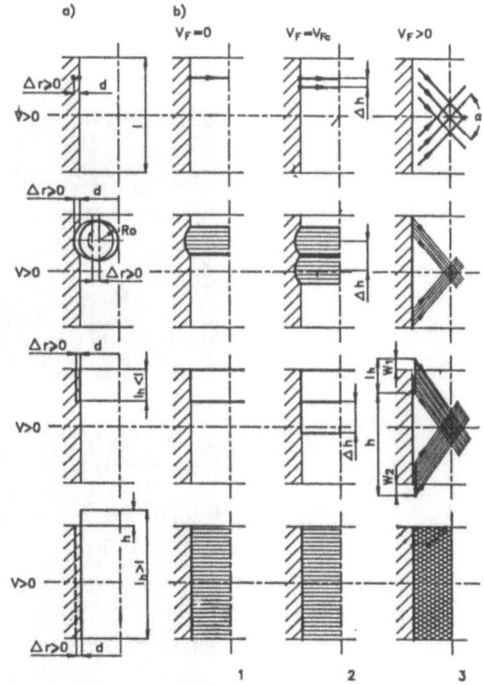

Figure 2. Basic methods of surface shaping of rotational cylinder with point, envelope and shaped tools in axial section of the cylinder (a) and the machining marks (b).

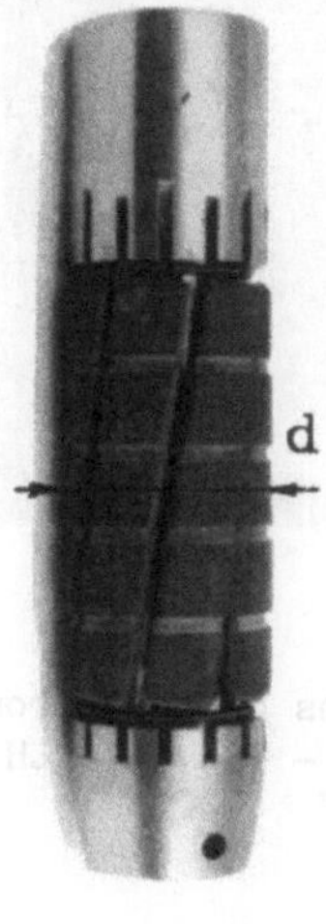

Figure 3. The tool enabling one direction longitudinal honing (LH), one-direction transverse honing (TH) and two-directional honing (TDH).

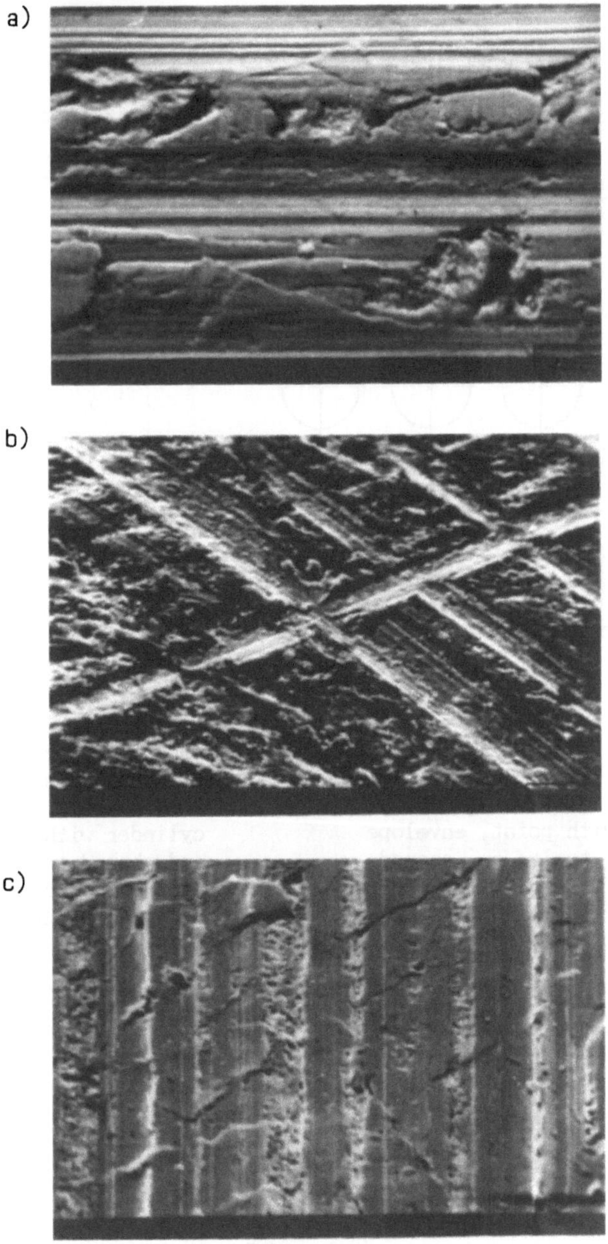

Figure 4. Typical photographs of cast iron cylinder surface after honing: a - TH, b - TDH, c - LH; mag. 400x.

assumed as one-direction honing and is shown in Figure 1 and Figure 2.

From Figure 1 and Figure 2 it is also evident that in case if:

$$l_h \geq l \quad \text{or} \quad l_r \geq \pi d, \tag{2}$$

where: l_r, l_h - length of shaping tools

l, d - dimensions of machined hole

then in order to shape a surface in the form of rotational cylinder only one speed $\bar{v}_f$ or $\bar{v}$ is necessary. If we presume that:

$$zl_r < \pi d, \quad zl_h < l \tag{3}$$

where: z - number of shaping tools

then both component speeds are necessary for machining.

In the case the point tool moves on shaped surface with relative speeds $\bar{v}$ and $\bar{v}_f$, determined as:

$$\bar{v} = \pi dn \tag{4}$$

$$\bar{v}_f = 2hf \tag{5}$$

where: $\bar{v}$ - peripheral speed,

$\bar{v}_f$ - feed speed,

d - diameter of machined hole,

n - rotational speed,

h - length of tool's step (stroke),

f - frequency of double steps of tool

then the resultant tool speed $\bar{v}_e$ is:

$$\bar{v}_e = \sqrt{\bar{v}^2 + \bar{v}_f^2} \tag{6}$$

Such kind of machining is called two-directional honing (TDH).

In Figure 3 there is shown an example of tool which enable one-direction longitudinal honing (LH), one-direction transverse honing (TH) and two-directional honing (TDH). Examples of surfaces obtained by means of such tool are shown in Figure 4.

INVESTIGATED FACTORS, RANGE OF THEIR VARIATIONS AND METHODS OF MEASUREMENTS

The assumed geometric factors (Cg) and material factors (Cm) of test samples result from the evaluation of a questionnaire relating to honed holes in Poland during the period 1975-70 [5]. According to this questionnaire, holes with d = 50H7 and l = 100 prevailed. Taking the material into consideration, 25% holes were made of steel, 74% of cast iron and 1% of other materials.

Macrogeometry of honed holes was established by measurements by means of measuring machine and numerical calculations. The shape deviations include the following errors: roundness deviation δr relative to mean circle, established in ten planes for every test sample (Figure 5) by means of 12-points measurement method on measuring machine [5,10]; cylindricity deviation δw in relation to mean cylinder, established on the basis of the above mentioned measurements and numerical calculations; axis straightness deviation (δp), established numerically on the basis of diameter of rotational cylinder circumscribed on the centers of average circles. As axis straightness deviation δp has assumed the maximum diameter of this cylinder.

The microgeometry of a surface after honing process was evaluated [11] in every hole at three planes (1;5 ÷ 6;10 - as shown in Figure 5) and in three points on circumference (spaced every 120°). For every hole mean values of Ra and Rmax parameters and linear load capacity of profile NL20 and NL50 were established. Then these values were averaged during repeated measurement for every association of investigated factors.

During construction experimental tests the controllable factors can be applied to the surface of honed cylinders, piston seals and working parameters chosen for particular kinematic pair. To these working parameters were included: Cg, Cm and Cd of tested cylinders, Cg, Cm and Cd of tested seals, test stand (Figure 6), characteristics and properties of hydraulic medium and its pressure, the method of honing process (TDH, LH, TH), kinetics of piston movements in cylinder, piston velocity v_t in cylinder, length of piston stroke, preclamping of seal, the way of placing and clamping the seal on the piston and piston itself in relation to the cylinder, working temperature of kinematic pair as well as stoppage and operation times of the kinematic pair.

The following factors can be assumed as disturbing factors: change of hydraulic medium properties, change of environment temperature, foreign magnetic fields, etc.

In our investigation the following were chosen as variable factors: method of honing (TDH, LH, TH), kind of cylinder material (steel, cast iron), type of cylinder seal (cast iron, rubber).

The following parameters, enabling evaluation of grindability and wear of honed surfaces, were chosen as output parameters: diameter increment δd of tested cylinders, giving evidence of the surface wear, mass decrement δm of seal rings, which characterizes the surface grindability, roughness of tested surface and medium leaks through the piston seal in the cylinder, reflecting the grinding-in of tested kinematic pair and boundary values for its correct performance.

The measurements of diameter increment δd of tested cylinders were made after demanded number of piston strokes x in cylinder; more frequently during the first working period of tested kinematic pair and less frequent after prolonged period of operation. The tests with rubber piston seal were stopped after $5 \cdot 10^4$ piston strokes because the mass decrements of sealing elements were so high (sometimes up to 50% of mass) so that leaks became excessive and difficult to measure. The diameter increments δd of test cylinders were measured at planes 2, 5÷6 and 9 (see Figure 5).

The measurement of the mass decrement δm of seal rings (in relation to number x of piston strokes) was carried out by triple weighing on analytical balance.

The roughness measurements of tested cylinders were performed in a function of number x (realized piston strokes) at the planes 2, 5÷6 and 9 (Figure 5) in two points on the cylinder periphery, spaced 90° to each

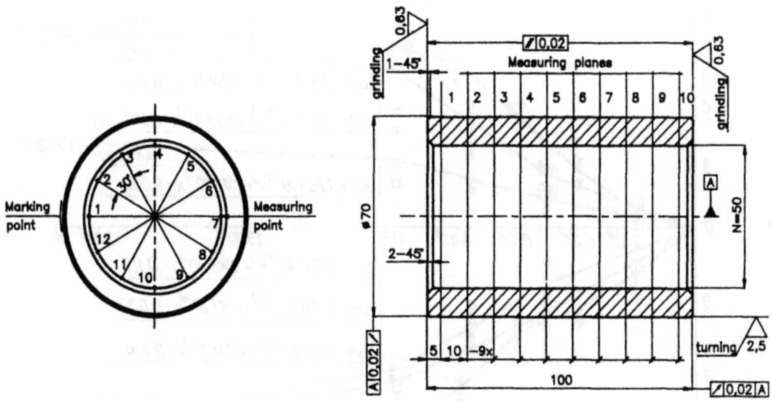

Figure 5. Geometrical features of the samples for construction investigations.

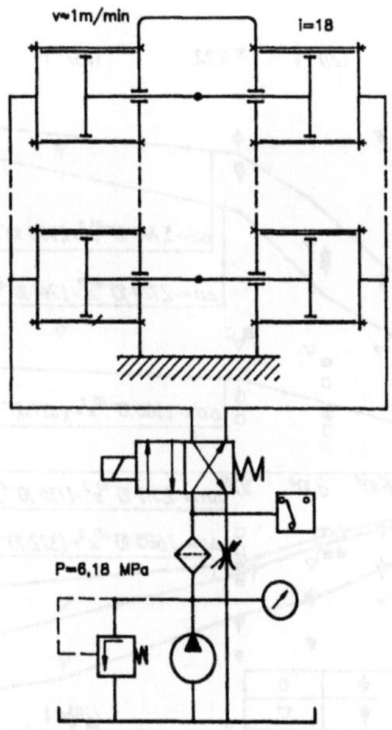

Figure 6. Stand for construction experimental tests - hydraulic diagram.

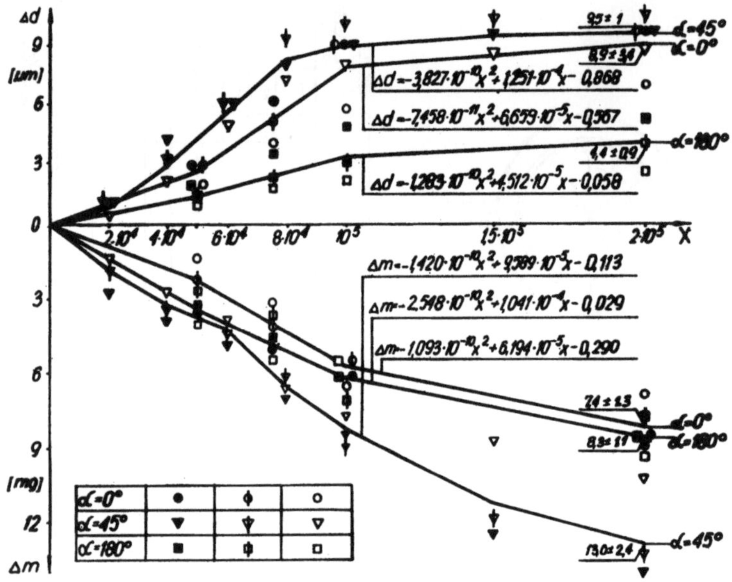

Figure 7. Diameter increment δd of steel cylinder and mass decrement δm of cast iron seal as a function of piston strokes (x) in relation to the directivity of machining marks α.

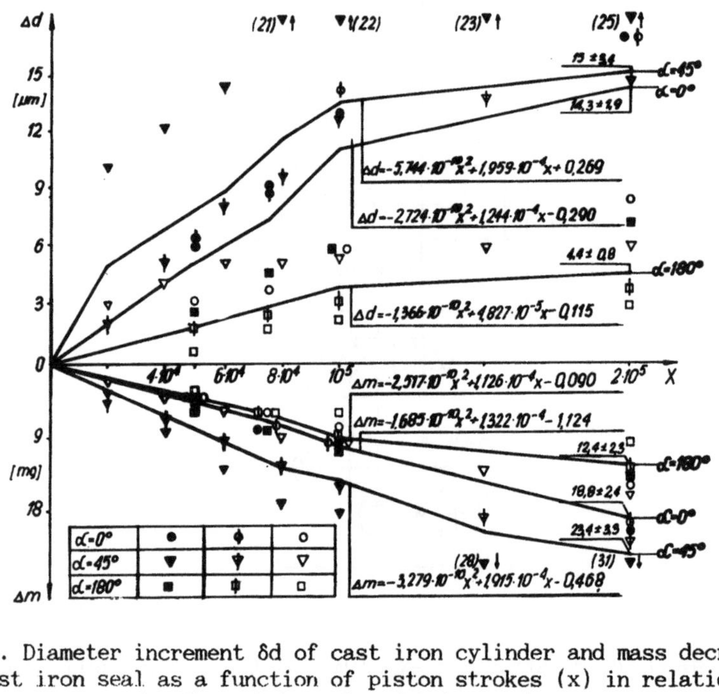

Figure 8. Diameter increment δd of cast iron cylinder and mass decrement δm of cast iron seal as a function of piston strokes (x) in relation to the directivity of machining marks α.

a)

b)

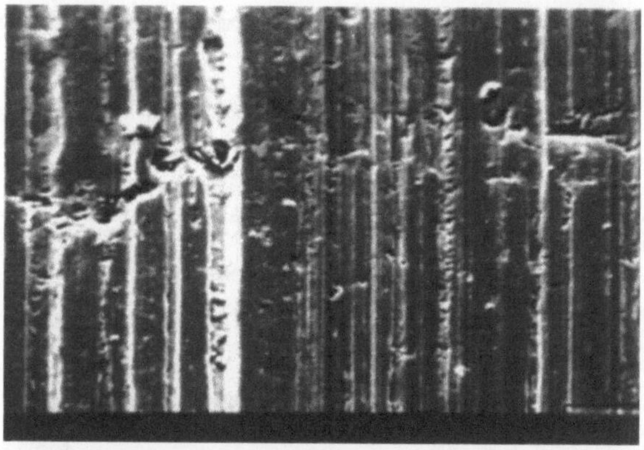

c)

Figure 9. Cast iron cylinder's surface machined through abrasive honing after operation with cast iron seal ($2 \cdot 10^5$ piston strokes): a - TH, b - TDH, c - LH; mag. 400x.

other. The surface of tested cylinders was investigated also by electron microscopy (Figure 9).

The measurements of leaks were carried out on relieved side of tested cylinder with the use of measuring cylinders. Dynamic leaks were measured for both types of seals and static ones only for the rubber seal.

Construction investigations were realized on the basis of the static, determined and complete test program with variable values marked as PS/DK-v. This program consists of N_k experiments:

$$N_k = r\prod_{k=1}^{k=i} n_k\,,\quad \text{dla}\ n_k = \text{const},\ r>1=\text{const}, \tag{7}$$

where: r - number of repeats (r = 3),
k - number of factors investigated (k = 4),
n_k - number of levels of investigated factors (3,3,2,2).

The inductive model was formulated in a form of the following relation:

$$y_0 = b_0 + \sum_{i=1}^{k} b_i \tilde{x}_i + \sum_{i<j}^{k} b_{ij}\tilde{x}_i\,\tilde{x}_j + \sum_{i=1}^{k} b_{ii}\tilde{x}_i^2\,, \tag{8}$$

where: y_n - n'th output factor,
b_0, b_i, b_{ij}, b_{ii} - regression coefficients,
k - number of tested factors,
$\tilde{x}_i, \tilde{x}_j$ - coded values of tested factors.

TEST RESULTS AND DISCUSSION

In this short paper only some test results are quoted. More extensive discussion of this problem can be found in other works [1÷11].

The diameter increment δd of steel cylinder mating with cast iron seal are shown in Figure 7a as a function of number x (piston strokes) for different honing methods (TDH, LH, TH). In Figure 8 there are shown the same relationships but for the cast iron cylinder. Both diagrams show also the mass decrements δm of seal in relation to δd.

The obtained results were then submitted to significance test of particular factor influence by using Snedecor's F-test. On the basis of this evaluation it is possible to ascertain that:

- in all investigated cases the number x of piston strokes has substantial effect on the diameter increment δd of cylinder and the mass decrement δm of the seal, what is self-evident and only confirms the correctness of carried out tests and calculations,
- intersection angle α of machining marks in all evaluated cases has essential effect on the mass decrement δm of the seal. It is noteworthy that the minimum values of δm are for α=180°,
- intersection angle α of machining marks has also essential influence on diameter increment δd of steel and cast iron cylinders mating with cast iron seals whereas has no essential effect in the case of the same cylinders mating with rubber seal. This last case can be explained with a very little increasing of diameter - many times less in comparison with cylinders mating with cast iron seals. Furthermore, for any kinematic pair examined, the diameter increment δd for the cylinders was the lowest for α=180°.

CONCLUSIONS

In all investigated cases the surface formed by using two-direction honing (most common now) were characterized by high or even higher grindability and wear in comparison with surfaces shaped by using one-direction longitudinal honing (LH). Such situation could be a result of still existing approach as to optimization of angle α only with taking into consideration the output of machining. Such approach proved to be insufficient. Also dynamic and static leaks on investigated cylinders showed the superiority of LH over TDH, which is described in detail in the above mentioned papers.

REFERENCES

1. Budzyński, A.F., Theoretical principles and technical factors of electrochemical honing. In Proceedings of ISEM6, Kraków, 1980.
2. Budzyński, A.F., Research on optimization of the process of electrochemical honing (ECH). In Proceedings of ISEM7, Birmingham, 1983.
3. Budzyński, A.F., Electrochemical honing - studies on the shape precision of holes being processed. In Proceedings of ISEM8, Moscow, 1986.
4. Budzyński, A.F., Theoretical and experimental research on the process of electrochemical honing (ECH). In Proceedings of ISEM9, Nagoya, 1989.
5. Budzyński, A.F., Theoretical and experimental research on the process of electrochemical honing (ECH). Assistant professor's dissertation, Technical University, Warsaw, 1982.
6. Karwowski, K., Method and the control system diminishing the rolling errors of holes undergoing honing. Doctor's dissertation, Technical University, Warsaw, 1982.
7. Kertyńska, H., Analysis of sliding properties of materials at cryogenic temperatures in thrust seal construction as an example. Doctor's dissertation, Technical University, Wrocław, 1979.
8. Orobiej, J., Examination of finishing processing parameters of the holes 0.1 ÷ 0.5mm in diameter. Doctor's dissertation, Technical University, Warsaw, 1984.
9. Seroka, St., Honing of holes making possible the obtaining of rolling area parallel to the axis of the geometrical structure. Doctor's dissertation, Technical University, Warsaw, 1982.
10. Wocianiec, R., Influence of certain parameters, especially of the tool shape, on the precision of shape and size of the holes undergoing honing. Doctor's dissertation, IOS, Kraków, 1982.
11. Zakościelny, St., Influence of the coarseness directivity and methods of production on wear and grindability of the structure undergoing honing, based on hydraulic servo-motors. Doctor's dissertation, Technical University, Gliwice, 1981.

Wear Resistant Surfaces with a Hardphase Gradient

H. Berns, A. Fischer and Ch. v. Nguyen

Ruhr-University Bochum
Bochum
Federal Republic of Germany

ABSTRACT

The boride content in a hardened steel matrix is gradually raised toward the surface to increase the wear resistance and avoid a step like change in properties from the stubstrate to a coating of constant hardphase content. Hardness, elastic modulus, bending strength, crack growth rate, wear resistance and thermal expansion were evaluated over the length of gradient.

1. INTRODUCTION

In the field of mineral extraction and further processing, machine parts and tools are protected by thick coatings of hard alloys in which hard phases like carbides and borides improve the abrasive wear resistance. The protective coating is usually deposited by open–arc welding of filler wire. Due to the turbulence within the weld pool the primary and eutectic hard phases are rather homogeneously distributed in each layer of a two or three pass cladding. This results in an abrupt change of properties from the substrate to the coating which is liable to peel off under stress or temperature cycling.
In order to achieve a continuous change in properties the hardphase content in the metal matrix must rise gradually from the interface to the surface of the coating. This can be brought about by thermal spraying, powder metallurgy (PM) or a combination of both. There is little literature available on the subject [1 to 4]. In the present study boride particles are embedded in a hardenable steel matrix by PM with 5 to 20 vol.% of hard particles. The aim is to get basic information on the wear resistance and the thermal as well as the mechanical properties along the gradient.

2. EXPERIMENTAL

Powders of steel and CrB_2 (Table 1) were mixed, hot isostatically pressed (HIP) and heat treated (Table 2). It was shown that a net like arrangement of CrB_2 particles around the steel grains is formed, if their size ratio is << 1 [5]. As this leads to poor toughness and wear resistance powders of similar size were used in the present study to gain a dispersion of CrB_2. Samples with constant contents of CrB_2 were prepared and others with a gradient in the CrB_2 concentration. After annealing the specimens were pre–machined and ground or polished after hardening and tempering.

The Young's modulus E and the fracture stress R_{bf} were measured in three point bending. For the evaluation of fracture toughness K_{Ic} and crack growth rate da/dN

CT specimens with or without a gradient were used, respectively (fig. 1). The abrasive wear resistance W_{ab}^{-1} was tested against grinding paper (fig. 2). The coefficient of thermal expansion was derived from constant CrB_2 specimens at 250 ^0C after tempering at 450 ^0C to avoid a superposition of carbide precipitation and retained austenite transformation. The Brinell macrohardness H_M was taken with a 2,5 mm WC–Co ball, the microhardness H_m with a Vickers indenter under a load of 0.49 N.

Light (LIMI) and scanning electron (SEM) microscopy as well as wave–length (WDX) and energy dispersive (EDX) X–ray analysis were applied to analyse the microstructure, the wear or the fracture surfaces. X–ray diffraction helped to identify hard phases and retained austenite (RA).

3. RESULTS

3.1 Microstructure

As shown in fig. 3 a the hard phase content in the hiped and hardened condition is about twice as high as the initial content of CrB_2 powder in the gradient. This is brought about by a reaction between CrB_2 and the steel powder. In fig. 4 a we distinguish the formation of shell–like zones around a CrB_2 particle and small hard phases precipitated within the steel matrix. EDX, WDX and X–ray diffraction lead to the schematic drawing of fig. 4 b. The partial transformation of CrB_2 to CrB releases boron to diffuse outward and combine with iron and 35 to 15 w% Cr to M_3B. The carbon migrates from the M_3B to the $M_{23}(B,C)_6$ area lowering the specific volume of this phase. At HIP–pressure this stabilizes the borocarbide, which shows the lowest Cr–content (~ 6 w%) and M/(B+C) ratio of the four hard phases present. The small precipitated phases are of a similar composition. There is, however, enough carbon left in the martensite to harden it sufficiently. Its average microhardness H_m amounts to 713 HV as compared to 730 HV for the pure steel.

3.2 Properties

The macrohardness H_M (fig. 3 b) is increased by the hardphase content. The elastic modules E is raised by hard phases while roughly obeying the inverse rule of mixture (fig. 3 f). In contrast the coefficient of thermal expansion α is lowered by an increasing hardphase content (fig. 3e). The resistance against grooving wear W_{ab}^{-1} is improved only if flint paper is used (fig. 3 c). The drop in bending strength R_{bf} towards the surface is more pronounced than that of fracture toughness K_{Ic} (fig. 3 g,h). The crack growth rate da/dN at constant ΔK is strongly depending on the ΔK level (fig. 3 d).

4. DISCUSSION

The CrB_2 and the steel powder are apparently not in equilibrium. At HIP temperature and under HIP pressure the system tends to lower its free energy by forming new phases. Due to insufficient information about the heat of reaction, the densities and the pressure influence it is not possible to evaluate ΔG. However, a look at the Fe–C–B–System [6] shows, that an alloy of the present steel with 15 vol. % CrB_2

would render a ratio C:B:M = 0,25:3:10 and fall into the γ + M_3 (B,C) + M_{23} $(B,C)_6$ region of the Fe–C–B phase diagram. This supports the results given in fig. 4. High Cr contents and a surplus of B apparently reduce the solubility of M_3 (B,C) for carbon.

The SiC particles are harder than all microstructural constituents of the PM materials. Therefore they cut the hard phases and the matrix as well (fig. 5 a). The wear resistance stays rather constant along the gradient. The remainders of CrB_2 in the hardphase center withstand the attack of Al_2O_3 particles. This is honoured by a slight improvement of W_{ab}^{-1} (fig. 5 b). The scratching by flint –being softer than all hard phases – is confined to the matrix areas (fig. 5 c). The spacing of the coarse hard phases, i.e. the free scratch length is reduced from ~ 200 μm at the interface to ~ 100 μm at the surface. As a result the wear resistance is increased by an order of magnitude. In the same direction the size of the coarse hard phases grows from 60 to 75 μm, which is considerably larger than the groove width. Thus, the coarse hard phases are effective barriers under the test conditions given. The precipitated small hard phase are too small (~ 2 μm) and are removed within the chips.

While coarse hard phases are benificial for the wear resistance they tend to fracture under a tensile stress. This crack initiation triggers the fracture in bending. As the size of the coarse hard phases decreases towards the substrate and as their spacing is growing the bending strength increases in this direction. The same is true for the fracture toughness but to a lesser degree. To explain this we consider a dispersion of hard phases in a metallic matrix (fig. 6). Loading a smooth three–point–bending or precracked CT specimen will eventually lead to a plastic zone within the matrix. As the cracking of hard phases or phase boundaries requires much less energy than fracturing the matrix, the interaction of the plastic zone size ω and the hardphase spacing $l_x = 2\, l_y$ plays an important role. Let us postulate that a coarse hard phase cracks, if it is touched by the plastic zone and then look at four cases.

a) $\omega >> l_x$: In three–point bending of smooth samples a number of hardphases will crack already at low hard phase contents. These microcracks will coalesce into a lager one, which triggers the fast fracture.

b) $\omega \geq l_x$: In K_{Ic} tests with pre–cracked CT specimens we estimate $\omega = (1-2\nu)^2\, K_I^2/\pi \cdot R_{p0,2}^2)$. The plastic zone is much smaller than in smooth bending but it still touches hard phases ahead of the main crack resulting in crack extension.

This also holds true for cyclic loading at high ΔK (26 as compared to K_{Ic} values of 28 to 33 MPa $\sqrt{m}$). Cracking of hard phases ahead of the main crack accelerates its growth. The growth rate, therefore, declines with the gradient toward the substrate.

c) $l_y < \omega < l_x$: We expect little crack growth by fracturing of hard phases ahead of the crack front. However, the plastic zone still touches coarse hard phases parallel to the main crack plane. This causes crack branching lowering the effective ΔK. The crack velocity is reduced by this type of crack tip shielding. This applies to ΔK = 18 MPa $\sqrt{m}$ and consequently da/dN is raised toward the substrate.

Of course, crack branching also occurs under the conditions of b) where cracks ahead govern the growth rate. At ΔK = 22 MPa $\sqrt{m}$ the effect of cracks, parallel and ahead of the main crack are in balance and the crack velocity is independent of the gradient.

d) $\omega << l_y$: Normal fatigue crack growth through the matrix is encountered.

5. CONCLUSIONS AND OUTLOOK

By mixing CrB_2 and tool steel powder PM – HIP materials with or without a gradient in hard phase content were manufactured. Their investigation rendered the following results:

1. An interaction of CrB_2 and the steel led to hard phases of twinfold content.
2. The elastic modulus increases with the gradient the coefficient of thermal expansion decreases while the macrohardness stays unchanged.
3. The hardphase content has little effect on the abrasive wear resistance as long as the abrading particles are harder (SiC, Al_2O_3). It is improved by an order of magnitude if the abrasive is softer than the hard phases (flint).
4. The fracture of smooth or pre–cracked specimens under single or cyclic loading depends on the ratio of plastic zone size and hardphase spacing. At certain levels of ΔK crack tip shielding by crack branching due to hard phases is encountered.

Future work will center on the investigation of coated specimens and components with steeper gradients of finer powder. This will influence the interdiffusion during HIP. Therefore, other hardphases will be considered, too. TiB_2 already failed [1]. Finer steel powders may prove an economic problem as the size of $< 100\ \mu m$ used in the present work is a standard one. To obtain a dispersion of hard particles their size ought to match that of the steelpowder.

REFERENCES

1. Holleck, H., Möglichkeiten und Grenzen einer gezielten Stoffauswahl für verschleißmindernde Hartstoffschichten. In Hartstoffschichten zur Verschleißminderung, ed. H. Fischmeister, H. Jehn, Deutsche Gesellschaft für Metallkunde, Oberursel, 1987, pp. 25–43

2. Hirano, T., Yamada, T., Teraki, J., Niino, M., Kumakawa, A., A study on a functionally gradient material. Design system for a thrust chamber. Proc. Japanese/German Seminar "Functional Gradient Materials", DLR, Cologne, May 1989, pp. 15–20

3. Bloschies, G., Fiber preform with ceramic surface layer. Proc. Japanese/German Seminar "Functional Gradient Materials", DLR, Cologne, May 1989, pp. 262

4. Begg, A.R., Brown, C.W., Charman, N.E., Graded structure composites, Britisch Petroleum, EP 260850, 23.3.1988

5. Fischer, A, van Nguyen, C., Härtbare PM–HIP Verbunde mit Boriden, Proc. 2nd Intern. Tooling Conf. "New Materials and Processes for Tooling", Bochum, Sept. 1989, pp. 223–237

6. Stadelmaier, H.H., Gregg, R.A., Die ternäre Phase $Fe_{23}B_3C_3$ im Dreistoffsystem Fe–C–B. Metall 17, 1963, 5, pp. 412–414

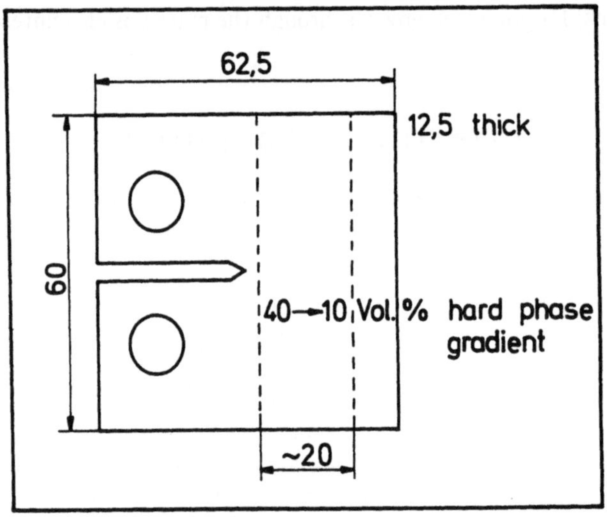

Figure 1. Compact tensile (CT) specimen

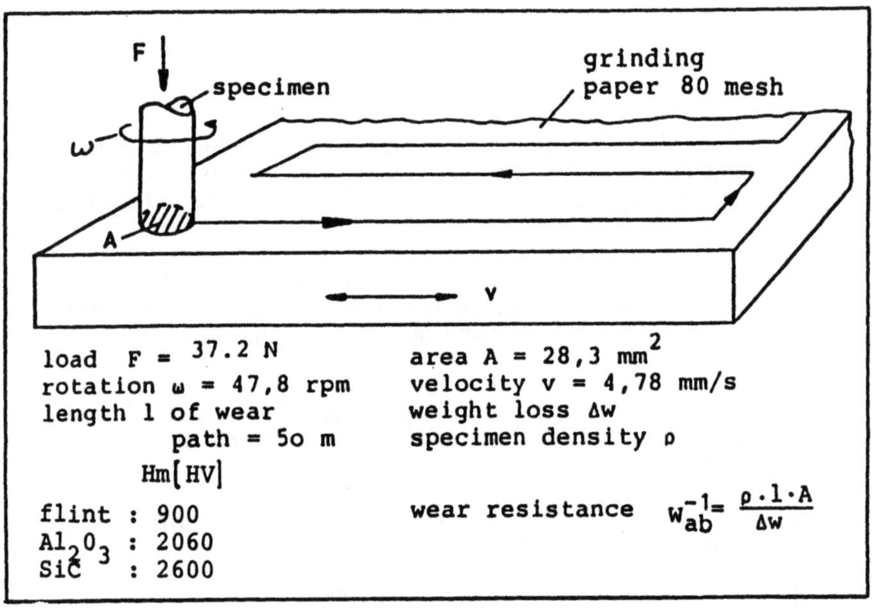

Figure 2. Abrasive wear test

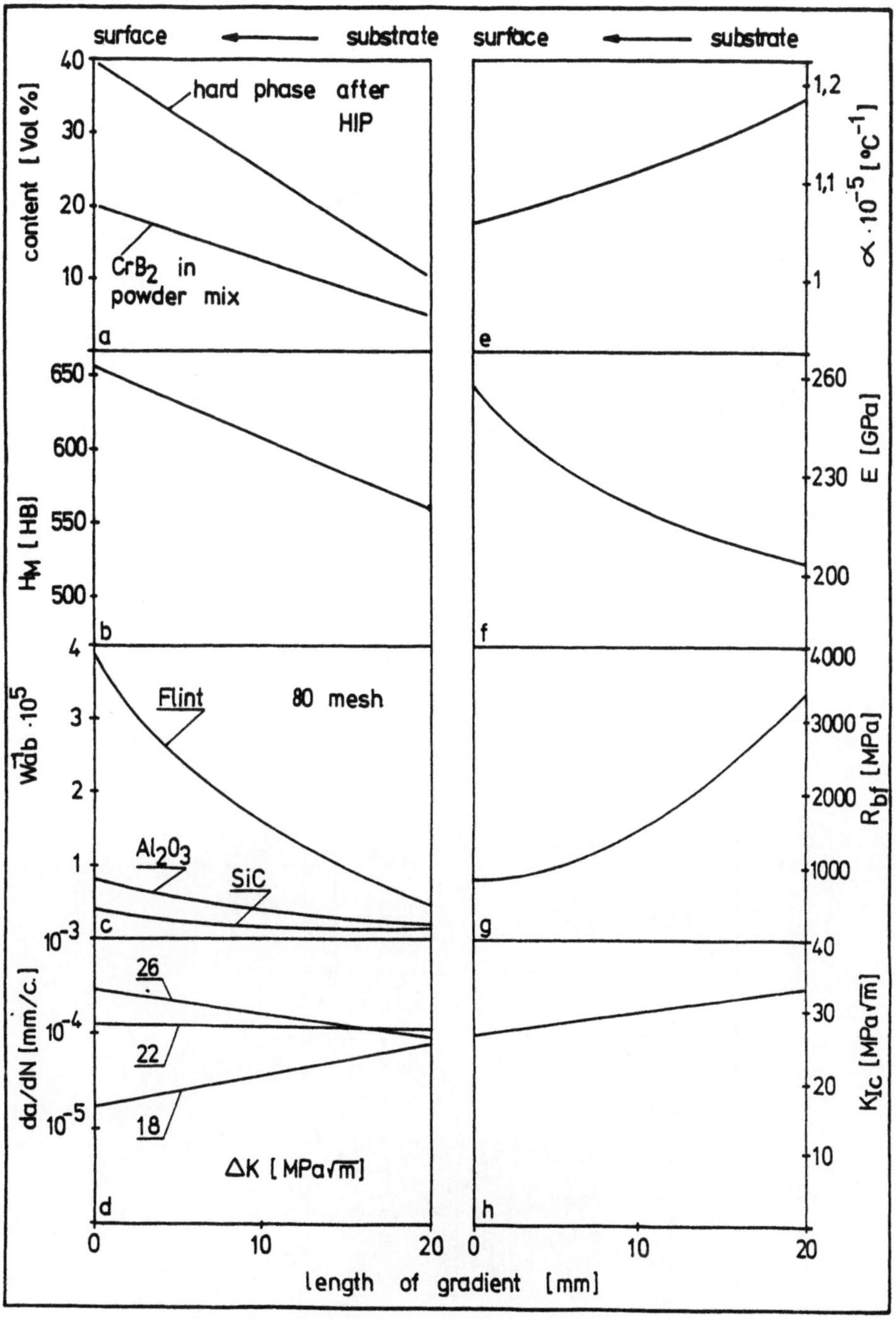

Figure 3. Test results of specimens which were hardened from 900 °C in oil and tempered at 200 °C (e. at 450 °C) for 2 hours; (left column: specimen with hard phase gradient; right column: specimens with constant hard phase content corresponding to a certain length of gradient)

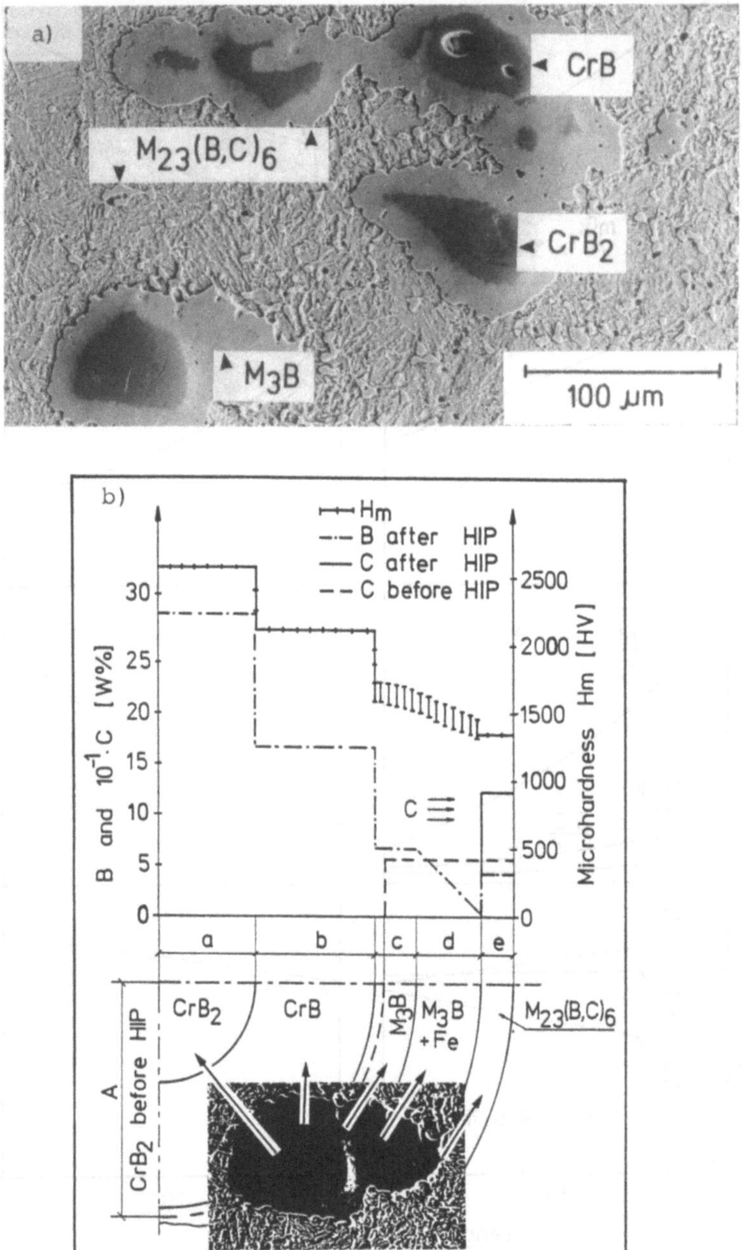

Figure 4. Microstructure at a content of 15 vol.% CrB_2, (a) micrograph; (b) schematic representation of phase analysis, chemical composition and microhardness

Figure 5. Wear grooves caused by (a) SiC, (b) Al_2O_3, and (c) flint

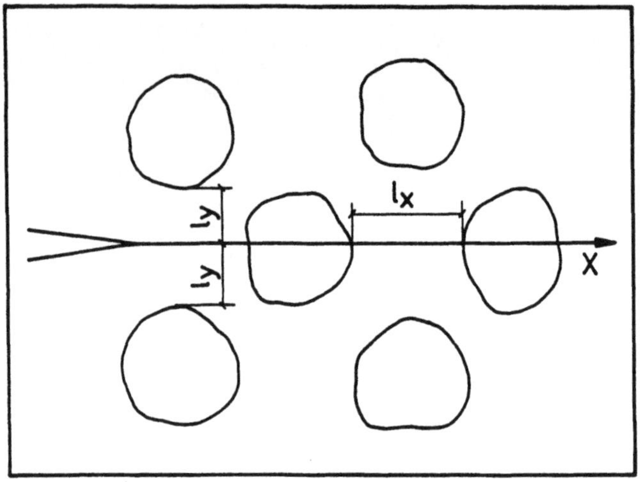

Figure 6. Schematic representation of a hard phase dispersion in a hard metal matrix with relevant distances l_x and l_y

Material	Volume Fraction in % of Hard Particles (Powder)	Hard Phases (Microstructure)	Hardness after Tempering H_M (HB)	H_m (HV0,05)
56NiCrMoV7 (< 100 µm)	0	0	573	Steelmatrix 730
+ CrB_2 (< 100 µm)	5	12	580	Steelmatrix 713
	7	18	587	Hard Phases
	10	21	610	CrB_2 2570, CrB 2100
	15	31	630	M_3B, 1400- 1800
	20	41	653	$M_{23}(B,C)_6$

Table 1: Powder Mixtures and Microstructures of PM-Materials.

Heat Treatment	Temperature in °C	Time in h	Cooling Medium
Annealing	700	10	Furnace
Hardening	900	0.5	Oil
Tempering	200	2	Air

Table 2: Heat Treatment Parameters

Surface Characterization and High-Temperature Adhesive Bonding of Carbon Fiber–Polyimide Matrix Composites

D. J. D. Moyer and J. P. Wightman

Virginia Polytechnic Institute and State University
Blacksburg Virginia USA

ABSTRACT

Carbon fiber-polyimide matrix composites were surface pretreated in numerous ways. Changes in the surface chemistry of the pretreated composites were determined through the use of X-ray photoelectron speectroscopy (XPS). The effect of surface pretreatment on the wettability of the composite surfaces was studied using contact angle measurements. Pretreated composite surfaces were bonded with a high temperature adhesive in both the single lap shear and the wedge configurations. Both types of joints were placed in various environments to determine the effect of surface pretreatment on the strength and durability of these adhesively-bonded composites. After the samples debonded, the failed surfaces were examined again using XPS to determine the locus of failure.

INTRODUCTION

Carbon fiber composites are being increasingly used as primary and secondary structures in commercial and military aircrafts and spacecrafts, where once only metal components were used. Many applications require the composite materials to be joined to itself or to metal. However, the effective use of composites may be limited by the reliability and efficiency of the joining methods used.

Adhesive bonding has several major advantages in comparison with mechanical fastening. Mechanical fasteners create areas of stress concentration due to uneven distribution of loads over large areas. In the specific case of composite materials, fastener holes severely weaken composites reducing the weight advantage of using composite materials and mechanical fasteners can be structurally inefficient due to low shear strengths of composite materials. Adhesive bonding significantly reduces the problems associated with mechanical fasteners stated above, while also yielding higher joint efficiency indexes (relative strength/weight of the joint region) and avoiding potential corrosion problems. If the full potential of adhesively bonded composite materials is to be reached, a thorough understanding of the factors which affect the adhesive bonding of composites must be obtained. In this paper attention will be focused on the role of surface pretreatment on adhesively bonded composites. Two recent reports (1,2) have focused on the adhesive bonding of thermoplastic composites following surface pretreatment of the composites.

Composite structures are made against surfaces from which they must be released after the matrix has cured. These surfaces are usually coated with release agents which may contain fluorinated polymers or silicons and these release agents may be transferred to the composite surface. It has been shown that contamination of the composite by release agents can cause reduction in the strength of adhesively-bonded composite joints (2,3) . To increase bond strength, various surface pretreatments have been used to remove the contamination, the most common being surface abrasion (4,5). Recently, the plasma pretreatment of polymers and solids has been a subject of considerable research interest. One specific advantage of the modification of materials by means of plasmas is that the process selectively alters the surface while leaving the bulk unaffected. In the present work, this plasma treatment was one of the techniques used to pretreat the composite surfaces.

The purpose of this study was to analyze the surfaces of a carbon fiber-polyimide matrix composite, before and after a series of surface pretreatments. Surface characterization included X-ray photoelectron spectroscopy (XPS) and contact angle measurements. Also, this study examined the effect of surface pretreatment on the strength and durability of high-temperature adhesively bonded composites placed in various environments. After samples debonded, further surface characterization was performed on the failed surfaces in order to best understand the cause of failure.

EXPERIMENTAL

Materials

A composite panel (thickness : 2.64 mm; fiber volume: 70%) obtained from personnel at NASA-LaRC was made from Celion 6000 graphite fibers with an NR 150B2 polyimide finish in a polyimide (LaRC-160) matrix and was used for the entire study. The 16 ply panel had a ply orientation of $[0_5/\pm 30/0]_s$. Details of the fabrication process have been previously reported (6).

The adhesive and primer used was LaRC-TPI made by Mitsui Toatsu Chemicals. LaRC-TPI adhesive tape was prepared by brush-coating a polyamic acid [7.5 wt% solids in diglyme for the first two coats and 24.1 wt% solids in diglyme for the following coats] onto 112 E-glass cloth with an A-1100 (γ-aminopropylsilane) finish prior to coating. The glass cloth (tightly mounted on a metal frame) was oven dried for 30 minutes at 100°C. The 0.1 mm (0.004 in) thick glass cloth served as a carrier for the adhesive as well as for bondline thickness control and an escape channel for solvent. Coatings of the polymer solution were applied to the glass cloth until a thickness of 0.25-0.30 mm (0.010-0.012 in.) was obtained. After application of each coat, the adhesive tape was air-dried for 1 hour, placed in a forced air oven for 1 hour at 100°C, 2 hours at 150°C and 3 hours at 175°C. The procedure used to prepare the tape was required to drive off the solvent and reaction product volatiles when converting the polyamic acid resin to the polyimide. Imidization of polyamic acids to polyimides generally occurs in the 160°C range, with the degree of conversion being a function of time and temperature (7).

The LaRC-TPI primer used was 20 wt% solids in diglyme. A thin coat of primer was applied with a brush to the composite samples immediately after pretreatment. Samples were then placed in an oven for 30 min at 60°C, 15 min at 100°C and 15 min for 150°C. Samples were allowed to cool to room temperature and then placed in a desiccator until bonding.

Surface Pretreatments: The composite samples were pretreated in nine different ways noted as 1) as-received, 2) methanol wash, 3) methanol wash and sulfuric acid soak, 4) gritblast and methanol wash, 5) methanol wash and 5 min ammonia plasma, 6) methanol wash and 5 min argon plasma, 7) methanol wash and 7 min argon plasma followed by 5 min ammonia plasma, 8) methanol wash and 5 min nitrogen plasma and 9) methanol wash followed by either 0.5, 1, 2, 5, 10 or 20 min oxygen plasma. The details of each pretreatment have been reported (8).

X-ray Photoelectron Spectroscopy (XPS): XPS analysis was performed on both pretreated composites and sections cut from debonded adherend surfaces. The as-received sample was analyzed using XPS and then analyzed again after it was methanol washed. All other samples were analyzed using XPS after the methanol wash procedure and then after the pretreatment. The oxygen plasma series was performed on the same sample with the noted time being the total time the sample had been exposed to the plasma. The details of the instrumentation and quantification of surface composition have been reported (9).

Contact Angle Measurements: Contact angle measurements were determined for all nine pretreatments stated above. 10 μl deionized water drops and 6 μl diglyme drops were placed on the composite surfaces by a syringe needle held approximately 3 mm above the surface. The contact angle of the drop was measured by a Ramé-Hart #100-00 115 NRL goniometer telescope. At least three drops for each liquid were measured for each pretreatment.

Lap Shear Test: Lap shear coupons (10.16 x 2.54 cm) were pretreated, primed and bonded with one layer of the LaRC-TPI adhesive tape. The lap shear coupons were placed in a jig set for 1.27 x 2.54 cm overlap. The jig was placed in a Carver Laboratory Model C press and heated from room temperature to 316°C (600°F) at 50 psi bonding pressure for 2 hours. The heat was turned off and the joints allowed to cool to room temperature under pressure. All lap shear joints were aged for 12 days in a desiccator before they were placed in a particular environment. After environmental aging lap shear bonds were pulled to break at room temperature on an Instron testing machine at a cross speed rate of 0.127 cm/min. The failure surfaces were studied by XPS to determine the locus of failure.

Wedge Test: Wedge coupons, (2.54 x 12.7 cm) were pretreated, primed and bonded with one layer of LaRC-TPI cut to the same dimensions as the wedge coupons leaving 1 cm clear at each end. All bonding conditions were identical to the lap shear bonding conditions stated above, except that no jig was used to prepare the wedge joints. A 0.38 cm thick wedge was slowly driven into one end of the wedge joint, with the use of a vise, causing an initial crack to propogate. The position of the initial crack was measured with a ruler to the nearest 1 mm and the joint was placed in one of three environments. The crack propagation with time was then determined periodically by removing the joint from the environment to measure the crack. The failure surfaces were examined by XPS to determine the locus of failure.

RESULTS AND DISCUSSION

Surface Composition

The XPS calculated atomic percentages are listed in Table 1 for each element in all samples before and after the specified surface pretreatment.

TABLE 1

XPS results of pretreated composites

Sample Number	Pretreatment	Atomic Concentration (%)* C	O	N	F
1	As-received	46.5	12.4	2.0	36.3
1	Methanol Wash	49.7	7.6	2.3	40.4
2	Methanol Wash	49.6	8.8	2.3	27.7
2	Sulfuric Acid Soak	50.3	10.7	2.5	34.2
3	Methanol Wash	51.5	6.2	2.1	39.6
3	Gritblast	72.6	7.8	6.1	13.5
4	Methanol Wash	50.9	9.0	2.4	36.8
4	0.5 Min. Oxygen Plasma	41.8	19.1	2.7	33.6
4	1 Min. Oxygen Plasma	42.4	21.2	2.8	30.2
4	2 min. Oxygen Plasma	46.2	31.5	4.2	12.9
4	5 Min. Oxygen Plasma	50.8	36.8	2.4	0.8
4	10 Min. Oxygen Plasma	48.4	36.6	3.6	0.7
4	20 Min. Oxygen Plasma	50.1	35.4	3.6	0
5	Methanol Wash	49.9	7.0	2.6	39.7
5	5 Min. Nitrogen Plasma	42.3	13.6	2.9	39.1
6	Methanol Wash	51.6	6.9	1.8	38.5
6	5 Min. Argon Plasma	54.0	22.4	4.4	15.7
7	Methanol Wash	52.5	7.8	1.6	38.1
7	5 Min. Ammonia Plasma	49.0	7.9	5.2	37.4
8	Methanol Wash	51.8	9.0	1.6	37.6
8	7 Min. Argon Plasma then 5 min Ammonia Plasma	63.5	17.1	10.7	5.5

*Balance due to Si, Na, S and Ca.

These results show that there is a large amount of fluorine (36 percent) on the as-received composite. By examination of the C 1s photopeak, this fluorine can be assigned specifically to fluorocarbon. The source of this fluorocarbon is the release cloth which is transferred to the composite panel during the fabrication of the composite panel. A similar result for polyphenylene sulfide composites has been reported by Bonazza (2). Table 1 also reveals that the amount of fluorine decreases with only the gritblast, 5 min argon, 7 min argon followed by 5 min ammonia and all oxygen plasma pretreatments. The amount of fluorine decreased as the time the sample was placed in the oxygen plasma increased. As previously mentioned Parker and

Waghorne (3) reported almost a linear relationship between the concentration of surface fluorine and both the joint strength and the proportion of interfacial failure.

Examination of the XPS C 1s photopeaks show that the 5 min argon, 5 min nitrogen, 7 min argon followed by 5 min ammonia and all oxygen plasma pretreatments increase the level of oxygenated carbon species. The 5 min ammonia and 7 min argon followed by 5 min ammonia plasma increase the amount of nitrogen present. The amount of oxygen increased as the time the sample was placed in the oxygen plasma increased, with a plateau occuring sometime after 2 minutes. The increase in oxygen for the argon and nitrogen plasma treatments may be attributed to the fact that the samples which are extremely reactive after plasma treatment were exposed to the atmosphere immediately after treatment.

Due to chemical reactions which might occur, increasing the number of carbon-oxygen and carbon-nitrogen functional groups may offer great potential for enhanced adhesive bonding with LaRC-TPI. For example, primary amines from either the LaRC-TPI primer of the ammonia plasma pretreated surface have the potential to react with ketone groups on the composite surface or the LaRC-TPI primer, respectively, to form ketamines (9).

The effect of surface pretreatment on interfacial contact angle measurements is shown in Figures 1 and 2. Because the composite surfaces

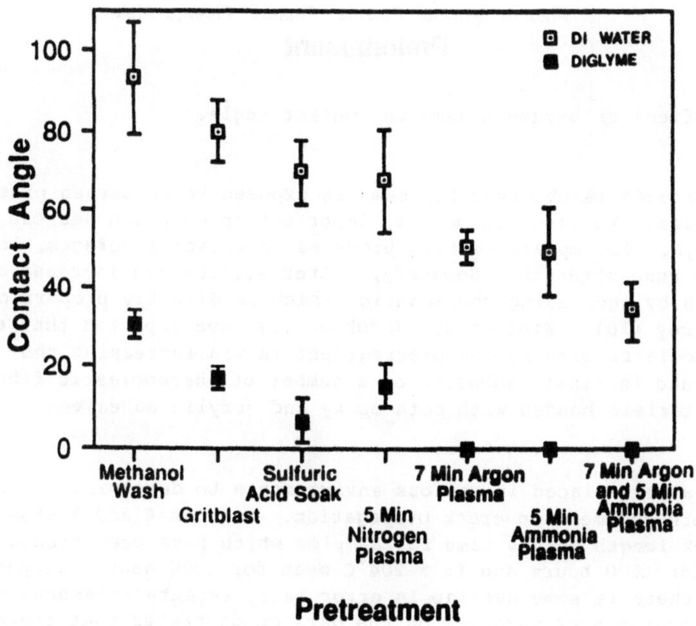

Figure 1. Effect of surface pretreatment on contact angle.

are not homogeneous, some scatter in the contact angle measurements was present. However, it can be clearly seen that the wettability of the pretreated composite surfaces increases due to a combination of a decrease in the amount of fluorocarbon and an increase in the carbon-oxygen and carbon-nitrogen functionality. Also an increase in wettability of the

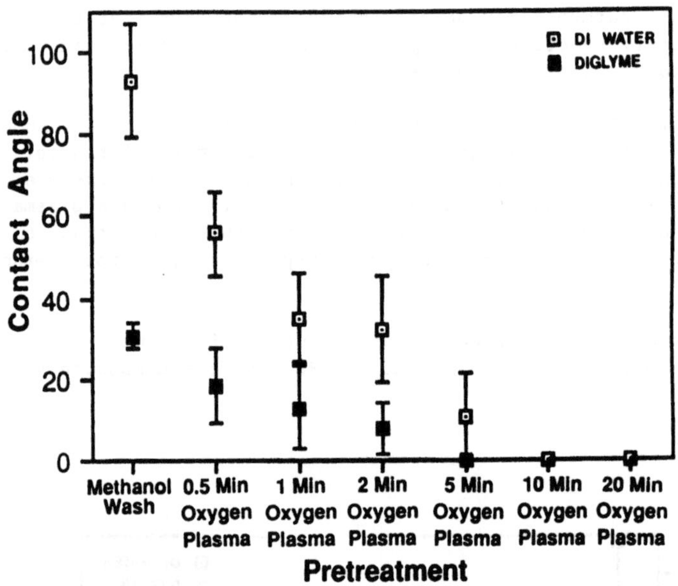

Figure 2. Effect of oxygen plasma on contact angle.

composite surfaces is observed for samples exposed to an oxygen plasma for increasing time. Wetting can be very important in adhesion and can affect it in two ways. Incomplete wetting produces interfacial defects, lowering the adhesive bond strength. Secondly, better wetting can increase adhesive bond strength by increasing the adhesion which is directly proportional to fracture energy (10). Kinloch and Kodokian (1) have reported that strong adhesion associated with corona pretreatment is via increasing the wettability and intrinsic adhesion of a number of thermoplastic fibre-composite materials bonded with both epoxy and acrylic adhesives.

Wedge Test

Wedge samples were placed in various environments to determine the effect of surface pretreatment on crack propagation. Figures 4 and 5 show the average crack length versus time for samples which have been placed in a desiccator for 1000 hours and in a 204°C oven for 1000 hours, respectively. Even though there is some overlap in error bars, repeated-measures analysis and one-way analysis of variance on the data demonstrates that there are definite statistical differences between the oxygen plasma pretreated surfaces in comparison to the methanol wash and gritblast wedge samples.

For all three environments including a hot wet-one, the oxygen plasma pretreatment was shown to be the best pretreatment, since the mean crack lengths of the oxygen plasma pretreated samples were always lower than for both gritblast and methanol wash wedge samples. Also there was a decrease in initial crack length for the samples which had been oxygen plasma pretreated for longer periods of time. In addition, it can clearly be seen that the error bars in Figures 3 and 4 are larger for the methanol wash and

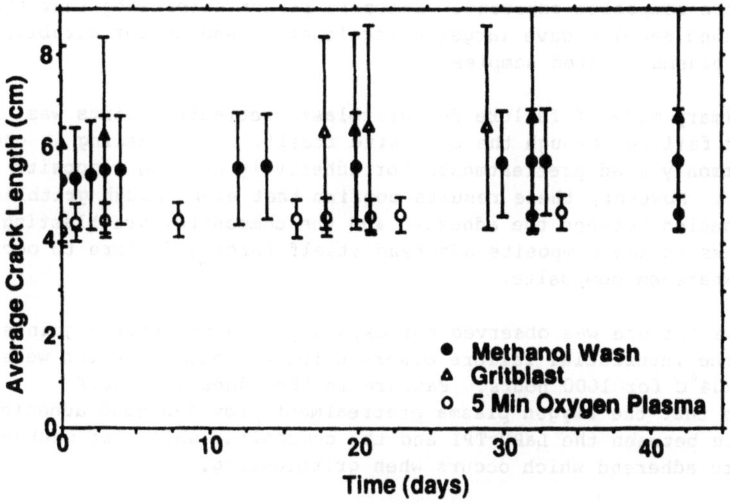

Figure 3. Effect of pretreatment on crack propagation (dessicator - 1000 hours.

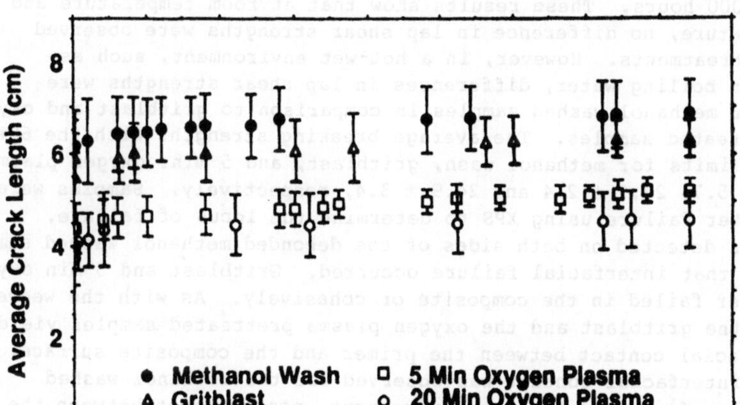

Figure 4. Effect of pretreatment on crack propagation (204°C - 1000 hours).

gritblast pretreated samples than for wedge samples which have been oxygen plasma pretreated. Therefore oxygen plasma pretreatment is considered to be the best surface pretreatment since reproducibility is very important when determining optimum surface pretreatments.

After the wedge samples had debonded, the failed surfaces were examined using XPS to determine cause of failure. Fluorine was detected on both sides of failed methanol washed samples. Therefore the primary mode of failure for the methanol washed samples was interfacial between the primer and the composite adherend. Hence it is not surprising that the methanol washed samples gave larger crack lengths, and poorer durability than oxygen plasma treated samples.

The primary mode of failure for gritblast pretreated joints was interlaminar failure through the composite itself. Gritblasting is one of the most commonly used pretreatments for adhesively bonding composite joints today. However, these results confirm that even though gritblasting improves adhesion between the adhesive and the composite, gritblasting also produces flaws in the composite adherend itself forcing failure to occur within the weakened composite.

Cohesive failure was observed for oxygen plasma pretreated joints, except for the interfacial failure observed for the samples which were exposed to 204°C for 1000 hours. Failure in the adhesive itself demonstrates that the oxygen plasma pretreatment provides good adhesion at the interface between the LaRC-TPI and the composite, while not weakening the composite adherend which occurs when gritblasting.

Lap shear joints were made from methanol wash, gritblast and 5 min oxygen plasma pretreated composites. The average breaking strengths with the 68% confidence limits for methanol wash, gritblast and 5 min oxygen plasma were 18.8 ± 4.3, 16.2 ± 2.1 and 23.5 ± 5.1 respectively for samples aged in a desiccator at room temperature for 1000 hours, and 10.2 ± 2.3, 11.5 ± 1.2, and 10.8 ± 3.4 respectively for samples environmentally aged at 204°C for 1000 hours. These results show that at room temperature and at high temperature, no difference in lap shear strengths were observed between pretreatments. However, in a hot-wet environment, such as immersion in boiling water, differences in lap shear strengths were observed for methanol washed samples in comparison to gritblast and oxygen plasma pretreated samples. The average breaking strengths with the 68% confidence limits for methanol wash, gritblast, and 5 min. oxygen plasma were 12.2 ± 5.7, 23.2 ± 2.4 and 24.9 ± 3.4, respectively. Samples were examined after failure using XPS to determine the locus of failure. Fluorine was detected on both sides of the debonded methanol washed samples which means that interfacial failure occurred. Gritblast and 5 min oxygen plasma either failed in the composite or cohesively. As with the wedge test, both the gritblast and the oxygen plasma pretreated samples yielded good interfacial contact between the primer and the composite surface, while poor interfacial contact was observed for the methanol washed samples. Therefore in a hot-wet environment, poor contact between the adhesive and the adherend will yield lower lap shear strengths than with samples which exhibit good interfacial contact.

ACKNOWLEDGEMENTS

The authors acknowledge the financial support of this research under the NASA Graduate Student Researchers Fellowship program. We thank T. L. St. Clair and D. J. Progar for helpful technical discussions, and Frank Cromer for training to operate the XPS. Additional support of this research by the Center for Adhesive and Sealant Science and the Virginia Institute of Material Systems is also acknowledged.

REFERENCES

1. Kinloch, A. J. and Kodokian, G. K. A., The adhesive bonding of thermoplastic composites, Final Report DAJA-45-86-C-0037, 1989, European Research Office, U. S. Army, London, England.

2. Bonazza, B. R., Adhesive bonding of polyphenylene sulfide advanced thermoplastic composites, Proc. 45th Annual SPI Conf., 1990, Washington, DC.

3. Parker, B. M. and Waghorne, R. M., Surface pretreatment of carbon fiber-reinforced composites for adhesive bonding, Composites, 1982, **13**, 280.

4. Crane, L. W., Hammermesh, C. L. and Maus, L., Surface treatment of cured epoxy graphite composites to improve adhesive bonding, SAMPE J., 1976, **12**, 6.

5. DeVilbiss, T. A., Messick, D. L., Progar, D. J. and Wightman, J. P., SEM/XPS analysis of fractured adhesively bonded graphite fibre-reinforced polyimide composites, Composites, 1985, **16**, 207.

6. DeVilbiss, T. A., Progar,D. J. and Wightman, J. P., SEM/XPS analysis of fractured adhesively bonded graphite fibre surface resin-rich/graphite fibre composites, Composites, 1988, **19**, 67.

7. Progar, D. J., Evaluation of a thermoplastic polyimide (422) for bondinng GR/PI composite, J. Adhesion Sci. Technol., 1988, **2**, 449.

8. Moyer, D. J. D. and Wightman, J. P., Characterization of surface pretreatment of carbon fiber/polyimide matrix composites, Surface and Interface Analysis 1989, **14**, 496.

9. Mohanty, D. K., Wu, S. D. and McGrath, J. E., Synthesis and characterization of ketamine functional poly(ether ether) ketone networks, Polymer Preprints, 1988, **29**, 352.

10 Wu, S., Polymer Interface and Adhesion, Marcel Dekker, Inc., New York, 1982, p 360.

Contact Angles on Imperfect Solid Surfaces

J. Gaydos, C. J. Budziak, D. Li and A. W. Neumann
University of Toronto
Toronto
Ontario
Canada

Abstract

The wetting behaviour of solids is governed by the Young equation which assumes that the surface of the solid is ideal. However, in practical engineering applications such as the spreading of adhesives or paints the wetting phenomena are more complicated than the Young equation would indicate. Real solids are both rough and chemically heterogeneous and these surface characteristics lead to non-uniform wetting. Non-uniform wetting of solids may be characterized in part by contact angle hysteresis or the difference in equilibrium contact angles which exists between adjacent heterogeneous patches on a solid. The presence of these patches, which vary in size, inhibits uniform spreading. In addition, the line of contact between the liquid and the solid may experience large contortions in the vicinity of the patch-patch boundary. When contortions are large an additional quantity, known as line tension, is apparent. When present, the line tension rounds-out the contortions, reduces the magnitude of the contact angle difference between adjacent patches and improves the likelihood of uniform spreading. Recent progress in the characterization of the dynamics of the contact line and of the influence of line tension on wetting will be discussed. Estimates of the maximum patch size, with a given line tension, for which spreading is not inhibited by contact angles differences between adjacent patches will also be presented.

Introduction

A large variety of phenomena involve the spreading of a liquid on a solid. These phenomena range from quite common effects like insect flotation on quiescent pools of water and cellular phagocytosis to important industrial processes which involve steam condensation, and the spreading of solder, paints, lubricants and inks. The essential feature in all of these capillary processes is that the system is arrayed so as to minimize its energy. In the next section we shall briefly discuss the equations and boundary conditions which are common to all of these equilibrium problems. A complete derivation of these relations on the basis of the stationary principle of the total grand canonical potential or free energy Ω_t is presented in the Appendix. Next, a brief discussion of current theoretical approaches to non-uniform or heterogeneous solid surfaces follows. Only systems of simple geometry and macroscopic size will be considered for simplicity. Finally, the paper ends with a description of recent progress in the measurement and description of these effects using digital image processing and axisymmetric drop shape analysis (ADSA).

Equilibrium Conditions

If a liquid in a vessel is to be in equilibrium the following conditions are physically necessary:

(1) The hydrostatic (or Poisson) condition

$$\nabla P^{l} = -\rho \nabla \phi \tag{1}$$

where P^l is the pressure of the bulk liquid phase, ρ is the density and ϕ is a conservative force-field potential.

(2) The Laplace equation of capillarity for the pressure change across a curved surface

$$P^{l} - P^{v} = \gamma^{lv} J \tag{2}$$

where P^v is the constant pressure in the fluid (or vapour) phase above the liquid, γ^{lv} is the surface tension and J is the mean curvature of the liquid-vapour surface.

(3) The Young equation of capillarity

$$\gamma^{lv} \cos\theta = \gamma^{sv} - \gamma^{sl} \tag{3}$$

which may be considered as a force balance in the plane of the solid between the adjacent surface tensions at the location of the three-phase line (see fig. 1). Under some conditions the system is either so small or the three-phase contact line L^{slv} so contorted that an additional effect along the boundary arises. The line tension, denoted by σ^{slv}, yields this additional effect. When present, the line tension which is a one-dimensional analog of the surface tension rounds-out the contortions or straightens the contact line. The boundary condition (3) is also modified [1] to include the effect of σ^{slv}

$$\gamma^{lv} \cos\theta + \sigma^{slv} \kappa_{gs} = \gamma^{sv} - \gamma^{sl} \tag{4}$$

where κ_{gs} is the geodesic curvature (see Appendix) of the three-phase contact line in the plane of the A^{sl} surface.

In what follows we shall assume that the A^{sl} surface may be represented by a function of the form

$$\xi(\mathbf{x}) = 0 \tag{5}$$

where $\mathbf{x}$ is the equation of the A^{lv} surface, $\xi(\mathbf{x}) > 0$ inside the solid boundary (i.e., in the direction of positive $\mathbf{n}^{sl}$) and $|\nabla\xi| \neq 0$ on surface A^{sl}. With the equation for the A^{lv} equilibrium surface expressed in the form $z = \zeta(x, y)$ equations (a11) and (a12) simplify to

$$div \frac{\nabla\zeta}{\sqrt{1+(\nabla\zeta)^2}} = \frac{\rho}{\gamma^{lv}}\phi + \frac{\lambda}{\gamma^{lv}} \tag{6}$$

and

$$\frac{-\zeta_x \xi_x - \zeta_y \xi_y + \xi_z}{\sqrt{1+(\nabla\zeta)^2}\ |\nabla\xi|} = \cos\theta \tag{7}$$

with $\xi(x, y, f(x,y)) = 0$ on the three-phase line. Finally, if the liquid is acted upon by an uniform gravitational field of intensity g, directed along the z-axis, the conservative force-field potential ϕ, which is assumed to be a known function, is represented by $\phi = gz$. Using the following notation

$$b = \frac{\rho g}{\gamma^{lv}} \quad ; \quad c = \frac{\lambda}{\gamma^{lv}} \tag{8}$$

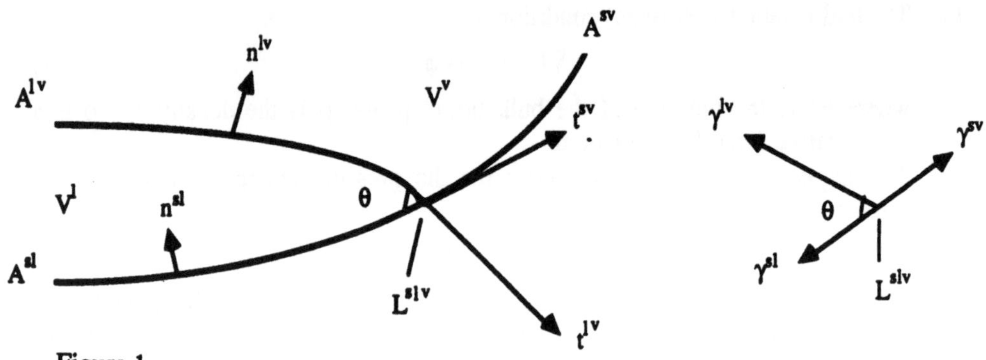

Figure 1

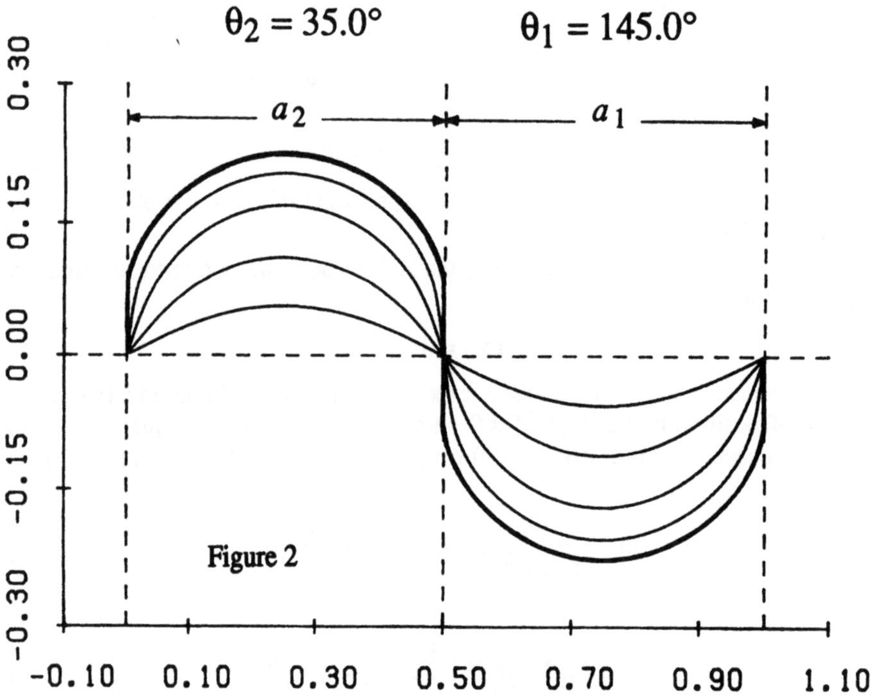

Figure 2

eq. (6) may be expressed as

$$div \frac{\nabla \zeta}{\sqrt{1+(\nabla \zeta)^2}} = b\zeta + c \tag{9}$$

If the system possesses a characteristic length l, which might correspond to the periodicity of a heterogeneous pattern for example, then the notation introduced in eq. (9) may be replaced by the dimensionless parameters

$$b_o = bl^2 = \frac{\rho g l^2}{\gamma^{lv}} \quad ; \quad c_o = cl = \frac{\lambda l}{\gamma^{lv}} \tag{10}$$

where b_o is commonly known as the Bond number.

When the Bond number is small (or $b \ll 1$) eq. (9) closely resembles the differential equation for a minimal surface [2], given by

$$\frac{(1+\zeta_y^2)\zeta_{xx} - 2\zeta_x \zeta_y \zeta_{xy} + (1+\zeta_x^2)\zeta_{yy}}{(1+\zeta_x^2+\zeta_y^2)^{3/2}} - b\zeta = c \tag{11}$$

For an axisymmetric system (with r representing the radius) eq. (9) simplifies to the ordinary differential equation [3]

$$\frac{1}{r}\frac{d}{dr}\left[\frac{r\zeta_r}{\sqrt{1+\zeta_r^2}}\right] = b\zeta + c \tag{12}$$

Analytical solutions to both (11) and (12) are not possible in general; however, for the case where $|\zeta_r| \ll 1$ such that the boundary condition (3) with $\cos\theta \approx 1$ permits eq. (12) to be linearized to yield

$$\zeta_{rr} + \frac{1}{r}\zeta_r - b\zeta = c \tag{13}$$

one may express the solution to this Bessel equation [4] which remains finite at $r = 0$ as

$$\zeta(r) = \begin{cases} \tilde{c}\, I_o(\sqrt{b}\, r) - \frac{c}{b} & \text{for } b > 0 \\ \tilde{c} + \frac{1}{4} c r^2 & \text{for } b = 0 \\ \tilde{c}\, J_o(\sqrt{|b|}\, r) - \frac{c}{b} & \text{for } b < 0 \end{cases} \tag{14}$$

where $\tilde{c}$ is an arbitrary constant and I_o and J_o are Bessel functions of the first kind. When the approximation $|\zeta_r| \ll 1$ is not valid a numerical routine has recently been developed to solve eq. (12) for the case of an axisymmetric sessile drop [5]. Required input parameters are b and V^l in addition to the three-phase contact line profile. This profile, which may be measured easily and precisely using recent developments in image processing [6], yields both r and $\kappa_{gs} = 1/r$.

On real solid surfaces these problems are enhanced by intrinsic characteristics of the surface which include chemical heterogeneity due to polycrystallinity or adsorption of chemical species, roughness and surface deformation. These influences have been discussed by a number of investigators [7-20] in connection with the experimental difficulty of verifying Young's equation (3). In practical situations the contact angle θ can achieve multiple values, however, in many cases the contact angle has been found to have two fairly reproducible angles. One angle θ_a corresponds to the case where the liquid slowly advances over the solid surface while the other angle θ_r is for a receding liquid. Contact angle hysteresis is defined simply as the difference, $\theta_a - \theta_r$. Obviously, the presence of this difference is at odds with eq. (3) which predicts one unique value for a given system. Early work by Wenzel [21] on physical roughness and Cassie [22] on heterogeneity attempted to find average properties for the surface. Recent studies [8,12,14,15,20] have considered 'weak fluctuations' in local interfacial energies, according to the prescription

$$h(\xi(\mathbf{x})=0)=\gamma^{sv}-\gamma^{sl}-\langle\gamma^{sv}-\gamma^{sl}\rangle \qquad (15)$$

where the angular brackets denote an area average, for relatively straight contact lines (i.e., $\kappa_{gs}=1$) and small contact angles (i.e., when either $|\nabla\zeta| \ll 1$ or $|\zeta_r| \ll 1$) with general surface geometries or patch structures. At the other extreme, the case of 'strong fluctuations' for the simple geometry of a stripwise planar wall has been solved analytically [7] when $b_o \ll 0$. In the latter case, both κ_{gs} and $|\nabla\zeta|$ achieve significant non-zero values close to the patch-patch or strip boundaries. Intermediary results [16,17] for the case of strong fluctuations on periodic solid structures of characteristic length l with $\kappa_{gs} \neq 0$ and $|\nabla\zeta| \ll 1$ such that eq. (11) may be written as

$$\zeta_{xx}+\zeta_{yy}-b\,\zeta=c \qquad (16)$$

have yielded estimates of the patch size below which hysteresis vanishes as $l \approx 2\,\mu m$ which is significantly larger than the 1 nm thickness of the liquid interface and slightly higher than the 0.1 μm lower limit estimated previously [7,13]. Thus, when the characteristic length l of the patch structure is extremely small, but non-zero one would expect hysteresis to vanish. Similar conclusions have also been reached by those researchers studying the problem in the weak fluctuation limit. In those situations where it is not correct (i.e., the strong fluctuation limit) to model the three-phase contact line as straight or nearly straight it becomes necessary to consider the influence of κ_{gs} and to replace boundary condition (3) with (4), which involves the line tension. Physically, the existence of even a small line tension (e.g., $\sigma^{slv} \approx 10^{-8}\,J/m$), when combined with a small radius of curvature (e.g., $\kappa_{gs} \approx 10^{+6}\,m$), will produce a discernible impact (as illustrated in fig. 2) on the contact angle in (4) and position of the contact line for virtually all liquids [23]. In fig. 2 only two solid sections or strips, of characteristic length $l=a_1+a_2$, are shown of a periodic stripwise wall. The periodic stripwise wall arrangement, which is discussed elsewhere [7,23], is used to model contact angle hysteresis and the influence of σ^{slv} upon this hysteresis. The surface of the solid wall is composed of two alternating strips of material which are sufficiently long that edge effects are not important. The alternating sections of the wall consist of homogeneous solid pieces which are joined together at a straight solid-solid or patch-patch boundary (the vertical dashed lines in fig. 2). Each solid section of the wall has an unique contact angle across its entire surface, but the contact angle between adjacent sections is unequal. Consequently, the three-phase contact line is very contorted in the vicinity of the solid-solid boundary. Curve 1 represents the computed shape of the contact line for a liquid on a solid surface when σ^{slv} is zero. In this case, the difference between the contact angles on adjacent surfaces is so large that the contact line even follows the solid-solid boundary over a portion of its path length. The behaviour of the contact line illustrated in curves 2 through 4 represents progressively larger values of the ratio $\sigma^{slv}/(\gamma^{lv} l)$ from 0.2×10^{-4} to 0.4×10^{-4}. If we consider the case of water at room temperature with $\gamma^{lv} \approx 0.072\, J/m^2$ in contact with a solid wall whose characteristic length $l \approx 0.001\, m$ then curve 4 represents the position of the contact line when $\sigma^{slv} \approx 0.3 \times 10^{-8}\, J/m$. As σ^{slv} is increased from zero the length of the contact line is reduced because each section of the contact line is pulled towards the position of the dashed horizontal line. In addition, the contact angle difference between adjacent sections is reduced. In the limit of infinitely large σ^{slv} the contact line would be a straight, horizontal line at the same level as the dashed horizontal line shown in fig. 2. Thus, in the strong fluctuation limit where the three-phase contact line between the liquid and the solid may experience large contortions (i.e., $\kappa_{gs} \gg 1$) in the vicinity of the patch-patch boundary, the line tension term $\sigma^{slv}\kappa_{gs}$ in boundary condition (4) will influence the local equilibrium value of the contact angle. As noted previously, the line tension rounds-out the contortions, reduces the magnitude of the contact angle hysteresis between adjacent patches and improves the likelihood of uniform spreading. Recent progress in this direction indicates that even in the strong fluctuation limit a patch size limit of 1 μm is not unreasonable [23] given the fact that σ^{slv} values as large as $10^{-6}\, J/m$ have been reported [24-26].

The determination of σ^{slv} is still in its infancy; one approach is to measure the drop size dependence of contact angles. In the case of an axisymmetric sessile drop sitting on a horizontal, smooth and homogeneous solid surface the three-phase contact line is circular so that κ_{gs} is constant and equal to $1/r$. Under these conditions, eq. (4) simplifies to

$$\gamma^{lv}\cos\theta + \frac{\sigma^{slv}}{r} = \gamma^{sv} - \gamma^{sl} \tag{17}$$

As $r \to \infty$, the contribution of the line tension term will become insignificant so that eq. (17) is equivalent to eq. (3) or

$$\gamma_{lv}\cos\theta_\infty = \gamma^{sv} - \gamma^{sl} \tag{18}$$

where θ_∞ indicates that the radius of the drop is extremely large. Eliminating $\gamma^{sv} - \gamma^{sl}$ from eq. (17) using eq. (18) yields the dependence of the contact angle on drop size or radius as

$$\cos\theta = \cos\theta_\infty - \frac{\sigma^{slv}}{\gamma^{lv} r} \tag{19}$$

Thus, on suitably prepared surfaces a plot of $\cos\theta$ versus $1/r$ gives an estimate of σ^{slv} from the slope if γ^{lv} is known.

Most recently the dependence of the contact angle on drop size has been investigated for dodecane and ethylene glycol drops on carefully prepared FC721, Zonyl FSC and DDOA surfaces [25]. The FC721 and Zonyl FSC surfaces were prepared by dip-coating fluorosurfactant FC721 and Zonyl FSC on molecularly smooth mica surfaces; DDOA surfaces were prepared by coating a Dimethyldioctadecylammonium (DDOA) monolayer on a mica surface by the Langmuir-Blodgett technique. The sessile drops were formed by slowly feeding liquid through a small hole in the solid substrate by a motor-driven syringe until the desired drop size was achieved. Three images were acquired and stored in the computer for later analysis by ADSA-P [27]. Essentially ADSA-P is a general numerical method which fits a set of profile coordinates which have been acquired using the digital image techniques described below to eq. (12). It does not require specific points along the profile and is not restricted to angles of any particular size. In addition to the contact angle θ, ADSA-P determines the interfacial tension γ^{lv}, volume V^l, surface area A^{sl}, radius of curvature and contact radius r. For all systems investigated the contact angle decreased from 3° to 5° with a change in contact radius from 1 to 5 mm. Beyond approximately 5 mm the contact angles tended to reach a final plateau value. This behaviour has also been observed with sessile drops of various n-alkanes on Teflon (FEP) [24]. In all these systems the results are consistent with a positive value of σ^{slv} of approximately 10^{-6} J/m.

Digital Image Analysis

Many techniques are available for the measurement of contact angles for solid-fluid-liquid systems [28]. The most accurate technique is based on the capillary rise of a liquid on a planar solid [28,29]. In this way, the measurement of a contact angle is reduced to the measurement of a length, which can be easily performed and with much higher accuracy and objectivity than the direct determination of an angle. The connection between the capillary rise h and the contact angle θ for a vertical, flat solid surface is obtained by integrating eq. (2) with appropriate boundary conditions to get

$$\sin\theta = 1 - \frac{1}{2}bh^2 \tag{20}$$

In addition to accuracy, further advantages include the ability to measure both advancing and receding as well as dynamic and static contact angles. As is well known, any imperfections in the solid surface will distort the linearity of the contact line. Since the contact line is observed directly, any irregularities are easily detected and can be quantitatively measured.

Conventionally, a cathetometer (i.e., a non-contact method of measuring length using a vertical measuring telescope) has been used to see the contact line and to measure h. Recently we have automated the technique and have utilized digital image analysis and processing to detect the contact line. The apparatus consists of three motorized, computer controlled translational stages which are used to move a horizontally mounted microscope fitted with a charge coupled device (CCD) scanner along the three orthogonal directions, x, y and z. The x, y and z-axis stages have a total travelling displacement of 5, 150, and 50 mm, respectively, and a resolution of 0.1 μm per step for the x and z-axis stages and 10 μm per step for the y-axis stage. The contact line is digitally recorded under high magnification, 160X with a small working field of 250 μm, using a digital video processor and the subsequent array of intensity levels is stored in computer memory. The working field is split or digitized into 480 x 512 pixels with an 8 bit resolution per pixel which permits 256 intensity or gray levels. Features in the image frames are distinguished from the background by locating the boundaries or edges of these images; the edge has a significant change in magnitude of the intensity level. The Sobel 3 x 3 edge detection operator is used to detect these changes and thus the edge. The pixels identifying the edge of the contact line are located by the computer to an overall uncertainty of ± 0.8 μm in the x-direction. Full details of the automation are given in reference [30]; the main advantages of the automation are that

(1) the accuracy of both the capillary rise and the contact angle measurement is increased by an order of magnitude. For example, eight consecutive contact angle measurements yield a confidence limit of $\pm 0.01°$ at the 95% level of confidence,

(2) detailed study of the contact line may be undertaken with relatively high accuracy such that objects as small as 0.8 μm are resolved, and

(3) with our driving mechanism very slow rates [31] of contact line movement, as low as $1 \times 10^{-7} m/s$, are attained.

At present a total length of only 250 μm, in the horizontal or x-direction, may be investigated at one time at this magnification. Longer sections of the contact line may be digitized by sequentially moving the microscope laterally. Hence, it is possible to reconstruct the complete line from the smaller working fields. The ability to perform this reconstruction is particularly advantageous when dealing with heterogeneous surfaces where the contact line is contorted. An example of a single field and a line shape is depicted in fig. 3. The depression in the line shape near the middle of the field is appoximately 20 pixels or 10 μm in depth at maximum. In connection with the stripwise wall mentioned above, the reconstruction ability permits an accurate measurement of the capillary rise and contact angle as well as the behaviour of the contact line at the boundary between the adjacent patches. Ultimately it is expected to yield σ^{slv} as described elsewhere [23].

Developments in digital image analysis have also been applied towards the detection of the curve L^{slv} around the perimeter of a sessile drop (see fig. 4) whose contact angle is small but not sufficiently small that solution (14) applies. The sessile drop arrangement is advantageous whenever one is restricted to only small amounts of liquid or to only a few square millimeters of substrate as is the case with many biological substrata.

If accuracy is an important factor in these measurements then a precise evaluation of the perimeter is necessary. The approach handles situations where the contact angle is small and it is difficult to acquire coordinates along the drop profile [5]. The modified program (ADSA-CD) does not require coordinates along the drop profile like ADSA-P; instead, the drop volume V^l, liquid surface tension γ^{lv} and the contact radius $\dot{r}$ are required as input data. The contact radius or drop perimeter curve is easily acquired by digitizing the drop image from above (when $\theta \leq 90°$) so that the flatness of the profile is not an impediment.

The relationship between contact angle, drop volume and contact radius of the drop's perimeter cannot be expressed in closed analytical form, but the numerical solution represented by ADSA-CD solves eq. (12) using a set of ordinary differential equations which are given elsewhere [5]. The key point is that digital image analysis permits one to assess how closely the A^{sl} surface approximates a circle. When this wetted region is circular a precise numerical solution using ADSA-CD is possible. Even when slightly imperfect, determination of the contact area, for example by image analysis, will allow determination of an effective contact diameter and hence an average contact angle from which a solution is possible. As an example, the contact angles of sessile drops of water on layers of T. thiooxidans bacteria cells were calculated using ADSA-CD over a range of times from essentially zero to 90 minutes

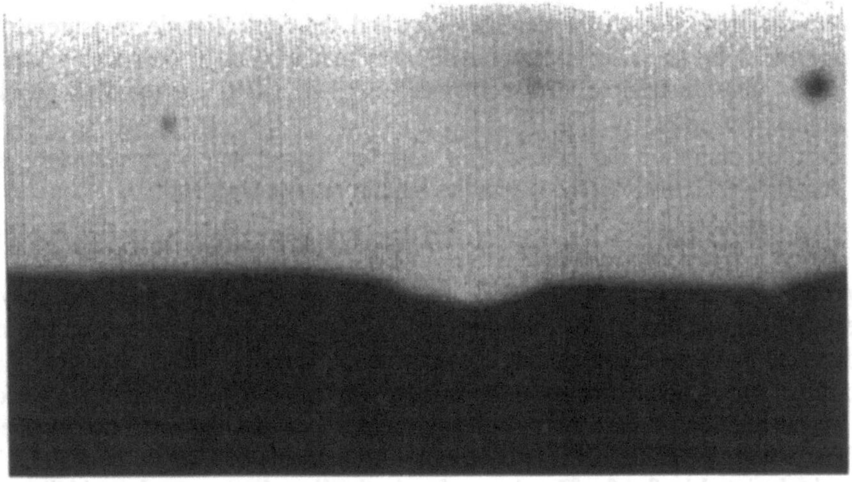

Figure 3 Actual image of a contorted contact line observed at 160X magnification on a rough, heterogeneous solid surface.

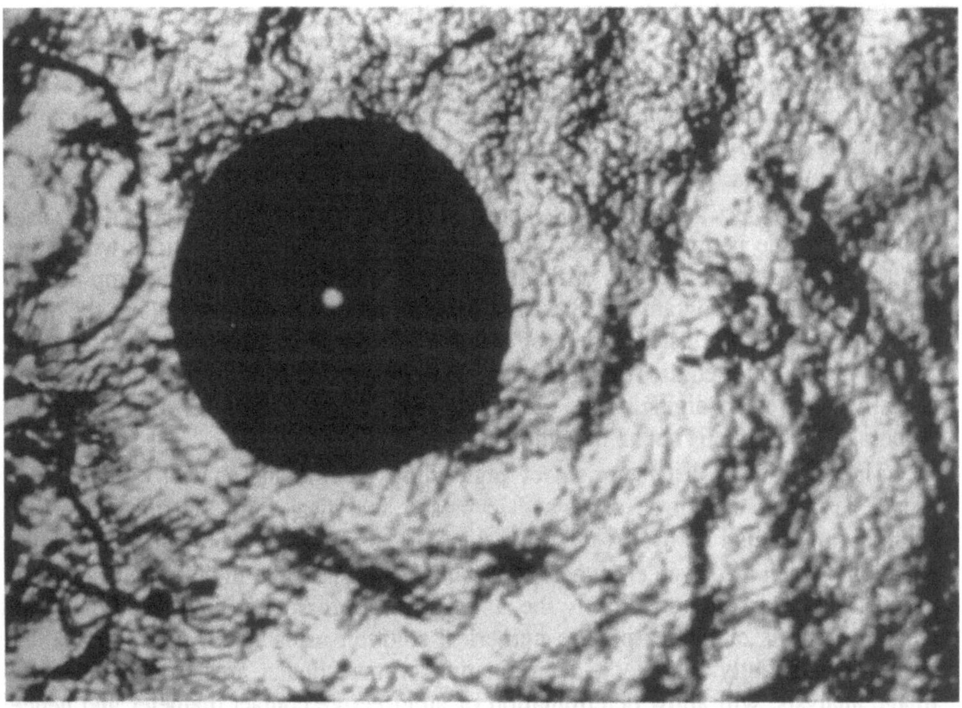

Figure 4 Image of a non-axisymmetric sessile drop of water on a thin layer of T. thiooxidans cells.

[32]. The average contact angle at the 95% level of confidence was 16.9 ± 0.9°. More drastic irregularities like those described elsewhere [20] are also subject to experimental study.

Conclusion

The image analysis and processing methodology allows the study of contact angle and wetting phenomena well outside the domain of studies done with ideal solid surfaces. For non-ideal surfaces exhibiting irregular contact lines, the local contact line variation can be observed under both static and dynamic conditions. High resolution on the order of ± 0.8 μm with the present apparatus permits heterogeneous surfaces with both a small minimum patch size and a small contact angle difference between adjacent patches to be studied. Slightly higher resolution is possible by increasing the magnification and using a short-wavelength laser. With these developments the actual experimental restriction on the study of the wetting properties of heterogeneous surfaces is connected with the preparation of controlled surfaces. However, future prospects on the manufacture of suitable surfaces is promising given recent developments in semiconductor microfabrication technology [33].

References

(1) Boruvka, L. and Neumann, A.W., *J. Chem. Phys.*, 1977, **66**, 5464.

(2) Struik, D.J., **Lectures on Classical Differential Geometry**, 2nd. ed., Addison-Wesley, London, 1961.

(3) Shanahan, M.E.R., *J. Chem. Soc. Faraday Trans. I*, 1982, **78**, 2701.

(4) Lebedev, N.N., **Special Functions & Their Applications**, (R.A. Silverman, trans.), Dover, New York, 1972.

(5) Skinner, F.W., Rotenberg, Y. and Neumann, A.W., *J. Colloid Interface Sci.*, 1989, **130**, 25.

(6) Cheng, P., Li, D., Boruvka, L., Rotenberg, Y. and Neumann, A.W., Automation of Axisymmetric Drop Shape Analysis for Measurements of Interfacial Tensions and Contact Angles. *Accepted for publication in Colloids and Surfaces*, 1989.

(7) Boruvka, L. and Neumann, A.W., *J. Colloid Interface Sci.*, 1978, **65**, 315.

(8) de Gennes, P.G., *Rev. Mod. Phys.*, 1985, **57**, 827.

(9) Eick, J.D., Good, R.J. and Neumann, A.W., *J. Colloid Interface Sci.*, 1975, **53**, 235.

(10) Fortes, M.A., *J. Colloid Interface Sci.*, 1984, **100**, 17.

(11) Huh, C. and Mason, S.G., *J. Colloid Interface Sci.*, 1977, **60**, 11.

(12) Joanny, J.F. and de Gennes, P.G., *J. Chem. Phys.*, 1984, **81**, 552.

(13) Neumann, A.W. and Good, R.J., *J. Colloid Interface Sci.*, 1972, **38**, 341.

(14) Pomeau, Y. and Vannimenus, J., *J. Colloid Interface Sci.*, 1985, **104**, 477.

(15) Robbins, M.O. and Joanny, J.F., *Europhys. Lett.*, 1987, **3**, 729.

(16) Schwartz, L.W. and Garoff, S., *J. Colloid Interface Sci.*, 1985, **106**, 422.

(17) Schwartz, L.W. and Garoff, S., *Langmuir*, 1985, **1**, 219.

(18) Shanahan, M.E.R., *J. Adhesion*, 1987, **20**, 261.
(19) Shanahan, M.E.R., *Revue Phys. Appl.*, 1988, **23**, 1031.
(20) Shanahan, M.E.R., *J. Phys. D*, 1989, **22**, 1128.
(21) Wenzel, R.N., *J. Phys. Chem.*, 1949, **53**, 1466.
(22) Cassie, A.B.D., *Diss. Faraday Soc.*, 1948, **3**, 11.
(23) Boruvka, L., Gaydos, J. and Neumann, A.W., A Novel Strategy for Determining Line Tension from the Shape of a ·Liquid Meniscus near a Stripwise Heterogeneous Wall. *Accepted for publication in Colloids and Surfaces*, 1989.
(24) Gaydos, J. and Neumann, A.W., *J. Colloid Interface Sci.*, 1987, **120**, 76.
(25) Li, D. and Neumann, A.W., Determination of Line Tension from the Drop Size Dependence of Contact Angles. *Accepted for publication in Colloids and Surfaces*, 1989.
(26) Wallace, J.A. and Schürch, S., *J. Colloid Interface Sci.*, 1988, **124**, 452.
(27) Rotenberg, Y., Boruvka, L. and Neumann, A.W., *J. Colloid Interface Sci.*, 1983, **93**, 169.
(28) Neumann, A.W. and Good, R.J., Techniques of Measuring Contact Angles. In **Surface and Colloid Science**, vol. 11, R.J. Good and R.R. Stromberg, eds., Plenum, New York, 1979, pp. 31-91.
(29) Neumann, A.W., *Adv. Colloid Interface Sci.*, 1974, **4**, 105.
(30) Budziak, C.J. and Neumann, A.W., Automation of the Capillary Rise Technique for Measuring Contact Angles. *Accepted for publication in Colloids and Surfaces*, 1989.
(31) Cain, J.B., Francis, D.W., Venter, R.D. and Neumann, A.W. *J. Colloid Interface Sci.*, 1983, **94**, 123.
(32) Duncan-Hewitt, W.C., Policova, Z., Cheng, P., Vargha-Butler, E.I., Neumann, A.W., *Colloids and Surfaces*, 1989, **42**, 391.
(33) Broers, A.N., *Proc. Roy. Soc. Lond.*, 1988, **A416**, 1.

Appendix

To derive relations (2)-(4) we must consider the expression for the total free energy Ω_t and its first variation $\delta\Omega_t$. The free energy for the system consisting of the liquid and the surrounding solid wall has the form

$$\Omega_t = \gamma^{sv} \, |A^{sv}| + \gamma^{sl} \, |A^{sl}| + \gamma^{lv} \, |A^{lv}| + \rho \iiint_{V^l} \phi \, dV^l \tag{a1}$$

where $|A^{ij}|$ represents the area of the ij^{th} surface and the integration is carried out over the volume of the liquid phase. It should be stressed that we have assumed *a priori* that the solid wall is ideal, rigid, isotropic, smooth, and chemically homogeneous. A necessary condition for this system to be in equilibrium is for the first variation $\delta\Omega_t$ to vanish for all allowed variations of the surface A^{lv} and contact line L^{slv}. If we treat the liquid phase as incompressible then the only allowed variations of A^{lv} and L^{slv} are those which leave the volume constant

$$\delta \iiint_{V^l} dV^l = 0 \tag{a2}$$

If each point on the surface A^{lv}, which is represented by a vector **x**, is subjected to a small displacement $\delta\mathbf{x}$ then the area changes according to

$$\delta\,|A^{lv}| = -\iint_{A^{lv}} J\,\mathbf{n}^{lv}\cdot\delta\mathbf{x}\,dA^{lv} + \int_{L^{slv}} \mathbf{t}^{lv}\cdot\delta\mathbf{x}\,dL^{slv} \tag{a3}$$

where $\mathbf{n}^{lv}$ and $\mathbf{t}^{lv}$ are unit vectors which are normal to A^{lv} (and directed away from V^l) and to L^{slv} (see figure 1). In addition,

$$\delta\,|A^{sl}| = -\delta\,|A^{sv}| = \int_{L^{slv}} \mathbf{t}^{sv}\cdot\delta\mathbf{x}\,dL^{slv} \tag{a4}$$

where $\mathbf{t}^{sv}$ is the unit normal vector to the three-phase contact line L^{slv}. Finally,

$$\delta\iiint_{V^l} \phi\,dV^l = \iint_{A^{lv}} \phi\,\mathbf{n}^{lv}\cdot\delta\mathbf{x}\,dA^{lv} \tag{a5}$$

Two additional conditions arise from physical considerations. To insure that the set of permissible variations does not permit any flow of liquid through the solid boundary we require

$$\mathbf{n}^{sl}\cdot\delta\mathbf{x} = 0 \tag{a6}$$

along the three-phase contact line. Also, since the vectors $\mathbf{t}^{sv}$, $\mathbf{t}^{lv}$, $\mathbf{n}^{sl}$ and $\mathbf{n}^{lv}$ are coplanar on the three-phase boundary

$$\mathbf{t}^{lv}\cdot\delta\mathbf{x} = (\mathbf{n}^{lv}\cdot\mathbf{n}^{sl})\,\mathbf{t}^{sv}\cdot\delta\mathbf{x} \tag{a7}$$

Combining these results yields

$$\delta\Omega_t = \iint_{A^{lv}} (-\gamma^{lv}J + \rho\phi)\mathbf{n}^{lv}\cdot\delta\mathbf{x}\,dA^{lv} + \int_{L^{slv}} \left[\gamma^{lv}(\mathbf{n}^{lv}\cdot\mathbf{n}^{sl}) + (\gamma^{sl} - \gamma^{sv})\right]\mathbf{t}^{sv}\cdot\delta\mathbf{x}\,dL^{slv} \tag{a8}$$

With the volume constraint, given by eq. (a2), written in the form

$$\iint_{A^{lv}} \mathbf{n}^{lv}\cdot\delta\mathbf{x}\,dA^{lv} = 0 \tag{a9}$$

the complete variation problem becomes

$$\delta\Omega_t + \lambda\iint_{A^{lv}} \mathbf{n}^{lv}\cdot\delta\mathbf{x}\,dA^{lv} = 0 \tag{a10}$$

where λ is a Lagrangian multiplier. The first variation $\delta\Omega_t$ with volume constraint (a9) is only satisfied if

$$-\gamma^{lv}J + \rho\phi + \lambda = 0 \tag{a11}$$

across the deformable surface A^{lv} and

$$\gamma^{lv}(\mathbf{n}^{lv}\cdot\mathbf{n}^{sl}) + (\gamma^{sl} - \gamma^{sv}) = 0 \tag{a12}$$

along the three-phase boundary. After the integrated form of eq. (1) (i.e., $P^l = -\rho\phi + \mathit{constant}$ in phase V^l) is substituted into eq. (2) and the Lagrangian multiplier is given the value $\lambda = P^v - \mathit{constant}$ one may see that eq. (a11) is identical to eq. (2). Equation (a12) is identical to eq. (3) since $\mathbf{n}^{lv}\cdot\mathbf{n}^{sl}$ is the cosine of the contact angle, θ, formed by the liquid at the solid boundary.

If the total system under consideration is small or the line of contact between the liquid and the solid experiences large contortions in the vicinity of the patch-patch boundary the energy expression in eq. (a1) is modified to include the energy of the linear phase, given by $\sigma^{slv}\,|L^{slv}|$. For this situation the equation for the equilibrium surface (i.e., eqs. (a11), (6) or (9)) is not changed, however, the boundary condition (i.e., eq. (a12)) is modified to

$$\gamma^{lv}(\mathbf{n}^{lv}\cdot\mathbf{n}^{sl}) + \sigma^{slv}\kappa_{gs} + (\gamma^{sl} - \gamma^{sv}) = 0. \tag{a13}$$

We may explain the meaning of κ_{gs} by recognizing that the curvature κ of the three-

phase contact line is defined as the rate of change of the contact line tangent vector $\mathbf{t}^{slv}$. Referring to fig. 1 we see that $\mathbf{t}^{slv} = \mathbf{n}^{lv} \times \mathbf{t}^{lv}$ (pointed into the page) so that $\kappa = -\mathbf{t}^{lv} \cdot \mathbf{t}^{slv\prime}$ where the $\prime$ denotes differentiation with respect to arc length along L^{slv}.

Engineering the Interphase in Reinforced Plastics

A. Khelawan* and M. R. Piggott
University of Toronto
Toronto
Ontario
Canada

ABSTRACT

Careful investigation of the mechanics of the interphase between fibres such as Kevlar, carbon and glass, and thermosetting polymers, strongly suggests that the interphase fails by brittle fracture rather than yielding. Works of fracture can be as low as 10 Jm^{-2}, so interphase toughening should improve the properties of fibre composites very significantly. To this end, glass fibre reinforced epoxies have been made with rubber interphases. Both reactive and non-reactive liquid and solid rubbers were investigated, and the viscoelastic properties of the rubbers were modified using fine solid particles. The results indicate that the work of fracture of the composites can be increased significantly, without any great loss in other mechanical properties. Neat rubbers were better overall than filled rubbers, and the best results were obtained when there was a chemical interaction between the rubber and the silane sizing on the fibres.

INTRODUCTION

Fibres for use as reinforcements are usually surface coated to protect the fibres from damage due to handling and to provide a bridging layer the - interphase - to the polymer to ensure good adhesion. Thus, silanes are used to coat glass fibres [1]. The silanes also make the fibre surface hydrophobic, so that glass reinforced plastics can be used with confidence in the presence of water.

A great deal of effort has been expended in trying to measure the properties of fibres interphases. Four methods are currently used, and these and the results obtained have recently been reviewed [2]. The values obtained for the interphase strength are sometimes unexpectedly high, i.e. higher than the strength of the adjacent polymer. This suggests that the stress criterion used to evaluate the data obtained from the tests may be inappropriate, and instead, a fracture criterion may be more suitable [3]. When this is used, the works of fracture are found to be very small [4].

---

* Sinclair & Valentine Co., 4590 Dufferin Street, Downsview, Ontario M3H 5S5

Fibre composites generally lack ductility and are not tough [5]. What toughness there is comes from processes taking place at the interface or in the interphase [6]. The biggest contributor to toughness is usually the frictional work of pulling the fibre out of the fracture surface in the polymer [7]. If the low work of interphase fracture could be increased, it should be able to contribute significantly to the overall energy absorbed in composite fracture, and thus have a direct effect on increasing fibre composite toughness.

One way of toughening the interphase is to use yielding surface layers, e.g. soft metals, and MoS_2 [8]. These were not entirely successful, but flexiblized epoxy surface layers [9] were found to increase the shear strength and off axis strength by 40% and 67% respectively. The use of viscous coatings increased toughness nearly two fold, but this caused severe loss in off-axis properties [10]. A silicone rubber coating was also found to give a two fold increase in toughness, but the flexural strength was decreased by 60-70% and the interlaminar shear strength by about 40% [11].

One could well expect that decreasing the mechanical strength of the interphase should lead to losses in off-axis properties of fibre composite, since Wadsworth and Spilling in 1968 [12] and Harris et al in 1971 [13] demonstrated very clearly that such losses might be expected. In this work, therefore, an attempt was made to use shear thinning materials as fibre coatings. These should give low interphase shear strengths at high rates of shear, as in fracture, but relatively high shear strengths at the low rates of shear appropriate to normal mechanical tests such as for shear and off axis strengths. Thus it was hoped to increase fracture work without a great loss in shear strength.

EXPERIMENTAL METHOD

E-glass fibres from Fiberglas Canada were coated from solutions containing a rubber and various amounts of fumed silica. Prior to the coating, the fibres were thoroughly washed with the solvent to remove all soluble sizing which would otherwise dissolve in the solution and affect the coating process. The rubbers used were as follows:

1. EPDM (Polysar Ltd) a terpolymer of ethylene, propylene, and ethyldiene norbornene, in the approximate ratio 50:46:4.
2. ATBN (B.F. Goodrich) an amine terminated butadiene acrylonitrile liquid rubber: grade Hycar 1300 x 21.
3. CTBN (B.F. Goodrich) a carboxyl terminated butadiene acrylonitrile liquid rubber. Three grades were used: Hycar 1300 x 9, 13, and 31.

Toluene was used as solvent for EPDM and methyl ethyl ketone for ATBN & CTBN. The fumed silica was Cabosil M5 (Cabot Corporation) and was used as received.

Fibre composites were made by pultrusion as described previously. [14] The fibre volume fraction was 33 ± 1%. The resin used was Shell epoxy EPON 815 with 14 parts triethylene tetramine. The composites were cured at 100°C for 18h.
The composites were notched and tested in the Izod machine using the reversed notch method. They were also tested for shear strength using the short beam flexure method.

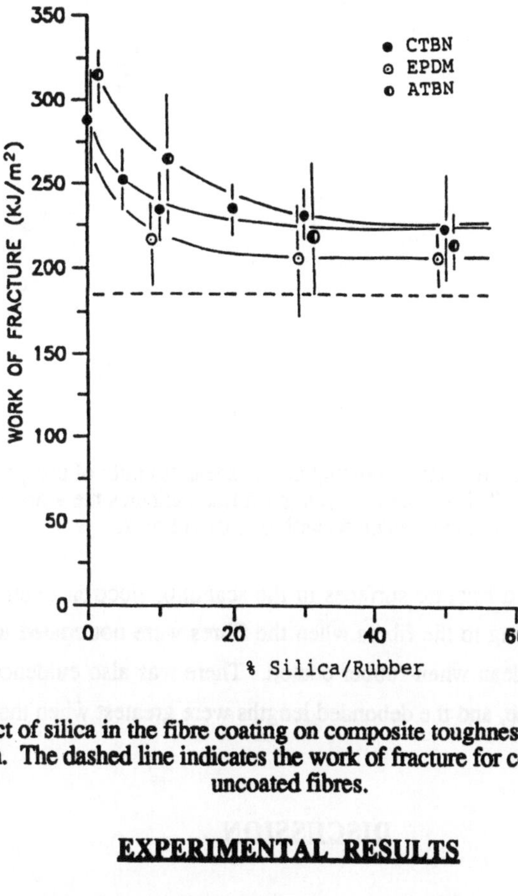

Figure 1. Effect of silica in the fibre coating on composite toughness. Coated from 1.5% rubber solution. The dashed line indicates the work of fracture for composites made with uncoated fibres.

EXPERIMENTAL RESULTS

Fig. 1 shows the effect of silica on the toughness of the composites, and fig. 2 shows the corresponding in-plane shear strengths. Both toughness and shear strength decreased. (The dashed line in fig. 2 is the shear strength of the composites made with fibres without a rubber or silica interphase)

The use of the rubbers without silica, on the other hand, did increase the toughness by 45-70%, fig. 3, with a loss in shear strength, although this was not very great for the CTBN coated fibres, fig. 4.

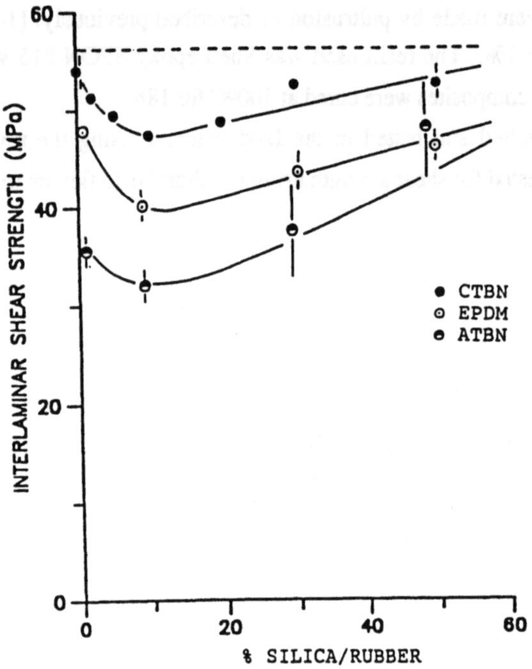

Figure 2. Effect of silica in the fibre coating on the shear strength of composites. Coated from solution containing 1.5% rubber. The dashed line indicates the work of fracture for composites made with uncoated fibres.

Examination of the fracture surfaces in the scanning electron microscope revealed pieces of polymer adhering to the fibres when the fibres were not coated with rubber, but came out more or less clean when rubber coated. There was also evidence of debonding below the fracture surface, and the debonded lengths were greatest when the toughness was greatest.

DISCUSSION

Silica can indeed increase the shear strength of the composite under certain conditions, namely, when it is used with ATBN rubber, and is used in large enough quantities, i.e. 30% or more as a proportion of the rubber. However, it reduces the toughness of the composite, as compared with that obtained when the neat rubbers were used. Overall there is no significant benefit from the use of silica since toughness was only increased about 8% relative to the value obtained with uncoated fibres, while shear strength was decreased by about the same amount ; see the dashed lines in figs 1 & 2.

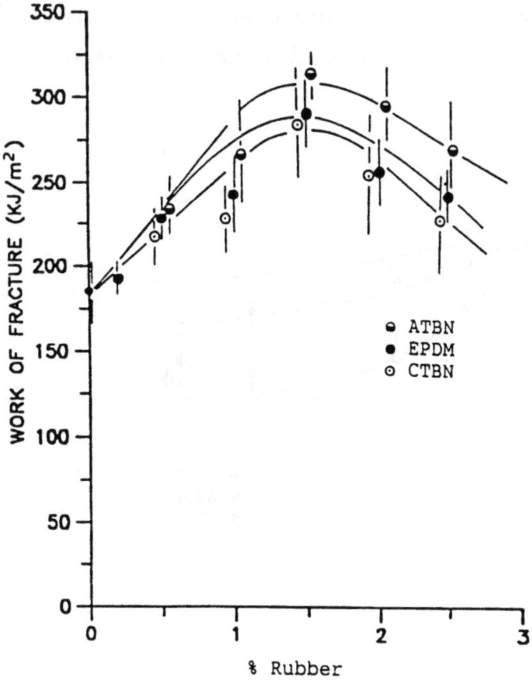

Figure 3. Work of fracture of pultrusions made with fibres coated with EPDM, ATBN and CTBN rubber.

The CTBN rubber without the silica does appear to have a beneficial effect overall, however. When coated from a 1.5% solution, an increase in toughness of about 35% could be obtained, with almost no loss of shear strength, see figs. 3 & 4. This improvement requires very little rubber; weight gain measurements on fibres coated with the rubber, before being used in the composite, indicate that the film thickness required for optimum toughening is only 75 nm.

The chemistry of the rubber is very important. The carboxyl terminated rubber can evidently bond quite well with the epoxy matrix, or the silane on the fibre surface, thus giving very little loss in shear strength. The amine terminated rubber presumably does not bond too well, since it reduces the interlaminar shear strength very significantly. This is probably because of poor interaction with the silane.

CONCLUSION

Fillers which make rubbers shear thinning do not improve the prospects for increasing the toughness of fibre composites without at the same time decreasing the shear strength. A carboxyl terminated liquid rubbber used alone, however, can be very beneficial and could well be worth exploiting commercially, since little is needed and it is not very expensive.

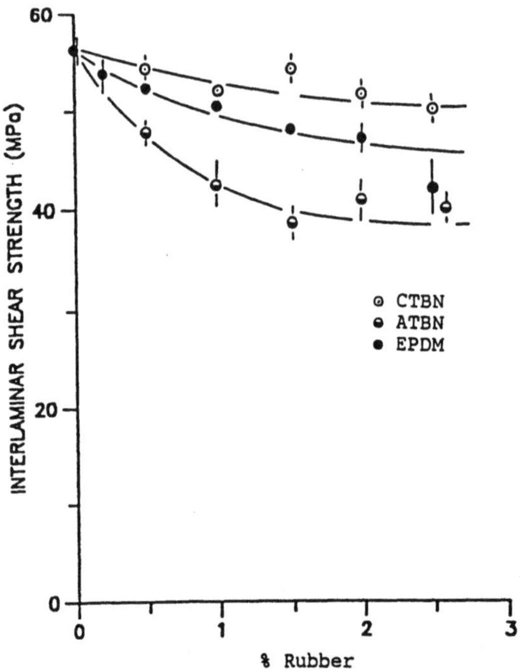

Figure 4. Interlaminar shear strength of pultrusions made with fibres coated with EPDM, ATBN and CTBN rubbers.

ACKNOWLEDGEMENTS

The authors are grateful to Polysar for funding support and supply of rubbers. They also gratefully acknolwedge support from NSERC (Canada) and URIF (Ontario) and from B.F. Goodrich for supply of rubber, and Fiberglas Canada for the fibres used.

REFERENCES

1. E.P. Plueddemann, Silane Coupling Agents (Plenum Press, N.Y., 1982), Chpt. 1.
2. M.R. Piggott, Advances in Composites (Eds. T.L. Vigo & B.J. Kinzig, American Chemical Soc. 1989) Chpt. 4.
3. M.R. Piggott, Comp. Sci. Tech. 30, 1987, p.295.
4. M.R. Piggott and M.M. Reboredo, Proc. 34th Int. SAMPE Symp. (Reno, Nevada,May 1989), 1913
5. M.R. Piggott, J. Mater Sci. 5, 1970, p. 669.
6. M.R. Piggott, J. Mater Sci. 9, 1974, p. 494.
7. G.A. Cooper and M.R. Piggott, Proc. ICF4 3, 1977, p. 303.
8. J.V. Mullin and V.F. Mazzio, J. Mech. Phys. Solids 20, 1972, p. 391.
9. L.D. Tryson and J.L. Kardos, 30th Ann. Conf. SPI, 1981, 2E.
10. N.H. Sung, T.J. Jones and N.P. Suh, J. Mater Sci. 12, 1977 p. 239.
11. N.L. Hancox and H. Wells, Fib. Sci. Tech. 10, 1977 p. 9.
12. N.J. Wadsworth and I. Spilling, Brit. J. Appl. Phys. 1, 1968 p. 1049.
13. B. Harris, P.W.R. Beaumont and E.M. De Ferran, J. Mater. Sci. 6, 1977 p. 238.
14. P.W.K. Lam and M.R. Piggott, J. Mater Sci .24, 1989 p. 4068.

Ferroelectric Thin Films from Metallo-organics: The Influence of Different Atmospheres during RTA Processing

Y. Shi, E. Sánchez-Carambot, W. Otaño, L. J. Martínez-Miranda and J. J. Santiago-Avilés

University of Pennsylvania
Philadelphia
Pennsylvania
USA

ABSTRACT

Barium Titanate ($BaTiO_3$) films have been deposited on platinum via spin-casting metallo-organic compounds and rapid thermal annealing (RTA). This method, called metallo-organic decomposition (MOD), produces tetragonal $BaTiO_3$ films with easily controllable stoichiometry as determined by x-ray diffraction (XRD) and Auger electron spectroscopy (AES). The precursors used were barium neodecanoate and titanium dimethoxy dineodecanoate. The film precursor is obtained by mixing the individual precursors in a nearly 1:1 molar ratio before processing. Variable thickness (0.1-3 μm) is achieved by spin-casting multiple layers of the films. The films were RTA processed to 650° for 30s or to 950° for 15s under controlled atmosphere (O_2, Ar, and partial vacuum). The films processed under O_2 showed better stoichiometry and purity than the ones processed under partial vacuum (10^{-3} Torr) as evidenced by XRD and AES. The carbon contamination from pyrolysis residues, prevalent in the partial vacuum processed films, fell below the instrument detection limit for the O_2 and Ar processed films. All the films are uniform and crack-free as shown in the scanning electron micrograph (SEM). Average grain size determined by SEM ranged from 0.01 μm to 0.5 μm, depending on the annealing condition.

INTRODUCTION

Barium titanate films are deposited on clean platinum substrates by metallo-organic decomposition as described in a previous article (1). After our early report on the use of RTA for MOD thin $BaTiO_3$ film deposition (2). Vest and coworkers reported in a series of articles (3-5) the use of MOD for thin film oxide deposition using conventional furnace annealing, The potential of enhancing purity and grain size by annealing in an oxidizing atmosphere was suggested by several workers (6). It was mostly claimed that an oxidizing ambient will provide the needed oxygen activity for the removal of carbon in the pyrolysis of the metallo-organic. In the early stages of $BaTiO_3$ nucleation and grain growth, if the system is envisioned as a $BaTiO_3$ filler surrounded by a carbon-rich amorphous binder, then removal of carbon is critical to the nuclei-nuclei interaction leading to grain growth.

Our aim is to try to validate this assertion by systematically annealing samples of the metallo-organic precursor deposited on Pt foils. The reactor utilized in processing these samples is an RTA system composed of a pill-box shaped vacuum chamber with a water-cooled base and a fused silica top. The system allows the coupling of radiant energy from quartz-halogen tungsten lamps to the sample. The reactor can also be operated as a flowing or static reactor by flushing or filling with suitable gases. Details of the RTA reactor have been published elsewhere (7).

Some of the characteristics of RTA vis a vis conventional furnace annealing are its speed and high throughput. Isothermal annealing events lasting 30 min at, lets say 800°C in conventional furnace annealing (CFA) can be reproduced by RTA in 5 to 30 sec with a maximum temperature not much in excess of CFA. Thermally activated events with slow kinetics are suppressed in RTA in such a way that interdiffusion is sometimes avoided. Since a minimal interaction between substrate and thin film overlayer is desired in our case, RTA is usually favored.

EXPERIMENTS

In our study, each sample consists of 10 metallo-organic precursor layers. After each layer is casted it is processed by RTA in a partial vacuum, a partial or total pressure of CVD grade oxygen and/or a partial or total pressure of CVD grade argon.

In the first experiment, all samples were annealed in one atmosphere of flowing O_2 except one, which was annealed in a partial vacuum of 10^{-3} Torr, at 650°C for 30s. The coating and heat processing was repeated 10 times. Four of the samples were further annealed to 950°C after the tenth coat application. The previous description is summarized in Table 1.

TABLE 1

Sample #	Temp.	Time	Atmosphere
0	prepared in vacuum, no second anneal		
1	no second anneal		
2	+ 950°C	15 s	O_2 (1 atm)
3	+ 950°C	15 s	O_2 (10 mTorr)
4	+ 950°C	15 s	Vacuum (10^{-3} Torr)
5	+ 950°C	30 s	O_2 (10 mTorr)

Table 1: Second Annealing Scheme for samples previously annealed at 650° for 30 s each coat.

In the second experiment, the samples were annealed at a higher temperature of 950°C for 15 s after each coating.One sample was annealed in rough vacuum (10-3 Torr), one in one atmosphere of O_2 and one in one atmosphere of Ar. Table 2 summarizes the annealing scheme for this second set of samples.

TABLE 2

Sample #	Temp.	Time	Atmosphere
6	950°C	15 s	O_2
7	950°C	15 s	Ar
8	950°C	15 s	Vacuum

Table 2: Annealing scheme for samples in second experiment.

RESULTS

The samples were characterized structurally and compositionally by XDR, Auger electron spectroscopy (AES) elemental surveys and depth profiles, and SEM.

The XDR results for the first group of samples show the presence of polycrystalline tetragonal $BaTiO_3$ and a minority phase, identified by indexing as $BaTiO_4$ (see Figure 1). The intensity of Ba_2TiO_4 peaks as compared to the $BaTiO_3$ peaks is larger for the vacuum processing. Evidence of grain growth with annealing time was observed (sample 5 with the longest annealing time, show the most intense $BaTiO_3$ peaks) but the appearance of minority phases other than Ba_2TiO_4 is also apparent.

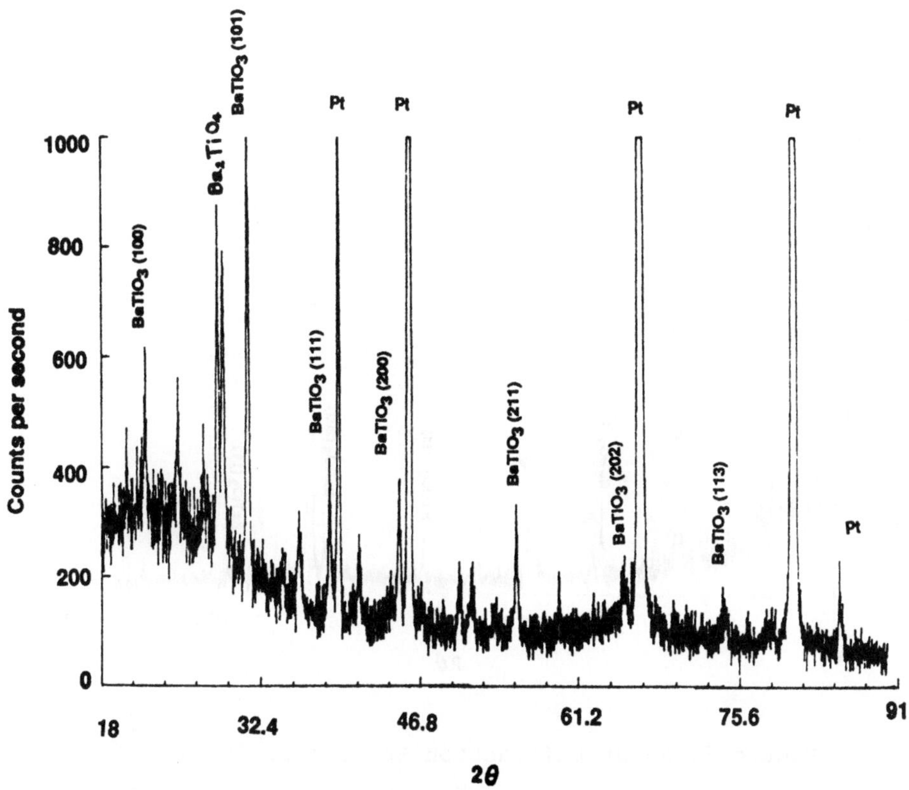

Figure 1: X-Ray Diffraction Pattern of Sample 5.

The AES results show the presence of carbon residues in the film interior (approximately 10 nm) only for the sample processed in vacuum at 650°C (sample zero). Sample 1 (processed at the same temperature but in the presence of O_2) shows no evidence of residual carbon in the film interior.

The SEM micrographs of the samples in the first group show a uniform featureless topography for samples 0 and 1. The samples exposed to a second annealing at 950°C show a typical ceramic grain structure with mean grain size of approximately 0.2 µm in diameter.

The samples in the second group, namely samples 6, 7 and 8, were all polycrystalline $BaTiO_3$ with no presence of a second phase or crystalline contamination as evidenced by XRD. An interesting feature in the XRD results is the presence of texture (increased ratio of (100)/(101) and (200)/(101) intensities) for the sample annealed in Ar as shown in Figure 2.

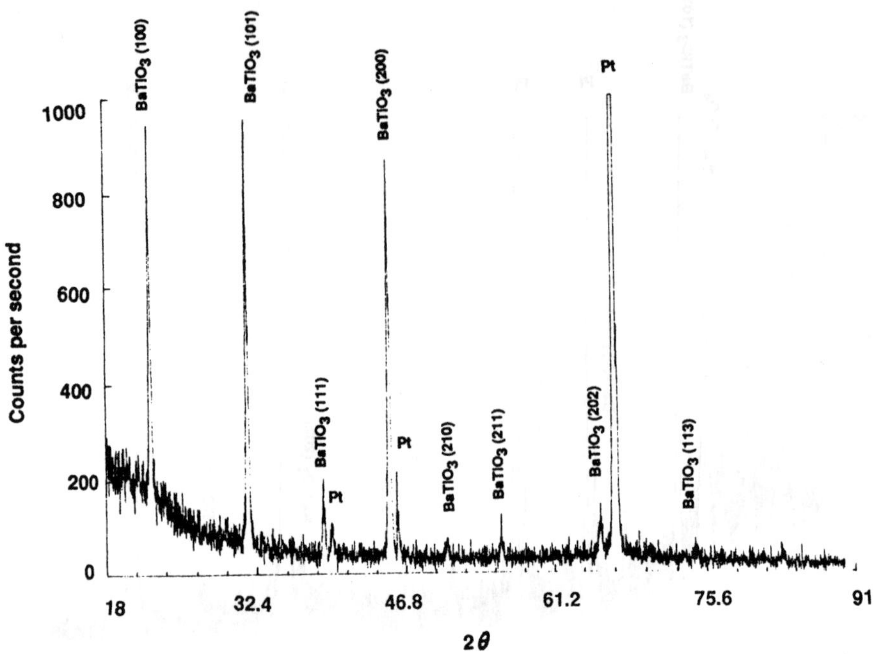

Figure 2: X-Ray Diffraction Pattern of Sample 7.

The AES measurements of the samples in the second group show that no carbon residues were present in the interior of the film after depth profiling. The only difference was the presence of a thin carbon layer on the surface for the samples processed in vacuum. Such a layer, if conducting, will impede the poling process by shorting the upper and lower electrode of the capacitor. The microstructure, as revealed by SEM, shows mean grain size in the vicinity of 0.2 μm. The sample processed in oxygen (sample 6) shows sintering and grain coalescence to a larger extent than the sample annealed in Ar or vacuum.

DISCUSSION AND CONCLUSION

We can point to several consistent features of our experimental results. The first is that the solid state reactions conducive to the formation of polycrystalline $BaTiO_3$ are not yet completed at 650°C, when the samples are processed via RTA. The results point to decomposition of Ba_2TiO_4. The second feature is that the development of grain size compatible with optimal ferroelectric properties (~0.1 μm or more) can only be achieved in RTA by annealing at high temperature (in our case, 950°C). The third feature is the consistent appearance of carbon-rich film on the samples processed in a partial vacuum. This is probably due to the absence of convection and/or reactions capable of oxidizing and/or transporting the carbon.

In addition, the formation of large grained (>0.1 μm) polycrystalline $BaTiO_3$ thin films from RTA-MOD is apparently insensitive to the ambient atmosphere except at low temperatures (650°C) where annealing in an atmosphere of O_2 enhances film purity.

REFERENCES

1. J.R. Hanrahan, E. Sánchez-Carambot, Q. Jiang, D. Berry and J.J. Santiago-Avilés, Proc. of 1st International Ceramic Science and Technology Congress, American Ceramic Society, Anaheim, CA (1989).
2. X.Y. Hung, J.J. Santiago-Avilés and D. Berry, Abstract to Alfreds Conference on Electronic Ceramics, Alfred, NY (1987).
3. R.W. Vest and J. Xu, Ferroelectrics, **93**, 21(1989).
4. R.W. Vest and J. Xu, IEEE Trans. UFFC (November 1988).
5. J.J. Xu, A.S. Shaikh and R.W. Vest, IEEE Trans. UFFC, **36**, 307(1989).
6. J.R. Hanrahan, E. Sánchez-Carambot, D. Berry, Q. Jiang and J.J. Santiago-Avilés, Thin Solid Films, in press.
7. J.J. Santiago-Avilés, C.S. Wei and J. Van der Spiegel, Materials Letters **2(6A)**, 477(1984).

Ultrasonic Verification of Surface Treatment Adequacy in Metal-to-Metal Adhesive Joints

P. Dickstein[1]

University of Toronto
Toronto
Ontario
Canada

ABSTRACT

Two classes of aluminum-to-aluminum adhesive bonds were ultrasonically tested by the pulse-echo technique. The two classes differed in the surface preparation of the adherends prior to bonding, resulting in different interfacial properties of the joints. These properties have a crucial effect on the long-term adhesive properties of the specimens.

Features extracted from the ultrasonic echo signals were found to be significantly sensitive to the interfacial characteristics of the joints. This sensitivity was verified by means of conventional statistical tests.

Based on the ultrasonic features data,the application of basic pattern recognition techniques enabled the classification of the joints according to the surface preparation of the adherends, and the evaluation of the upper bound for the probability of mis-classification of the specimens. The results imply the possibility of ultrasonic adhesive, rather then cohesive, bond classification.

INTRODUCTION

Traditional methods of joining two engineering materials together, such as welding and riveting, have serious limitations. Riveting can lead to galvanic corrosion between two metals, and also causes large stress concentrations due to the non-uniform distribution of the load; extra material is needed in the joining area to prevent failure at the stress raisers, and this leads to a substantial increase in weight of the final structure. Welding brings in a host

1- Permanent mailing address: P.O.Box 11871, Tel Aviv, Israel 61166.

of problems ranging from cracking in the heat affected zone to porosity, slag inclusions and complications in nondestructive testing (NDT); its high expense is also a negative factor.

One of the main alternatives for joining materials together is the use of adhesive bonding. Although the restrictions concerned with welding and riveting are not generally encountered when using adhesives, the latter have their own set of limitations, such as the pronounced sensitivity of breaking strength to the surface preparation of the adherends, proper polymerization of the adhesive, and environmental conditions during manufacture and service. The variety of parameters that affect a bond's quality imposes heavy demands on quality control measures and nondestructive testing techniques.

Several NDT methods are recommended for the inspection of adhesive joints [1-3]. In addition to methods for detecting localized flaws, voids, or delaminations, there also exist commercial bondtesters, which may determine the cohesive quality of the bulk adhesive layer. However, there currently does not exist a reliable method for determination of the adhesive characteristics. Experience has shown that the surface treatment of the adherends prior to bonding is a major parameter in establishing a joint's long-term resistance to adhesive failure [4,5]. It is a nondestructive characterization of adherend surface condition in a bonded joint that is the topic of this study.

One of the more promising nondestructive methods for inspecting adhesive bonds is ultrasound [3,6]. However, a variety of bond parameters can be inferred from the characteristics of sound propagation. Among them are: bulk material properties (elastic modulus, attenuation, density, etc.), geometrical properties (structural dimensions, discontinuities), and properties dependent on environmental factors (absorbed moisture, temperature). In light of these numerous factors, the quantitative ultrasonic determination of a specific microstructural or morphological state such as surface condition of the adherends, independent of all other states is not simple and requires complicated procedures.

One approach for the quantitative acquisition of discriminatory information for the classification of samples according to their state is the application of basic pattern recognition techniques and cluster analysis to features derived from the ultrasonic echos. Several ultrasonic experimental studies on adhesive joints that used some basic concepts from the pattern recognition field [7-11] have been published, and a theoretical procedure for the application of pattern recognition techniques to the characterization of the material state via analytical quantitative ultrasonics was outlined [11,12].

In this work, two classes of aluminum-to-aluminum adhesive bonds are ultrasonically characterized; the two classes represent different methods for pretreatment of the adherend surfaces with significant implications for the long-term durability of the bonds. The significant difference between the ultrasonic features of the two classes, verifiied by means of statistical tests, enables the use of mathematical techniques to assign specimens to the appropriate group and to evaluate the probability of misclassification.

MATERIALS AND METHODS

The Specimens

Aluminum-to-aluminum bonded plate specimens were prepared, following the ASTM D1002-72 standard for the strength properties of adhesives in shear by tension loading (metal to metal). The two aluminum alloy adherends were Al-2043 T3, each 1.62 *mm* thick. The final specimen geometry is shown in Fig.1.

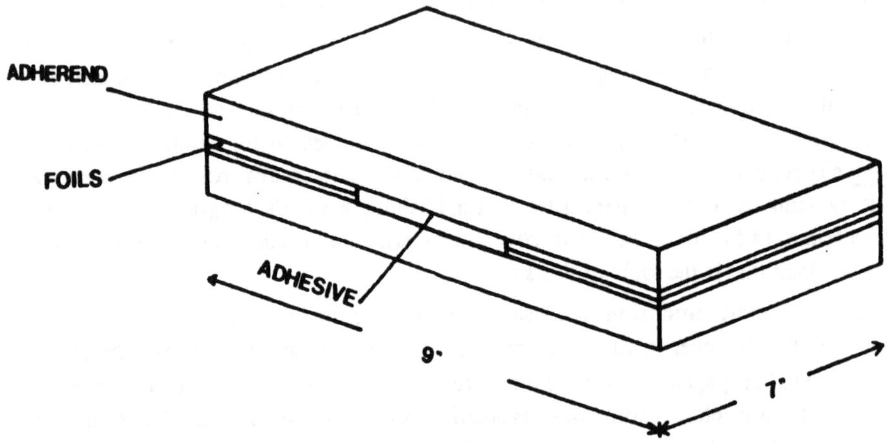

Figure 1: The specimen geometry

Two different series of specimens were produced, by applying different methods of surface pretreatment to the adherends prior to bonding:

1. Unsealed chromic acid anodization (CAA), according to the MIL-A-8625D standard.
2. Chromate conversion coating treatment (CCT), according to the MIL-C-5541 standard. The solution used was Chemotas No. 720, manufactured under licence from Metallgesellschaft.

In preparing the joints the supported film epoxy adhesive FM-73 was used, of 0.38*mm* nominal thickness, and $0.42 \pm 0.02 kg/m^2$ density. The adhesive was polymerized at $120^0 C$ in a hot press for 90 min at a pressure of 35*Psi*, following the manufactor's recommendations.

An epoxy-phenolic based primer BR-127, manufactured by Bloomingdale Department of American Cynamid, was applied to the CAA adherends by brushing to a thickness of approximately 2.5*μm*. The primer layer was air dried for 30 min and then polymerized at $120^0 C$ for 30min.

Each of the two series of bonded specimens featured five different thicknesses of the adhesive layer. These thicknesses were controlled by foils during the manufacturing process, resulting in adhesive layers of nominal thickness *W*=0.0, 0.05, 0.1, 0.2 and 0.3 *mm*.

The Experiment

The specimens were immersed in an ultrasonic immersion tank with an X-Y scanning bridge. A transducer of 1*MHz* nominal central frequency was driven by a Panametrics 5052 PRX100 pulser-receiver. This choice of the probe's frequency was based on the results of a previous study [11], implying that unless a very high frequency probe can be used, which is capable of resolving the interface structure, the probe's frequency should be in the range of the lower resonant modes of the joints. The gated ultrasound echo signals were digitized by a Tektronix 7912 A/D programmable digitizer, with a sampling rate of 1 *GHz*. The digitized signal were processed and stored in a Digital PDP 11/34 computer.

A minimum of 10 measurements of the interfacial echo signal were taken corresponding to each of the five adhesive-layer thicknesses, for the two series of specimens. These signals were then used for the data analyses and processing.

In addition, a test for the cohesive properties of the specimens from the two series was carried out by means of a Fokker bond-tester. No significant difference between the two series was observed in this test as expected, since the two series were prepared using the same adhesive and the same curing-cycle [13].

DATA PROCESSING AND RESULTS

Extraction of ultrasonic features

Two features were extracted from each of the ultrasonic echos obtained from the specimens. These features were:

1. f_1- a parameter derived from the higher order crossing analysis (HOC) [14]. This parameter is used to measure the degree of similarity between two signals. Throughout this work each ultrasonic echo was compared to white noise.
   Let Z(t) designate an ultrasonic echo signal of interest, and Y(t) a signal of white noise. Each signal is digitized to produce two sets of N points, {Z} and {Y}; the average zero-offset is subtracted from each set yielding a mean of zero in both {Z} and {Y}. An intermediate parameter d_j is defined for each signal [14]:

$$d_j = \begin{cases} T_0, & j=0 \\ T_j - T_{j-1}, & j=1,2, \cdots h-1 \\ (N-1) - T_{j-1}, & j=h \end{cases} \tag{1}$$

   where T_i is the number of crossings of the time-axis by the ith drivative of the ultrasonic signal. For this work, a value of 16 was used for h. Going to larger values of h does not appreciably alter the value of f_1, as the very high order derivatives of $Z(t)$ approach white noise. The value of f_1 is then given by:

$$f_1 = \sum_{j=0}^{j=h} \frac{(d_{jZ} - d_{jY})^2}{d_{jY}} \tag{2}$$

   Note that there are other candidates for the reference signal Y(t) besides white noise, e.g.: the ultrasound echo signal from an aluminum sheet, or the echo signal from an "ideal" bond.

2. f_2- the mean frequency of the power spectrum $S(f)$, obtained from a Fourier-transform of the time-domain ultrasonic signal. The value of f_2 is given by:

$$f_2 = \frac{\int_0^\infty f \cdot S(f) \cdot df}{\int_0^\infty S(f) \cdot df} \tag{3}$$

This parameter is related to the frequency-dependant attenuation of the ultrasonic signals when travelling through the joints.

These two features were selected for this study, since they were found to be the most sensitive to the interfacial conditions of adhesive joints, out of additional numerous features studied in a previous work [11,13]. The sensitivity of f_1 may be explained by the sound-scattering from highly non-uniform oxide layers, and that of f_2 by the damping of particular modes of vibration by this layer [11].

A mapping procedure of the features values, after normalization, in the plane defined by f_1 and f_2, is shown in Fig.2. This procedure yields the formation of two separable clusters, each cluster corresponds to a different series of specimens.

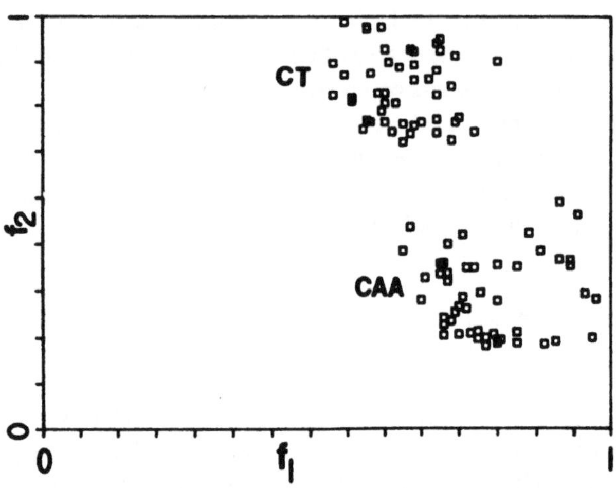

Figure 2: Separable clusters formation in the mapping procedure

Check for statistical significance

The apparent sensitivity of f_1 and f_2 to the interfacial condition in the sample joints was verified applying the conventional statistical ANOVA test.

The null hypothesis that the mean of the f_1 values obtained from the CAA specimens is not significantly different from that obtained from the CCT specimens was rejected with a 99% confidence level. The same result, regarding the mean values, was obtained for f_2.

The null hypothesis that the variance of the f_1 values obtained from the CAA specimens is not significantly different from that obtained from the CCT specimens was accepted. The same hypothesis, regarding the f_2 values, was rejected with a 90% confidence level.

In Fig.3, the Gaussian distribution of the f_1 values obtained from the CAA specimens is demonstrated, as well as the distribution of the f_1 values from the CCT specimens.

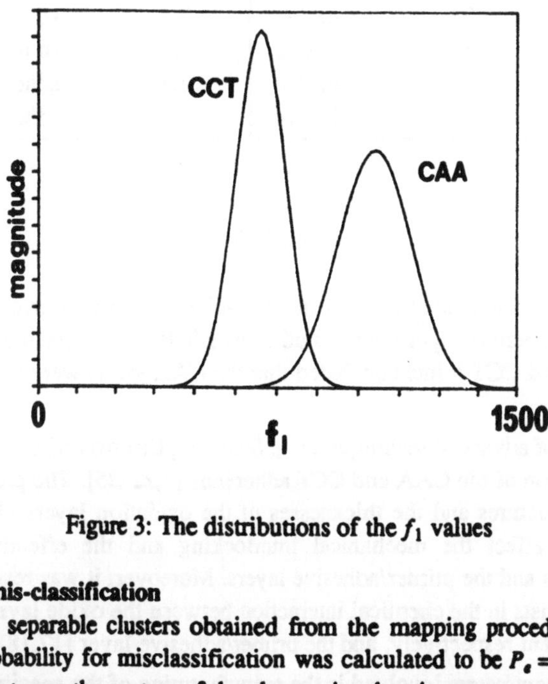

Figure 3: The distributions of the f_1 values

Probability of mis-classification

Referring to the separable clusters obtained from the mapping procedure (see Fig.2), the upper-bound probability for misclassification was calculated to be $P_e = 10\%$. In calculating P_e the distance between the centers of the clusters is taken into account, as well as the distribution of the points within each of the clusters. The procedure for computing the value of P_e is rather complicated and it is described in detail elsewhere [11].

Note that P_e represents the fraction of production samples which would be misclassified in the long term, assuming that the production samples corresponding to each series have a distribution similar to that of the training samples used in this work.

Destructive testing of the joints

The bonded samples (Fig.1) were cut into five equal strips, which were then subjected to shear-strength destructive testing according to the ASTM-D3165-73 standard. The ultimate-loads, as well as the modes of failure are presented in Table 1. The crucial effect of the surface pretreatment on the quality of the joints is evident: the CAA specimens have proper adhesive properties, so that the adhesive strength is larger than the cohesive strength, and the failure occurs within the adhesive layer. The CCT specimens have poor adhesive properties, so that the adhesive strength is smaller than the cohesive strength, and the failure occurs at the interface between the adhesive layer and the adherends.

Table 1.
The mean breaking-loads (B.L) and mode of failure (M.O.F) of the joints

	CAA		CCT	
W*mm*	B.L N/mm^2	M.O.F	B.L N/mm^2	M.O.F
0.0	3.6	cohesive	2.2	adhesive
0.05	3.37	cohesive	2.4	adhesive
0.1	3.5	cohesive	2.37	adhesive
0.2	3.3	cohesive	2.3	adhesive
0.3	3.24	cohesive	2.4	adhesive

DISCUSSION

The advantage of CAA adherends as a substrate for adhesive bonding over CCT adherends is well established. Numerous experimental studies which dealt with the initial strengths and durabilities of CAA and CCT joints concluded that the CAA joints were significantly superior [4,5,15-21].

The application of advanced techniques, e.g. Scanning Electron Microscopy (SEM) has enabled the examination of the CAA and CCT adherends [4,22-25]. The pronounced differences between the structures and the thicknesses of the oxidation layers obtained with the two treatments may affect the mechanical interlocking and the effective contact area between the adherends and the primer/adhesive layers. Moreover, it was revealed that a pronounced difference exists in the chemical interaction between the oxide layers formed in the CAA and CCT treatment respectively, and the primer/adhesive layer [17,18].

As all the other parameters involved in the manufacturing of the specimens (e.g. curing cycle) were kept unchanged, the sensitivity of the ultrasonic features should be due to the differences in the oxide layers formed following the CAA and CCT treatments. A difficulty arises from the fact that the ultrasonic wave-length is of the order of 6.32*mm*, which is much larger than the largest observed thickness of the oxidataion layer, approximately 0.005*mm*. However, there exist some modelling of the interface characteristics, referenced by Stone [3], which imply that it should be possible to predict differences in ultrasonic responses from significantly different pretreatment processes. Stone also mentions an unpublished work in which a distinction was obtained between the ultrasonic responses from an etched surface and one with additional CAA.

At least two independent studies reported recently about a partly closed structure [22] of the oxide layer, or the formation of a macroscopic planar surface [23] in intimate contact with the adhesive obtained on CAA adherends. This phenomenom, to a less extent, was found also on Sulforic Acid Anodized (SAA) adherends, but not with any other treatment. It is possible that this planar surface, not existing on the CCT adherends, contributed to the differences in the ultrasonic features obtained in this study, by affecting the Rayleigh scattering.

A less pronounced sensitivity of ultrasonic features to the interfacial condition may be expected when ultrasonically testing joints having surface pretreatment other than CAA. Further research is needed to clarify this point.

Note that the structure and thickness of the oxidation layer can be affected by parameters such as the current density applied in the anodizing process, or the detailed composition of the adherend alloy (e.g. Al-2024-T3 or Al-6061). Thus, any comparison of observations from different studies should be done very carefully.

It should be kept in mind that the results of this study, as well as the consequent conclusions, relate to the particular experimental set up of this work. Adhesive joints can be examined with several other ultrasonic techniques differing in the mode of propagation of the ultrasonic waves and their frequencies. These techniques, referenced elsewhere [11], were not addressed in this study.

SUMMARY

1. Ultrasonic features were found to be significantly sensitive to the interfacial conditions in joints of which the adherends underwent CAA and CCT treatments.
2. Based on the sensitivity of the ultrasonic features, adhesive joints can be classified by means of basic pattern recognition techniques, and the upper bound for the probability of misclassification can be evaluated.
3. The oxide layer formed on the CAA adherends has a unique closed structure. This special structure may be responsible, in part, for the ultrasonic sensitivity. Further investigation is required to verify whether this sensitivity is apparent when testing joints having other surface pretreatments.

ACKNOWLEDGEMENT

The experimental work of this study and part of the data processing was performed by the author while with Prof. Y. Segal, Department of Nuclear Engineering, Technion- Israel Institute of Technology. Part of the data processing was performed while with Prof. A.N. Sinclair, Departmet of Mechanical Engineering, University of Toronto. The advice of Prof. J.K. Spelt, Department of Mechanical Engineering, University of Toronto, is appreciated. Thanks to Mrs. S. Dickstein for her contribution to the statistical analysis of the experimental data.

REFERENCES

1. R.D. Adams and P. Cawley, A review of defect types and nondestructive testing techniques for composites and bonded joints, NDT Int., **21**, No. 4, pp. 208-222, 1988.
2. R.J. Schliekelman, Quality control of adhesive bonded joints, in Structural Adhesives in Engineering, IMECHE 1986-86, Mechanical Engineering Publications Limited, London ,1986.
3. D.E.W. Stone, Non-destructive methods of characterizing the strength of adhesive bonded joints- a review, Report No. 86058, British Royal Aircraft Establishment, 1986.
4. S.R.. Hartshorn, The durability of structural adhesive joints, in Structural Adhesives, Chemistry and Technology, ed. S.R. Hartshorn, p. 347, Plenum Press, New York ,1986.
5. W.G. Brockman, The importance of surface pretreatment prior to bonding, in Structural Adhesives in Engineering, IMECHE 1986-86, Mechanical Engineering Publications Limited, London ,1986.
6. D.W. Fitting and L. Adler, Ultrasonic Spectral Analysis for Nondestructive Evaluation, Plenum Press, New York ,1981.
7. J.L. Rose, M.J. Avioli and R. Bilgram, A feasibility study on the nondestructive evaluation of an adhesively bonded metal to metal bond: An ultrasonic pulse echo approach, British J. of NDT, pp. 67-71, March 1983.
8. P.L. Flynn, Cohesive bond strength prediction for adhesive joints, J. Testing and Evaluation, **7**, No. 3, p. 168, 1979.
9. G. Alers, P.L. Flynn and J. Buckley, Ultrasonic techniques for measuring the strength of adhesive bonds, Materials Evaluation, p. 77, April 1977.
10. R.S. Williams and P.E. Zwicke, Assessment of adhesive properties using pattern recognition analysis of ultrasonic NDE data, Materials Evaluation, **40**, p. 312, March 1982.
11. P.A. Dickstein, Y.Segal, E.Segal and A.N. Sinclair, Statistical pattern recognition techniques: a sample problem of ultrasonic determination of interfacial weakness in adhesive joints, J. NDE, **8** ,No.1,pp. 27-35, 1989.
12. J.H. Williams and S.S. Lee, Pattern recognition characterization of micromechanical and morphological materials states via analytical quantitative ultrasonics, in Materials Analysis by Ultrasonics: Metals, Ceramics, Composites, ed. A. Vary, p. 187, Nayes Data Corporation, New-Jersey, 1987.
13. P. Dickstein, Spectral Analysis of Ultrasound from Multi-Layered Media, D.Sc Thesis, Technion-Israel Institute of Technology, January 1988.
14. B. Kedem, Spectral analysis and discrimination by zero-crossings, Proc. IEEE, **74**, No. 11,p. 1477, 1986.
15. J.D. Minford, Permanence of adhesive-bonded aluminum joints, in Adhesives in

Manufacturing , ed. G.L. Schneberger, Marcel Dekker Inc., New-York, pp. 567-601, 1983.

16. W. Brockmann, O.D. Hennemann and H. Kollek, Failure mechanisms in the boundary layer zone of metal/polymer systems, in Adhesive Joints: Formation, Characteristics and Testing , ed. K.L. Mittal, Plenum press, pp. 469-484, 1982.
17. A.E. Yaniv, N.Fin, H. Dodiuk and I.E. Klein, Oxide treatments of Al-1100 for adhesive bonding surface characterization, Appl. Surface Sci. , **20** , pp. 538-548, 1985.
18. H. Dodiuk, A.E. Yaniv, I.E. Klein, N. Fin and L. Drori, Chemical interaction in the system aluminum-oxide-primer-adhesive, Appl. Surface Sci. **25** ,pp. 137-153, 1986.
19. J.D. Minford, Effect of surface preparation on stressed aluminum joints in corrosive saltwater exposure, Adhesive Age , pp. 36-41, 1980.
20. P. Poole and J.F. Watts, Effect of alloy composition and surface pretreatment on the durability of adhesive bonded aluminium alloy joints, Int. J. Adhesives. Adhesions , **5** , No. 1, pp. 33-39, January 1985.
21. J.D. Minford, Joint durability studies with abraded, etched, coated and anodized aluminum adherends, in Adhesive Joints: Formation, Characteristics and Testing , ed. K.L. Mittal, Plenum press, pp. 485-501, 1982.
22. R.J. Davies and A.J. Kinloch, The surface characterization and adhesive bonding of aluminium, in Adhesion 13 , ed. K.W. Allen, Elsevier Applied Science, London, pp. 8-22, 1989.
23. J.A. Bishop, E.K. Kim, G.E. Thompson and G.C. Wood, Effects of the environment on bonded aluminum joints: an examination by electron microscopy, in Adhesion 13 , ed. K.W. Allen, Elsevier Applied Science, London, pp. 201-220, 1989.
24. W. Brockmann, O.D. Hennemann, H.Kollek and C. Matz, Adhesion in bonded aluminium joints for aircraft construction, Int. J. Adhesives and Adhesions , **6** , No. 3, pp. 115-143, July 1986.
25. J. Bishop, T. Smith, G.E. Thompson and G.C. Wood, Examination of the adhesive-adherend interface in the aluminum-aluminum bonded joint by electron microscopy, in Aluminum Surface Treatment Technology , eds. R.S Alwitt and G.E. Thompson, The Electrochemical Society, Pennington, NJ., 1986.

The Unidirectional Tensile Characteristics of Ion-Nitrided and Ion-Carbonitrided Surface Layer of 40CrNiMo

Guan Delin, Yu Zhiwei, Zhang Hongtien* and Yang Lieyu

Institute of Materials and Technology
Dalian People's Republic of China

**Institute of Metal Research*
Shenyang People's Republic of China

ABSTRACT

The mechanical behavior of ion-nitrided and ion-carbonitrided surface layers of 40CrNiMo has been studied by acoustic emission and SEM. It shows that the unidirectional tensile yield strength of materials can be raised by the above-montioned surface treatment. The plastic deformation of the surface layer occurs before the low yield point or after according to the difference of residual stress distribution. The formation of surface deformation streaks and their influence on the surface layer splitting are discussed.

INTRODUCTION

The wear resistance of workpieces can be improved by the ion bombardment treatment owing to hard penetration and diffusion layers. A good deal of research about the technology has made a considerable advances during the last twenty years. In consideration of service security, the mechanical behavior of the surface layer, particularly the fatigue, call for more and more attention.

Differing from the surface mechanical hardening treatments such as shot peening or sandblasting, the ion bombardment improves mechanical properties by modifing the chemical compositions and structures of the surface layer. The parameters having influence on mechanical properties depend on the concentration and existent forms of the elements, the depth of penetration and diffusion layers, the structure, as well as the residual stress arising therefrom.

In order to determine the effects of various parameters on the mechanical behavior of the surface layer, it's feasible to divide various deformations into several simple elementary types. The study of these elementary deformations is conductive to understanding the mechanical behavior of the surface layer under complicated conditions.

The present paper aims to brief the unidirectional tensile characteristics of ion-nitrided and ion-carbonitrided surface layers of 40CrNiMo.

MATERIALS AND METHODS

The chemical compositions of the test materials are as follows (wt.%):

C	Cr	Ni	Mo	Si	Fe
0.47	0.92	1.43	0.11	0.22	remnant

The specimens in the form of 6.35mm diameter and 55mm gauge length were machined from hot rolled Ø20 bars, and then hardened by oil quenching at 1123K and tempered at 923K for 2.5h, giving them an average hardness of HRc25.

The ion bombardment treatment was carried out in a model LDZ-15A ion bombardment installation, which gave the specimens a surface layer consisting of the compound layer and the transition layer which contain $\gamma'(Fe_4N)$ and ε

($Fe_{2-3}N$). The depth of the compound layer formed by ion-nitriding and ion-carbonitriding remains 12μm.

The tensile test was carried out on the Shimadzu-AG500A tensile machine with a displacement velocity of 0.3mm/min of the cross beam at ambient temperature. The deformation process was monitored by an accoustic emission system with a central frequency of 100 KHz, a bandspread of 40KHz, a threshold level of 0.5V and a gain of 100dB. In order to display the evolution of the surface layer during the tensile deformation, the acoustic emission spectra of the untreated specimen was modified to zero.

The axial residual stress along the radial direction was measured by a model MSF-2M X ray diffractometer with K_α Cr target and (211) diffraction, ψ was fixed at 0°, 15°, 30° and 45°. To attain the different measurement depths, various polishing operations were conducted by means of an electrolytic process.

Finally, the surface of the specimens was observed by SEM.

RESULTS

1. Acoustic emission during the unidirectional tensile process

The stress evolution and acoustic emission according to the deformation are given in Fig.1, which show a little increment of yield stress for both ion-nitriding and ion-carbonitriding in comparison with the untreated specimens (- 605MPa).

Below the low yield point, the tensile process of the ion-nitrided specimens does not accompany the acoustic emission phenomenon; when the low yield point is reached, a little amount abrupt acoustic emission appears. Considerable amount of acoustic emission occur from an extension ΔL = 0.2mm, and decrease slowly at ΔL = 0.3mm.

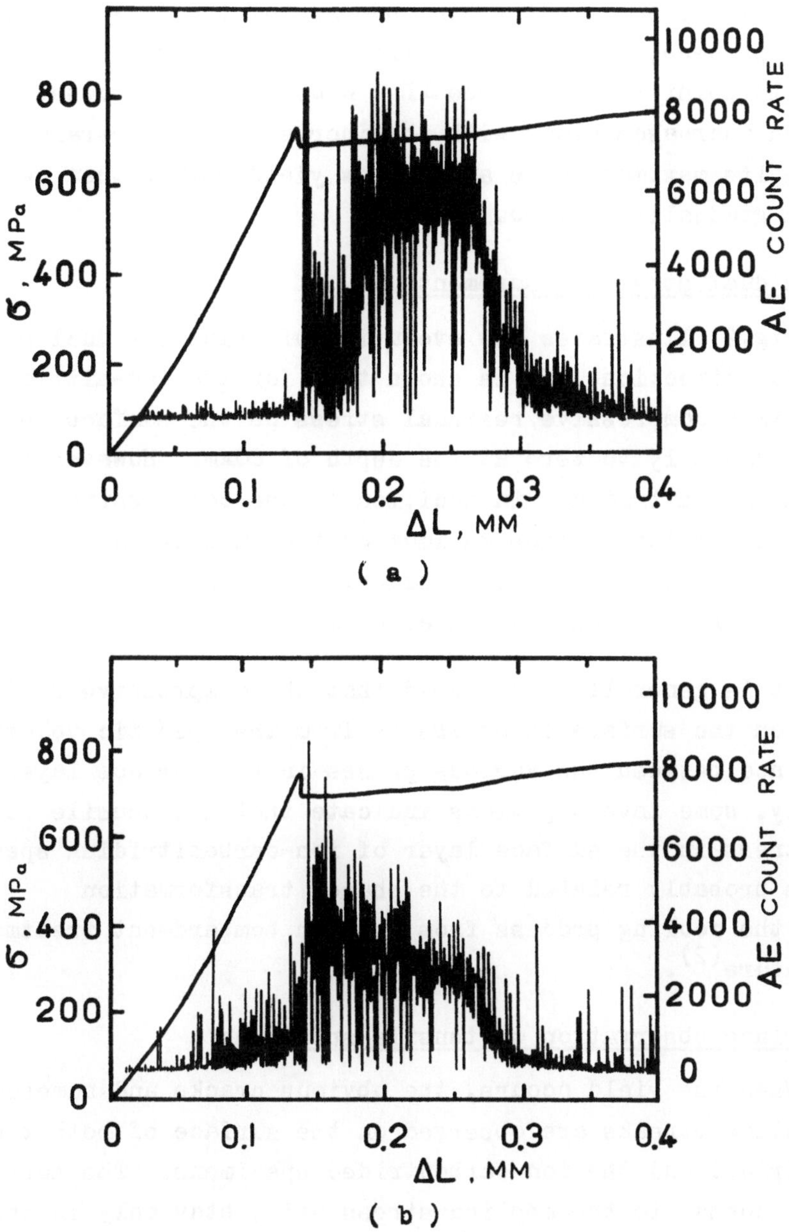

Fig.1 The variation of stress and acoustic emission count rate with extension for ion-nitriding and ion-carbonitriding
(a) ion-nitriding (b) ion-carbonitriding

Unlike ion-nitriding, in the case of the ion-carbonitriding process, the acoustic emission occurs even below the low yield point (exactly from ΔL = 0.05mm). The accoustic emission increases along with the increase of the strain, and reaches its maximum value at the low yield point, then decreases gradually to zero.

2. Residual stress measurement

Fig.2 illustrates the evolution of axial residual stress in radial direction. It is shown that for the ion-nitriding the maximum compressive residual stress on the surface decreases radially to zero at the depth of 80μm. However for the ion-carbonitriding, in addition to the compressive residual stress which remains maximum on the surface, a tensile residual stress zone appears inside. The maximum tensile residual stress reachs 400MPa at 45μm.

It is generally considered that the compressive residual stress on the surface layer issues from the specific volume difference between the various phases in the surface layer(1). Recently, some investigations indicate that the tensile residual stress in the surface layer of ion-carbonitriding specimens is probably related to the phases transformation during the cooling process from the ion bombardment treatment temperature(2).

3. Surface observation of tensile specimens

When the yield occurs, the obvious cracks and numerous deformation streaks are observed on the surface of both the ion-nitrided and the ion-carbontrided specimens. The surface cracks, normal to the applied stress axis, stay only in the compound layer, some of which are linked together with the deformation streaks. Moreover, below the yield point the deformation streaks exist merely on the surface of the ion-carbonitrided specimens, but not on the ion-nitrided specimens. Evidently, in the case of carbonitriding the acoustic emission below the low yield point is related to the deformation streaks; the acoustic emission after the yield is caused

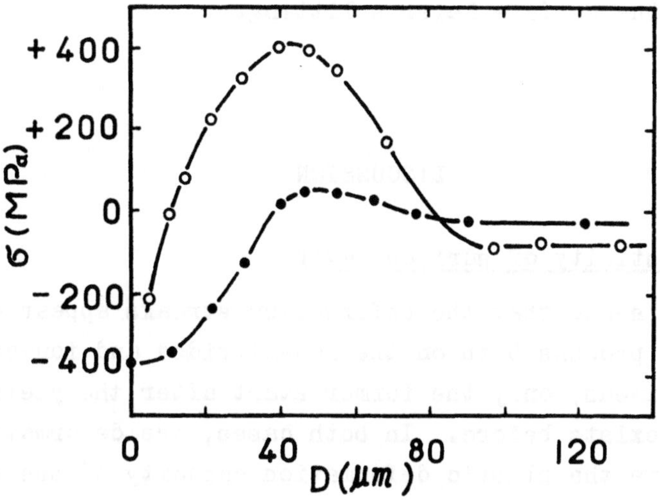

Fig.2 The residual stress distribution after ion bombardment treatment
• ion-nitriding o ion-carbonitriding

(a)

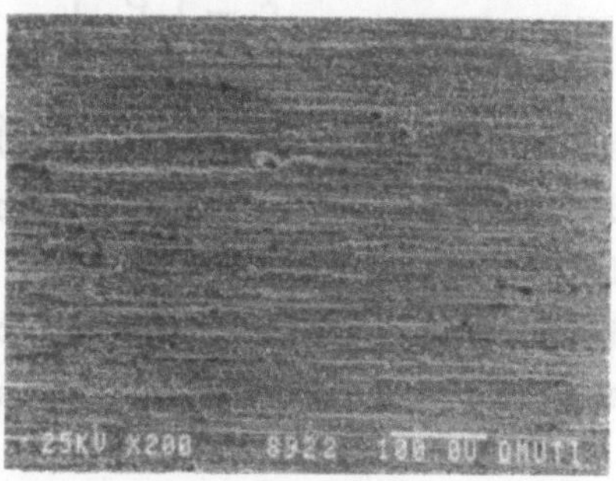

(b)

Fig.3 The surface morphology below yield
(a) ion-nitriding (b) ion-carbonitriding

chiefly by the surface layer splitting.

DISCUSSION

1. The plasticity of surface layer

It is shown that the deformation streaks appear during the tensile process both on the ion-nitrided and ion-carbonitrided specimens; only the former start after the yield and the latter exists before. In both cases, the deformation streaks prove the plastic deformation capacity of the surface layers.

Owing to the existence of the tensile residual stress zone in the surface layer of the carbonitrided specimens, it's possible that the local plastic deformation occurs before the macro-yield as a result of superimposure of the residual stress and the applied stress. The local plastic deformation can be determined as:

$$\varepsilon = b \rho_m L \qquad (1)$$

where b is Burgers vector, ρ_m is mobile dislocations density and L means free path. It should be mentioned that the acoustic emission appears at the applied stress level 200MPa, which, added to the maximum tensile residual stress, equals the yield strength of materials.

Some investigation on the relation between the acoustic emission and deformation or creaks indicate that the high energy state arising from the local stress concentration can be translated to the low energy state by the plastic deformation or cracks, and release strain energy. The elastic wave manifests as acoustic emission[3]. The energy derived from the plastic deformation can be expressed as:

$$E \propto \rho_m \bar{v}^2 L^2 \qquad (2)$$

where $\bar{v}$ is the average velocity of dislocations. Then from (1) and (2)

$$E \propto \frac{1}{b} \varepsilon L \bar{v}^2 \tag{3}$$

Assuming that L and $\bar{v}$ are constant, the acoustic emission energy increases with the plastic deformation.

If the acoustic emission energy was divided into two parts as the plastic deformation Ep and the surface layer splitting Ec, then

$$E = E_p + E_c \tag{4}$$

Obviously, according to preceding results the acoustic emission count rate corresponding to Ep is less than 4000. In the carbonitriding process, Ep is released as the acoustic emission below the low yield point. As soon as the yield occurs, the surface layer splits and the acoustic emission energy becomes Es. As for the nitriding process, owing to the non-existence of the tensile residual stress zone in the nitrided surface layer, the local plastic deformation cannot occur before the yield as in the carbonitriding process. There are, of course, no surface deformation streaks and the correspondent acoustic emission. Beyond the yield point, an incubation is discovered between $\Delta L = 0.15$mm and $\Delta L = 0.17$mm, where the acoustic emission count rate is nearly less than 4000, corresponding to the plastic deformation of the surface layer, and immediately the surface layer splits.

2. <u>The strength of the surface layer</u>

For both the nitriding and the carbonitriding, the plastic deformation occurs on the surface layer after the local or total yield, and then the surface layer splits. That is to say the surface cracks are related to the surface deformation streaks. Suppose that the yield strength of the surface layer and matrix are respectively σ_{ss} and σ_{sm} , their volume percentages are Vs and Vm, then the yield strength of the specimen is given by

$$\sigma_{St} = \sigma_{ss} V_s + \sigma_{sm} V_m + f(\sigma_{ss}, \sigma_{sm}) \quad (5)$$

Since $\sigma_{ss} > \sigma_{sm}$, therefore the yield strength of the ion bombarded specimen is 20-70MPa more than the untreated. σ_{St} depends not only on the yield strength, but also on the depth of the surface layer. As noted, the increment of the fatigue strength for smooth specimens treated by nitriding has nothing to do with the depth of the compound layer, but according to the fact that the increment of the fatigue strength has to be below a critical strain(4), the thick compound layer means obviously higher critical strain for the improvement of fatigue.

In addition, it should be pointed out that $f(\sigma_{ss}, \sigma_{sm})$ in (5) expresses the stress concentration effect of the matrix to surface la yer during the matrix yields. σ_{St} depends also on the concordance of yield strength between the matrix and surface layer. Certainly, an appropriate yield strength of the matrix is necessary in order that the specimens procure a maximum yield strength.

CONCLUSION

1. The ion bombardment treatment can increase the unidirectional tensile yield strength of 40CrNiMo. The increment of yield strength depends on the yield strength of the surface layer, and its concordance with the matrix.

2. In the nitriding process, the axial residual stress is compressive, which decreases radially. Before the yield, there are no deformation streaks on the surface layer, and beyond the yield point the plastic deformation occurs rapidely on the surface layer, then splits the surface layer.

3. In the carbonitriding process, in addition to the compressive residual stress on the surface layer, there is a tensile residual stress zone. The plastic deformation of the surface layer occurs before the yield, forming deformation streaks. After the yield, the surface layer splits at once.

REFERENCES

1. Noyan, I.C. and Cohen, I.B., The nature of residual stress and its measurement. In Residual Stress and Stress Relation, ed. Eric Kula and Volken Weiss, Plunum Press, New York, 1982, pp. 2-4.

2. Xu Xiaolei, Hei Zukun and Yang Lieyu, The study on the structures of ion-carbonitriding surface layer of 40CrNiMo, Annual Progress Report, Dalian Maritime University, Dalian, October 1989.

3. Zhu Zuming, Mu Zaiqin, Wang Shuangquan and Yin Huazi, Acoustic emission during plastic deformation and microcrecking process of -Ti. Acta Metallurgica Sinica, 1982, vol. 18, No. 3, pp. 272.

4. Guan Delin, Yu Zhiwei and Yang Lieyu, Fatigue behavior of 40CrNiMo treated by plasma nitro carburizing method, STRUC ENG -89, Los Angeles, 9-11 August 1989.

Friction and Metal Transfer in Sliding Aluminum against Ion-Implanted Steel

P. C. Nautiyal

Indian Institute of Petroleum
Dehardrun India

and J. A. Schey

University of Waterloo
Waterloo Ontario
Canada

ABSTRACT

Workpiece-lubricant-die interactions were studied in twist-compression experiments with 6061-T6 aluminium sliding against bare and ion-implanted steel anvils. From the magnitude and trends in coefficient of friction and from the extent of metal transfer and retransfer it could be concluded that, for optimum lubrication under predominantly boundary conditions, both contact surfaces should allow the formation of adsorbed/reacted films with the boundary additives. This condition was reliably satisfied with bare steel. Rapidly increasing friction and metal transfer indicated that a TiC surface interfered with the adsorption of stearyl alcohol and a TiN surface interfered with both stearic acid and stearyl alcohol.

INTRODUCTION

In the plastic working of aluminum and its alloys, metal transfer to the die results in the formation of die pickup which may develop into a more or less continuous die coating. The extent of pickup and the thickness and stability of the coating affect the surface quality of the product, and excessive pickup or an unstable coating may force a termination of the process. Metal transfer under dry (unlubricated) conditions has been repeatedly investigated (see, for example, Chen and Rigney [1]), but little is known about metal transfer in the presence of lubricants. Within a larger effort centering on basic material-process-lubricant interactions, the authors already reported on the effect of lubricants on aluminum transfer against a tool steel surface [2]. The present paper deals with friction and aluminum transfer against ion-implanted surfaces.

Implantation of Ti and C ions into tool steel surfaces has been reported to yield benefits in some instances, as for punches used in aluminum can body makers. The effects of changing the surface composition by ion implantation or by various methods of surface coating cannot be considered in isolation but must be dealt with as an integral part of the tribosystem [3]. In particular, there is the possibility that a coated surface may interfere with or prevent rather than promote the adsorption/reaction processes necessary for the functioning of boundary and EP additives. There is some evidence of this scattered in the literature. For example, Sioshansi and Au [4] found higher friction, but reduced wear, in lubricated sliding of 4140 and O6 steel against Ti plus C implanted 52100 steel. Gabriel et al [5] reported that an ion-plated TiN surface reduced the wear of a non-plated countersurface when a neat mineral oil was used but actually increased wear when an EP agent (ZnDTP) was added to the lubricant. Kohser et al [6] observed that vapor deposition of a Sn layer followed by N implantation increased the wear life of a heat-treated steel only when a fatty acid capable of forming a tin soap was used. In the work of Tan et al [7] the scuffing resistance of steel balls increased with a S additive on a CVD TiN coating yet a P additive was more effective on a bare steel surface. Schey and Newnham [8] reported that a stearic acid film, which was always effective when applied to steel dies, broke down when boronized dies were in contact with steel specimens but survived when the specimen was made of aluminum. Obviously, lubricant interaction with both surfaces was necessary for steel but interaction with the specimen was sufficient for aluminum.

In view of the sparse knowledge on lubricant/surface interactions, the present work aimed at studying the effects of ion implantation on friction and aluminum transfer in the presence of typical boundary additives.

MATERIALS AND EXPERIMENTAL METHODS

The workpiece material was represented by 6061-T6 aluminum alloy specimens machined from bars of 25.4 mm diameter. One end was bored to 19.05 mm I.D. to give an annular contact area, with the end face finely turned to 0.6 μm CLA roughness, measured transverse to the turning marks. The die (tool) material was represented by flat, rectangular anvils of AISI O1 die steel at HRC 62; the test surfaces were lapped to give a nondirectional surface finish of 0.1 to 0.15 μm CLA roughness. Ion implantation was performed by Spire Corporation, Bedford, Mass., to the conditions shown in Table 1.

Table 1 Conditions of Ion Implantation

Ion beam	Energy, keV	Dose, 10^{17} ions/cm^2
B	60	2
N	60	4
Ti/C	120/60	2 / 1
Ti/N	120/60	2 / 1

In the twist-compression test used throughout this program, the annular end face of the specimen is rotated at 10 rpm against the flat anvil, under an applied pressure of 90 MPa. Upon initial approach, a squeeze film is established, the thickness of which was estimated at 7, 18, and 45 μm for the predominantly paraffinic base mineral oils of 4.5, 30, and 285 mm^2/s viscosity at 38 °C [2]. For the sake of brevity, these will be referred to as light, medium, and heavy oils. In the course of continuing rotation, the lubricant is gradually squeezed out and conditions shift towards the predominantly boundary lubrication regime, as shown by the disappearance of the turning marks. Because of the severe conditions imposed, the test is eminently suitable for exploring the effects of additives. In this instance, boundary additives were restricted to 1% stearic acid and 5% stearyl alcohol in the base oils. In some previously reported experiments [9] the shorter-chain lauric acid and lauryl alcohol and the unsaturated oleic acid were ineffective even at 5% concentration, as was stearyl alcohol at 1% concentration. Among EP additives, TCP at 5% concentration was ineffective but chlorine (in the form of chlorinated paraffin) gave a noticeable improvement at 5% Cl concentration. None of these additives were further investigated here.

While the coefficient of friction and, particularly, its variation for the duration of a test are sensitive measures of the effectiveness of lubricants, metal transfer to the anvil and retransfer to the specimen give valuable indications of the ability of the lubricant/coating system to prevent undesirable die pickup and coating development. Therefore, anvil and specimen surfaces were investigated by optical microscopy, surface roughness measurement, and SEM. In previous work [9], anvils implanted with B and N behaved similarly to unimplanted surfaces, therefore, the present investigation concentrated on Ti plus C and Ti plus N (for brevity, TiC and TiN) implanted anvils. Previous work had also shown that continuous rotation leads to very severe conditions with significant tool pickup on implanted surfaces, even in the presence of additives. Therefore, in the present work experiments were conducted with interrupted rotation: At the end of one revolution, the anvil was dropped, the lubricant film was allowed to spread over the anvil surface, and then the anvil was raised and kept in contact for a further revolution. The procedure was repeated until ten revolutions were accumulated. This procedure should be more typical of metalworking processes where the tool surface is relubricated between successive workpieces, as in an aluminum can bodymaker.

RESULTS AND DISCUSSION

To put the results in perspective, it is necessary to review some results obtained with continuous rotation [9]. With neat mineral oils (Fig. 1a), the effect of viscosity is clearly evident. Even though scatter between individual tests (shown by symbols) is relatively large, it is obvious that implanted surfaces are no better than bare steel. Indeed, TiN caused friction to rise steeply from the first to the twentieth revolutions even with the heavy oil. The addition of 1% stearic acid significantly improved the situation on bare steel, even with the light oil (Fig. 1b). The TiC- and, in particular, the TiN-implanted surface made matters worse, suggesting that adsorbtion/reaction on both specimen and anvil surfaces was necessary for maintaining a low interfacial shear strength. Much the same applies to 5% stearyl alcohol (Fig. 1c). On bare steel it was as effective as the acid; with the light oil it was even more effective than 1% stearic acid on both bare and implanted surfaces, although implantation again impaired performance.

Variations in friction reflected events at the interface. When friction dropped in the course of rotation, as with the neat light oil (Fig. 1a top), metal-to-metal contact during the first revolution resulted in heavily smeared transfer particles covering some 10% of the contact zone on the anvil, and particles retransferred to the specimen covered some 20% of the annular end face. By the tenth revolution, some 50% of the anvil and 20% of the specimen were covered with dull particles, with a concomitant

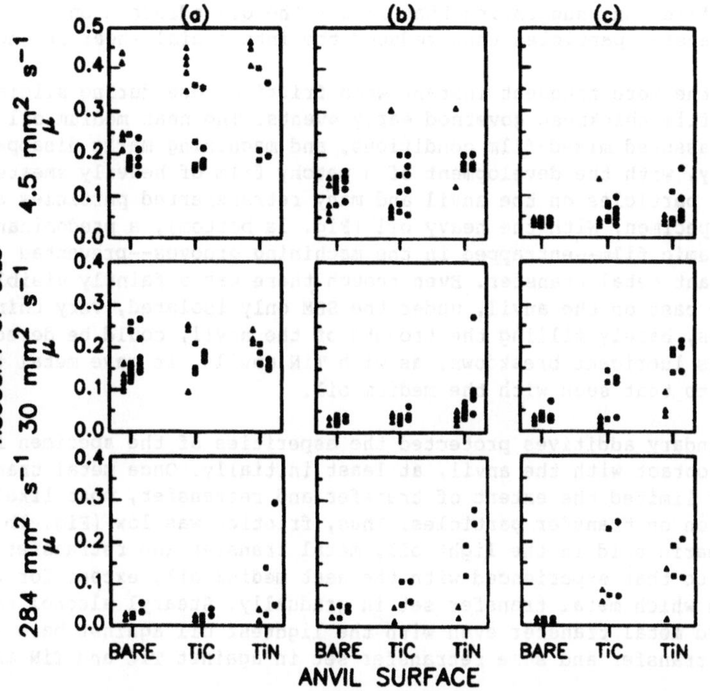

Fig. 1 Variation in the coefficient of friction in twist-compression testing of 6061-T6 Al alloy workpieces against bare steel anvils with (a) neat mineral oils and with (b) 1% stearic acid and (c) 5% stearyl alcohol as additives. Continuous rotation: ▲ 1st; ■ 10th; ● 20th revolution.

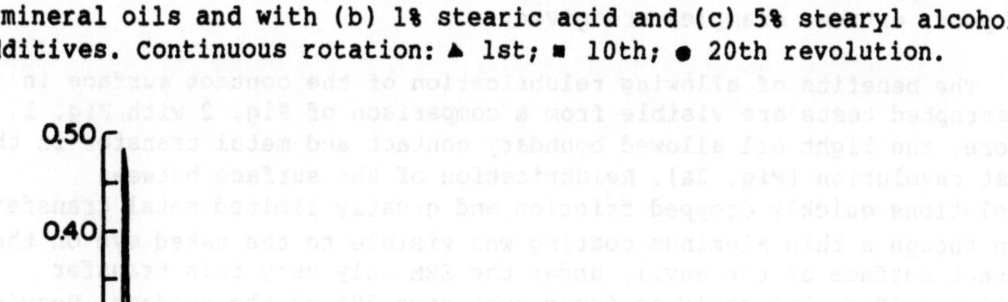

Fig. 2 Variation of friction in twist-compression testing with interrupted rotation against bare steel anvils with neat (a) light, (b) medium, and (c) heavy base mineral oils.

drop in friction, suggesting that it was the oil adsorbed on the transfer and retransfer particles that reduced the interfacial shear strength.

In the more frequent instant when friction rose during sliding, initial film thickness governed early events. The neat medium oil (Fig. 1a, center) assured mixed-film conditions, and machining marks disappeared only gradually, with the development of a patchy film of heavily smeared transfer particles on the anvil and many retransferred particles embedded in the specimen. With the heavy oil (Fig. 1a bottom), a predominantly hydrodynamic film--entrapped in the machining grooves--prevented significant metal transfer. Even though there was a faintly visible aluminum cast on the anvil, under the SEM only isolated, very thin transfer particles, barely filling the troughs of the anvil, could be detected. When there was lubricant breakdown, as with TiN anvils, it gave metal transfer similar to that seen with the medium oil.

Boundary additives protected the asperities of the specimen from direct contact with the anvil, at least initially. Once metal transfer set in, they limited the extent of transfer and retransfer, most likely by adsorption on transfer particles. Thus, friction was low (Fig. 1b and 1c); with stearic acid in the light oil, metal transfer and retransfer were similar to that experienced with the neat medium oil, except for the TiN anvil on which metal transfer set in gradually. Stearyl alcohol totally prevented metal transfer even with the lightest oil against bare steel, but gradual transfer and some retransfer set in against TiC and TiN anvils.

In reporting on the present work, typical torque traces are shown in Fig. 2 and Figs. 5 - 8 to illustrate trends. When replicate tests gave highly reproducible results, only one trace is shown; when reproducibility was poor, extreme examples are given.

The benefits of allowing relubrication of the contact surface in interrupted tests are visible from a comparison of Fig. 2 with Fig. 1. As before, the light oil allowed boundary contact and metal transfer in the first revolution (Fig. 2a). Relubrication of the surface between revolutions quickly dropped friction and greatly limited metal transfer. Even though a thin aluminum coating was visible to the naked eye on the contact surface of the anvil, under the SEM only very thin transfer particles (Fig. 3a) could be found over some 30% of the surface. Heavier pickup was present only at the edges of the contact zone (Fig. 3b) where the lubricant film was the thinnest. Considering the minimal pickup on the anvil, surprisingly heavy damage was inflicted on the specimen which showed evidence of gauging and patches of heavily deformed retransfer particles (Fig. 3c). Thus, compared to the tests with continuous rotation, interrupted rotation allowed the low-viscosity lubricant to minimize metal transfer to the anvil but without substantially reducing damage to the workpiece.

With the medium oil, continuous rotation resulted in an increase of friction and metal transfer (Fig. 1a). In contrast, interrupted rotation gave gradually decreasing friction (Fig. 2b), and at the end of the tenth revolution a steady-state low-friction condition was attained. Under the SEM, a faint aluminum cast on the anvil was similar to that earlier found with the heavy oil. Transfer particles barely filled the troughs of the anvil surface (Fig. 4a) and the specimen surface was free of backtransfer. Indeed, remnants of machining marks were clearly evident (Fig. 4b).

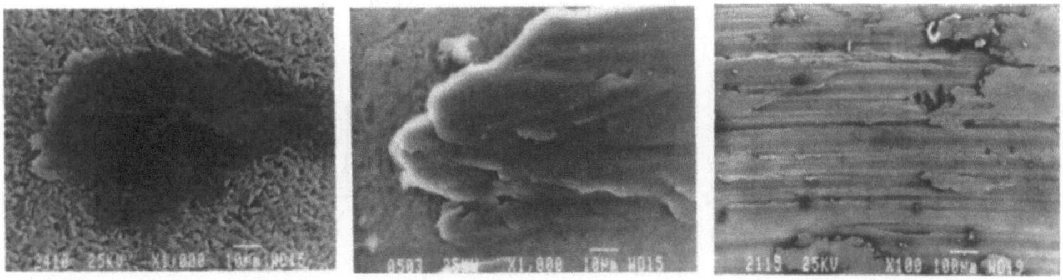

Fig. 3 Interrupted twist-compression tests with light oil on bare steel anvil. SEM photographs after 10 revolutions: (a) center and (b) edge of contact area on anvil, (c) specimen showing grooving and back transfer.

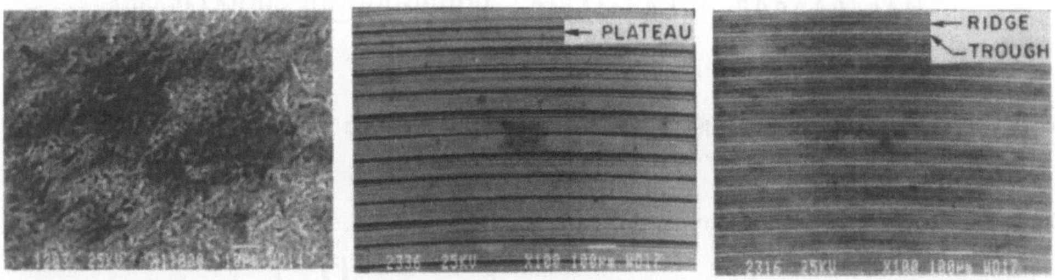

Fig. 4 Interrupted twist-compression tests on bare steel anvil: (a) center of contact area on anvil and (b) on specimen with medium oil; (c) specimen with heavy oil.

With the heavy oil, friction was extremely low (Fig. 2c) and metal transfer was practically absent, with only a faint aluminum cast at the edge of the contact zone. The asperities of the machined grooves of the specimen were barely deformed (Fig. 4c).

Since the medium oil was most sensitive to changes in test conditions, the effects of ion implantation were first explored with this oil. With the neat oil, there was practically no difference between bare and TiC anvils (Fig. 5a and b), but TiN already showed some adverse effects (Fig. 5c). It should be noted that ion-implantation does not change surface topography, hence surface composition must be responsible for the poorer results. This was also the case with 1% stearic acid in the oil (Fig. 6). With bare anvils, friction was consistently low and metal transfer limited (even less than that in Fig. 4a). Friction stayed at a low value also on TiC, but tests on TiN had poor reproducibility and frequently gave high friction and metal transfer.

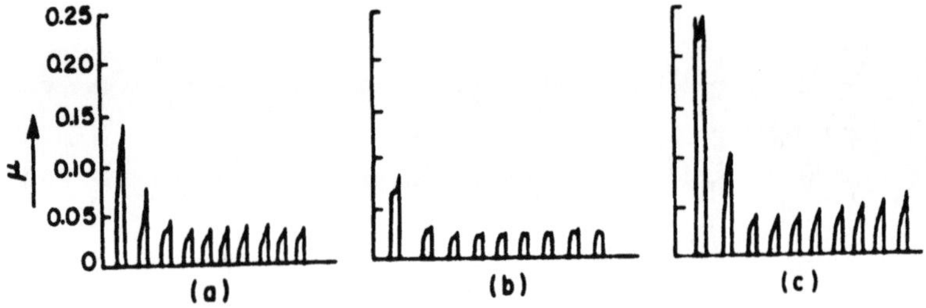

Fig. 5 Variation of friction in interrupted twist-compression testing with neat medium oil against (a) bare, (b) TiC, and (c) TiN anvils.

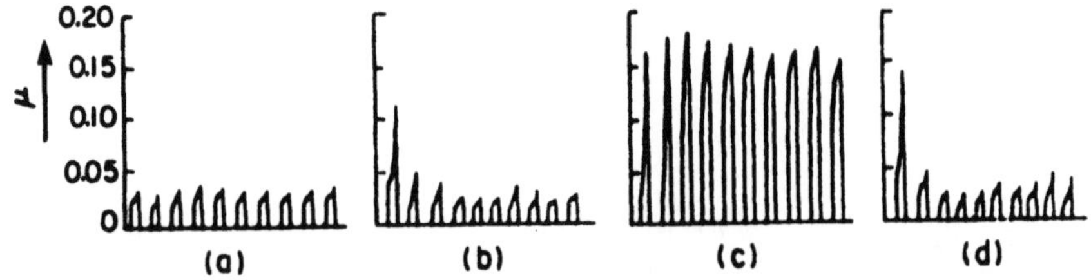

Fig. 6 As Fig. 5, but with 1% stearic acid in medium oil: (a) bare, (b) TiC, (c) and (d) TiN anvils.

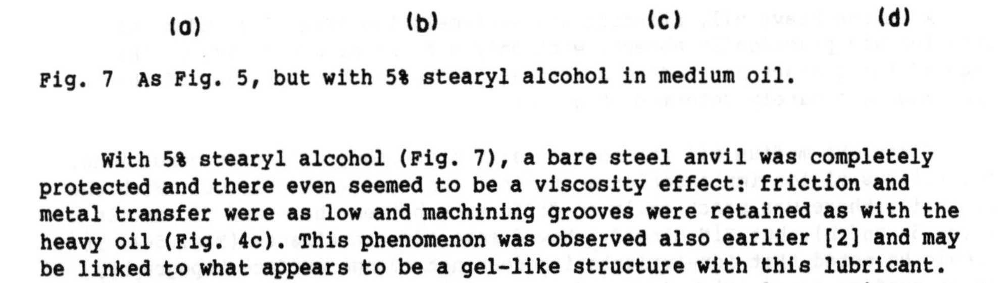

Fig. 7 As Fig. 5, but with 5% stearyl alcohol in medium oil.

With 5% stearyl alcohol (Fig. 7), a bare steel anvil was completely protected and there even seemed to be a viscosity effect: friction and metal transfer were as low and machining grooves were retained as with the heavy oil (Fig. 4c). This phenomenon was observed also earlier [2] and may be linked to what appears to be a gel-like structure with this lubricant. Unexpectedly, lubrication broke down not only on TiN but also on TiC, as evidenced by rising friction, moderate transfer to the anvil, and severe damage and back transfer to the specimen (similar to that in Fig. 3c, but with less grooving under the generally smaller backtransfer particles).

The possible adverse effect of implantation was explored also with the heavy oil (Fig. 8). Evidently, under the predominantly hydrodynamic conditions, the boundary additives prevent all transfer and keep friction very low with the bare anvil (Fig. 8, row a). Interrupted rotation now

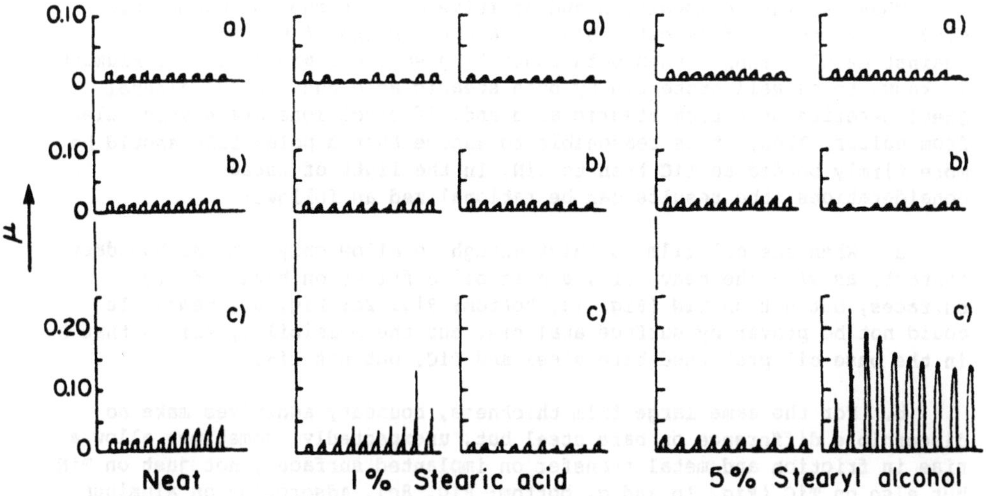

Fig. 8 Variation of friction in interrupted twist-compression testing with heavy oil against (a) bare, (b) TiC, and (c) TiN anvils.

allowed similarly favorable conditions to develop with TiC anvils (Fig. 8, row b). However, TiN again interfered with the action of stearyl alcohol, as seen from the poor reproducibility, occasional high friction (Fig. 8, row c) and accompanying severe damage to the specimen.

GENERAL DISCUSSION AND CONCLUSIONS

In interpreting the results, it is necessary to consider the events at the interface. As long as oil films are thick enough to provide total separation of surfaces, neither surface composition nor boundary additives play any role. Under the severe conditions imposed by the twist-compression test, such situation did not exist even with interrupted rotation. As soon as there is some boundary contact, boundary and/or EP additives are needed to protect asperities. Even though the base oils used here are highly refined, they contain minor amounts of sulfur (0.06, 0.07, and 0.20% in the light, medium, and heavy oils, respectively). If interface temperatures are high enough, sulfur is an effective EP agent on steel.

The major additive effects in this instance are attributable to boundary additives. It is generally accepted that on nonreactive surfaces they form physically adsorbed, oriented films, which are desorbed at the melting point of the acid. When applied to reactive surfaces, a soap of higher melting point forms, thus extending protection of asperities to higher temperatures [10]. It has also been found that, for the same chain length and saturation, a higher percentage of alcohol is needed to provide the same protection as acid. The ternary Ti-C-Fe or Ti-N-Fe alloy surface layer of implanted anvils should have characteristics similar to that of TiC and TiN ceramics. No information could be found on the adsorption of boundary additives on such surfaces, nevertheless, it is reasonable to assume that the more stable TiN would be less reactive than TiC, and both would be much less reactive than steel and, particularly, aluminum.

Thus, surface composition and additive effects should become more evident as the oil film gets thinner. Additives should better protect against metal transfer when both contacting surfaces are reactive. Aluminum is known to be well protected by both stearic acid and stearyl alcohol. Steel benefits more from stearic acid and, if conditions are severe, also from sulfur. Also, it is reasonable to assume that a polar film should be more firmly bonded to TiC than to TiN. In the light of these considerations, the results can be rationalized as follows:

a. When the oil film is thick enough to allow only minimal boundary contact, as with the heavy oil, a neat oil suffices on bare and TiC surfaces, but not on TiN (Fig. 1a, bottom; Fig. 2c; Fig. 8c, neat). It could not be proven by surface analysis, but the possibility exists that S in the base oil protected bare steel and TiC, but not TiN.

b. For the same large film thickness, boundary additives make no discernible difference on bare steel but, unexpectedly, sometimes allow a rise in friction and metal transfer on implanted surfaces, not just on TiN but also on TiC (Fig. 1b and c, bottom; Fig. 8c). Adsorption on aluminum should take place in all cases, and this is reflected in low initial friction. However, as the oil film thins out, lack of adsorption on TiN and, to a lesser extent, on TiC allows metal transfer and increased friction. It is conceivable that, in a competition for surface sites, S prevented the action of stearyl alcohol on TiC (Fig. 1c, bottom). Poor reproducibility of results indicates that chance events may determine the initiation of metal transfer.

c. When the initial film is thinner and contains no boundary additives, as with the neat medium oil, initially high friction further increases if the film thins out on sliding (Fig. 1a, center). The benefits of boundary additives are clearly evident in lower friction and metal transfer if both surfaces are reactive, but the benefit is lost on TiN and, with stearyl alcohol, also on TiC (Fig. 1b and c, center).

d. For the same moderate initial film thickness, replenishment of the film results in reduced friction with a neat oil (Fig. 2b; Fig. 5) but friction again rises on TiN with stearic acid (Fig. 6c and d) and on both TiC and TiN with stearyl alcohol (Fig. 7b-d).

e. When boundary contact prevails from the beginning, as with the light oil, friction is high and is governed by the shear strength of oil films adsorbed on the transfer and retransfer surfaces. Thus, it is much reduced by boundary additives on reactive surfaces (bare steel anvils in Fig. 1b and c vs. Fig. 1a and Fig. 2a). With TiN anvils the benefits are primarily attributable to adsorption on the aluminum specimen; with TiC there is perhaps also adsorption on the anvil (Fig. 1b and c, top). With stearyl alcohol, friction is as low with the light as with the medium oil, possibly because of the gel-like structure previously observed.

The present experiments gave no information on wear. It is entirely possible that the high hardness of implanted surfaces is of great benefit in situations where lubrication minimizes metal transfer and friction, and wear of the die is the process-limiting factor. It is also possible that additives other than fatty acids or alcohols are effective even on implanted surfaces.

ACKNOWLEDGMENTS

The authors are grateful to the Manufacturing Research Corporation of Ontario and the Natural Sciences and Engineering Research Council of Canada for funding this work.

REFERENCES

1. Chen, L.H. and Rigney, D.A., Transfer During Unlubricated Sliding Wear of Selected Metal Systems. Wear, 1985, 105, 47-61.

2. Nautiyal, P.C. and Schey, J.A., Transfer of Aluminum to Steel in Sliding Contact: Effects of Lubricant. Submitted to J. of Tribology.

3. Schey, J.A., Tribology in Metalworking: Friction, Lubrication and Wear, ASM, Metals Park, OH, 1983, pp. 115-118.

4. Sioshansi, P. and Au, J.J., Improvements in Sliding Wear for Bearing-Grade Steel Implanted with Titanium and Carbon, Mater. Sci. Eng., 1985, 69, 161-166.

5. Gabriel, H.M., Schmidt, F., and Kloos, K.H., Influence of Additive-Treated Lubricants on the Wear Behaviour of TiN Ion-Plated Structural Parts. In Proc. 4th Eurotrib, Elsevier, London, 1985, Vol.1, Paper 1-4.5.

6. Kohser, R.A., Hale, E.B., and O'Keefe, M.J., A Surface Layer-Lubricant Interaction and Its Effect on the Wear Resistance of Steel. In Tribological Mechanisms and Wear Problems in Materials, ASM International, Materials Park, OH, 1987, 111-119.

7. Tan, Z.H., Wang, Y.Z., and Yu, M.S., The Scuffing Resistance of TiN Coating under Lubricated Condition, In Proc. 5th Eurotrib, Espoo, Finland, 1989, Vol. 2, 378-383.

8. Schey, J.A., and Newnham, J.A., Effect of Die Composition on Lubrication in Metalworking. Lubric. Eng., 1970, 26, 129-137.

9. Nautiyal, P.C., and Schey, J.A., Effects of Ion-Implanted Surfaces in Lubricated Sliding of Aluminum on Steel. In Proc. Nat. Conf. Industrial Tribology, Trivardrum, India, 1989, pp. 111-13.1 to 111-13.4.

10. Ref. 3, pp. 50-58.

The Effect of Ion Implantation on the Fretting Fatigue Properties of Titanium Alloy Fan and Compressor Blades

J. E. Elder, R. Pishva

Hawker Siddeley Canada Inc.
Mississauga Ontario Canada

W. Doswell

Carleton University
Ottawa Ontario Canada

G. Weatherly

University of Toronto
Toronto Ontario Canada

and W. Wallace

National Aeronautical Establishment
Ottawa Ontario Canada

ABSTRACT

The feasibility of applying ion implantation to produce substantial improvement in the wear/fretting fatigue resistance of titanium alloys used as gas turbine fan and compressor blade material has been studied. Detailed microscopy and mechanical test results have been used to analyze the various mechanisms contributing to wear and fretting fatigue damage in titanium alloys. The manner in which ion implantation might modify the normal wear/fretting fatigue response of these alloys is discussed in detail, along with the factors to be considered in choosing optimum parameters.

INTRODUCTION

Severe fretting fatigue has been observed on the dovetail surfaces of aero gas turbine engine fan and compressor blades. The blade material is Ti-6Al-4V. The dovetail fretting fatigue damage develops as a result of low amplitude slip and rubbing between the dovetail surface and the inner surface of the disc posts. If severe fretting fatigue is allowed to continue, premature fatigue crack initiation may occur which may result in premature fatigue failure. An in-flight blade fatigue failure would be catastrophic.

Ion implantation has been identified in a number of earlier studies (1,2) as being a viable technique for improving the wear, fatigue and fretting fatigue properties of Ti-6Al-4V. Ion implantation is a technique with a particularly good potential for improving the fretting fatigue resistance of the dovetail surface of fan and compressor blades because it has been successful in improving both the tribological and fatigue properties of titanium alloys (1,2). Note that many of the commercially used anti fret coatings applicable to titanium alloys may improve fretting resistance, however, they reduce the fatigue strength of the material. Accordingly, the use of these coatings for improving the fretting fatigue strength of titanium alloys is questionable.

In an earlier part of this investigation, the ion implantation parameters which appear to have a high potential for increasing the fretting fatigue strength of Ti-6Al-4V were identified (1). From these results three different ion implantation treatments were selected for experimentation. These include:

1. Nitrogen - peak concentration greater than 48.7×10^{21} N/cm^3,
2. Carbon - peak concentration greater than 16.0×10^{21} C/cm^3,
3. Nitrogen plus carbon (dual implant)

The experimental procedure included both an analysis of the microstructure of the implanted layers using transmission electron microscopy and an evaluation of the fretting fatigue strength of ion implanted Ti-6Al-4V. It is hoped that a correlation may be developed between the microstructure of the implanted layer and the fretting fatigue properties. The results of these analyses which have been obtained to date are presented.

TRANSMISSION ELECTRON MICROSCOPY OF ION IMPLANTED Ti-6Al-4V

Transmission electron microscopy has been carried out on nitrogen, carbon and carbon plus nitrogen implanted Ti-6Al-4V to characterize the implanted layer microstructure. For each implantation treatment it appears that the surface structure transformed completely from titanium to titanium carbide or nitride. Furthermore, the titanium carbide or nitride appears to have formed with an epitaxial relationship to the titanium. With the exception of the thin foils which received the dual N^+ plus C^+ implant, both cubic Ti(C,N) and hexagonal close packed (α-Ti) diffraction patterns were observed. Tetragonal Ti_2N was also observed in the nitrogen implanted material. Only cubic diffraction patterns were obtained from the foils which received the dual C^+ plus N^+ implantation. Note that these foils received the highest dose of all the samples examined.

Figures 1a, b and c give diffraction patterns of the nitrogen implanted Ti-6Al-4V. Many of the diffraction patterns were sharp, while other patterns showed clear evidence of twinning. Figure 1a is the [100] pattern from Ti_2N and shows both the (020) TiN and the (002) Ti_2N reflections. Figure 1b shows superimposed diffraction patterns of TiN and Ti. The orientation of the TiN with respect to the Ti is also shown in Figure 1b i.e. (0001) Ti parallel to (111) TiN and $[11\bar{2}0]$ Ti parallel to $[1\bar{1}0]$ TiN. The diffraction pattern of Figure 1c shows two orientations of TiN in a twin relationship. These patterns come from <011> TiN orientations. Figure 2 illustrates a dark field micrograph of the nitrogen implanted Ti-6Al-4V showing one orientation of TiN. The twinning is quite evident in this micrograph. It is likely that the twinning is associated with relief of the transformation strains for Ti to TiN.

Note that the structures presented in Figures 1 and 2 are typical of single crystal patterns. These foils were given extremely high doses and as a result it would be expected that the diffraction patterns would reveal severe lattice damage. The results presented here suggest that during the implantation treatment the foils attained a high enough temperature to cause annealing. The foils were approximately 3mm thick when they were implanted and were affixed to an Al substrate using conductive paint. Further analysis is required to confirm that these structures would also be obtained in a bulk sample using the same implant conditions.

THE EFFECT OF ION IMPLANTATION ON THE FRETTING FATIGUE STRENGTH OF Ti-6Al-4V

Fretting Fatigue Testing Procedure

The fretting fatigue test facility used in this investigation included a fretting fatigue test apparatus and specimen (Figure 3), a MTS 250KN servo hydraulic test frame, MTS 810 material test system and a Hewlett Packard Data Acquisition/Control Unit (Model 3497A). This facility is located at the National Research Council, Structures and Materials Lab.. The fretting fatigue test specimen is a fretting bridge arrangement which provides a relatively inexpensive means of inducing fretting and has been used by other research groups (3,4). The fretting fatigue test apparatus is enclosed in a chamber such that testing could be carried out in dry air as humidity effects are known to influence fretting wear. During testing, compressed air was passed through desiccants before being bled into the chamber. Testing was carried out at room temperature to model the service conditions of fan blades.

Prior to testing, the fatigue specimens were polished to a one micron finish. With the exception of the base line fatigue and fretting fatigue specimens, all fatigue specimens were ion implanted on the surfaces of the gauge length. Base line fatigue tests were conducted by simply subjecting the fatigue specimen to an axial cyclic load in the absence of the fretting bridges. Fretting fatigue tests were carried out by clamping two fretting bridges to the gauge length of the fatigue samples with a constant normal load. The fretting bridge pads were ground to a 600 grit level and polished to a nine micron level. The fretting bridge material was Ti-5Al-2Sn-2Zr-4Mo-4Cr.

Fretting fatigue tests were carried out under a normal load of 50 MPa and cyclic fatigue load of 560 MPa. The R ratio used was 0.25 with a frequency of 4 Hertz. At this stage of the investigation only a carbon implantation treatment has been evaluated.

Fretting Fatigue Test Results

The two baseline fatigue tests which were carried out were discontinued after the specimens had accumulated one million cycles. Two baseline fretting fatigue tests were carried out on polished unimplanted Ti-6Al-4V and the specimens lasted 39,370 and 52,434 cycles respectively before failing. Only two carbon implanted specimens have been tested as of yet. These samples lasted 114,438 and 94,670 cycles respectively before failing. Note that width measurements of wear scars on the failed specimens and fretting bridges revealed that the slip amplitude was on the order of 5-7 microns during testing.

Scanning electron microscopy of the fretting wear scars revealed the fretting damage on the polished and the ion implanted specimens to be very similar. Figure 4 illustrates the typical fretting damage observed on the fatigue specimens. In all cases, the extent of the damage tended to differ along the edge of the wear scar compared to the middle. The middle region of the wear scars typically took on a smeared, debris free appearance whereas, the outer edges were typically rough and debris covered.

It was noted for the case of the unimplanted specimens, the cracks which had propagated to failure had initiated from within the wear scars at areas which were in close proximity to the interface between undamaged and fretting damaged material (Figure 5a). The crack propagation route on the wear surface was found to follow very close to the outer limits of the wear scars which corresponded to the interface between the smeared middle region and the debris covered edge region. For the case of the carbon implanted specimens, cracks had initiated from areas close to the boundary between damaged/undamaged material but in the middle region of the wear scar (Figure 5b). In general, it appears that a semi-elliptical crack propagated from the initiation site, in a direction perpendicular to the surface. It appears that final specimen failure was the result of overload along shear planes, after the crack covered approximately 60 - 70% of the specimen cross-section.

THE FRETTING DAMAGE PROCESS

It appears that the mechanisms of fretting in both the unimplanted and the carbon implanted specimens include adhesion, oxidation, delamination and abrasion. During the initial stages of fretting it is likely that adhesion predominated. The debris generated as a result of adhesion may readily oxidize and subsequently cause abrasion. Oxide debris may also have been formed as a result of delamination of platelets from the surface oxide layer. This mechanism of debris generation was likely predominant in the later stages of fretting. Figure 6 illustrates the typical near surface cracking which results in platelet delamination. Once a platelet becomes loose, it is free to abrade and smear the fretting wear surface giving the middle region of the wear scar its typically smeared appearance. As fretting continues it is probable that the large platelets become fragmented into fine debris which works its way to the edges of the wear scar. This explains the rather rough debris covered appearance near the edges of the wear scar. Figure 7 illustrates the typical appearance of material degraded through delamination, smearing and abrasion.

The crack initiation site for the unimplanted specimens occurred at the interface between the edge and middle region of the wear scar. The fretting degradation process which occurs along the outer edges of the wear scars is enhanced by the edges of the fretting bridges which cause a stress concentration. A stress concentration also likely results from the localized entrapment of debris in the edge region. It should also be noted that the crack initiation sites are also in close proximity to the boundary between damaged material and undamaged material. High stress concentrations are also known to exist at these areas because it is an interface between material under tensile and alternating shear stresses (5). Therefore, for the case of the unimplanted specimens, crack initiation appears to be motivated by fretting bridge edge effects, the presence of an excessive amount of debris and stress concentrations which develop at the interface between the material undergoing fretting and the undamaged material.

The two carbon implanted specimens had crack initiation sites close to the boundary between fretting damaged/undamaged material. However, the initiation sites did not occur within the interface of the edge and middle regions but in the smeared middle region alone. The likely explanation for this different crack initiation site is that the carbon ion implantation treatment has reduced or delayed the generation of debris within the wear scar.

The results of the TEM investigation confirmed that the titanium in the implanted layer had been completely transformed to titanium carbide. Apart from providing a harder surface layer, the titanium carbide may have some chemical influence on the fretting process. A hardened surface layer will directly serve to reduce the amount of debris within the wear scar and accordingly, less debris will be trapped under the edges of the fretting bridges resulting in a lower stress concentration. In addition, the rate of fretting damage accumulation would be slower owing to the smaller amount of debris. Therefore the edge effects likely were moderated by the carbon implantation treatment.

As a result, the crack initiation site for the carbon implanted specimens was shifted from the interface of the edge and middle region to the middle region. Further investigation is required to determine if the TiC layer had a chemical effect on the fretting process. Such effects as a change in the oxidation kinetics during fretting and a change in the coefficient of friction are possible.

CONCLUSIONS

1. Ion implantation has been identified as being a viable technique for deterring the development of fretting fatigue damage on Ti-6Al-4V. Therefore, it is likely that it may be useful for deterring the development of fretting fatigue damage on the dovetail surfaces of Ti-6Al-4V aero gas turbine engine fan and compressor blades.

2. Mechanical testing revealed that fretting causes a definite reduction in the fatigue life of polished Ti-6Al-4V specimens. Carbon implantation was found to increase the fretting fatigue life of Ti-6Al-4V by at least a factor of two. Note that only two fretting fatigue tests have been carried out at present on the carbon implanted material. Further testing is planned to improve the statistics of these results.

3. Differences in the mechanism of the fretting degradation process between unimplanted and carbon implanted Ti-6Al-4V specimens was not apparent. However, the carbon implantation was found to alter the fretting fatigue crack initiation process. The increase in the fretting fatigue life by carbon implantation was associated with a delaying in the generation of debris and subsequent damage. This perhaps was the result of a surface hardening effect. The microstructural analysis of the carbon implanted layer suggested that a fully carburized layer was formed as a result of the implantation. Therefore, it is expected that this layer would have a much greater surface hardness than that of the unimplanted material. Ultra-micro hardness tests are planned to confirm this.

4. Transmission electron microscopy has revealed that several different stoichiometries of titanium nitrides and carbides were formed as a result of implanting with nitrogen and carbon respectively. It is hoped that towards the end of this investigation a correlation may be developed for the various implantation treatments between the microstructure of the implanted layer and the fretting fatigue improvement which results.

Figure 1a. Diffraction pattern from nitrogen implanted Ti-6Al-4V illustrating the [100] pattern form Ti_2N showing the TiN (020) and Ti_2N (002) reflections.

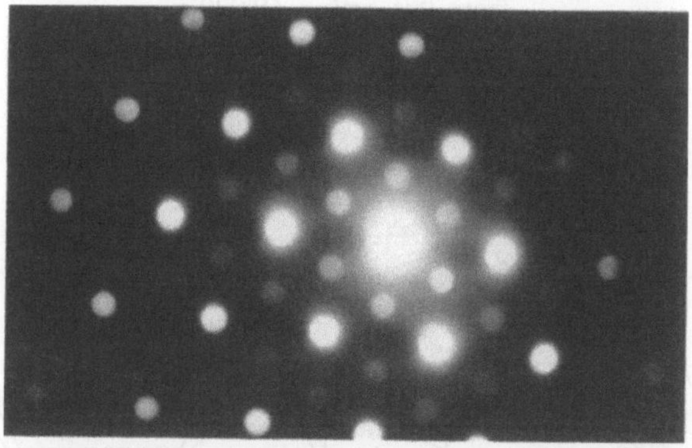

Figure 1b. Diffraction pattern from nitrogen implanted Ti-6Al-4V illustrating superimposed patterns of TiN and Ti with (0001) Ti // (111) TiN and [11$\bar{2}$0] Ti // [1$\bar{1}$0] TiN.

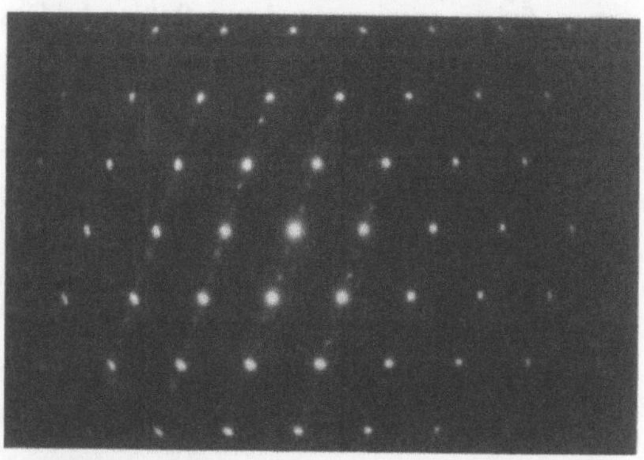

Figure 1c. Diffraction pattern from nitrogen implanted Ti-6Al-4V illustrating two orientations of TiN in twinned relationship. These patterns come from <011> TiN orientations.

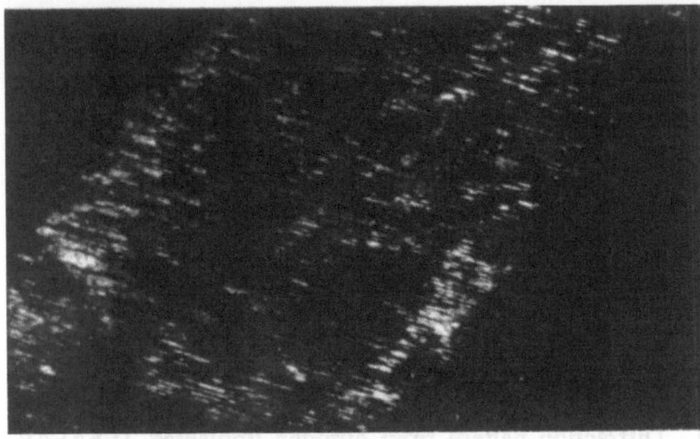

Figure 2. Dark field micrograph of nitrogen implanted Ti-6Al-4V showing one orientation of TiN. The diffraction pattern in figure 1c was taken from this grain. 60,000 x

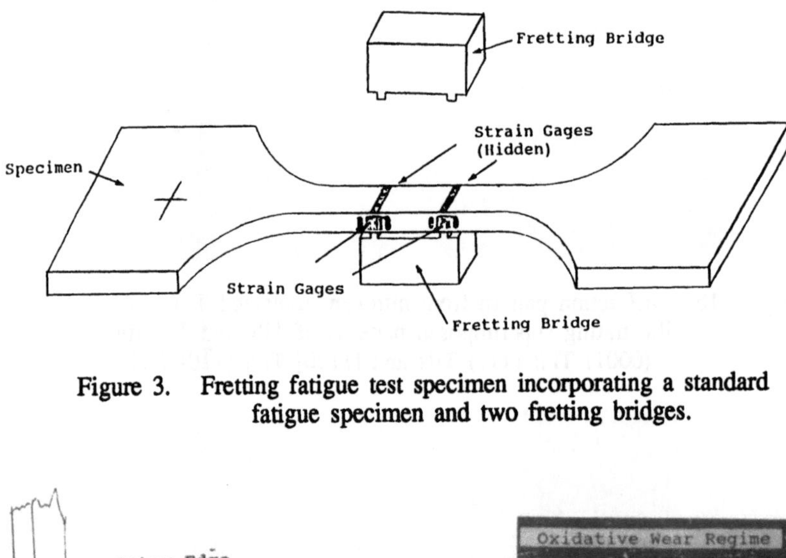

Figure 3. Fretting fatigue test specimen incorporating a standard fatigue specimen and two fretting bridges.

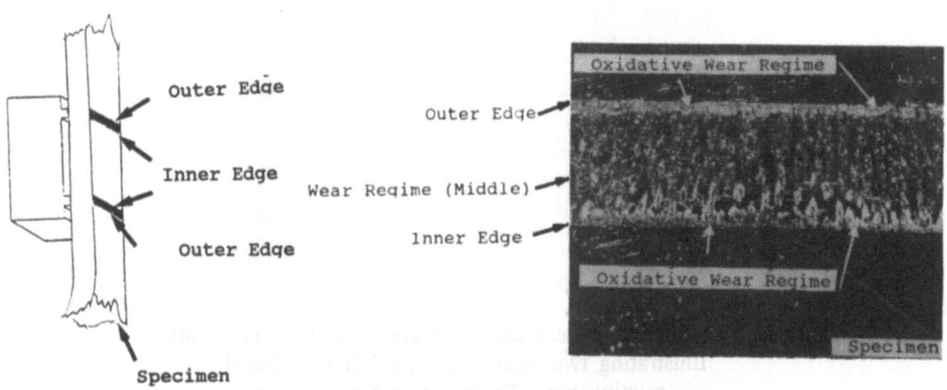

Figure 4. The typical appearance of a fretting wear scar introduced on a fatigue test specimen.

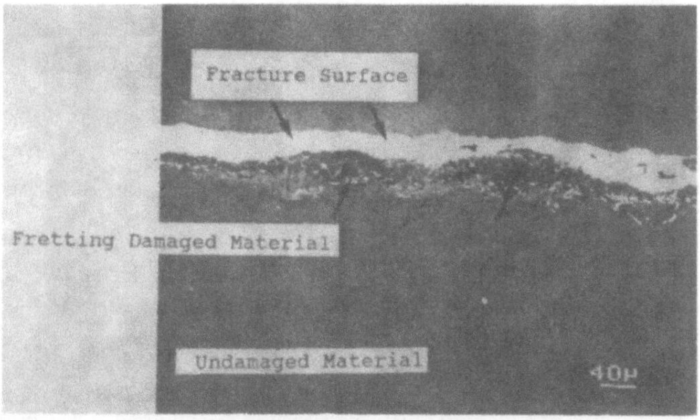

Figure 5a. Crack propagation route near damaged/undamaged material for polished, unimplanted Ti-6Al-4V fretting fatigue test specimens.

Figure 5b. Crack propagation route on wear scar surface for carbon implanted Ti-6Al-4V fretting fatigue test specimens.

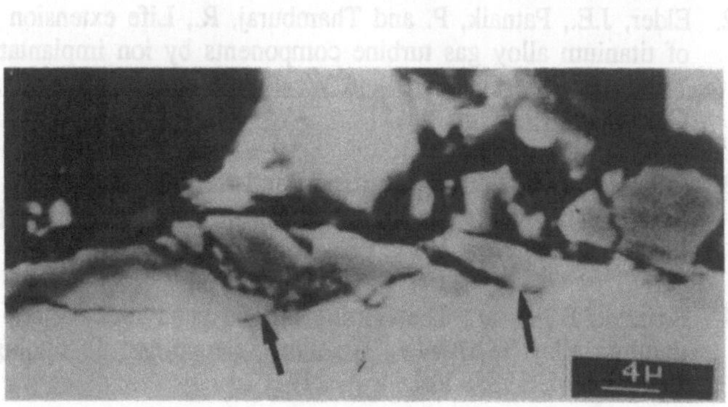

Figure 6. Near surface cracking which results in platelet delamination.

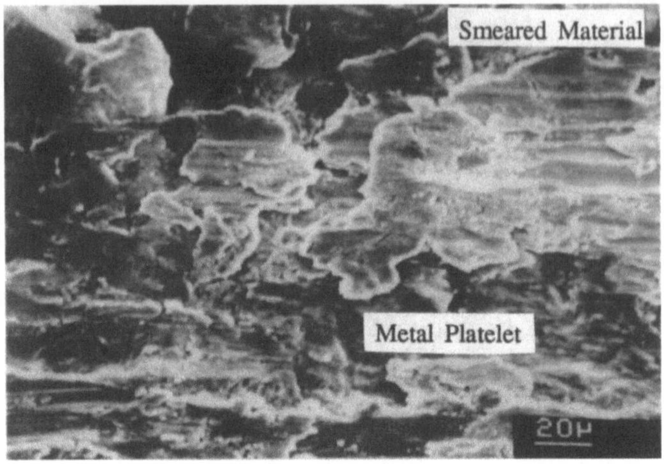

Figure 7. Typical appearance of material degraded through delamination, smearing and abrasion.

ACKNOWLEDGEMENTS

The authors of this paper would like to acknowledge CRAD (Department of National Defence Canada) for their financial support for this work.

REFERENCES

1. Elder, J.E., Thamburaj, R. and Patnaik, P.C., Optimising ion implantation conditions for improving wear, fatigue and fretting fatigue of Ti-6Al-4V. International Materials Reviews, Vol. 33, No. 6, 1988, pp. 289-313.

2. Elder, J.E., Patnaik, P. and Thamburaj, R., Life extension of titanium alloy gas turbine components by ion implantation, Tribological Mechanisms and Wear Problems in Materials, ASM, Cincinnati, 1987, pp. 11-21.

3. Duquette, D.J., Fretting corrosion and fretting fatigue, Surface Treatments for Improved Performance and Properties, Plenum Press, New York, ch. 10, p. 173.

4. Broszeit, E., et. al., The fretting fatigue behaviour of the titanium alloy Ti-6Al-4V, Titanium Science and Technology, 1984, DGM, 1985, Vol. 4, p. 2171.

5. Hoeppner, D.W., Environmental effects in fretting fatigue, Fretting Fatigue, Applied Science Publishers Ltd, London, Ch. 6, 1981.

Martensitic Transformation in 304 Stainless Steel after Implantation with Neon

J. Y. Xu, B. J. van Brussel, J. Noordhuis, P. M. Bronsveld and J. Th. M. de Hosson

University of Groningen
Groningen
The Netherlands

ABSTRACT

X-ray diffraction on neon implanted AISI 304 austenitic stainless steel revealed both a tensile strain in the implanted layer which increases with implantation dose and which is responsible for a martensitic transformation at a critical dose of $2.3*10^{17} Ne^{+}/cm^{2}$, and a three times smaller compressive strain in the underlying substrate which can be explained by the existence of neon bubbles at high pressure.

INTRODUCTION

Austenitic stainless steels represent a class of materials widely used owing to their good ductility and corrosion resistance. However, when exposed to temperatures in the range 400-900 Celsius they are susceptible to intergranular corrosion caused by the precipitation of Cr-rich carbides at grain boundaries. It has been shown that a surface melting procedure can be used to restore the corrosion resistance by dissolving the Cr carbides [1].
Another aspect of these particles is their influence on wear properties. Many successful wear-resisting materials consist of particles of a hard phase dispersed in a more ductile matrix. Such dispersion can be prepared by laser melting.

In a study aimed at improving the wear properties of common steels by laser melting a considerable tensile stress was measured in the laser treated surface. This tensile stress state detrimentally influences the crack growth velocity. Ion implantation, introducing compressive stresses, might neutralize this effect [2]. In line with the established classical techniques of nitriding and nitrocarburizing nitrogen would seem to be the most likely candidate for implantation [3]. Unfortunately, a brittle layer of ε-$Fe_2(C,N)_{1-x}$ was formed at a dose of $3*10^{17}$ N^+/cm^2 exhibiting a deleterious effect on the wear performance [4]. In general, implantation with an inert gas is not very effective in reducing wear rates. However, neon implantation into laser-melted steel turns out to reduce the wear rate substantially as a result of a conversion of residual stresses into compressive ones. Transmission electron micrographs gave clear evidence of the formation of a large number of bubbles, of Orowan looping and of pinning of dislocations by these bubbles [2].

Moreover, some austenitic stainless steels have a metastable fcc crystal structure from which stress-induced martensite [5] is formed easily leading to another compressive stress contribution. In figure 1 a so-called Schaeffler diagram is depicted which indicates the metastable character of several types of steels [6]. Also noble gas implants have been found to induce a martensitic transformation [7]. The transformation might be of practical importance since the conventional work hardening of this steel is due to the stress induced formation of martensite.

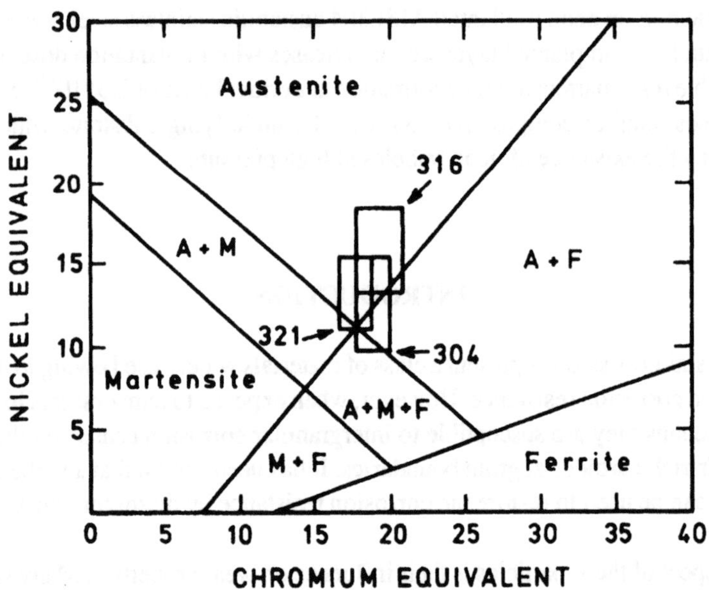

Figure 1 Schaeffler diagram showing the phase distributions in stainless steels. Indicated are the composition regions for AISI 304, 316 and 321 stainless steels.

In the following we will report on a X-ray diffraction study of neon implanted AISI 304 stainless steel with low austenitic stability in order to determine the critical dose for the fcc to bcc transformation.

EXPERIMENTAL DESCRIPTION

As sample material AISI 304 stainless steel was chosen with a composition as indicated in table I. From a 10 mm diameter commercially available round bar a piece of 10 mm length was cut and reduced in thickness by mixed rolling from 10 to 5 mm thickness. This flat piece was subsequently cut in five smaller pieces which were ground by hand, solution annealed at 1063 Celsius for half an hour in a silica ampoule filled with argon and water-quenched. Finally the surface was polished electrochemically and implanted with doses up to $4*10^{17}$ Ne^{+}/cm^{2}.

Implantations were carried out using an Extrion 200kV - 200 µA machine. The neon was implanted with an energy of 110 keV and a current close to 60 µA keeping the temperature of the sample fixed at 100 degrees centigrade. For a dose of $2.5*10^{16}$ ions/cm^{2} the projected range is 80 nm with a deviation of 15 nm as calculated according to the so-called TRIM model [8,9]. One should realize that the range is shifted somewhat towards the surface as a function of dose because of the sputtering effects.

TABLE 1

AISI 304 chemical compositions in wt%

Cr	Ni	C	Mn	Mo	Cu	V
18.33	9.40	0.038	1.40	0.27	0.220	0.067
Co	**Sn**	**P**	**S**	**Si**	**B**	**Fe**
0.17	0.013	0.032	0.013	0.48	0.001	bal.

X-ray scans were performed with a Philips powder diffraction set-up including a graphite monochromator in the outgoing beam and a xenon-filled proportional detector. A MicroVax 2000 was used for the automatic data handling program. A copper tube was used at 40 kV and 25 mA. The polished surfaces of a set of parallel samples were implanted, sliced off, disc punched, and thinned from the back in order to study the microstructure using a JEOL 200 kV transmission electron microscope.

Crucial in this experiment was to have the implantation depth and the X-ray penetration depth of the same order of magnitude. The fraction Gx of the total diffracted X-ray intensity contributed by a layer of thickness X is given by the formula [10]:

$$X = K_x \sin\theta / 2\mu$$

with

$$K_x = \ln[(1/(1-G_x))]$$

where μ/ρ is the mass absorption coefficient in cm^2/g and θ the Bragg angle. The penetration depth X (at G_x = 95%) of the copper beam into AISI 304 at 2 θ = 147 degrees corresponding to the 420 fcc peak is 6.6 µm taking μ/ρ = 278 cm^2/g. The penetration depth X (also at G_x = 95%) of a chromium beam into AISI 304 at 2 θ = 126.8 degrees corresponding to the 220 fcc peak is 16 µm taking μ/ρ = 106. From such considerations we came to the conclusion that a copper tube should be used. As our data are concerned with a 0.1 µm surface layer it is good to realize that the fraction G_x of the diffracted intensity for a 0.1 µm thick layer using CuKα is 10% at 2 θ = 50.8 degrees, the diffraction angle of the 200 fcc peak.

RESULTS

In figure 2 we have put together three X-ray diffraction scans taken on three different samples of AISI 304 stainless steel. Figure 2a corresponds to a sample before implantation, figure 2b and 2c to samples which were implanted with doses of $2.1*10^{17}$ and $4*10^{17}$ ions/cm^2, respectively. Figure 2a contains sharp 111 and 200 fcc peaks. Texture could be experienced by rotating the sample around its surface normal. It was thus possible to reduce the 111 peak considerably as is clearly shown in the figure. In figure 2b a strong satellite peak is present on the low angle side of the 200 fcc peak. This is an indication of an expanded fcc phase [11] which, judging from the halfwidth is microcrystalline. The main feature in figure 2c is the development of a 110 bcc peak. Moreover, the satellite of the 200 fcc peak has collapsed. When aligning the three curves in one figure a shift is encountered of the 111 and 200 with respect to the previous positions. One could determine the shift of these peaks with respect to a higher angle reflection say the 331 fcc peak. In this case due to a higher diffraction angle the penetration depth into the sample will be two times bigger than for the 200 peak and the effect of the implanted layer will be obscured. We have monitored the position of the 331 fcc peak and found no effect as a function of implantation dose.

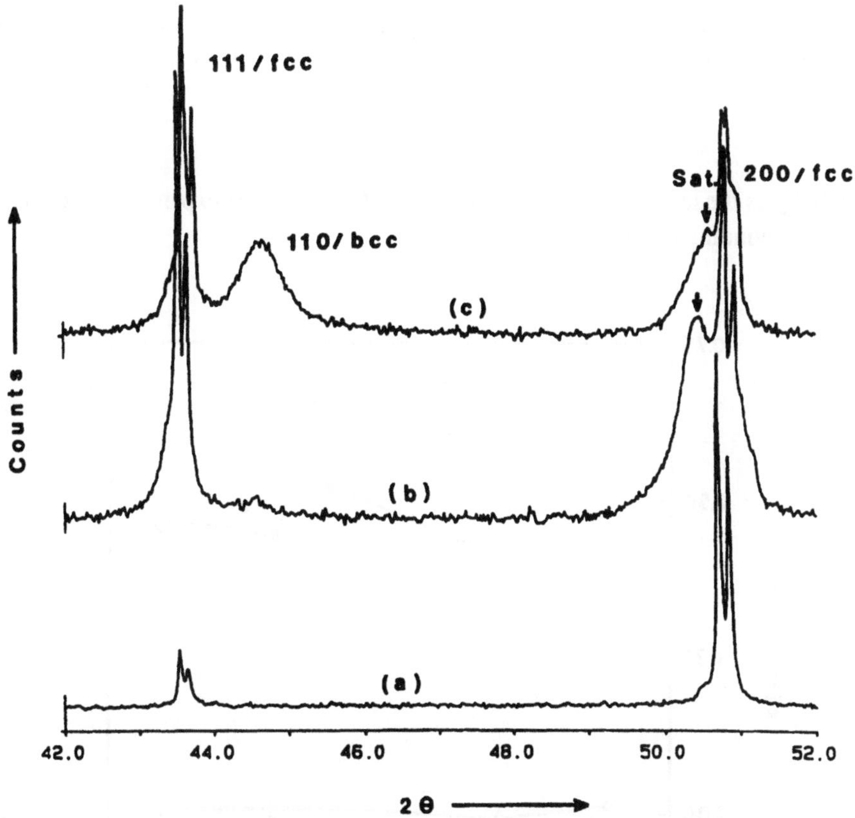

Figure 2 X-ray powder diffraction scans : CuKα radiation, 40 kV /25 mA, stepsize 0.02° 2θ, counting time 10-20 sec, DS=1°, RS=0.2mm, SS=1°. a) before implantation, b) after impl. with $2.1*10^{17}$ ions/cm^2, c) after implantation with $4*10^{17}$ ions/cm^2.

In figure 3 we have plotted the strain in % calculated from the expression:

$$\varepsilon_{331}= 100*(d_{331}-d_{331}^{\circ})/d_{331}^{\circ}$$

versus the logarithm of the dose with d_{331}° the lattice parameter of the 331 fcc peak before implantation. From this plot it is obvious that the 331 peak position is independent of dose. It also means that the lattice constant of the deeper lying substrate for different samples is the same so that one can compare the lattice constant for a layer closer to the surface, e.g. corresponding to the 200 peak, as a function of implantation. The corresponding strain component is represented by:

$$\varepsilon_{200}=100*(d_{200}-d_{200}^{\circ})/d_{200}^{\circ}$$

with d_{200}° the lattice parameter of the 200 fcc peak before implantation. We have done similar data collection for the 200 satellite peak according to the expression:

$$\varepsilon_{sat} = 100*(d_{sat} - d_{200})/d_{200}$$

in which d_{200} is the lattice parameter of the substrate and d_{sat} the expanded lattice parameter corresponding to the implanted layer.

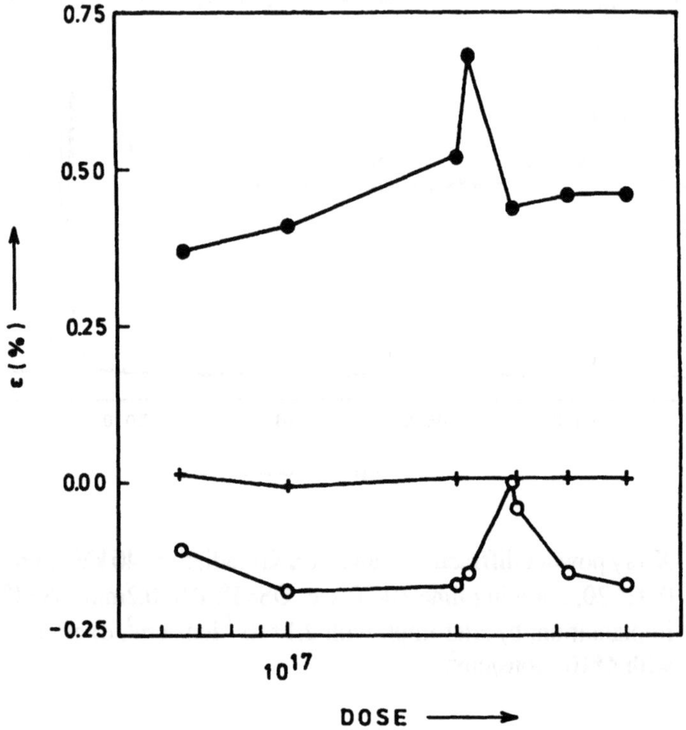

Figure 3 Strain vs implantation dose (110 keV Ne^+, 60 μA, 100 Celsius) calculated from the 331 (+) , the 200 (O) and the satellite of the 200 (●) diffraction peaks.

From figure 3 we conclude that in between 2.1 and 2.5 $*10^{17}$ ions/cm^2 a transformation takes place.

In figure 4 we show a bright and dark field TEM micrograph of a sample which was implanted with 50 keV neon ions on one side up to a dose of $3*10^{17}$ ions/cm^2 and thinned electrochemically from the backside. As diffraction vector part of the bcc ring was chosen as depicted in the inset which also shows a considerable amount of texture. The bright patches in figure 4a are the neon bubbles and in figure 4b the bcc precipitates.

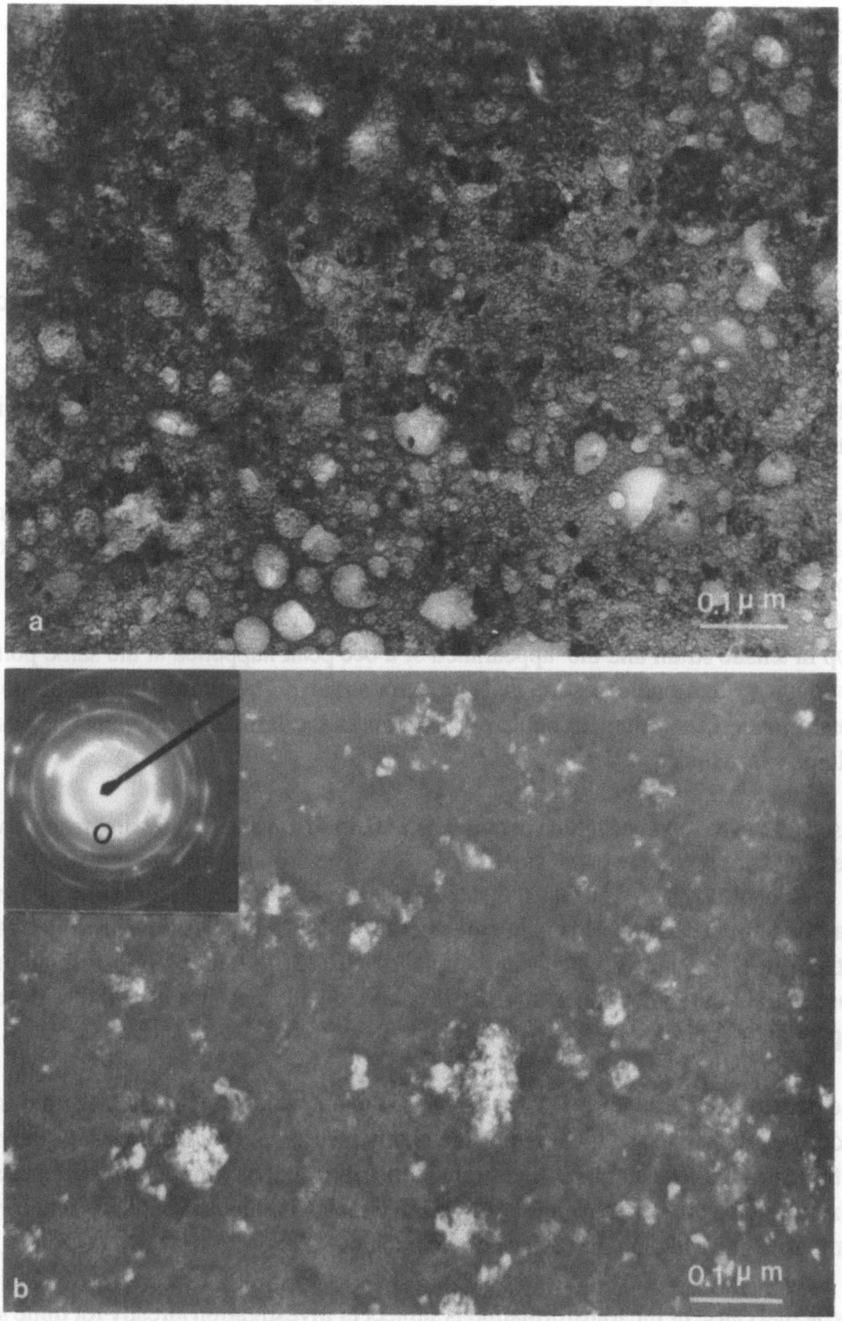

Figure 4 **Bright (a) and dark (b) field transmission electron micrographs of an implanted backside thinned sample showing neon bubbles and bcc precipitates. The black circle on the inset indicates the part of the diffracted intensity in the reciprocal lattice used for imaging.**

It is difficult to pinpoint the bcc patches as being directly connected to the neon bubbles. To improve upon this result, avoiding specifically the two-dimensional projection, one would like to try stereo microscopy but by rotation around a matrix axis chosen as stereo axis it is not possible to image the same set of precipitates.

DISCUSSION

Vardiman and Singer [12] studying the reverse reaction of bcc transforming into fcc of AISI 304 under influence of Ni^+, N^+ and Ne^+ implantation state that neon lacks any capacity to stabilize austenite. This is a good starting point for our discussion. Our X-ray results show that Ne^+ implantation favors the fcc to bcc transformation with a critical dose between $2.1*10^{17}$ and $2.5*10^{17}$ ions/cm^2.

Looking at the satellite contribution in figure 3 we see that initially a positive or tensile strain is introduced which increases steadily from 0.37% to a value of 0.68% corresponding to a dose of $2.1*10^{17}$ ions/cm^2. At a dose of $2.5*10^{17}$ ions/cm^2 the strain is down to a value of 0.44% while a bcc peak indicates the fcc to bcc transformation. The drop in tensile strain and the appearance of a bcc peak happened at the same point in the graph so that one can state that the extra strain is favourable for the fcc to bcc transformation according to one's expectation as the transformation encompasses a volume expansion.

Based on the fcc and bcc lattice parameters of 0.3590 nm and 0.2867 nm, respectively, one calculates a volume expansion $\delta V/V \approx 2\%$. The drop in strain of only 0.24% may be explanable by the fact that the satellite peak with its large half-width is from a heavily deformed layer. This is in agreement with TEM observations in which rather outspoken layers are reported after implantation [13]. After the transformation the tensile strain curve continues more or less horizontally because at a higher dose the implantation saturates.

The negative or compressive strain of the 200 main peak is related to the strain in the underlying substrate and is caused by the highly pressurized neon bubbles. At the point of transformation the compressive strain in the substrate has temporarily been relaxed to zero strain as the effect of the pressurized bubbles is shielded by the newly formed martensite nucleated on the surface of the bubbles.

The actual cause of the transformation is subject of investigation already for quite some time. More and more one realizes that the fcc structure is a metastable one which easily can be transformed. Kajiwara [14] investigated the role of dislocations and grain boundaries in the martensitic nucleation process. He took into account three kinds of deformation:

- the lattice deformation which generates a martensitic lattice from the austenitic one

- the lattice invariant deformation which produces an interface with no average distortion between the two phases (i.e. an invariant habit plane)

- the deformation accommodating the shape strain of the martensite with a possible tendency of self-accommodation.

He came to the conclusion based on a thorough experimental investigation of Fe alloys with a large amount of Ni that dislocations themselves do not act as favourable nucleation sites for the martensitic transformation. Certain grainboundaries are the real nuclei formers, while dislocations are necessary for accomodation. It is plausible that whenever the plastic accommodation is impeded the transformation is suppressed. This can be lack of dislocations, too small a grain size or work hardening. The yield point of a material is then directly related to the ease with which the alloy transforms martensitically. As a matter of fact it is this idea of a low yield point which made us choose this type of sample.

A nice example of this impedement is described by Noordhuis and De Hosson [15]. Their TEM results have shown bcc precipitates in samples which were implanted below the critical dose but were given an additional anneal treatment. They also observed preferred nucleation on grainboundaries. By taking into account the shear stress, the chemical free energy, the interfacial energy and the resultant strain energy they show that the shear field of an inert gas bubble can induce a local martensitic transformation, provided the bubble size and pressure are large enough. The heavily deformed implantation layer is hampering the TEM observations so that it is difficult to link the bcc precipitates unequivocally to the neon bubbles.

CONCLUSION

Implantation of 110 keV Ne^+ into AISI 304 stainless steel produces a layer which is heavily deformed but which produces a satellite peak in a X-ray diffraction scan on the low angle side of the 200 fcc peak. The shift of this peak to a lower diffraction angle represents an expansion of the lattice and favours the fcc to bcc transformation. The 200 peak itself shifts slightly to a higher diffraction angle under implantation indicating a compression in the underlying substrate probably caused by the highly pressurized neon bubbles. The actual nucleation sites for the martensitic transformation in our samples are likely to be neon bubble surfaces.

ACKNOWLEDGEMENTS

This work is part of the research program of the Foundation for Fundamental Research on Matter (FOM-Utrecht) and has been made possible by financial support from The Netherlands Organization for Research (NWO-The Hague), The Ministry of Foreign Affairs of The Netherlands and the Academia Sinica of The People's Republic of China.

REFERENCES

[1] A. La Barbera, S. Martelli, A. Mignone, F. Pierdominici, M. Vittori, G. Lulli, Proc. First Europ. Conf. Laser Treatment of Materials, Bad Nauheim, FRG, 1986, p. 179.

[2] H. De Beurs, J.A. Hovius, J.Th.M. De Hosson, Acta Metall., 1988, **36**, 3123.

[3] H. Dimigen, K. Kobs, R. Leutenecker, H. Ryssel, P. Eichinger, Mat. Sci. Eng., 1985, **69**, 181.

[4] H. De Beurs, J.Th.M. De Hosson, Appl. Phys. Lett., 1988, **53**, 663.

[5] Z. Nishiyama, Martensitic Transformations, Acad. Press, New York, 1978.

[6] A. Schaeffler, Met. Prog., 1969, **56**, 680.

[7] E. Johnson, In: Ion Implantation, 1988, ed. F. Wöhlbier, Defect and Diffusion Forum 57/58, 241, Trans Tech Publ., Switzerland, 1988, p. 241.

[8] J. Biersack, L. Haggmark, Nucl. Instr. Meth., 1980, **174**, 257.

[9] J. Ziegler, J. Biersack, U. Littmark, The Stopping and Range of Ions in Solids, Vol. 1, Pergamon Press, New York, 1984.

[10] B.D. Cullity, Elements of X-ray Diffraction, Acad. Press, New York, 1977.

[11] P. Rozenak, D. Eliezer, Acta Metall., 1987, **35**, 2329.

[12] R. Vardiman, I. Singer, Materials Lett., 1983, **2**, 150.

[13] A. Johansen, E. Johnson, L. Sarholt-Kristensen, S. Steenstrup, E. Gerritsen, C.J.M. Denissen, H. Keetels, J. Politiek, N. Hayashi, I. Sakamoto, First European Conf. on Accelerators, in Appl. Res. and Tech., Frankfurt, 1989.

[14] S. Kajiwara, Metall. Trans., 1986, **17A**, 1693.

[15] J. Noordhuis, J.Th.M. De Hosson, in press.

Elevated Temperature Ion Implantation: A New Direction for Surface Engineering

F. J. Worzala, J. R. Conrad, R. A. Dodd, M. Madapura and K. Sridharan

University of Wisconsin—Madison
Madison Wisconsin USA

ABSTRACT

The advantages of ion implantation as a method for surface treatment are substantial. Improvements in hardness, wear and corrosion resistance of many steels have been obtained through relatively short time exposure to a flux of interstitial ions such as carbon or nitrogen. However, the depth of the implanted layer is extremely small after conventional implantation, generally in the range of 1000Å or less. For many applications, such as abrasive and corrosion wear, these layers are of insufficient thickness to provide long term improvement.

In order to achieve increased layer thickness, ion implantation with nitrogen has been carried out at 550°C. Using the Plasma Source Ion Implantation (PSII) device, specimens of several different alloys were implanted at elevated temperatures, using the ion flux as the heating source.

A number of commercial alloys were investigated to study the effectiveness of elevated temperature nitrogen ion PSII implantation in creating engineered surfaces having superior properties. These included several alloy steels and several nickel-based superalloys. Detailed evaluation of

metallography, hardness and wear have been conducted. Selected alloys were examined using transmission electron microscopy and scanning auger microscopy. Support for this research was provided by NSF Grant DMC-8712461 and ECS-8314488, and by DAAL03-89-K-0048.

INTRODUCTION

Over the past two decades the use of ion implantation has been promoted as a significant method for modifying the surface of materials (3). A number of industrial applications of the process have been reported (1,2). The major advantages of ion implantation are: (a) a wide variety of atoms can be implanted, (b) because of the high energy of the implanted ions, entry into the surface is not controlled by thermodynamic reaction rates (c) the implanted layer is contiguous with the workpiece, hence is integrally bonded and (d) implantation is generally conducted at low temperatures, such that thermal degradation of the microstructure and surface mechanical properties do not occur.

A major disadvantage of the process is the fact that most implanters are line-of-sight machines which require elaborate target manipulation arrangements to achieve a uniform dose on all surfaces. In addition, for non-planar targets, masking (undesirable from a contamination standpoint) is sometimes required to minimize the 'retained dose' problem (3). The masking requirement introduces more fixturing, in addition to manipulation fixturing, further complicating the process. An innovative, cost-effective approach that circumvents the above problems, called the Plasma Source Ion Implantation (PSII), has recently been developed (4). In PSII the target is placed directly in a plasma source and is pulse biased to a high negative potential relative to the chamber walls. Ions are accelerated across the plasma sheath, normal to the target surface and regardless of the contour of the target, all surfaces are uniformly implanted. The PSII Process has received considerable industrial attention and its effectiveness in improving surface properties of commercial alloys has been demonstrated (5,6).

The use of nitrogen ion implantation to improve wear resistance of a variety of metallic alloy systems has been widely reported (7). Best results have been obtained in ferrous alloys, where strong nitride formers are present as alloy additions. Tool and die steels containing Cr, Nb, Ti and V show significant improvement, most likely as a result of the formation of fine, metastable, nitride precipitates (8). Post implantation annealing leads to the formation of more stable nitrides, which can result in further property improvement.

One of the major limitations of ion implantation is the very shallow depth of penetration achieved. Ions having typical energies in the range of 50-90 keV, produce a hardened region which is only 60-100 nanometers deep. For most engineering applications experiencing wear, this level of hardened zone thickness is insufficient. This is especially true for applications which experience abrasive and/or adhesive wear. For this reason, ion implantation has not supplanted carburizing and nitriding for such applications as gears and bearings. However, by implanting at elevated temperatures, not only can more stable precipitates be formed, but the hardened layer thickness can be increased significantly. The combined effects of thermal and irradiation enhanced diffusion can produce thick, hardened layers in relatively short times. The present work describes the results of high temperature nitrogen implantation of several alloys having engineering importance.

EXPERIMENTAL

The alloys studied in this investigation include two Fe-Ni based superalloys, Incoloy 908 and 909, and several steels, including AISI 4340 and AISI 410 stainless. The compositions and heat treat condition of these alloys are presented in Table I. Implantation was conducted using the PSII process. Nitrogen atoms having an energy of 50 keV were implanted to a dose of 3×10^{18} atom/cm^2 at an average current of 50 μA/cm^2. The implantation time was three hours, and the steady state sample temperature was approximately 550°C. The chamber was

TABLE 1
Chemical Composition (wt.%) of Alloys Investigated

Alloy	Fe	Ni	Co	Cr	Al	Ti	Nb+Ta	Mn	C	Mo
908	40.6	49.8	-	3.8	1.0	1.6	3.0	-	.01	-
909	41.4	38.9	12.7	-	.06	1.6	4.9	-	.01	-
410	86.6	-	-	12.3	-	-	-	1.0	.15	-
4340	96.6	1.95	-	.82	-	-	-	-	.4	.25

operated at a pressure of $4x10^{-4}$ Torr.

Analysis of nitrogen profiles of implanted specimens were performed by Auger electron spectroscopy. Composition profiles to depths less than 500Å were obtained by argon ion sputtering. At greater depths, nitrogen composition was determined by mechanical polishing, followed by Auger analysis or by electron probe microanalysis. The structural characteristics of the implanted specimens were studied by conventional metallography and transmission electron microscopy. Standard techniques of specimen preparation were employed. X-ray diffraction was used to determine the phases present and to measure lattice parameters of the implanted layer so as to determine the existence of residual stresses.

Specimens were evaluated by means of hardness and wear testing. Knoop microhardness measurements were made with 50 and 100 gram loads, to obtain hardness profiles across the implanted and diffused layer. A pin-on-disk tester was used to characterize the wear of implanted specimens. Samples were secured on a rotating platform and a diamond tipped indenter was pressed onto the sample with a specified load. The sample was rotated at 50 rpm for specified times, with an oil lubricant applied continuously. The loads varied from 250 to 1000 grams. The wear track depth was measured by means of an alpha-step profilometer.

RESULTS AND DISCUSSION

The effects of elevated temperature implantation on microstructure are readily apparent in the metallographic cross-sections of implanted specimens. Figure 1 presents typical microstructures of two of the alloys investigated, INCO 908 and AISI-410 stainless. Here it can be seen that the etched layers of about 15-20μm, representing the nitrogen diffusion zone, are visible.

The nitrogen concentration profiles across the diffusion zone for the INCO specimens are presented in Figure 2. A peak in the nitrogen concentration is observed at the 250-500 Å range, which represents the penetration depth of the energetic nitrogen ions. At greater depths a plateau can be observed up to distances of about 5μm, after which a further decrease in nitrogen level is observed. The concentration of nitrogen in the plateau region varies between 9 and 12%, being higher for alloy 908. For the 410 stainless steel, (Figure 3) a similar behavior was observed, with the plateau at 10.5% but extending to a depth of only 3μm, before dropping off. The peak concentration near the surface is not shown in Figure 3, because the analysis was made on the cross-section using a scanning Auger microscope. A somewhat different profile was observed for AISI 4340, as seen in Figure 3. Here we see data obtained by electron microprobe, and verified by Auger analysis. The peak concentrations are much lower, in the neighborhood of only 0.8%. This level drops off quickly, reaching background at depths of 80 microns. Microscopic evidence of an etching effect could also be seen to the same approximate depth.

Comparing the nitrogen profiles of the four specimens, it is apparent that alloy 4340 is behaving differently than the other alloys. The peak concentrations are much lower and fewer of the implanted ions are retained, even though diffusion to significant depths can be observed. This behavior is believed to arise because of the low level of nitride formers in 4340 (∿1% Cr), as compared to that of the other alloys.

The microhardness profiles of the implanted specimens

TABLE 2
Hardness Values for Alloys Investigated
Average Knoop, 100 Gram Load

Alloy	Before Implanted	Implanted Surface	Subsurface
908	460	1000	500
909	400	900	490
410	320	990	320
4340	690	580	350

reflect the nitrogen levels observed, as shown in Figure 4a. Hardness at the surface reached a values which were 200% higher than the unimplanted material. This was also true for the 410 stainless specimen, but not for 4340. Here a decrease in hardness was observed, after implantation of a quenched specimen, as seen in Table 2. These differences in hardness resulting from implantation are again attributed to the lack of nitride formers in alloy 4340, and the martensite tempering which occurred during implantation.

The results of x-ray diffraction analyses indicated the presence of chromium nitrides in the implanted regions of alloys 908 and 410. No such evidence was found in the 909 or 4340 alloys. In all cases, peak broadening was observed for implanted specimens, indicating the presence of a residual compressive stress.

Several of the alloys were examined by TEM. In the 908 alloy, the implanted specimens were found to contain a high density of fine particles, identified as chromium nitrides. The 909 alloy, on the other hand, contained particles too small to be identified by electron diffraction. These could be iron nitrides, or perhaps nitrides of the refractory precipitation hardening elements niobium and titanium.

The pin-on-disc wear test results indicated that substantial improvement in wear resistance can be achieved through elevated temperature ion implantation. A typical wear curve for the 410 stainless alloy is shown in Figure 4b. Included in this figure is the curve for ion nitrided 410, for comparison. This specimen was nitrided for 5 hours at 550°C. It can be seen that the wear characteristics of both the nitrided and implanted specimens have considerably better properties than the untreated control. Both the 908 and 909 alloys showed similar improvement, while the 4340 alloy exhibited only modest change.

The micrometallurgical reactions that occur at the surface of a specimen during ion implantation are complex, especially when diffusion can occur at elevated temperatures. The presence of nitrogen levels in excess of 10 atom percent will promote the formation of nitrides, if a strong nitride former is present. This was shown to be the case for the two alloys containing high concentrations of chromium, namely the 908 and 410 alloys. But it appears that in order to be most effective, the nitride forming element should be in solid solution. Alloy 909 contains no chrome but does contain titanium, niobium and tantalum, all strong nitride formers. However, in the aged condition a large percentage of these alloying elements are incorporated into the γ' intermetallic compound precipitates. Therefore, they are not readily available for reacting with the nitrogen ions. Consequently a lower hardness was observed for the 909 alloy. The 4340 alloy, which contains only 1% chromium was not hardened appreciably by ion implantation. Rather, because it was implanted in the as-quenched condition, the alloy was softened. What little chromium nitride that was formed was insufficient to counteract the tempering that occurred during the elevated temperature implantation. It can be seen from these results that not only is the alloy composition important in ion implantation, but metallurgical structure is also critical. In order for nitride particles to form and provide optimum wear resistance, the nitride forming element should be in solution. There also seems to be some evidence to indicate that outward diffusion can occur, if no strong nitride formers are available. A significant amount of the nitrogen implanted into the 4340 specimen appears to have been lost in this manner.

CONCLUSIONS

By means of elevated temperature nitrogen ion implantation, a hardened layer of substantial thickness can be achieved in relatively short times. The characteristics of this layer depend upon the composition of the implanted material and to a certain extent its microstructure. Alloys containing nitride forming elements in solution were improved substantially through the formation of stable nitrides in the diffusion zone. This was shown to be true for alloys 908 and 410, where significant hardness and wear improvements were observed. Less pronounced improvements were observed in alloy 909, where nitride formers were tied up as intermetallic compounds. An enhancement of hardness was observed, presumably as a result of nitride formers in solution prior to implantation. The precipitates formed during ion implantation of the 909 alloy were too small to be identified, but it is assumed they are nitrides of the refractory elements titanium, niobium and/or tantalum. Alloys which contain low percentages of nitride formers, such as alloy 4340, do not respond appreciably to elevated temperature nitrogen ion implantation especially when they are in the hardened condition. The effects of rapid diffusion of nitrogen and tempering of the martensite resulted in an overall softening of the microstructure. However a diffusion zone of approximately 80 microns was observed which had peak hardness levels in the range of 600 Knoop.

It is believed that the elevated temperature implantation of nitrogen by means of PSII has significant industrial potential. For critical applications where thick, hardened layers are required, the process has several distinct advantages over nitriding. These include: 1) the absence of a brittle white layer, 2) better surface finish, and 3) shorter treatment times. However, based on the results of this study, it is imperative that close consideration be given to the alloy constituents and microstructure of the materials to be implanted.

REFERENCES

1. P. D. Townsend, J. C. Kelly and N. E. W. Hartley, Ion Implantation and Their Applications, Academic Press, London, 1976.
2. G. Dearnaley, Surface Engineering, Eds., R. Kossowsky and S. C. Singhal, NATO ASI Series, p. 125, 1984.
3. F. A. Smidt, B. D. Sartwell, Nucl. Instrum. and Meth., 86, p. 70, 1985.
4. J. R. Conrad, J. L. Radtke, R. A. Dodd, F. J. Worzala, N. C. Tran, J. App. Phys., 62(11), p. 4591, 1987.
5. F. J. Worzala, R. A. Dodd, J. R. Conrad and J. L. Radtke, Proc. Int. Conf. on Tool Materials, sponsored by Uddeholm Research Foundation, Chicago, IL, Sept. 1987.
6. J. R. Conrad, R. A. Dodd, F. J. Worzala, X. Qiu and R. S. Post, Proc. ASM Conf. on Ion Implantation and Plasma Assisted Processes for Industrial Applications, Atlanta, GA, May, 1988.
7. G. Dearnaley and N. E. W. Hartley, Thin Solid Films, 54, p. 215, 1978.
8. G. K. Hubler, Metastable Materials Formation by Ion Implantation, Eds., S. T. Picraux, W. J. Cohyke, Proc. Mat. Res. Soc., 27, Boston, MA, p. 341, 1981.

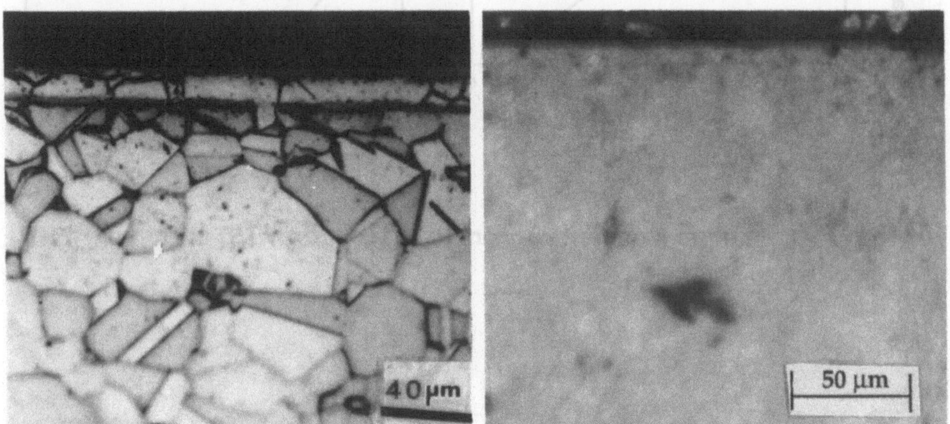

(a) (b)

Fig. 1 Typical Microstructures After Implantation, Depicting Diffusion Zone. a) Alloy 908, b) AISI 410.

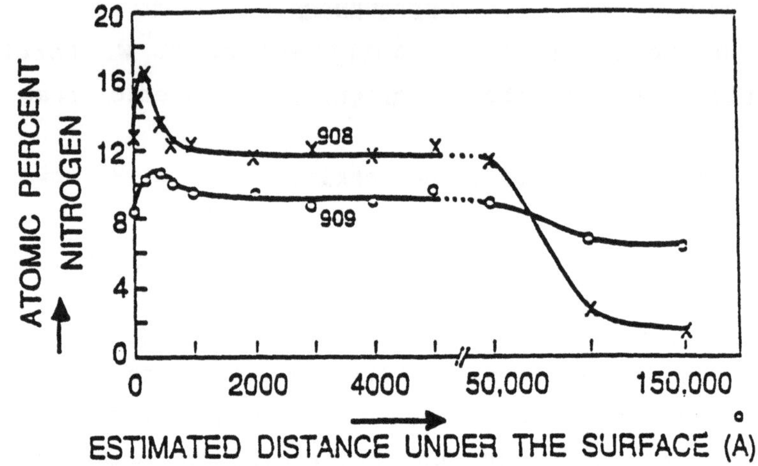

Fig. 2 Nitrogen Concentration Profiles of Alloys 908 and 909.

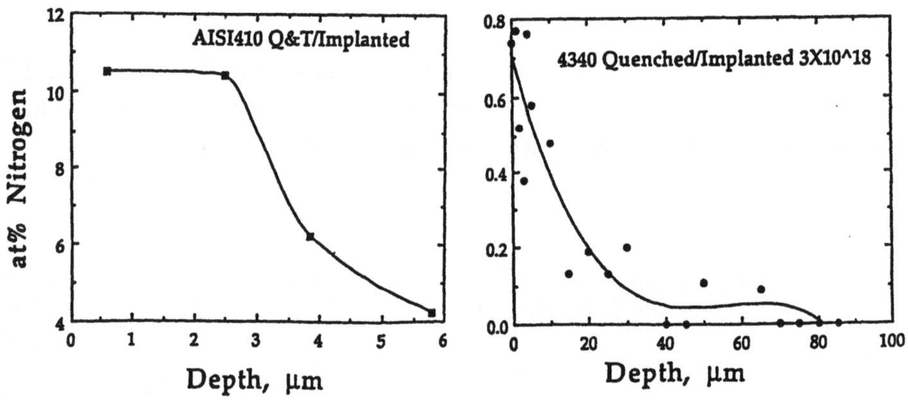

Fig. 3 Nitrogen Concentration Profiles of AISI 410 and 4340.

A)

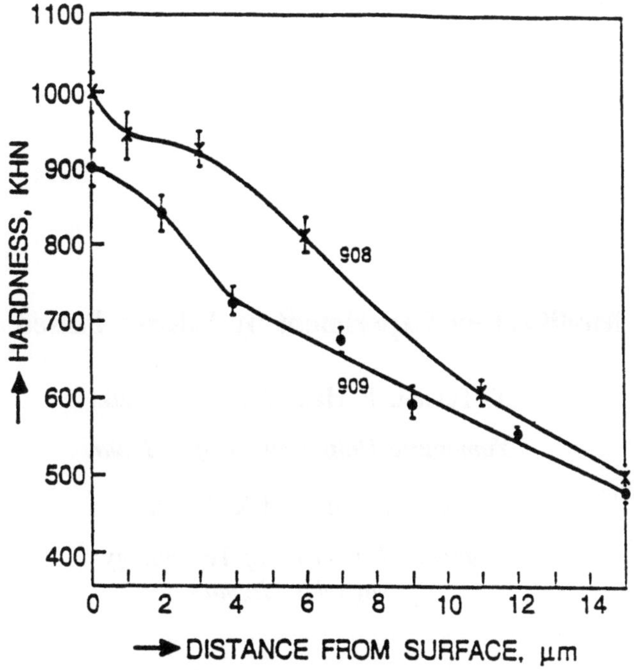

B)

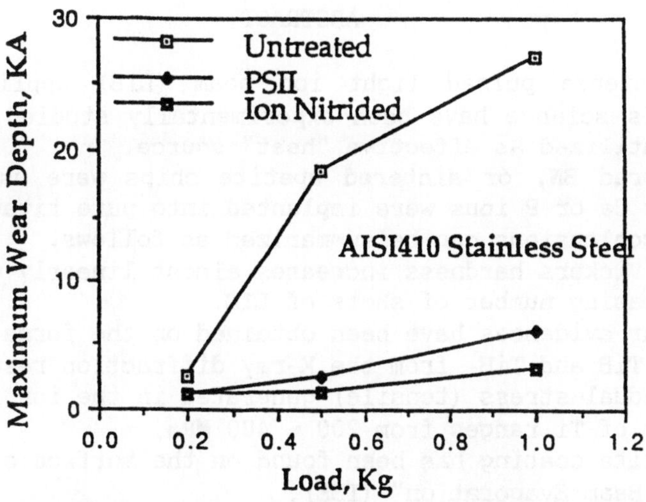

Fig. 4 a) Hardness of Alloys 908 and 909 Across Implanted Zone.
b) Wear Behavior of AISI 410 Stainless.

Surface Modification Experiment by Intense Pulsed Ion Beam

T. Takaai, T. Hatsushika, T. Suzuki

Yamanashi University Kofu Japan

K. Masugata and K. Yatsui

Nagaoka University of Technology
Nagaoka Japan

ABSTRACT

An intense pulsed light ion beam (LIB) applications in materials science have been experimentally studied, where the LIB is utilized as effective "heat" source.

Sintered BN, or sintered apatite chips were used; B or N ions, or Ca or P ions were implanted into pure titanium.

The conlcusions can be summarized as follows.

(1) The Vickers hardness increases almost linearly with increasing number of shots of LIB.
(2) Clear evidences have been obtained on the formation of TiN, TiB and TiB_2 from the X-ray diffraction results.
(3) Residual stress (tensile) generated in the ion implanted layer of Ti ranges from 200 ~ 400 MPa.
(4) Apatite coating has been found on the surface of Ti by "Ion Beam Evaporation" (IBE).

INTRODUCTION

Conventional, steady-state charged ion beams have been successfuly applied in the R & D of materials such as in semicnoductor processing, welding, and lithography. On the other hand, intense pulsed light ion beam (LIB), which significanlty differs from the conventional ion beams, have been recently considered for various applications particualy in materials science (1). If the LIB is irradiated onto the materials, the surface will be quickly heated to produce high density plasma. Using such a high power density "ablation" plasma, we expect quick depostion of various thin films, which has been named as "Ion Beam Evaporation" (IBE) (2). In fact, we have succeeded in the preparation such as of ZnS, YBaCuO, ITO, $BaTiO_3$, and so on. For this case, the LIB is utilized as the "heat" source since the pulse width is short compared with the conduction time of heat, and the phenomenon can be considered as "adiabatic". Since the quick heating cycle can be applied of the order of $\sim 10^6$ K/s, the LIB can be utilized for the various surface modifications, high-energy ion implantation, ion beam annealing, and so on. Furthermore, the ablation plasma produces very high pressure of more than 10^5 bar. A strong shcok wave is thus produced, by which the characteristics of the materials will be changed such as the wear speed (3).

In this paper, some of these experimental studies will be presented.

EXPERIMENTAL SETUP

The experiment has been carried out on the "ETIGO-I" (4) and "ETIGO-II" (5), pulse power machine in the Nagaoka Univeristy of Technology. Figure 1 illustrates the cross sectional view of ETIGO-II. Basically, it consists of Marx generator, intermediate capacitor, pulse forming line, and ion diode. Specifically, they produce the pulse power output of 1.2 MV, 240 kA, 50 ns, and 14 kJ for ETIGO-I and 3 MV, 460 kA, 50 ns, and 70 kJ for ETIGO-II.

Figure 2 illustrates the ion diode. As the ion diode, we have normally used a magnetically insulated diode (MID), where a transverse magnetic field is applied to suppress the motion of electrons. As the flashboard anode, we used 1) sintered BN (thickness = 1 mm, purity = 95 %), and or 2) sintered apatite chips. Several kinds of ions such as 1) B or N ions, or 2) Ca or P ions were used for the surface modification experiment. In the ion implantation experiment, a target of titanium (1 mm thick) was placed around the geometrically focused point.

Figure 3 shows typical waveforms of the MID. Typically, the MID was operated at beam energy of 0.6 ~ 1 MeV, ion current density of 1 ~ 5 kA/cm², and beam power density of

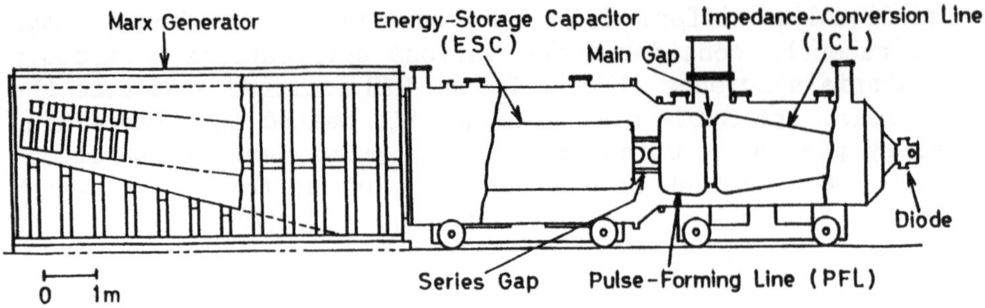

Fig. 1 Cross-sectional view of intense pulse-power generator, ETIGO-II, at the Nagaoka Univ. of Technology.

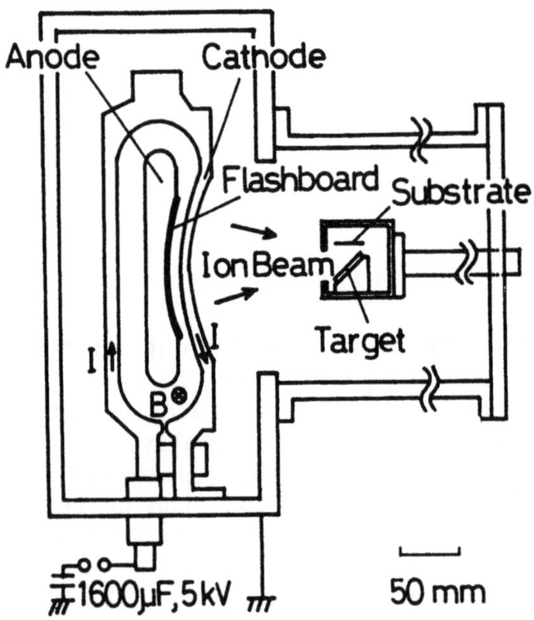

Fig. 2 Cross-sectional view of MID. The cathode works as a one-turn theta-pinch coil, producing transverse magnetic field for the insulation of electrons to reach the anode. Ions extracted from the MID focus toward the geometrically focusing point. Gap length between anode and cathode is 10 mm. For the ion implantation, where BN is used as the flashboard anode, titanium target is placed at z = 145 mm from the anode. In the apatie experiment, apatite is used as the flashboard anode, and is compared with the IBE experiment, where plastic and apatite is used as anode and target, respectively.

several GW/cm^2. High density "ablation" plasma will be produced by the irradiation of high power density LIB.

EXPERIMENTAL RESULTS AND DISCUSSIONS

Surface Modification of Pure Titanium

Figure 4 shows Vickers hardness of the titanium plotted against the number of shots, where boron nitride was used as the flashboard anode (6). As seen from Fig. 4, hardness increases with increasing the number of shots by a factor of ~ 60 % by the irradiation of LIB (beam energy ~ 750 keV, ion-current density ~ 3 kA/cm^2). It is widely known (7) that the hardness of pure titanium is strongly affected by constituting interstitial solid solution with oxygen, nitogen and carbon. Boron, however, is found to be beyond the Hume-Rotherry's atom size fit range (8) and has no solid solution with titanium. Hence, the increase in the hardness observed above can be considered by solutioning of excess nitrogen. In other words, nitrogen is not only present as an interstitial solute in pure titanium, but as a coherent precipitation of TiN which is isotropic with NaCl in its crystal strucure.

Figure 5 shows an X-ray diffraction pattern of Ti after being irradiated by 100 shots (1). Certain peaks corresponding to TiN and TiB_2 are clearly observed, from which we find the production of intermetallic compounds of TiN and TiB_2.

Pulsed heating and cooling of the surface take place simultaneously with the implantation. Generation of the thermal stress will be found to exist as a residual stress of the surface layer ranging from 200 to 400 MPa. From the above consideration, we have studied the residual stress against the number of shots. A typical results is shown in Fig. 6. In Fig. 6, we find the presence of tensile stress, but do not observe noticeable effects on the tensile residual stress by the the change of the number of shots. From Fig. 6, furthermore, the tensile residual stress changes into the compressive stress by being annealed for 30 min at 973 K. The compressive stress, on the other hand, increases with increasing the number of shots, for which reason we do not understood.

Figure 7 shows the difference of the interplanar spacings of Ti with the number of shots as a parameter. Here, the positive slope means the shrinkage of the unit cell of Ti.

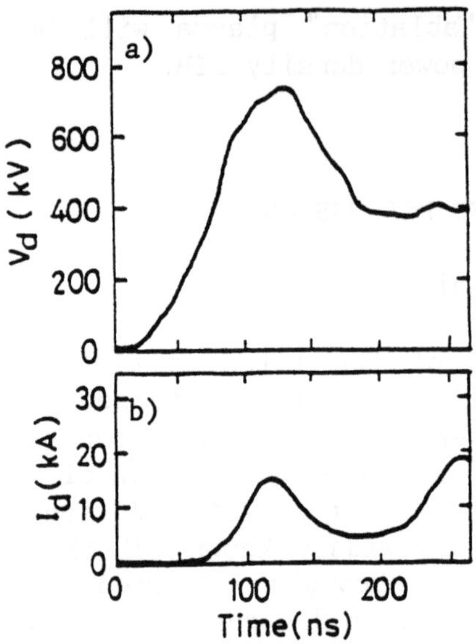

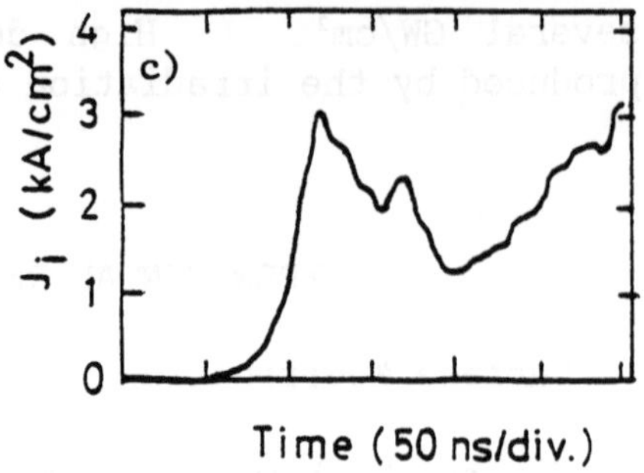

Fig. 3 Typical waveforms of a) diode voltage, b) diode current, and c) ion current density (at z = 145 mm).

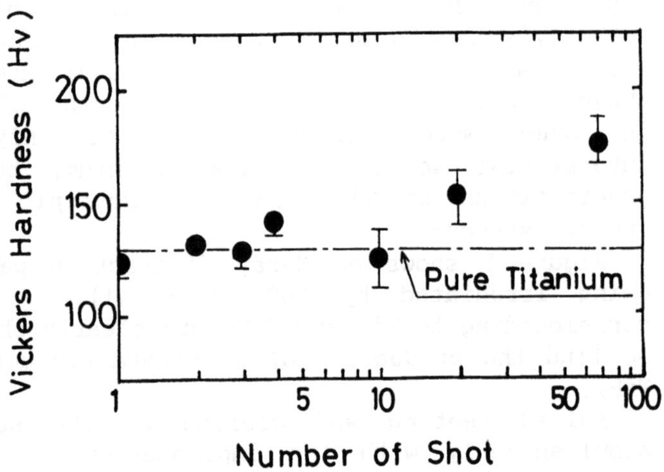

Fig. 4 Vickers hardness vs number of shots.

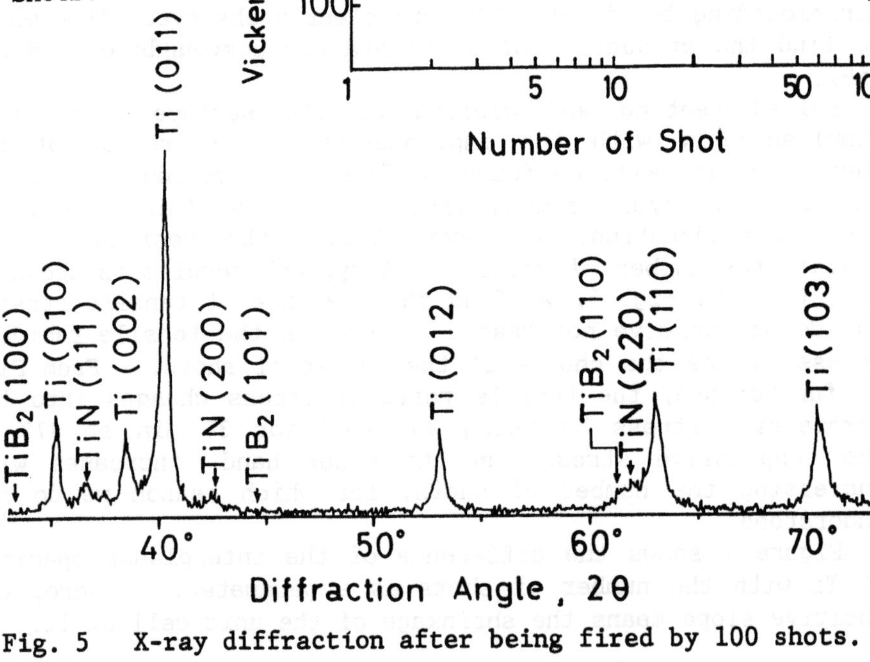

Fig. 5 X-ray diffraction after being fired by 100 shots.

Fig. 6 Residual stress vs number of shots.

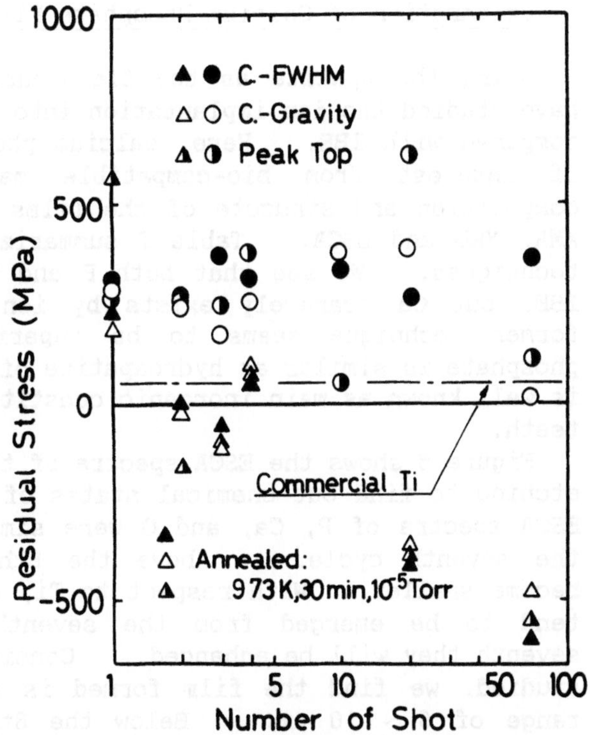

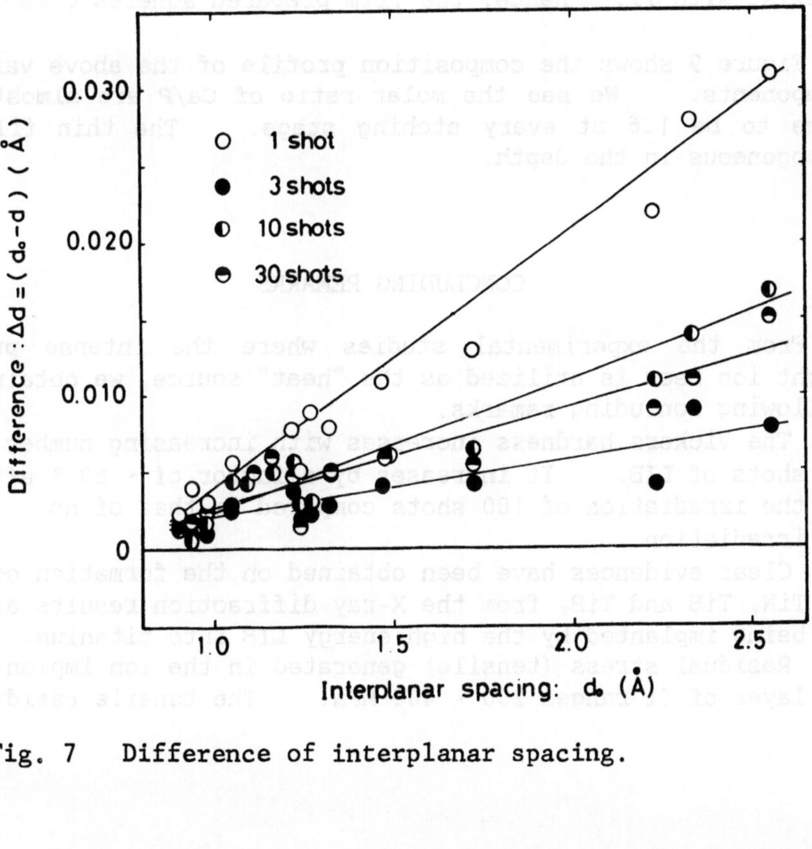

Fig. 7 Difference of interplanar spacing.

Preparation of Calcium Phosphate Films on Titanium

Using the apatite as the flashboard anode of the MID, we have studied the ion implantation into Ti, which has also been compared with IBE. Here, calcium phosphate films on Ti are of interest from bio-compatible materials. Chemical composition and strucute of the films formed were studied by XMA, XRD and ESCA. Table I summarizes the results of both techniques. We see that both P and Ca can be implanted by IBE, but Ca scarecely exists by ion implantation. The former technique seems to be superior since the calcium phosphate is similar as hydroapatite of $Ca_{10}(PO_4)_6(OH)_2$, which is well known as main inorganic constituent of human bones and teeth.

Figure 8 shows the ESCA spectra of the film by using Ar ion etching to find out chemical states of the components. The ESCA spectra of P, Ca, and O were similar from the first to the seventh cycle. Above the 8th cycle, however, they become smaller. With respect to Ti, on the other hand, they tend to be emerged from the seventh layer, and over the seventh they will be enhanced. Considering the etching time studied, we find the film formed is relatively thin in the range of 6 ~ 10 μm. Below the 8th layer, P, Ca, and O coexist with Ti. Hence, the film prepared adheres closely to Ti.

Figure 9 shows the composition profile of the above various components. We see the molar ratio of Ca/P are almost the same to be 1.6 at every etching stage. The thin film is homogeneous in the depth.

CONCLUDING REMARKS

From the experimental studies where the intense pulsed light ion beam is utilized as the "heat" source, we obtain the following conluding remarks.

(1) The Vickers hardness increases with increasing number of shots of LIB. It increases by a factor of ~ 60 % after the irradiation of 100 shots compared to that of no irradiation.

(2) Clear evidences have been obtained on the formation of TiN, TiB and TiB_2 from the X-ray diffraction results after being implanted by the high energy LIB into titanium.

(3) Residual stress (tensile) generated in the ion implanted layer of Ti ranges 200 ~ 400 MPa. The tensile residual

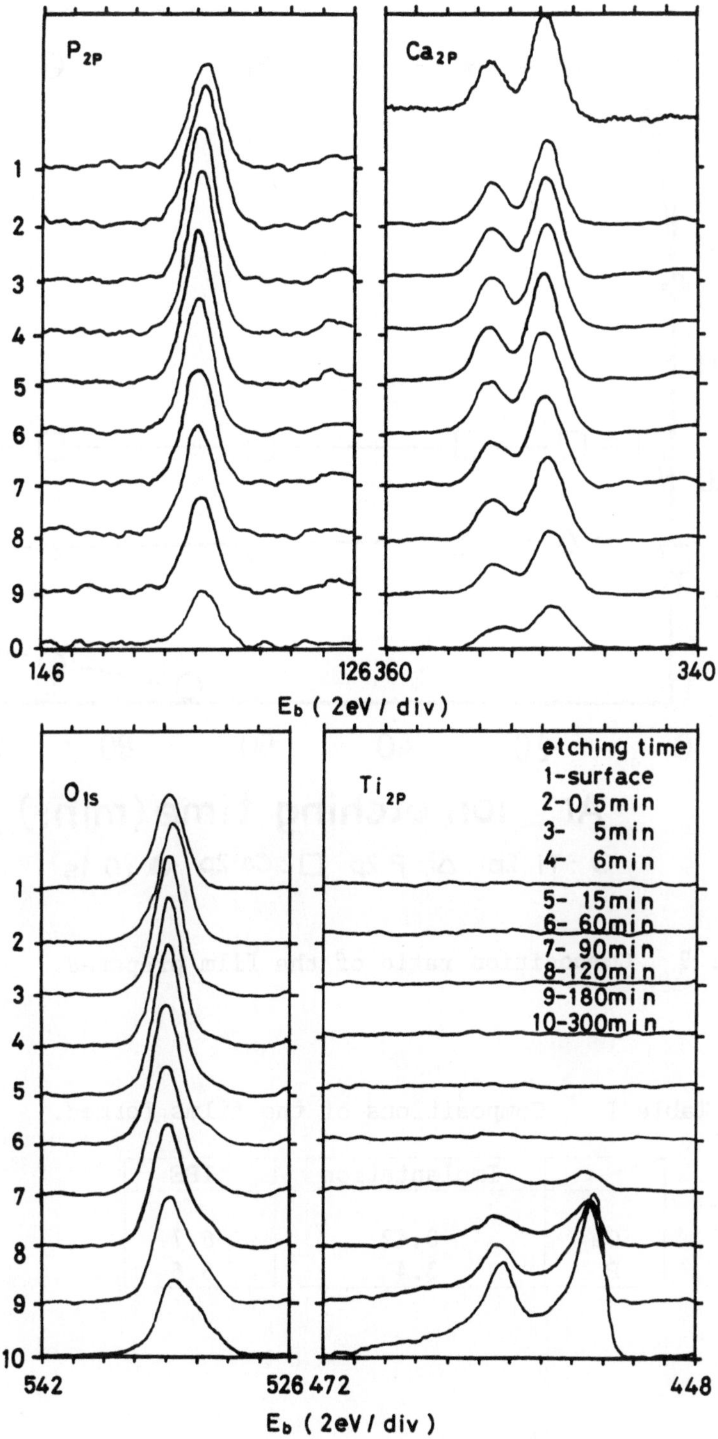

Fig. 8 ESCA spectra of the film prepared.

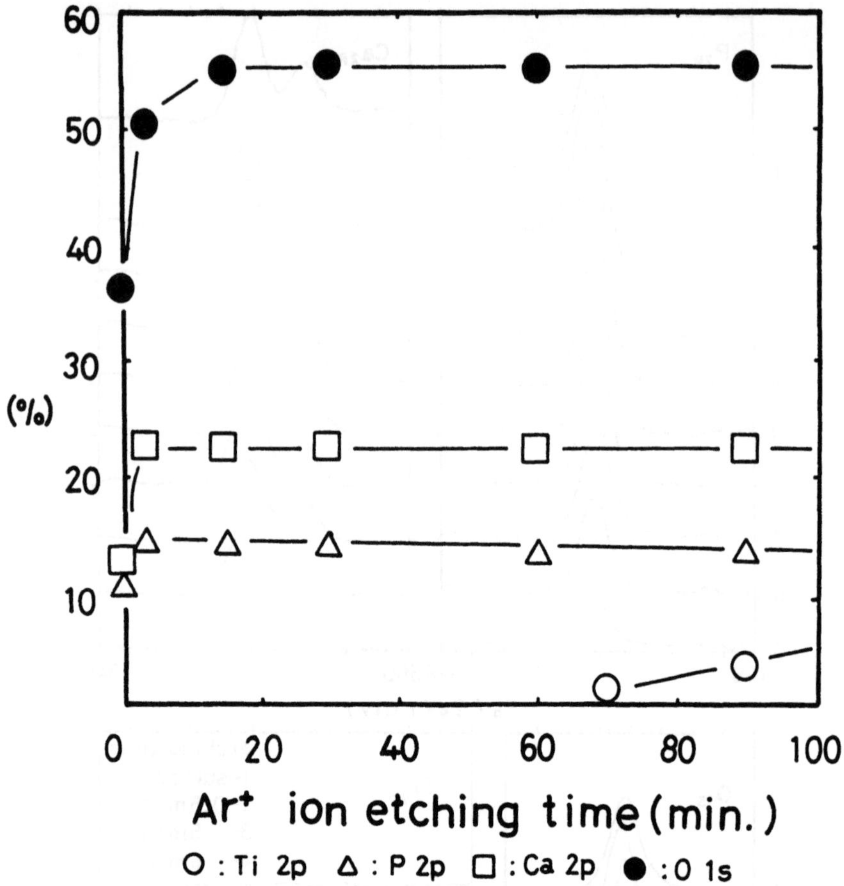

Fig. 9 Composition ratio of the film prepared.

Table I. Compositions of the films formed.

	Implantation	IBE
Ca	0.42	8.7
P	3.4	5.6

stress changes into the compressive stress by after being annealed.

(4) Apatite coating has been found on the surface of Ti by the irradiation of LIB by "Ion Beam Evaporation" (IBE). The film thickness prepared is estimated to be 6 ~ 10 μm. The film seems to be homogeneous.

ACKNOWLEDGMENTS

Experimental support given by Dr. Yutaka Shimotori of Nagaoka Univeristy of Technology and Mr. Toyomasa Honda of Yamanashi University are highly appreciated.

REFERENCES

1. K. Yatsui: Laser and Particle Beams 7 (1989) 773.
2. Y. Shimotori et al.: Jpn. J. Appl. Phys. 28, 468 (1989).
3. A. D. Pogrebjak et al.: Phys. Lett. A129 (1987) 259.
4. K. Yatsui et al.: Laser and Particle Beams 3 (1985) 119.
5. A. Tokuchi et al.: in Proc. 2nd Int'l Symp. on ICF Res. by High-Power Particle Beams, Nagaoka, ed. K. Yatsui (1986), p. 430,
6. K. Yatsui et al.: in Proc. 7th Int'l Conf. on High-Power Particle Beams, Karlsruhe, ed. by W. Bauer et al. (1988), p. 522.
7. R. I. Jaffee, H. R. Ogden, and D. J. Maykuth: Trans. Amer. Inst. Min. (Metall.) Engrs., 181 (1950) 1261.
8. W. Hume-Rothery: "The structure of Metals and Alloys" (Inst. of Metals, 1950), p. 60.

PSII Nitrogen Implanted M2 Tool Steel for Wear Resistance in Wood Machining Tools

S. Raebel, F. J. Worzala and J. R. Conrad

University of Wisconsin—Madison
Madison
Wisconsin
USA

ABSTRACT

PM M2 saw blade inserts in the as-sintered or quenched and tempered conditions were ion implanted with nitrogen using Plasma Source Ion Implantation. The implanted inserts were subjected to a simulated sawing test to rank their performance against unimplanted M2 and stellite, a material used in the forest products industry for wood machining. Based on the results of this investigation, nitrogen ion implantation appears to be a promising method for reducing the wear of wood machining tools.

INTRODUCTION

The wear experienced by saw blade inserts is the major problem encountered in saw blade maintenance in the forest products industry. Saw blade inserts are exposed to a corrosive, erosive environment due to high temperatures at the cutting edge of the tip and the presence of vaporized moisture in the wood. Teeth made from hard materials such as carbide and stellite must be reground at frequent intervals. When the material being sawed is plywood, fiberboard, or particleboard, the presence of polymeric glues further reduces tool life. Saw tooth wear is an expensive problem for the forest products industry. Valuable tooth material and production time are lost due to the erosion of teeth and subsequent regrinding. The worn tips also produce an inferior wood surface finish requiring planing, an additional processing step resulting in wood loss and production time loss.

Many methods for reducing the wear of saw blade tips and other wood machining tools have been investigated. One of the more significant methods of wear reduction has been the surface modification of saw blade tips with wear resistant materials such as chromium, tungsten, and titanium carbide, and diamond. A number of researchers have found that surface modification of saw blade tips has been successful in reducing the wear of wood machining tools [1-5]. Some researchers have investigated the possibility of surface modification of one face of the saw blade tip to produce a self-sharpening effect as a result of one face of the tooth wearing away less quickly than the others [3-5].

In this investigation, another method of surface modification, Plasma Source Ion Implantation, or PSII, has been investigated as a potential means of reducing wear in circular saw blade inserts. Ion implantation of nitrogen, boron, or carbon into steel has been found to enhance resistance to wear, corrosion, and fatigue failure [6-8]. PSII is an ion implantation process in which the substrate is placed directly into the ion source and pulsed to a high negative potential relative to the chamber walls, thus forming a plasma sheath enveloping the target. Ions are accelerated normal to the target surfaces across the plasma sheath and embed into the target. PSII has been found to increase the wear resistance of steels and other materials for many applications, including punches, dies, and prosthetic implants [8-10].

SUBSTRATE PREPARATION

The saw tooth inserts implanted in this investigation were fabricated by SSI Technologies using the powder metallurgy process. SSI Ultimet M2 powder blended with 0.75% Acrawax C lubricant was compacted at 50 tsi into a green compact, then vacuum sintered for 45 minutes at 1254°C. Some of the specimens were then austenitized at 1100°C for one hour, quenched, and tempered twice at 540°C for two hours.

IMPLANTATION

The rake faces of the saw tooth inserts were ground and polished with 1 micron diamond paste to a smooth finish. Saw tooth inserts in the as-sintered and quenched and tempered conditions were implanted with nitrogen in two batches: Batch I was implanted at an energy of 50 keV and a dose of 3E17 atoms/cm^2 for 460 minutes and Batch II was implanted at an energy of 40 keV and a dose of 3E18 atoms/cm^2 for 394 minutes. The substrate/implantation conditions of the saw tooth inserts investigated are listed in Table 1.

Specimens in Batch I developed a gold surface layer during implantation, while Batch II specimens had a cloudy dark appearance. Scanning auger microscopy was utilized to determine the compositions of the surfaces of the two batches and to obtain the nitrogen profile from surface of the specimen into the substrate. Oxygen, in addition to carbon, nitrogen, and iron, was present at the surfaces of the gold colored specimens, indicating an oxide contamination layer. This layer was completely removed after 200 Angstroms of sputter cleaning. Nitrogen concentration was profiled from

TABLE 1
Substrate/implantation parameters for the four specimens tested.

Batch I	Batch II
Specimen 1:	Specimen 3:
Substrate: as-sintered	Substrate: as-sintered
Energy: 50 keV	Energy: 40 keV
Dose: 3E17 atoms/cm^2	Dose: 3E18 atoms/cm^2
Time: 460 minutes	Time: 394 minutes
Specimen 2:	Specimen 4:
Substrate: quenched and tempered	Substrate: quenched and tempered
Energy: 50 keV	Energy: 40 keV
Dose: 3E17 atoms/cm^2	Dose: 3E18 atoms/cm^2
Time: 460 minutes	Time: 394 minutes

the surface of the saw tooth into the substrate for each specimen. These profiles are shown in Figure 1. For Batch I specimens, the nitrogen showed a peak concentration from the surface to approximately 600 Angstroms into the substrate, and decreased quickly thereafter. Batch II specimens showed a nitrogen content that was high at the surface and decreased steadily to a depth of approximately 1200 Angstroms.

MICROHARDNESS

Knoop microhardness measurements were taken at the surfaces of the specimens and into the substrates to determine if local hardening had taken place. Knoop microhardness readings were taken with a 10 g load applied for 10 seconds. The specimens exhibited a two phase structure on the surface consisting of a light phase and a darker matrix phase as illustrated in Figure 2, and hardness measurements were taken of each phase. The substrate measurements were taken in cross section. Batch I specimens showed a very slight overall increase of the dark surface phase microhardness over the substrate microhardnesses, and light phase readings were high, indicating a carbide phase. For specimen 1, light phase readings were in the 1125-1400 range, dark phase readings were in the 700-875 range, and substrate microhardnesses ranged from 600-800. For specimen 2, light phase readings ranged from 1050-1350, dark phase readings ranged from 700-850, and the substrate measured 600-800. Batch II specimens showed a slightly lower hardness at the surface than in the substrate, indicating perhaps slight surface overheating, depletion of carbon in the dark phase, or radiation damage. The light phase measurements

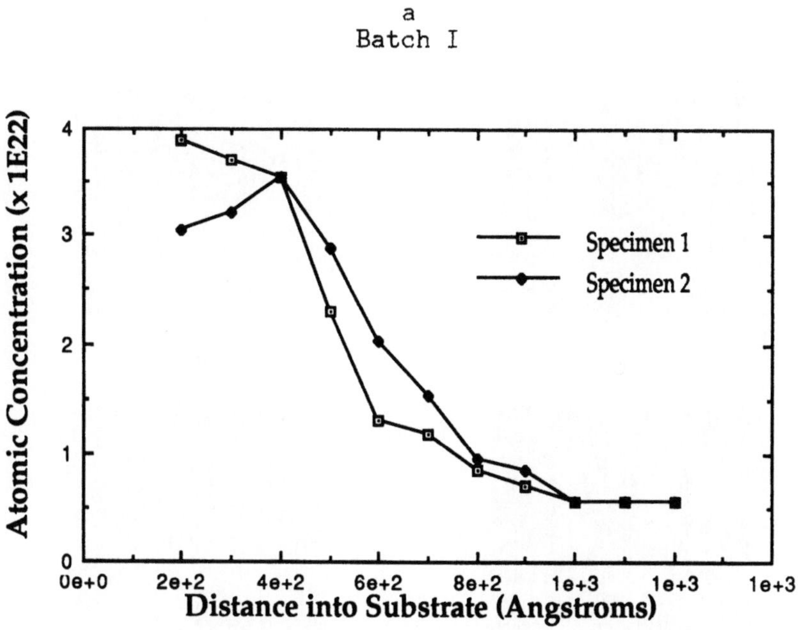

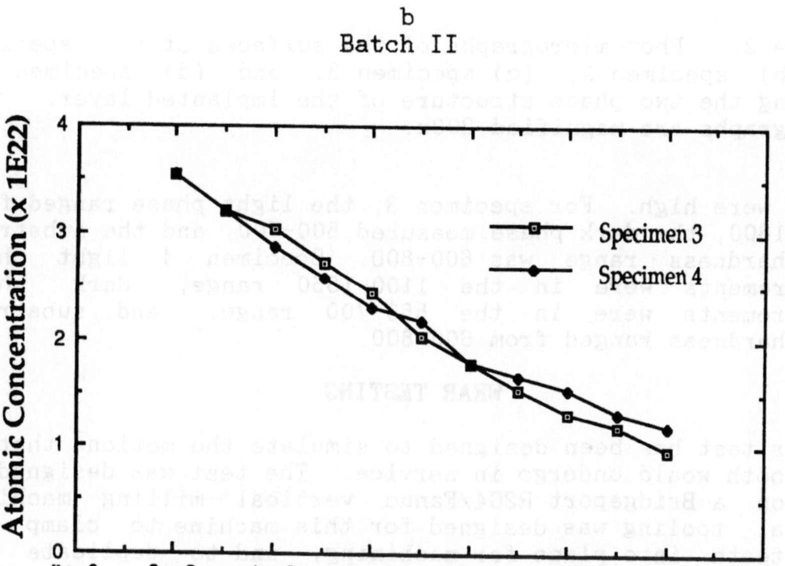

Figure 1. Atomic concentrations of implanted nitrogen versus depth into the saw tooth substrate for (a) specimens 1 and 2, and (b) specimens 3 and 4.

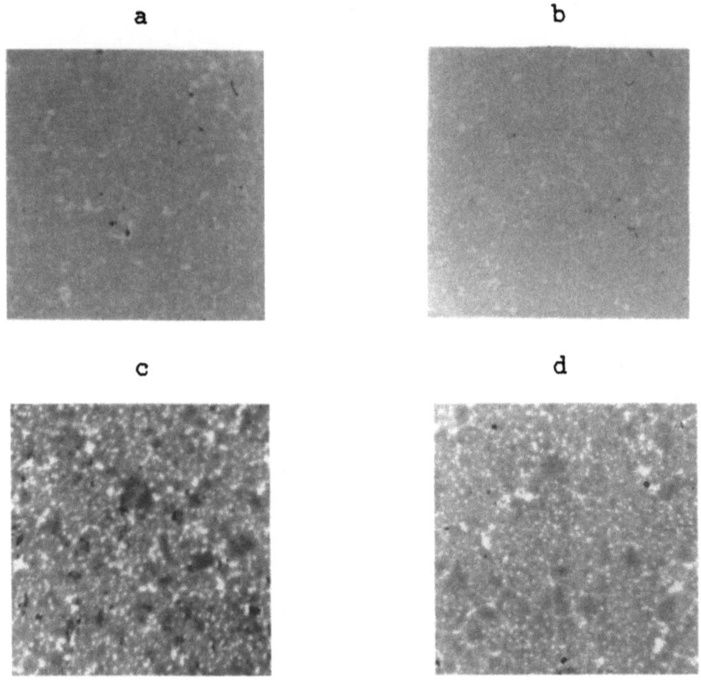

Figure 2. Photomicrographs of the surfaces of (a) specimen 1, (b) specimen 2, (c) specimen 3, and (d) specimen 4, showing the two phase structure of the implanted layer. All photographs are magnified 200x.

again were high. For specimen 3, the light phase ranged from 1300-1600, the dark phase measured 500-700, and the substrate microhardness range was 600-800. Specimen 4 light phase measurements were in the 1100-1550 range, dark phase measurements were in the 550-700 range, and substrate microhardness ranged from 600-800.

WEAR TESTING

A wear test has been designed to simulate the motions that a saw tooth would undergo in service. The test was designed to run on a Bridgeport R2G4/Fanuc vertical milling machine. Special tooling was designed for this machine to clamp the saw teeth into place for machining, and to duplicate the angles at which the teeth would contact the wood. The mill was programmed to make 120 17 inch passes through a 4" x 4" x 24" particleboard log which was clamped lengthwise onto the mill bed. Speeds and feeds were determined from reference sources on wood machining [11,12], and also from the maximum capabilities of the mill. The parameters of the test are listed in Table 2.

TABLE 2
Bridgeport Mill wear test parameters:

Spindle speed:	4200 rpm
Tool diameter:	3 inches
Number of inserts in tool:	2
Distance sawed per run:	170 feet
Forward feed rate:	236 inches per minute
Wood sawed:	particleboard

Before testing, saw teeth were ground to a sharp point at the cutting edge of the insert. Two saw teeth from each of the implanted specimens, along with unimplanted M2 tool steel in the as-sintered and the quenched and tempered conditions, and stellite, which is a material currently used in the forest products industry for this purpose, were subjected to four runs each of the wood machining test. The tip rounding of the rake and clearance faces of the tooth was measured for each set of specimens after each test run, and plotted versus distance sawed. These plots are pictured in Figure 3.

Specimen 4 performed better than stellite in the wear tests and much better than the unimplanted quenched and tempered M2 tool steel. An example of the cutting edges of these tips after testing is shown in Figure 4. Specimen 1 also performed better than stellite and much better than unimplanted as-sintered M2 tool steel. Specimens 2 and 3, however, did not perform as well as either the stellite or the unimplanted M2. Examination of photographs taken of the cutting edge of the insert after grinding but before testing showed a discoloration of the metal close to the tip, indicating that overheating had taken place during the grinding stage. Excessive heating may cause nitrogen induced defects to become unstable and atoms to diffuse away from the surface, rendering the layer ineffective [8]. These specimens are currently being reground for retesting.

CONCLUSIONS

Saw tooth inserts fabricated from M2 tool steel in the as-sintered and quenched and tempered conditions were implanted with nitrogen in two batches and evaluated as potential wood machining tools.

Specimens implanted in Batch I showed a nitrogen profile which is high at the surface and decreases quickly after 600 Angstroms. The microhardnesses of the dark phase of the surfaces of the specimens exceeded substrate hardnesses slightly. Specimens implanted in Batch II showed a nitrogen profile which was high at the surface and decreased steadily to 1200 Angstroms, and a surface dark phase microhardness which was slightly below the substrate microhardness. All specimens showed a light carbide phase at the surface which

a
As-sintered M2

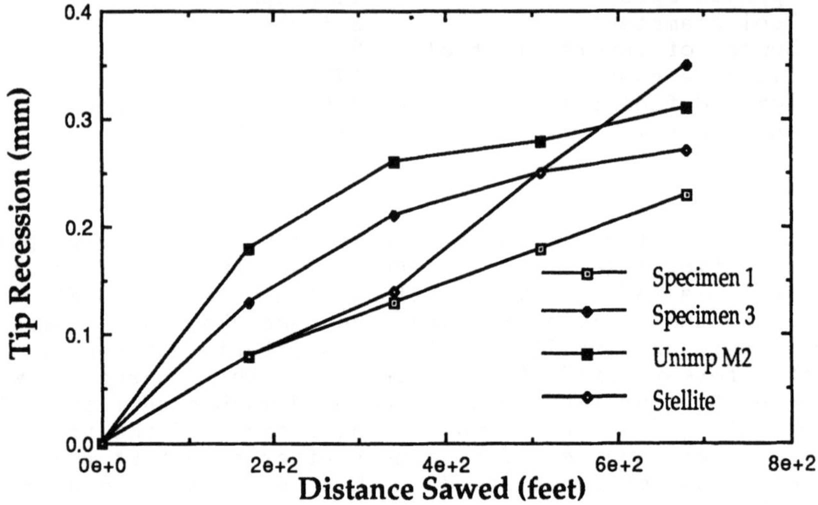

b
Quenched and Tempered M2

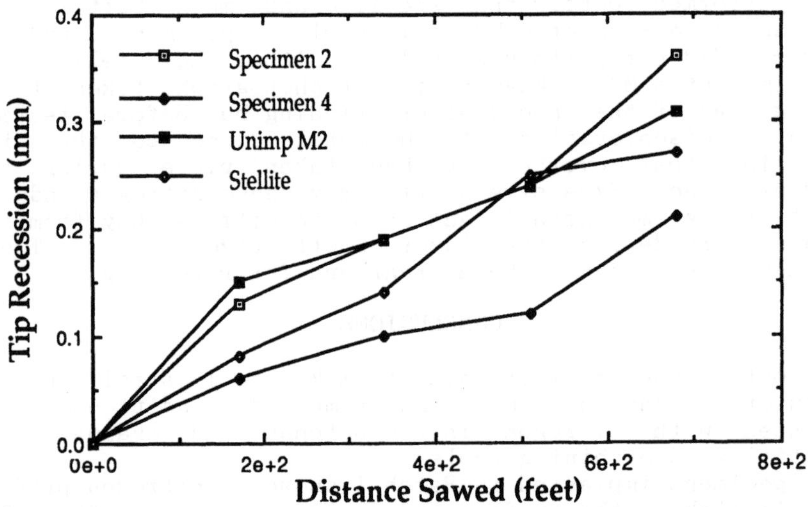

Figures 3a and 3b. Saw tooth rake face tip recession versus distance sawed in feet. Figures 3a and 3b show the recession of implanted M2, unimplanted M2, and stellite for the as-sintered substrates and quenched and tempered substrates, respectively.

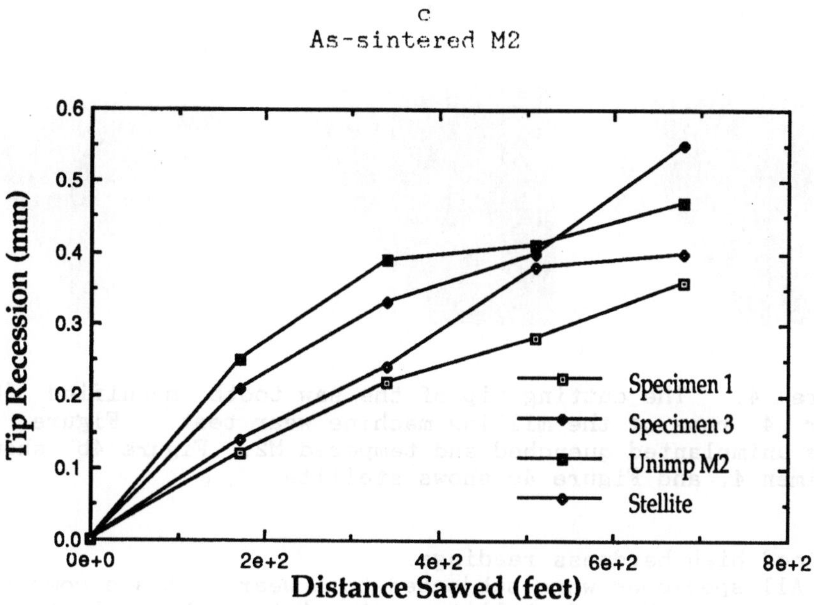

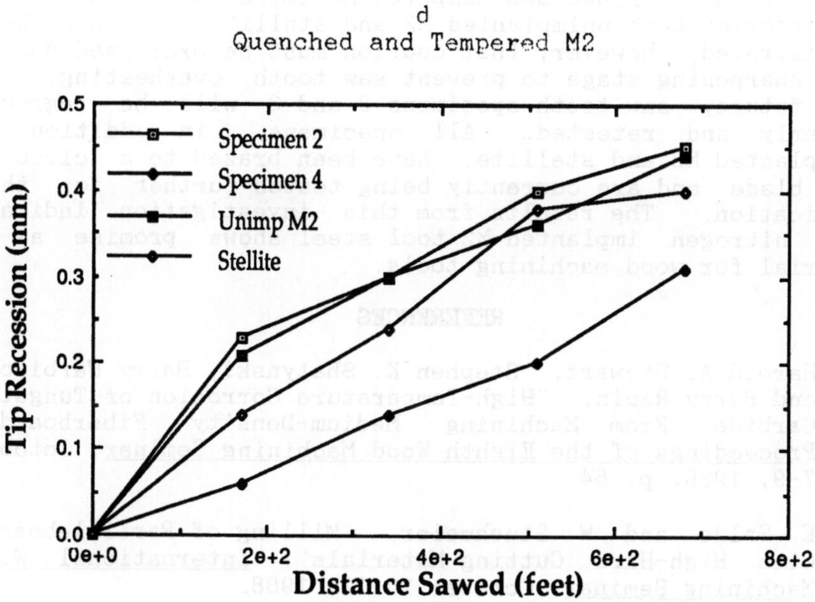

Figures 3c and 3d. Saw tooth clearance face tip recession versus distance sawed in feet. Figures 3c and 3d show the recession of implanted M2, unimplanted M2, and stellite for the as-sintered and quenched and tempered substrates, respectively.

a b c

Figure 4. The cutting tip of the saw tooth magnified 50x after 4 runs of the milling machine wear test. Figure 4a shows unimplanted quenched and tempered M2, Figure 4b shows specimen 4, and Figure 4c shows stellite.

produced high hardness readings.

All specimens were subjected to a wear test and compared to unimplanted M2 and stellite. As-sintered M2 implanted in Batch I and quenched and tempered M2 implanted in Batch II outperformed both unimplanted M2 and stellite. It has been demonstrated, however, that caution must be exercised during the sharpening stage to prevent saw tooth overheating. In the future, saw tooth specimens 2 and 3 will be reground properly and retested. All specimens, in addition to unimplanted M2 and stellite, have been brazed to a circular saw blade and are currently being tested further for this application. The results from this investigation indicate that nitrogen implanted M2 tool steel shows promise as a material for wood machining tools.

REFERENCES

1. Harold A. Stewart, Stephen K. Shatynski, Barry Harbison, and Barry Rabin, "High-Temperature Corrosion of Tungsten Carbide From Machining Medium-Density Fiberboard", Proceedings of the Eighth Wood Machining Seminar, October 7-9, 1985, p. 54.

2. E. Salje and W. Stuehmeier, "Milling of Particleboards with High-Hard Cutting Materials", International Wood Machining Seminar, October 10-12, 1988.

3. Eberhard D. Kirbach, "Saw Tips With Self-Sharpening Characteristics", Proceedings of the Eighth Wood Machining Seminar, October 7-9, 1985, p. 196.

4. F. Worzala and J. Danielson, "Recent Advancements in Materials for Wood Machining Applications", International Wood Machining Seminar, October 10-12, 1988.

5. Yo S. Song and Frank J. Worzala, "Coated Powder Metal Substrates for Wood Machining Applications", Report to SSI Technologies, October 28, 1988.

6. G. Dearnaley, "Fundamentals of Ion Implantation", Surface Modifications and Coatings, American Society for Metals, Metals Park, OH, 1986, pp. 399-408.

7. James D. Destefani, "Ion Implantation Update", Advanced Materials and Processes, 134, (4), October 1988, pp. 39-43.

8. Frank J. Worzala, Richard A. Dodd, John R. Conrad, and Jeffrey L. Radtke, "Ion Implantation of Tools and Punches - Laboratory Curiosity or Commercial Process?", Proceedings of the International Conference on Tool Materials, Uddeholm Research Foundation, September 28 - October 1, 1987, Chicago, IL.

9. J. R. Conrad, R. A. Dodd, and F. J. Worzala, "Plasma Source Ion Implantation: A New Manufacturing Process for Surface Modification of Materials".

10. J. R. Conrad, R. A. Dodd, F. J. Worzala, and X. Qiu, "Plasma Source Ion Implantation: A New, Cost-Effective, Non-Line-of-Sight Technique for Ion Implantation of Materials", Surface and Coatings Technology, 36, 1988, pp. 927-937.

11. Richard W. Perkins, "Woodcutting Tools and Machines", Mark's Standard Handbook for Mechanical Engineers, 8th Ed., McGraw-Hill Book Company, New York, 1978, pp. 13-80 - 13-84.

12. Erik Oberg, Franklin D. Jones, and Holbrook L. Horton, Machinery's Handbook 22nd Ed., Industrial Press Inc., New York, 1984, p. 2186.

Plasma Nitriding and Nitrocarburizing of Austempered S.G. Cast Iron

R. Priestner and S. Korichi

University of Manchester Institute of Science and Technology
Manchester
UK

ABSTRACT

The effects of varying austempering parameters on the response of S.G. cast iron to plasma nitriding and nitrocarburizing was studied. The ideal austempered microstructure of bainite and austenite yielded poor compound layers and the shallowest case depth. A microstructure consisting wholly of bainite obtained by austempering at too high a temperature yielded the best surface-engineered product. Adding carburizing potential to the plasma reduced the thickness of compound layers and case depth. The results were explained by the different distributions of cementite arising when the differently austempered microstructures were tempered on heating to the nitriding temperature. It is concluded that S.G. iron destined to be plasma nitrided should be austempered at higher than normal temperatures, and that nitriding should be preferred to nitrocarburizing.

INTRODUCTION

Austempered spheroidal graphite (S.G.) cast irons display combinations of strength and ductility which allow them to compete with forged steel in a number of engineering and agricultural applications. Their properties derive from melt processing, which spheroidizes the graphite, and from heat-treatment processing in the solid state, which optimizes the microstructure of the metallic matrix. The austempered product is, thus, a composite, in which the ductility of the matrix is not compromised by the second constituent, spheroidal graphite.

In Europe a number of grey iron castings, such as gear and valve casings, are often nitrided to improve their corrosion resistance. Heating to nitriding temperatures for treatments of short duration has little effect on the relatively stable matrix of conventional flake irons, nor on mechanical properties since they are already relatively brittle. Nor is there a general demand for surface hardening treatments, since hard surfaces can be more economically produced in white or mixed irons by control of compositional and casting parameters.

The story is quite different for austempered S.G. irons. Firstly, the field in which they compete would be greatly widened by effective and economic surface engineering. Secondly, the matrix microstructure obtained by austempering is thermally unstable, and is destroyed completely by tempering at the nitriding temperature. Nevertheless, a surface engineering process which yields a hard, adherent, corrosion resistant surface layer on a case of matching hardness and useful depth would be of value, provided core properties were not sacrificed entirely. In this paper a study has been made of the influence of prior matrix microstructure, obtained by variation in austempering parameters, on the response of S.G. iron to plasma nitriding and nitrocarburizing.

A review of the microstructural response of S.G. irons to austempering cannot be given here. However, some information is needed in order to understand the procedures and results of the research. Austempering consists, basically, of austenitizing the matrix at temperatures, above 850°C, followed by quenching to the austempering temperature, T_A, holding for a controlled time, then cooling to room temperature. The concentration and distribution of carbon and alloying elements controls the transformation to bainite during austempering, which generally proceeds in two stages. Three characteristic temperature regimes then exist in which different microstructures are developed.

(i) At low austempering temperatures the first stage of bainite formation slows virtually to a stop well before complete transformation is achieved. Some of the carbon rejected from the bainitic ferrite diffuses in solution into the remaining austenite. The austenite is thus partly stabilised against the martensite transformation on cooling to room temperature, but part of it does transform to martensite.

(ii) At intermediate austempering temperature, the first-stage reaction produces more bainite before slowing to a virtual stop, and very little martensite is formed on cooling to room temperature. The dual-phase, ferrite-plus-austenite matrix has high ductility and is the optimum microstructure sought by austempering. After a long time at an intermediate temperature the remaining austenite precipitates carbides, which allows the second stage of the bainite reaction to proceed to completion.

(iii) At high austempering temperatures the second stage bainite reaction follows quickly on the heels of the first, producing a microstructure wholly of bainitic ferrite and carbide.

These decomposition routes may, of course, produce a spectrum of possible microstructures, especially when controlled cooling is used instead of isothermal austempering.

MATERIALS AND METHODS

The compositions of the materials are listed in Table 1.

TABLE 1

Compositions of the cast irons, Wt%

	C	Si	Mn	S	P	Ni
Cast 1	3.6	2.4	0.31	0.023	0.024	-
Cast 2	4.05	2.1	0.19	0.01	0.01	1.14

Plate specimens 25 mm x 20 mm in area and 8 mm thick were machined from keel-blocks. Specimens were austenitized for one hour at 870°C to 1050°C in argon. They were quenched into a salt bath for austempering and then cooled in air. For microstructural and X-ray diffraction analysis, specimens were ground on their centre-plane. Specimens destined for plasma treatment were ground to a standard finish on all surfaces. X-ray diffraction spectra were obtained from surface compound layers without further grinding. Differential scanning calorimetry (DSC) was done at a scan rate of 0.33 Ks^{-1}. Carbide particles were extracted by etching in 40% Nital. Washed and dried filtrates were dispersed on glass slips for X-ray diffraction analysis.

Plasma nitriding and nitrocarburizing treatments were of 4 hours duration at 570°C in a commercial plant. Two atmospheres were used:

Treatment J - 50/50 nitrogen/hydrogen

Treatment K - 49/1/50 nitrogen/methane/hydrogen.

The gas pressure and voltage of operation were determined by the mass and surface area of the charge. Temperature was monitored and controlled by thermocouples inserted in dummy specimens.

RESULTS

Austempered microstructures

Early experimentation led to the selection of 300°C, 380°C nd 440°C as austempering temperatures which, after austenitizing at 900°C, characterized the three regimes of austempering behaviour described in the Introduction. The experiments described here are largely restricted to treatments at these temperatures.

Figure 1 illustrates the growth of bainitic ferrite and Figure 2 shows X-ray diffraction spectra at stages during austempering at 380°C. After short austempering times martensite (α') formed on cooling. The martensite was progressively replaced by ferrite (α) and austenite (γ) as carbon rejected from growing bainitic ferrite stabilized the austenite against martensite formation. Very little martensite formed after austempering for longer than about 60 minutes. At longer times than 20 hours the austenite decomposed to second-stage bainite.

The fractional austenite content was estimated from integrated peak intensities, and its carbon content was estimated from the change in lattice parameter, indicated by the shift in the $(200)_\gamma$ peak in Figure 2. These data are shown in Figure 3. The maximum retained austenite fraction was lower after austempering at 300°C than at 380°C, because much of the austenite transformed to martensite on cooling. At 440°C the second stage bainite reaction quickly followed the first reaction, so eliminating retained austenite after less than 2 hours at temperature.

The dependence of hardness on austempering time is illustrated in Figure 4. The high hardnesses after short times at temperature are due to so little bainite having formed that much of the austenite transformed to martensite on cooling. The high final hardness after austempering at 300°C is partly due to the persistence of martensite in the microstructure. The second stage of bainite formation at 440°C had little effect on hardness.

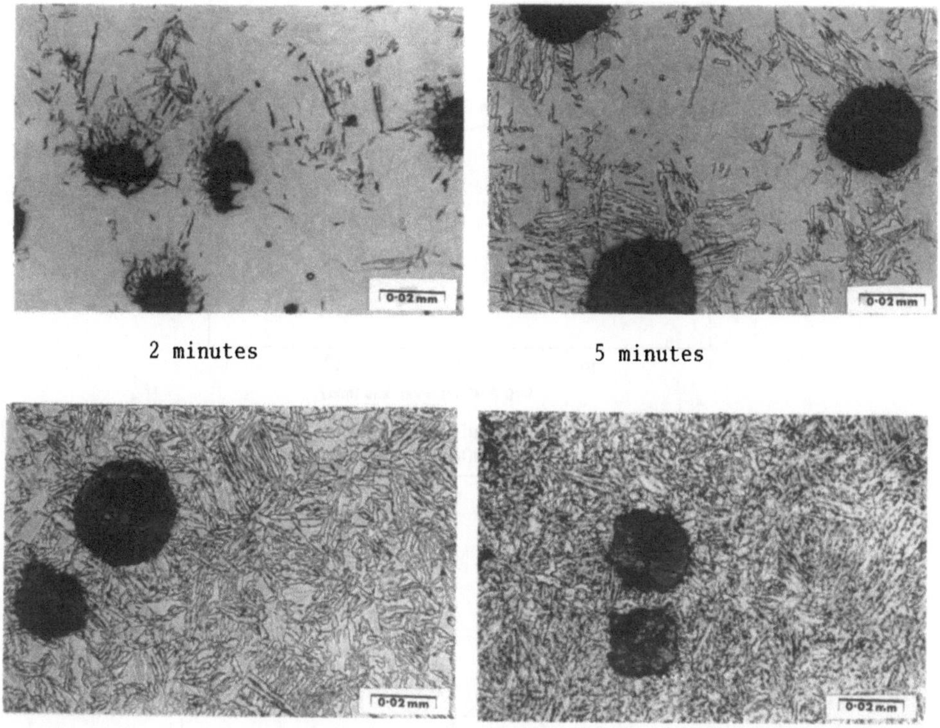

2 minutes   5 minutes

1 hour   marker = 0.02 mm   120 hours

Figure 1. Austempered at 380°C, Cast 1 austenitized at 900°C.

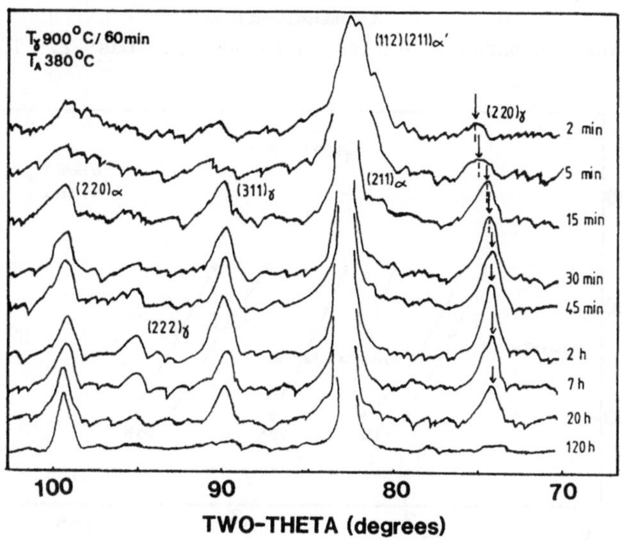

Figure 2. X-ray diffraction spectra after austempering at 380°C
Cast 1 austenitized at 900°C.

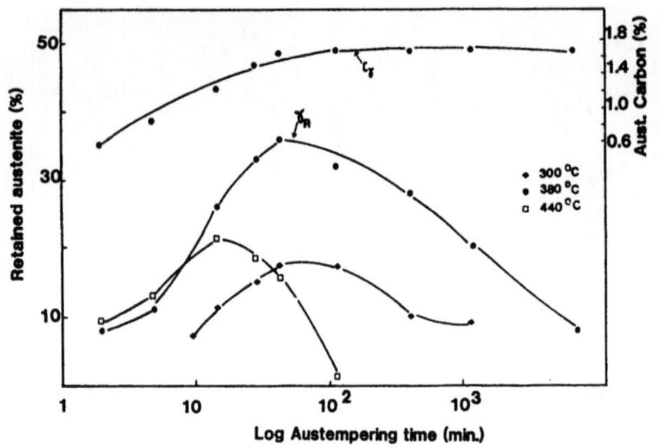

Figure 3. Retained austenite, γ_R, and austenite carbon conc., C_γ, at 300°C and 380°C. Cast 2, T_γ = 900°C.

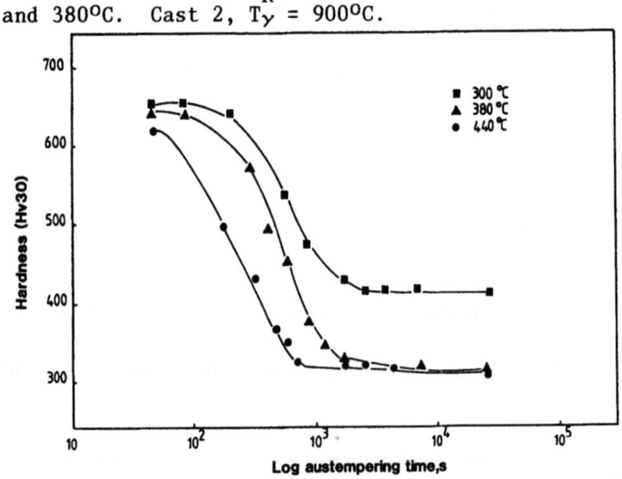

Figure 4. Change in hardness during austempering. Cast 2. T_γ = 900°C.

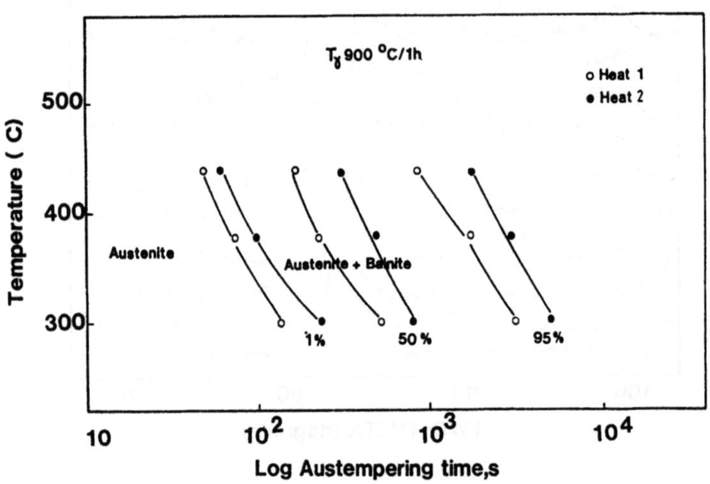

Figure 5. Isothermal TTT diagram for 1st bainite reaction. T_γ = 900°C.

The stability, with respect to martensite formation, of the austenite present at the end of the first-stage bainite reaction at 380°C was due to its increased carbon content, the carbon having been rejected from the bainitic ferrite as it formed. The relative instability of the austenite at 300°C was puzzling, since its carbon content achieved virtually the same values as at 380°C. However, the following suggests that more carbide precipitated during the bainite transformation at 300°C than at 380°C. X-ray diffraction analysis of extracted particles showed ϵ carbides to be present after austempering to the end of the first bainite reaction at both temperatures. Many more extractions were required after the 380°C treatment, in order to obtain a sufficient volume of extract for analysis, than after 300°C.

X-ray spectra from powders extracted during the second-stage bainite reaction displayed a large number of diffraction peaks, some of which coincided with silicon carbide, some with cementite, but neither phase could be unambiguously identified.

Figure 5 shows the partial, isothermal, transformation diagram deduced from metallographic, hardness and X-ray diffraction data. Although the alloyed cast transformed more slowly than the unalloyed cast, at equivalent stages the microstructures were identical in the two materials.

Tempering of austempered microstructures

Figure 6 shows some results from differential scanning calorimetry for specimens austempered at the three characteristic temperatures. The large exothermic events near 780K (~510°C) in specimens austempered at 300°C and 380°C were caused by the transformation of retained austenite to ferrite and carbide. In the specimen austempered at 300°C, this was preceded by tempering of the martensite present in that specimen. The large exothermic event did not occur in the specimen austempered at 440°C, because it was already fully transformed by the second-stage bainite reaction. A small event above 760K (~490°C) was probably the transformation of silicon carbide to cementite. Metallography, X-ray diffraction, and the results in Figure 6 clearly showed that all the austempered microstructures transformed completely to ferrite and cementite on heating to the plasma treatment temperature of 570°C.

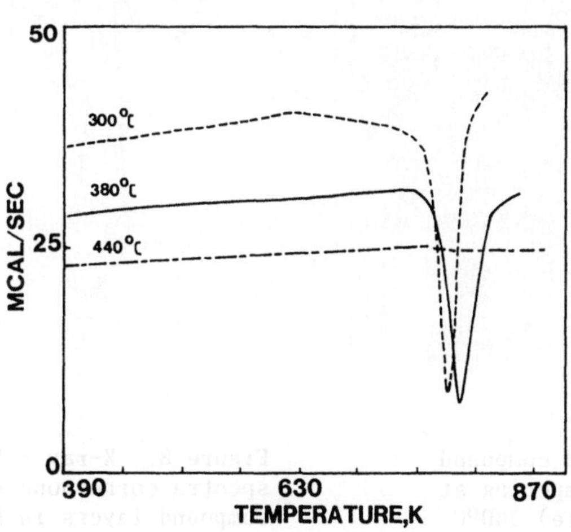

Figure 6. DSC heat change in specimens austempered 2 hours at characteristic temperatures. Cast 2, austenitized at 900°C.

Plasma nitriding: Treatment J

Figure 7 compares the microstructures of the compound layers obtained by nitriding specimens austempered at each of the three characteristic temperatures. Figures 8 and 9 show the X-ray diffraction spectra and hardness profiles for the same samples. They show that prior austempered microstructure had a profound influence on the result obtained by nitriding. Austempering at 380°C yielded the thinnest, least compact compound layer and the shallowest case depth. Both ϵ and γ' phases were present in all the compound layers, but cementite was also present in the compound layer on specimens austempered at 380°C. The presence of cementite was associated with softer, as well as thinner, compound layers. The most compact compound layer was obtained after austempering at 440°C. This fully bainitic microstructure also yielded the deepest case and the highest fraction of ϵ in the compound layer.

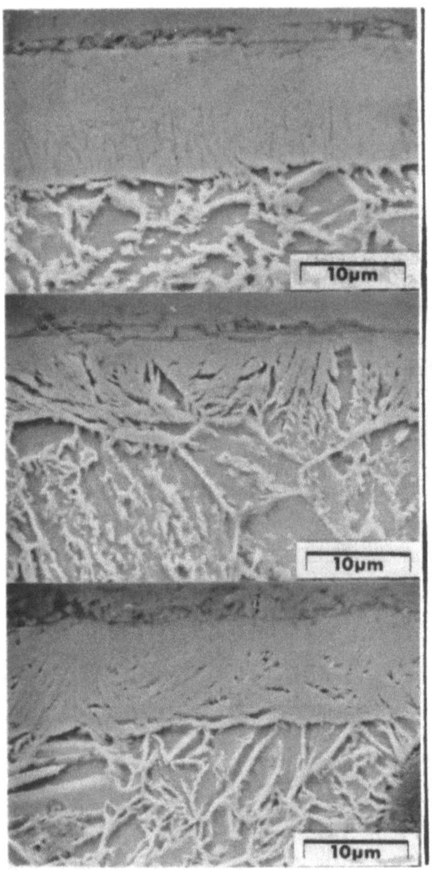

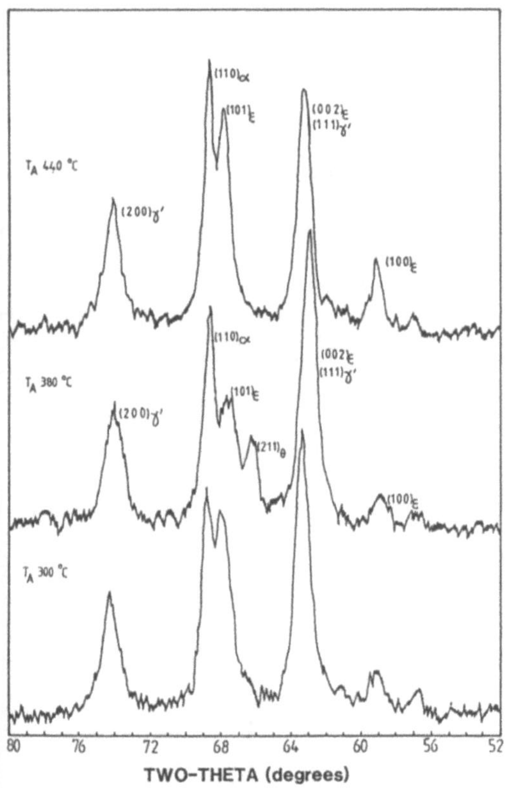

Figure 7. Nitrided compound layers after austempering at (top) 440°C, (middle) 380°C (bottom) 300°C, for 2 hrs. Austenitized at 900°C. Cast 2.

Figure 8. X-ray diffraction spectra correspondening to compound layers in Fig. 7.

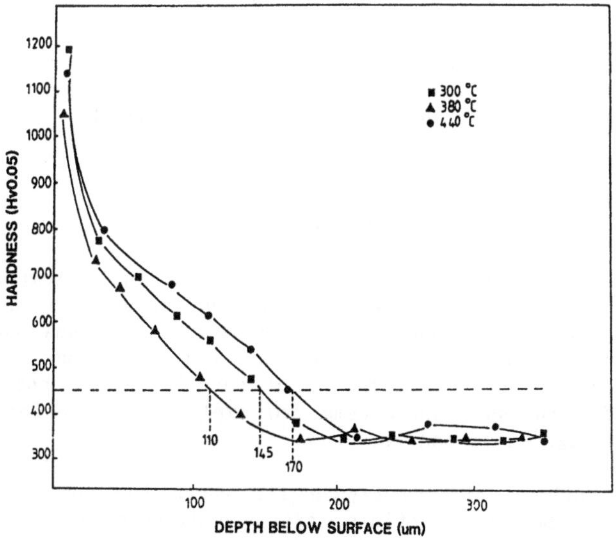

Figure 9. Hardness profiles, treatment J after 2hr. austempering. Cast 2.

The carbon needed to stabilize the ϵ nitrocarbide during nitriding was derived from the substrate, as also occurs in the nitriding of high carbon steels. The effect of varying the total carbon content of the substrate matrix by varying the austenitizing temperature was, therefore, also determined, and some results are shown in Figure 10. The increased supply of carbon increased compound layer thickness upto T_{γ} = 1000°C, but decreased case depth.

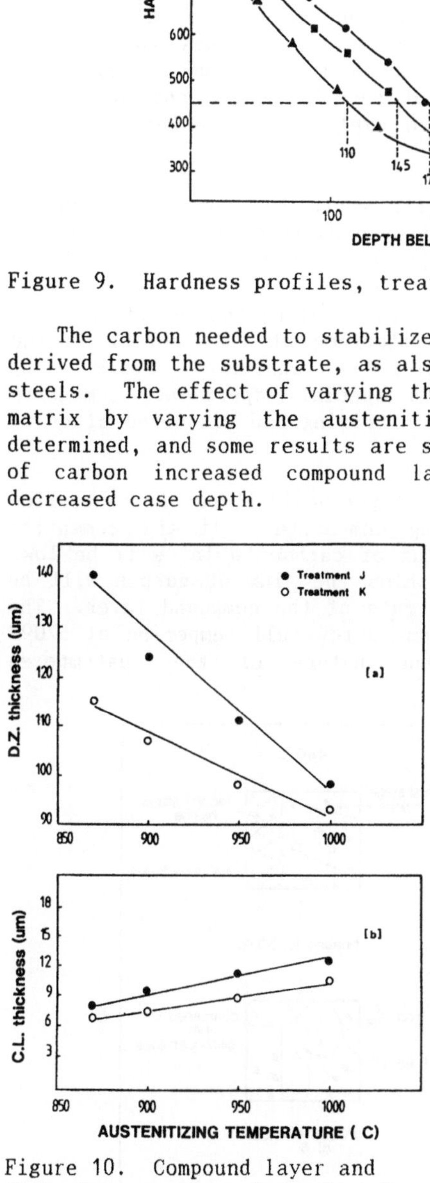

Figure 10. Compound layer and diffusion zone (case) depths after austempering 45 min. at 380°C. Cast 1.

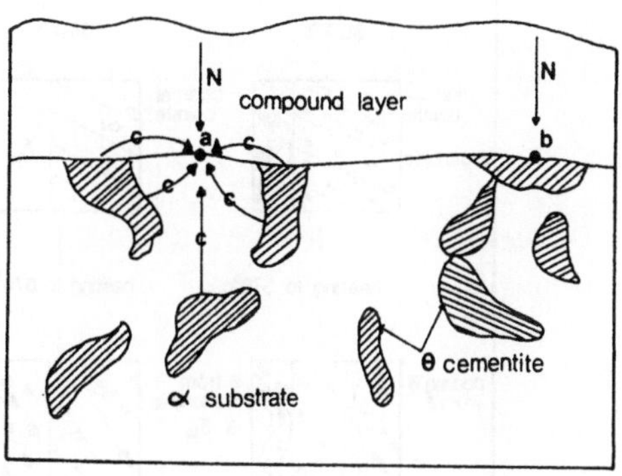

Figure 11. Schematic illustration of growth of compound layer by dissolution of cementite.

Plasma nitrocarburizing: Treatment K

As in nitriding, the thinnest, least compact compound layers were obtained after austempering at 380°C, when they also contained cementite and were softer. The presence of carburizing potential in the plasma reduced both the compound layer thickness and the case depth, Figure 10.

DISCUSSION

Since the austempered microstructures all became fully tempered during heating to the nitriding temperature, the strong influence of austempering conditions on nitriding response was surprising. It was particularly significant that the microstructure developed for optimum strength and ductility yielded the worst combination of case depth and compound layer constitution and hardness. An attempt to explain these effects is made with the help of Figures 11 and 12.

At a point such as 'b' in Figure 11, where the compound layer is contiguous with a cementite particle, nitrogen diffuses from the compound layer into the cementite and transforms it to ϵ nitrocarbide ahead of the general interface. This releases carbon to diffuse through the ferrite of the substrate, or through the compound layer, or along the interface, to a point such as 'a', thus advancing the interface at that point. Large cementite particles might not be completely transformed before being entrained in the compound layer. In a special experiment, cementite present in the compound layer at 4 hours nitriding had dissolved after 16 hours.

At 'a' the growth rate of the compound layer would be partly controlled by the flux of carbon from transforming cementite. If the cementite particles are large and far apart the flux of carbon to 'a' will be low. With a uniform, fine dispersion of particles the flux of carbon will be higher, giving a uniform and high growth rate of the compound layer. The distribution of cementite particles present after full tempering at 570°C must, therefore, have depended on the nature of the austempered microstructure.

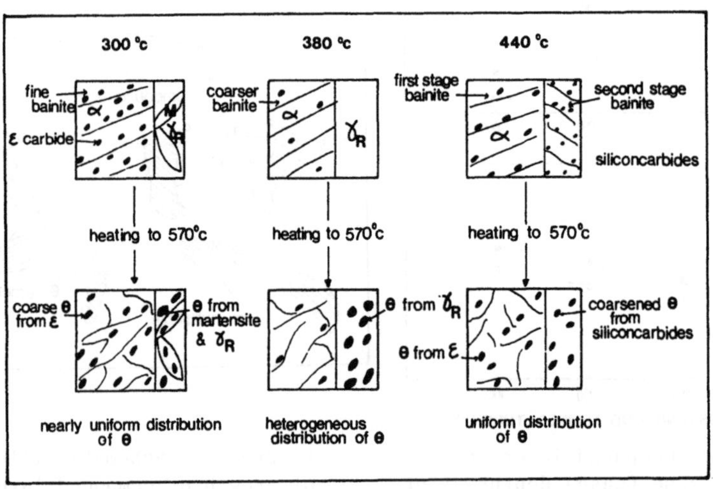

Figure 12. Effect of structure on carbide distribution after tempering.

How this may have arisen is described schematically in Figure 12. Consider first the microstructure resulting from austempering at 380°C, bainite and retained austenite. The DSC results showed that the retained austenite transformed at about 510°C, directly to ferrite and cementite. At that temperature this cementite would be very coarse compared with the other bainitic transformation products. The ϵ carbide present in the first-stage bainite would also transform to coarser cementite. However, since much carbon had been segregated to the retained austenite, the final distribution of cementite would be heterogeneous, and more of it would be of the coarsest variety obtained from decomposition of the retained austenite. After austempering at 300°C, tempering of the martensite and first-stage bainite would yield relatively finely dispersed carbides. The retained austenite, which had been partitioned by the martensite transformation into small, discrete volumes might yield coarse cementite particles on tempering, but the distribution of particle should be expected to be finer and less heterogeneous after tempering than for specimens austempered at 380°C. Finally, the material austempered at 440°C was fully bainitic before tempering, and transformation of ϵ and silicon carbides to cementite within the two bainites would be expected to give the most uniformly distributed cementite particles in the tempered structure.

The heterogeneous distribution of cementite obtained by tempering material austempered at 380°C, with groups of large particles being generated from the retained austenite volumes, is, therefore, thought to be responsible for the low rate of growth of the compound layer, its lower content of ϵ nitrocarbide, its non-uniform penetration into the substrate, and the entrainment of cementite. The shallower case depth arises due to the balance that is needed betwen the rate at which nitrogen can replace carbon in cementite ahead of the interface, and the rate at which carbon so released can diffuse to the interface.

During nitrocarburizing, the compound layer would form initially from carbon and nitrogen in the plasma. This would reduce the driving force for carbon to diffuse in the substrate from cementite particles to the compound layer. After an initial rapid start, the rate of growth of the compound layer would be slower than during nitriding. Also, since less carbon would be drawn from cementite ahead of the interface, its conversion to nitrocarbide would be slower, leading to a lower case depth, as shown in Figure 10.

CONCLUSIONS

The microstructure of bainite and retained austenite that leads to good combinations of strength and ductility in austempered S.G. cast iron also leads to the least satisfactory response to plasma nitriding. Isothermal austempering at a higher temperature than normal to complete both stages of the bainite reaction led to a harder, thicker compound layer with a higher ϵ content and no cementite, and to a greater case depth. This is probably due to a more uniform distribution of carbides in the tempered microstructure, which might also be expected to confer better ductility in the core. There seems to be no reason to nitrocarburize such irons, instead of simply nitriding them.

BCIRA and Blandburgh-Nemo are thanked for supplying material and plasma treatments, respectively.

Techniques for Improved Surface Integrity of Aerospace Aluminum Alloys

W. Wallace, M. D. Raizenne

National Aeronautical Establishment
Ottawa Ontario Canada

and A. Ankara

Middle East Technical University
Ankara Turkey

ABSTRACT

The optimum performance of aerospace components and structures usually depends on the achievement of high strength, high fatigue resistance and resistance to many forms of surface degradation including wear, fretting, corrosion and stress corrosion cracking. In the case of high strength aluminum alloys, trade-offs are often made between strength and corrosion resistance, such as those brought about by standard heat treatments T651 or T7351. The Retrogression and Reaging Process, first described by Cina et al. has shown that near surface properties can be modified by novel heat treatment procedures to obtain strength levels equivalent to the T651 temper together with stress corrosion resistance equivalent to that of the T7351 temper. This success has stimulated other work to investigate derivative heat treatments that might provide similar benefits in thick section parts. This paper reviews these new approaches to heat treatment and provides examples of the effects of these treatments on the properties of thick section parts with major dimensions up to 76 mm. The effects of processing on the microstructure conductivity, resistivity, yield strength, hardness, fatigue, corrosion fatigue and stress corrosion cracking of typical 7000 series alloys are described.

INTRODUCTION

During 1981-85 one of the present authors was involved in an AGARD-NATO study of corrosion related problems in aircraft structures. A handbook of over fifty case histories of corrosion related failures was published, that included information on causes, mechanisms and corrective actions[1]. The case histories were submitted by national aerospace and defence laboratories of the NATO nations and by far the most commonly reported problem was stress corrosion cracking, closely followed by pitting, intergranular corrosion, exfoliation, and corrosion fatigue and most of these failures involved aluminum alloys. Notwithstanding the fact that considerable progress has been made in the development of aluminum alloys and tempers providing high strength, toughness and stress corrosion resistance, the 200 case histories collected showed that there is still a large inventory of equipment in service employing stress and exfoliation corrosion prone aluminum alloys. It is not altogether surprising that there has been renewed interest in the development of heat treatments to produce corrosion resistant tempers, and particularly in treatments that can be applied retroactively to fully manufactured and service exposed parts.

RETROGRESSION AND REAGING (RRA)

The RRA process was first described by Cina et al.[2,3] as a process that could be applied to

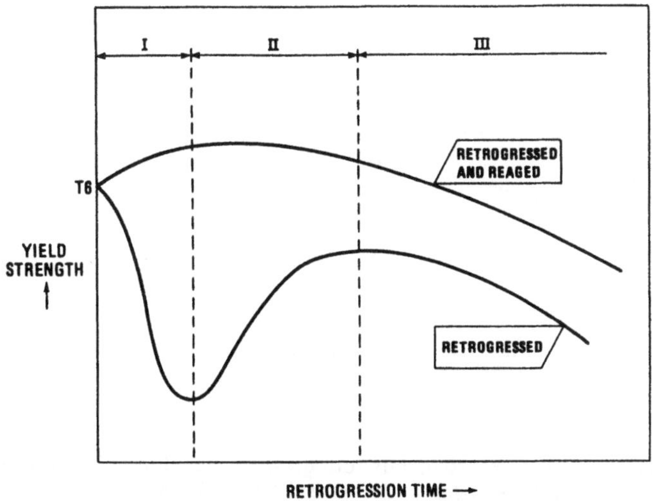

Figure 1. Schematic of the retrogression and reaging process.

7075 material in a T6 condition to achieve stress corrosion resistance equivalent to the T73 temper without loss of strength. Retrogression involves short time exposure to high temperature, typically 1 to 100 seconds at temperatures of 200-280°C. During such treatment, hardness or yield strength first fall to a minimum before increasing again to a secondary peak, while continued treatment causes a further decrease in strength as illustrated in Fig. 1. Thus there are three distinct stages to the retrogression process.

During the first few seconds of retrogression, electrical resistivity increases, while conductivity decreases towards a minimum beyond which it increases with time towards values typical of the T73 temper (2,4). Figure 2 shows the change of electrical conductivity for 7475-T6 aluminum during retrogression at 220°C for times up to ten minutes, together with corresponding data after reaging. The important observation by Cina was that strength and hardness could be restored by reaging using the original T6 conditions, that is 16-24 hours at 120°C, and this additional heat treatment would continue to increase conductivity and stress corrosion resistance towards T73 values. According to Cina the optimum processing involved reaging from the initial minimum, that is the end of Stage I in the retrogression hardness curve, and since this involves very short times, the process appeared most suitable for very thin section parts, typically 1-6 mm, or for surface processing of thicker section parts.

Wallace et al (5-7) noticed that reaging often restores strengths to levels higher than the starting T6 or T651 values, especially when carried out from the end of Stage I. They investigated the use of longer retrogression times and lower temperatures so as to establish the longest times that would still allow strength to be restored on reaging to the T6 or T651 starting values. Typically, reaging was carried out at the end of Stage II of the retrogression curve or longer, and retrogression temperatures were usually in the range 180-240°. In this way the T6RRA condition would provide the highest conductivity, stress corrosion resistance and exfoliation resistance, closely meeting T73 values, while still achieving original or minimum T6 strength. Retrogression times employed were typically 6 minutes to 3 hours at 220° to 180° respectively in silicone oil baths, and these longer times allowed improved properties to be achieved through thicker section parts up to 25 mm. Kaneko (8) and Brown (9) described the use of molten salt or Wood's metal baths to achieve similar results, while induction heating has been used by Cina et al. (1,2) and at the Middle Eastern Technical University, Turkey, to process thick section products up to 75 mm diameter or thickness.

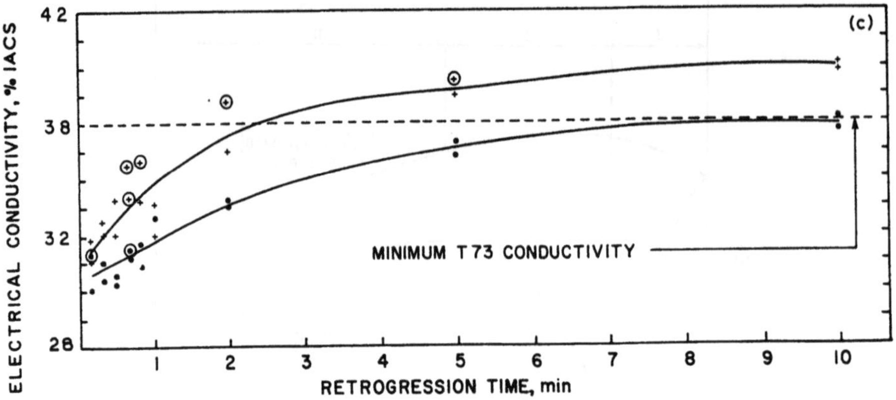

Figure 2. Electrical conductivity versus retrogression time for 7475 aluminum after retrogression at 220°C and after RRA (5).

Stress corrosion and exfoliation corrosion resistance

The beneficial effects of RRA on resistance to exfoliation and stress corrosion cracking have been confirmed in numerous studies using the ASTM Standard G34-79, EXCO test, and ASTM G38-73 C-ring test respectively (2,3,10,11). Figure 3 shows the effect of retrogression time at 220° on the resistance to stress corrosion crack growth of 7075-T6 RRA, and compares crack growth rates with those for 7075-T651 and 7075-T7351 materials. These results were obtained using bolt loaded double cantilever beam specimens, tested by intermittent wetting with 3.5% acqueous NaCl solution following the method described by Hyatt (12). For 6 minutes retrogression the crack growth rates are below about 6×10^{-10} m/s, while for12 minutes retrogression the growth rates are below 10^{-10} m/s and very similar to 7075-T7351. Similar results are shown in Figure 4 for 7475

Table 1 **Heat treatments applied to double-cantilever-beam (DCB) specimens for stress-corrosion tests, electrical conductivities expected from previous results, actual conductivities, and 0·2% yield strengths obtained from specimens machined from broken DCB specimens**

Specimen	Heat-treatment conditions	Expected conductivity, % IACS	Actual conductivity, %IACS	0·2% yield strength, MN m^{-2}
2	As-received (T7351) condition	. . .	42·2	405·8
3	'T6-type' condition (Ref. 1)*	33–34	33·6	490·5
4	'T6' + 3 h at 180°C + 24 h at 120°C	39–40	40·9	498·6
5	'T6' + 30 min at 200°C + 24 h at 120°C	39–40	40·2	510·2
6	'T6' + 6 min at 220°C + 24 h at 120°C	39–40	40·2	519·5
7	'T6' + 2·5 min at 240°C + 24 h at 120°C	40–41	40·6	488·2
8†	'T6' + 1 min at 260°C + 24 h at 120°C	38–40	43·7	403·5
9†	'T6' + 10 s at 280°C + 24 h at 120°C	38–39	44·7	302·6

*30 min at 447°C + 24 h at 120°C.
†'Over-retrogressed'.

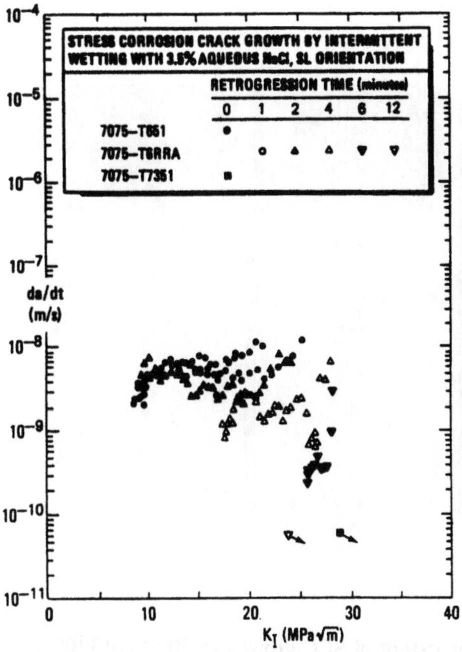

Figure 3. The effects of retrogression time at 220°C on the stress corrosion crack growth rates of 7075-T651 aluminum (13).

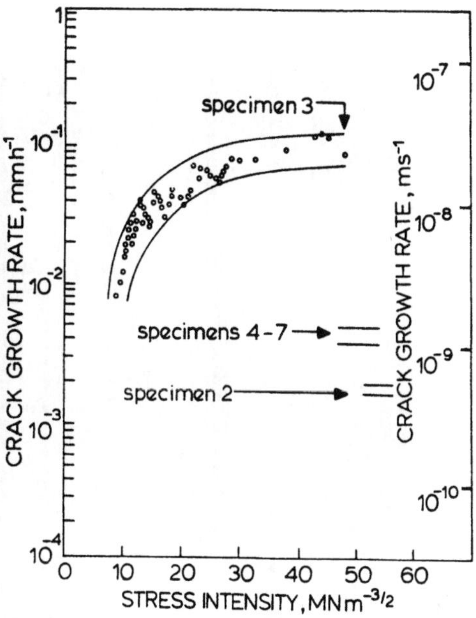

Figure 4. Stress corrosion crack growth rates for the 7475 aluminum tempers listed in Table 1 (7)).

alloy in the T6, T6RRA and T7351 conditions. The specimens examined in Fig. 4 were heat treated according to the conditions shown in Table 1, where values of electrical conductivity and yield strength are also reported. Specimens 8 and 9 gave stress corrosion resistance at least equivalent to T7351 material, but were overaged. Specimens 4 to 7 gave yield strengths very close to, if not higher than T6 material, and yet produced plateau crack growth rates fifty to sixty times slower. For one specimen (#6), this improved stress corrosion resistance was obtained together with a yield strength 6% higher than the same material in the T6 condition. The extent of stress corrosion crack growth in these specimens after about 700 hours of testing is shown in Figure 5.

Fatigue and corrosion fatigue

Notwithstanding the beneficial effects of RRA on stress corrosion resistance and exfoliation resistance, its effects on fatigue and corrosion fatigue have not been widely studied. The results reported here were obtained from a collaborative study between the NAE-NRC and Ecole Polytechnique - Montreal (13,14).

Fatigue crack growth rates were determined in dry argon and flowing 3.5% aqueous NaCl solution using 12.7 mm thick compact tension (CT) specimens conforming to ASTM Standard E647-83. Specimens were loaded in the short transverse direction to produce crack growth in the longitudinal direction, and crack lengths were determined automatically using a compliance technique (15). Measurements of crack closure loads were made periodically and crack growth rates were plotted against $\Delta K_{apparent}$ and against $\Delta K_{effective} = K_{max} - K_{closure}$ (13-15).

Figure 6 shows the effects of a corrosive environment and compares fatigue crack growth rates for 7075-T651, T6RRA, and T7351 in argon and in 3.5% NaCl at R=0.1, and at

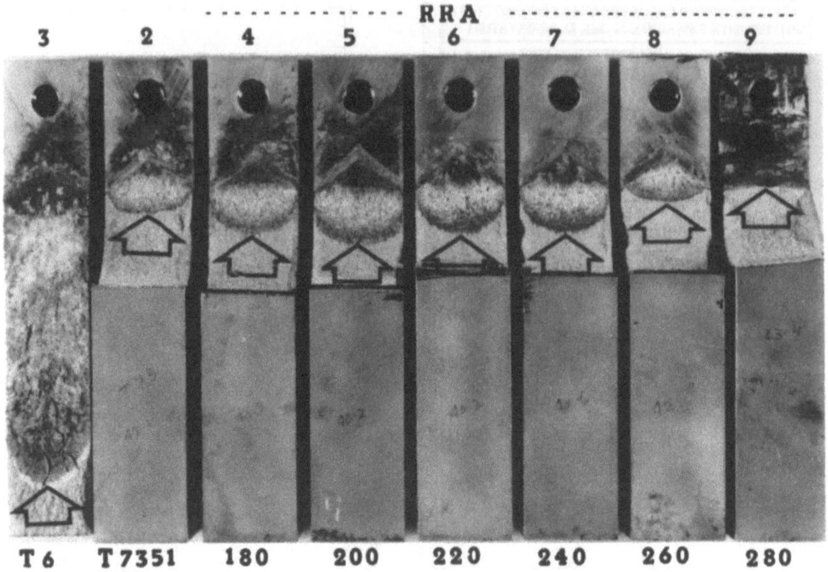

Figure 5. DCB specimens illustrating the extent of SCC growth in 700 hours for the T6RRA specimens of Table 1.

a frequency of 20Hz. Crack growth rates for the three tempers are similar under similar testing conditions, and a strong effect of environment is seen which increases as stress intensity range ΔK decreases. The effects of crack closure are almost negligible for tests in argon, however in 3.5% NaCl the effects are quite large, and the apparent threshold stress intensities for the non-corrected data are in the range 2.5 to 3.5 MPa√m, compared to less than 2 MPa√m for the corrected data. At the slowest growth rates recorded in 3.5% NaCl, (~10^{-9}m/cycle), 7075-T651 appears to be slightly more resistant to crack growth than T6RRA or T7351 material at this frequency (20Hz).

Figure 7 shows the effects of frequency and compares the fatigue crack growth rates for the three alloy tempers at 2Hz and 20Hz in 3.5% NaCl solution. The data all falls within a reasonably narrow scatter band, with the 2Hz data falling towards the high side of the scatter band at high values of ΔK, and towards low side of the scatter band at low values of ΔK. This effect is not obviously due to thicker corrosion products and increased closure because the effect is still seen after correcting for closure.

Figure 8 shows the effects of load ratio, and compares data obtained at R=0.1 with data obtained at R=0.5, all in 3.5% NaCl solution at 2Hz. When the data are plotted against ΔK, two distinct scatter bands are apparent with data for R=0.1 indicating slower growth rates than for R=0.5 at comparable values of ΔK. However when the data are corrected for crack closure the two data sets tend to merge to form a single broad scatter band. Thus the apparent differences are due to the wedging effects of the thick corrosion products and increased crack closure that occurs in corrosive solutions at R=0.1. Differences in fatigue crack growth rates between the three alloy tempers are negligible, at intermediate to high values of $\Delta K_{effective}$ at this frequency (2 Hz), although there are indications that the T6RRA and T7351 material are more resistant to crack growth at the lowest values of ΔK employed

MECHANISMS

It is generally agreed that the initial loss of strength during retrogression (Stage I) is due to the dissolution of strengthening agents. Park and Ardell argued that since η'-$MgZn_2$ is the

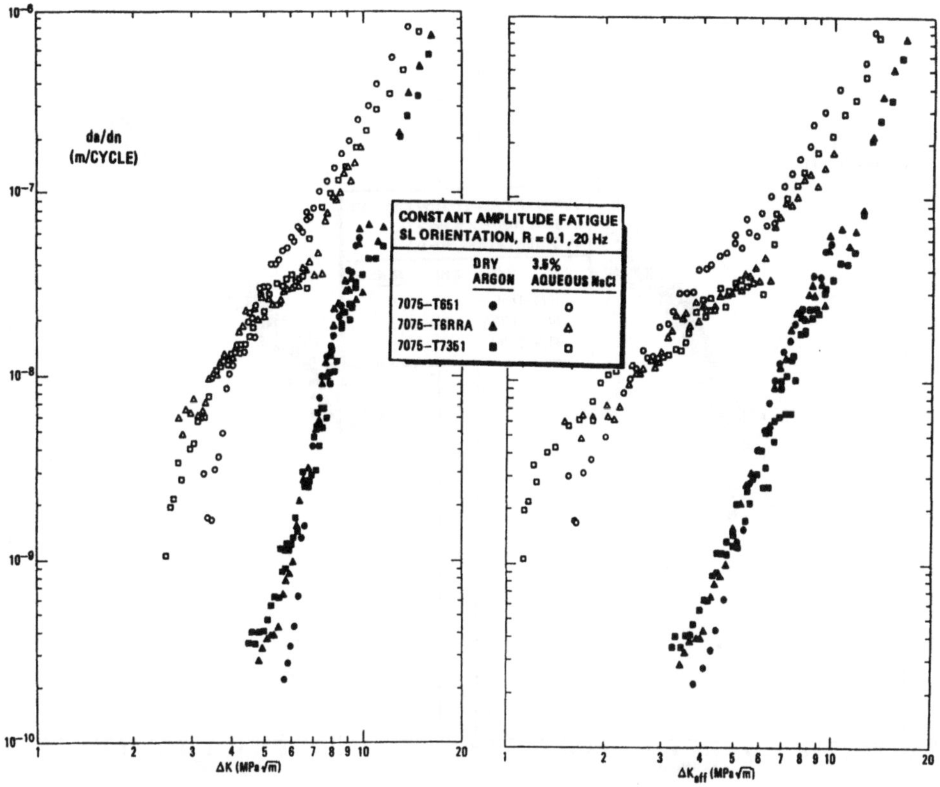

Figure 6. The effects of environment on fatigue crack growth rates in 7075-T651, and T7351 aluminum (14).

main strengthening phase in 7075-T6, it is likely to be the finer particles of this phase that dissolve (16,17). Danh et al. (18) found that the volume fraction and size of the η ' and η precipitates remained largely unchanged during Stage I, but increased rapidly during stages II and III, see Figure 9. They suggested therefore that G.P. zones present in the T6 material, and which could not be imaged directly in their study, were probably dissolving during stage I, and that the recovery of strength during stage II was due to the growth of the remaining η'.

Three explanations have been proposed for the improved SCC resistance of RRA material. Tallanker and Cina (19) have suggested that dislocations, produced during the quenching and stress relieving operations of the T651 temper, are responsible for the poor stress corrosion resistance in this material. Accordingly, the stress fields associated with the dislocations act in concert with the applied stress to promote cracking, while the dislocations also serve as sites for the concentration of hydrogen and hence promote hydrogen embrittlement. The improved properties of RRA and T73 materials were believed to be due to the lowering of dislocation densities in these tempers. A criticism of this explanation is that there is very little difference in SCC behaviour between T6 and T651 material, and between T73 and T7351 material, where dislocation densities would be expected to differ because of the cold stretching operations. As indicated in Figure 10, dislocation densities are not always high in T6, T6RRA or T73 materials which have not been cold stretched.

It has also been widely recognized (10,11,16,17,19,20) that RRA and T73 treatments lead to significant increases in the size and spacing of grain boundary precipitates

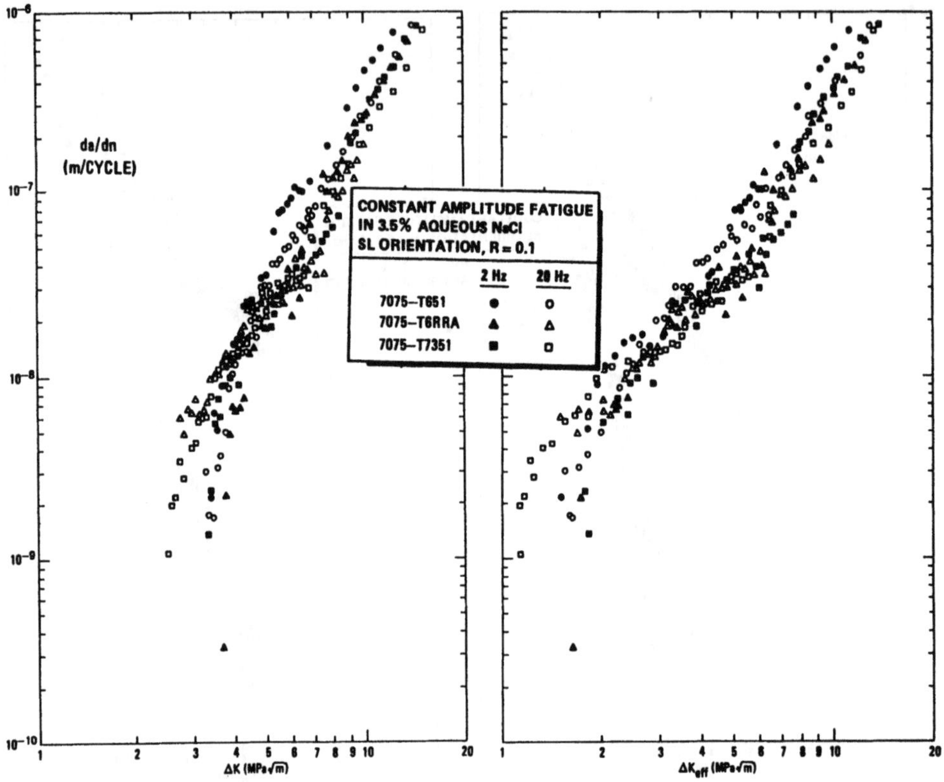

Figure 7. The effects of frequency on fatigue crack growth rates in 7075-T651, T6RRA and T7351 aluminum (14).

compared to T6, as indicated in Fig. 10, and following an argument first proposed by Christodoulou and Flower (21), Rajan et al. (20) suggested that the larger precipitates contribute to improved stress corrosion resistance by acting as more effective trapping sites for hydrogen. Accordingly, hydrogen produced by hydrolysis at the crack tip diffuses along grain boundaries until it meets an incoherent particle interface where it condenses to form a small molecular gas bubble, thereby lowering the hydrogen concentration in solid solution. The evidence for such a mechanism is presented in several papers (20-22). These arguments have been extended by Thompson et al. (10,11) who have shown that less hydrogen is liberated during the corrosion of T6RRA material than with T6 material, and they also suggested that the larger grain boundary particles may reduce the concentration of magnesium in solution at grain boundaries while the hydrogen bubbles may serve to blunt a growing stress corrosion crack.

CONCLUSIONS

1. Retrogression and reaging treatments applied using silicone oil baths, molten metal or molten salt baths, or induction heating can be used to produce high strength and stress corrosion resistance in 7000 series aluminum alloys.

2. Treatments applied to fully heat treated 7475-T6 or T651 material, and typically involving retrogression for 3 hours at 180°C followed by reaging for 16-24 hours at 120°C, produce strength levels higher than the starting material together with resistance to stress corrosion cracking very similar to T7351 material.

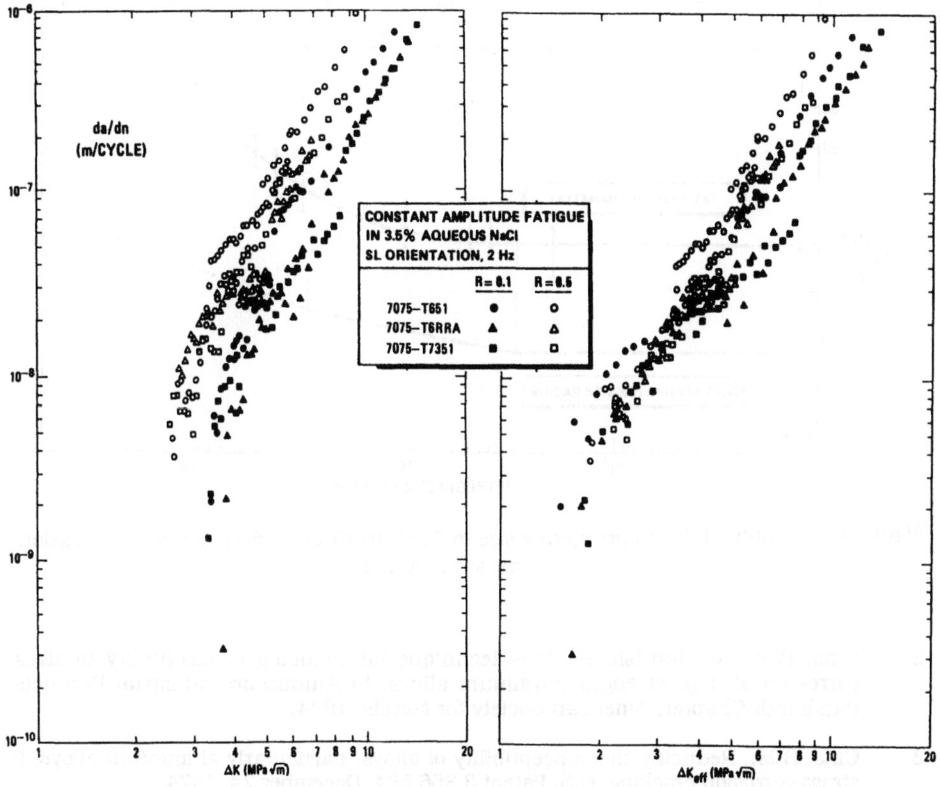

Figure 8. The effect of load ratio on fatigue crack growth rates in 7075-T651, T6RRA and T7351 aluminum (14).

3. These treatments have very little effect on fatigue or corrosion fatigue crack growth rates, however the T6RRA and T73 tempers appear to be slightly more resistant to fatigue crack growth in the threshold region at low frequencies in corrosive solutions.

ACKNOWLEDGEMENTS

The work described here has been carried out largely by graduate students at Carleton University, Ottawa, Ecole Polytechnique, Montreal, and the Middle Eastern Technical University, Ankara, Turkey. Their diligence and perseverance are gratefully acknowledged. The work was carried out under NAE-NRC Project 07317, Aerospace Metallic Materials, and thanks are due to the National Research Council Canada for permission to publish. One of the authors (A.A.) was assisted by the Advisory Group for Aerospace Research and Development (AGARD-NATO) under the Support To Nations Program, and this assistance is gratefully acknowledged.

REFERENCES

1. Wallace, W., Hoeppner, D.W., and Kandachar, P.V., AGARD Corrosion Handbook Volume 1, Aircraft Corrosion: Causes and Case Histories, AGARD-AG-278 Volume 1, July 1985.

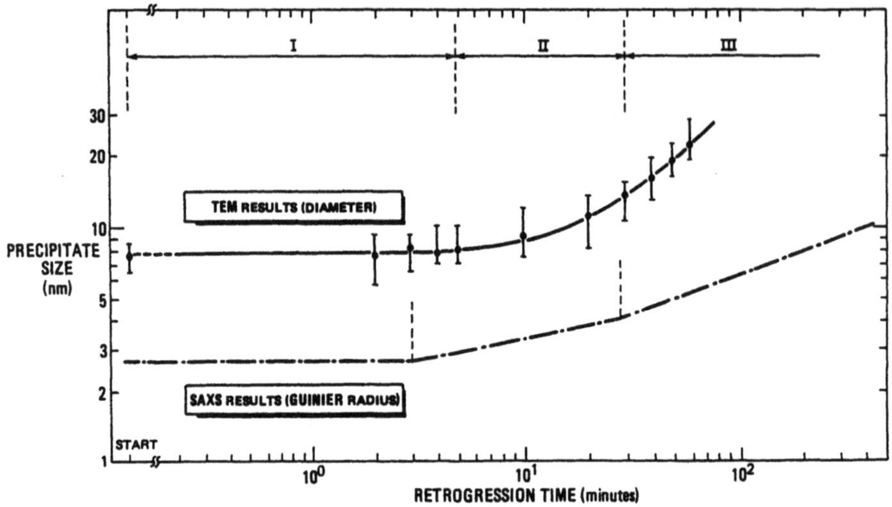

Figure 9. $MgZn_2$ (η' + η) precipitate size in 7075-T6RRA as a function of retrogression time at 220°C (18).

2. Cina, B.M. and Ranish, B., New technique for reducing susceptibility to stress corrosion of high strength aluminum alloys, in Aluminum Industrial Products, Pittsburgh Chapter, American Society for Metals, 1974.

3. Cina, B.M., Reducing the susceptibility of alloys, particularly aluminium alloys, to stress corrosion cracking, U.S. Patent 3,856,584, December 24, 1974.

4. Ankara, A., Durlu, N., and Wallace, W., Kinetics of retrogression and reaging of 7075 aluminum alloy, Turkish Journal of Engineering and Environment, 1988, 12, 3, 302-309.

5. Islam, M.U. and Wallace, W., Retrogression and reaging response of 7475 aluminium alloy, Metals Technology, 1983, 10, 386-392.

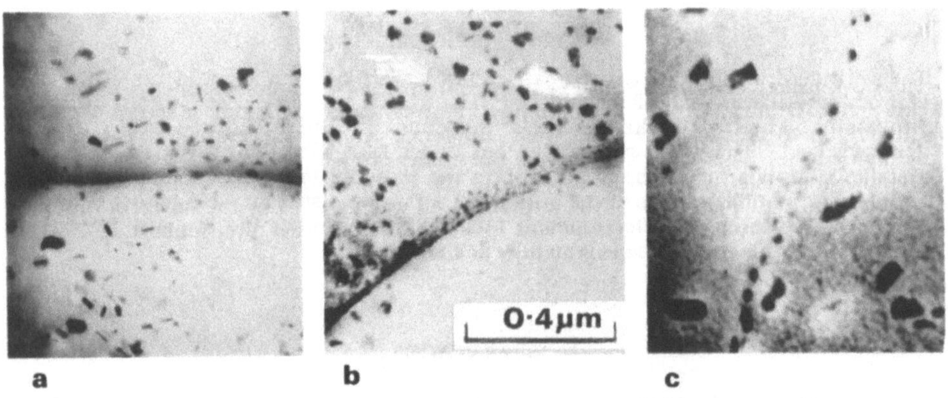

Figure 10. TEM observations on a) 7075-T6, b) 7075-T6RRA and c) 7075-T73.

6. Wallace, W., Beddoes, J.C. and deMalherbe, M.C., A new approach to the problem of stress corrosion cracking in 7075-T6 aluminum, Canadian Aeronautics and Space Journal, 1981, 27 (3), 222-232.

7. Islam, M.U. and Wallace, W., Stress corrosion crack growth behaviour of 7475-T6 retrogressed and reaged aluminium alloy, Metals Technology, 1984, 11, 320-322.

8. Kaneko, R.S., RRA: Solution for Stress Corrosion Problems with T6 Temper Aluminum, Metal Progress, April 1980, 117, 41-43.

9. Brown, M.H., Producing combined high strength and high corrosion resistance in Al-Zn-Mg-Cu Alloys, U.S. Patent 4,832,758, May 23, 1989.

10. Thompson, J.J., Tankins, E.S. and Agarwala, V.S., A heat treatment for reducing corrosion and stress corrosion cracking susceptibilities in 7XXX Al Alloys, Paper No. 204, Conf. Corrosion 86, Nat. Assoc. of Corr. Eng., Houston, TX, Mar. 17-21, 1986.

11. Thompson, J.J., Tankins, E.S., and Agarwala, V.S., Stress corrosion cracking resistant heat treatment for 7000 series aluminum alloys - RRA, Naval Air Development Center, Warminster, PA, 1987, unpublished work.

12. Hyatt, M.V., Use of precracked specimens in stress corrosion testing of high strength aluminum alloys, Corrosion, 1970, 26, 11, 487-503.

13. Baldantoni, A., Corrosion sous tension et fatigue-corrosion des materiaux structuraux aeronautiques, thesis submitted in partial fulfillment of the requirement of Master of Applied Science, Department of Metallurgical Engineering, Ecole Polytechnique, University of Montreal, June 1985.

14. Baldantoni, A., Wallace, W., Raizenne, M.D. and Dickson, J.Y., The NRC contribution to the FACT programme, in AGARD Report No. 713, The Fatigue in Aircraft Corrosion Testing (FACT) Programme, February 1989.

15. Raizenne, M.D., KRAKS - an MTS Basic 770 computer program for automated fatigue crack growth rate testing, NAE-NRC Report LTR-ST-1538, February 1986.

16. Park, J.K. and Ardell, A.J., Effect of retrogression and reaging treatments on the Microstructure of Al 7075-T651, Met. Trans., 1984, 15A, 1531-1543.

17. Park, J.K. and Ardell, A.J., Precipitate microstructure of peak-aged 7075 Al, Scripta Met., 1988, 22, 1115-1119.

18. Danh, N.G., Rajan, K., and Wallace, W., A TEM study of microstructural changes during retrogression and reaging in 7075 aluminum, Met. Trans., 1983, 14A, 1843-1849.

19. Talianker, M. and Cina, B., Retrogression and reaging and the role of dislocations in the stress corrosion of 7000 - type aluminum alloys, Met. Trans., 1989, 20A, 2087-2092.

20. Rajan, K., Wallace, W. and Beddoes, J.C., Microstructural study of a high strength stress corrosion resistant 7075 aluminum alloy, J. Mat. Sci, 1982, 17, 2817-2824.

21. Christodoulou, L. and Flower, H.M., Hydrogen embrittlement and trapping in Al-6%Zn-3%Mg, Acta Met., 1980, 28, 481-487.

22. Scamans, G.M., Hydrogen bubbles in embrittled Al-Zn-Mg alloys, J. Mat. Sci., 1978, 13, 27-36.

Duplex Hardening: An Advanced Surface Engineering Technique for Helicopter Transmissions

D. P. Davies

Westland Helicopters Ltd
Yeovil
UK

ABSTRACT

In order to reduce weight, improve reliability and increase safety, in addition to providing enhanced oil loss tolerance and performance improvement, significant changes in the design of future helicopter transmission are envisaged. Such designs are likely to place unacceptable demands on conventional gear steels, thereby restricting further opportunities for advancement in transmission development. Improvements can be made by changing to one of the newly developed temper resistant carburising steels, but the indications are that in order to achieve a fundamental advance in performance it will be necessary to change both the material and the heat treatment process. This paper summarises the properties attainable with a novel surface engineering technique, known as 'Duplex Hardening' and discusses its potential and suitability in the processing of high performance gears.

INTRODUCTION

Future Design Requirements

High performance gears in the modern helicopter are required to operate at ever increasing torque and speed. The net consequence of this is that a significant increase in temperature at the contact surface can be expected which can only be partially alleviated by improved and heavier lubrication cooling systems. It is likely that in future designs the main gearbox input speeds will be as high as 25,000 rpm and the normal operating temperature (currently 90°C) will be up to 150°C. Furthermore, it is anticipated that the next generation of military helicopter gearbox will have to cater for extra oil loss tolerance (i.e. gear steels must have the capability to withstand once only excursions to 300°C), as a result of battle damage or due to the failure of a critical component.

Future Material Requirements

It will be appreciated that steels will need to have improved fatigue properties at elevated temperatures to cope with the anticipated rise in (a) gearbox operating temperature and (b) torque. Furthermore, it should also be noted that the risk of scuffing, which is associated with the generation of very localised high temperatures (measured at approximately 200°C at gear tooth contact) is above the tempering temperature of conventional carburising steels even during normal operation. Consequently, the limited temperature capability of conventional carburising steels may influence normal operation and failure modes to a greater extent than has been realised and hence steels will need to have greater temper resistance to cope with the anticipated rise in temperature at the tooth contact.

GEAR MATERIALS

Conventional Carburised Steels

By raising the tempering temperature of conventional carburising steels to achieve the desired oil loss tolerance, a lower surface hardness, impaired wear resistance, a reduction in favourable residual compressive stress and a corresponding drop in fatigue performance would result (1). In addition, the surface contact phenomena which depends on material hardness would also be adversely affected by increasing the tempering temperature.

Temper Resistant Carburised Steels

The limited elevated temperature performance of conventional carburising steels, led to the development of a number of temper resistant carburising steels. The aim of these developments has been to retain as many as possible of the advantages currently exhibited by conventional steels, whilst making sufficient improvements to meet the highest temperature requirement (i.e. up to 300°C).

Evaluation of such steels by various investigators (2) have shown that while some are capable of meeting the oil loss tolerance requirement, improvements in performance are only minimal when compared to conventional carburising steels. The desire to achieve a substantial increase in performance, in addition to meeting the most demanding of temperature requirements prompted Westland Helicopters to evaluate a novel surface engineering process known as 'Duplex Hardening'.

DUPLEX HARDENING

Duplex hardening is a surface engineering technique that involves nitriding on top of a carburised surface, hence the stability of the carburised case is of paramount importance. The development of the temper resistant carburising steels has meant that the stability of the carburised case can be maintained beyond 500°C via secondary hardening, while the presence of nitride forming alloying elements such as Mo, V, Cr and W has ensured that these steels should also respond to nitriding treatments. Work carried out at Westland Helicopters (3) on Vasco X-2M has confirmed this assumption (Figure 1). However before this process could be used for transmission components a number of its basic properties needed to be characterised and evaluated.

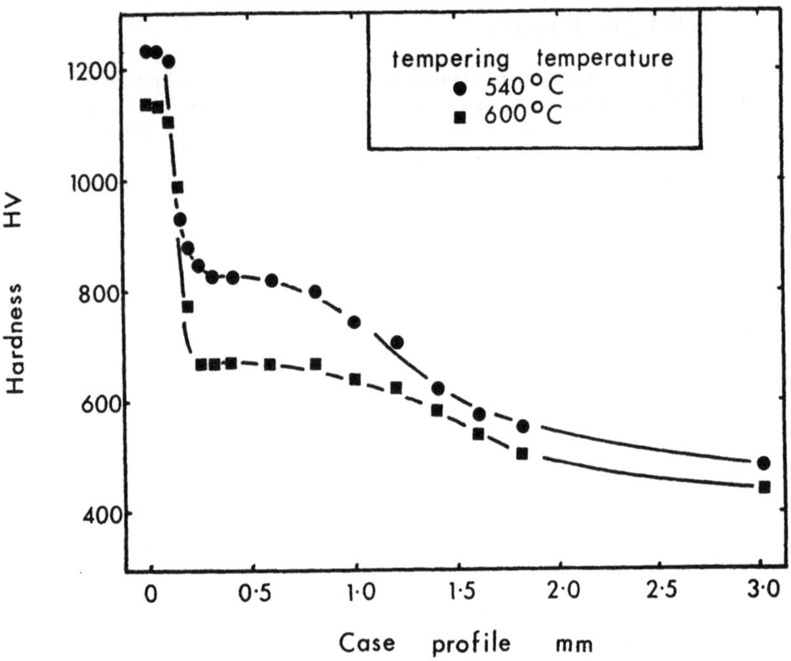

FIGURE 1 Duplex Hardening Case Profile of Vasco X-2M (Gas Nitrided)

NITRIDING PROCESS

Gas Nitriding

Gas nitriding is usually carried out between 480-550°C (Westland standard practice is 520°C). Consequently, trials were carried out to determine the effect that tempering temperature had on case hardness/ nitriding response of carburised Vasco X-2M. From Figure 2 it can be seen that by raising the tempering temperature after carburising, both the carburised and nitrided case hardnesses fall. This phenomenon is well documented and is due to the fact that by raising the tempering temperature more stable secondary hardening precipitates are formed which essentially remove the available alloying elements from the matrix. Moreover, once coarsening of the alloy carbides has occurred the nitriding response will be further retarded due to the alloy carbides being large and difficult to dissolve. Consequently, with carburised Vasco X-2M, tempering beyond 600°C would be likely to produce an unacceptable surface hardness and exacerbate the possibility of 'case crushing'.

Plasma Nitriding

Although the Westland programme was primarily designed to evaluate the performance of duplex hardening, consideration of manufacturing difficulties and especially distortion could not be overlooked, since dimensional changes could not be corrected by grinding, because of the shallowness of the nitrided case. Plasma nitriding was therefore evaluated, because substantial reductions in distortion have been claimed, but also because if offers far greater flexibility, especially with regard to nitriding temperature.

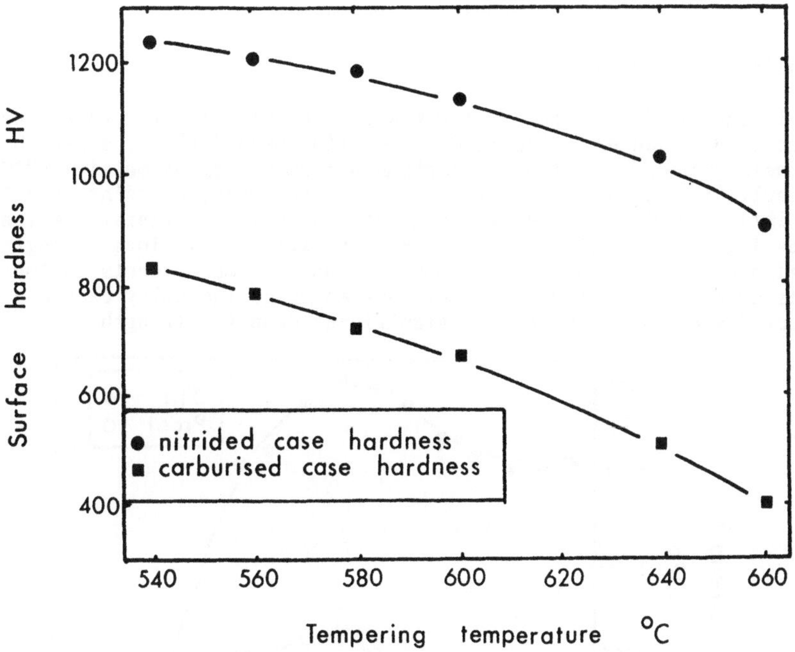

FIGURE 2 Effect of Tempering Temperature on Duplex Hardening Response

Acceptable case profiles were produced by plasma nitriding carburised Vasco X-2M in the range 400-450°C (Figure 3).

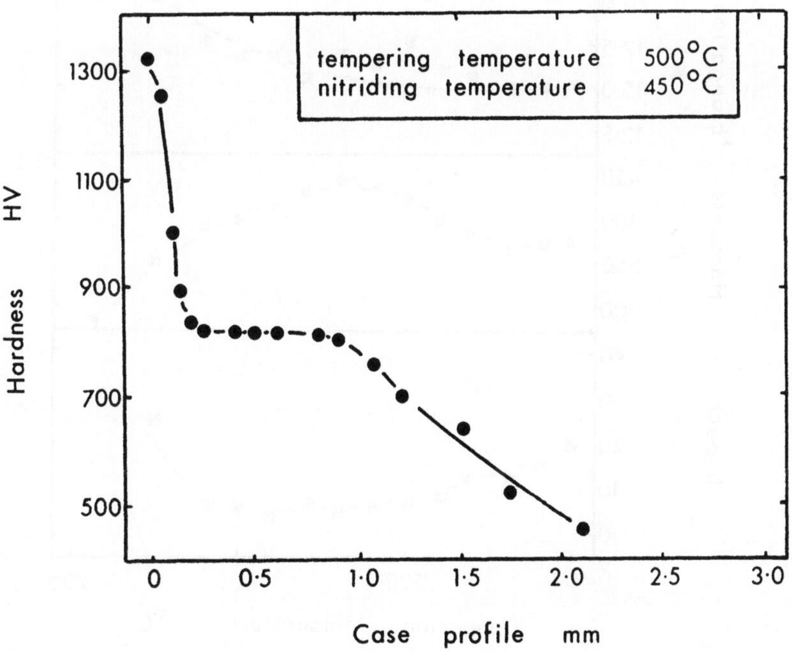

FIGURE 3 Duplex Hardened Case Profile of Vasco X-2M (Plasma Nitrided)

MECHANICAL PROPERTIES

Tensile/Impact Properties

The influence that tempering temperature has on the core properties of Vasco X-2M is shown in Figure 4. Tempering beyond 450°C, strength and hardness gradually increase, reaching a peak at approximately 540°C, as a result of alloy carbide precipitation. An important point to note is the influence that secondary hardening has on impact resistance i.e. tempering at 500-560°C will produce a condition of maximum strength but it also produce a state of minimum toughness. Improvements in impact resistance are only realised once coarsening of the alloy carbides has occurred and this results in a significant drop in strength.

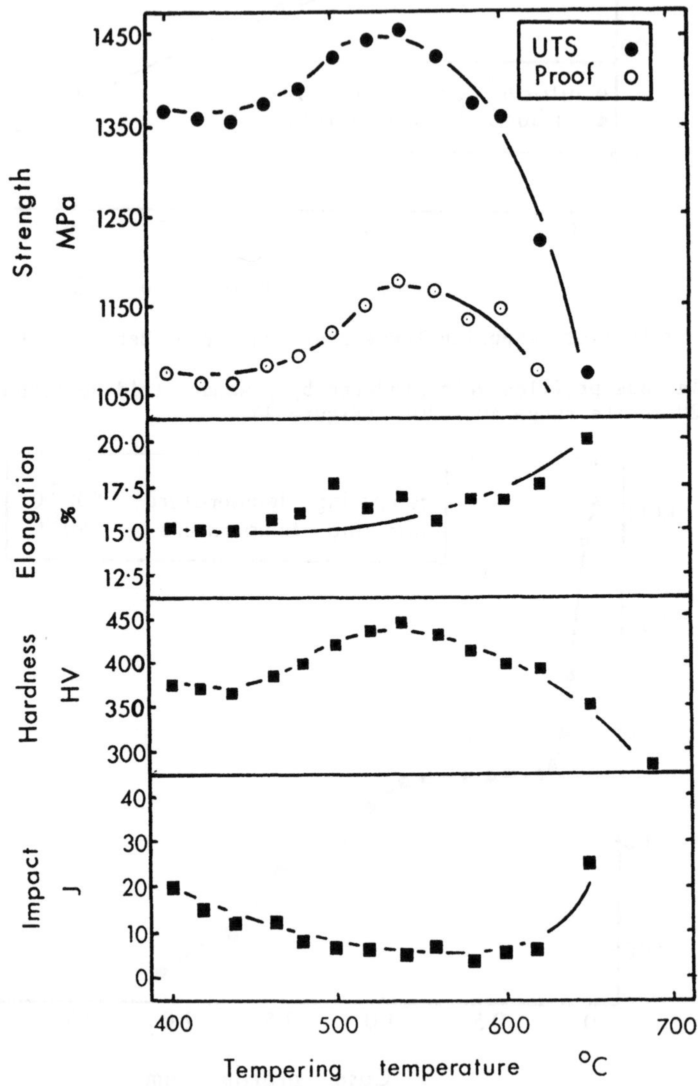

FIGURE 4 Influence of Tempering Temperature on the Mechanical Properties of Vasco X-2M

Fracture Toughness Properties

Traditionally the use of intrinsically tough materials necessary to engender a degree of damage tolerance combined with the application of a safe life design philosophy, has mitigated against any systematic study of fracture properties of gear steels for helicopter transmissions. However, as duplex hardening would require the steel to be tempered at or near the secondary hardening region a reduction in toughness would have to be judged against this current standard.

Fracture toughness measurements were made using compact tensile specimens, in accordance with BS 5447/ASTM E399. Testing was carried out at both room and elevated temperature, to more fully characterise the toughness properties of the steel and more closely represent the proposed gearbox running temperature of 150°C.

It was apparent from the room temperature results (Figure 5) that by increasing the tempering temperature beyond 250°C, reductions in fracture toughness were apparent. The loss in toughness was considered essentially microstructural, i.e. by tempering certain steels above 250°C tempered martensite embrittlement occurs, due to the production of cementite films at martensite plate boundaries which substantially weaken the matrix. Tempered martensite embrittlement is usually avoided by tempering above 350-400°C. Unfortunately, in the case of Vasco X-2M a further fall in toughness was experienced. This phenomenon is well known, and seems to be due to the embrittlement introduced by virtue of very high strengths obtained from the extremely fine alloy carbides, precipitated during secondary hardening. It is interesting to note that substantial improvements in toughness will only occur once coarsening of the alloy carbides has occurred at 650°C.

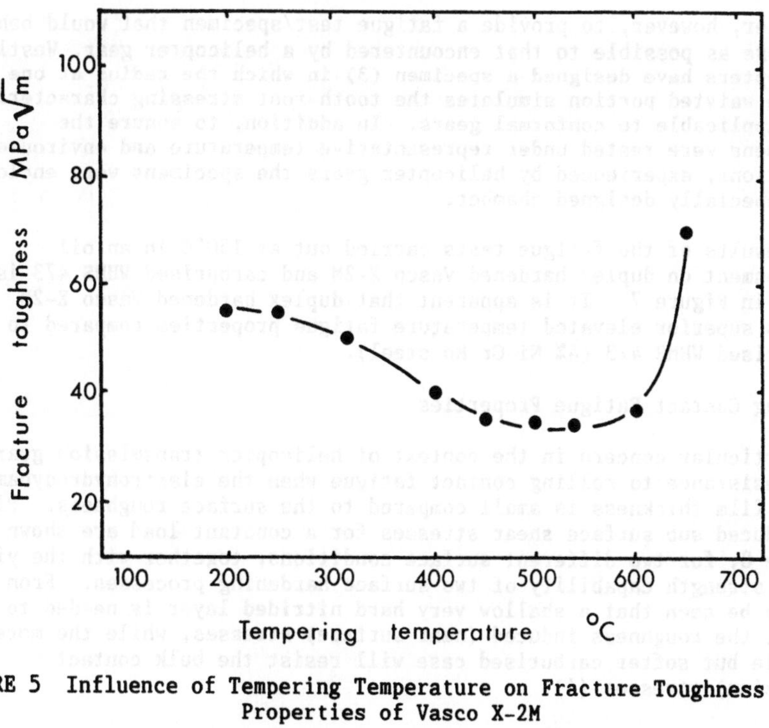

FIGURE 5 Influence of Tempering Temperature on Fracture Toughness Properties of Vasco X-2M

While helicopter transmissions may be required to endure cold soak/start conditions below -30°C, gears spend the majority of their working life at elevated temperature (i.e. gearbox running temperatures). Consequently, fracture toughness measurements were performed at 150°C, in order to determine the influence of temperature on fracture resistance. From Table 1 it can be seen that by raising the test temperature the fracture toughness value of Vasco X-2M tempered at 540°C was raised from 35 MPa$\sqrt{m}$ at room temperature to 55 MPa$\sqrt{m}$ at 150°C. This improvement was characterised by a change in fracture, from cleavage to a mixed mode of cleavage and microvoid coalescence.

TABLE 1

Effect of Test Temperature on the Fracture Toughness Properties of Vasco X-2M

TEST TEMPERATURE °C	FRACTURE TOUGHNESS K_{IC} MPa $\sqrt{m}$
20°C	35
150°C	55

Fatigue Properties

Unnotched rotating bend fatigue tests were performed on duplex hardened and carburised specimens of Vasco X-2M. The results (Figure 6) show that duplex hardening improves the fatigue endurance by approximately 25% over its carburised equivalent.

In order, however, to provide a fatigue test/specimen that would behave as close as possible to that encountered by a helicopter gear, Westland Helicopters have designed a specimen (3) in which the radius at one end of the waisted portion simulates the tooth-root stressing characteristics applicable to conformal gears. In addition, to ensure the specimens were tested under representative temperature and environmental conditions, experienced by helicopter gears the specimens were enclosed in a specially designed chamber.

The results of the fatigue tests carried out at 150°C in an oil environment on duplex hardened Vasco X-2M and carburised WHMS 473 is shown in Figure 7. It is apparent that duplex hardened Vasco X-2M offers superior elevated temperature fatigue properties compared to carburised WHMS 473 (4% Ni-Cr-Mo steel).

Rolling Contact Fatigue Properties

Of particular concern in the context of helicopter transmission gears is the resistance to rolling contact fatigue when the elastrohydrodynamic (EHD) film thickness is small compared to the surface roughness. Plots of induced sub surface shear stresses for a constant load are shown in Figure 8, for two different surface conditions, together with the yield shear strength capability of two surface hardening processes. From this it can be seen that a shallow very hard nitrided layer is needed to resist the roughness induced (near surface) stresses, while the more ductile but softer carburised case will resist the bulk contact (Hertzian) stresses (4).

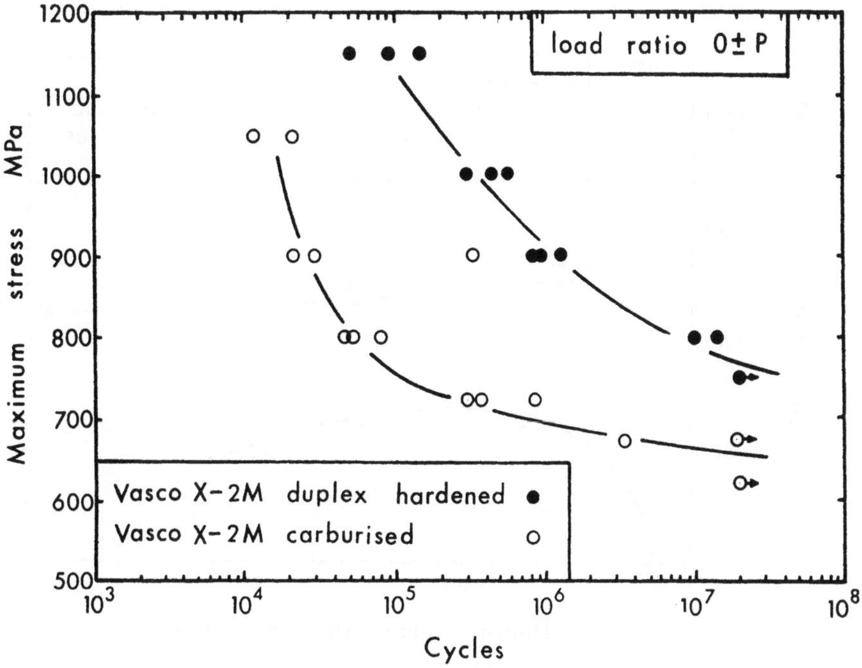

FIGURE 6 Effect of Duplex Hardening on Fatigue Performance of Vasco X-2M

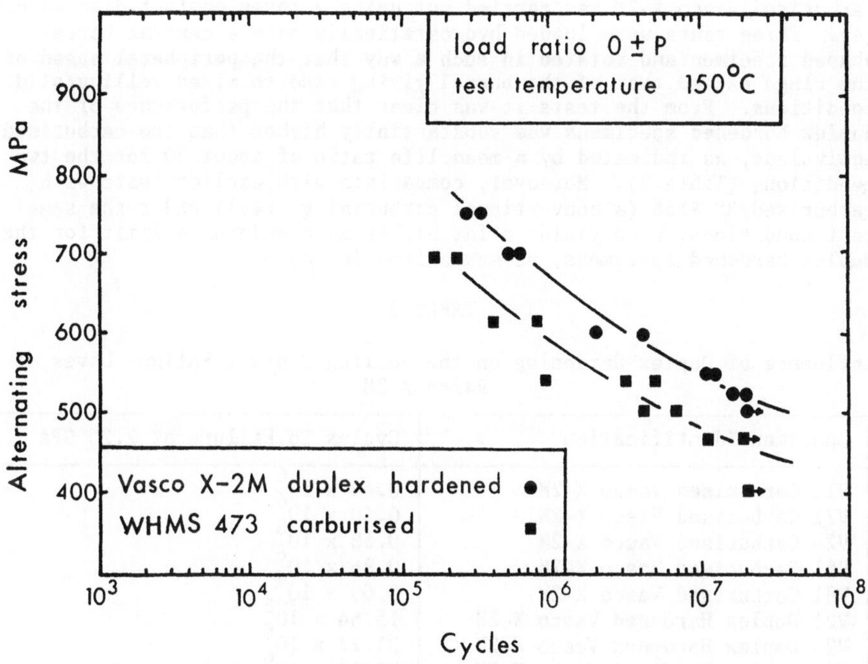

FIGURE 7 Influence of Duplex Hardening on Fatigue Performance at Elevated Temperature

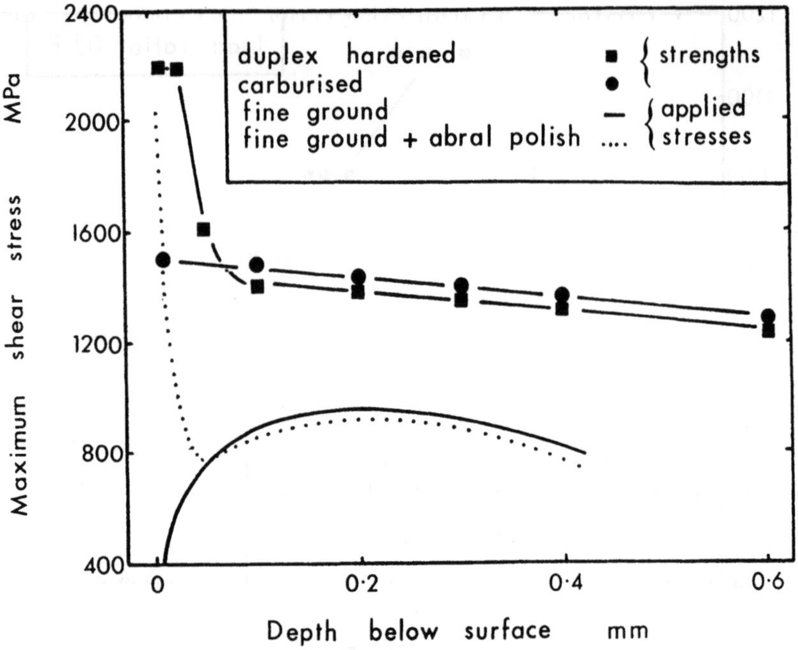

FIGURE 8 Influence of surface roughness/yield stress on principal shear stresses

Rolling contact fatigue evaluation of duplex hardened Vasco X-2M and carburised Vasco X-2M was carried out using a three contact disc machine (4). Three rings were loaded hydrostatically onto a central barrel shaped specimen and rotated in such a way that the peripheral speed of the rings exceed that of the barrel giving rise to mixed rolling/sliding conditions. From the tests it was clear that the performance of the duplex hardened specimens was substantially higher than the carburised equivalent, as indicated by a mean life ratio of about 30 for the two conditions (Table 2). Moreover, comparison with earlier tests with carburised BS S156 (a conventional carburising steel) under the same test conditions, also yields a far higher mean endurance limit for the duplex hardened specimens, as summarised in Table 3.

TABLE 2

Influence of Duplex Hardening on the Rolling Contact Fatigue Lives of Vasco X-2M

Specimen Identification	Cycles To Failure at 2.25 GPa
V11 Carburised Vasco X-2M	0.44×10^6
V71 Carburised Vasco X-2M	0.60×10^6
V24 Carburised Vasco X-2M	0.68×10^6
V51 Carburised Vasco X-2M	0.84×10^6
V81 Carburised Vasco X-2M	3.69×10^6
V21 Duplex Hardened Vasco X-2M	15.64×10^6
V23 Duplex Hardened Vasco X-2M	31.27×10^6
V44 Duplex Hardened Vasco X-2M	31.58×10^6 non failure
V72 Duplex Hardened Vasco X-2M	38.16×10^6 non failure

TABLE 3

Influence of Duplex Hardening on the Endurance Fatigue Limit

Specimen Identification	Endurance Limit at 10^8 Cycles
Vasco X-2M Carburised	1.15 GPa
BS S156 Carburised	1.13 GPa
Vasco X-2M Duplex Hardened	1.87 GPa

The relatively low endurance limits given in Table 3 are due to specimen configuration i.e. the edge chamfer inducing a stress concentration.

DISCUSSIONS AND CONCLUSIONS

It is clear that a significant benefit can be obtained from the use of duplex hardening as a surface engineering treatment for helicopter gears, in that both axial and rolling contact fatigue properties are superior when compared to both conventional and temper resistant carburised steels.

However, the steel chosen for this demonstration, into the feasibility and benefits of duplex hardening (ie. Vasco X-2M), has certain limitations with respect to toughness and nitriding response. It is believed that other temper resistant steels may offer a better overall property balance. Moreover, the practicality and reproducability of the proposed treatment in a gear manufacturing context has yet to be proven.

ACKNOWLEDGEMENTS

The author is indebted to Westland Helicopters Limited for permission to publish this paper. Thanks are also due to G Carr and P J Small for their invaluable assistance with the experimental work and to B C Gittos, N L Bottrell, R Sanders and A V Olver for their support.

REFERENCES

1. Davies D P 'Gear Materials in Helicopters Transmissions'
   Metals and Materials, July 1986, p342

2. Thomas W A and Patel P R 'V-22 Osprey - The Commitment to an Advanced Carburised Grade Gear Steel'
   Conference Proceedings : Carburising Processing and Performance
   Lakewood, Colorado, 12-14 July 1989

3. Davies D P and Gittos B C 'Gear Steels for Future Helicopter Transmissions'
   Proceedings I Mech E, Vol 203, May 1989

4. Olver A V 'Fatigue in Helicopter Transmissions'
   Westland Helicopters Research Paper 758, Aug. 1989

Low Cycle Fatigue Fracture Behaviour of Conventional and Damage Tolerant Microstructures of Inconel 718 at 650°C

Hong Bande, Yi Xiao

Harbin Institute of Technology
Harbin People's Republic of China

P. Au, A. K. Koul and W. Wallace

National Aeronautical Establishment
Ottawa Ontario Canada

ABSTRACT

The effects of microstructure and strain amplitude on low cycle fatigue lives and fracture characteristics of Inconel 718 were studied. All the tests were conducted in air at 650°C Results show that the conventional microstructure had only marginally longer fatigue lives than the damage tolerant microstructure irrespective of the applied strain amplitude. The fracture surface and fracture profile observations demonstrate that the resistance to intergranular cracking was improved by the serrated grain boundaries in the damage tolerant microstructure but the concomitant precipitation of large δ along serrated grain boundaries promoted the formation of microvoids and decohesion at the δ plate-matrix interfaces during cyclic straining. The impact of the two microstructures on LCF crack initiation life is also discussed.

INTRODUCTION

Inconel 718 is a high strength, weldable wrought nickel base superalloy which was developed for use at temperatures up to about 700°C. This material is used extensively in the aerospace, nuclear and petrochemical industries where high strength, low cycle fatigue resistance and corrosion resistance at elevated temperatures are important design considerations.

The physical metallurgy of Inconel 718 is now firmly established and has been described by several authors [1-3]. It is well known that the alloy is strengthened through the precipitation of the γ" phase (bct DO_{22} structure) during aging. The γ" forms as coherent disk-shaped precipitates with an orientation relationship of (001) γ"//{001} γ and <100> γ"//<100>γ with the matrix. After aging, this alloy also forms a small amount of spherical γ' phase with an ordered fcc ($L1_2$) structure. These precipitates dissolve between 843 to 871°C. However, γ" transforms to orthorhombic δ phase (Ni_3Nb) after exposure to temperatures above

815°C and becomes incoherent with the matrix. The δ phase contributes little to the strength of the alloy. Besides these precipitates, the grain size and grain boundary morphology have considerable effect on the material's elevated temperature fatigue behaviour. Therefore, attempts were made to design heat treatments for Inconel 718 to improve its fatigue and crack propagation resistance without sacrificing other mechanical properties [4, 5]. A damage tolerant microstructure has recently been developed for this alloy [5]. This microstructure is characterized by slightly coarser grains than the conventional microstructure and contains heavily serrated grain boundaries. The damage tolerant microstructure was shown to possess better high temperature fatigue crack growth resistance than the conventional fine grained microstructure with little loss in low cycle fatigue life as a function of total strain range. However, this microstructure lowers the low cycle fatigue (LCF) life as a function of plastic strain range. The objective of the present study was, therefore, to investigate this aspect in greater detail, especially the influence of the two microstructures on the LCF fracture behaviour of Inconel 718 at 650°C.

EXPERIMENTAL PROCEDURE

Material

The material used in this study was procured in the form of hot rolled bars 22 mm in diameter. All bars were taken from the same material heat and its chemical composition is given in Table 1.

TABLE 1

Composition of Inconel 718 used in the present study

C	Mn	Si	Cr	Ni	Co	Mo	Nb+Ta	Ti	Al	B	S	P	Cu	Fe
0.03	0.11	1.12	18.3	53.2	0.53	2.99	5.17	0.99	0.42	0.003	0.002	0.01	0.06	Bal.

As received material had been stress relieved at 954°C for one hour at the rolling mill and was subjected to two different heat treatment schedules to obtain microstructures that varied in grain size, grain boundary and matrix precipitate morphologies. Table 2 lists the heat treatments employed, the grain sizes produced and the material designation adopted in this paper. One batch of material in Table 2 was given the conventional heat treatment for Inconel 718. An annealing temperature of 954°C dissolves the γ' and γ'' phases and maintains the fine grain size that is developed during sub-δ phase solvus wrought processing; followed by standard double aging at 718°C and 621°C. Another batch of material was subjected to a modified heat treatment schedule to produce the damage tolerant microstructure, Table 2. A solutioning temperature of 1032°C followed by furnace cooling produced a coarse grain size. Direct aging at 843°C precipitated a large amount of plate-like δ phase at the grain boundaries and to a limited extent within the grain interiors, and the grain boundaries assumed a serrated morphology. Reheating to 926°C dissolved the overaged γ' and γ" precipitates formed during furnace cooling to 843°C and subsequent double aging at 718°C and 621°C produces fine γ' precipitates.

TABLE 2

Heat treatments and material designation

Material designation	Heat treatment	Average grain size
Conventional	954°C/1h/AC+718°C/8h → (2°C/min) 621°C/8h/AC	ASTM 11
Damage Tolerant	1032°C/1h → (3°C/min) 843°C/4h/AC+926°C/1h → (3°C/min) 718°C/8h → (2°C/min) 621°C/8h/AC	ASTM 4

Specimen and Test Procedure

ASTM E606 axial fatigue specimens were manufactured from heat treated materials to a diameter of 7.62 mm in the test section and with a minimum gauge length of 33.02 mm. After fabrication, the test section of each specimen was polished in the axial direction with successively finer grit emery papers (grade 320, 400 and 600) to remove circumferential machine marks.

Fully reversed, constant amplitude low cycle fatigue tests were performed in laboratory air at 650°C in a closed loop electrohydraulic testing system under total axial strain control. The strain ranges employed were typically 2, 1.5, 1.25 and 1 percent. In the case of damage tolerant material, additional tests were performed at lower strain ranges (0.83 and 0.65 percent). A triangular waveform with a constant strain rate of 0.002 per second was used and specimens were loaded in tension first. Specimen gauge section was heated by radio frequency induction and the test temperature was monitored and controlled by an infrared radiation pyrometer. Details on the capability of the testing facility are given in Reference 6.

An x-y recorder was used to obtain cyclic total axial strain versus load plots, which were converted to engineering stress-strain hysteresis loops using the specimen cross sectional area. Specimen failure; Nf, was defined by the number of cycles for a 5 percent drop in the stabilized tensile stress value. After testing, specimens were prepared for optical metallography, SEM and TEM examination. An immersion etchant consisting of one part H_2O_2, two parts H_2O and three parts HCl was found to be effective for optical microscopy.

EXPERIMENTAL RESULTS

Microstructures

Figure 1 shows the typical microstructures of the two materials. The samples were obtained from the grip ends of the LCF specimens and oriented to view along transverse sections. As can be seen in Figure 1 a, the conventional microstructure consists of fine grains and the grain boundaries are mainly planar and decorated by plate-like δ phase (Ni_3Nb). However, the damage tolerant microstructure (Figure 1 b) possesses much coarser grains. A large number of δ plates intersect the grain boundaries at rather obtuse angles, giving the boundaries a serrated appearance. Furthermore, many large δ plates also precipitate along noncoherent twin boundaries within the grains.

Transmission electron microscopy reveals that the γ' in both microstructures is present as ellipsoidal, disk shaped precipitates lying on (001). Dark field imaging in <001> matrix zone axis orientation reveals that the γ' disks in the conventional microstructure are fine and distributed more densely than those in the damage tolerant microstructure, Fig. 1 (c) and (d). The γ' disks are approximately 27 nm in diameter and 12 nm in thickness in the conventional microstructure, whereas the γ' discs in the damage tolerant microstructure reveal a bimodal distribution where the finer γ' discs are similar in size to the conventional microstructure but the larger discs are approximately 51 nm in diameter and 20 nm in thickness. The larger γ' discs in the damage tolerant microstructure are believed to have formed during furnace cooling from 926°C to 718°C.

LCF Strain-Life Curves

Constant amplitude strain-life fatigue curves obtained for the two materials tested at 650°C are shown in Figure 2. The curves are presented in cycles to failure versus total and plastic strain ranges. As can be seen for the strain range values considered, the fine grained

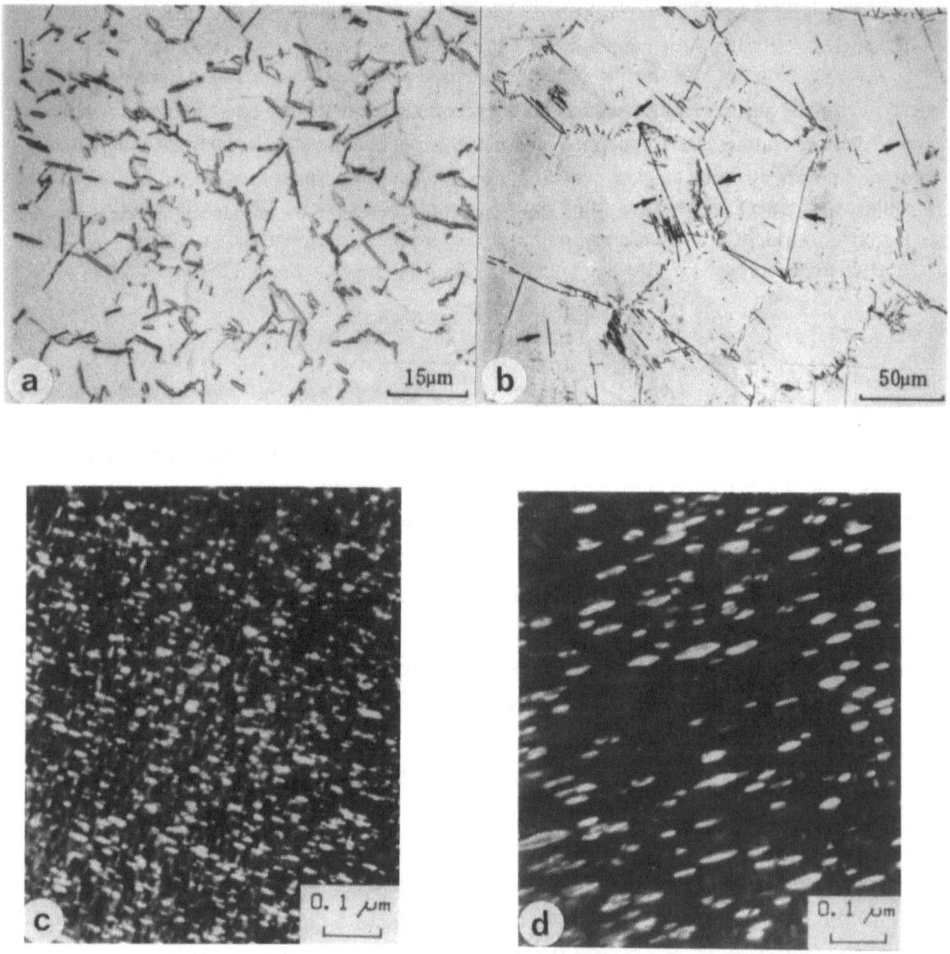

Figure 1. Optical micrographs showing the grain structure and γ'' distribution in experimental Inconel 718 alloy. (a) conventional grain structure, (b) damage tolerant grain structure, (c) conventional γ'' distribution and (d) damage tolerant γ'' distribution. Note: the arrows in Fig. 1(b) indicate the precipitation of δ-plates along noncoherent twin boundary ledges.

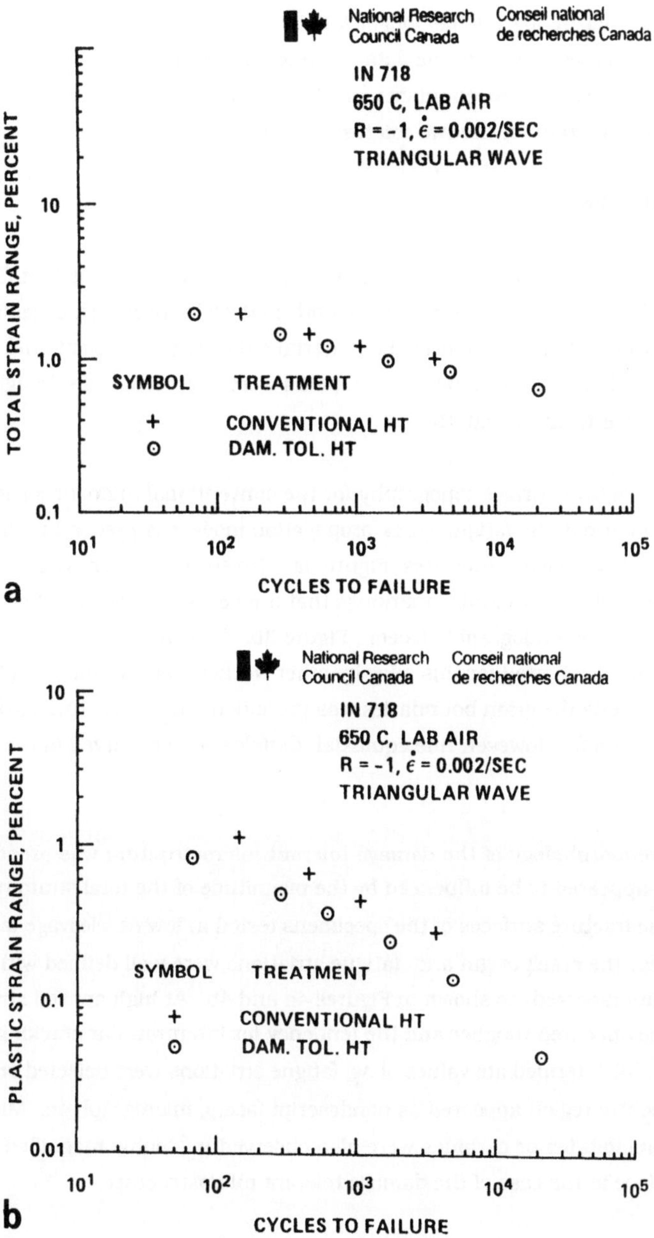

Figures 2. **LCF strain-life curves for conventional and damage tolerant microstructures of Inconcel 718 alloy at 650°C (a) cycles to initiation versus total strain range and (b) cycles to crack initiation versus plastic strain range.**

conventional microstructure always has longer fatigue lives than those of the coarse grained damage tolerant microstructure when the data is plotted as a function of plastic strain range. These observations indicate the mechanisms for plastic deformation and fracture during LCF testing may be quite different for the two microstructures.

Fracture Surface Morphology

Scanning electron microscopy of specimen fracture surfaces reveals that fatigue crack initiation was predominantly transgranular for both materials, although occasionally intergranular initiation was also noted in the material subjected to conventional heat treatment. It is also observed that fatigue crack paths were microscopically "rougher" in the coarse grained damage tolerant material

The typical fracture surface topography for the conventional microstructure tested at 650°C is shown in Figure 3. Its fatigue crack propagation mode was predominantly intergranular with a few cleavage fractures, Figure 3a . No striations were visible on the fracture surface, but a series of parallel markings that are believed to be slip offsets were superimposed on some crystallographic facets , Figure 3b. Similar fracture surface morphologies have been reported for this alloy by other workers [7]. As shown in Figure 3, secondary cracking along the grain boundaries was prominent on the fracture surface, but no micropores were observed. However, fine equiaxial dimples were observed in the overload area.

The fracture morphology of the damage tolerant microstructure was predominantly transgranular and appeared to be influenced by the magnitude of the total strain range applied. For instance, on the fracture surfaces of the specimens tested at low $\Delta\varepsilon$, cleavage facets were usually present near the crack origin and fatigue striations were well defined with plenty of secondary cracks interspersed, as shown in Figures 4a and 4b. At high applied strain ranges the fracture surface appeared rougher and the tendency for intergranular cracking increased, Figures 4c and 4d . At intermediate values of $\Delta\varepsilon$, fatigue striations were detected on a few facets but most of the fracture region appeared as nondescript facets, micrograph 4c. Micropores originating from precipitates or carbides were also observed in highly magnified micrographs of the fracture surface in the case of the damage tolerant microstructure.

DISCUSSION

In the present study, the coarse grained damage tolerant microstructure had serrated grain boundaries instead of the planar boundaries of the fine grained conventional microstructure. The difference in grain size and grain boundary morphology changed the fracture mode from mainly intergranular for the conventional microstructure to a

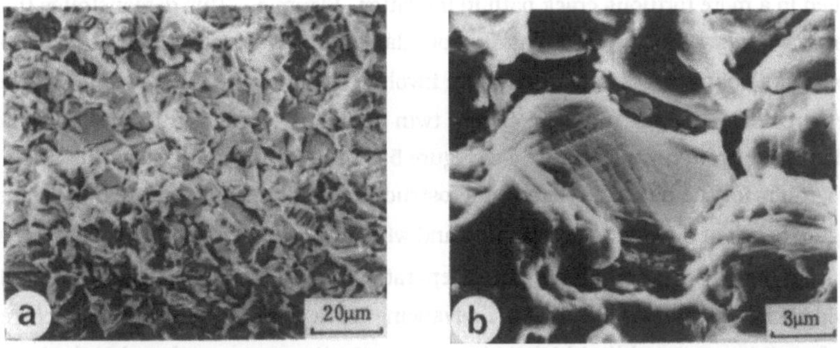

Figure 3. Scanning electron micrographs of fracture surface of the material subjected to conventional heat treatment and tested at a total strain range $\Delta\varepsilon$ = 1.02 percent at 650°C; Nf = 3,800.

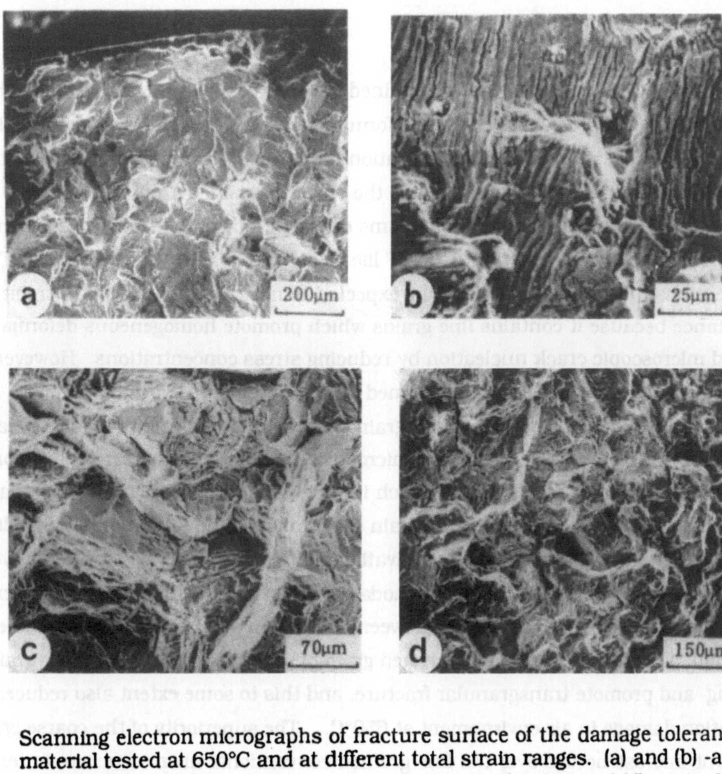

Figure 4. Scanning electron micrographs of fracture surface of the damage tolerant material tested at 650°C and at different total strain ranges. (a) and (b) -at $\Delta\varepsilon$=0.65 pct and Nf=21,500, (c) -at $\Delta\varepsilon$=1.0 pct and Nf=1,800 and (d) -at $\Delta\varepsilon$=1.25 pct and Nf=650.

predominantly transgranular fracture for the damage tolerant microstructure and also resulted in a more tortuous crack path in the latter, Figure 5. This demonstrates that the serrated boundaries improved the grain boundary cracking resistance of Inconel 718 alloy. A concomitant change in the microstructure involved the presence of abundant twins and large δ plates precipitated along the noncoherent twin boundary ledges within the grain interiors in the damage tolerant microstructure (see Figure 5 b and 1 b). Figure 6 shows the SEM micrographs for the damage tolerant microstructure that were taken from the same section as in Figure 5, but at a higher magnification and with a better contrast of the δ phase. As can be seen from the two micrographs, δ plates separated from the matrix during strain cycling and formed elongated dimples ahead of the advancing crack and also beneath the fracture surfaces. Therefore, it is suggested that higher stress concentrations were produced at the interface of δ plate-matrix during cyclic straining. It promoted premature formation of microvoids and decohesion along δ plate-twin boundaries as indicated by arrows in Figure 6 b. In the conventional microstructure, however, the shorter length of δ plates (usually less than 10μm) allowed the matrix to accommodate larger cyclic strains before decohesion of the δ plate-matrix interface took place.

The LCF crack initiation life, defined in this work by a 5% load drop in the stabilized tensile stress value, corresponds to the formation of a 1.5 to 2 mm long crack in Inconel 718 material [6]. Therefore, the crack initiation life in reality comprises the life to microscopic crack initiation and the life to propagate the crack up to a length of 1.5 to 2 mm. It follows that the deformation and fracture mechanisms operative during both microscopic crack initiation and crack propagation contribute to LCF life as defined in a 5% stress drop test. The conventional microstructure would be expected to have a superior fatigue crack initiation resistance because it contains fine grains which promote homogeneous deformation and retard microscopic crack nucleation by reducing stress concentrations. However, in terms of fatigue crack propagation, the fine grained material would be expected to have a lower fatigue crack propagation resistance because grain boundary sliding occurs readily ahead of the crack tip. In contrast, the damage tolerant microstructure contains coarser grains along with a serrated grain boundary structure which improve the intergranular crack propagation resistance of the material. A coarse grain size promotes the formation of long slip bands which also reduce the tendency for activating additional slip systems on a per cycle basis. This is because long slip bands can accommodate a larger number of dislocation reversals than short slip bands since the distance between dislocation pile-up obstacles is greater [10-12]. In addition, a coarse grain size and serrated grain boundary structure suppress grain boundary sliding and promote transgranular fracture, and this to some extent also reduces the impact of oxidation damage in air environment at 650°C. The superiority of the coarse grained damage tolerant microstructure over the fine grained conventional microstructure in terms of fatigue crack propagation resistance under plane strain linear elastic conditions has previously been

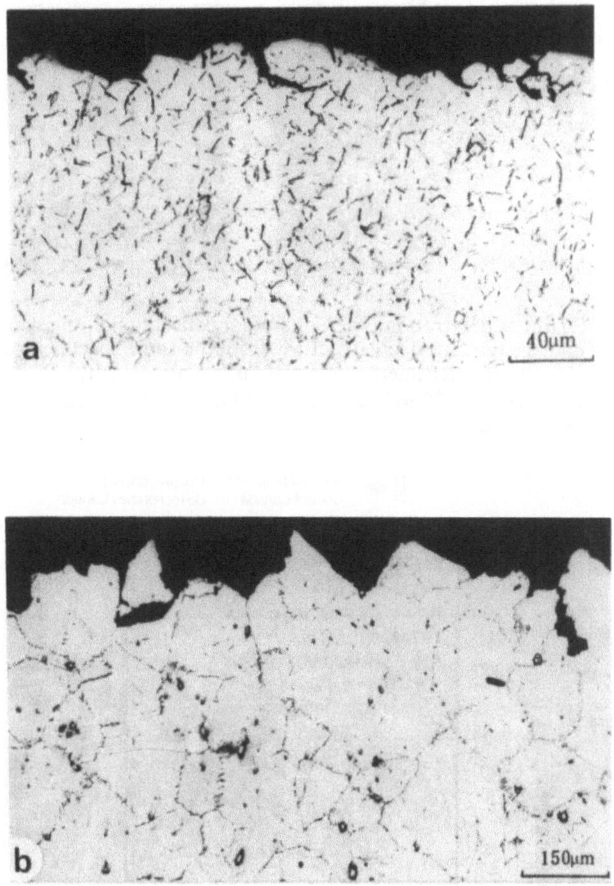

Figure 5. Profile of typical fatigue crack path for LCF specimens tested at 650°C and $\Delta\varepsilon$ of 1.0 pct. (a) conventional microstructure; (b) damage tolerant microstructure.

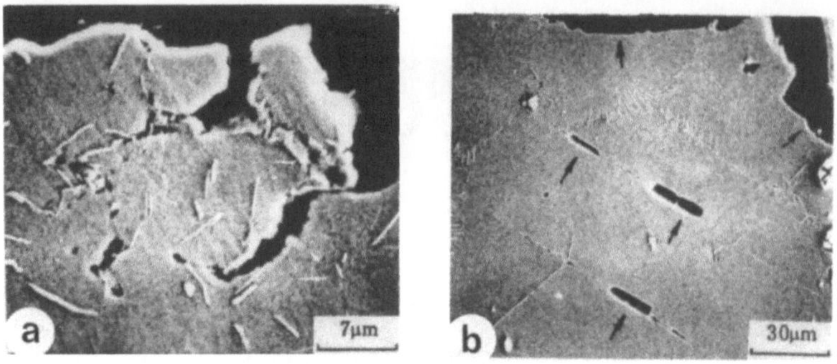

Figure 6. Scanning electron micrographs of (a) conventional microstructure and (b) damage tolerant microstructure, from the same specimens as in Figure 5, showing the detrimental effect of large δ plates on the nucleation of microvoids. The arrows in (b) indicate the cracking path and elongated dimples along δ plate-twin boundaries.

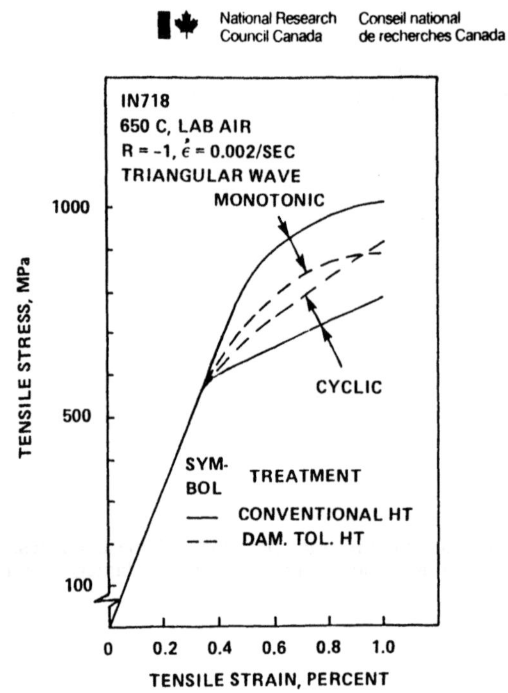

Figure 7 Tensile and cyclic stress-strain behaviour of the two microstructures of Inconel 718 at 650°C.

discussed in detail elsewhere [4, 5]. However, some investigators have shown that up to 70% of the LCF life of conventional microstructure IN-718 tested at room and elevated temperatures can be consumed during microscopic fatigue crack initiation [8, 9]. Therefore, under strain controlled LCF testing conditions, the conventional microstructure has somewhat longer total fatigue lives (as defined by a 5% load drop) than the damage tolerant microstructure in spite of the superior fatigue crack growth resistance of the latter.

As regards the differences in the LCF lives as a function of plastic strain range between the two microstructures, Fig. 2b, these results simply indicate that for a given total strain range, the stabilized hysteresis loop is considerably wider in the case of the conventional microstructure and the plastic strain constitutes a larger fraction of the total strain. As a result, the data points in Fig. 2b are shifted upwards along the Y-axis. The widening of the stabilized hysteresis loop for the material subject to the conventional heat-treatment can be directly attributed to the severe cyclic softening of the material at 650°C whereas the material subjected to the damage tolerant heat treatment only reveals marginal cyclic softening, Fig. 7. Cyclic softening of the material is considered deleterious because the load bearing capacity of the material is reduced with cyclic strain accumulation. The cyclic softening of the conventional microstructure is most probably related to the presence of fine γ' discs which can lose coherency through interaction with dislocation or be sheared by dislocations during cyclic straining [13-16]. In contrast, the damage tolerant microstructure contains a bimodal γ' distribution where larger γ' precipitates cannot be sheared easily and the loss of coherency is not as severe as for fine γ' precipitates. As a result, the cyclic softening is not as noticeable in the damage tolerant material as it is in the conventional microstructures.

CONCLUSIONS

The microscopic crack initiation generally occurred in a transgranular mode in both conventional and damage tolerant microstructures of Inconel 718 alloy tested under LCF conditions at 650°C. In the conventional fine grained material, the fatigue fracture mode was predominantly intergranular accompanied by extensive secondary cracking. In contrast, the coarse grained damage tolerant material revealed mainly transgranular fracture with well defined fatigue striations. The differences in the fracture mode are caused by the differences in the grain size and grain boundary morphology of the two materials where grain boundary sliding and environmental damage is suppressed in the case of the damage tolerant material and this also improves its resistance to intergranular cracking. However, large δ plates precipitated along the noncoherent twin boundaries of the damage tolerant material promoted the formation of microvoids and the decohesion of the δ plate matrix interfaces both ahead of the advancing crack front and beneath the fracture surface.

Strain controlled isothermal LCF test results indicate that the conventional fine grained microstructures had marginally longer LCF crack initiation lives (defined by a 5% drop in the stabilized tensile stress value) than the coarse grained damage tolerant microstructure for applied total strain values in the range of 1 to 2%. It is suggested that LCF crack initiation life during a 5% stress drop test is governed by both microscopic crack initiation and crack propagation. While the conventional microstructure has superior microscopic crack initiation resistance, the intergranular crack propagation resistance is considerably superior in the case of the damage tolerant microstructure. As a result, there is negligible difference in the LCF life of the two microstructures as a function of total strain range. The conventional microstructure also undergoes severe cyclic softening which is considered deleterious for turbine disc applications.

REFERENCES

1. Barker, J.D., Krueger, D.D. and Chang, D.R., Thermomechanical processing of Inconel 718 and effect on properties. Proc. of the Nicholas Grant Symposium, June 1985, pp. 125-137.

2. Oblak, J.M., Paulonis, D.F. and Duvall, D.S., Coherency strengthening in Ni-base alloys hardened by DO_{22} γ' precipitates. Met. Trans., 1974, 5, pp. 143-153.

3. Hall, R.C., The Metallurgy of Alloy 718. J. Basic Engineering, Sept. 1967, pp. 511-516.

4. Thamburaj, R., Terada, T., Koul, A.K., Wallace, W. and M.C. de Malherbe, The influence of microstructure and environment upon elevated temperature crack growth rates in Inconel 718, Proc. Conf. on 'High Temperature Creep,' Tokyo, Japan, ASME/JSME, 1986, pp. 275-287.

5. Koul, A.K., Au, P., Bellinger, N., Thamburaj, R., Wallace, W., and Immarigeon, J-P. Development of a damage tolerant microstructure for Inconel 718 turbine disc material, Superalloys 88 book, Ed., S. Reichman et al., Met. Soc. of AIME, 1988, pp. 3-12.

6. Au, P., Development of an isothermal low cycle fatigue testing facility. NRCC, NAE LTR-ST-1601, September 1986.

7. Mills, W.J. and James, L.A., Effect of heat-treatment upon the fatigue- crack growth behaviour of Alloy 718 weldments - part II: Microscopic behaviour. J. Eng. Mater. Technol., 1985, 107, pp. 41-47.

8. Fournier, D. and Pineau, A., Low cycle fatigue behaviour of Inconel 718 at 198K and 823K. Met. Trans., 1977, 8A, pp. 1095-1105.

9. Sanders, T.H., Frishmuth, R.E. and Embley, G.T., Temperature dependent deformation mechanisms of Alloy 718 in low cycle fatigue. Met. Trans., 1981, 12A, pp. 1003-1010.

10. James, L.A., The effect of grain size upon the fatigue-crack propagation behaviour of Alloy 718 under hold-time cycling at elevated temperature. Engineering Fracture Mechanics, 1986, 25, pp. 305-314.

11. Krueger, D.D., Antolovich, S.D. and Van Stone, R.H., Effect of grain size and precipitate size on the fatigue crack growth behaviour of Alloy 718 at 427°C. Met. Trans. 1987, 18A, pp. 1431-1449.

12. Pédron, J.P. and Pineau, A., The effect of microstructure and environment on the crack growth behaviour of Inconel 718 alloy at 650°C under fatigue, creep and combined loading. Mat. Sci. Eng., 1982, 56, pp. 143-156.

13. Lerch, B.A., Jayaraman, N. and Antolovich, S.D., A study of fatigue damage mechanisms in Waspaloy from 25 to 800°C, Mat. Sci and Eng., 1984, 66, pp. 151-166.

14. Fournier, D. and Pineau, A., Low cycle fatigue behaviour of Inconel 718 at 298K and 823K, Met Trans, 1977, 8A, pp. 1977-1095.

15. Rao, K.B.S., Schiffers, H., Schuster, S. and Nickel, H., Influrence of time and temperature dependent processes on strain controlled LCF behaviour of Alloy 617, Metall. Trans., 1988, 19A, pp. 359-371.

16. Hwang, S.K., Lee, H.N. and Yoon, B.H., Mechanism of cyclic softening and fracture of a Ni-base γ' strengthened alloy under LCF, Metall. Trans., 1989, 20A, pp. 1793-2801.

Corrosion and Wear Protection of Compressor Blades by Aluminum Coatings and a New Method for Testing their Erosion Resistance

Kh. G. Schmitt-Thomas and Th. Happle

Technische Universität München
München
Federal Republic of Germany

Abstract

Aluminum coatings on compressor blades guarantee a very good cathodic corrosion protection provided they are not damaged. In operation, however, the additional erosive stress - caused by particles in the intake air of the compressor - damages coatings and considerably reduces the protection against corrosion. Aiming at an improvement of the operating behaviour of compressor blades under such complex stress conditions, it is investigated, how various coating systems behave under combined erosive and corrosive stress condition. In order to find out the erosion stability of new and precorroded coating systems an erosion testing method has been developed. The following paper deals with the most important influencing factors of erosion (blasting wear) and their presentation in the new erosion testing method taking into account the experiences and results being available up to now.

1. Introduction

The compressor blades of stationary gas turbines and jet engines are usually protected by coatings against corrosion. The coatings should avoid the failure of the blades by corrosion fatigue. In operation of the compressor the coating systems will be damaged by blasting wear (erosion by particles). To assure a sufficient corrosion protection by local damages of the coatings, anodic coatings will be preferentially applied.

Inorganically bound aluminum coatings - consisting of aluminum pigments, embedded in an inorganic matrix (chromat/phosphat-compounds) - are applied as compressor blade coatings and assure in the undamaged condition a good corrosion resistance. The application of the aluminum coatings under corrosive and erosive conditions required informations about the erosion behavior of the aluminum coatings and the classification of the coatings with regard to their erosion resistance.

For this reason originally coated specimens - with inorganically bound aluminum coatings - and the same coatings after erosion stress are investigated in a corrosive environment to find out the influence of erosive degradation on the protection against corrosion. Furthermore different systems of coatings had been compared with respect of their corrosive and erosive behaviour. An erosion test method was developed in order to determine the erosion resistance of the coatings more easily, reproducible and in a manner that approximates actual operation. The basic method of the erosion test is to move the samples by a rotational motion through a solid particles fluidized bed. The fluidized bed contains the abrasive particles which are kept in suspension by a gas stream.

2. Composition of inorganically bound aluminum coatings

The inorganically bound aluminum coatings (metal-ceramic aluminum coatings) are inorganic coatings with a matrix of components of chromats and phosphats containing aluminum pigments in a dispersion (fig. 1). The coatings are composed of an aluminum pigmented basic layer and a surface coating (sealing). To get a coating free of pores the basic layer is applied in two or more spraying steps. The dispersion consisting of a mixture of chromium-phosphoric acid is sprayed on the surface. The basic layer is burned in at 350°C. Due to the glass-like compounds of polyphosphats, wrapping the aluminum pigments, this coating is not electro-conducting. To get the electric conductivity for the cathodic protection the basic

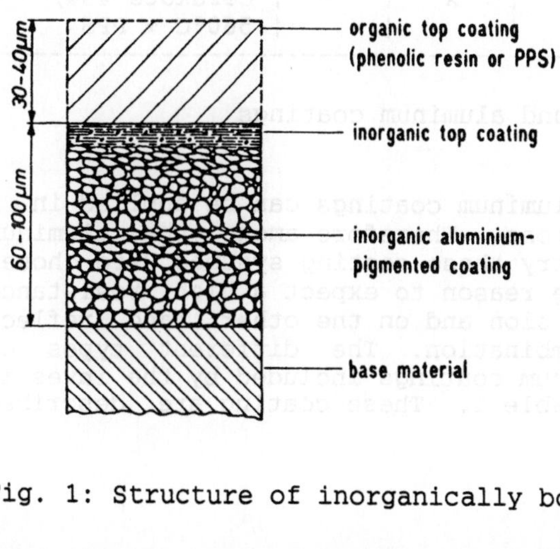

Fig. 1: Structure of inorganically bound aluminum coatings

layer is shotpeened with glass balls. By local destroying of the glassy wrap of the aluminum particles the electric conductivity will be achieved. An other possibility to achieve the electric conductivity is a sintering process of the particles at a burn-in temperature of about 550°C. The selection of the processes depends from the material of the components. On applying high temperatures for treatment always the possibility of a deformation of the components is inherent.

To improve the corrosion and erosion resistance the aluminum pigmented basic coating is sealed with an inorganic or an organic top coating. The inorganic sealing is similar to the basic coating matrix without containing aluminum pigments. The organic top coating in use are phenolic resins or thermoplastics, e.g. polyphenylensulfid. Compared with the basic coatings the surface roughness of the sealing is less. On this way precipitation of corrosive particles and pollution of the surface is diminished.

Group	top coating			burning-in temperature, shot-peening		inhib. pigments	coatings
	in-org.	no	org.	350°C, yes	550°C, no		
	1. figure:			2. figure:		3.fg.	
	1	0	2	1	2	1	
110	X	-	-	X	-	-	P484 + P100, P484 + P201, Sermetel 725, Sermetel 1141
111	X	-	-	X	-	X	Sermetel EF-1
220	-	-	X	-	X	-	Cerakote 484/ 560°C + PPS

Table 1: Inorganically bound aluminum coatings

The inorganically bound aluminum coatings can be applied in a lot of parameter variations. Therefore among the aluminum coatings offered by industry these coating systmes were chosen which on the one hand gavę reason to expect a high resistance against corrosion and erosion and on the other hand reflect the possibilities of combination. The different types of inorganically bound aluminum coatings included by the investigation are presentet in table 1. These coating are described in the literature /1-10/.

3. Blasting wear (erosive) testing method

A blast wear testing method was been designed corresponding to the mechanisms of degradation under blasting wear of erosive conditions of compressor blades /1,2,8-10/. The basic idea of the testing method is that the specimens, which are tested, pass a solid fluidized bed by a rotational movement (fig. 2).

The fluidized bed contains the abrasive particles which are kept floating by a gas stream. The continuity in the fluidized bed is attained by gathering the abrasive particles in a basin. These particles are then led back into the gas stream that enters at the bottom of the basin. The size of the particles is kept constant by sifting out the smaller ones prior to their getting back into the gas stream.

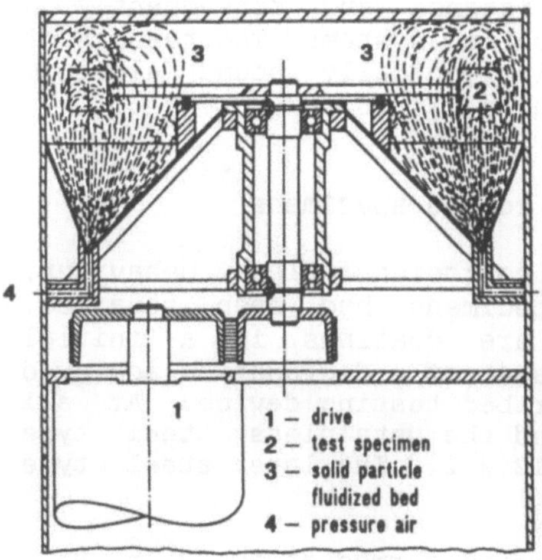

a) scheme

b) erosion tester

Fig. 2: Erosion testing method using fluidized solids bed

The sample holders were mounted at an angle of 90° to each other at the outer end of the flywheel which is connected to the driving motor by a shaft and a driving belt. This enables the testing of up to four samples so that in one test run equal conditions for one or more main samples and the reference samples are guaranteed. The particle impact angle may be adjusted from surface parallel streaming to vertical particle impact. By adjusting the angles of gradient of the sample holders the impact angle of the particles can be preset continuously variable or on a screen with a step adjustment of 15°. The sample holders flanged to the flywheel can be exchanged so that different sample forms or components can be tested with suitable sample holders.

With the erosion tester in use impact velocities can be steplessly adjusted between 20 to 100 m/s. However much higher impact velocities can be set - the way they occur in reality. In the blast wear test the kind of abrasives as well as the size of particles, the distribution of size and the mass of abrasives may be variable (according to the requirements). The density of the fluidized bed is adjustable trough the amount of abrasive particles in the feed-in and collecting basin. An extension of the erosion tester for combined testing erosion and corrosion is intended.

The erosion behavior of inorganically bound aluminum coatings was investigated by determining mass loss and reduction of the coating thickness through blasting wear in relation to the testing time. The erosion tests showed that the aluminum coatings are sensitive to blasting wear - especially to micro-cutting and micro-ploughing processes by flat impact angles. The results of the erosion tests provided a quantitative distribution of the erosion resistance and consequently a definite classification of the coating systems. The results of the basical erosion tests at inorganically bound aluminum coatings are given in the literature /1,2,8-10/.

4. Corrosion fatigue behavior of coated specimens

To compare the influence on the corrosion fatigue behaviour, the condition of the coated specimens had been regarded. Subject of the investigation are coatings in a initial condition and after different stages of a degradation achieved by erosive stresses at the discribed testing device. At all cases for basic material served the stainless steel type X20Cr13 and the type X10CrNiMoV12 2 2. The later steel type shows the higher strength.

The corrosion fatigue tests were performed at a 22 % NaCl solution (pH7) and at a temperature of 80°C. During the test, the medium was aerated. The specimens were loaded in an alternating tensile range with an R-value of 0.1 (ratio of lower to the upper stress) at a frequency of 50 Hz.

Under fatigue load at a corrosive environment the steels type X20Cr13 and X10CrNiMoV12 2 2 show a remarkable decrease of the fatigue strength. At the solution of 22 % NaCl (pH7) and a temperature of 80°C the fatigue strength of uncoated X20Cr13 is settled with 180 N/mm^2 (upper stress; without corrosive conditions: 640 N/mm^2). The fatigue strength of the uncoated steel type X10CrNiMoV12 2 2 amounts to 160 N/mm^2. Each value is rated at 3×10^7 cycles.

The corrosion fatigue of the coated steel X20Cr13 at corosive conditions is settled in a narrow scattering band of 550 to 580 N/mm^2 after application of coatings of the group 110 and 220 (fig. 3 and 4). Compared with the above mentioned coating groups the coating of the group 111 applied on the steel

X20Cr13 is rated with higher corrosion fatigue strengths of 630 N/mm². Accordingly to the higher fatigue strength of the X10CrNiMoV12 2 2 the coated specimens of this material are also settled in a higher domaine of corrosion fatigue. In this respect the results differ from the corrosion fatigue tests of the uncoated X20Cr13 and X10CrNiMoV12 2 2. In uncoated condition for the later steel type the corrosion fatigue strength shows with 160 N/mm² the lower value. Corrosion fatigue strength of the X10CrNiMoV12 2 2 is raised from 160 to 640 N/mm² by applying a coating of group 110. On applying the coating of group 111 this value is rated with 720 N/mm².

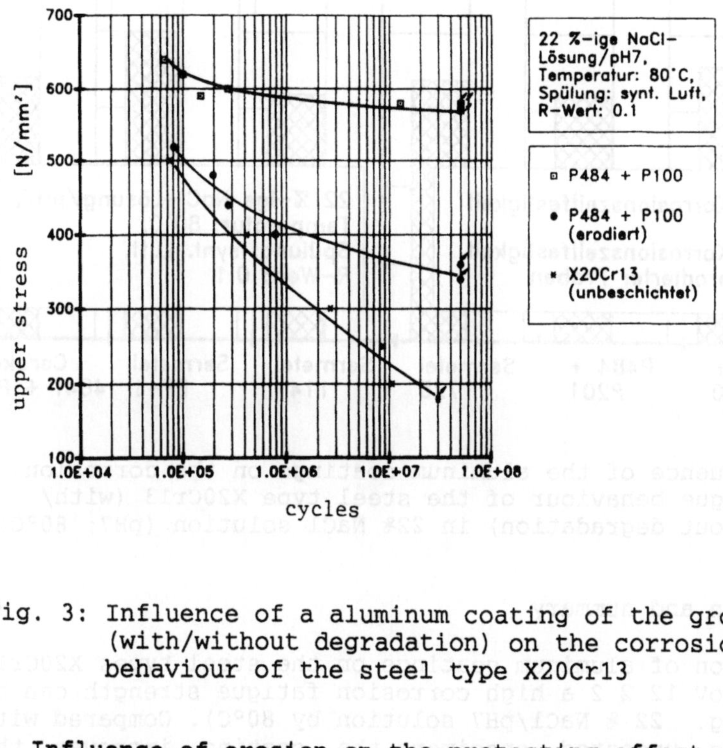

Fig. 3: Influence of a aluminum coating of the group 110 (with/without degradation) on the corrosion fatigue behaviour of the steel type X20Cr13

5. Influence of erosion on the protecting effect of aluminum coatings

To invastigate the influence of degradation of aluminum coatings by wear on corrosion fatigue behaviour, tests had been performed with coated specimens of X20Cr13 after erosion tests by the described testing device. The parameters of the erosive stresses were as follows: medium corundum, impact velocity 60 m/s and angle of attack 90°C (direct impact).

The tests were continued up to the complete degradation of the coatings. The corrosion fatigue conditions are 22 % NaCl solution (pH7) at a temperature of 80°C. Compared with coated specimens at initial condition after degradation by wear corrosion fatigue strength degreases from 550-630 N/mm² to 320-400 N/mm² depending from the type of coating (fig. 3 and 4). At all cases the corrosion fatigue strength of the specimens of partially degrated coating is settled far above the corresponding value of uncoated specimens of X20Cr13 with 180 N/mm².

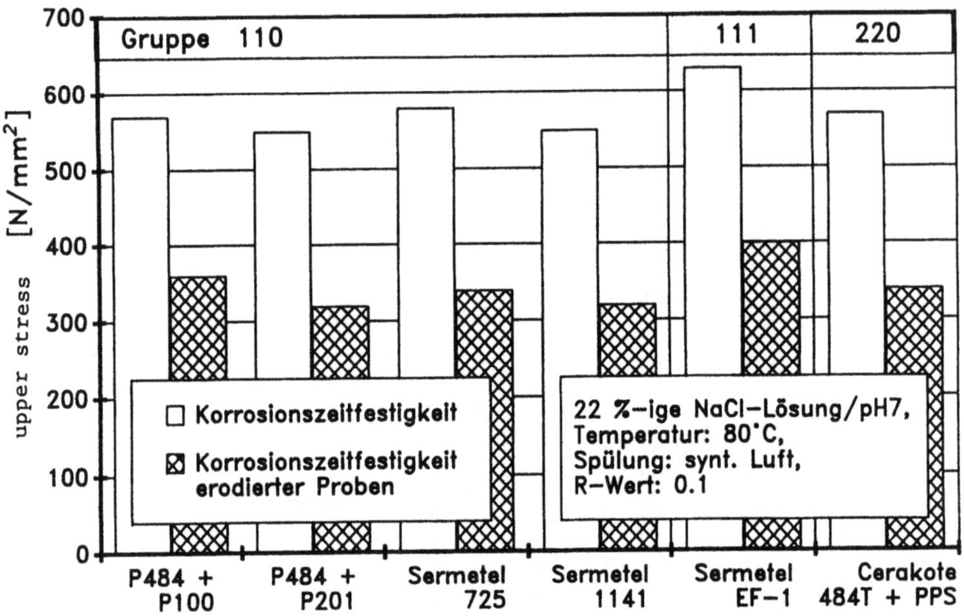

Fig. 4: Influence of the aluminum coatings on the corrosion fatigue behaviour of the steel type X20Cr13 (with/ without degradation) in 22% NaCl solution (pH7; 80°C)

6. Discussion and summary

By application of aluminum coatings on the steel types X20Cr13 and X10CrNiMoV 12 2 2 a high corrosion fatigue strength can be achieved (e.g. 22 % NaCl/pH7 solution by 80°C). Compared with corresponding uncoated specimens the coating improves the corrosion fatigue strength about 300 % by 5×10^7 cycles. The corrosion fatigue strength of the coated material is settled within a narrow scatter band. After partial degradation by erosion the range of qualifications of the coatings remains about the same range as without damage however the level of corrosion fatigue strength generally decreases. On the failing of inorganically bound aluminum coatings pitting corrosion is mainly causal. At high concentration of chlorine ions pitting corrosion is settled within the domaine of the free potential of corrosion. An improvement of pitting corrosion can be achieved by inhibitors, e.g. strontium chromates. Pitting corrosion however cannot completely be prohibed.

The erosion behavior of inorganically bound aluminum coatings was investigated. The erosion tests showed that the aluminum coatings are sensitive to blasting wear - especially to micro-cutting and micro-ploughing processes by flat impact angles. The results of the erosion tests provided a quantitative distribution of the erosion resistance and consequently a definite classification of the coating systems /1,2,8-10/.

7. Literatur

/1/ Schmitt-Thomas, Kh. G., Th. Happle:
VDI-Z Special Ingenieur-Werkstoffe März'88 (1988), 24/29
/2/ Happle, Th.: Magazin Neue Werkstoffe 1/88 (1988), S.37/41
/3/ Puchinger, F.: Werkstoffe u. Korrosion 27 (1976), 707/710
/4/ Kolkman, H. J., A. J. A. Mom: ASME Paper 84-GT-255 (1984)
/5/ Simon, H.: Metalloberfläche 39 (1985), 321/324
/6/ McMordie, B. G., M. F. Mosser: ASME Paper 86-GT-290 (1986)
/7/ Mosser, M. F.: SAE, technical paper 880879 (1988)
/8/ Schmitt-Thomas, Kh. G., Th. Happle:
In: Werkstoffprüfung 1988, DVM e.V. (1988), 305/312
/9/ Schmitt-Thomas, Kh. G. und Th. Happle:
Proceedings of the ASM - World Material Congress, Chicago, USA, Sept. 24-30, 1988, S.23/37
/10/Schmitt-Thomas, Kh. G. und Th. Happle:
Proceedings of 9th European Congress on Corrosion, Vol. II: Energy Generation EG-207, Utrecht, Oct. 2-6, 1989+

Effects of Surface Finish on the Solid State Welding of High Strength Aircraft and Aerospace Aluminium Alloys

H. M. Tensi and M. Wittmann

Technische Universität München
München
Federal Republic of Germany

ABSTRACT

This paper reports some effects of different surface preparation methods on the bond quality of diffusion welded Aluminium alloy sheets. The preparation methods included grinding and steel brushing as well as "RF sputter cleaning" with a subsequent sputter deposition of Cu, Ag or Au.

Surface and depth profile analyses were carried out by means of Auger-Electron-Spectroscopy (AES) and Glow Discharge Source (GDS) Plasma-Emission-Spectroscopy in order to verify the surface characteristics of the sheets to be bonded.
For the welding experiments a DB-Hot-Press with a vaccum chamber has been developed especially for the Diffusion Bonding of high strength Al-sheets. The bonding process is characterized especially by the application of the weld pressure during the heating phase and by a controlled cooling phase.

Mechanical tests were carried out by shear testing and peel testing at room-temperature and SPF-temperature (516°C). Additionally metallographic and fractographic methods were used for a further characterization of the joints. The results show that basic material properties can be obtained in the solid state welding by an optimum surface treatment and optimized bonding parameters.

INTRODUCTION

The combined application of Diffusion Bonding and Superplastic Forming is highly attractive for the manufacture of aircraft structural-parts. The advantages of SPF are based on the extrem plasticity of the materials resulting in high material utilization, reduced part count and low assembly costs. As a complementary joining process Diffusion Bonding exhibits some considerable benefits compared with conventional joining techniques like riveting, fusion welding and adhesive bonding (1).

Although the joint formation in the Diffusion Bonding process is up to now not described by a uniform theory it is generally recognized that a bond is formed when the surfaces are brought together within the range of interatomic forces. This involves both intimate contact between the surfaces to be bonded and the removal of surface barriers. The geometric and the physical state of mating surfaces is therefore of a high importance for the quality of the bond. The geometry of a surface is described by its roughness and waviness. The physical state concerns above all the type and the thickness of surface contaminants like oxides and the degree of plastic deformation. Additionally the metal type itself determines the diffusion weldability (2).

For the Diffusion Bonding of Aluminium the main problem is the oxide layer which forms again in oxidizing atmosphere immediately after its removal. The Aluminium oxide is very compact and stable and therefore prevents pure metal to metal contacts and diffusional processes which are necessary for a joint for-mation (3). So a successful production of DB-joints of Aluminium alloys requires surface preparation methods which either avoid the formation of an oxide layer or which cause a fragmentation of the oxide during the welding process itself. The disruption of oxide layers necessarily requires a certain degree of bulk deformation for surface extension and extrusion of material through gaps in the fractured surface layer.

EXPERIMENTAL PROCEDURE

Materials and preparation methods

Sheets of the high strength Aluminium alloy AA 7475 and the Al-Li alloys 2091/8090 were used for the DB-experiments. Table 1 shows the nominal weight composition of the tested alloys.

TABLE 1
Nominal weight composition of examined Al-alloys

alloy sheet	composition in wt %					
	Zn	Cu	Mg	Li	Zr	Cr
7475/2mm	5.2-6.2	1.2-1.9	1.9-2.6	/	/	0.18-0.25
2091/1.7mm	/	1.8-2.5	1.1-1.9	1.7-2.3	0.04-0.16	/
8090/1.7mm	/	1.0-1.4	0.5-0.9	2.3-2.6	0.08-0.16	/

Blanks of 25 x 100 mm with the 100 mm length parallel to the sheet rolling direction were diffusion bonded in the self made vaccum hot press described later. Prior to bonding the surfaces of the specimens were prepared by different methods.

The examined surface treatments were:

1. Grinding with SiC emery papers of different roughness and cleaning with ethanol,
2. Chemical cleaning followed by scratch brushing with rotating brushes,
3. Mechanical cleaning by grinding followed by "RF ion etching" and a subsequent sputter coating with Copper, Silver or Gold.

The sheets were diffusion bonded immediately after finishing the surface preparation by grinding or scratch brushing. The sputter-coated specimens which were prepared in groups of about 30 pieces had to be stored up in a vaccum container during the time between sputter-coating and Diffusion Bonding.

Surface analyses

Surface analyses were accomplished in order to classify the physical state of the surfaces to be bonded or to find explanations for differences in the bond quality due to different preparation methods. For this purpose Auger-Electron-Spectroscopy (AES) and GDS-Plasma-Emission-Spectroscopy were available. Both methods were used for depth profile analyses.

The results from the surface analyses revealed some interesting characteristics due to the different surface states.

Figure 1 represents the monitored depth profile of a ground and uncoated AA 7475 surface. The plot indicates that in air the natural oxide layer was already established traceably within an hour. The oxide layer contains Carbon which was probably brought onto the surface by cleaning with ethanol.

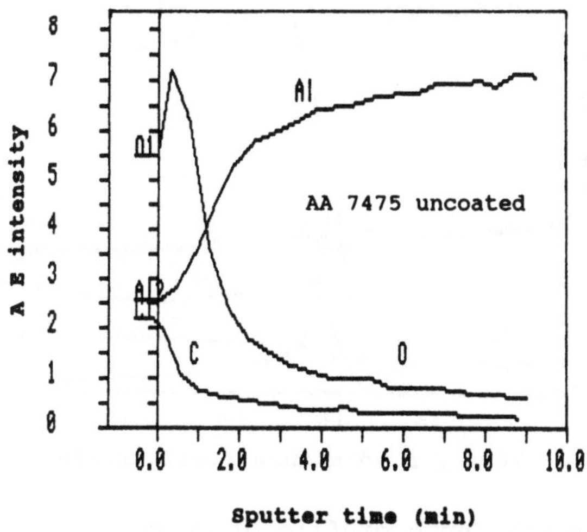

Figure 1. AES-Depth profile of a ground AA 7475 surface, exposed to air for about one hour

The depth profile of a Copper coated AlLi- (2091) sheet (Figure 2) shows that the oxide layer on the Al-surface was removed completely by "RF ion etching" before the Copper atoms were deposited on the clean surface. On the Copper surface nevertheless an oxide layer has formed similar to that on the pure Aluminium. The thickness of the oxide layers could not be determined exactly because the etching rates are only known approximately. Their size may be in the range of 10 nm in both cases. On the Copper surface also C and H were detected probably due to cleaning the Cu-surface with ethanol before the analysis was started.

The Silver coated specimens did not show any oxide film on the surface. Surprisingly in the interface between the sputterlayer and the Aluminium surface a considerable amount of oxygen could be analysed. The concentration profile presented in Fig.3 indicates, that up-hill diffusion of Oxygen through the Ag-layer must have occured provided that the surface was cleaned in the same good quality as in the case of Cu-coating. The result is the formation of Al_2O_3 and LiO near the interface. This configuration appears detrimental for the production of a good bond which has been confirmed by mechanical tests.

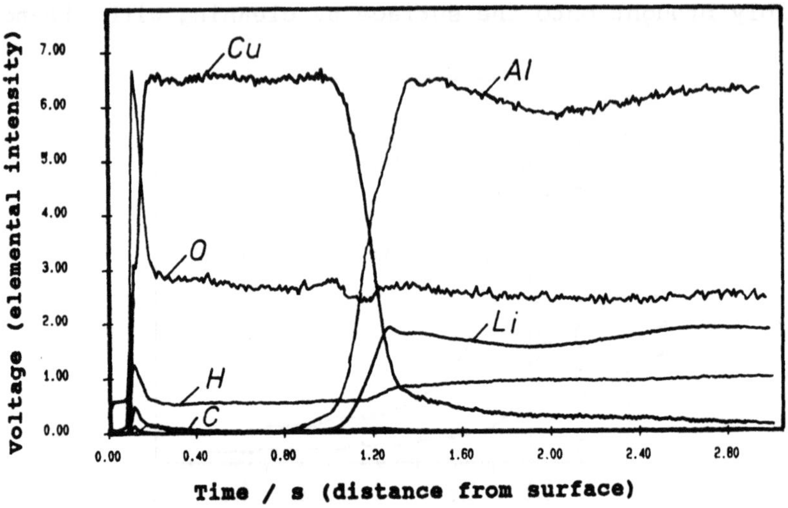

Figure 2. Elemental depth profile of Cu-coated (d=1000 Å) Al-Li 2091 using GDS-Plasma-Emission-Spectroscopy; ground surface

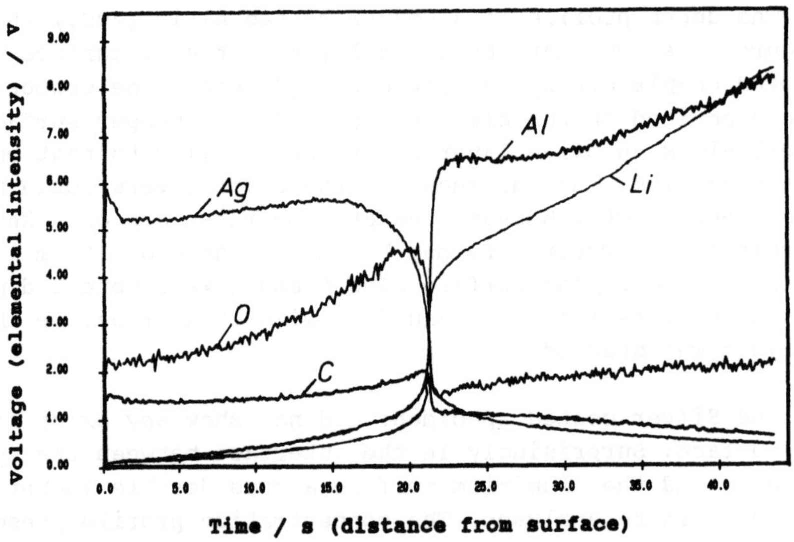

Figure 3. Elemental depth profile of Ag-coated (d=1100Å) Al-Li 8090 using GDS-Plasma-Emission-Spectroscopy; surface scratch brushed

Welding process and testing of joints
The experiments were carried out in a DB-Hot-Press with a vaccum chamber which had been developed especially for the diffusion bonding of high strength Aluminium sheets. The bonding process is characterized by the application of the weld pressure from the beginning of the heating phase and a defined temperature control in the cooling phase. This pilot plant even provides the option of quenching the specimens from bonding temperature. The welding temperature was about 510°C.

The bond quality was investigated by overlap shear tests at room temperature and peel-tests at room temperature and SPF temperature (516°C). Additionally metallographic and fractographic methods were employed for a further characterization of the joints (see Fig.7).

RESULTS FROM MECHANICAL TESTS

Influence of surface preparation on the shear strength
The effect of surface coating on the bond shear strength was measured and compared with the results from uncoated specimens. The range of clad layer thickness was 300 - 3000 Å for Cu, 1000 - 3400 Å for Ag and 1000 Å for Au. The uncoated specimens were ground with SiC emery papers of different grade in order to investigate the influence of the surface roughness on the bond quality.

Also the effect of a post bondig artifical ageing was investigated. This did not need an extra solution treatment because the bonds were quenched in cold water.

The shear strength data of diffusion bonded AA 7475 sheets are plotted in Figure 4. They show that the best results were obtained in the uncoated and artifically aged condition. In this case the shear strength values are in the range between 320 and 410 MPa. Surprisingly the shear strengths of Cu- and Ag- coated bonds were lower. Au-coated specimens showed even worse quality. Many Au-coated specimens failed during the preparation before they could be tested. Obviously artifical ageing increases the shear strength independant from the surface treatment. The influence of the interlayer thickness seems to be negligible. The influence of the surface roughness of Al-Li 2091 sheets on the shear strength is illustrated in Figure 5. Increasing surface roughness results in increasing shear strengths.

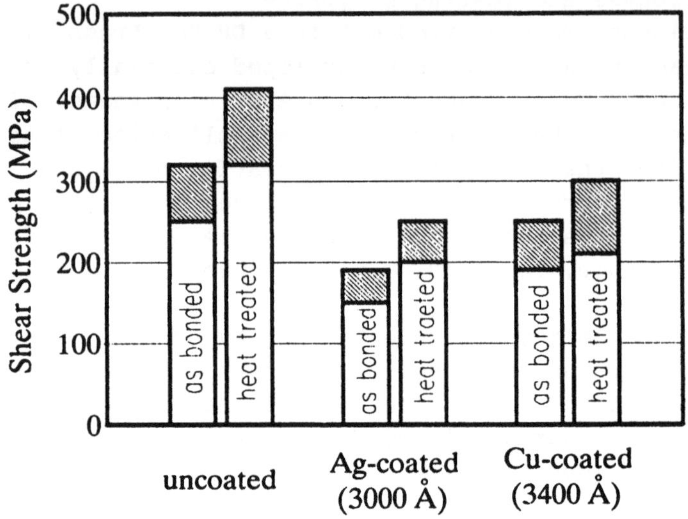

Figure 4. Shear strengths of diffusion bonded AA 7475 sheets at room temperature for different surface finishes

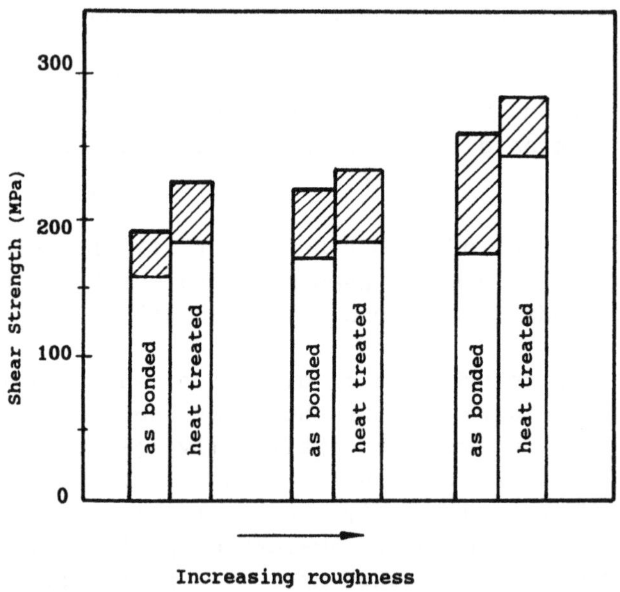

Figure 5. Influence of the surface roughness on the shear strength of diffusion bonded 2091 alloy sheets (4)

Influence of surface preparation on the peel strength

Peel tests were accomplished at room temperature and under SPF conditions (516°C). The influences of the preparation methods, grinding, scratch brushing and sputter coating were investigated. The results are presented in Figure 6a and b. They exhibit considerable differences in the bond strengths for both test conditions.

At room temperature Cu coated bonds showed relatively low peel strengths of 25 - 50 N/mm. The peel strengths of ground specimens (grade 1000) were in the range of 45 - 83 N/mm. The best results however were achieved with brushed surfaces. Peel strengths of 120 - 150 N/mm were measured. At SPF-temperature the same tendency was found. The peel strengths of sputter coated bonds (1 - 1,5 N/mm) were totally inacceptable. Medium values in the order of 2,5 - 3,8 N/mm were measured for ground surfaces. Excellent peel strengths could again be produced when the surfaces were scratch brushed. Values of 8 - 11 N/mm were obtained in this case. This makes the subsequent SPF of diffusion bonded sheets without any mechanical support of the DB-zone feasible (4).

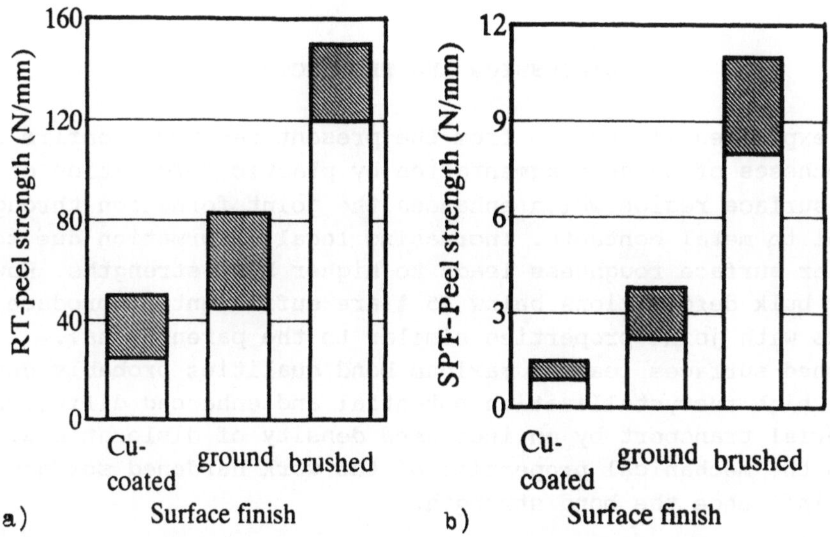

Figure 6. Effect of surface finish on the peel strength of AA 7475 sheets: a) at room temperature
b) at SPF temperature

Metallographic sections of well bonded specimens exhibit large areas where the bond interface cannot be localized, and where the bond line can merely be detected by connecting some of the rare voids which still exist (see Fig. 7a). SEM-fractographs of shear tested bonds indicate a ductile fracture behaviour similar to the parent material (Fig. 7b)

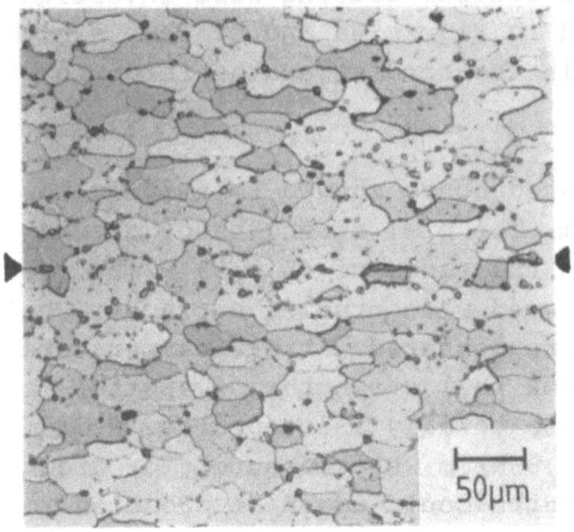

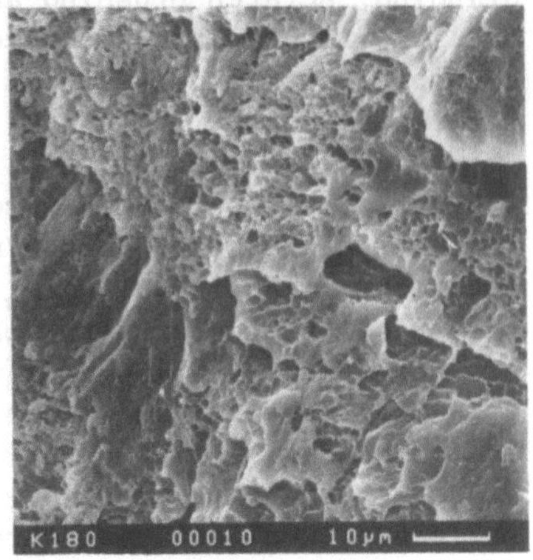

a) b)

Figure 7. a) Microstructure of well bonded AA 7475 sheets
b) SEM fractograph of a shear tested 7475 bond

DISCUSSION AND PROSPECTS

The experimental results from the present research confirm the hypotheses of oxide fragmentation by plastic deformation of the surface region which enhances the joint formation through metal to metal contacts. Increasing local deformation due to a higher surface roughness leads to higher bond strengths. However bulk deformations below 15 % are sufficient to produce bonds with joint properties similar to the parent metal. Brushed surfaces lead to maximum bond qualities probably due to a high recrystallization potential and enhanced diffusional material transport by an increased density of dislocations. Also the mechanical properties of the work hardened surfaces may influence the bond strength.

Sputtered interlayers seem unfavourable for the diffusion bonding of Aluminium alloys although the sputter materials themselves are considered easy to be bonded. The moderately stable oxides which are formed on Silver and Copper surfaces are considered as advantageous properties. Copper is additionally known for its ability to assimilate surface contaminants in the base metal.

The reason for the poor weldability of Ag-coated Aluminium sheets may however be found in the depth profile plotted in

Figure 3. Probably the deposited Silver layer was not dense enough to prevent the migration of oxygen through Silver towards the Aluminium margin. There an interfacial oxide layer could form which is responsible for the low bond strengths. Probably the same effect causes the poor bonding of Au-coated sheets.

Besides the above described positive properties of Copper the fact that it forms thick soft oxides is detrimental to diffusion bonding because they tend to smear along the weld interface. Only small areas are therefore available for welding in this case.

In summary the results exhibit that diffusion bonded joints can be produced in a quality which is required for the combined use with Superplastic Forming. Uncoated sheets bonded immediately after the surface finish exhibit promising bond strengths. If the transfer of the encouraging results from the laboratory status to the production level is possible the manufacture of diffusion bonded and superplasticly formed Al-components can be expected. Certainly a lot of research and development is still necessary before riveted parts will be replaced by DB/SPF parts in an extensive range.

REFERENCES

1. Winkler, P.J., Diffusion Bonding and Superplastic Forming, two complementary manufacturing techniques. Proceedings of Int. Conf. on Superplasticity and Superplastic Forming, Blaine, USA, Aug. 1988
2. Jellison, J.L., The role of surface contaminants in the solid-state welding of materials. Treatise on clean surface technology, volume 1, Plenum Press, New York, London, 1988, p.205-234
3. Kazakov, N.F., Diffusion Bonding of materials. Pergamon Press, Frankfurt, 1981
4. Tensi, H.M., Wittmann, M., Friedrich, H.E., Schmidt,J.J., Diffusion bonding of high strength aircraft and aerospace aluminium alloys. Light weight alloys for aerospace applications, Proc. of a TMS-symposium, Las Vegas, USA, 1989

The authors whish to thank SPECTRUMA-Instruments, Kirchheim, FRG, for the generous support of this research.

Surface Modifications of Zirconium Alloys by Laser Glazing

K. F. Amouzouvi, L. J. Clegg and R. C. Styles

Whiteshell Nuclear Research Establishment
Pinawa
Manitoba
Canada

ABSTRACT

Surface modification of Zr-2.5Nb and Zircaloy-2 materials by laser glazing has been studied. The effects of laser glazing parameters and conditions on the surface contamination and surface finish, the microstructure, the microhardness and the tensile properties have been characterized using light microscopy, scanning Auger microscopy, transmission and scanning electron microscopy, and microhardness and tensile tests. A smooth surface finish, free of contamination, could be produced if the laser glazing was performed at high scan rates (i.e., 1.92 m/s traverse speed) in a protective atmosphere of argon, with helium as a cover gas. The resulting microstructure was found to consist of a uniform, fine, α'HCP martensite in both Zr-2.5Nb and Zircaloy-2 alloys, although very fine intermetallic particles could be observed in the Zircaloy-2. Post-glaze ageing resulted in precipitation of very fine ellipsoidal particles of Nb-rich β-phase particles in the Zr-2.5Nb and in finer and additional intermetallic particles in the Zircaloy-2. Based on these results, the potential of laser glazing to improve the corrosion resistance of Zr-2.5Nb and Zircaloy-2 pressure tube materials is discussed. Laser glazing is expected to improve the corrosion resistance of Zircaloy-2 material, while Zr-2.5Nb will require post-glaze ageing before optimum corrosion resistance is achieved.

INTRODUCTION

Hydrogen can have a deleterious effect on the tensile and fracture properties of zirconium and its alloys. Hydrogen is produced and picked up during aqueous corrosion of zirconium alloys. Surface modifications, to produce within the surface layers, metallurgical structures with much improved corrosion resistance may minimize hydrogen pickup. Previous investigators [1-3] have shown that the rate of corrosion of zirconium alloys is strongly dependent on the microstructure.

Laser irradiation has been used to modify the surface microstructure of Zircaloy-2 and -4 tubing without changing the microstructure of the bulk of the tubing [5,6]. The procedure involved laser heating or melting of the tube surface and β-quenching from natural cooling of the tube. This treatment resulted in a metallurgical structure (small grains with a fine distribution of intermetallic particles) with enhanced corrosion resistance. A laser-glazing process [6] has also been used on Zircaloy-4 sheet for nuclear fuel cladding [7] to reduce its susceptibility to stress-corrosion cracking.

The present investigation studied the effect of laser glazing on the microstructure of surface layers on specimens from Zr-2.5Nb and Zircaloy-2 pressure tubes. The microstructures, surface topography and surface contamination resulting from the glazing process were characterized. The effect of the laser-glazing process on the mechanical properties (strength and ductility) was also determined. In addition, the influence of post-glaze heat treatments on the resulting microstructures was studied. This paper describes and discusses the results obtained in the present study. Based on these results, the potential of laser glazing to improve the corrosion resistance of pressure tube material is evaluated.

MATERIALS AND METHODS

The materials used in this study were from standard cold-worked pressure tubes made from Zr-2.5Nb (2.5 wt% Nb, 0.6 wt% O_2, Zr balance) and Zircaloy-2 (1.5 wt% Sn, 0.14 wt% Fe, 0.12 wt% Cr, 0.06 wt% Ni, 0.7 wt% O_2, Zr balance). Coupons (25.4 mm x 25.4 mm x 4.0 mm) and tensile specimens were prepared from flattened pressure tube.

Laser glazing was carried out at the Illinois Institute of Technology Research Institute (IITRI) by rapidly scanning the surface of the specimens with a beam from a Continuous-Wave (CW), multikilowatt, AVCO HPL-8 CO_2 laser. Powers between 6 and 10 kW were used, which correspond to average power densities of 5×10^7 to 8.5×10^7 W/mm^2. Laser beam traverse speeds of 0.48 m/s, 0.53 m/s, 0.96 m/s and 1.92 m/s, and specimen translation speeds between 8.3×10^{-5} m/s and 1.5×10^{-4} m/s were used in this study. With these power densities and laser-beam traverse speeds, surface melting followed by fast quenching to the cold and solid substrate was obtained. Laser glazing was performed with helium as a cover gas in either air or a protective atmosphere of argon (enclosed chamber filled with argon).

The microstructures of the surface layers of the glazed specimens were characterized by transmission electron microscopy (TEM). Microhardness characterizations of the glazed layers were also performed. Scanning Auger Microanalysis (SAM) was used to investigate whether surface contamination occurred during the laser-glazing process and to determine the extent of such contamination. Light microscopy and scanning electron microscopy (SEM) examinations were performed to determine the effect of laser glazing on the surface topography.

Tensile tests were carried out using an Instron testing machine to determine the effect of laser glazing on the elongation at fracture and tensile strength (yield stress at 0.2% strain, and ultimate tensile stress). The tensile specimens were glazed on both of the largest flat surfaces. A nominal strain rate of 7.7×10^{-3} s^{-1} was used during the tensile testing.

RESULTS

Macroscopic Appearance of Laser-Glazed Specimens

Visual examination of laser-glazed specimens revealed that specimens glazed in air, with helium cover gas, exhibited some discoloration (gold or brown coloured appearance), whereas specimens glazed in a protective atmosphere of argon, with the helium cover gas, did not exhibit any discoloration. The surfaces of the latter specimens were shiny and silvery. The laser-glazed surfaces exhibited a rough surface finish (Figure 1). This surface topography is typical of laser-glazed surfaces processed with CW lasers. The rough surface finish of laser-glazed specimens is mostly attributed to surface rippling effects [8]. These ripples, as revealed by light microscopy and SEM, are shown in Figure 1.

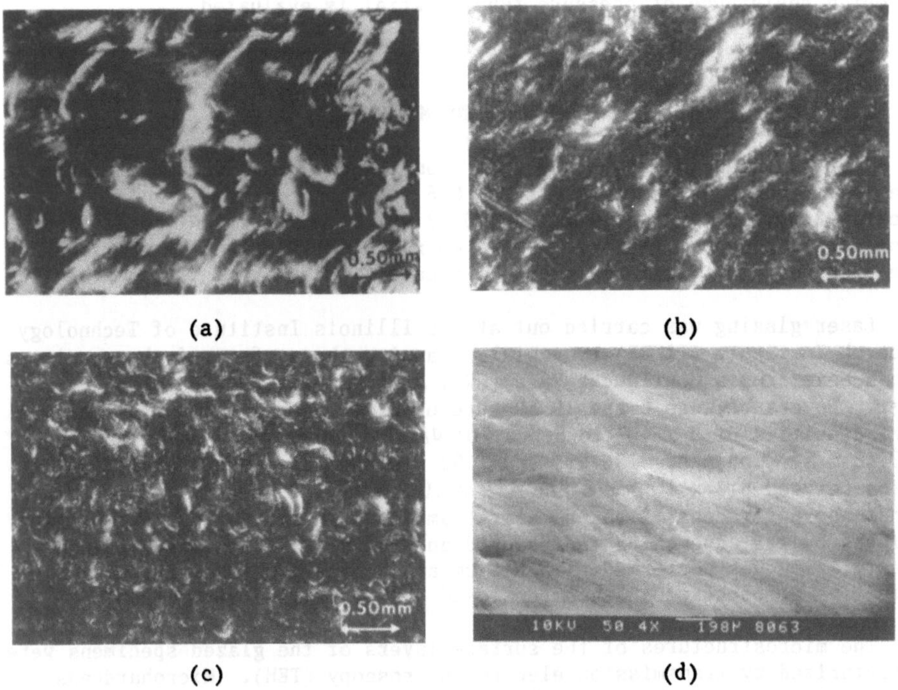

Figure 1. Optical and scanning electron micrographs showing the effect of laser scanning rates on the surface finish of laser-glazed Zr-2.5Nb specimens at 8-kW laser power, translation speed of 1.4×10^{-4} m/s and traverse speeds of (a) 0.53 m/s, (b) 0.95 m/s and (c) 1.91 m/s. (d) Rippled surface finish (same glazing conditions as in (a)) as revealed by SEM.

SAM Analysis of Surfaces of Laser-Glazed Zr-2.5Nb Specimens

An Auger depth profile (Figure 2(a)) recorded from a silver-coloured specimen after 1-min sputter cleaning revealed a surface oxide, about 10 nm in thickness.

Auger spectrum (Figure 2(b)) recorded at a sputtered depth of 40 nm revealed mainly metallic Zr, with a small amount of oxygen. SAM analyses performed at different locations on the cross section of the silver-coloured specimen at 25 μm, 60 μm and mid-thickness (≃2 mm) showed much lower concentrations (indistinguishable from background) of oxygen than that observed at 40 nm, Figure 2(b).

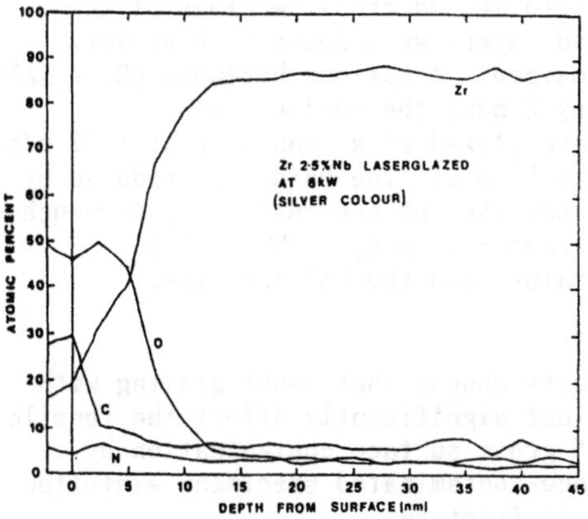

(a)

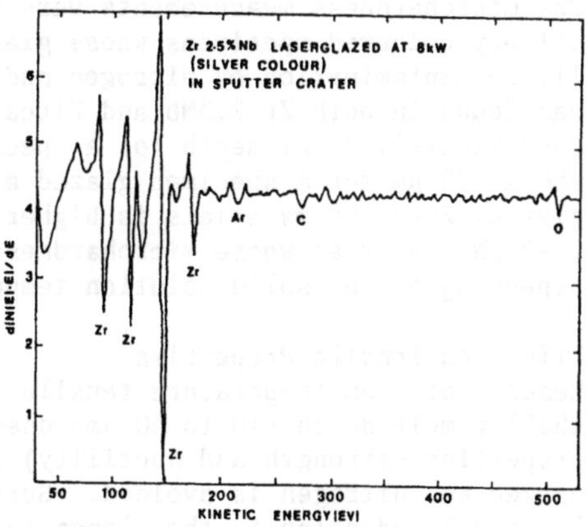

(b)

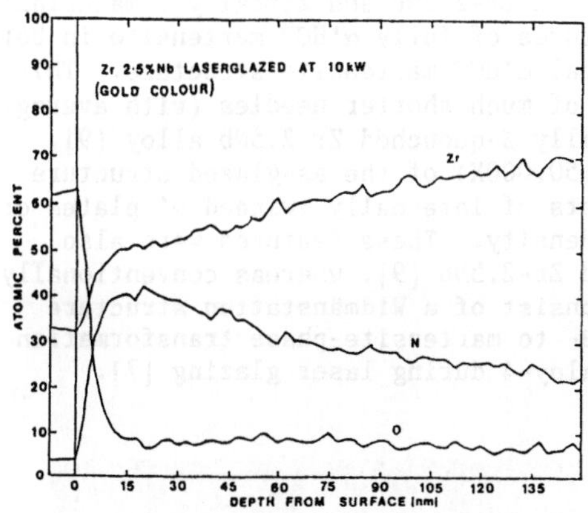

(c)

Figure 2. (a) Auger depth profile of laser-glazed specimens (shiny and silvery). (b) Auger spectrum of specimen in (a) but recorded at 40-nm depth. (c) Auger depth profile of laser-glazed specimen (gold coloured).

An Auger depth profile (Figure 2(c)) recorded from the gold-coloured specimen revealed a different profile, compared with that of the silver-coloured specimens. The contaminated surface layer was composed of ZrO_2 at the sample surface, but the underlying layer contained more nitrogen than oxygen. While the concentration of oxygen steeply decreased away from the free surface (estimated oxide thickness less than 10 nm), a significant and abrupt increase in the nitrogen concentration occurred up to 18-nm depth. The nitrogen concentration then decreased slowly with depth. Auger spectra recorded on the cross-section indicated a depth of nitrogen- and oxygen-contaminated layers of the order of 1 μm, with the nitrogen concentration tailing off slowly into the bulk over a glazed depth of several micrometres. The melt depth ranged from 10 to 75 μm, depending on the laser-beam traverse speed.

Microhardness Characterizations

The microhardness measurements were performed on cross sections of the silvery coloured specimens whose glazed layers were found to have very little contamination by nitrogen and oxygen. A maximum hardness ($H_v \simeq 323$) was found in both Zr-2.5Nb and Zircaloy-2 near the surface, at approximately 25-μm depth for a specimen glazed at a scan rate of 1.92 m/s, and at 75 μm for a specimen glazed at 0.53 m/s. The hardness produced by glazing Zr-2.5Nb materials is higher than that of conventionally β-quenched Zr-2.5Nb material whose microhardness ranges from $H_v \simeq 250$ to 275, depending on the solid solution temperature and the holding time.

Effect on Tensile Properties

Results of room temperature tensile tests showed that laser glazing with shallow melt depth (10 to 50 μm) does not significantly affect the tensile properties (strength and ductility) if gross surface contamination by oxygen and nitrogen is avoided. Surface-contaminated specimens exhibited 30 to 35% reduction in the elongation at fracture.

TEM Examinations of As-Glazed and Post-Glazed Heat-Treated Specimens

TEM examinations of the glazed layers in Zr-2.5Nb and Zircaloy-2 materials revealed fine and uniform microstructures of fully α'HCP martensite in both materials. Figure 3(a) shows a typical α'HCP martensite structure. This structure is much finer and consists of much shorter needles (with average length of ≃ 2 μm) than in conventionally β-quenched Zr-2.5Nb alloy [9]. Examinations at high magnification (150,000X) of the as-glazed structure revealed that the α'martensite consists of internally twinned α' plates or grains with a very high dislocation density. These features were also observed in conventionally β-quenched Zr-2.5Nb [9], whereas conventionally β-quenched Zircaloy-2 was found to consist of a Widmänstatten structure with plates of α-zirconium [10]. A β- to martensite-phase transformation was also previously observed in Zircaloy-4 during laser glazing [7].

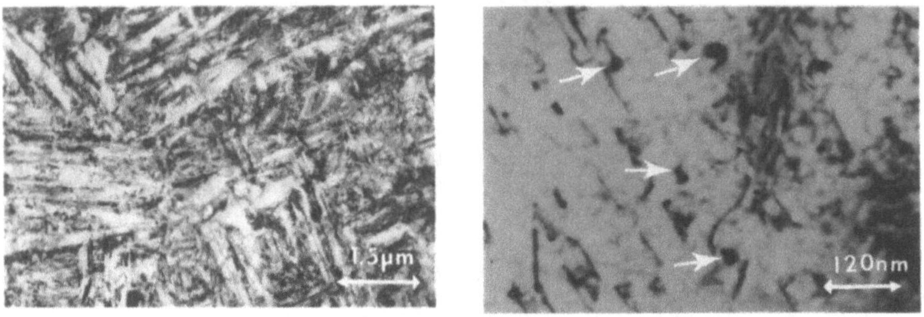

(a) (b)

Figure 3. Transmission electron micrographs showing (a) a typical α'martensite structure in an as-glazed Zr-2.5Nb specimen and (b) fine intermetallic particles (indicated by arrows) in an as-glazed Zircaloy-2 specimen.

TEM examination at high magnification of as-glazed Zircaloy-2 material revealed very fine (≤15 nm) second-phase particles (Figure 3(b).

These fine precipitates were much smaller than those observed in the as-received Zircaloy-2 material and are believed to be Zr-Fe-Cr-Ni intermetallic particles, usually observed in as-received Zircaloy-2.

Fine ellipsoidal β-phase precipitates, 5.5 to 29 nm in size (major axis) were observed in recrystallized grains (Figure 4(a,b)) in glazed Zr-2.5Nb specimens that were aged at 673 K or 723 K for 24 h or longer. Unlike the conventionally β-quenched Zr-2.5Nb materials, where precipitation of the particles occurs preferentially at twin boundaries, precipitates in laser-glazed specimens were mostly observed in recrystallized grains (Figure 4(a) to (c)). Ageing of glazed Zr-2.5Nb at 763 K for 24 h to 90 h resulted in a similar precipitation behaviour observed at 673 K and 723 K, except that higher density and slightly coarser (9 to 55 nm) precipitates were observed. Precipitation in the glazed Zr-2.5Nb aged at 763 K for 90 h occurred within the α'martensitec plates (Figure 4(d)) and

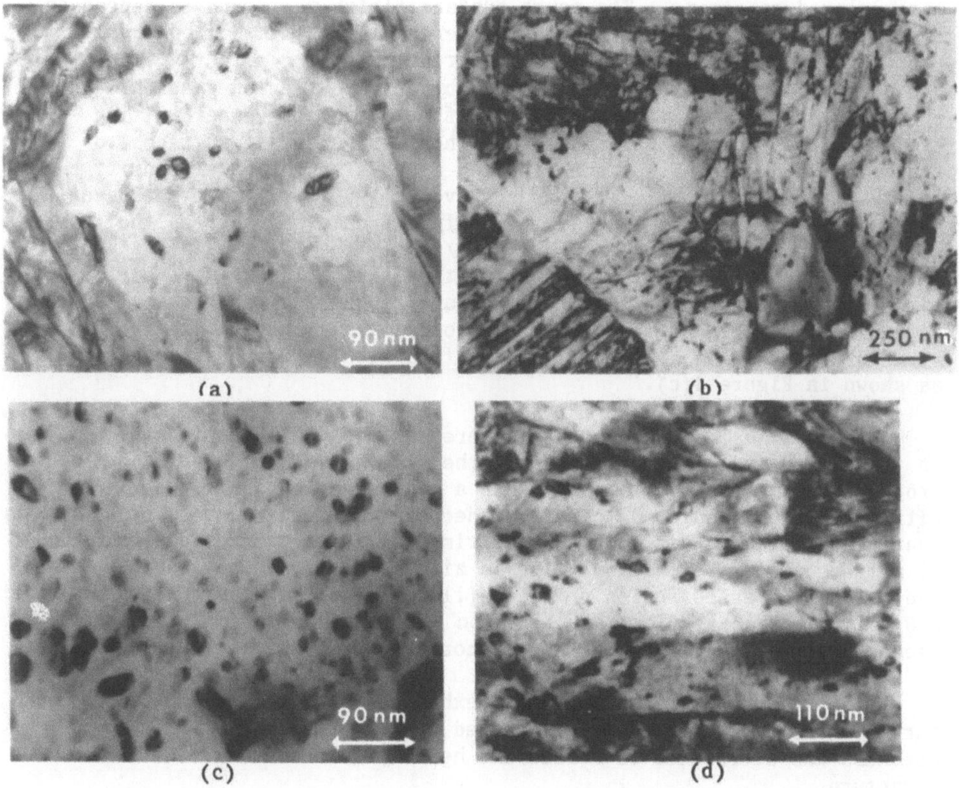

Figure 4. Transmission electron micrographs showing fine ellipsoidal Nb-enriched particles in (a) a recrystallized grain of glazed Zr-2.5Nb specimen aged at 673 K for 24 h, (b) recrystallized grains of glazed Zr-2.5Nb specimens aged at 723 K for 24 h (note absence of particles within martensite plates adjacent to recrystallized grains), (c) a recrystallized grain of a glazed specimen aged at 763 K for 24 h and (d) within martensitic plates of a glazed specimen aged at 763 K for 90 h.

within recrystallized grains. Energy dispersive X-ray (EDX) analyses of the precipitates obtained after ageing at 673 K to 763 K showed that these precipitates were niobium-enriched.

Ageing of glazed Zircaloy-2 specimens at 723 K for 24 h did not result in significant coarsening of the intermetallic particles initially present in the as-glazed specimens. Instead, a higher density of much finer precipitates was produced. The intermetallic particles observed in as-glazed or glazed and heat-treated specimens were at least 10 to 20 times smaller than those observed in conventionally β-quenched Zircaloy-2 materials.

DISCUSSION

Optical and SEM examinations revealed a rough surface finish on all the laser-glazed specimens. The rough surface finish was found to depend on the coarseness of tracks produced by overlapping passes during the glazing process and was mostly attributed to surface rippling effects. The ripples are due to liquid flow, mostly induced by surface-tension gradients during the laser surface melting. The liquid flow during the laser melting creates differences in height between the first and last surfaces to freeze. The degree of surface rippling depends on the power density but is mostly affected by the laser-beam scan velocity, i.e., traverse speed [8]. As shown in Figure 1(a) to (c), the surface became smoother as the traverse speed increased from 0.53 m/s to 1.91 m/s. If the laser-beam scanning velocity exceeded a critical velocity, the liquid did not have sufficient time to form ripples and therefore surface roughening due to rippling can be avoided [8]. Thus, it is possible to eliminate or minimize the ripple structure and obtain a laser-glazed surface with a smooth surface finish, as shown in Figure 1(c).

SAM analyses have shown that severe surface contaminations by nitrogen of the glazed surface could occur if the glazing was not carried out in a protective atmosphere of argon. Such a contamination, particularly by nitrogen, which was found to persist deep in the glazed layers of specimens that were glazed in air, could be detrimental to the corrosion performance. If nitrogen is present in solution in zirconium, it can be absorbed into the oxide film as corrosion proceeds [11]. The presence of nitrogen in the ZrO_2 lattice increases oxygen diffusion (by introducing additional mobile vacancies) and hence produces higher corrosion rates [11,12].

Laser-glazed Zr-2.5Nb specimens exhibited a higher (17 to 30%) hardness than conventionally β-quenched Zr-2.5Nb material. This higher hardness was attributed to a fine and heavily deformed α'HCP martensitic structure.

The lower ductility exhibited by contaminated specimens was attributed to the presence of a brittle nitrogen-enriched (ZrN) surface layer. Cracking of the ZrN surface layers during testing probably resulted in the reduction of the total elongation to fracture.

The finer α'martensite microstructure compared with conventionally β-quenched Zr-2.5Nb was due to the high cooling rate during laser glazing, which resulted in very fine prior β-grains. Laser glazing resulted in β- to α'martensite phase transformation in Zircaloy-2, in agreement with previous results [7]. In contrast, conventionally β-quenched Zircaloy-2 was found to consist of a Widmänstatten structure, with platelets of α-zirconium [11]. This observation and the present results suggest that a much higher cooling rate (such as that achieved with laser glazing) was required for a martensitic reaction to occur in the Zircaloys, probably because of the presence of tin, which is an α-phase stabilizer. The rapid cooling rates in laser glazing (10^4 to 10^5 K/s) were insufficient to suppress the precipitation of intermetallic particles in Zircaloy-2. Instead, a significant refinement of these intermetallic particles occurred.

In addition to precipitation of Nb-enriched β-phase particles in Zr-2.5Nb material and additional intermetallic particles in Zircaloy-2, post-glaze ageing resulted in partial recrystallization of the α'martensite structure at temperatures as low as 673 K. This propensity of glazed specimens to recrystallize at a low temperature was attributed to the presence of a high density of twins and dislocations, similar to what is observed in heavily deformed materials. The low recrystallization temperature is consistent with well-established recrystallization mechanisms, where the recrystallization temperature decreases as prior deformation is increased, because severe deformation increases the number of sites at which recrystallization nucleation can occur. Lesurf [3] found that, in conventionally β-quenched Zr-2.5Nb, an additional 30% cold work was necessary before recrystallization could occur after ageing treatment for 24 h at 773 K. This would indicate that laser glazing produces a more heavily deformed structure than conventional β-quenching.

POTENTIAL OF LASER GLAZING TO IMPROVE CORROSION RESISTANCE OF PRESSURE TUBE MATERIALS

In Zr-2.5Nb material, the rapid cooling (10^4 to 10^5 K/s) of laser glazing produced a fine α'HCP martensite supersaturated with niobium, similar to conventionally β-quenched Zr-2.5Nb. Such a microstructure in conventionally β-quenched Zr-2.5Nb was previously found [13,14] to exhibit a poor corrosion resistance. However, in conventionally β-quenched Zr-2.5Nb, this disadvantage was largely overcome by ageing treatments [3,13,16] at 773 K for 24 h or by a combination of ageing and cold working, which resulted in precipitations of β-Nb particles. Precipitation of β-Nb particles reduces the niobium supersaturation of the α'martensite and is accompanied by α' to equilibrium α-zirconium phase transformation. Optimum corrosion resistance has been achieved when the Nb concentration was reduced to the equilibrium concentration in α-Zr, i.e., 1.08 at.% Nb [14,15]. Thus, a laser-glazed Zr-2.5Nb component requires a post-glaze ageing treatment to maximize its corrosion resistance.

Laser glazing did not prevent the precipitation of intermetallic particles in Zircaloy-2. Instead, a significant refinement of the intermetallics has occurred. One of the current hypothesis on the "breakaway" oxidation phenomenon observed in Zircaloys, which involves a transition of the oxide from a protective to a non-protective state, is based on the presence of coarse second-phase intermetallic particles in the alloy. The breakaway results from the cracking in the oxide [17-19] at sites of local heterogeneities, i.e., coarse intermetallic particles, where the oxidation growth stress (due to the high Pilling-Bedworth ratio (1.56)) may be concentrated. The onset of the "breakaway" oxidation is determined to a large extent by the size and distribution of the intermetallic particles present in the material. Glazing of Zircaloy-2, which resulted in a significant refinement of the intermetallic particles, should improve the oxidation resistance of Zircaloys. This conclusion has been corroborated by results of previous investigations [10], where conventional β-quenching has resulted in an improved corrosion resistance, provided the cooling rate through the ($\alpha + \beta$) phase exceeded 50 to 80 K/min. With glazing, cooling rates of 10^4 to 10^5 K/s were achieved, i.e., 4 to 5 orders of magnitude higher than those obtainable in the conventional β-quenching. Thus, unlike Zr-2.5Nb, laser glazing could be used to improve the corrosion resistance of Zircaloys without any post-glaze heat-treatments, providing surface contamination is avoided.

CONCLUSIONS

Laser glazing of Zr-2.5Nb and Zircaloy-2 materials resulted in a substantial modification of the surface in both alloys. Some laser-glazed specimens processed without sufficient inert gas shielding exhibited macroscopically observable contamination by nitrogen and oxygen. Surface contamination was minimized by glazing in a protective atmosphere of argon. High scanning rates (1.92 m/s) during the glazing treatment resulted in the best surface finish. The melted zone consisted of fine α'HCP martensite in both Zr-2.5Nb and Zircaloy-2 materials. In Zr-2.5Nb, the α'martensite was supersaturated with niobium and required ageing at 673 K, 723 K, and 773 K for 24 to 90 h to precipitate niobium-enriched β-phase particles. In Zircaloy-2, the rapid cooling associated with the glazing did not prevent the precipitation of intermetallic particles, rather it produced a significant refinement of the intermetallic particles. Laser glazing is expected to result in improved corrosion resistance of Zircaloy-2, compared with that of as-received material. Zr-2.5Nb will require post-glaze ageing to maximize its corrosion resistance. Laser glazing with a shallow melt depth does not affect the tensile properties of Zr-2.5Nb material if surface contamination by nitrogen is avoided.

ACKNOWLEDGEMENTS

The authors are indebted to Dr. R. Dutton, for suggesting work on laser surface treatment and for his encouragement throughout the course of this study. We are grateful to Dr. W.H. Hocking and to Mr. J. Betteridge for carrying out the SAM analyses, and to Dr. C.D. Cann and Dr. R. Ploc for their comments on the manuscript and helpful discussions. We also thank Dr. C.E. Coleman for his useful comments and suggestions, and to the secretaries L. Graham and G. Oliveira for typing the manuscript.

REFERENCES

1. Cox, B. J. Electrochem. Soc., 1961, 24, 24-30.

2. Urbanic, V.F., Lesurf, J.E. and Johnson Jr., A.B. Corrosion, 1975, 31, 15-20.

3. Lesurf, J.E. In Applications-Related Phenomena in Zirconium and its Alloys: Proc., ASTM STP 458, 1968, pp. 286-300.

4. Anthony, T.R. and Cline, H.E. US Patent 4,294,631 (1981).

5. Sabol, G.P., McDonald, S.G., Nurminen, J.I. and Jacobsen, W.A. In Zirconium in the Nuclear Industry, Proc. Seventh Int. Symp., eds., R.B. Adamson and L.F.P. Van Swam, ASTM STP 939, 1987, pp. 168-186.

6. Breinan, E.M., Kear, B.H. and Banas, C.M. Physics Today, 1976, 29(11), 44-50.

7. Snow, D.B. and Breinan, E.M. In Applications of Lasers in Materials Processing, proceedings of ASM Conference, Metals Park, 1979, pp. 229-243.

8. Anthony, T.R. and Cline, H.E. J. App. Phys. 1977, 48(9) 3888-3894.

9. Banerjee, S. and Krishnan, R. Acta Met., 1971, 19, 1317-1336.

10. Anderson, T. and Vesterlund, G. In Zirconium in the Nuclear Industry, Proc. Fifth Int. Conf. ed., D.G. Franklin, ASTM STP 754, 1982, pp. 75-95.

11. Thomas, D.E. In Metallurgy of Zirconium, eds, B. Lustman and F. Kerze, Jr., McGraw-Hill, New York, 1955.

12. Kass, S., Grozier, J.D. and Shubert, F.L. Corrosion, 1964, 20, 350t-360t.

13. Urbanic, V.F. and Gilbert, R.W. Presented at IAEA Technical Committee Meeting on Fundamental Aspects of Corrosion of Zirconium-Based Alloys for Water Reactor Environments, Portland, Oregon, 11-19 September 1989.

14. Klepfer, H.H. J. Nucl. Mat., 1963, 9, 65-76.

15. Williams, C.D. and Ells, C.E. Phil. Mag., 1968, 18, 763-772.

16. Cox, B. Atomic Energy of Canada Limited Report, AECL-5610, 1976.

17. Dawson, J.K., Long, G., Seddon, W.E. and White, J.F. J. Nucl. Mat., 1968, 25, 179-200.

18. Bradhurst, D.H. and Heuer, P.M. J. Nucl. Mat., 1970, 37, 35-47.

19. Ahmed, J. and Keys, L.H. J. Less-Common Metals, 1975, 39, 99-107.

Laser Shock Hardening of Welded Maraging Steel

G. Banaš, F. V. Lawrence Jr, J. M. Rigsbee

University of Illinois at Urbana–Champaign
Urbana
Illinois
USA

and H. E. Elsayed-Ali

University of Rochester
Rochester
New York
USA

ABSTRACT

The effect of laser-induced high intensity stress waves on the hardness, residual stresses, and fatigue resistance at the heat affected zone of welded 18 Ni (250) maraging steel was investigated. Laser shock processing increased the hardness, induced compressive residual stresses, and increased fatigue strength of weldments.

INTRODUCTION

The feasibility of utilizing laser-induced shock waves to modify material properties was demonstrated by several researchers [1 - 7]. Pressure measurements showed that laser-induced shock waves could produce peak compressive surface stresses up to several GPa [8]. These compressive stresses rapidly decreased in intensity as the shock wave propagated into the interior of the metal. In order to obtain that magnitude of peak stress, a laser power density of several times 10^9 W/cm^2 with nonosecond pulse duration was required [8]. If the stored elastic distortional energy due to the laser generated shock wave was greater than or equal to that necessary for yielding, plastic deformation would occur resulting in compressive residual stresses within the material. Residual stresses have been proven to increase the fatigue resistance of materials due to the mean stress effect [9].

The initial impact of the laser pulse energy is analogous to a high deformation rate explosive shock wave deforming the material. The subsequent stress wave generated by the laser pulse can give rise to elastic and plastic deformation, formation and rearrangement of dislocations, and formation of vacancies and vacancy clusters leading to microporosity [10 -

12]. The above-mentioned effects are important from the metallurgical viewpoint since they can alter the microstructural and mechanical properties of the laser irradiated metal. Beyond a threshold energy density for a particular material, melting and crater formation are possible. If melting is involved, then the subsequent ultra-rapid solidification can produce metallic glasses, ultrafine grain structures, metastable phases, and a large concentration of vacancies [11].

In our study, we fatigue tested both as-welded and laser-shock-hardened specimens and determined hardness and residual stress changes in the heat affected zone (HAZ) of the laser treated 18 Ni (250) maraging steel weldments.

TABLE 1
Chemical composition of 18 Ni (250) maraging steel.[a, b]

C%	Mn%	P%	S%	Si%	Ni%	Cr%	Mo%	Cu%	Ti%	Co%	Al%
0.02	0.01	0.008	0.007	0.08	18.29	0.03	5.40	0.04	0.34	7.31	0.01

[a] Chemical composition analysis was done by CHICAGO SPECTRO SERVICE LABORATORY, Inc.
[b] Iron -- balance.

TABLE 2
Mechanical properties of 18 Ni (250) maraging steel.[a]

Monotonic properties		
Elastic modulus, GPa	E	186
Hardness, HRC		50-52
Yield strength (0.2%), MPa	S_y	1,700
Ultimate tensile strength, MPa	S_u	1,600
Reduction in area	%RA	55
True fracture strength, MPa	σ_f	2,146
True fracture ductility	ε_f	0.80
Strength coefficient, MPa	K	2,160
Strain hardening exponent	n	0.03
Cyclic properties		
Fatigue ductility coefficient	ε_f'	0.80
Fatigue ductility exponent	c	-0.61
Fatigue strength coefficient, MPa	σ_f'	2,232
Fatigue strength exponent	b	-0.063
Cyclic strength coefficient, MPa	K'	2,282
Cyclic strain hardening exponent	n'	0.10
Cyclic yield strength, MPa	S_y'	1,245
Transition fatigue life, reversals	$2N_t$	330
Propagation properties		
Fracture toughness, $MPa\sqrt{m}$ (R = 0)	K_C	330

[a] All mechanical properties were obtained from experiments [13] and calculations [14].

EXPERIMENTAL PROCEDURE

A 18 Ni (250) maraging steel was used in this study which had the chemical composition given in Table 1. A filler metal of a similar composition was used during welding. The monotonic, cyclic, and fatigue properties after aging of this steel are given in Table 2.

The dimensions and geometry of the specimens used in this study are shown in Fig. 1. The button-shaped bead-on-plate was made using gas tungsten arc (GTA) welding with parameters given in Table 3. The specimens were coated with a thin layer (0.1 mm) of black paint and were placed under a 3.5 mm of water during laser irradiation. The thin layer of black paint provided a sacrificial material with a low heat of vaporization that enhanced laser coupling and, to some degree, protected the surface of the specimen from melting and vaporization. The water overlay acted to confine the vaporized sacrificial material and was previously found to be effective in increasing the amplitude of the shock wave [15].

TABLE 3
GTA welding parameters

Plate thickness, mm	4.15
Electrode diameter, mm	2.3
Voltage, V	24
Current, A	150
Heat input, kJ/mm	4.5
Travel speed, mm/s	1.7
Shielding gas	Argon
Gas flow rate, m^3/h	0.51
Gas cup size, mm	12.7

Experiments were performed using a laser consisting of a 100 MHz Nd:YAG (λ = 1.06 μm) oscillator followed by a Pockels cell for pulse selection. The oscillator pulse was injected into a regenerative amplifier capable of providing 0.5 mJ pulses with 100 ps pulse width and up to 1 kHz repetition rate. These pulses were cleaned using a switch-out Pockels cell. A double-pass amplifier operating at 8 Hz repetition rate was used to amplify the laser pulses to the 30 mJ level. The specimens were irradiated with 20 mJ pulses (100 ps pulse width) at 8 Hz repetition rate. Since the laser beam was focussed to 0.1 mm spot diameter at the specimen, a technique of pulse overlapping was used in order to completely cover the weldment zone (~28 mm^2); see Fig. 2. A computer controlled X-Y table was used to move the water tank containing the specimen in the required directions while processing.

Specimens were fatigue tested in a 20 kip servo-hydraulic material testing system (MTS) machine with self-aligning grips using a sine wave form, load control, a zero-to-max. stress cycle (stress ratio is equal zero), and a test frequency of 15 Hz. Two Micro-measurements strain gauges (6.4 mm length and 120 Ω resistivity) were attached to both sides of each specimen using M-bond 200 adhesive (cyanocryle) to monitor bending stresses. A proper distance (one-third of the plate thickness) from the weld toe was maintained so that the monitored strain would be independent of the weld toe stress concentration and would approximate the nominal surface strain. The hardness traverse of both unshocked and shocked samples was measured using Leitz microhardness tester with Knoop indenter and a load of 2 N. The longitudinal (axial) surface residual stresses were measured using an X-ray residual stress apparatus. The irradiated spot was 2.5 mm diameter with an X-ray surface

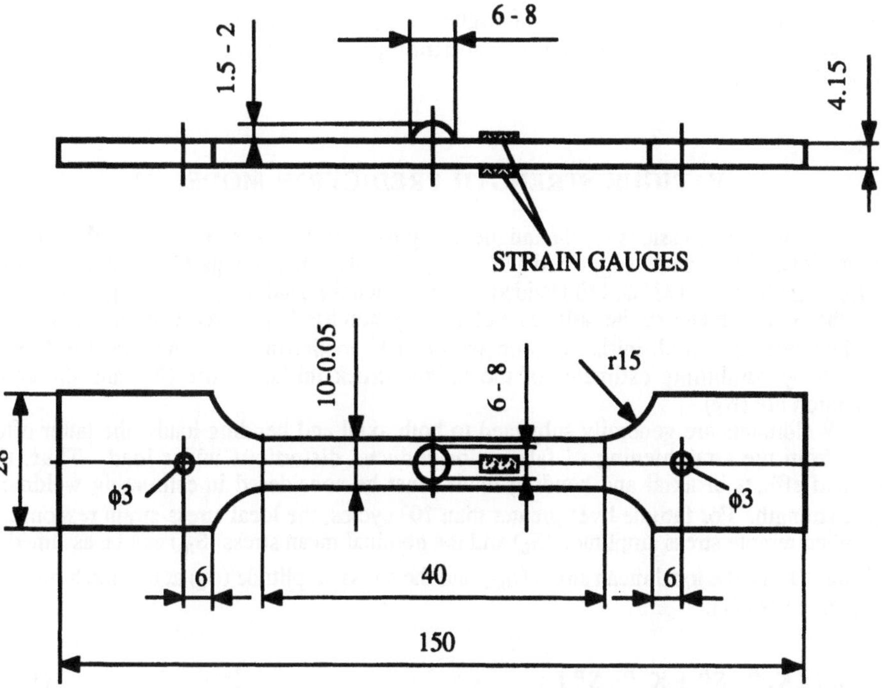

Figure 1. Fatigue test specimen with button-shaped beat-on-plate.

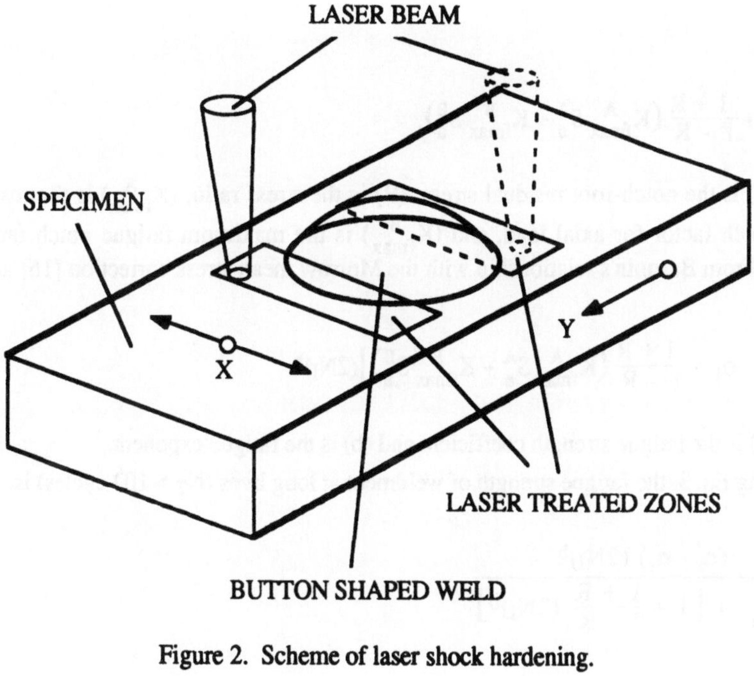

Figure 2. Scheme of laser shock hardening.

penetration depth less than 5 µm. Fatigue strength predictions of laser-shocked and unshocked weldments were made using initiation-propagation model (I-P) [16 - 22] and compared to those of experimental.

FATIGUE STRENGTH PREDICTION MODEL

Empirical expressions for the fatigue strength of weldments subjected to either axial or bending loads have been suggested [18 - 21, 23, 24]. A propagation model has been developed by Maddox [23] and El Haddad [24] in which the total fatigue life (N_T) is predicted solely through estimates of the fatigue crack propagation life (N_P). Lawrence et al., [16 - 21] have developed a hybrid, initiation-propagation (I-P) model in which the total life (N_T) is predicted by combining estimates of the fatigue crack initiation life (N_I) and the crack propagation life (N_P).

Weldments are generally subjected to both axial and bending loads; the latter often results from the straightening of fabrication induced distortions under load. Thus, the combined effects of axial and bending loads must be considered in estimating weldment fatigue strength. For fatigue lives greater than 10^5 cycles, the local stress-strain response to the applied remote stress amplitude (S_a) and the nominal mean stress (S_m) can be assumed to be elastic. Thus, the local mean stress (σ_m) and the stress amplitude (σ_a) at the notch root can be expressed as [21]:

$$\sigma_m = \sigma_r + \left(K_{fmax}^{A} S_m^{A} + K_{fmax}^{B} S_m^{B}\right) \tag{1}$$

since $S_m = \left(\frac{1+R}{1-R}\right) S_a$

thus,

$$\sigma_m = \sigma_r + \frac{1+R}{1-R}\left(K_{fmax}^{A} S_a^{A} + K_{fmax}^{B} S_a^{B}\right), \tag{2}$$

where (σ_r) is the notch-root residual stress, (R) is the stress ratio, (K_{fmax}^{A}) is the maximum fatigue notch factor for axial load, and (K_{fmax}^{B}) is the maximum fatigue notch factor for bending. From Basquin's relationship with the Morrow mean stress correction [16] and Eqs. 1 and 2;

$$\sigma_a = \left[\sigma_f' - \sigma_r - \frac{1+R}{1-R}\left(K_{fmax}^{A} S_a^{A} + K_{fmax}^{B} S_a^{B}\right)\right](2N_I)^b \tag{3}$$

where (σ_f') is the fatigue strength coefficient and (b) is the fatigue exponent.
Rearranging Eq. 3, the fatigue strength of weldment at long lives ($N_T > 10^5$ cycles) is:

$$S_a = \frac{(\sigma_f' - \sigma_r)\,(2N_I)^b}{K_{fmax}^{eff} + \left[1 + \frac{1+R}{1-R}\,(2N_I)^b\right]} \tag{4}$$

where:

$$K_{fmax}^{eff} = \left(1 - \frac{S_a^B}{S_a^A + S_a^B}\right) K_{fmax}^{A} + \frac{S_a^B}{S_a^A + S_a^B} K_{fmax}^{B} \tag{5}$$

Fatigue stress concentration factor and residual stresses

The diagram of cross-section of button shaped bead-on-plate weldment is shown in Fig. 3. The (K_{fmax}) concept [21] was used to estimate the effect of stress concentration factor on the fatigue strength. Peterson's equation correctly interrelates the fatigue notch and elastic stress concentration factors:

$$K_f = 1 + \frac{K_t - 1}{1 + \frac{a}{r}} \tag{6}$$

where (K_t) is the elastic stress concentration factor for a given notch geometry, (r) is the notch-root-radius, (K_f) is the fatigue notch factor, and (a) is the Peterson's material parameter. The (K_{fmax}) occurs at the notch radii numerically equal to Peterson's parameter (a) ($r_c = a$):

$$K_{fmax} = 1 + \frac{K_t - 1}{2} \tag{7}$$

Elastic stress concentration factor (K_t) was determined using finite element analysis considering both tensile and bending stresses.

The residual stresses at the weld toe may be simulated as shown in Fig. 3. A remote stress of (σ_r/K_f) is then applied to produce a local stress equal to the residual stress (σ_r). The normally applied remote stress range (ΔS) is then added to give ($\Delta S + \sigma_r/K_f$) which is equivalent to the remote stress range (ΔS) in Eq. 8.

$$\frac{(K_f \Delta S)^2}{E} = \Delta\sigma\Delta\varepsilon \tag{8}$$

where : ($\Delta\sigma$) is the local stress, and ($\Delta\varepsilon$) is the local strain. The local and remote stress-strain behavior can then be bounded by Neuber's rule (Eq. 8) as shown in Fig. 3. The first reversal (half of cycle) is bound by:

$$\frac{\left(K_f \frac{\Delta S}{1 - R} + \sigma_r\right)^2}{E} = \Delta\sigma\Delta\varepsilon \tag{9}$$

Fatigue properties

Both parameters of Basquin's equation, (σ_f') and (b) from Eq. 4, are related to tensile strength and can be estimated from hardness measurements. McMahon [22] assumed a linear relationship between (σ_f') and hardness:

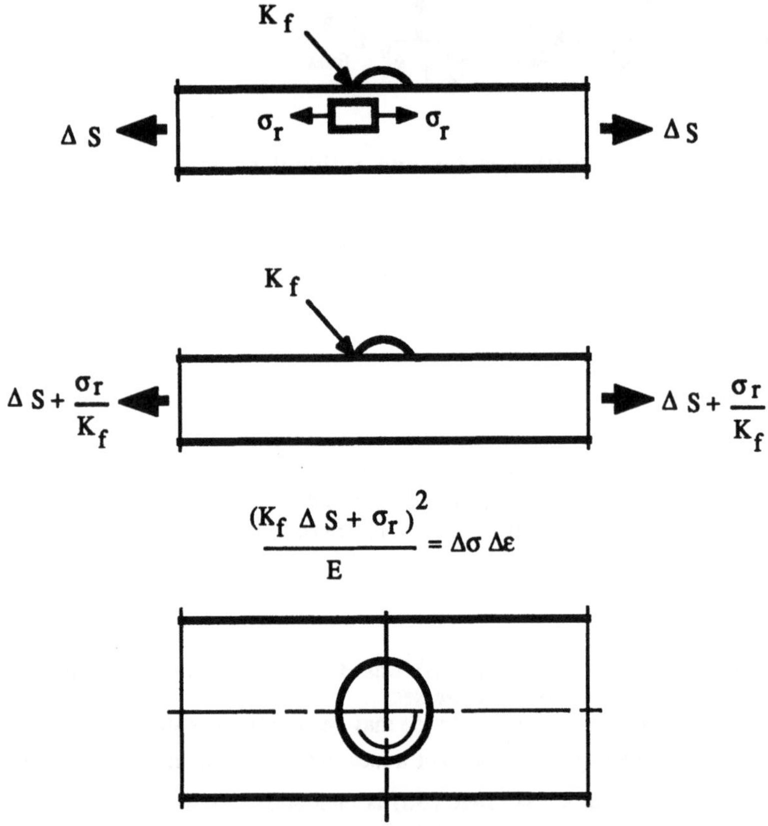

Figure 3. Mechanical simulation of residual stresses at the wled toe by elastic superposition and Neuber's rule.

$$\sigma_f' = 3.3\ \mathrm{DPH} + 370 = 0.95\ S_u + 400 \quad \mathrm{(MPa)} \tag{10}$$

where (DPH) is the diamond pyramid hardness (Vickers) and (S_u) is the ultimate tensile strength.

There is a general trend of (b) decreasing in absolute value with increasing hardness [22]. A relationship commonly used to relate (b) to the hardness (or ultimate strength) is:

$$b = -\frac{1}{6}\log\left(\frac{2\sigma_f'}{S_u}\right) = -\frac{1}{6}\log\left(\frac{2\mathrm{DPH} + 1380}{\mathrm{DPH}}\right) \tag{11}$$

The exponent (b) is defined by the fatigue strength coefficient (σ_f') at one reversal and the fatigue strength at 10^6 reversals [22].

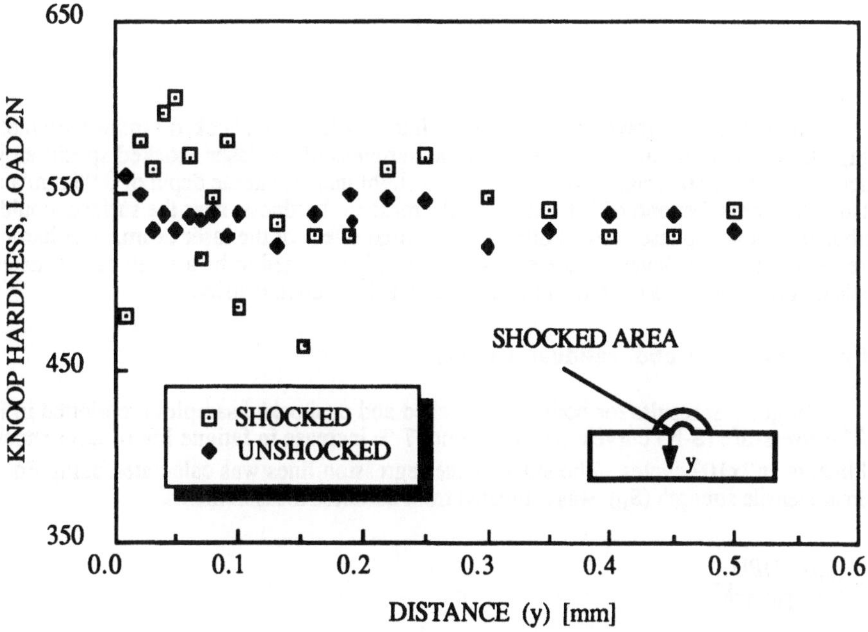

Figure 4 Knoop hardness traverse measured in (y) direction from the weld toe surface for laser-shocked and unshocked specimens.

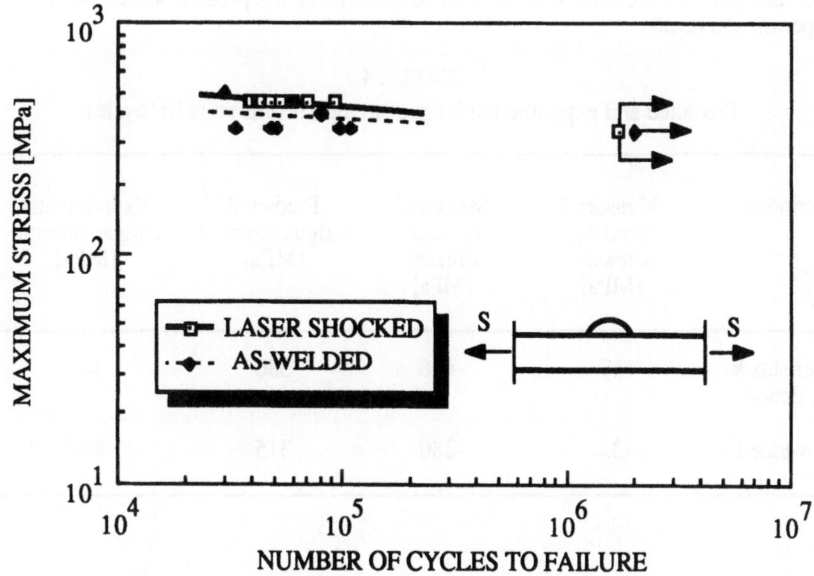

Figure 5. S-N curves for as-welded and laser-shock-hardened specimens; R = 0, f = 15 Hz, tests performed in air at room temperature.

RESULTS

Hardness

Knoop hardness traverse results for both unshocked and shocked samples are plotted in Fig. 4. The data indicate a decrease in the hardness of the laser shocked specimen just under the surface (0.01 mm from the surface), a slight increase at the depth of 0.05 mm and a constant hardness beyond of 0.1 mm. The decrease in hardness, near the surface, could be attributed to a new phase formed due to the thermal effect of the laser beam. The hardness increase in the deeper layers of the shocked sample is assumed to be a result of the material plastic deformation and a consequent increase in the dislocation density.

Fatigue resistance and residual stresses

Fatigue test results for both laser shocked and unshocked samples are plotted in Fig. 5. The stress-life (S-N) curves indicate about 17 % increase in fatigue life of laser shocked weldments at $2x10^6$ cycles. The slope of the regression lines was calculated using Eq. 11. Ultimate tensile strength (S_u) was estimated from hardness measurements:

$$S_u = S_u^{BM} \frac{DPH}{DPH^{BM}} \tag{10}$$

where (S_u^{BM}) is the ultimate tensile strength of the base metal, and (DPH^{BM}) is the hardness of the base metal.

The fatigue strength of both laser-shocked and as-welded samples were calculated using Eq. 4. Results of those calculations are listed in Table 4. Despite the small number of fatigue data points, particularly for as-welded specimens, the predictions are consistent with the experimental results.

TABLE 4

Predicted and experimental fatigue strength results at $2x10^6$ cycles.

Description	Measured bending stresses [MPa]	Measured residual stresses [MPa]	Predicted fatigue strength [MPa]	Experimental fatigue strength [MPa]
Laser shock hardened	17	-416	380	346
As-welded	32	-280	315	296

CONCLUSIONS

1. Laser shock hardening with 10^{12} W/cm^2 power density altered the hardness and residual stresses in the HAZ of welded 18 Ni (250) maraging steel to the depth of 0.05 mm.

2. Laser shock hardening increased the fatigue strength at 2×10^6 cycles by 17 % for the 18 Ni (250) weldments.

ACKNOWLEDGMENTS

We would like to thank Dr. W. Donaldson[a)], Mr. L. Kingsley[b)], and Dr. D. Wake[c)] for their assistance. This program was supported by the Materials Processing Consortium of the University of Illinois at Urbana-Champaign. Partial support for laser facilities was provided by the sponsors of the Laser Fusion Feasibility Project at the Laboratory for Laser Energetics.

REFERENCES

1. S. A. Metz and F. A. Smidt, Jr., Appl. Phys. Lett. 19, **207** (1971).
2. F. Osterman, Metall. Trans. 2, **2897** (1971).
3. B. P. Fairand, B. A. Wilcox, W. J. Gallagher, and D. N. Williams, Jr. Appl. Phys. 43, **3893** (1972).
4. T. R. Hsu, J. Testing and Evaluation 1, 457 (1973).
5. B. P. Fairand, A. H. Clauer, R. G. Jung, and B. A. Wilcox, Appl. Phys. Lett. 25, **431** (1974).
6. A. H. Clauer, B. P. Fairand, and B. A. Wilcox, Metall. Trans. A8a, **119** (1977).
7. A. H. Clauer, B. P. Fairand, and B. A. Wilcox, Metall. Trans. A8a, **1872** (1977).
8. B. P. Fairand, A. H. Clauer, Jr. Appl. Phys. 50, 1497 (1979).
9. R. W. Landgraf, "Effect Of Mean Stress on The Fatigue Behavior of a Hard Steel." M. S. Thesis University of Illinois at Urbana- Champaign, (1966).
10. O. T. Inal and L. E. Murr, Jr. Appl. Phys. 49, **2427** (1978).
11. K. Mukherjee, T. H. Kim, and W. T. Walter, Proc. of Symp. on Laser in Metallurgy, Physical Metallurgy and Solidification Comities of the Metallurgical Society of AIME, Chicago, 137 Feb. (1981).
12. C. S. Coffey, Jr. Appl. Phys. 62, **2727** (1987).
13. G. Banaś and J. Lunarski, Report on "Badania wplywu umocnienia na wytrzymalosc zlaczy spawanych ze stali N18K9M5TPr, Politechnika Rzeszowska, Poland (1983).
14. D. F. Socie, M. R. Mitchel, and E. M. Caulfield, FCP Report No. 26, College of Engineering, University of Illinois at Urbana-Champaign (1978).
15. B. P. Fairand and A. H. Clauer, Opt. Commun. 18, **588** (1976).

---

a), b) Laboratory for Laser Energetics, University of Rochester
c) University of Illinois

16. F. V. Lawrence, Jr., R. J. Mattos, Y. Higashida, and J. D. Burk, ASTM STP 648, **134** (1978).

17. F. V. Lawrence, Jr., N. J. Ho, and P. K. Mazumdar, Amm. Rev. Mater. Sci. 11, **401** (1981).

18. Y. Tobe and F. V. Lawrence, Jr., Welding Journal, 56, **266s** (1975).

19. J. McMahon and F. V. Lawrence, Jr., FCP Report No. 131, College of Engineering, University of Illinois at Urbana-Champaign (1986).

20. F. V. Lawrence, Jr., Proc. of the Conf. on Fatigue of Welded Constructions, Paper No. 47, The Welding Institute, Abington Hall, Cambridge (1987).

21. J. -Y. Yung and F. V. Lawrence, Jr., Fatigue Fract. Engng Mater. Struct. 8, **223** (1985).

22. J. McMahon, "Predicting Fatigue Properties Through Hardness Measurements." M. S. Thesis, University of Illinois at Urbana-Champaign (1984).

23. S. J. Maddox, Welding Journal, 53, **401** (1974).

24. M. H. El Haddad, T. H. Topper, and I. F. C. Smith, Jr. Testing and Evaluation, 8, **301** (1980).

Laser Surface Heat Treatment Used for Induction of Residual Stresses into Circular Saw Blades for Better Working Behaviour

H. K. Tönshoff and A. R. Rosenthal

Laser Zentrum Hannover eV
Hannover
Federal Republic of Germany

ABSTRACT

Residual stresses in workpieces and tools are the object of investigations since several years. Industries and instituts of research and development are interested to change the working behaviour of parts by generating tensile or compressive residual stresses in the area near the surface.

Residual stresses could be induced by laser radiation in the tensile as well as in the compressive range. Measurements such as hardness, x-ray diffraction, polish cross-section and thermography were used to indicate these stresses. Different parameters of the laser process causes variable residual stresses in the steel sheets. This paper gives some examples and explains advantages and disadvantages of stress-input by laser surface heat treatment.

INTRODUCTION

Recently laser surface heat treatment plays a growing part in production engineering. Meanwhile methods like laser cutting and laser welding are established technologies, but surface treatment by laser is the youngest field of laser material processing. Figure 1 shows the production system of laser heat treatment under German Industrial Standard (DIN) No. 8580. Lattice structure of material will be rearranged by the high intensity of the laser beam, so that features of material can be changed.

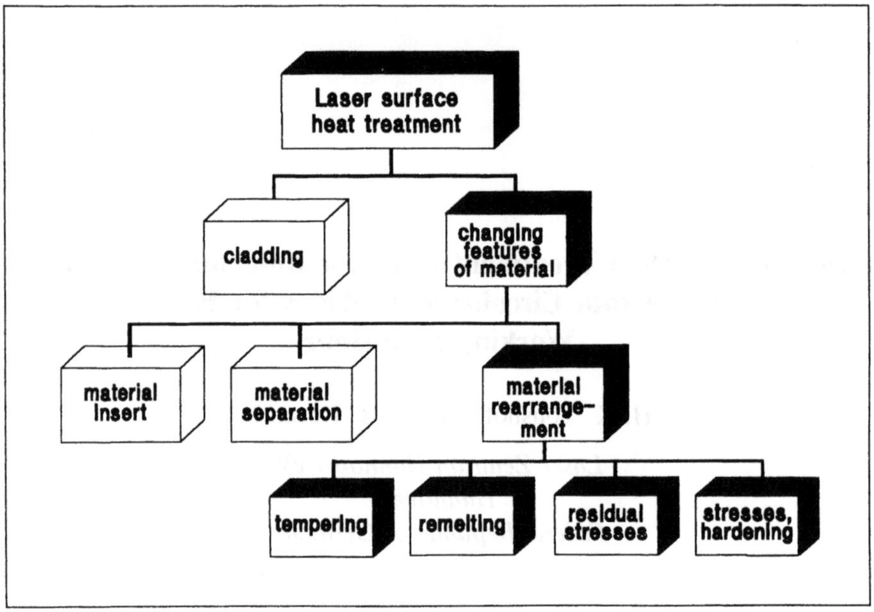

Figure 1. Laser production system (DIN 8580)

STATE OF THE ART

Especially residual stresses in the compressive or tensile range influence the working behaviour of tools and workpieces. They are used to prestress disklike rotating tools, for example saw blades, to increase their flexural stiffness and eigenfrequencies.

It exists no necessity to cool the cutting edge area of the blade in process using wood and metall as workpiece. This means that a temperature drop is set during the blade-workpiece interaction above the tool cross-section which extends from higher values at the tooth area to lower values at the center of the circular saw blade [3,5]. Together with load tensions generated by cutting and feed power components the thermal load results in tangential heat compressive stresses with about 100 MPa (figure 2)[1,4].

These requirements entail an increase of the run-out [7,10], an enlargement of the resulting kerf and a setback in surface quality of the cut. Therefore a stress balance between residual stresses and load tensions of the tool in process were demanded to get smaller vibrational amplitudes [2,13,8,9]. To reach this goal laser material processing as a new technology will be applied in competition to traditional techniques shown in figure 3.

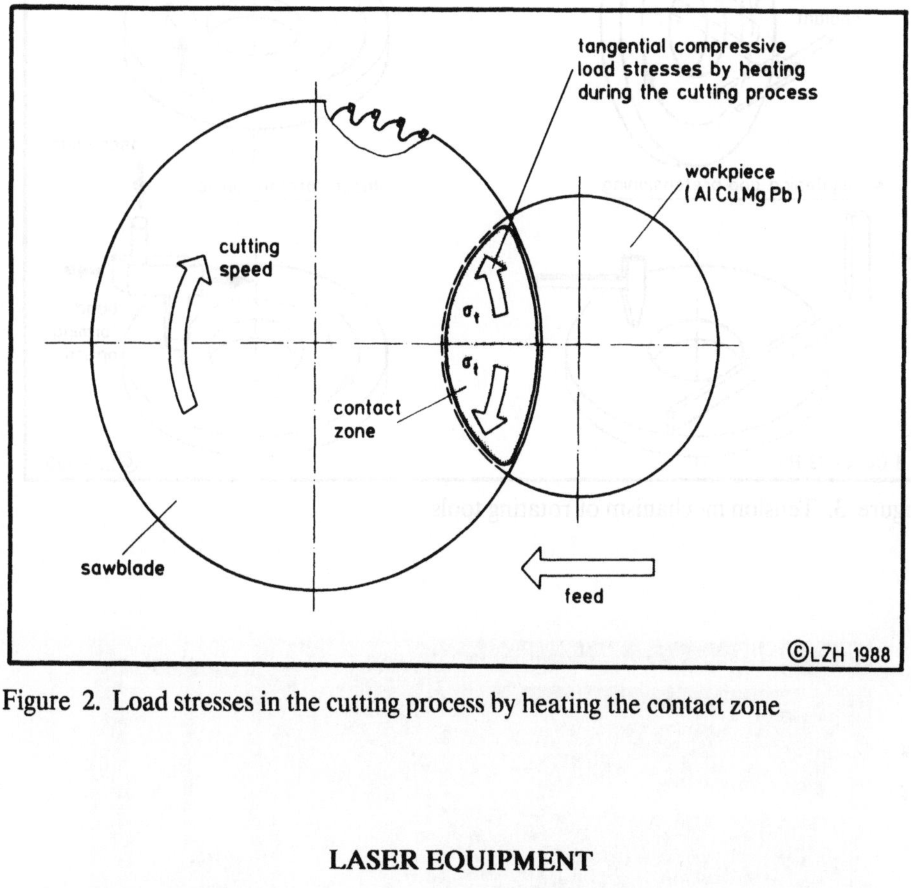

Figure 2. Load stresses in the cutting process by heating the contact zone

LASER EQUIPMENT

An 1500 W fast axial flowed CO_2-laser coupled with a surface treatment work station was used for the tests and depicted in figure 4. The optical system for material treatment based on Mo-coated copper mirrors. An integrator shapes the circular intensity distribution of the laser beam to a square-law beam with a cross-section of 5x5 mm. This beam profile was generated by an integration of multiple reflection in a square tube based on the principle of fibre transmission. Therefore it was possible to get a homogeneous laser beam intensity distribution in the beam cross section. A diagrammatic representation of the integrator optics shows figure 5.

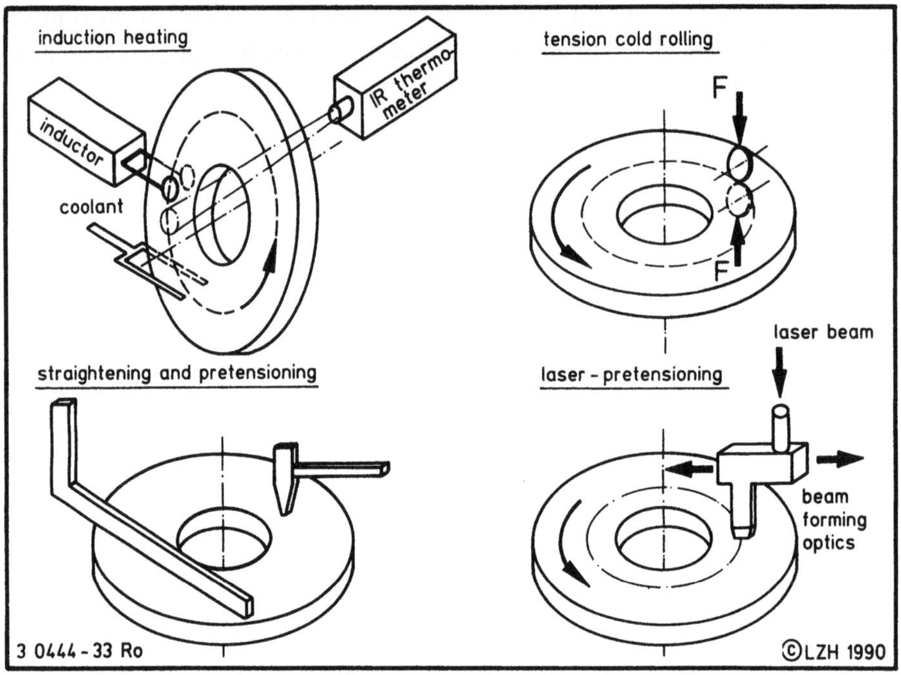

Figure 3. Tension mechanism of rotating tools

Figure 4. CO_2-Laser with surface treatment testing stand

Nitrogen flowed through the optical system on the surface of the specimen to protect the mirrors and prevent oxidation around the working beam spot. The gas pressure amount 0.15 MPa. The surfaces of the blades were coated with graphite to increase the absorption of the laser beam for better incoupling into the material. A thickmess of the layer between 8 and 10 μm could be measured. The graphite coating has no disturbing influence to the result of the laser treatment. A computer numerical control moves the workpiece under the beam in x- and y-axis with a fixed speed rate to generate the laser track geometry.

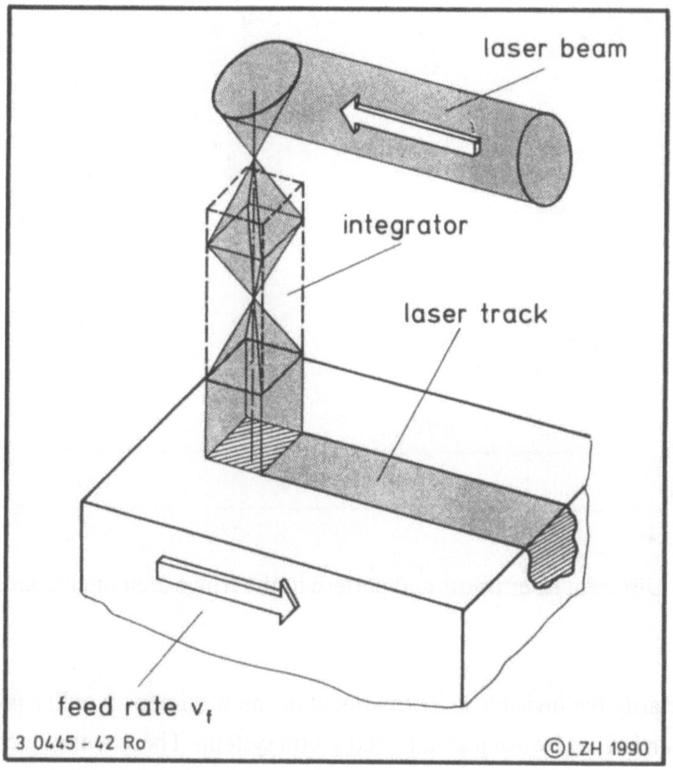

Figure 5. Copper mirror optical system (integrator)

RESEARCH

For the induction of residual stresses several laser track pattern are produced on the testing stand by CO_2-laser radiation. Mainly the fringe area of the saw blades was treated to minimize vibrations of the teeth during their mesh. Some significant track pattern are illustrated in figure 6. The left single-track will be used in comparison with the mechanical method of cold rolling. Four coaxial tracks in the middle of figure 6 increase the pretensioned area on the blade surface. The meander-track on the right side makes possible a 2-dimensional state of stress in radial and tangential direction.

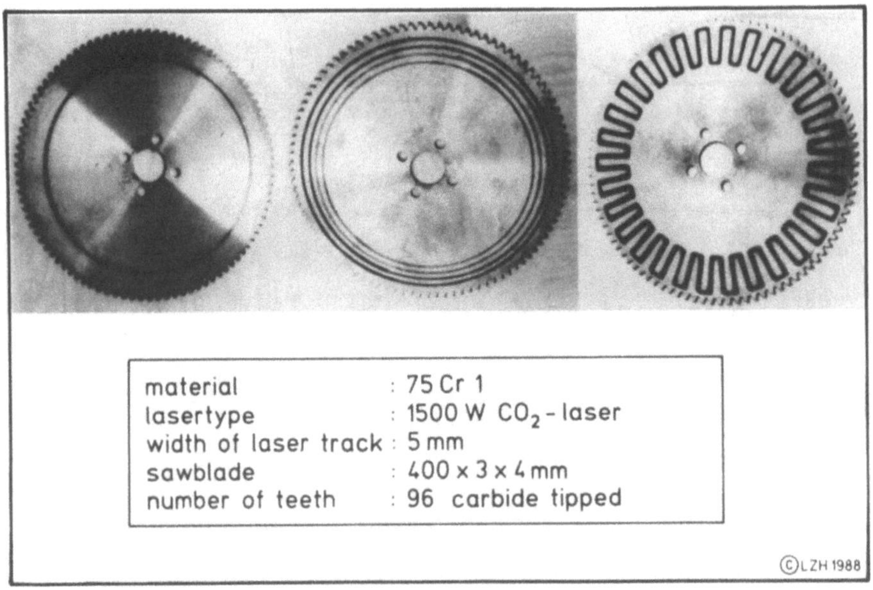

Figure 6. Different laser tracks and pattern in the fringe area of circular saw blades

To clarify the invisible laser treatment of the specimen an online process monitoring was carried out by using an infrared vision system. Thereby the steep temperature gradient can be visualized (figure 7). The smaller hot spot shows the heated lower part of the integrator optics while the large hot spot represents the heat affected zone on the specimen surface.

STRESS MEASUREMENT

By slow rotating steel saw blades the compensation of heat compressive stresses released by the cutting process is important during the saw blade-workpiece interaction. Therefore tangential tensile stresses must be placed into the sawblade in the teeth area. For the nondestructive measurement of these tangential stresses near the surface the x-ray diffraction technique was applied.

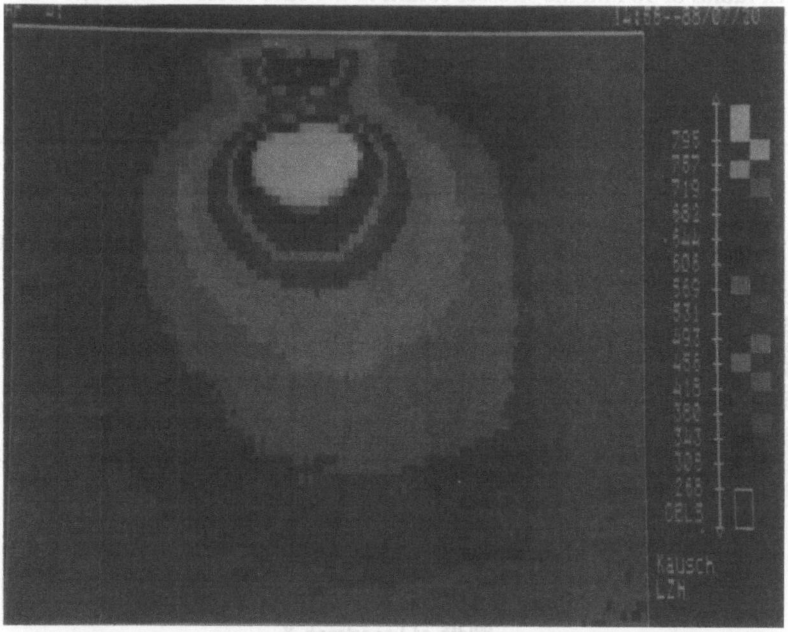

Figure 7. Temperature profile during laser hardening (specimen surface)

To make clear the change of stress in the laser track measuring points were laid across the track. To get some more informations about the influence of heat treatment on the material structure additionally surface hardness was measured across the track. The measuring results of residual stresses and surface hardness of the laser track is shown in figure 8.

If special parameters of laser power P_L and feed speed v_f are be chosen the specimen is hardened and the heating of the material over A_{c3} changes the structure of the material to martensite. The expansion of the martensite volume causes high tangential compressive stresses in the middle of the track. Different parameters of the laser process causes a process temperature below A_{c3}. The heat expansion with following cooling contraction leads to the desired tensile residual stresses (figure 8) [6,11,12].

In comparision with traditional pretension methods laser treatment inducing very high tensile residual stresses on the surface of the saw blades with high temporal and local dissolving of the laser beam (figure 9). In spite of the small difference between the track radius of 10 mm the stresses return to the base stress level, which was measured before laser treatment.

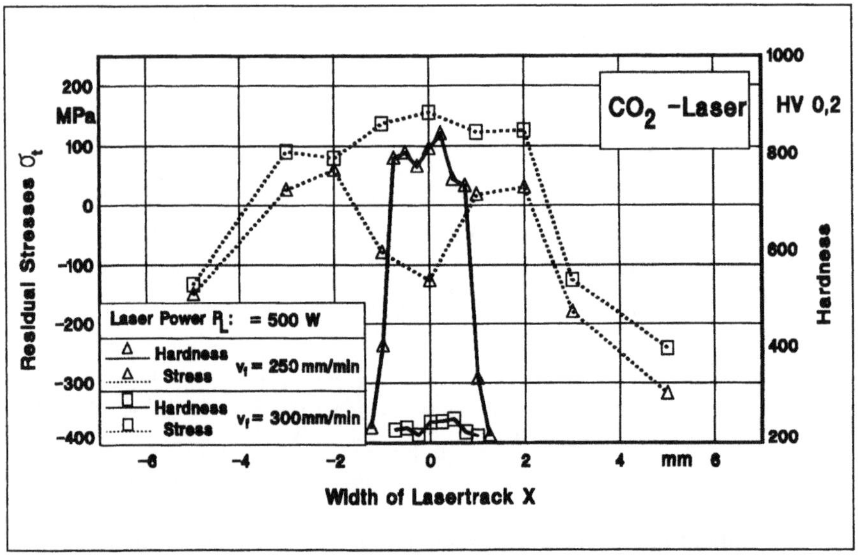

Figure 8. Tangential residual stresses and hardness of a circular saw blade by CO_2-laser heat treatment

Trial cuts were carried out on a circular saw and lead to characteristic results for cutting quality and tool behaviour. Reference blades, which were straighten but not prestressed, are used for comparison and valuation the test data of the different pretensioned saw blades.

First the difference between the gap and the cutting edge width of the carbide tips was measured. Figure 10 shows a comparison of the reference tool and a laser pretensioned saw blade cutting aluminum alloy ALCuMgPb (meander track). The tooth feed f_z was varied to increase cutting power, i.e. the external forces increase blade vibration by stimulation of eigenfrequencies. It can be observed, that vibrations grows with greater values of the feed f. The same figure makes clear, if conicity of the gap reduces the quality of the cut. The lasered blade minimize gap width as well as conicity of the gap.

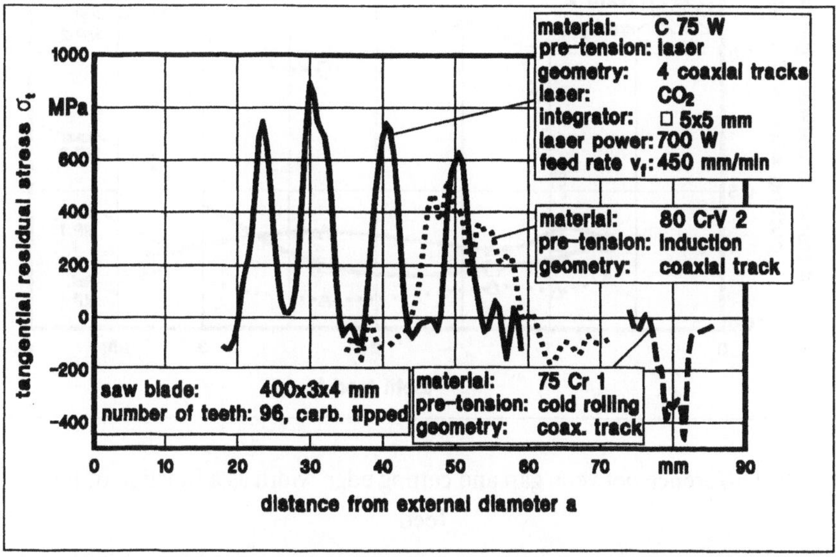

Figure 9. Tangential residual stresses in saw blades induced by different pre-tensioning methods

Figure 11 describes the second measurement during a cut with growing tooth feed f_z. The roughness of the cut surface is an indication of the vibration during the cut. The curve of the reference blade is very different from the one of the lasered blade. Trough laser heat treatment a good damping of the vibrations can be reached, so that the roughness don't exceed the value of 20 μm. The trial cut of the reference blade had to be interrupted to prevent a tool damage.

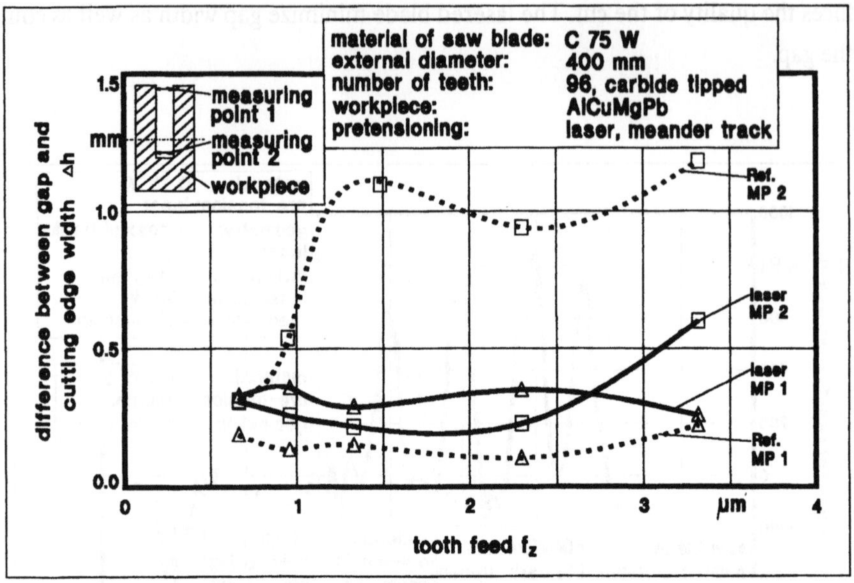

Figure 10. Difference between gap and cutting edge width as a function of tooth feed

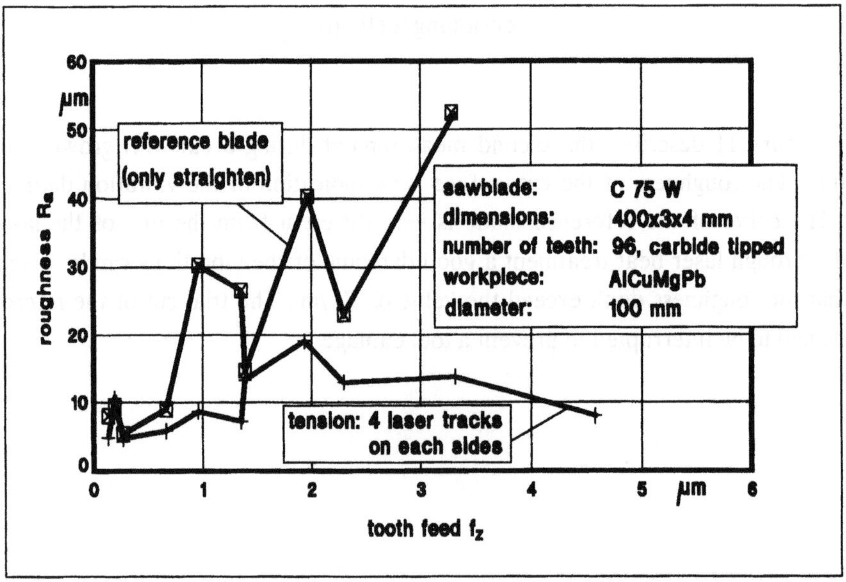

Figure 11. Roughness of the cut workpiece surface

CONCLUSION

The investigations make clear, that laser heat treatment is a good method for inducing residual stresses in parts. Especially disklike rotating tools can be pretensioned by high tensile stresses resulting in improving the working behaviour of the tools. The high thermal and local dissolving is a great advantage to generate complex track geometries. Laser machines are CN-controlled and working on a high level of automation. Therefore bottlenecks in manufactoring plants caused by manual labor, which depends on the experience of skilled workers, can be avoided. This leads to higher reproducibility and productivity in production. The disadvantage is the high investment for laser machines. Shift work and different treatments like laser cutting, laser pretensioning and laser straightening on the same machine may increase the economy of operations and reduce the costs.

REFERENCES

1. Hackenberg, P.: Spannungen in mechanisch und thermisch vorgespannten Kreissägeblättern. Dr.-Ing. Diss. RWTH Aachen 1974.

2. Huber, H.: Residual stresses in circular saws introduced by mechanical and thermal means. Proceedings of the 5. Wood Machining Seminar. University of California, Richmond 1977.

3. Mote, C.D. jr. und S. Holoyen: The temperature distribution in circular saws during cutting. Norsk Treteknisk Institut. Mitteilung Nr. 49, 1976.

4. Münz, U.V.: Vorspannungszustand und Arbeitsverhalten von Kreissägeblättern. Holz als Roh- und Werkstoff 36. 1978, 9, S. 345-352.

5. Nieh, L.T. und C.D. Mote jr.: Vibration and Stability in thermally stressed rotating disks. Experimental Mechanics 1975 S. 258-264.

6. Rosenthal, A., Meyer, C. und V. Bödecker: 2. Teil: Oberflächenhärten mit dem kW-Festkörperlaser. Industrie Anzeiger Nr. 53/54. Konradin Verlag 1988.

7. Schajer, G.S. und C.D. Mote jr.: Analysis of optimal roll tensioning for circular saw stability. Diss. University of California, Forest Products Laboratory (1982).

8. Scherger, A.: Schwingungs- und Geräuschuntersuchungen an Trennscheiben mit metallischem Kern. Diss. Universität Hannover (1981).

9. Schulze, R.: Trennschleifen von Hartgestein. Diss. Universität Hannover (1980).

10. Tönshoff, H.K. und J. Jendryschik: Dynamical Behaviour of Disklike Rotating Tools. International Journal computers & structures Vol.21 Nov. 1/2 1985.

11. Tönshoff, H.K.; A. Rosenthal: Verbesserung des Arbeitsverhaltens von scheibenförmigen Trennwerkzeugen durch Laserstrahlbehandlung. Blech Rohre Profile Jg. 35, (1988) Nr.10.

12. Tönshoff, H.K., A. Rosenthal: Laser heat treatment for pre-tensioning disklike rotating tools. Proc. of International Conference on Residual Stresses II, Nancy Nov. 1988. Elsevier Applied Science Publishers LTD, Essex 1989.

13. Westphal, R.: Geräuschminderung beim Kreissägen von Leichtmetall in Fertigungsbetrieben. Diss. Universität Hannover (1980).

Residual Stress Evaluation of Materials Manufactured by High-Energy Process

J. F. Flavenot and J. Lu

CETIM
Senlis
France

ABSTRACT

This paper presents several applications of the step by step hole drillling method for measuring residual stress distribution introduced in different components manufactured by the high energy process.

The first application concerns a NiCrAlY plasma sprayed coating. The influence of air plasma spraying, subatmospheric plasma spraying and post-treatment on the residual stress distribution is studied; the application of laser in mechanical engineering is also presented in the case of photonic peening(laser induced stress waves) and welding respectively for an aluminium alloy and a titanium alloy.

INTRODUCTION

During these last years the development of high energy process in manufacturing has been increased for replacing more traditional manufacturing process. The increase of quality of the manufactured components requested by the users has produced development of new technologies to improve their surface properties; high temperature coatings are now applied to many gas turbine components which are submitted to high stresses and degradation of the surface by oxidation, hot corrosion and erosion [1]. The present current pratice is to apply protective coatings based on MCrAlY compositions to extend the life and reliability of theses components. Plasma spraying is often used for manufacturing these coatings. On the other way thermal barrier are now used to increase the efficiency of car engine or diesel engine. This process is now a well established technique of forming metallic and ceramic coatings to protect the components against the corrosion. The mechanical or adhesion properties of these coatings depends on the quality of this coating (compositon, porosity), but also on the residual stresses produced by the process [2],[3]. For this reason, it is very important to check the residual stress distribution in these coatings in order to optimize their mechanical properties.

Laser surface treatment and laser welding are other high energy process the applications of which have recently increased.

For surface treatment laser is used to create new surfaces with new alloy systems or more simply for surface hardening after heating. These treatments are aimed to improve friction, wear or fatigue properties [4].

A more recent use of laser for surface treatment is the "photonic peening" which consist of the creation of shock waves on the surface of the metals. These shock waves produce a superficial plastic deformation and compressive residual stress which increase the fatigue strength of the component as in the case of a shot peening. But with a great advantage : the surface roughness is not modified. In this case a high energy pulsed laser beam can generate local pressure pulses of up to 6 to 10 GPa [5] without heating the surface.

Laser is of course, now very well used for welding of sheet metal or heavy section components. In this case, laser beam are very similar to electron beam and produces a very high gradient of temperatures and a very high gradient of residual stresses.

For these different high energy laser processes the knowledge of the produced residual stresses allows to explain the mechanisms which appear during the treatment and the modification of the mechanical properties.

THE PRINCIPLE OF THE INCREMENTAL HOLE DRILLING METHOD

The hole-drilling method involves monitoring the change in strains when a hole is drilled into a residually stressed component. Measurement is performed by means of a special three-element strain gauge rosette. These strain measurements can be then related to the original residual stress in the analysed sample at the hole location .

To obtain the gradient of the residual stress with depth , the hole is drilled in steps. For each hole, of depth "z", the surface strains $\varepsilon_i(z)$ are measured. Once the hole drilling is complete, the residual principal stresses, $\sigma_1(z)$ and $\sigma_2(z)$ can be calculated from equations involving measured values of $\varepsilon_i(z)$ and the calculated correlation coefficients Ain and Bin . The determination of these theoretical coefficients represents the main problem in the hole-drilling method. G.S.Schajer has carried out [13] an extensive review of the mathematical formulation for the stress calculation procedure, and has considered the following four established methods for estimating non-uniform residual stress fields from relaxation strain data : the incremental strain method [6] , the average stress method [7], the power series method [8], the integral methods [9–12].

The integral method can be considered as the most "fundamental" of the four . CETIM uses this method for different kinds of materials and components . The theoretical approach may be found in references ([9]-[13])

CETIM has developed special finite element software (CATROR) for calculating A_{in} and B_{in} , details of which are given in Ref.[12]. The program is applicable for all isotropic plate materials , such as thin and thick plate. For multilayer materials, other additional assumptions than for normal materials must be made. It is assumed that nodal displacements are continuous at the interface of the different layers. The only input data necessary is that of the plate geometry, the depths of the drilling increments , the thickness of each layer and their elastic constants , and the strain gauge rosette geometry .

In the case of residual stress measurement in coatings a great difficulty is the knowledge of the elastic constant of the coatings.

The experimental procedure is very simple and consist in drilling the hole step by step. For each step of drilling the strains are measured in the three strain gauges used. The difficulty of the experimental procedure is to find the best drilling conditions which do not introduce heating or parasite residual stresses. For each case , these otimized drilling conditions are defined using preliminary tests and metallographic checking of the hole geometry and surface intergrity of the hole.

(A) NiCrAlY COATING CASE

This coating is manufactured by **TURBOMECA** using two methods: low pressure plasma spraying and air plasma spraying.

The coating material is based on NiCrAlY composition and the substrate is a nickel alloy (NC15K10DAT).

The residual stress was measured on the samples with and without the post-heat treatment.Table 1 shows the condition of the post-heat treatment.

Coating method	Heat treatment	Coating thickness
Air plasma spraying	1060°C / 2 hours	1mm
	without	0.5mm
Low pressure plasma spraying	1080°C / 4 hours	1mm
	without	0.5mm

Table 1 : Heat treatment used on the experimental samples.

Figure 1 shows the results obtained for the coating produced by air plasma spraying, with and without heat treatment. Tensile residual stresses were observed in the coating before heat treatment, and compressive residual stresses were observed after heat treatment. In sample which is not subjected to heat treatment, the level of the tensile stress is very high and this can be a source of microcrack.

Figure 2 shows the results obtained for the coating produced by low pressure plasma-spraying, with and without heat treatment.

It is interesting to note that in both cases ((1) and (2), fig.2), the residual stresses are compressive in nature, indicating that the effect of heat treatment is not as important as in the case of air plasma-spraying.

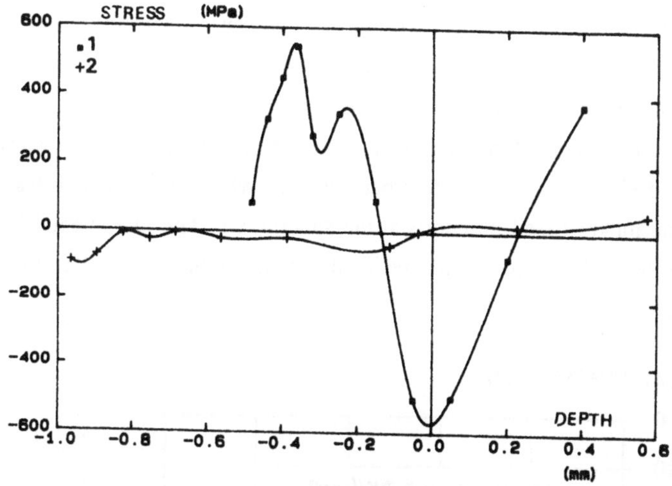

Fig. 1 Residual stress distribution obtained in the sample produced by air plasma spraying - without heat treatment (1) and with heat treatment (2) .

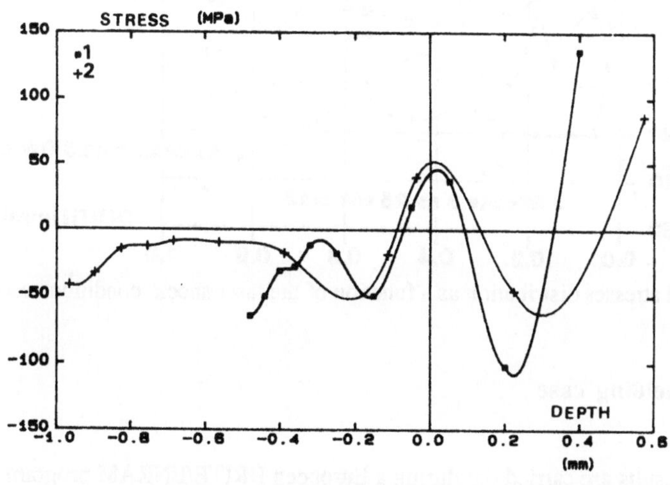

Fig. 2 Residual stress distribution obtained in the sample produced by low-pressure spraying -without (1) and (2) with heat treatment.

(B) Laser shocked aluminium alloy

The samples were manufactured by PSA and Ecole Polytechnique [5]. The aluminium alloy (AS7G grade) treated by the laser shock was studied by CETIM. Fig.3 shows the results obtained by the incremental hole drilling method. It can be seen that for this case, the residual stresses distribution does not modify when the pulse duration charge between 3 ns and 25 ns.

But the residual stresses increase with an increasing number of laser shocks. For all cases, the residual stresses of compression are observed up to a depth of 1mm. The prestressed layers are more important in comparison with a shot peening treatment. In this case, the hole drilling method is choosed because of the microstructure of the aluminium alloy. It was difficult to use the X-ray diffraction method. Another case for which the X-ray diffraction has a limitation is the nickel based monocristalline superalloys. The incremental hole drilling method has been used to evaluate the residual stresses induced by different laser shock treatments [14].

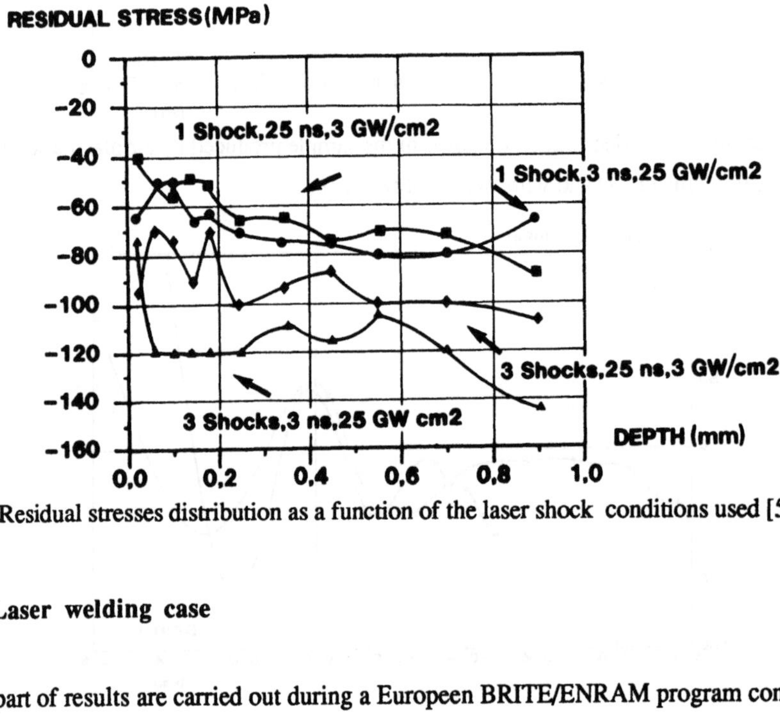

Fig.3 Residual stresses distribution as a function of the laser shock conditions used [5]

(C) Laser welding case

This part of results are carried out during a Europeen BRITE/ENRAM program concerning laser sheet metal welding.The example shown is a Ti sample welded by AEROSPATIALE FRANCE using a high power CO_2 laser. The X-ray diffraction method and the hole drilling method have been combinated to evalute respectively the surface residual stresses and the in depth residual stresses.

The measurements by the X-ray method were carried out on the center of the weld, at 2 and 4mm (center of three X-ray spot) from the center of the weld bead. The residual stresses on the two sides (A and B) of the welded part were measured. For the case of the incremental hole drilling method, the hole is drilled by a milling machine using a tungsten carbide drill.

The hole diameter is 2mm. On side A, three measurements were carried out, at the center of the weld bead, at 2 and 4mm from the center of the weld bead. To have a minimum effect due to relaxation of residual stresses during drilling, the locations of measurements were shared out along the weld bead(fig.4).

The results of measurements are shown in fig.5 and in fig.6. Because of the diameters of the X-ray spot and of the hole (2mm), and the stress gradient near the weld bead, the results can not be considered as punctual. In fact, a mean stress on the analyzed zone was measured. Following observation can be made: very high tensile residual stresses were measured in the parallel direction of the weld bead on the two sides of the sheet at the center of the weld bead. The level of the stresses decreases quickly as a function of the distance from the center of the weld bead. The residual stress are maximum at the center through the thickness of the sheet. In the perpendicular direction of the weld bead, the residual stresses are much smaller.

Figure 4 Hole locations along the weld bead on the sheet welded by laser

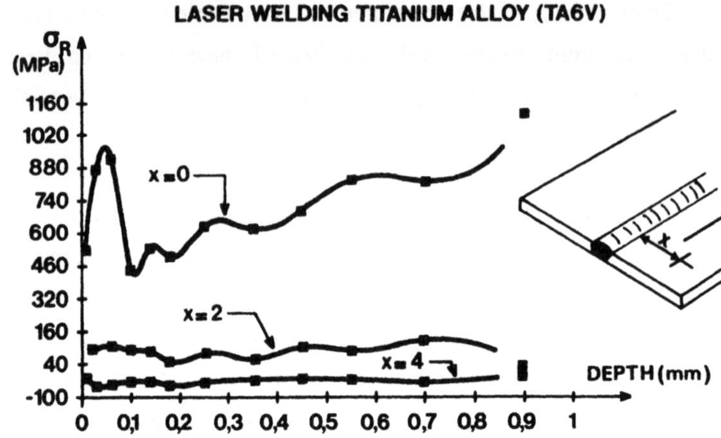

Figure 5 Residual stresses obtained in the direction parallel to the weld bead

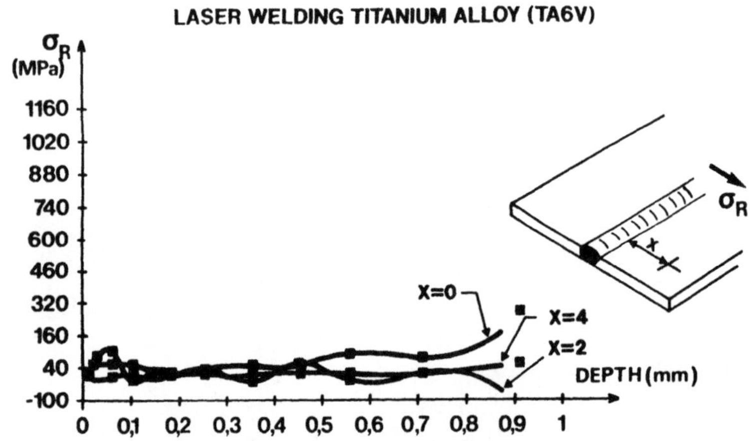

Figure 6 Residual stresses obtained in the direction perpendicular to the weld bead

CONCLUSION

The purpose of this paper has been to show the procedure and applications of residual stress distribution measurement by the incremental hole drilling method.

A finite element program has been developed to calculate the calibration coefficients for the incremental hole drilling method in the case of multi-layer materials. The main difficulty in this calculation is that of finding the elastic modulus of each coating layer material - the elastic modulus of a sprayed structure is a function of the structure's porosity and of its mechanically interlocked nature, which vary with spray parameters. This necessitates the development of methods for accurately measuring the elastic modulus of each layer.

The study presented here shows the suitability of the modified hole drilling method for controlling the residual stress levels in materials manufactured by high-energy process. An advantage of the method is its speed- only two or three hours are necessary to evaluate the full stress gradient, thus opening up the possibility of using the method for process optimization.

The study has considered four types of application-plasma sprayed NiCrAlY, laser shock treated aluminium alloy and laser welded titanium alloy. The method has, however, also been used to measure residual stress in ceramic coating materials [15] , stellite coating and electron beam welded part.

REFERENCE

[1] Sivakumar, R., and Mordike, B.L., Laser melting of plasma sprayed NICoCrAlY coatings. Surface Engineering 1987, 3, N.4, pp299-307.

[2] Dründling, H.W., Schneider K., and Singheiser, L., Mechanical properties of coated systems, Materials Science and Engineering, 1987(88), pp177-189.

[3] Hobbs, M., Thermal barrier coatings for diesel engines, Surfacing Journal, 1985, 16, pp101-108.

[4] Proceedings of the fourth International colloquium on welding and melting by eletron and laser beams, Cannes, September 1988, Edited by CEN/Saclay France

[5] Fournier, J., Generation of shock waves by high-energy pulsed laser, Mechanical and metallurgycal analysis , Ph.D thesis, Ecole Polytechnique,1989

[6] Kelsey, R.A. , Measuring Non-Uniform Residual Stresses by the Hole-Drilling Method, Proceeding SESA , Vol.14 , No.1 , 1956, 181-194 .

[7] Nickola, W.E. , Pratical Subsurface Residual Stres Evaluation By The Hole-Drilling Method, Proceeding of the Spring Conference On Experimental Mechanics , New Orleans , June 8-13 , 1986 , 47-58 .Society for experimental mechanics .

[8] Schajer, G.S. , Application of finite element calculations to residual stress measurements , ASME , Journal Of Engineering Materials and Technology , Vol.103 , No.2 , 1981 , 157-163 .

[9] Bijak-Zochowski, M. , A semidestructive method of measuring residual stresses , VDI-Berichte , Vol.313, 1978 , 469-476 .

[10] Niku-Lari, A. , Lu, J. , Flavenot, J.F. , Measurement of residual stress distribution by the incremental hole-drilling method , Experimental Mechanics , Vol.25, No.6 , 1985 , 175-185 .

[11] Flaman, M.T. , and Manning, B.H. , Determination of residual stress variation with depth by the hole-drilling method , Experimental Mechanics , Vol.25 , No.9 , 1985 , 205-207

[12] Lu, J. , Niku-Lari, A. , Flavenot, J.F. , Récents développments de la mesure des contraintes residuelles par perçage incremental , Matériaux et Techniques , Décembre 1985 , 709-718 .

[13] Schajer, G.S. , Measurement of Nonuniforme Residual Stresses Using the Hole-Drilling Method . Part 1 - Stress calculation Procedures , ASME , Journal of Engineering Materials and Technology , Vol 110 , Oct 1988 , pp338-349

[14] Forget, P., Jeandin, M., Lu, J., Contraintes résiduelles engendrées par choc laser sur des superalliages aéronautique, To be published in Proceedings of Journées Nationales sur les nouvelles technologies, 1990.

[15] Lu, J., Flavenot, J.F., The effect of the finishing treatments on the residual stresses distribution in advanced coating materials, Proceedings of the 1989 SEM Conference on Experimental Mechanics, June 1989, Boston, pp927-934

Finite Element Analysis of Main Crack–Microcrack Interaction under Mode-I Loading

S. A. Meguid, S. X. Gong and P. E. Gaultier

University of Toronto
Toronto
Ontario
Canada

ABSTRACT

A finite element study is made of the interaction between a main crack and symmetric microcracks under plane stress conditions. Specifically, our attention is devoted to evaluating the stress intensity factor at the main crack in the presence of (i) two symmetric microcracks, and (ii) four symmetric microcracks. The work was further extended to examine a case which involves main crack-microcrack-microcrack interaction. The findings of the work were also compared with an analytical model.

1. MOTIVATION

Considerable efforts are currently being expended on the computation of the effects of near tip microcracking on main crack propagation in ceramic materials [1-2] and hydrogenated and outgassed austenitic stainless steels [3-4]. The main interest in these problems stems from the possible toughening effect introduced by the presence of these microcracks near the tip of a main crack. In view of the complexity of this current interaction problem closed-form solutions are not feasible, and most previous studies are confined to approximate analytical expressions.

In these investigations attention was devoted to evaluating the stress intensity factor at the main crack in the presence of (i) two symmetric microcracks and (ii) four symmetric microcracks in close proximity to the main crack. Furthermore, the work compares the finite element results with an analytical model which is based on the work of Gong and Horii [5]. The work was further extended to examine the effects of microcrack-microcrack interaction upon main crack stress intensity factor.

2. ANALYTICAL APPROACH

The analytical approach, utilizes the superposition principle and the complex potentials of Muskhelishvili [6]. As shown by Muskhelishvili the stress components σ_x, σ_y and τ_{xy} and displacement components u_x, u_y can be expressed in terms of two complex potentials $\Phi(z)$, $\psi(z)$, as follows:

$$\begin{aligned} \sigma_x + \sigma_y &= 2[\Phi'(z) + \overline{\Phi'(z)}\,] \\ \sigma_y - \sigma_x + 2i\tau_{xy} &= 2[\bar{z}\Phi''(z) + \psi'(z)\,] \\ 2G(u_x + iu_y) &= \kappa\,\Phi(z) - z\overline{\Phi'(z)} - \overline{\psi(z)} \\ z = x + iy \ \text{ and } \ i &= \sqrt{-1}, \end{aligned} \tag{1}$$

To simplify the analytical analysis, the main crack is assumed to be semi-infinite, which corresponds to the assumption that the length of the main crack ($2a$) is much larger than the length of the microcrack ($2c$) and the distance between the main crack and the microcrack (d). Under these conditions the undisturbed stress field (ie., in the absence of microcracks) is simply given by the singular near-tip stress field with the applied stress intensity factors K_I and K_{II}. The solution to the present problem is obtained by the superposition of three subproblems, each of which contains one single crack; as depicted in Fig. 1.

In subproblem 1, the main crack is subjected to the applied stress intensity factors K_I and K_{II} of mode I and mode II. The undisturbed stress field is given by the singular near-tip stress field. In this case, Muskhelishvili's stress functions for subproblem 1 are

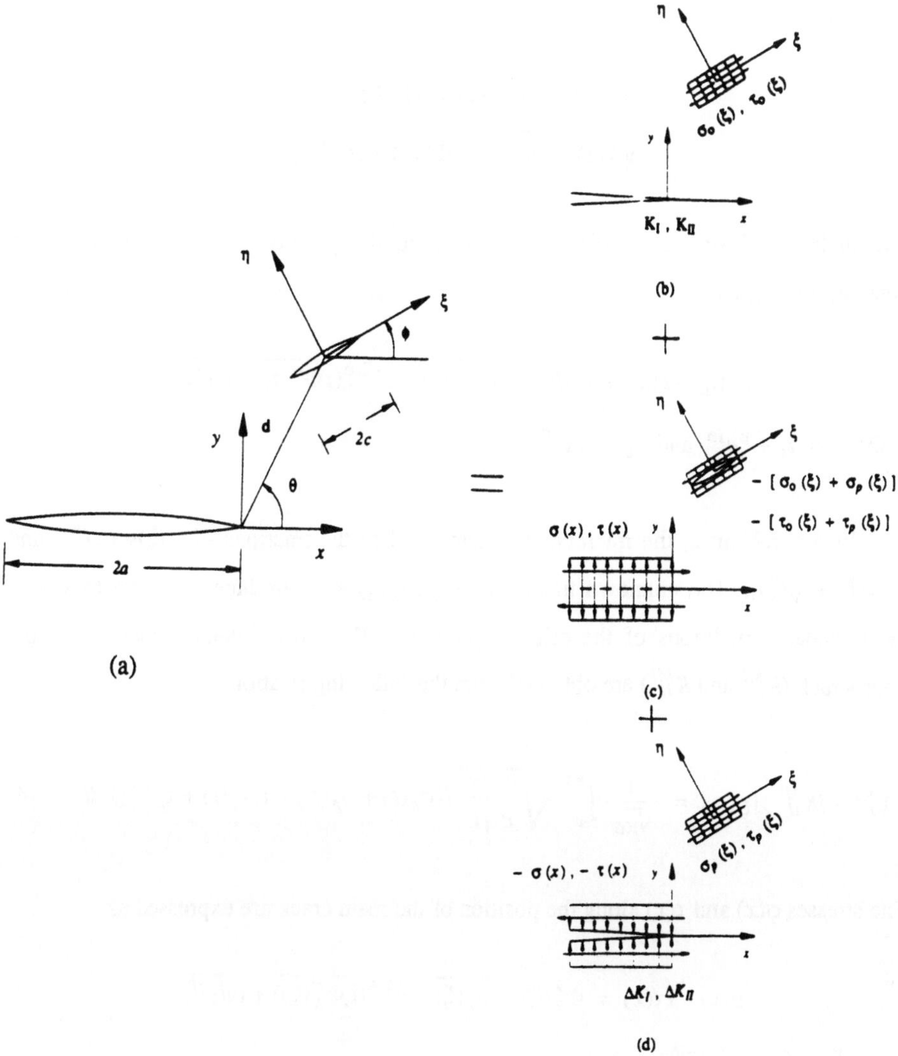

Fig.1 The superposition principle:
(a) the original problem
(b) subproblem 1
(c) subproblem 2
(d) subproblem 3

given by:

$$\Phi_1'(z) = (K_I - iK_{II})/(2\sqrt{2\pi z}),$$
$$\psi_1'(z) = \overline{\Phi_1(\bar{z})} - \Phi_1'(z) - z\Phi_1''(z). \tag{2}$$

The undisturbed stresses $\sigma_o(\xi)$ and $\tau_o(\xi)$ along the position of the microcrack are obtained from (2) as:

$$\sigma_o(\xi) - i\tau_o(\xi) = \Phi_1'(z) + \overline{\Phi_1'(z)} + e^{-2i\phi}(z\overline{\Phi''(z)} + \overline{\psi_1'(z)}) \tag{3}$$

with $z = z_c + \xi e^{i\phi}$ and $z_c = de^{i\theta}$.

In subproblem 2, the microcrack is subjected to the tractions $-[\sigma_o(\xi)+\sigma_p(\xi)]$ and $-[\tau_o(\xi)+\tau_p(\xi)]$. The unknown tractions $\sigma_p(\xi)$, $\tau_p(\xi)$ are introduced in order to satisfy the boundary conditions of the original problem. The stress intensity factors for the microcrack (K_I^{MI} and K_{II}^{MI}) are obtained from the following relation:

$$(K_I^{MI} - iK_{II}^{MI})|_{\xi=\pm c} = \frac{1}{\sqrt{\pi c}} \int_{-c}^{+c} \sqrt{\frac{c \pm t}{c \mp t}} \{[\sigma_o(t) + \sigma_p(t)] - i[\tau_o(t) + \tau_p(t)]\}\, dt \tag{4}$$

The stresses $\sigma(x)$ and $\tau(x)$ along the position of the main crack are expressed as:

$$\sigma(x) - i\tau(x) = \Phi_2'(\zeta) + \overline{\Phi_2'(\zeta)} + e^{2i\phi}(\zeta\overline{\Phi_2''(\zeta)}) + (\overline{\psi_2'(\zeta)}) \tag{5}$$

with $\zeta = (x - z_c)\, e^{-i\phi}$.

In subproblem 3, the main crack is loaded by the tractions $\sigma(x)$ and $\tau(x)$ of the magnitude determined by (5) but with an opposite sign. The stress intensity factors for the main crack (K_I^{MA}, K_{II}^{MA}) are obtained using

$$\Delta K_I - i\Delta K_{II} = \sqrt{\frac{2}{\pi}} \int_0^{\infty} \frac{1}{\sqrt{t}} [\sigma(-t) - i\tau(-t)]\, dt. \tag{6}$$

Now we consider the superposition of the three subproblems. The traction-free condition along the main crack is automatically satisfied, while traction-free condition of the original problem for the microcrack leads to the appropriate consistency equation. The stress intensity factors of the main crack for the original problem are obtained as follows:

$$\begin{Bmatrix} K_I^{MA} \\ K_{II}^{MA} \end{Bmatrix} = \begin{Bmatrix} K_I \\ K_{II} \end{Bmatrix} + \begin{Bmatrix} \Delta K_I \\ \Delta K_{II} \end{Bmatrix}. \tag{7}$$

3. FINITE ELEMENT MODELLING

In view of the symmetric nature of the different systems investigated, it was only necessary to discretize one quarter of each system for the finite element analysis using eight-noded parabolic elements. The following dimensions were selected for the models: $a/d = 10$; $c/d = 1/4, 1/2$ and 1 for different values of θ and ϕ.

Two approaches were adopted in the determination of the stress intensity factor at the main crack, K_I^{MA}. The first utilized the stress field ahead of the main crack-tip and involves correlating the finite element nodal point stresses σ_y^* with the well known crack-tip stress equations [7]. The second utilizes the displacement field directly behind the main crack-tip. Using least square regression analysis, it was possible to evaluate the stress intensity factor at the main crack K_I^{MA} for the different studied cases. K_I^{MA} was then normalized by the undisturbed stress intensity factor K_I, which was again determined from the finite element analysis.

4. ANALYSIS OF RESULTS AND DISCUSSIONS

A detailed analysis of numerous configurations is provided in reference [8]. Here we only present representative cases which evolved from the earlier work [8].

In view of the fact that main crack-microcrack interaction is influenced by location (c/d and θ), orientation (ϕ) and number of microcracks (n), it is important to investigate the effect of these variables upon the stress intensity factor and the stress field at the main crack. The appropriate dimensions for main and microcracks were selected to enable the comparison with the zeroth order solutions.

Two representative cases illustrating the effects of microcrack location upon the interaction problem are provided in Figs. 2 and 3. Fig. 2(a) shows the deformed geometry associated with mode-I deformation, while Fig. 2(b) shows the associated von Mises stress trajectories for symmetric microcracks with $c/d = 1/4$; $\theta = 45^o$; $\phi = 90^o$. The finite element results of the normalized stress intensity factor for the main crack K_I^{MA}/K_I was computed to be 0.960 which compares well with the analytical results of the zeroth order solution of 0.987. Similarly, Figs. 3(a) and 3(b) show the deformed geometry and the Mises trajectories for the case where $c/d = 1/4$; $\theta = 90^o$; $\phi = 0^o$. In this case, the finite element result of the normalized stress intensity factor K_I^{MA}/K_I was 0.954 while the zeroth order solution was 0.953. Both these results indicate that the tip of the main crack experiences some form of stress shielding.

An interesting configuration would be to include the above microcracks into one system and thus enabling the examination of (i) the effects of multiple microcracks, (ii) the interaction between microcracks, and (iii) the resultant effect on the stress trajectories at the main crack. Fig. 4(a) shows the deformed geometry for $c/d = 1/4$ and with the following combined configuration:

microcrack (a): $\theta = 45^o$, $\phi = 90^o$ and

microcrack (b): $\theta = 90^o$, $\phi = 0^o$.

The deformed geometry of Fig. 4(a) confirms the lack of deformation of microcrack (a) as it is aligned with the applied far-field stress, while microcrack (b) is subjected to pure opening mode type of deformation. The stress trajectories of Fig. 4(b) confirm the above deformation pattern and show a limited stress gradient at and near microcrack (a). It is also interesting to note that the stress field at the main crack is vastly different from that observed with the undisturbed case, Fig. 4(c). The presence of microcracks (a) and (b) resulted in a combined shielding effect at the tip of the main crack. This is evidenced by the finite element results of the normalized stress intensity factor at the main crack K_I^{MA}/K_I which is 0.958 which compares well with an analytical value of 0.940.

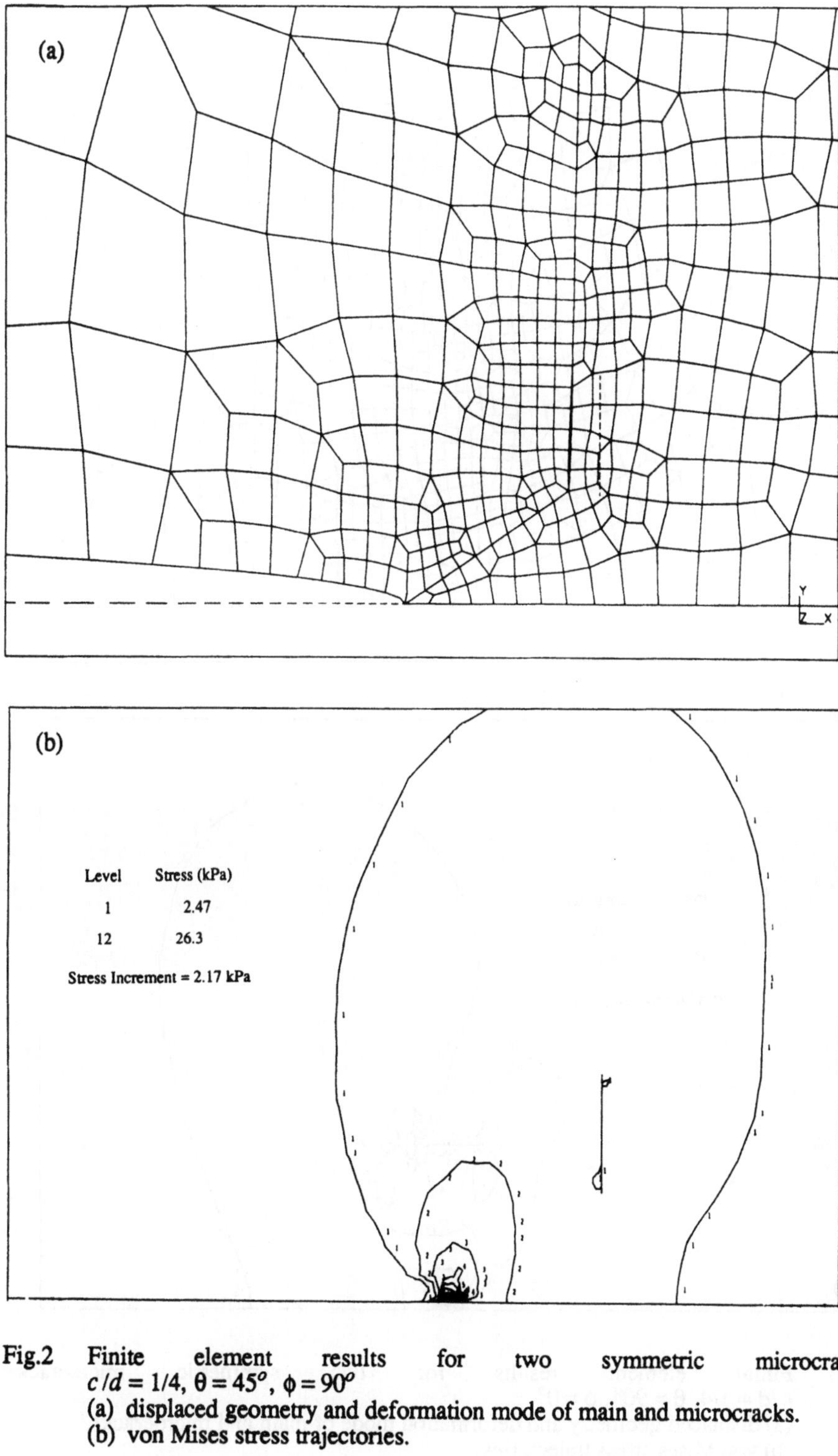

Fig.2 Finite element results for two symmetric microcracks: $c/d = 1/4$, $\theta = 45^o$, $\phi = 90^o$
(a) displaced geometry and deformation mode of main and microcracks.
(b) von Mises stress trajectories.

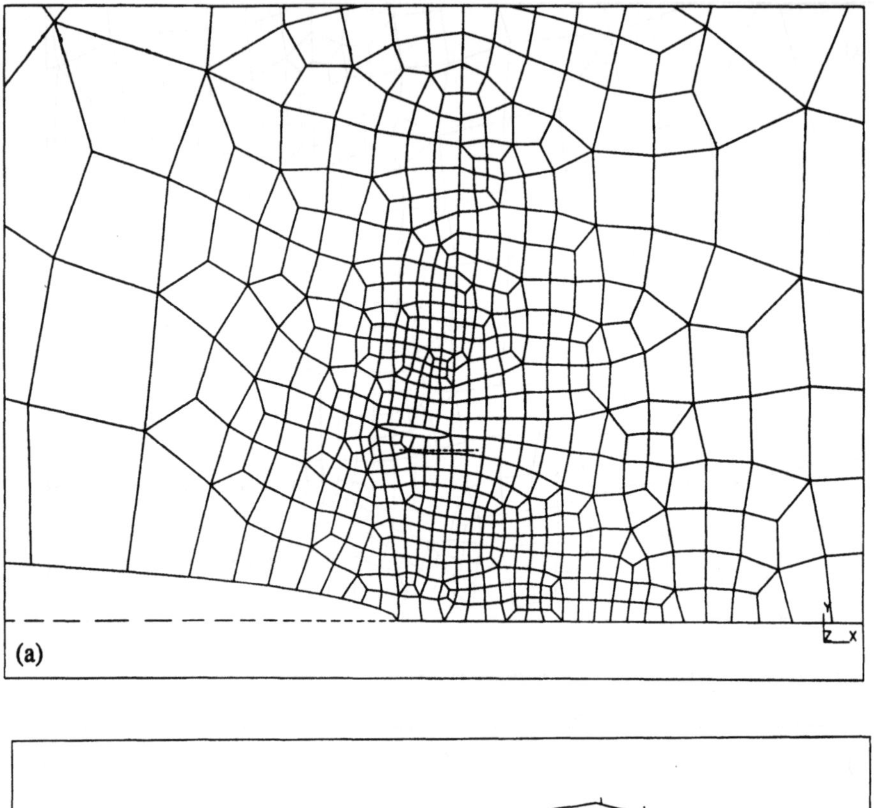

Fig.3 Finite element results for two symmetric microcracks: $c/d = 1/4$, $\theta = 90^o$, $\phi = 0^o$
(a) displaced geometry and deformation mode of main and microcracks.
(b) von Mises stress trajectories

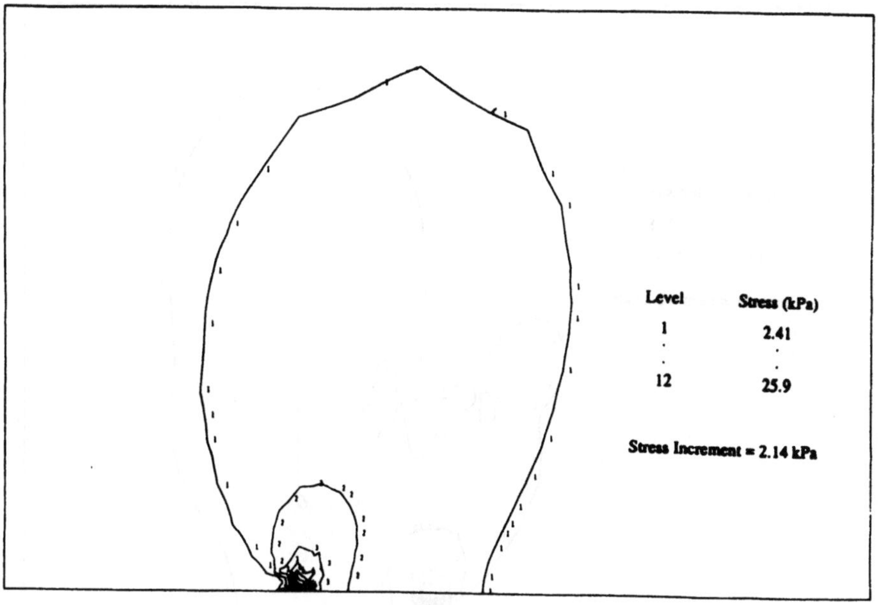

Fig.3 (c) von Mises stress trajectories for a main crack in the absence of microcracks.

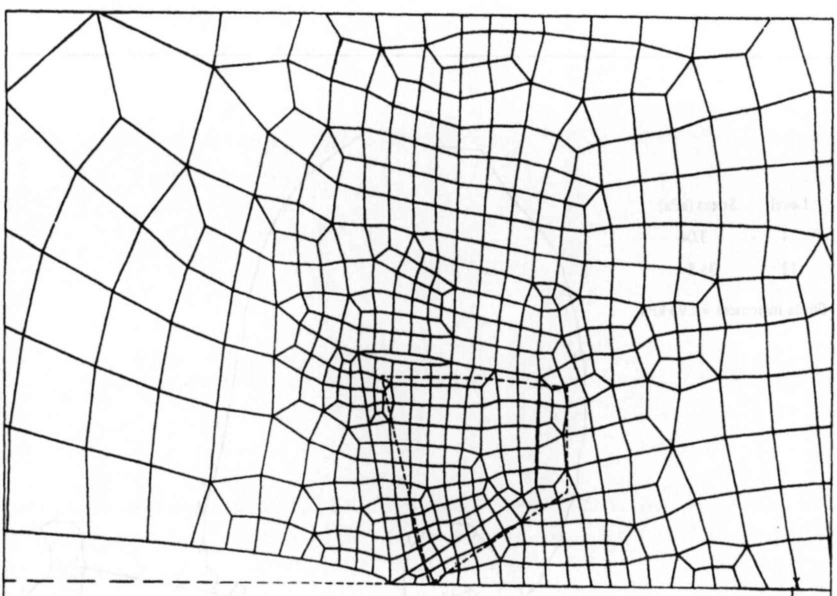

Fig.4 Finite element results for four symmetric microcracks: $c/d = 1/4$, $\theta = 45^o$, $\phi = 90^o$ and $c/d = 1/4$, $\theta = 90^o$, $\phi = 0^o$
(a) displaced geometry and deformation mode of main and microcracks.

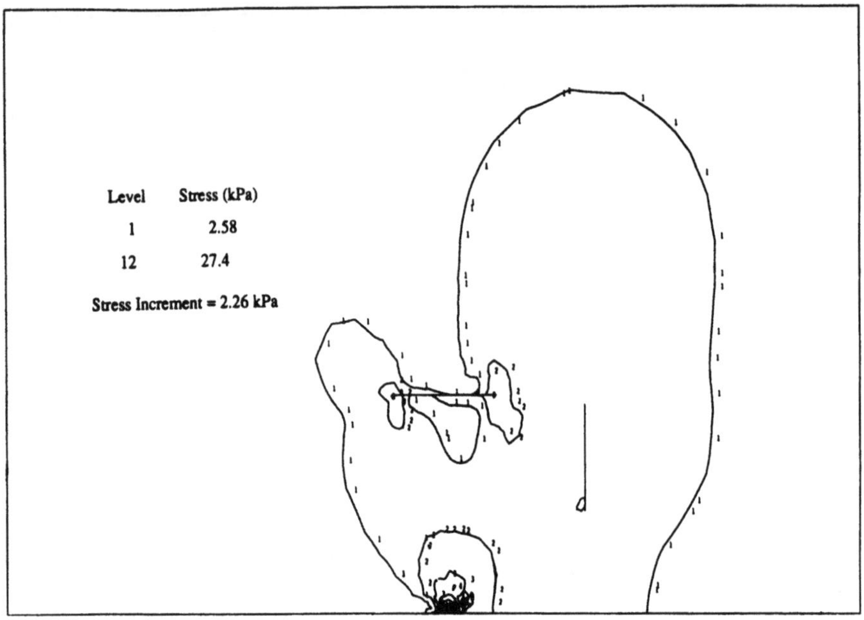

Fig.4 (b) von Mises stress trajectories.

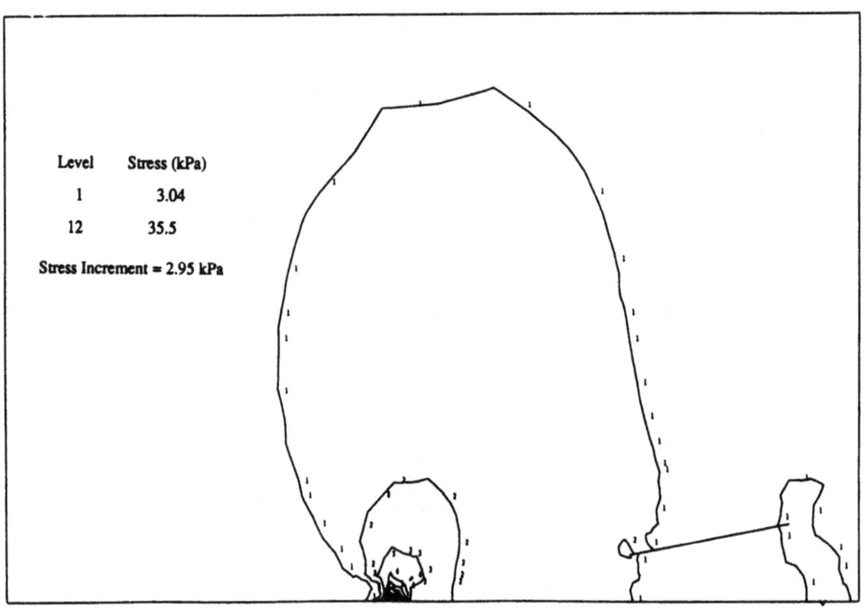

Fig.5 von Mises stress trajectories for two symmetric microcracks: $c/d = 1/4$, $\theta = 10^o$, $\phi = 10^o$

The stress trajectories also indicate that limited interaction exists between microcracks (a) and (b). This is contrary to Fig. 5, which provides a good example of main crack-microcrack-microcrack interaction. In this case, $c/d = 1/4$, $\theta = 10^o$, and $\phi = 10^o$ for two symmetric microcracks. In this configuration, the zeroth order approximation of the analytical solution should break down as it does not allow for such interaction. However, both the finite element and the zeroth order solution gave comparable values to K_I^{MA}/K_I. In this case, the finite element prediction was 1.015, while the zeroth order estimate was 1.016. This can be explained by the inappropriate selection of microcracks location ($c/d = 1/4$). Further details regarding this aspect of the work can be obtained in [8].

CONCLUSIONS

A comprehensive finite element study is made of the interaction problem between a main crack and symmetric microcracks. Three cases were considered. The first examined the effect of two symmetric microcracks and the second of four symmetric microcracks upon the stress intensity factor at the main crack-tip, K_I^{MA}/K_I. The results revealed shielding of K_I^{MA}/K_I in two and four symmetric microcracks configurations. The study was further extended to examine the effect of main crack-microcrack-microcrack interaction upon K_I^{MA}/K_I. It is important to note that the analytical method is incapable of providing the stress trajectories in the regions of main and microcracks. This is important as it enables a complete and an intuitive understanding of the interaction problem. Since the analytical model is only adaptable to a semi-infinite crack in an infinite body, the finite element method would prove more apt to compare with future experimental models.

ACKNOWLEDGEMENTS

The financial support provided by the Natural Sciences and Engineering Research Council and Energy Mines and Resources of Canada is gratefully acknowledged.

REFERENCES

[1] N. Claussen, J. Stub and R.F. Pabst, Effect of induced microcracking on the fracture toughness of ceramics, Bull. Am. Ceram. Soc., 56, pp 559-562 (1977).

[2] A.G. Evans and K.T. Faber, Crack-growth resistance of microcracking brittle materials, J. Am. Ceram. Soc., 67, pp 255-260 (1984).

[3] P. E. V. de Miranda and R. Pascual, Initial stages of hydrogen induced short fatigue crack propagation, Proc. of the Conference on the Behaviour of Short Fatigue Cracks, Ed. K. J. Miller and E. R. de Los Rios, Mechanical Engineering publications, London (1986).

[4] N. Narita, C. J. Altstetter and H. K. Birnbaum, Hydrogenated phase transformations in austenitic stainless steels, Metal. Trans., 13A, pp 1355-1365 (1982).

[5] S. X. Gong and H. Horii, General solution to the problem of microcracks near the tip of a main crack, J. Mech. Phys. Solids, 37, pp 27-46, (1989).

[6] N.I. Muskhelishvili, Some Basic Problems of the Mathematical Theory of Elasticity, translated from Russia by J.R.M. Radok, Noordhoff, Leiden, (1953).

[7] S.A. Meguid, Engineering Fracture Mechanics, Elsevier Applied Science Publishers, London, England, (1989).

[8] S.A. Meguid, P.E. Gaultier and S. X. Gong, A comparison between analytical and finite element analysis of main crack-microcracks interaction, submitted to Engng. Fracture Mechanics, 1990.

The Effect of Surface Treatments on the Fretting Wear of an Aluminium Alloy (RR58)/Steel (BS 970 080 M40) Couple

R. B. Waterhouse
University of Nottingham
Nottingham
UK

ABSTRACT

Fretting experiments were carried out using a cross-cylinder arrangement. Tests were run for up to 10^6 cycles. Coefficients of friction were continuously monitored and wear measurements were made by carrying out a profilometer survey on the resultant fretting scars on both specimens. The steel specimens were untreated, but the aluminium specimens were tested in the following conditions: untreated, MoS_2 antifret coating, shot-peened, and shot-peened and hard anodised.

Coefficients of friction were initially low, rising to 0.5 to 0.7 after 10^6 cycles, except for the shot-peened and anodised coating which ranged between 0.8 and 1.05, eventually falling back to 0.8 at 10^6 cycles. The wear on the aluminium alloy was greatest on the MoS_2 coated specimens being twice that for the untreated surface. Wear of the steel surface was highest against the antifret coating and lowest against the shot-peened surface. The recommended treatment to prevent fretting damage in this type of contact is shot-peening of the aluminium alloy surface.

INTRODUCTION

When two surfaces in contact are vibrated so that they experience tangential, oscillatory, relative movement of small amplitude (<75μm) surface damage can occur with the formation of wear debris, which may be able to escape from the contact or, depending on the circumstances, may become compacted between the surfaces. This constitutes the processes of fretting wear and fretting corrosion. If the movement is the result of one of the members of the contact being cyclically stressed, then fatigue cracks may be initiated leading to eventual fatigue failure.

One of the methods of minimising the effect of fretting is by surface treatment [1,2]. Harder surfaces are more resistant to damage and some methods rely on this, e.g. work hardening by shot-peening or surface rolling, or diffusion processes, e.g. carburising or nitriding of steel surfaces. A further series of methods depend on the application of a suitable coating. The coating may have various functions, e.g. lowering the coefficient of friction, preventing metal-to-metal contact, taking up the movement within the coating (or insert) if of low elastic modulus. However, many coating treatments tend to reduce the fatigue strength of materials, e.g. in the present case anodising is known to produce reductions of fatigue strength in aluminium alloys of up to 65% [3]. To counteract this, shot-peening is frequently applied as a pre-treatment since it induces a compressive residual stress in the surface which inhibits the propagation of fatigue cracks and raises the fatigue strength. The surface roughening also has a slight beneficial effect where fretting-fatigue is concerned [4]. The combination of shot-peening followed by anodising has been shown to improve the fatigue strength of aluminium alloys [5].

The present investigation is concerned with certain of these treatments applied to an aluminium alloy surface in contact with an untreated steel surface.

MATERIALS AND METHODS

The specimens both of the aluminium alloy and the steel were cylindrical 12.5mm dia. and 19mm long. The nominal composition and mechanical properties of the materials are shown in Table I.

TABLE I
Composition and mechanical properties of the materials

Al alloy RR58:	Al 2Cu 1.5Mg 1Fe 1Ni
Steel BS 970 080 M40:	0.4C 0.8Mn

Material	YS or 0.2% PS MPa	UTS MPa	Elongation %	Hardness H_v
RR58	340	400	8	140
BS 970 080 M40	260	525	16	269

The treatments applied were:

(a) antifret coating of MoS_2 applied and stoved at 190°C for 2h. Coating thickness 13 to 18μm.

(b) shot-peened with glass beads to an intensity of 6 on the Almen A scale.
(c) shot-peened as in (b) and hard anodised, to MOD specification DEF 151 Type 3 to a thickness of 25μm.

The steel samples were untreated. All the specimens were ultrasonically degreased in acetone before testing.

The fretting rig is shown in Fig. 1. The fretting couple was a crossed-cylinder arrangement. The steel cylinder formed the lower member and the aluminium alloy cylinder the upper members. The movement was parallel to the axis of the steel cylinder and hence, at right angles to the axis of the aluminium alloy cylinder. The fretting movement was produced by an electromagnetic vibrator mounted in a trunnion to which is connected a goose-necked arm which holds the upper moving specimen. The normal load between the specimens was adjusted by fixing the position of the vibrator in the trunnion from the readings of the appropriate strain gauges on the arm. Other strain gauges give readings of the tangential force so that the coefficient of friction can be determined. This is recorded at regular intervals during a test.

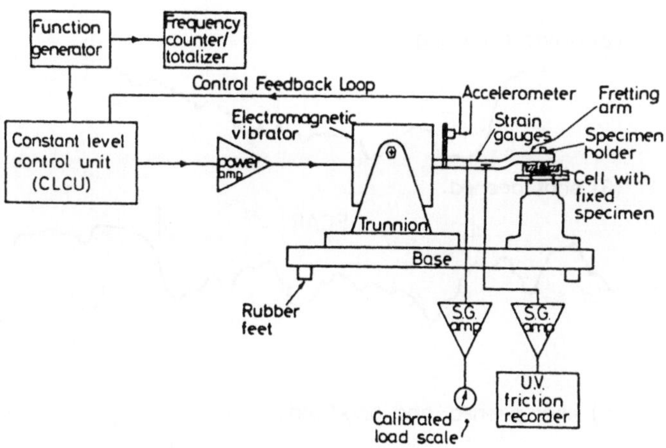

Fig. 1. Fretting rig.

The mechanical variables were held constant as follows:

normal load	3N
peak to peak amplitude	40μm
frequency	50Hz

Duplicate tests were run for 10^3, 10^4, 10^5 and 10^6 cycles.

At the end of each test, loose debris was removed by ultrasonic cleaning, the specimen was examined in the SEM using EDAX to monitor whether there was material transfer from one surface to the other, i.e. Al on the steel and Fe on the aluminium alloy, and photographs taken in every case.

The scars were then surveyed with a Talysurf profilometer at 22µm intervals. From the traces the area on each trace below the original surface was determined and hence the volume of material removed could be determined. The shot-peened surfaces were so rough that it was not possible to gain quantitative measurements of the wear which appeared to be slight. Fig. 2 shows typical traces over the centre of the scars for each type of aluminium alloy specimen, after 10^6 cycles of fretting.

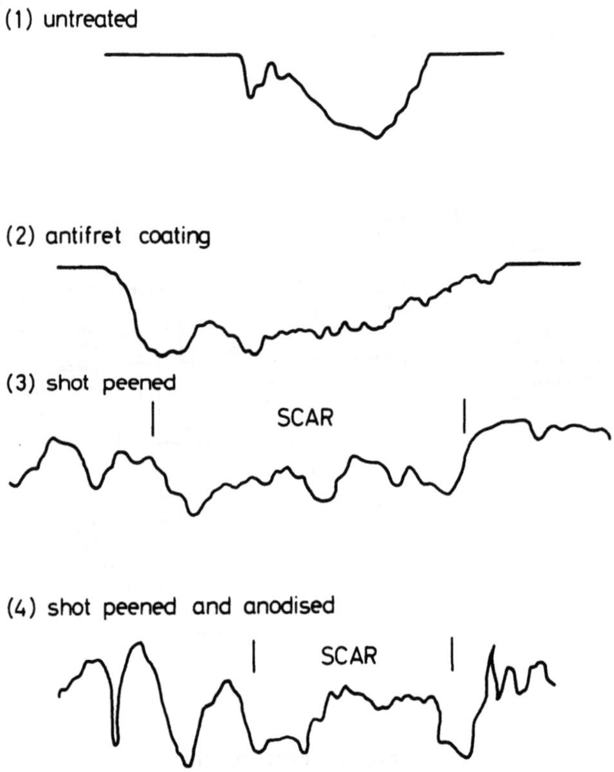

The vertical magnification in 1, 2 and 3 is X1000, in 4 it is X2000.

The horizontal magnification is X100.

Fig. 2. Typical profilometer traces across scars, 10^6 cycles.

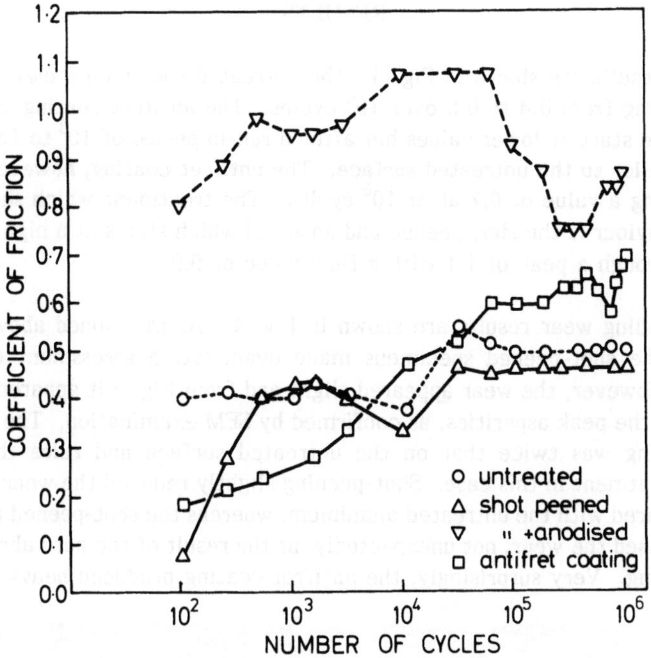

Fig. 3. Coefficient of friction vs. no. of cycles.

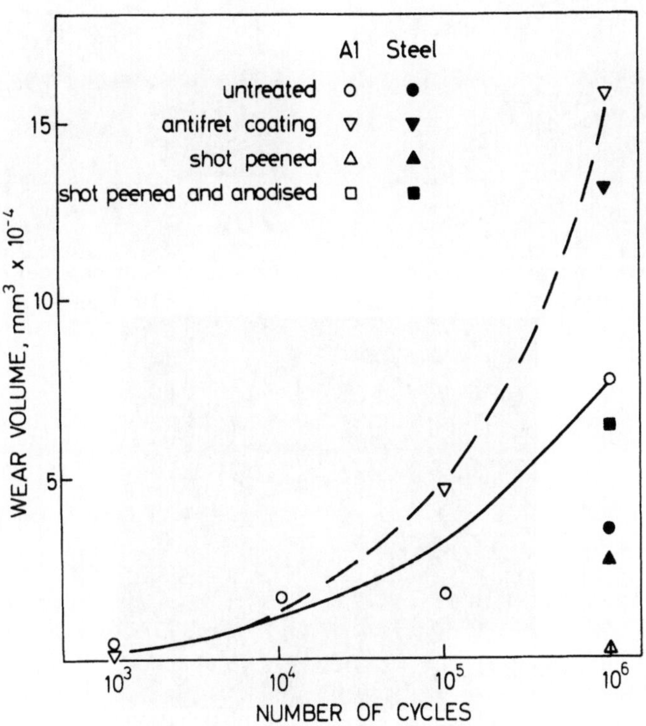

Fig. 4. Wear vs. no. of cycles.

RESULTS

The friction results are shown in Fig. 3. The untreated specimen shows little change in friction rising from 0.4 to 0.5 over 10^6 cycles. The antifret coating and the shot-peened surface start at lower values but after a run-in period of 10^3 to 10^4 cycles the friction is similar to the untreated surface. The antifret coating, however, continues to rise reaching a value of 0.7 after 10^6 cycles. The treatment which shows entirely different behaviour is the shot-peened and anodised which starts at a high value of 0.8 and passes through a peak of 1.1 with a final value of 0.9.

The corresponding wear results are shown in Fig. 4. As mentioned above the rough surfaces of the shot-peened specimens made quantitative assessment of the wear impossible. However, the wear appeared slight and from Fig. 2 it apparently involved some wear of the peak asperities, as confirmed by SEM examination. The wear on the antifret coating was twice that on the untreated surface and rules this out as a protective treatment in this case. Shot-peening slightly reduced the wear on the steel surface compared with the untreated aluminium, whereas the shot-peened and anodised surface increased the wear, not unexpectedly, as the result of the hard abrasive nature of the anodising. Very surprisingly, the antifret coating produced heavy wear on the steel surface.

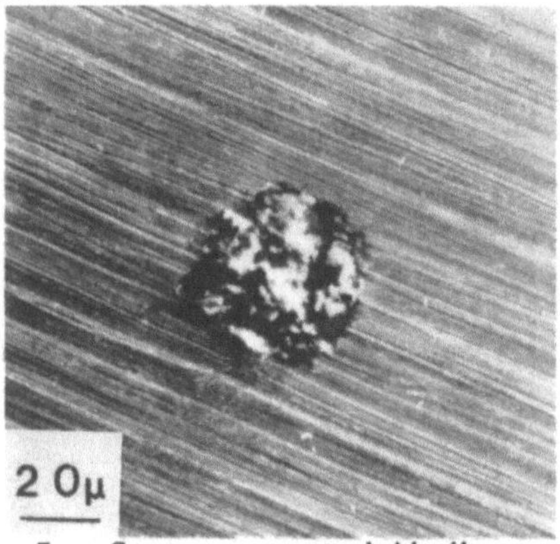

Fig. 5a. Scar on untreated Al alloy, 10^5 cycles

Fig. 5b. Scar on steel counterface, 10^5 cycles.

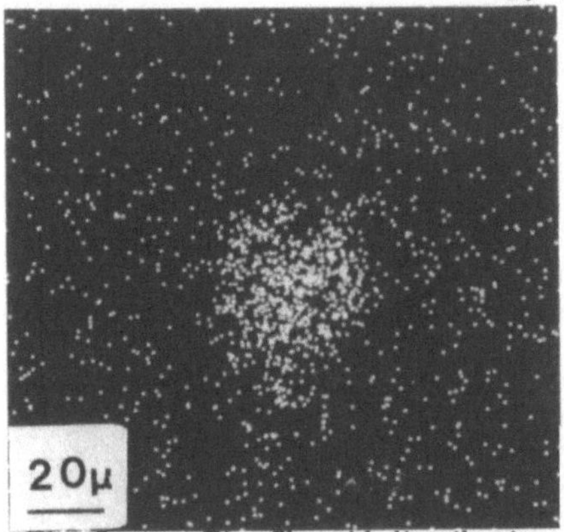

Fig. 5c. as Fig. 5b - Al distribution.

All the specimens were examined in the scanning electron microscope (SEM) and were also subjected to EDAX to determine the extent of transfer of material from one surface to the other. With the untreated aluminium, scars were detected after 10^3 cycles; the scars on the alloy and the steel appeared very similar at 10^5 cycles, Figs. 5a and 5b. There is evidence of charging particularly on the aluminium specimen, due to the poor conductivity of Al_2O_3. Little iron is transferred to the aluminium surface but some aluminium is transferred to the steel surface, Fig. 5c. The scars on the antifret coating and steel counter surface are considerably larger. At 10^5 cycles it appeared as a black area, Fig. 6a, but at 10^6 cycles the coating appeared to be broken through with some charging in the central area and a surrounding black fringe, Fig. 6b. Again, there was transfer of aluminium to the steel surface.

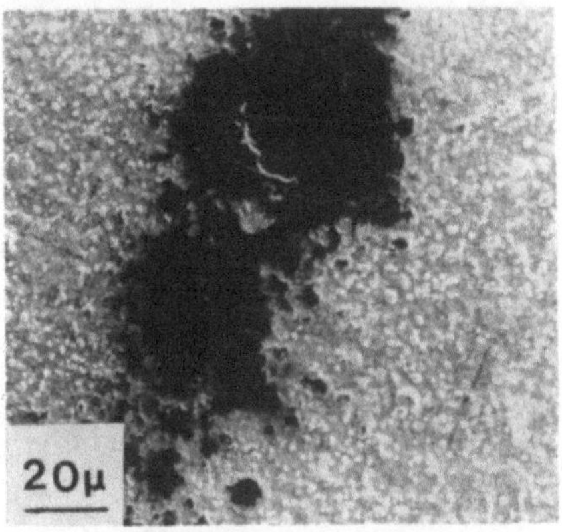

Fig. 6a. Scar on antifret coating, 10^5 cycles

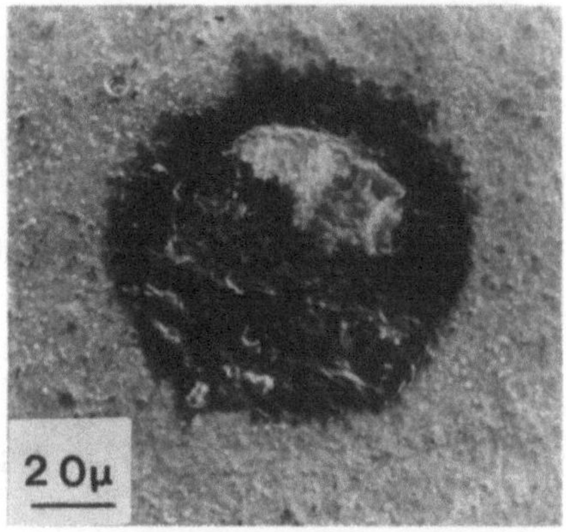

Fig. 6b. Scar on antifret coating, 10^6 cycles.

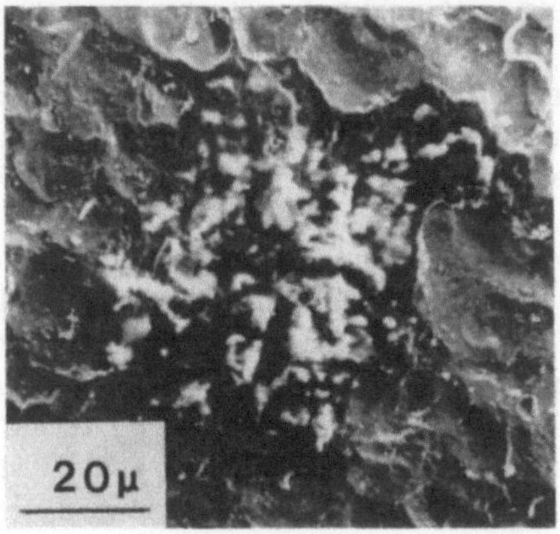

Fig. 7. Scar on shot-peened Al alloy, 10^5 cycles.

On the shot-peened aluminium surface a scar was visible after 10^3 cycles. At 10^5 cycles oxide had formed on the tops of the asperities with pronounced charging up, Fig. 7. Some transfer of aluminium to the steel surface was observed. With the shot-peened and anodised surface the scar was not detected until 10^4 cycles where it appeared as a smooth, polished area. This was more pronounced at 10^5 cycles, Fig. 8a. In contrast to the other surfaces, there was considerable transfer of iron to the aluminium surface, Fig. 8b. The scar on the steel surface became apparent after 10^3 cycles and was very pronounced at 10^6 cycles but no transfer of aluminium was detected.

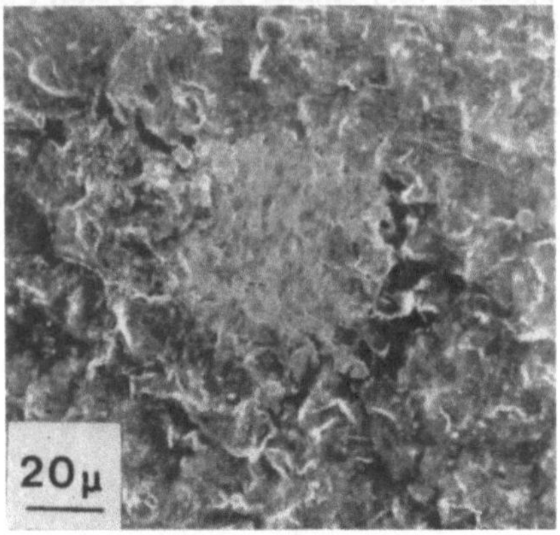

Fig. 8a. Scar on shot-peened and anodised Al alloy, 10^5 cycles.

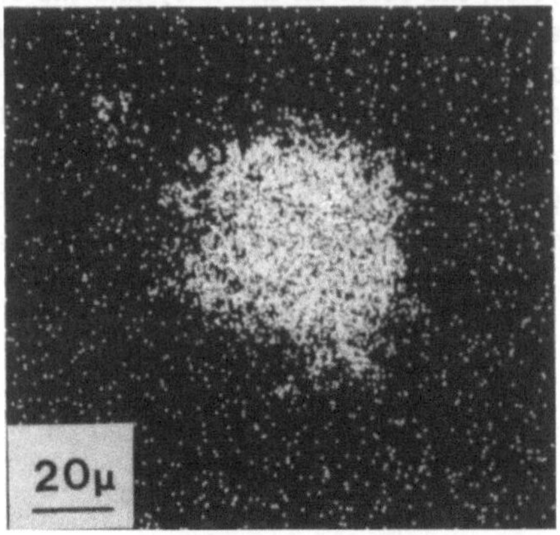

Fig. 8b. as Fig. 8a - Fe distribution.

DISCUSSION

The surprising result from this investigation is the very heavy wear suffered by the antifret coating, Fig. 4, being almost twice that of the untreated aluminium alloy. The depth of the scars after 10^6 cycles is about 22μm which is comparable with thickness of the coating (nominally 13 to 18μm). It therefore appears that most of the wear was borne by the coating and this is substantiated by the very small amount of Fe transfer and this only on one of the 10^6 cycle specimens. Some concern was expressed that the coating treatment may result in over-ageing of the alloy. Hardness tests were carried out on the untreated and the coated specimens with the following results: untreated alloy 139 Hv, antifret coated specimen 148 Hv. It appears that the mechanical properties have not been adversely affected. The coefficient of friction rises steadily throughout the tests to 0.65 - 0.70, Fig. 3, in contrast to the untreated specimens, which have a constant coefficient of friction of 0.4 up to 10^4 cycles and then a rise to 0.5, at which it remains constant for the remainder of the test. The higher coefficient of friction of the coated specimens suggests that the coating is steadily worn away although EDAX photographs of iron distribution on the steel specimens and aluminium distribution on the aluminium specimens do not indicate this.

Shot-peening has previously been shown to reduce the coefficient of friction and wear in the fretting of mild steel surfaces [6]. The hardening of the surface contributes to this, and also the rough surface enables any loose debris that does form to be accommodated in the hollows rather than being trapped in the contact areas where it may contribute to further damage since the oxide is harder than the metal. The transfer of material from one surface to the other is, in this case, from the aluminium alloy to the steel, with none in the opposite direction. The oxide on aluminium is alumina which has a hardness of ca. 2000 H_v and it is likely that it could become embedded in the steel surface. This also applies to the untreated aluminium alloy. Comparison of the aluminium distribution on the steel samples produced by the untreated and shot-peened surfaces indicates that they are about the same. The coefficients of friction of the two surfaces also are very similar after 5 x 10^4 cycles, Fig. 3.

The notable feature of the shot-peened and anodised surfaces is the very high coefficient of friction in the early stages, but still remaining high at the end of the test 0.8 to 0.9. High coefficients of friction are typical in fretting sintered alumina against steel surfaces. Fretting sintered Al_2O_3 against a cold drawn 0.64C steel (H_v 412) resulted in a coefficient of friction of 0.7 to 0.8 [7] whereas the same material against a bearing steel (H_v 715) gave an even higher friction of 0.9 to 1.0 [8]. The significant observation in the case of the anodised coating is the very high transfer of iron to the specimen, Fig. 8b. It appears to be highly compacted debris from the steel. No aluminium was detected on the steel specimens. It could be that the hard anodised coating on the rough surface which shows fairly sharp ridges, Fig. 7, is machining away the steel.

The question has been asked whether a harder steel would be beneficial in this particular contact. Since the hardness of Al_2O_3 is so much greater than anything attainable with even tool steels, it seems unlikely that this would have much effect.

It must be remembered that the geometry of the specimens used in this investigation allows debris to escape comparatively easily, whereas in a flat on flat contact, the debris is more easily trapped and may contribute to the damage [9].

CONCLUSIONS

The following conclusions can be drawn:

(1) The coefficients of friction of the untreated and shot-peened aluminium alloy/steel couple are very similar at 0.45 to 0.5 and are constant after 5 x 10^4 cycles.

(2) The coefficient of friction of the antifret coating rises steadily reaching a value of 0.65 to 0.70 at 10^6 cycles and could still be increasing. The wear volume is about twice that on the untreated specimens. The depth of the pits on the antifret coating are roughly comparable with the thickness of the coating after 10^6 cycles. Very little, if any, material transfer occurs on the antifret coating and the steel.

(3) Of the untreated and shot-peened surfaces there is a little transfer of Fe to the Al surface, but more Al is transferred to the steel surfaces.

(4) The shot-peened and anodised coating shows the highest coefficient of friction between 0.8 and 0.9 after 10^6 cycles, comparable with the friction of Al_2O_3 against steel. Considerable transfer of Fe occurs to the aluminium surface.

(5) The wear of the shot-peened surfaces, particularly when anodised, appears minimal. However, the high coefficient of friction of the anodised surface and the heavy wear of the steel counterfaace, makes this treatment inadvisable. Plain shot-peening appears to be the best method of reducing fretting damage in these circumstances.

ACKNOWLEDGEMENT

The author is grateful to Napier Turbochargers Ltd., Lincoln, for financing this work.

REFERENCES

1. Beard, J., The rational selection of palliatives for avoidance of fretting, Proc. Int. Conf. on Tribology, 1-3 July, London, 1987, Instn. Mech. Engrs., 1, 311-319.

2. Chivers, T.C. and Gordelier, S.C., Fretting fatigue palliatives: some comparative experiments, Wear, 1984, 96, 153-175.

3. Beitel, G.A. and Bowles, C., Influence of anodic layers on fatigue-crack initiation in aluminium, Met. Sci. J., 1971, 5, 85-91.

4. Leadbeater, G., Noble, B. and Waterhouse, R.B., The fatigue of an aluminium alloy produced by fretting on a shot-peened surface, Advances in Fracture Research, Proc. 6th Int. Conf. on Fracture, Pergamon, 1984, 3, 2125-2132.

5. Emeric, D.A., Levine, S. and Washburn, K., The effect of surface coatings on the fatigue strength of aluminium alloys, Report No. 2339, US Army Mobility Equipment Research and Development Command, Fort Belvoir, VA 22060, USA, Sept. 1981.

6. Bergman, C.A., Cobb, R.C. and Waterhouse, R.B., The effect of shot-peening on the friction and wear of a carbon steel and an austenitic stainless steel in fretting conditions, Wear of Materials 1987, ASME New York, 33-37.

7. Kuno, M., Waterhouse, R.B. and Pearson, B.R., Fretting wear in sintered alumina- and tungsten carbide cermet-metal couples, Wear of Materials 1987, ASME New York, 371-379.

8. Kuno, M., Fretting damage of high carbon chromium bearing steel, PhD thesis, University of Nottingham, 1988, p. 64.

9. Kuno, M. and Waterhouse, R.B., The effect of oscillatory direction on fretting wear under crossed cylinder contact conditions, Proc. Conf. Eurotrib, 12-15 June 1989, Helsinki, Finland, 3, 30-35.

Degradation of a Biomedical Polyurethane: Surface Integrity and Fatigue Effects

M. M. Sinnott and D. W. Hoeppner

University of Utah
Salt Lake City
Utah
USA

ABSTRACT

Polyurethanes degrade under a variety of conditions. In cardiovascular applications of polymeric materials emphasis is placed on elastomers that have extended fatigue lives. The effect of surface integrity on the performance properties of Biomer®, a segmented poly(etherurethaneurea) used on many blood contacting surfaces, has been investigated using tensile tests at room temperature, and biaxial fatigue tests at body temperature. No significant differences other than ultimate tensile strength (UTS) were noted in the stress-strain curves of the specimens with varying surfaces, however preliminary results from fatigue tests indicate a decreasing fatigue life with increasing surface roughness.

This study reports on specimen analysis prior to and following fatigue testing in an attempt to describe mechanisms for mechanical degradation related to differences in surface integrity and fatigue loading. Analysis methods used include Fourier transform infrared spectroscopy (FTIR), and scanning electron microscopy (SEM). FTIR indicated degradation of the polymer occurred related to cyclic stresses applied to the material. All fatigued specimens showed signs of degradation regardless of their surface roughness. Surface evaluation using SEM indicated the presence of micropores prior to and following cyclic loading of the specimens. Micropores seen following cyclic loading were larger and in some instances more numerous. Microcracks were observed propagating between micropores in a typical fatigue failure mechanism.

INTRODUCTION

Biomedical components, such as those used in circulatory assist devices, are often subjected to a duty cycle in which the stress is continuously varying. Many materials experience failure due to such cyclic loading at stresses far below stresses that cause failure during static loading.

This phenomena is known as fatigue, and propagation of surface initiated fatigue cracks can lead to premature catastrophic failure.

Fatigue failure occurs through a process of nucleation, and propagation of cracks. In the propagation phase the fatigue crack grows a small amount during each loading cycle until final failure occurs. Since cracks often start on the surfaces of components, the surface integrity may be an important factor in the fatigue life of the component.

The roughness of a surface is the finely spaced irregularities that are produced by the cutting action of tool edges and abrasive grains on surfaces that are machined. Surface roughness is defined as the deviation of the topography of a surface from an ideal atomically smooth planar surface. It is measured as the average (arithmetical) deviation from the mean line of the surface profile, and is expressed in micrometers (μm) or microinches (μin). [1]

Semi-crystalline polymers are essentially two phase systems consisting of spherulitic clusters of crystals embedded in an amorphous continuum. So called "segmented polyurethanes" belong to this family of polymers and it is generally accepted that their properties are due primarily to the phase separation of soft and hard segments leading to the formation of hard-segment domains which are dispersed in the "rubbery" polyether matrix. Their strength results from the entanglement of the polymer chains and from the presence of the crystalline regions which reinforce the weaker amorphous regions. If the molecular structure is degraded by chain scission then the important properties of the polymer will be lost.

MATERIALS AND METHODS

Films of solution grade Biomer® (supplied by Ethicon, Inc. as a 30 weight % solution and diluted to a 20 weight % solution in dimethylacetimide (DMAC)) were cast on plates with varying surface finishes of 2 μin, 63 μin, 125 μin, and 250 μin. Thin sheets of Biomer with different surface finishes were obtained since the mold surface of the sheets reflected the surface of the plate upon which it was cast. The thicknesses of the films were from 0.007 to 0.009 inches.

Tensile and fatigue specimens were cut from thin films using standard dies and an arbor press. Specimen geometries and treatment prior to testing has been reported previously.[2]

Tensile tests were conducted using an Instron Model 1000 Tensile Machine according to ASTM-D-882: Standard Test Methods for Tensile properties of Thin Plastic Sheeting. The specimen geometry was a standard "dumbbell" shape.

A fatigue test protocol was developed with reference to ASTM-D-671-78: Standard Tests Methods for the Flexural Fatigue of Plastics by Constant Amplitude of Force since no standard method for fatigue testing of thin polymer films could be found. The developed protocol outlines the determination of the effect of repetition of the same magnitude of inflation pressure on thin sheets of viscoelastic polymers. The fatigue test apparatus and methods have been previously described.[2] The fatigue specimen geometry was a thin disc that was clamped at the edges and inflated under a given pressure to a hemispherical shape.

In preparation for Fourier transform infrared spectroscopy (FTIR) a small portion from each of the fatigued and non-fatigued ("virgin") specimens was dissolved in DMAC and solution cast directly onto clean glass plates. The films were dried in a convection oven at 50°C for 24 hours to remove residual solvent. Film thicknesses were estimated to be approximately 0.004 inches using a dead weight dial micrometer.

A NIC 1200S Fourier transform infrared spectrometer (Nicolet Instrument Corp.) was used to obtain transmission spectra of the thin polymer films over the 400-4600 cm^{-1} wavelength region. Fifty scans per sample were taken and averaged to provide the final spectra with a resolution of 2 cm^{-1}. The fatigued sample spectra were subtracted from the virgin spectra in order to investigate the effects of surface finish and fatigue on the molecular structure of the material. A difference spectra was plotted for each surface roughness over the 1000-3400 cm^{-1} wavelength region. All data were processed using the standard software provided with the instrument. No smoothing procedures were used when plotting the spectra.

Prior to Scanning Electron Microscopy (SEM) evaluation a small section from each fatigued specimen (both mold and air sides) was cut from the center of the specimen as this was the area that sustained the highest stresses during testing. A similar section was cut from a sheet of unstressed material for each surface finish. The sample sections were mounted on aluminum stubs, and were sputtered with AuPd using a Protractor Vacuum Etch/Sputter Machine set at 5 mtorr vacuum, and 10 mamps, for approximately four minutes. The thin coating of AuPd that was deposited on the surface of the samples was adequate to provide protection of the polymer surface from electron beam bombardment without obscuring surface detail.

SEM examination was performed using an SS-40 Scanning Electron Microscope (International Scientific, Inc.) at 50, 500, 1000, and 2000X.

RESULTS AND DISCUSSION

Tensile Tests

Average stress-strain curves for the specimens grouped according to surface finishes were plotted together as a means of comparison. The results shown in Figure 1 are typical of an uncrosslinked polyurethane. Elongation at fracture is between 720 to 750% which is in close agreement with other investigators [3]; and ultimate tensile strength (UTS) is between 5900 and 6600 psi, also within the range of reported values.[4]

Variations in the moduli at 100%, 300%, and 500% elongation as well as the variations in UTS between groups were analyzed using the Student "t" test for unpaired samples. The 100% modulus is indicative of material stiffness, whereas the 300 %, 500% and UTS are related to mechanical integrity of the material. A 95% ($p < 0.05$) level of statistical confidence was utilized as the criterion for significance. No significant differences were noted between any of the sample 100%, 300% or 500% moduli. There was no significant difference in the UTS of the 2 μin compared to the 63 μin specimens; or the 125 μin compared to the 250 μin specimens. There was a small but significant difference in the UTS of all other combinations compared. This result indicates that the surface finish of the material above a specified roughness may affect its UTS.

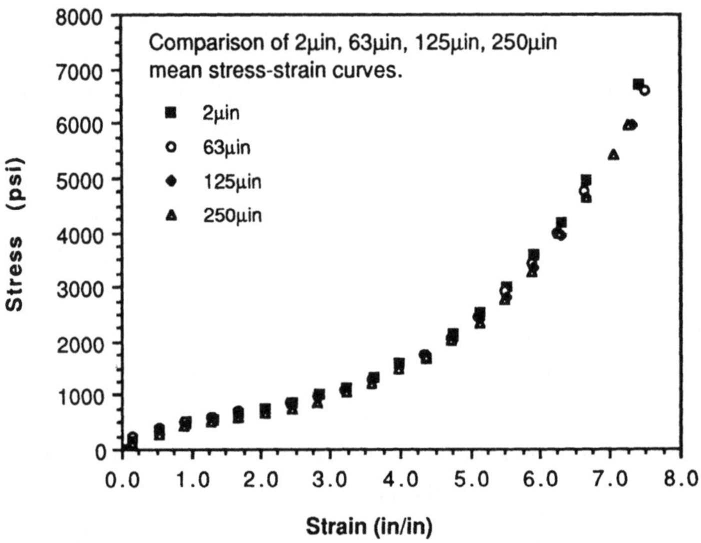

Figure 1. Comparison of mean stress-strain curves.

Changes in the surface roughness will effect stress concentrations due to changes in cross section and geometry. Stress concentrations may enhance localized events such as surface crack nucleation and propagation, which in turn leads to failure of the roughest specimens at a lower UTS. Properties indicative of bulk characteristics (i.e., the moduli at 100%, 300%, and 500% elongation) remain unchanged between groups.

Fatigue Tests

Fatigue tests were run in excess of 4 months. Cycles to failure were recorded for those specimens that failed; unfailed specimens were suspended following 20,886,600 cycles. The results of the test are shown in Figure 2. All 250 μin specimens failed early in the testing (at 33,211; 37,290; and 42,487 cycles). One 2μin and one 125 μin specimen also failed (at

1,116,025 and 2,810,585 cycles respectively). All other specimens survived until suspended from the test.

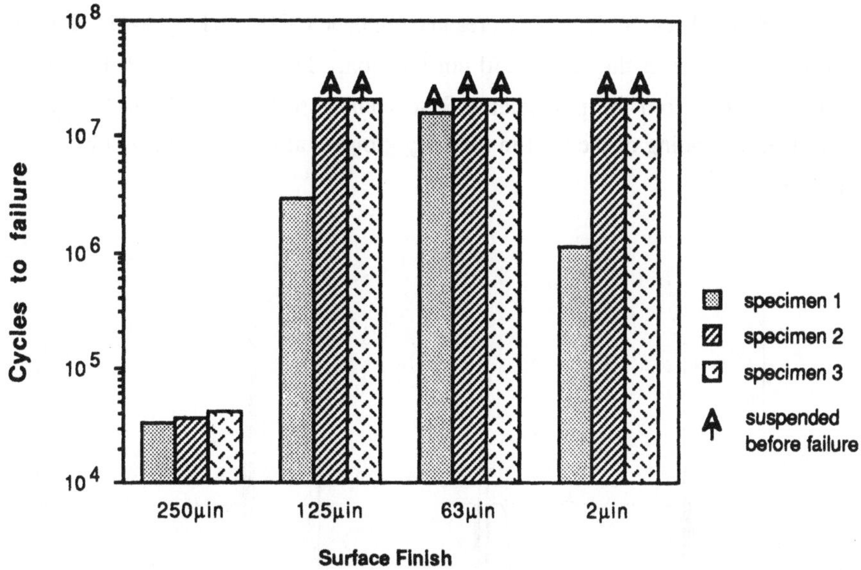

Figure 2. Cycles to failure for Biomer specimens with various surface finishes.

FTIR Spectroscopy

Difference IR spectra for the specimens grouped according to surface finishes were plotted. Peak assignments indicating possible polymer degradation are presented in Table 1. [5,6]

Wavenumber (cm^{-1})	Relative intensity*	Assignments
2938	VS	υ_a (C-H) in CH_2
2852	VS	υ_s (C-H) in CH_2
1540-1525	S	υ (C-N) + δ (N-H) amide II
1225	S	υ (C-N) + δ (N-H) amide III
1110	VS	υ (C-O-C) aliphatic ether
1080	S	υ (C=O-C) in hard segment

* Relative intensity based on sample at room temperature. S = strong; VS = very strong. υ = stretching; υ_a = asymmetric stretching; υ_s = symmetric stretching; δ = bending.

TABLE 1. FTIR band assignments seem in difference spectra.

Figure 3 shows the difference spectra for each surface roughness. Peaks on the spectra indicate a relative decrease in the molecular structure associated with absorption at that wavenumber.

Urea and urethane C-N stretching and N-H bending amide II modes absorb in the 1525-1540 cm^{-1} region. Similarly urea and urethane C-N stretching and N-H bending amide III modes absorb in the 1200-1250 cm^{-1} region. The C-O-C stretching vibration of the aliphatic ether from the polyether soft segment absorbs at 1110 cm^{-1}. Symmetric and asymmetric bending of the C-H in the CH_2 group is absorbed at 2852 cm^{-1} and 2938 cm^{-1} respectively.

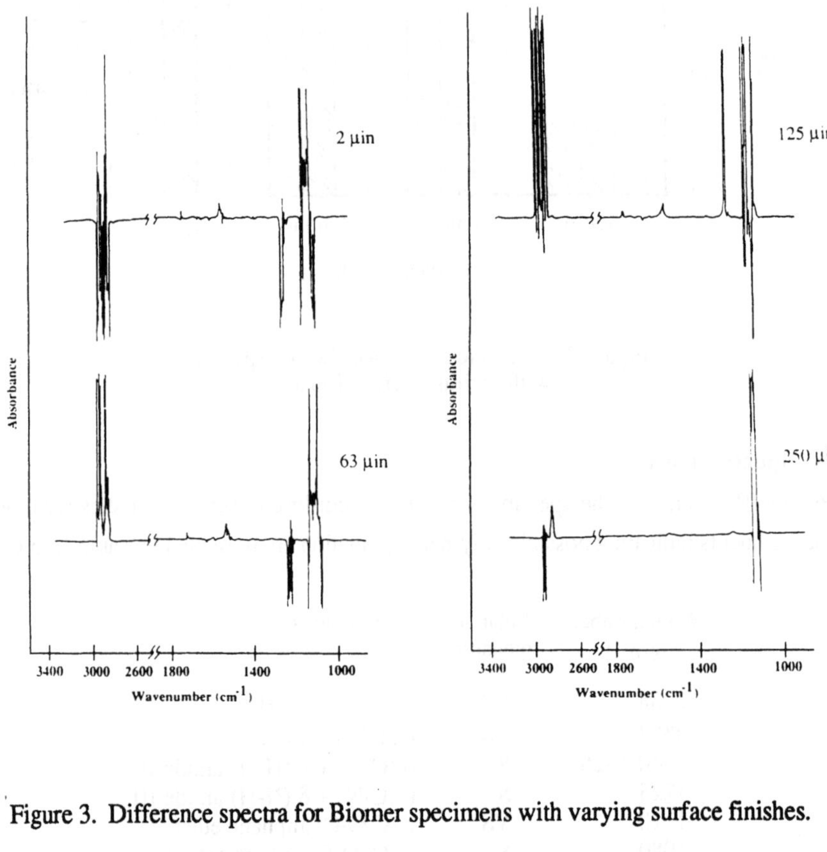

Figure 3. Difference spectra for Biomer specimens with varying surface finishes.

According to Marchant et. al. [5] degradation of a poly(etherurethaneurea) should lead to a relative decrease in CH_2-O, C-O-C, C-N and possibly C=O, and N-H absorptions. This pattern was generally observed for all the fatigued specimens. The presence of stresses seen in the fatigue loading of components may accelerate their degradation regardless of their surface finish.

Scanning Electron Microscopy

Surface analysis was performed using SEM on the mold and air side of both fatigued and unfatigued specimens. All specimen surfaces, prior to or following cyclic loading, showed discontinuities in the form of micropores. Micropores tended to be larger in diameter on the specimens that received cyclic loading. Figures 4 and 5 illustrate this difference. In some

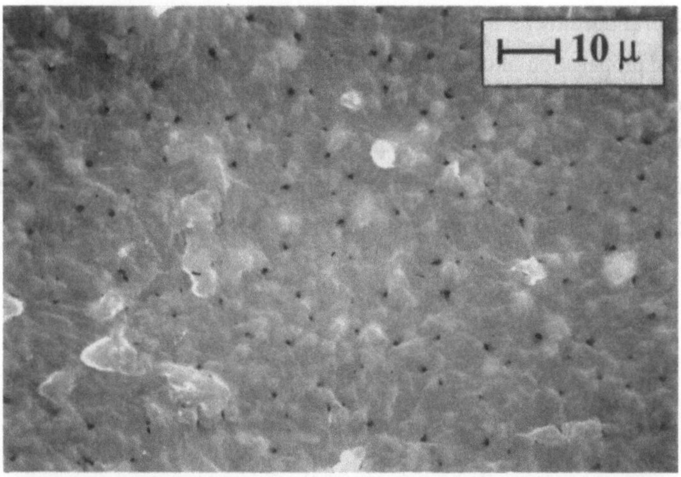

Figure 4. Surface of 2 µin specimen, air side, prior to cyclic loading.

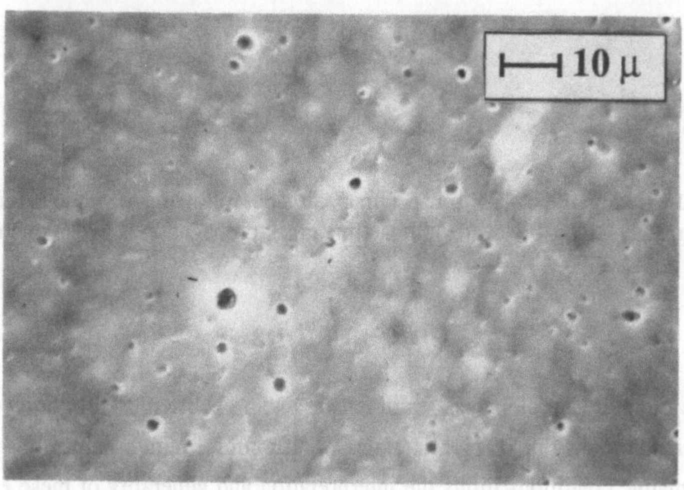

Figure 5. Surface of 2 µin specimen, air side, following to cyclic loading (approximately 1.11 x 10^6 cycles).

instances the pores were more numerous. Fatigue loading appears to "stretch" pores to a larger size and may cause subsurface pores to rupture thereby increasing the number of pores on the surface of some specimens.

Surface cracks in implanted polyurethane samples have been reported to start preferentially at small "defects" such as microbubbles.[7] Other authors have reported the presence of microcracks in the blood diaphragms of artificial hearts even after very short pumping durations (i.e.,within two weeks).[8]

In several specimens it appeared that fatigue failure occurred through a process of nucleation, and propagation of cracks. Cracks on the surfaces of the fatigued specimens in this study not only started at micropores but were also seen to propagate between pores. (See Figure 6) Areas containing numerous parallel cracks similar to crazing were also noted on some fatigued specimens. Since the cracks appeared to propagate from pore to pore, the integrity of the surface may be an important factor in the fatigue life of the component.

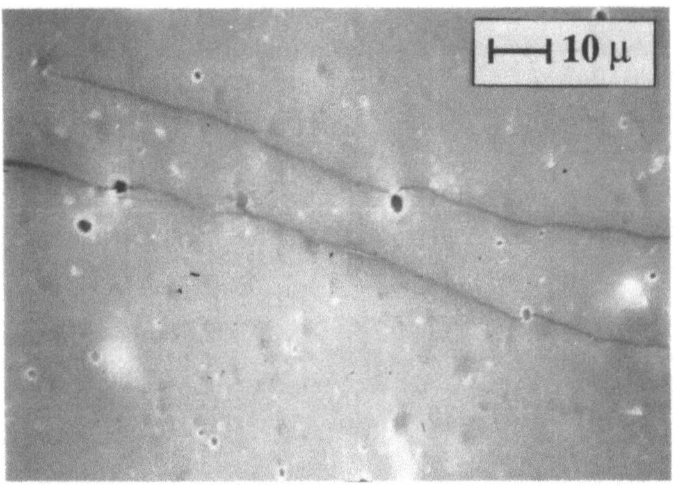

Figure 6. Surface of 125 μin specimen, air side, following to cyclic loading (approximately 2.81 x 10^6 cycles) illustrating crack propagation between micropores.

CONCLUSIONS

In the literature simple stress-strain tensile testing is often used to draw conclusions regarding dynamic behavior of biomaterials. The noninformative nature of static results applied to dynamic conditions limits the interpretation and generality of these experiments. This is not to say that tensile tests are not worthwhile but only that conclusions made based on their results should be applied with caution. For cyclic loading conditions dynamic fatigue testing should be done.

Results based on the tensile and fatigue tests done in this study indicate that mold surface finishes of at least 63 μin should be specified to maximize UTS and fatigue life of the material. It may be necessary to perform fatigue tests of longer duration in order to better characterize the fatigue properties of this material.

FTIR indicated degradation of the polymer occurred related to mechanical stress applied to the material. All fatigued specimens showed signs of degradation regardless of their surface roughness.

Surface evaluation using SEM indicated the presence of micropores prior to and following cyclic loading of the specimens. Micropores seen following cyclic loading were larger and in some instances more numerous. Microcracks were observed propagating between micropores in a typical fatigue failure mechanism.

This work was supported in part by a Biomedical Research Support Grant awarded by the National Institutes of Health, Grant no. S07 RR07092

REFERENCES

1. ASTM Committee on Terminology: Compilation of ASTM Standard Definitions, (sixth edition). Philadelphia, American Society for Testing and Materials, 1986.

2. Sinnott, M.M., Hoeppner, D.W., Romney, E., Dew, P.A., "Effects of Surface Integrity on the Fatigue Life of Thin Flexing Membranes", Trans. Am. Soc. Artif. Intern. Organs, 1989, **35**, 687-690.

3. Phua, S.K., Castillo, E., Anderson, J.M., Hiltner, A., "Biodegradation of a Polyurethane In Vitro", Journal of Biomedical Materials Research, 1987, **21**, 231-246.

4. Ethicon, Inc., Somerville, N.J., Technical Product Information.

5. Marchant, R.E., Zhao, Q., Anderson, J.M., Hiltner, A., "Degradation of a Poly(etherurethaneurea) Elastomer. An Infrared and XPS Study", Polymer, 1987, **28**, 2032-2039.

6. Lelah, M.D., Cooper, S.L., Polyurethanes in Medicine, CRC Press Inc, Boca Raton, Florida, 1986, 100.

7. Lemm, W., "Biodegradation of Polyurethanes", in Polyurethanes in Biomedical Engineering, edited by H. Planck, G. Egbers, and I. Syré, Elsevier Science Publishers B.V., Amsterdam, 1984, 103-108.

8. Hennig, E., Bucherl, E.S., "Mineralization of Circulatory Devices Made of Polymers", in Polyurethanes in Biomedical Engineering, edited by H. Planck, G. Egbers, and I. Syré, Elsevier Science Publishers B.V., Amsterdam, 1984, 109-134.

Microstructural and Surface Effects on Fatigue of Ti-6Al-4V Alloy

S. B. Young and W. J. Bratina

University of Toronto
Toronto
Ontario
Canada

ABSTRACT

Ti-6Al-4V was used to study some fundamental aspects of fatigue. Although, a variety of microstructures and surface conditions were compared, the majority of work was performed on smooth specimens in the beta annealed condition. Additional tests were made on specimens coated partially with spherical titanium particles and on standard FM specimens. An extensive SEM survey of surfaces was performed to elucidate the characteristics and growth behaviour of small surface cracks (microcracks and short cracks). Concepts of crack arrest and of an apparent fatigue limit were advanced. A comparison between individual segments of macrocracks and short cracks was proposed; mechanisms responsible for fatigue crack behaviour were tentatively deduced.

INTRODUCTION

Apart from their great industrial importance, it appears that titanium alloys represent a very suitable material for the study of fundamental aspects of fatigue due to their great variability of microstructures and surface conditions. For Ti-6Al-4V alloy, the most used commercial titanium alloy, appropriate heat treatments can result in microstructural conditions which range from fine equiaxed hcp alpha phase with intergranular beta phase (commercially called mill annealed structure) to drastically different coarse lamellar colonies of alpha within large prior-beta grains (beta annealed structure). Extensive observations of microstructure and dislocation configurations have been made by TEM and dislocation activities at the alpha-beta interface have been well studied [1,2].

In S/N testing, mill annealed titanium is significantly superior to the beta annealed condition, with respective high cycle fatigue strengths of about 620 and

500 MPa. Surface conditions and material microstructure strongly affect S/N behaviour via crack initiation mechanisms. Exactly opposite trends are claimed for fatigue crack growth (FCG) and fatigue crack propagation (FCP) rates, these being significantly lower, in the neighbourhood of one order of magnitude, for the beta annealed than the mill annealed morphology, particularly at intermediate ΔK values [3-5].

The following topics will be addressed in this paper:

1. Small surface cracks, ranging from microcracks to short cracks, and their behaviour as a statistical function of local stress condition and microstructure.
2. The occurrence of non-propagating cracks, in contrast to propagating cracks; mechanisms leading to a temporary or permanent crack arrest and the associated phenomenon of fatigue limit.
3. The apparent erratic growth behaviour of short cracks when attempts are made to apply da/dN versus ΔK format.
4. Modifications of surface conditions by the introduction of stress raisers on the surface by sintering of spherical titanium particles of various discrete diameters; this technique has commercial application in the biomaterials field.
5. Some additional studies on FCG and FCP rates as a function of the fracture mechanics (FM) parameter ΔK will also be presented. An attempt will be made to apply the information obtained from fatigue fracture surfaces to short crack behaviour.

Optical and electron microscopy observations will be emphasised, particularly as regards mechanisms responsible for crack behaviour.

MATERIALS AND METHODS

This paper presents mostly the results of S/N testing of various microstructures and surface conditions but complementary studies using FM techniques were also made. All specimens were tested in laboratory air at ambient temperature and pressure.

Materials: This work was performed in conjunction with a larger ongoing biomaterials project; hence, specimen substrate alloys met ASTM specification F 136 for surgical grade extra low interstitial (ELI) Ti-6Al-4V and coating particles were either of commercial purity titanium (ASTM F 67) or also Ti-6Al-4V. Three microstructural conditions were examined: mill annealed, beta annealed, and martensitic. The physical metallurgy of Ti and its alloys can be understood by considering the alpha-beta allotropic transformation (882°C for pure Ti) from hcp to bcc crystal structure; and the metal interstitial element content [6, 7]. Ti-6Al-4V is a two-phase alpha-beta alloy most commonly fabricated by thermomechanical processing and solution treating in the mill annealed condition [5-8], in which the substrate material was received. If not hot wrought but brought slowly from above the beta-transus to room temperature the beta-annealed alpha lamellar structure is produced. Faster cooling, for example by Ar or oil quenching, results in the basketweave martensitic microstructure [6, 9].

S/N Testing: Two rotating bending fatigue specimen variations were employed. Standard hourglass waisted specimens (ASTM standard E 466) were mostly used; the second specimen geometry was tapered in design to permit a uniform stress distribution along their gauge length. Whereas for the hourglass specimens fatigue is essentially stress induced, predominantly defect induced fatigue is to be expected for the tapered specimen type, with its larger uniformly stressed surface area. Test frequency was 50 Hz and R was -1. Practices outlined in ASTM E 739 were followed to generate and analyse the S/N data. In any rotating-bending fatigue specimen the highest stresses are to be found on the outer surface; this stress gradient is acceptable since fatigue cracks, in most cases, originate at free surfaces irrelevant of the type of loading.

FCG and FCP Studies: In some complementary experiments, notched FM specimens conforming with ASTM standard E 647 were employed; fatiguing was performed axially on an MTS loadframe at R = 0.1 and frequencies of 1-10 Hz; da/dN versus ΔK results were generated. This specimen format tends to localize fatigue cracking to the notch region. In fact, a single well-defined crack represents an ideal case and deviations may render an ASTM test invalid or at least questionable. In this work, the tortuousity of an actual crack path, the event of multiple crack formations, and crack branching were of main interest.

Surface modifications: Stress raisers in the form of sintered spherical particles (nominally 100 μm in diameter) were sintered to the smooth surfaces of some rotating bending specimens. For biomaterials applications, particles are normally applied in two complete layers [10, 11]; in this work however, some specimens were incompletely coated such that particles were isolated or in small separated groups on the surface. This permitted control and examination of crack initiation sites. Surface micro-geometry was also varied by applying particles of different diameters to other specimens (approximately 20 and 350 μm).

RESULTS

Microcracks and Short Cracks: What constitutes a microcrack or a short crack is somewhat arbitrary. For example, according to Suresh and Ritchie [12] "cracks are defined as being short when their length is small compared to relevant microstructural dimensions (a continuum mechanics limitation), when their length is small compared to the scale of local plasticity (an LEFM limitation) and when they are merely physically small (i.e. crack length ≤ 500 to 1000 μm)." For this paper, small cracks are divided into microcracks, at dimensions less than and approaching microstructural features, and short cracks, which are longer but still small from a FM perspective. Small cracks may grow to become macrocracks, which then follow a conventional da/dN versus ΔK pattern.

On free surfaces, initiated microcracks varying from one to tens of micrometres in length were seen in a variety of microstructures. In the martensitic morphology, alpha needles appeared to be particularly susceptible to crack growth (Figure 1); microstructural sensitivity of fatigue cracks was also very evident in the beta annealed condition, where cracks seemed to develop and grow along microstructural features, sometimes intimately, for example preferentially along colony boundaries of alpha lamellae (Figure 2). In other cases, however, cracks

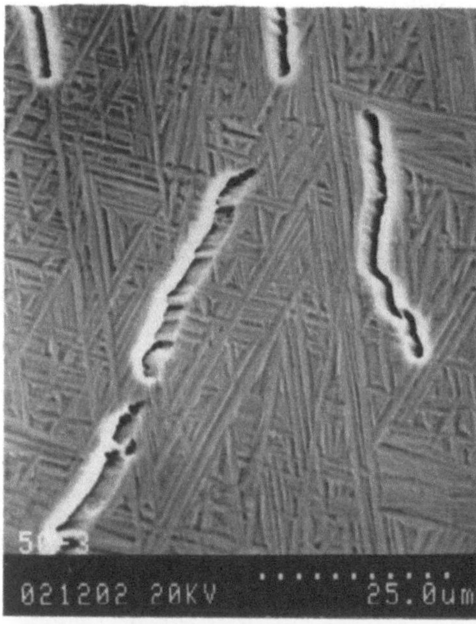

Figure 1: SEM micrograph of microcrack growth along alpha-beta interface in martensitic microstructure.

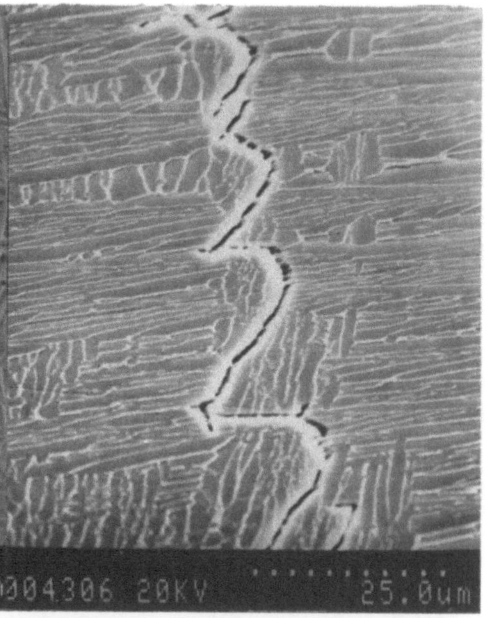

Figure 2: SEM micrograph of small crack path along colony boundaries in beta annealed microstructure.

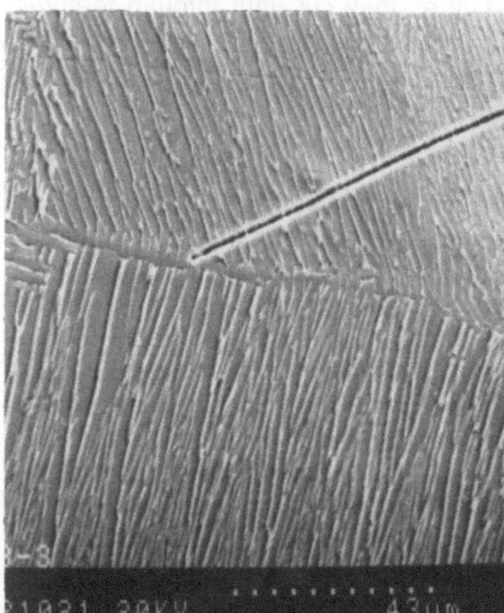

Figure 3: Microcrack growth through grain in beta annealed microstructure.

Figure 4: Grain boundary cracking; unfailed specimen: 10^7 cycles, 250 MPa.

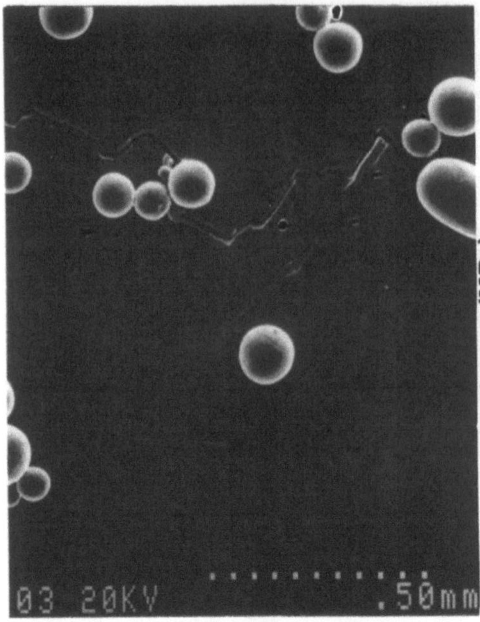

Figure 5: Crack failure route in partially coated specimen, note branching.

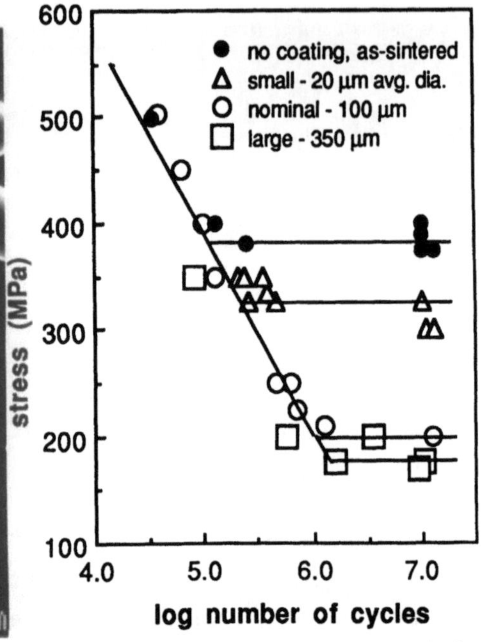

Figure 6: S/N curves for beta annealed material with surface modifications.

appeared to ignore microstructure completely as they cut right through features within beta grains (Figure 3). The beta annealed structure also appeared susceptible to short cracking up to about 100 µm along prior-beta grain boundaries, until seemingly halted by crack branching (Figure 4; note also the thermal etching running almost vertically on the left-most grain and the microcracks contained within that grain).

Surface modifications: Stress raisers significantly affected fatigue performance: in general, microcracks and short cracks were observed preferentially at sintered particles. Their occurrence, however, appeared to be a statistical event based on the function of local microstructure and stress environment. In failed beta annealed specimens, cracks tended to have formed through stress raisers (Figure 5), however in a rather haphazard fashion, perhaps by coalescence. Crack branching and secondary cracking were readily visible, apparently dependent upon a combination of local surface geometry and surface microstructure.

S/N testing: All microstructural and surface variations demonstrated an apparent fatigue limit, below which stress no failures were recorded. For uncoated specimens, the mill annealed material with a ground surface exhibits a fatigue limit of 620 MPa [10]; whereas the beta annealed condition with as-sintered surfaces performed to 380; once coated with 100 µm particles the value dropped to 200 MPa (Figure 6). Coatings with larger particles resulted in poorer fatigue performance of 175 MPa; on the other hand, the small particle coating significantly modified surface geometry, raising the endurance limit to 325 MPa.

FCG and FCP as Represented by FM Methodology: The progress of fatigue cracks plotted in a da/dN versus ΔK manner is not of direct concern in this paper. Numerous scanning electron micrographs of fractured surfaces from threshold and near-threshold regions have been taken, similar to previous work [2, 3, 13, 14]; however, they will not be reproduced here for lack of space. Also, in our laboratory, many determinations of da/dN versus ΔK have confirmed lower FCP rates for the beta annealed material than the mill annealed condition [3].

DISCUSSION

As indicated in the introduction, microstructure does affect fatigue performance of titanium alloy; however, surface treatment also significantly influences fatigue behaviour due to residual stresses and surface roughness influences. Industrially, specific surface treatments such as grinding and shot-peening are used to ensure reliable engineered surfaces.

In a previous study of Ti-6Al-4V, crack initiation plus short crack growth to 1 mm in surface length typically consumed at least 95% of total fatigue life [15]. Our own results on 316L stainless steel (fcc structure), HSLA steels (bcc), and mill annealed Ti-6Al-4V (hcp plus bcc) confirm this: on the basis of fatigue striation counts employing high resolution SEM and replica TEM techniques, macrocrack (Paris region) growth was found to represent less than 15% of total fatigue life [2, 14]. Considering that the initiation stage and the final fracture amount to less than 1% of total life, it follows that the remaining time must be spent in the threshold and near-threshold small crack region.

Recognising the importance surface condition plays in combination with observations that macrocrack propagation life consumes a very small fraction of total fatigue life, it appears logical that FM researchers have shifted their attention towards analysis of small surface cracks; thus approaching areas of interest to metal physicists. Small cracks have been observed to propagate at unusually high rates of growth at levels of stress intensity well below macrocrack thresholds. Thus, it is apparent that long crack mechanics do not accurately interpret small crack behaviour [16-19]. It has been proposed that small cracks grow anomalously fast only when their plastic zone size is smaller than the significant microstructural element size, at this scale a crack is too short to experience a material as homogeneous and continuum assumptions do not apply [18].

Fatigue Crack Initiation and Microcracks

The term "fatigue crack initiation" is not well defined but certainly includes the to- and fro-motion of dislocations in a specimen subjected to a cyclic stress, leading eventually to cyclic slip along favourable slip planes and leading further to more or less irreversible microcracks. Decades ago metallographers made detailed studies of the surface appearance of various metals and alloys subjected to fatigue stresses and early observations of slip bands, persistent slip bands (PSB), extrusions, incipient cracks, propagating and nonpropagating cracks have been examined [20]. PSB's have been studied extensively in fcc metals and to a lesser extent in bcc iron and iron-base alloys.

Basically, the current concept of fatigue cracking is that crack initiation occurs quickly and relatively easily and is hence unavoidable [17]. These very tiny cracks are similar in length to many microstructural features. Once formed, cracks may propagate quickly and readily for short distances along microstructurally influenced routes until barriers are encountered which impede propagation; a crack thus increases in length until another, stronger, microstructural barrier is encountered. An existing small crack is thus stable and nonpropagating unless its stress field is sufficient to overcome microstructural barriers; if no barriers of sufficient strength exist, the crack becomes conventionally unstable and propagates into a macrocrack regime.

Microcracks as shown in Figures 1 and 2 seem to be predominantly affected by microstructure; particularly Figure 2 shows a crack trying to accommodate its progress closely with microstructure. However, the crack shown in Figure 3 is appears primarily stress controlled, the microstructure playing only a minor role; yet, the subgrain boundary appears to act as an efficient, temporary or permanent barrier. At this stage no branching was observed although some crack fragmentation was seen, as in Figure 2. All these cracks are surface cracks, therefore subject to oxidation if in air; that is the cracks are irreversible. Microstructural barriers such as interfaces between alpha and beta phases may be relatively easily overcome (Figures 1 and 3) or by-passed (Figure 2). With continuous stress cycling there is no reason why cracks should not continue to grow as these minor microstructural barriers do not seem to be able to arrest them permanently. Other microstructural barriers such as grain boundaries appear to arrest cracks more effectively (Figure 3). This may be however deceiving, a particular crack may hesitate at the boundary for 10, 100, 1000, or many more cycles and then continue its progress either through an adjacent grain or along the grain boundary.

The fatigue process is statistical. To accurately interpret the phenomenon, a large number of SEM and other observations is required; the few photographs in this paper may not be exactly representative. It would appear that many microcracks, perhaps the great majority, do sooner or later arrest. Others may coalesce, becoming a short crack and then either arrest or propagate into a macrocrack which may or may not be a specific crack leading to final fracture.

Short Cracks

Although metallurgists became familiar with short cracks decades ago [20], it remained for FM researchers to do so in a quantitative manner by applying the da/dN versus ΔK analysis format. When plotted in this manner the behaviour of short cracks appears erratic. They occur at stress levels below the threshold, their growth rates are larger or smaller than predicted by near-threshold analysis, and occasionally they more or less arrest [13, 21-23].

High growth rates and the nonpropagating nature of short cracks will now be addressed. Large FCG rates appear to be directly associated with measurements at a microscale [23]; their accuracy being well beyond that of the usual da/dN versus ΔK technique. It is suggested that monitoring of an actual macrocrack path on a microscale in the FCG and FCP regions may yield most of the relevant data. As mentioned before, the FCP rate for mill annealed alloy is claimed to be higher than for beta annealed material; if da/dN is assumed to represent actual crack velocity per cycle, this cannot be correct. For a designer however the velocity component of crack path along a straight line is of major interests and this is in fact what is represented by da/dN measurements. In this engineering respect, beta annealed alloy is a very suitable material. The actual crack path [24] especially in beta annealed alloy is however greatly disturbed, being characterized by an unusual tortuousity, frequent kinks, deviations, temporary arrests, extreme branching, etc. Macrocrack path is therefore much longer if measured along its actual route and hence the crack velocity is much higher, perhaps about one order of magnitude higher.

It appears that the tortuous macrocrack path behaviour in beta annealed material mirrors the characteristics of short cracks exactly; the individual segments of this long path can be compared to short cracks. Individual straight line segments would experience significantly higher velocities than the velocities obtained by the averaging of macrocrack values used to plot da/dN versus ΔK values in a sigmoidal curve. In fact by plotting beta annealed alloy data in this manner, scatter appears excessive, whereas the mill annealed alloy with more homogeneous structure and therefore less tortuousity graphs with tolerable scatter. Briefly, short cracks and individual segments of macrocracks are indeed too short to experience a material as homogeneous; they thus respond stochastically to local microstructures and stress condition. This is in agreement with growth rate minima observed by Wagner and Lutjering on the same microstructure [21].

The unusually rapid progress of short cracks can sometimes reverse and result in a propagating crack becoming nonpropagating. Various mechanisms for crack arrest can be envisaged: strong microstructural barriers, reduced stress intensity at crack tips due to crack branching, plastic deformation at the crack tip combined with work hardening and strain ageing, etc. There are indications of fatigue limit behaviour in Ti-6Al-4V (Figure 6). However, a direct comparison with the fatigue limit phenomena of alpha iron-carbon alloys may be difficult to make. The fatigue limit in ferritic carbon steel could be considered, in the extreme, as an ideal equilibrium, where the generation of dislocations is balanced by their immobilisation due to dislocation-dislocation interactions, but primarily by dislocation-interstitial interactions, specifically carbon atoms. A large body of experimental evidence seems to confirm strain ageing as an important mechanism for fatigue limit in iron-carbon systems [20]. However, with the acceptance of the existence of microcracks and short cracks at fatigue stress levels [25], modifications of this thesis must be introduced. A complete arrest can be associated with crack branching, where the ΔK's of individual branches are reduced drastically and the progress of plastic deformation at the branch tips is effectively blocked by combined work hardening and strain ageing. In this manner, the mechanism of strain ageing remains still operative, although confined to the plastically deformed regions at the tips of individual crack branches. In a ferritic carbon steel there is an optimum of fatigue properties at approximately 220°C; this may be difficult to reconcile with the presence of cracks, as they would strongly oxidize and consequently be irreversible.

In Ti alloys there is no real mechanism analogous to strain ageing of iron-carbon alloys. What was observed was occasional branching of short cracks associated obviously with a drastic decrease of the original ΔK value (Figure 4). The energy dissipation associated with plastic deformation occurs and partial immobilisation of dislocations takes place by dislocation-dislocation interactions, i.e. via work hardening.

Little information on microcrack or short crack behaviour could be obtained from mill annealed specimen fractographs in the threshold or near-threshold region; as the fractures were indistinct and too oxidised. Basically we know that deformation must be via cyclic slip mechanisms but this was not experimentally evident. This is in contrast to work done on Ti-8.6Al and Ti-6Al-4V alloys with equiaxed grains [13]; slip plane facets were observed in Ti-8.6Al alloy with a grain size of about 100 μm. Conversely, in our study of beta annealed Ti-6Al-4V alloy, large featureless facets persisted almost to the final stage; fracture surfaces of this kind are no doubt associated with a rapidly advancing crack.

Surface Condition Modifications

To explain the transition in crack length from micro- to short to macro-crack, the mechanism of crack coalescence is proposed. It was apparent from the specimens coated with particles that cracking sought stress raisers such that particle-to-particle bridges formed during propagation (Figure 5). However, finite element analyses have shown that for the particle/sinterneck/substrate geometry the stress concentration resulting from the notch-effect extends too insignificantly beyond the particle to permit an advancing crack to "see" the stress raiser (perhaps 30 μm for a 100 μm particle) [26]. Coalescence of small cracks and macrocracks is an explanation of the behaviour observed.

CONCLUSIONS

The fatigue behaviour of Ti-6Al-4V was studied employing mostly smooth specimens in the beta annealed condition.

1. Surface fatigue cracking is the statistical phenomenon resulting from the local environment of microstructure, stress, and surface condition.
2. Small surface cracks (microcracks and short cracks) are influenced by microstructure and microstructural barriers which include: alpha-beta interfaces, alpha colonies, subgrain and prior-beta grain boundaries.
3. Small stable, nonpropagating, fatigue cracks (approximately 20 to 100 μm in length) were observed on the surfaces of specimens tested below their fatigue limits. Various mechanisms leading to eventual crack arrest were scrutinized.
4. Growth of small cracks can be better understood by considering the behaviour of individual segments of long cracks, and vice versa.
5. Fine sintered particles were found to act as stress raisers on a local scale; coalescence appeared to be a mechanism of crack growth.

ACKNOWLEDGEMENTS

The authors thank: S. Yue (particularly for Figs 1 to 3), R.M. Pilliar, E. Schoen, N.D.G. Mountford, and G.C. Weatherly; and acknowledge the support of the Natural Sciences and Engineering Research Council of Canada and the Ontario Centre for Materials Research.

REFERENCES

1. Brown, LM, Ogin, SL, Role of internal stresses in the nucleation of fatigue cracks. Fundamentals of Deformation and Fracture, ed. BA Billby, KJ Miller, JR Willis, Cambridge Univ., 1985, 501-528.
2. Yue, S, personal communication, McGill University.
3. Schoen, E, The effect of microstructure and environment on fatigue crack propagation in ELI Ti-6Al-4V. MASc Thesis, Univ. Toronto, 1987.
4. Rosenberg, HW, Chesnutt, JC, Margolin, H, Fracture properties of titanium alloys. Applications of Fracture Mechanics, ed. JE Campbell, WW Gerberich, JH Underwood, ASM, Metals Park OH, 1982, 213-252.
5. Microstructure Fracture Toughness and Fatigue Crack Growth Rate in Titanium Alloys, ed. AK Chakrabarti, JC Chesnutt, TMS, Warrendale, PA, 1987.
6. Collings, EW, The Physical Metallurgy of Titanium Alloys, ASM, Metals Park, OH, 1984.
7. Smith, WF, Structure and Properties of Engineering Alloys, McGraw-Hill, New York, 1981.
8. Jaffee, RI, Lutjering, G, Effect of microstructure and loading condition on fatigue of Ti-6Al-4V alloy. Microst. Fract. Tough. and Fatigue Crack Growth Rate in Titanium Alloys, 1987, 193-208.
9. Pilliar, RM, Weatherly, GC, Developments in implant alloys. CRC Critical Review of Biocompatibility, 1986, **4**, 371-403.
10. Yue, S, Pilliar, RM, Weatherly GC, The fatigue strength of porous-coated Ti-6% Al-4% V implant alloy. J Biomedical Materials Research, 1984, **18**, 1043-1058.
11. Young, SB, Fatigue of Porous Coated Titanium Implant Alloy, MASc Thesis, Univ. Toronto, 1989.
12. Suresh, S, Ritchie, RO, Near-threshold fatigue crack propagation: a perspective on the role of crack closure. Fatigue Crack Growth Threshold Concepts, ed. D Davidson, S Suresh, TMS-AIME, Warrendale, PA, 1984, 227-261.
13. Gerdes, C, Gysler, A, Lutjering, G, Propagation of small surface cracks in Ti-alloys. ibid, 465-78.
14. Bratina, WJ, Yue, S, Fatigue crack growth: a metallurgist's point of view. Time-Dependent Fracture, ed AS Krausz, Martinus Nijhoff, The Hague, 1985, 27-42.
15. Yoder, GR, Cooley, LA, Crooker, TW, Observations on a fracture mechanics approach to fatigue crack initiation in Ti-6Al-4V. Fracture Mechanics: 16th Symposium, ASTM, 1983, STP 868, 392-405.
16. Brown, MW, Interfaces between short, long, and non-propagating cracks. The Behaviour of Short Fatigue Cracks, ed. KJ Miller and ER de los Rios, EGF Pub. 1, London, 1986, 423-439.
17. Miller, KJ, Initiation and growth rates of short fatigue cracks. Fundamentals of Deformation and Fracture, 1985, 477-500.
18. Chan, KS, Lankford, J, The role of microstructural dissimilitude in the fatigue and fracture of small cracks. Acta Met, 1988, **36**, 193-206.
19. Morris, WL, James, MR, Small cracks and the transition to long cracks, MRS Bulletin, 1989, **8**, 18-24.
20. Thompson, N, Wadsworth, NJ, Metal Fatigue. Advances in Physics, 1958, **7**, 72-169.
21. Wagner, L, Lutjering, G, Microstructural influence on propagation behavior of short cracks in an (α+β) Ti alloy. Z Metallkde, 1987, **78**, 369-375.
22. Brown, CW, Taylor, D, The effects of texture and grain size on short fatigue crack growth rates in Ti-6Al-4V. Fatigue Crack Growth Threshold Concepts, 1984, 433-446.
23. Lankford, J, Davidson, DL, Near-threshold crack tip strain and crack opening for large and small fatigue cracks. ibid, 447-463.
24. Gray, GT, Thompson, AW, Williams JC, The effect of microstructure on fatigue crack path and crack propagation rate. ibid, 131-143.
25. Miller, KJ, Mohamed, HJ, de los Rios, ER, Fatigue damage accumulation above and below the fatigue limit. The Behaviour of Short Fatigue Cracks, 1986, 491-512.
26. Wolfarth, D, Filiaggi, M, Ducheyne, P, Parametric analysis of interfacial stress concentrations in porous coated implants, J Biomedical Materials Research, 1990, in press.

Corrosion and Wear Characteristics of Nickel Aluminide, NiAl, in NaCl Medium

F. A. Elrefaie, H. A. Ahmed, M. M. I. Ahmed

Cairo University
Giza Egypt

and K. E. Mohamed

Atomic Energy Authority
Inchas Egypt

ABSTRACT

The corrosion behaviour of nickel aluminide, Ni-Al, in 3% NaCl aqueous solution is investigated using electrochemical techniques. The wear resistance of this alloy in the abrasive 3% NaCl + fine sand medium is also studied. The cathodic and anodic regions of this aluminide are determined in the temperature range 25-65°C. The corrosion behaviour of NiAl is controlled by Al-forming compounds. The corrosion potentials as well as the polarization resistance are obtained. The electrochemical measurements indicated also that this phase shows no active-passive transition in the 3% NaCl solution and suffers passivating film breakdown. Over the investigated temperature range, NiAl is susceptable to pitting. Optical micrography supported the occurrence of pits. The wear resistance of this material is proved to be higher than the Ni-5% Al γ solid solution.

KEYWORDS

NiAl aluminide, corrosion, wear, NaCl aquous solution.

INTRODUCTION

Nickel-superalloys are used as structural materials. These alloys are often protected by nickel aluminide coatings NiAl, Ni_3Al and Ni_2Al_3 intermetallic phases [1-4] to improve oxidation as well as wear resistance [5-13]. However, the corrosion behaviour of these phases in humid and aqueous media has not been systematically studied. Therefore, this work is designed to investigate the corrosion aspects of NiAl phase in 3% NaCl solution at different temperature; the wear resistance of this phase is also determined in the abrasive 3% NaCl + fine sand aqueous medium.

EXPERIMENTAL

The intermetallic compound NiAl was prepared by electric arc melting of nickel (99.99 wt% pure) and 99.99% purity aluminium under argon atmosphere. The melt is then casted as cylinder, 10 cm long and 1 cm diameter, in water-cooled copper hearth. Discs, 0.1 cm thick and 1 cm diameter, were then sectioned from the casted rod by using spark cutter. The disc samples were then annealed at 1200°C for 24 h under flowing argon atmosphere then metallographically polished finishing with 1μ diameter paste. X-ray diffractometry and atomic absorption analyses indicated that the stoichiometric NiAl phase is the only existing phase in these samples.

The corrosion characteristics of the NiAl intermetallic phase have been obtained by using the standardized potentiodynamic procedure. A Wenking potentioscan Model Pos. 73 in parallel with a Philips, P.M-8120, X-Y recorder is used in this investigation. The scanning range was set from -500 mV to +500 mV versus saturated calomel electrode (SCE) while the scanning rate was 1 mV/sec.

The wear test was conducted using a magnetic stirrer apparatus. The testing medium was 3% NaCl + 5% sand abrasive medium. The kinetic behaviour is determined by weighing the sample before and after testing for different periods of time. Optical metallography is used to examine the surface microstructure of tested samples.

RESULTS AND DISCUSSION

The potentiodynamic polarization curves for the NiAl intermetallic phase in 3% NaCl solution at temperature in the range 25°C to 65°C are obtained; these curves are illustrated in Figs. 1-3. It is clear from these figures that the corrosion behaviour of the NiAl phase under these conditions is characterized by the existence of passivating film breakdown phenomenon at potentials higher than the free corrosion potential, E_{corr}, by ≈ 80-100 mV.

The corrosion behaviour of pure Ni and Al is previously studied [3,4]. Nickel shows active-passive transition while aluminium corrosion behaviour is similar to the NiAl intermetallic phase, i.e. active-passive transition is not observed while passivating film breakdown occurs. This indicates that the active anodic behaviour is controlled by aluminium in the intermetallic phase.

The values of the electrochemical parameters obtained during the potentiodynamic tests for samples subjected to different temperatures are given in Table 1.

It is clear from this table that upon increasing the temperature, the free corrosion potential and the breakdown potential of the passive layer increases, while the polarization resistance decreases. The polarization resistance decreases with temperature clearly indicates that the corrosion process is initially controlled by thermally activated process.

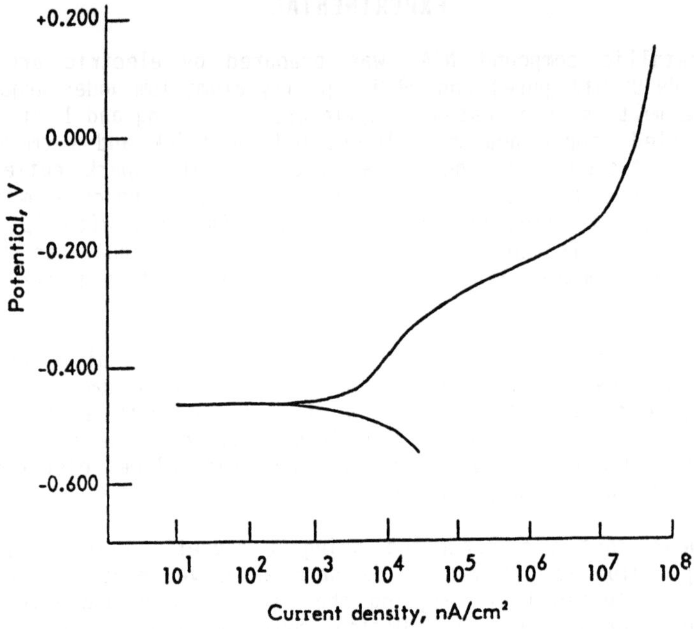

Figure 1. Potentiodynamic polarization curve for NiAl intermetallic phase in 3% NaCl aqueous solution at 25°C.

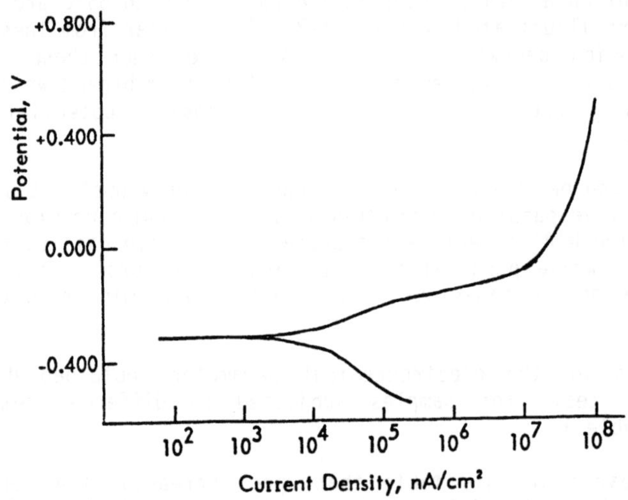

Figure 2. Potentiodynamic polarization curve for the NiAl intermetallic phase in 3% NaCl aqueous solution at 40°C.

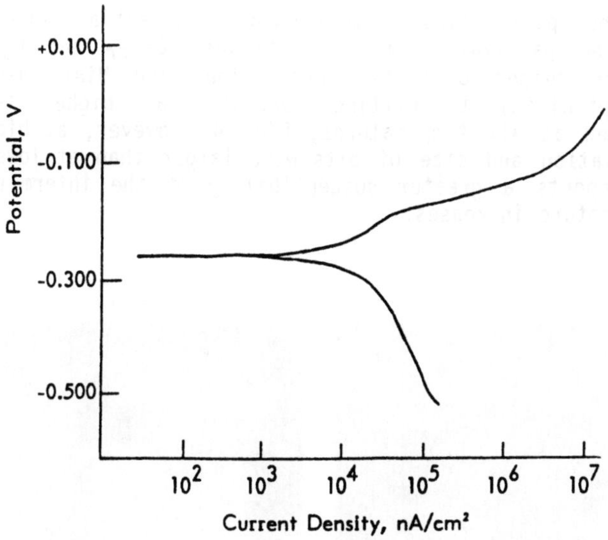

Figure 3. Potentiodynamic polarization curve for the NiAl intermetallic phase in 3% NaCl aqueous solution at 65°C.

TABLE 1
Effect of temperature on the corrosion parameters of NiAl intermetallic phase in 3% NaCl aqueous solution:
E_{corr} = free corrosion potential; R_p = polarization resistance;
E_p = passivating film breakdown potential.

Temperature (°C)	E_{corr} (mV)	R_p (ohm.m²)	E_p (mV)
25	-460	6.4×10^3	-300
40	-315	3.7×10^3	-200
65	-260	2.0×10^3	-180

The free corrosion potential increase with temperature may be explained on the basis of less stability of formed passive phases as the temperature increases; thus, the immunity range of the intermetallic phase increases.

Though the passivating film breakdown potential increases with temperature, the passivation interval between E_{corr} and E_p decreases with increasing temperature; this means that the NiAl phase shows a greater susceptibility to pitting corrosion at higher temperature. Pits were formed at all temperatures, Fig. 4; however, at high temperature the population and size of pits were larger than at lower temperature which supports a greater susceptibility of the intermetallic NiAl phase as temperature increases.

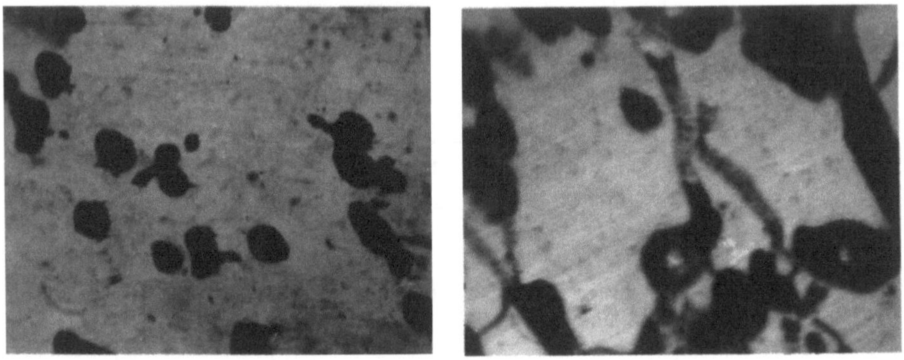

(a) (b)

Figure. 4 Pitting morphology formed at: a) 25°C (100X), b) 65°C (100X).

The wear resistance curve of the NiAl intermetallic phase in 3% NaCl and 5% fine sand medium is shown in the kinetic curve, weight loss versus time, of Fig. 5. On the same figure the wear resistance of as casted Ni-5% Al γ solid solution phase is illustrated. Figure 6 illustrates the wear resistance of NiAl as compared to heat treated solid solution phase. It is clear from both figures that the wear resistance of the NiAl intermetallic phase is better than casted Ni-5% γ solid solution phase. The NiAl phase has shown similar wear resistance as Ni-5% Al γ solid solution phase annealed at 900°C.

CONCLUSIONS

The corrosion behaviour of NiAl intermetallic phase in 3% NaCl aquous solution over the temperature range 25°C < T < 65°C is controlled by aluminium-phase formation and dissolution. This phase shows no active-passive transition behaviour; since passivating film breakdown occurs. The intermetallic NiAl phase is susceptable to pitting corrosion in the 3% NaCl solution at all temperatures; this susceptibility increases with temperature. The NiAl phase shows higher wear resistance in the abrasive 3% NaCl + 5% fine sand medium than the Ni-5% Al γ solid solution phase.

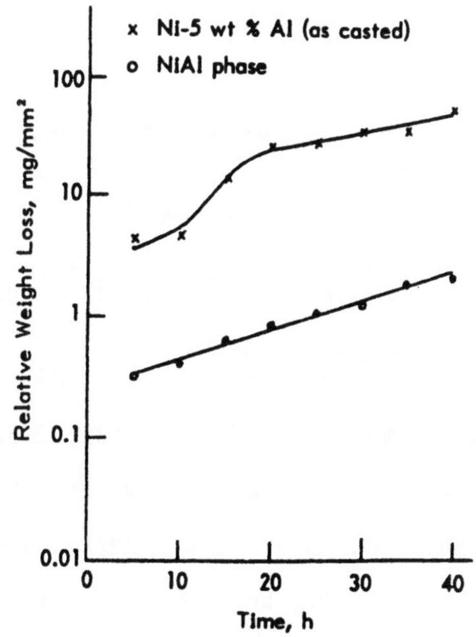

Figure 5. The relative weight loss of the NiAl intermetallic phase as compared to the as-casted Ni-5 wt% γ solid solution in the agressive 3% NaCl + 5% fine sand aqueous medium at 25°C.

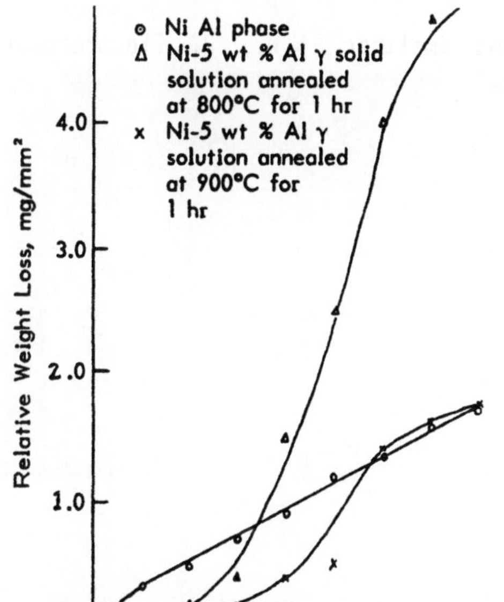

Figure 6. The relative weight loss of NiAl phase as compared to heat treated Ni-5 wt % Al γ solid solution in the agressive 3% NaCl + 5% fine sand aqueous medium at 25°C.

REFERENCES

1. Sims, C.T., J. Met., 1986, **18**, 1119.

2. Oblak, J.M., Doherty, J.E., Giamei, A.F., and Kear, B.H., Met. Trans., 1974, **5**, 1252.

3. Fleetwood, M.J., J. Inst. Met., 1970, **98**, 1.

4. Gresaffe, S.J. in "The Superalloys", Sims, C.T. and Hagel, W.C., Editors, Chap. 12, John Wiley and Sons, New York, 1972.

5. Ahmed, M.M.I., El-Raghy, L.S., and Elrefaie, F.A., IITT-International Conf. on Projection of Plasma Spraying, Paris, 1988.

6. Pettit, F.S., Trans. Met. Soc. AIME, 1967, **239**, 1296.

7. Wolf, J.S. and Evans, E.B., Corrosion, 1962, **18**, 129t.

8. Sttat, F.H. and Wood, G.C., Corros. Sci., 1977, **17**, 647.

9. Wood, G.C. and Sttat, F.H., Br. Corros. J., 1971, **6**, 247.

10. Hagel, W.C., Corrosion, 1965, **21**, 316.

11. Smialek, J.L., Met. Trans., 1978, **9A**, 309.

12. Kuenzly, J.D. and Douglass, D.L., Oxid. Met., 1974, **8**, 139.

13. Hendam, H.M. and Smeltzer, W.W., J. Electrochemical Soc., 1980, **127**, 1630.

The Study of Anticorrosion Technique for Natural Gas Processing Plants

Wang Su Yun

Daqing Petroleum Administration Bureau
Daqing People's Republic of China

and Gui Jing Lian

Daqing Petroleum Institute
Anda People's Republic of China

ABSTRACT

Comprehensive anticorrosion technique of sacrificial anode, zinc rich anticorrosion paint and the inhibitor CT2-1, was used in E-502 water cooling heat exchangers of 500# booster-type in a mild cooling state in natural gas processing plants. This enabled better working conditions with the corrosion rate decreasing and the service life of the exchangers increasing by 3-4 times.

THE EXISTING PROBLEM

Daqing Natural Gas Co. found that the E-502 water cooling heat exchangers of the booster-type mild cooling plant had been seriously corroded when it was used in the natural gas processing. In the exchangers, ground water and natural gas are used as a tube passes medium and a shell passes one, their compositions, properties and components are shown in table 1 and table 2 separately. The exchanger E-502 is built up with three heat exchangers (Ø553×L6861) connected in series, its function is to cool down the temperature of natural gas being passed from 155~160℃ to 16℃ about through the exchange process. The working conditions of the exchanger are shown in table 3.

TABLE 1

The composition and properties of the water

pH	total hardness mg N/L	total alkality mg N/L	specific conductance μs/cm	Cl^- mg N/L	HCO_3^- mg N/L	SO_4^{--} mg N/L	Ca^+ mg N/L	Mg^{++} mg N/L	soluted oxygen mg /L
7.63	3.0	4.90	200	1.12	9.0	0.44	1.90	1.10	10.14

TABLE 2

The component of the gas

component	He	H_2S	CO_2	O_2	N_2	CH_4	C_2H_4	C_3H_8	C_4H_{10}	C_5H_{12}
content(%)	0.0032	0.315	0.1153	trace	0.9428	74.3719	7.6916	8.9694	4.1888	1.3613

TABLE 3

Working condition of E-502

parameter / medium / process		tube passes	shell passes
		ground water	natural gas
flow rate(kg/h)		53000	19000
temperature(℃)	in	12	155～160
	out	40～43	16
pressure(MPa)	in	0.4	1.6
	out	0.4	1.6

On the seriously corroded places in the exchangers, for example, the corrosion rate of the inner and outer surfaces of the tube bundle could be come up to 0.7～0.8mm/y[1] and 2～3 mm/y[1] respectively. Especially, it is a very serious problem for the plant about 70% of the total corrosion products is FeS [2] which will be directly entered into the stream along with the processing medium, so that the normal operation will be seriously impaired. The thickness of settlements of the corrosion products resulting from about one year operation of the 500# plant could be reached up to 20 cm on the bottom of the light hydrocarbon storage tank D-509. The settlements could plug the pipeline and seriously influence the result of the hydrocarbon recovery.

Thus it can be seen that the corrosion mentioned above, including the tube plate, is serious. It is very serious consequences, a lot of $Fe(OH)_3$ of corrosion products would be settled near by the expanded joint or welded joint places of the tubes with tube plate(see Fig. 1). The service life of the exchanger depending upon the service life of its tube bundle will be only one or one and half year if it had not been protected with anappropriate anticorrosion method. In the worst condition, the tube bundle would be penetrated after operation in eight months. It is obvious that the anticorrosion technique must be used in order to increase the service life of the exchangers for the booster-type mild cooling plant.

Figure 1 The corrosion state of the tube plate

DETERMINATION OF THE ANTICORROSION TECHNIQUE AND ITS SIMULATED TEST

The tube bundle in the heat exchanger E-502 is made of low-carbon steel pipe(ASTM A334-74 Gr 1, $C \leqslant 0.30\%$), its wall thickness is 2mm, outside diameter is 21mm. By metallographic analysis, it is a ferrite-pearlite structure(see Fig. 2)which is composed of ferrite grain and dispersed cluster structure of pearlite and there are some cementite (Fe_3C) educed out from the grain boundary. The outside surface of the tubes is corrosed by H_2S existed in natural gas being processed. It is clear from Fig. 2 that the corrosion process of cluster structure of pearlite on the outside surface of the tubes will take precedence over all others, and then passes through the ferrite grain boundary to the deeper cluster structure of pearlite. When there are some cracks(see the arrow of Fig. 2)corrosed on the boundary, the cluster structure of pearlite will be in contact with the medium, which will make the corrosion further and further, finally the ferrite grain will be peeled off.

Because of that, an oil soluble adsorption-type organic inhibitor CT2-1 is used to protect the outside surface of the tube bundle, and the physical properties of inhibitor CT2-1 is shown in table 4. According to the ASTM-G15—76 standard definition about the corrosion test terms, the inhibiton can be defined as a chemical matter or a compound one, which can protect or slow down the corrosion when it as a proper concentration or a form exists in a kind of medium . [3]

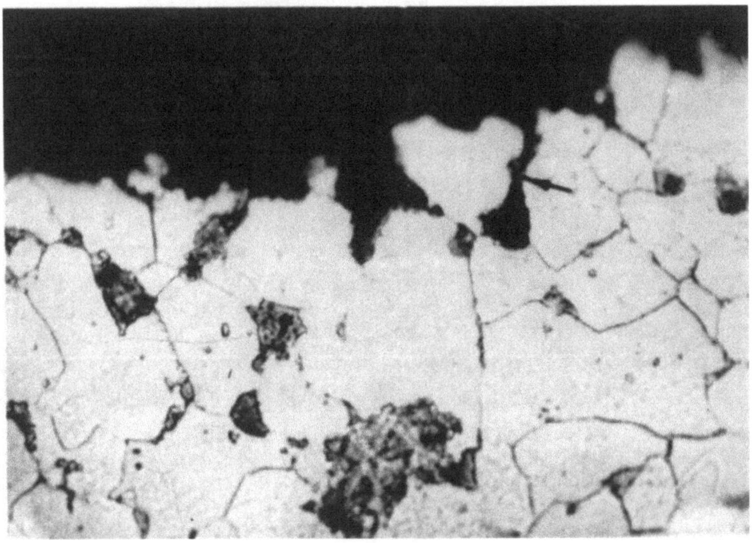

Figure 2 The metallographic structure of the tube bundle steel and corrosion of its outside surface

TABLE 4

The physical properties of the inhibitor CT2-1

colour	state	specific gravity	solubility
brown	thick fluid	0. 95～0. 97	hydrocarbon and alcohol

The corrosion coupon test was carried out in the inlet line of the E-502 exchangers. The result of that is shown in table 5, which shows a good anticorrosional effect by using the inhibitor CT2-1.

TABLE 5

The test data of the corrosion coupon

coupon size (L×W×T)	inhibitor	corrosion time (hour)	corrosion weight loss (g)
40×12×12	no	1200	1.443
40×12×12	yes	1392	0.096

This organic inhibitor to be an adsorption type is a chemical compound consisted of organo-amine and organic matters, which can be adsorbed on the metal surface to form an invisible film in molecular thickness that can block up the reaction of both the cathode and the anode. The CT2-1 would be concentrated at the interface between the metal and the electrolytic solution. Thus the inhibitor can effectively inhibite the corrosion even though its concentration is extremely low, and its application has a practical commercial sense.

In the physical and chemical adsorptional process of the inhibitor CT2-1, it has two functions, one is to make the energy state of the metal surface tend to be stable to block up the energy in the corrosion reaction, another is to form an oil wet protective film on the metal surface. The CT2-1 is a cathodic inhibitive type one that will adsorb cations to form an onium ion with hydrogen ions in acidic water solution, which can be adsorbed as a monomolecular film on the metal surface (see Fig. 3).

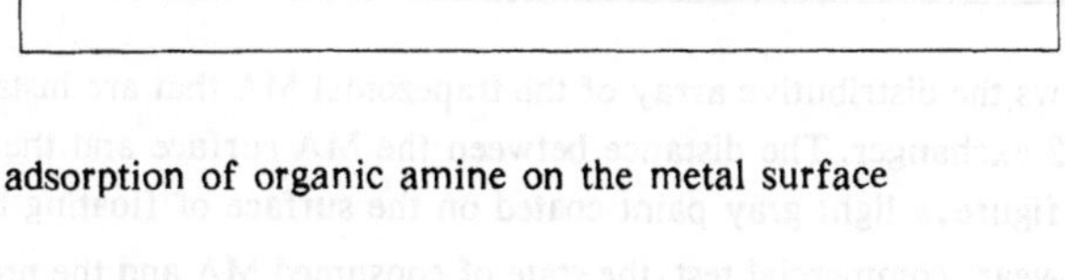

Figure 3 The adsorption of organic amine on the metal surface

When this kind of cations are adsorbed on the surface of cathode, the hydrogen ions in the acidic medium can hardly approach to the metal surface for being reduced, which can increase the overpotential of the hydrogen ions reaction to slow down the corrosion.

In addition, a high pure magnesium is used as sacrificial anode to carry out the cathodic protection in the water phase process, which is first tested in a testing heat exchanger. Fig. 4 shows the installation position of a magnesium anode(here after called "MA").

Figure 4 The MA installation in the testing heat exchanger

Considering the more expensive cost of the metal magnesium, a zinc-rich paint is used as a primer to the inner surface of the tube bundle and the tube plate in order to keep the magnesium consumption on the protected unit area in an economical limit. The formulation of the paint is shown in table 6.

TABLE 6

The formulation of the zinc-rich paint

component	zinc powder	water glass	water	lead oxide	1% alga acid-sodium
weight rate	100	20	12	2	30

Fig. 5 shows the distributive array of the trapezoidal MA that are installed on the floating head of the E-502 exchanger. The distance between the MA surface and the tube plate is about 25mm. From the figure, a light gray paint coated on the surface of floating head can be seen.

After two-year commercial test, the state of consumed MA and the protected tube plate is shown in Fig. 6 and Fig. 7 respectively.

Figure 5. The distributive array of the MA for the floating head of exchanger E-502

Figure 6 The consumed MA

Figure 7 The good state of the protected tube plate and the tube bundle of E-502

THE DESIGN AND CALCULATION OF THE MA

After measuring the electrochemical properties of the MA, the amount and the service life of the MA have been calculated, their results are shown as below:

Protected area of the exchangers:	237	cm^2
Potential difference between the tube end and the MA:	0.35	V
Current produced from the MA:	142	mA
Weight of a piece of the MA:	0.742	Kg
Number of the MA:	10	pieces
Service life of the MA:	3.7	years
Consumption of the MA:	6.3	kg

THE CALCULATION RESULT OF FLOW RESISTANCE FOR MA

The differential pressure of one unit of the E-502 is increased in 0. 09% after the installation of the MA.

THE RESULT OF THE COMMERCIAL TEST

The shell passes

No.	protective condition	test time (hour)	average weight loss (g)	corrosion depth (mm/y)	inhibiting rate (%)	surface state
1a	adding inhibitor	1126	0. 1932	0. 06	95. 3	bright
2	no adding inhibitor		1. 7960	0. 62		homogeneous pitting corrosion
3a	adding inhibitor		0. 1699	0. 057	95. 7	bright
4	no adding inhibitor		1. 5210	0. 53		homogeneous pitting corrosion
5a	adding inhibitor	720	0. 0932	0. 049	96. 7	bright
6	no adding inhibitor		0. 9755	0. 52		homogeneous pitting corrosion

The tube passes

item	consumption of the MA (%/month)	thickness of the zinc-rich paint (μm)	Max. corrosion rate (mil/y)	protectivity (%)	service life of the exchangers (year)	heat transfer coefficient ($W/m^2 \cdot K$)
no protected	—	—	31. 49	—	0. 8～1. 5	206. 96
protected	2. 1	120～150	1. 90	93～94	6～8	329. 58

CONCLUSION

1. If adding 6kg/month of the inhibitor CT2-1 into the inlet of the shell passes of E-502, the inhibiting rate could be higher than 95%, the yearly corrosion depth could be cut down from 15.7mil/y to 1.97mil/y. (The processing capacity of the plant is 300000~400000Nm^3/d, the content of H_2S is 70~150ppm, the content of CO_2 is 12ppm, the amount of water condensation is 3t/d).
2. If using the zinc-rich paint and the cathodic protection to protect the tube passes of the E-502, the corrosion rate of the tube bundle could be decreased from 11.8~15.7mil/y to 0.048~1.90mil/y, and the protectivity could come up to 93~94%.
3. Because of the effective protection, the exchangers have not been scalled for a working period of one year, which keeps in the heat transfer coefficient not to be changed.
4. The yearly economical efficiency of one exchanger could be reached to RMB 562000(about US$ 138723)after applicating the anticorrosion technique described above.

REFERENCES

1. The natural gas corrosion' s damage and losses. The Natural Gas Co. Publisher, November 1986.
2. The analysing report about the glycol fouling in the natural gas processing system. The Lan Zhou Chemi-physical Research Institute of China Scientific Institute Publisher, May 1984.
3. The theory and application of the metal corrosion, p. 263. The Chemical Engineering Publishing house, 1984.

Shot Peening—Current Trends, Future Prospects

J. J. Daly

Metal Improvement Company
Paramus
New Jersey
USA

The shot peening process is sparingly used compared to other processes such as heat treating, plating, anodizing and various coatings.

A suggested reason for this is that it is impossible to evaluate what has been done to a part which has been shot peened without destructively destroying the part. To date there is no nondestructive method available to us for evaluating residual stress profiles of shot to peened components. Therefore, we must rely on the controls currently available to us as well as the personnel and reputation of the company or department which performs the shot peening process. Additionally, we have seen the process become apparently over complicated when in essence it is a simple process. Its low cost makes it cost effective. It is adaptable to most all geometrically piece part shapes and it is relatively easy to apply. Over complication of the process leads to unfamiliarity and that leads to resistance to accept the process. In fact, the process was always simple to apply, where its influence on fatigue performance was complicated to understand and predict. Today the apparent over complication of the process stems from a much better grasp of its mechanics.

Although the approach remains simple, a more sophisticated approach to parameter selection and control now enables us to obtain optimum improvement and repeatability.

Evaluation of the shot peening process is best performed under controlled operating conditions in the laboratory or the field on actual shot peened hardware compared to unpeened hardware. In many cases, this is not possible due to the size of the part, the cost of the test, or the time frame required for evaluation. Alternatively, and in most cases, fatigue test specimens are used in lieu of parts and the type of testing or the specimen used in many cases I have seen does not represent the actual loading on the part or the surface finish of the peened part. In this case, fatigue test results may be erratic, inconclusive and in many cases give erroneous data. As an example, we know that shot peening does not show up well on high stress low cycle fatigue. When accelerated tests are performed by increasing the stress level above the design operating stress level, the S and N curves in many cases show that both the shot peened and non-shot peened curves meet near the yield point and in many cases will cross over yielding data which is inaccurate for comparison of shot peened versus non-shot peened components. Shot peening has proven to be of great value in extending fatigue life of components which loaded in bending or torsion are to a lesser extent in axial loading. Fatigue tests are conducted in axial loading when in actual practice bending or torsional loading causes fatigue failure. Additionally, fatigue specimens in many cases are erroneously polished and do not represent the actual surface finish of the component. Most finished machine components have a final surface finish of 125RMS compared to polished specimens having a finish of between 8 and 32RMS. Fatigue cracks primarily initiate at notches, and in many cases notches from machining. In reviewing fatigue test data where axial tests are performed on smooth versus shot peened specimens, one can draw the erroneous conclusion that super finishing yields better fatigue results than shot peening for all fatigue applications.

In pure axial loading, the entire cross section of the part is loaded in tension. The compressive stresses induced on the surface of the part by shot peening are balanced by subsurface tensile stresses which under loading may become undesirable causing premature subsurface failure. In axial loading conditions low shot peening intensities are recommended which yield depths of compressive stress that do not exceed 5% of the cross section of the part.

Higher intensity shot peening, yielding up to 20% of the part in compression have proven beneficial for bending and torsion applications as well as minimizing fatigue failures from foreign object damage and corrosion pitting. Subsequent shot peening to a lower intensity with a small shot size or polishing further improves fatigue life by bringing the maximum value of compressive stress closer to the surface as well as providing an improved surface finish. However, this improvement is offset with additional costs and economics then becomes a factor.

I believe that we who recognize the value of shot peening have to do more to address the resistance of those who question the reliability of the shot peening process because of their concerns over the lack of tangible data concerning controls. The advent over the past five years of computer monitored shot peening machines and the introduction of fluorescent Dyescan media for measuring coverage through the Peenscan process certainly is a step in the right direction to give industry this assurance.

This brings us to the subject of coverage on a given component versus coverage on the Almen strip. It is incorrect to correlate a direct relationship between coverage on a component and coverage on an Almen strip. The only direct relationship would be if the component is of the same alloy, the same hardness as the Almen strip. Otherwise there is no correlation. Coverage can only be measured by visual inspection or the use of fluorescent media, and the Almen strip is solely an instrument to be utilized for the purpose of measuring the kinetic energy, more commonly called intensity, and to control this kinetic energy from component to component.

There are too many shot peening callouts of improper shot size and intensity on components which were established either through standardization within the particular company, or it is specified due to the fact it worked well in the past on other components. When erroneous callouts are specified, maximum fatigue life benefits are sacrificed and in many cases fatigue life is reduced. It's for this reason that it is recommended to employ the following steps before specifying the shot size and intensity.

<u>Step 1</u>: <u>Determine the Proper Depth of Compression</u>. This depends on the type of loads to which the component is exposed, such as axial torsion, bending, stress corrosion,

fretting fatigue, corrosion fatigue, and also will the component be subject to mechanical damage in service. Remember the normal rule of thumb is that no more than 5% of the total cross-section of the area of the component to be shot peened should be in compression to allow for equalization of tensile stresses in the core to compensate for the compressive stresses at or near the surface. This 5% rule of thumb applies particularly to thin sections and axial loading. There are exceptions to the 5% rule, depending upon the thickness of the section and whether or not the part is subjected to mechanical damage, fretting fatigue, or corrosion fatigue, in which case a deeper depth of compression should be considered to give optimum fatigue results. I recommend, however, that no more than a maximum of 20% of the cross-section of the component to be shot peened should be in compression. In order to determine the relationship between intensity and depth of compression for different materials at different hardness levels, many residual stress curves have been developed to be used for such references. Work is progressing by Metal Improvement Company to establish a computer program which will alleviate and simplify this research process.

Step 2: Media, consisting of size, material, hardness and uniformity. The shot size has to be small enough to get into the radii or changes in sections. Size also plays an important role in surface finish requirements. As an example, the larger the ball for a fixed intensity, the better will be the surface finish. The trade-off is cost since a larger ball will take longer to obtain full coverage than a smaller ball.

Next is the material of the media. For iron contamination concerns, there are many nonferrous media materials available today -- ceramic, glass, stainless steel. Next is hardness of the media. It is a proven fact, and enough tests have been run to indicate that the hardness of the media should be as hard or harder than the component to be shot peened. One area where this has proven most successful is shot peening of case hardened carburized gears where hard shot, 55 Rc and above, produces a much higher magnitude of compressive stress than the use of softer shot.

Next is the uniformity of the media, which is important to obtain good surface finish and avoid surface damage. This is accomplished with automatic mechanical shot separators, an integral part of the machines, readily available on the market today.

Step 3: Kinetic Energy, more commonly called Intensity, which is controlled by utilization of the Almen strips located in fatigue critical areas, the locations of which should be specified by design or structures engineers. The Almen strip, developed by John Almen back in the early 1930's, is the most effective, inexpensive, practical method to measure intensity. As intensity is directly related to depth of compression, its control provides assurance that the requested depth is achieved. Though also directly related to surface finishing, the selection of optimum intensity for some materials may call for a compromise between maximum depth of compression and surface roughness.

Step 4: Surface Finish. To obtain optimum surface finish at a fixed intensity, the larger the shot, the better the finish. Frequently "dual shot peening" is used to obtain better finish as well as improvement in fatigue life. Dual shot peening is the use of a larger shot, following with smaller shot size at a lower intensity. This not only improves surface finish but brings the peak compressive stress closer to the surface of the material being shot peened.

Step 5: Coverage. This can be accomplished by visual inspection utilizing a 10-power magnifying glass. However, on many components such as case hardened carburized gears and parts of complex configurations and internal surfaces, visual inspection and the use of a boroscope are expensive, time consuming and ineffective. For this reason, the Peenscan method of inspection was developed utilizing fluorescent tracer liquid which exhibits a distinct change in color under a U.V. light depending upon different coverage rates obtained.

Step 6: Equipment. Great progress has been made in the last five years in the utilization of computer monitored shot peening machines with full automation and shot flow sensors. These computer monitored machines provide a distinct computer readout on a printer indicating that all of the parameters in the machine are controlled and within required specifications. I predict that within the next ten years most of the mechanical shot peening machines used in industry today will become obsolete and industry will solely depend on computer monitored machines.

In summary, we must not only continue to emphasize and educate the metalworking industry worldwide on the benefits, versatility and low-cost advantages of the shot peening process, but must also contribute time and effort to the education within the lecture halls on shot peening and its benefits.

Shot Peen Controls and Aircraft Engine Requirements

P. G. Bailey and J. M. Whalen

GE Aircraft Engines
Evendale
Ohio
USA

New shot peening requirements are emerging which are driven by aircraft engine customers. These include the military services and commercial air framers. The interest is in enhanced life engine components produced through quantified, verifiable benefits obtained by controlled, repeatable shot peen processes. These new requirements have the potential to greatly change the way peening is defined, and used in aircraft engine applications.

Today, the typical benefit that peening provides the aircraft engine manufacturer is fatigue life improvement. Peening benefits low and high cycle fatigue and retards crack growth. However, these benefits are not quantified, and are seldom used directly in calculating the component life. Peening is also used today to inhibit stress corrosion cracking, reduce fretting of contact surfaces, and to minimize the effects of machining which are typically not quantified.

Aircraft engine repair and overhaul facilities use peening to restore the repaired hardware to overhaul requirements. These requirements are often the same requirements that the aircraft engine OEM places on the new components.

Shot peening has progressed from manual controls and indirect process measurements in the 1970's to robotic nozzle motions and micro-processor controls in the 1980's. Required for the 1990's are quantitative benefits to be achieved through process enhancements and direct process measurements.

Shot peening technology may be divided into three chronological phases as shown in Figure 1. Phase one was the typical practice of shot peening from the 1970's into the 1980's. In this phase, the benefits of peening are not quantified, and the equipment has manual controls. All the nozzle positioning is done manually, and the operator monitors the process. In Typical Practice peening, the application of the process is measured indirectly with an Almen strip for compressive stress profile, and visually for coverage.

The next state of peening development is state-of-the-art (SOA) peening which is in use but not universally in practice. The benefits are still qualitatively defined as they were in the Typical Practice peening, but today the equipment is NC or microprocessor controlled. The key parameters of shot flow and air pressure are controlled and monitored by the machine controller. In robotic machines a manipulator sets all the nozzle positions very accurately and repeatably. In other systems, nozzle fixture motions are microprocessor controlled but without NC feedback. The same indirect process verification techniques used in typical practical peening are used in today's SOA peening.

Robotic shot peening represents the best approach for the future in shot peening equipment. This equipment offers improved process control and consistency for emerging needs which may require more precise control of impingement angles and local area coverage.

State-of-the-art in peening verification does not include any effective non-destructive method for measuring compressive residual stress. X-ray diffraction measures compressive stress only on the outer surfaces. Stress profiles are only obtained by destructive sectioning.

The next phase of peening may be labelled Enhanced Peening. This is the next generation of peening for the 1990's and beyond. In this phase of peening the benefits will be quantified with component specific life testing. The same robotic and/or microprocessor controlled equipment will be used as in SOA, but there will be several processor enhancements added to raise the level of process assurance. Direct non-destructive process verification techniques will be used to measure compressive stresses, and coverage.

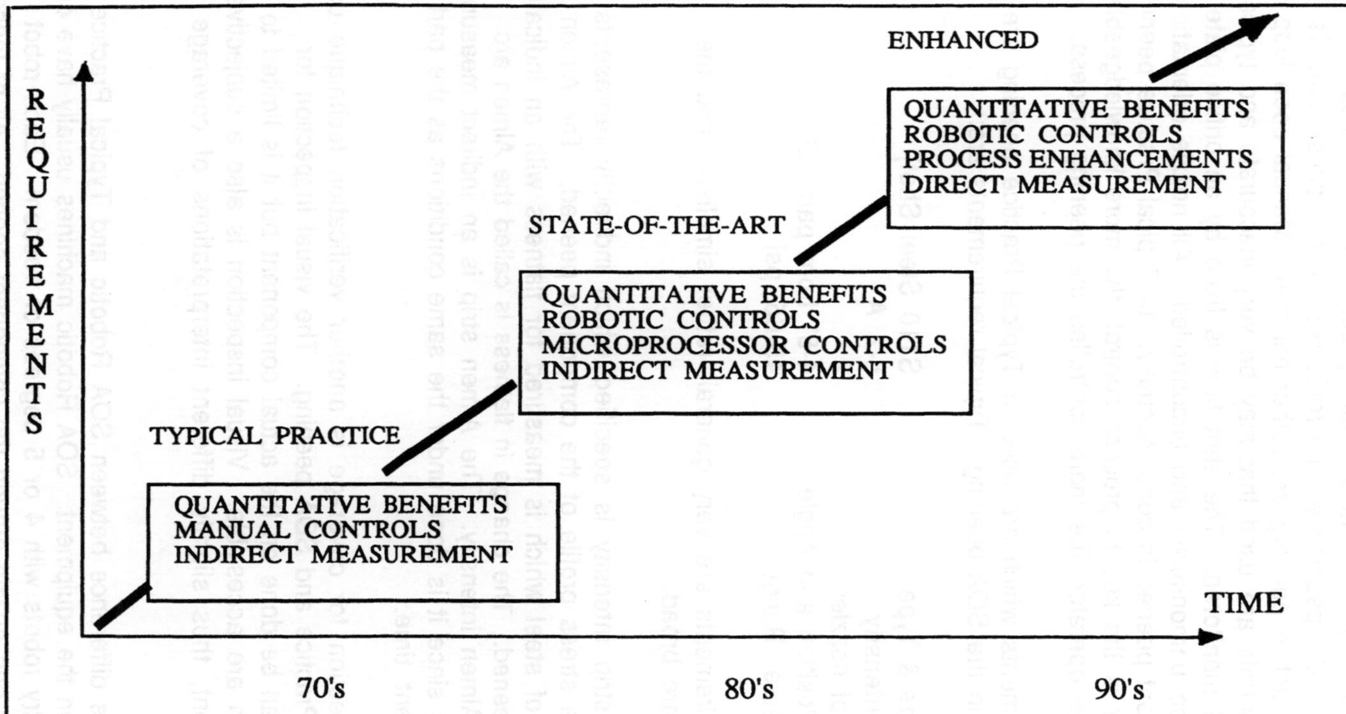

FIGURE 1: PHASES OF PEENING

In order to describe Enhanced Peening, some specifics about Typical Practice and SOA peening will be reviewed. The process controls used in Typical Practice peening are rather minimal. Manual open loop air pressure controls are used that may be very inaccurate and typically are not well monitored. The shot flow is fixed by an orifice plate, which is also unmonitored and uncontrolled. All nozzle orientation to the component peened is done manually. In Typical Practice peening, the operator is the key to process control, the more knowledgeable and alert the operator, the more controlled the peening process.

The requirements which are used in Typical Practice peening are still today used in the SOA peening. Typical requirements are:

Media Size & Type	S110 Steel Shot]
Almen Intensity	5-7 A
Number of nozzles	2
Nozzle Position and Angle	4-5" from part 45°
Air Pressure Range	45-65 psi

These requirements are very general. rather simplistic, and the tolerances are broad.

An Almen strip intensity is specified which indirectly represents the compressive stress profile of the component peened, The Almen strip is a piece of steel which is measured for flatness with an indicator, and then peened. The change in flatness is called the Almen arc height, or Almen intensity. The Almen strip is an indirect measure of the process since it is done under the same conditions as the part but at a different time.

Visual inspection for coverage is another verification technique used in Typical Practice and SOA peening. The visual inspection for coverage can be done on the actual component but it is limited to areas which are accessible. Visual inspection is also a subjective measurement, thus slightly different interpretations of coverage are probable.

The obvious difference between SOA Robotic and Typical Practice peening is in the equipment. SOA Robotic machines usually have one or two gantry robots with 4 or 5 degrees of freedom. Each robot manipulates one nozzle around the component peened. All of these nozzle movements are controlled by a CNC machine controller and a part program. Along with the robotic nozzle manipulation, the SOA equipment has closed loop air pressure and shot flow controllers. These controllers set and monitor their respective parameters. When specified tolerance bands are exceeded, they trigger machine shutdowns.

This SOA robotic equipment has significant advantages over its predecessor, the Typical Practice machine. These advantages are:

° Reduced nozzle setup time since part programs store all the nozzle positions electronically

° Improved process consistency due to the nozzle positioning repeatability, and closed loop control of key parameters, and

° Lends itself to automation when coupled with the right material handling, part fixturing, and cell software.

SOA microprocessor controlled multi-nozzle machines also have distinct process control advantages over Typical Practice machines when coupled with dedicated nozzle and part fixturing and malfunction shutdowns

° Reduced setup time
° Improved process consistency
Assurance of peening integrity via the malfunctioning shutdowns

Quantitative benefits through Enhanced shot peening will require advances in several areas. Direct verification of stress profiles and peening coverage will be required. Definition and control of peening damage is another requirement. Increased assurance and control of the process itself is a necessity also.

These new requirements have three implications for the shot peening process. First, the benefits of peening will need to be quantifiably defined. Component life testing will be required to determine the benefit that peening provides for each component. Secondly the result of the peening process will need to be measured non-destructively for both compressive stresses and coverage. Finally, the peening process needs to be better controlled and proven to repeatable to raise the level of quality assurance.

Enhanced peening processes needed to meet these new requirements will most likely utilize the current SOA robotic peening equipment, with the addition of several process enhancements. All of these enhancements should integrate into current SOA equipment.

The first enhancement is the establishment of a shot velocity sensor. Shot velocity is a critical parameter which contributes directly into the compressive stress profile of the component peened. The shot velocity sensor would detect nozzle wear as well as incorrect air pressure or shot mass flow rate. This device is necessary to assure that the process is consistent and repeatable.

Another needed process enhancement is adaptive control technology for the key parameters of shot velocity and coverage. Adaptive control technology will eliminate the effects of outside influences on the outcome of the peening process, improving process consistency.

Process modelling is another process enhancement needed. Process modelling would incorporate the hardware geometry as well as the process parameters and allow for process simulation. The simulation would be run to assure that uniform compressive stress is achieved over the entire component, including difficult areas, such as bolt holes, and dovetail slots. Process modelling would add assurance that the component was peened uniformly and also aid in process development. Less time would be required to develop part programs with the help of a process model.

The final area which needs development before Enhanced Peening can be implemented is process verification. A non-destructive direct measurement system is needed for both compressive stress profile and coverage. These techniques must measure the hardware directly in all areas. A new NDE technique is needed to measure peening damage. However, before damage can be measured, it must first be defined. All of the NDE methods will need to quantify the peening applied to the component.

An important point to remember is that all of the process enhancements discussed should integrate directly into existing SOA peening equipment. Also the new Enhanced Peening process will offer additional opportunities to reduce costs and improve productivity over the current SOA equipment. These savings will result from improved verification techniques introduced. Almen intensity checks may no longer be required. This would eliminate the time required to make these checks which often takes as long as the component peening time to complete.

Improved quality shot is another imperative for Enhanced shot peening. Conditioned cut wire shot is probably the most likely candidate.

In summary, new shot peening requirements are emerging that are driven by interest in enhanced part life. Recent progress in process controls has heightened that interest. Primary areas for process development are in the areas of shot quality, shot delivery control and non-destructive evaluation.

Fatigue Fracture and Residual Stress Relaxation in Shot-Peened Components

D. W. Hammond

British Aerospace plc
Filton UK

and S. A. Meguid

University of Toronto
Toronto Ontario Canada

ABSTRACT

This work is devoted to highlighting the beneficial effects of compressive residual stresses, their relaxation and the associated fatigue crack initiation and propagation in two contrasting materials. Accordingly, rotating bend fatigue tests were conducted on essentially un-notched specimens made from 7075 T7351 aluminium zinc alloy and 080M40 medium carbon steel. The tests involved assessing the life of un-peened, peened, and re-peened specimens where re-peening was applied after the exhaustion of a proportion of the specimens anticipated fatigue life. Whilst the steel demonstrated a very significant recovery of fatigue life associated with the re-peening treatment, this was not the case for the aluminium even though peening had proved highly beneficial. Residual stress measurements and fractographic examination were used to elucidate this discrepancy.

INTRODUCTION

It is now well established that residual stresses are present in nearly all engineering components and that they may profoundly influence the fatigue life of such components. The establishment of current techniques for residual stress measurement makes it possible to evaluate and hence take advantage of or allow for the influence of residual stresses. Some success has been encountered in the consideration of welding residual stresses and their influence on fatigue strength [1]. Attempts to evaluate peening in a similar way are less prevelant in the literature, but some work exists on the relaxation of peening residual stresses, [2], and on the effectiveness of periodic re-peening, in further extending the fatigue life of critical components [3], [4]. This study is concerned with establishing the way in which shot peening influences the fatigue fracture process and how peening and re-peening partially fatigued components can be best exploited.

MATERIALS AND METHODS

The selection of the materials tested reflected the desire to test alloys typical of those in current usage. This, together with their differing defect distribution, crystal structure and strengthening mechanisms prompted the current choice. The 7075 aluminium zinc alloy was in the T7351 condition, i.e., overaged for optimum stress corrosion resistance, and uniaxially stretched to control residual stresses. It possessed an essentially equiaxial grain structure with a grain size of 70μm and a significant volume fraction of strings of inter-metallic particles ranging in size from 10 - 25μm. Its yield strength was 450MPa and its ultimate was 530MPa. The 080M40 medium carbon steel, with manganese, was in a slow cooled condition, possessing a microstructure of coarse pearlite and ferrite. Its grain size was 100μm and it contained strings of rounded manganese inclusions of around 10μm.

The experimental programme was conducted using a specially built four-point rotating bend fatigue test rig which was capable of testing specimens of large diameter. It was calibrated using slip rings and a strain gauged specimen. Testing was performed at 50Hz in ambient conditions and the specimen gauge diameter was 12mm. The stress concentrator used was a 35mm radius and the overall diameter of the specimen was 34mm giving a value of stress concentration factor K_t near 1.1.

Peening was performed in a direct air pressure single nozzel system using a shot of 600 to 800μm diameter and a Hv hardness of 590-690 MPa. The peening conditions created Almen arc heights of 250μm 'A' for the aluminium and 400μm 'A' for the steel.

The fatigue testing program set out firstly to establish the life versus applied stress range relationships for the two materials in the peened and unpeened condition, and secondly to more closely examine the regime corresponding to $10^5 - 10^6$ cycles to failure. Such a regime is between the extremes of L.C.F. where rapid relaxation might be anticipated, and H.C.F., where relaxation may or may not be present. It also corresponds to a number of industrial applications, such as pressure vessels and landing gear assemblies.

RESULTS AND DISCUSSION

Figures 1(a) and 1(b) show the overall fatigue life of the aluminium and steel specimens as a function of the applied cyclic load. It may be noted that the improvements due to peening in the $10^5 - 10^6$ cycles to failure (N_f) regime is more significant in the aluminium alloy. This may be accounted for by the fact that this regime corresponds to applied stress of $\pm 0.9\sigma_y$ in the case of the steel rather than $\pm 0.5\sigma_y$ in the case of the aluminium. The improvement in the peened life of the steel at applied stresses beyond the yield stress indicates that work hardening of the exposed surface layers plays a significant role in this situation.

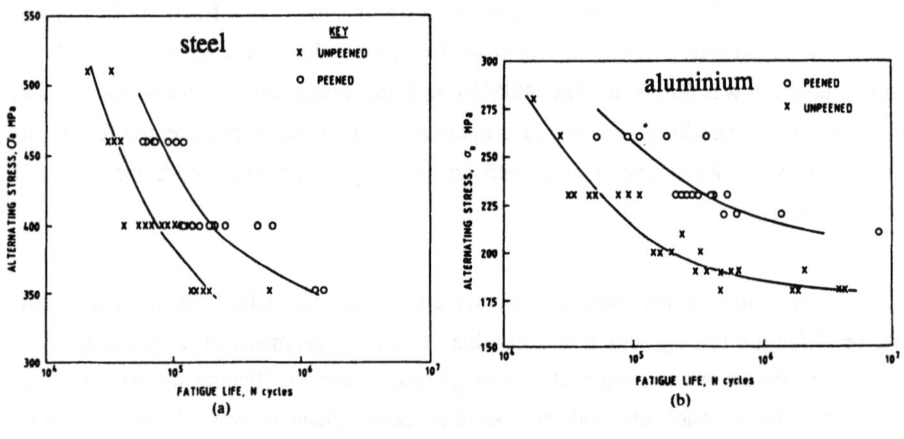

Figure 1. S-N. Curves for both materials in the un-peened and peened conditions

Figure 2 shows the response of the specimens to a re-peening treatment performed on peened specimens which had been fatigued to a known proportion of their fatigue life. It can be seen that whilst the steel specimens show essentially complete recovery of their peened life after re-peening, the aluminium ones are austensively not affected.

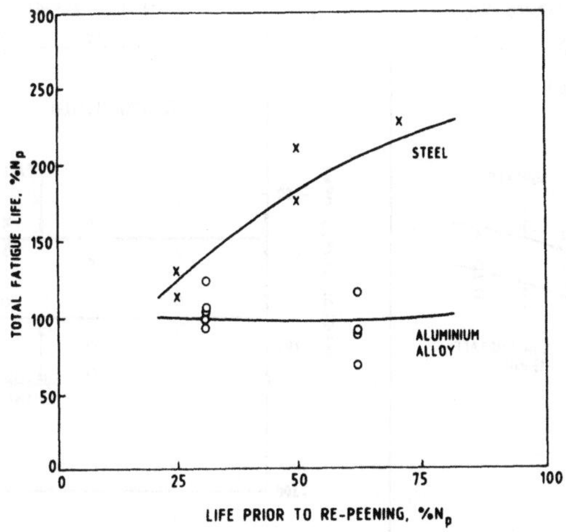

Figure 2. Total fatigue life as a function of cycles elapsed prior to re-peening

Figure 3 depicts the residual stress both at the surface and in the sub-surface regions in peened specimens subjected to cyclic loading to a known proportion of their anticipated life. Both of these results and those depicted in Figure 2 relate to applied loads of ±230 MPa for the aluminium and ±400 MPa for the steel. It can be seen that the residual stresses in the steel specimens relax more considerably than the fatigue strength improvement associated with peening. The aluminium specimens, however, did not show this trend. A consideration of the superposition of the peak compression load, and the residual compressive stress initially present in the steel also demonstrates the significance of work hardening effects.

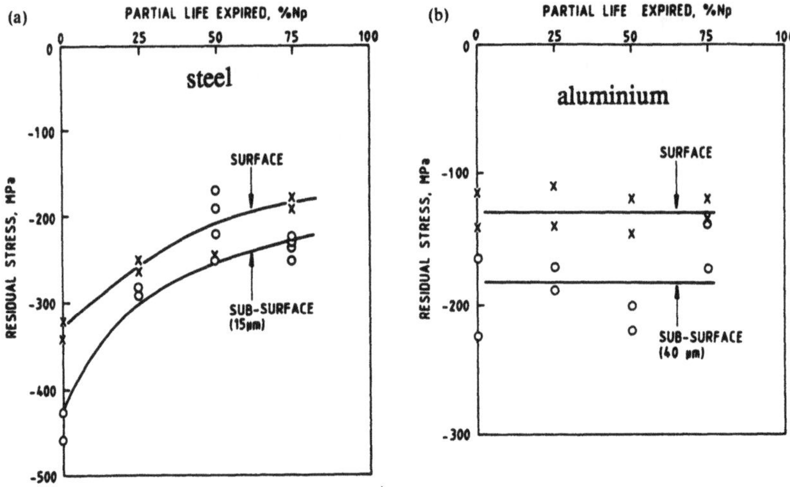

Figure 3. Relaxation of residual stress at and near the surface as a function of used life

Figure 4 depicts parts of a fracture surface typical of those encountered in the peened aluminium specimens. This sub-surface crack initiation was not observed in the un-peened specimens. This accounts for the observation that little or no residual stress relaxation was observed in the peened aluminium prior to failure.

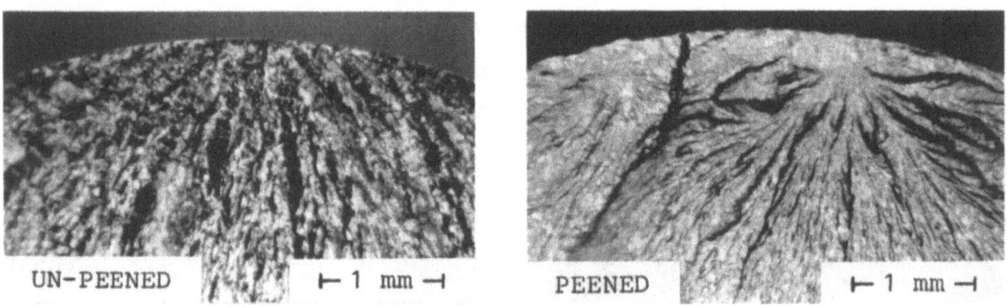

Figure 4. Fracture surfaces typical of those encountered in the aluminium in the un-peened and peened conditions

CONCLUSIONS

The present work was devoted to examining the fatigue fracture behaviour of 7075 T7351 aluminium alloy and 080M40 medium carbon steel in the presence of shot peening residual stresses. The findings may be summarized as follows:

(i) The peening treatment enhanced the fatigue life of both materials appreciably. This is due to the presence of compressive residual stresses in the vicinity of the surface, though there exists a large discrepancy between the magnitude of the residual stress and the fatigue strength improvement it yields.

(ii) The re-peening of partially fatigued specimens showed a marked effect on the steel specimens but not on the aluminium ones. This may, in part, relate to the greater degree of residual stress relaxation observed in the steel specimens prior to fracture.

(iii) The peening treatment promoted sub-surface crack initiation in the aluminium alloy. This was not observed in the steel. Clearly the development of sub-surface cracks will influence the effectiveness of the re-peening process as well as explains the lack of surface stress relaxation prior to failure in the aluminium alloy.

ACKNOWLEDGEMENTS

The authors wish to acknowledge the Natural Sciences and Engineering Research Council of Canada and the Science and Engineering Research Council of Great Britain for their support of this project. Thanks are also due to Gillian Smith of NEI Parsons, Newcastle, who performed some residual stress measurements.

REFERENCES

1. Meguid, S.A. and Chee., The effect of peening and re-peening upon partially-fatigued components, J. Mech. Wkg Tech., 1983, 8 pp. 129-146.

2. Hammond, D.W., The mechanics of peening and re-peening partially-fatigued components. M.Phil. Thesis, Cranfield Institute of Technology, 1986.

3. Glinka, G., Effect of residual stresses on fatigue crack growth in steel weldments under constant and variable loads. In Fracture Mechanics, ASTM STP 677, American Society for Testing and Materials, 1916 Race Street, Philaedelphia, 1979, pp. 198-214.

4. James, M.R., The relaxation of residual stresses during fatigue. In Residual stress and Stress, Relaxation, ed. E Kula and Volker Weiss, Plenum Publishing Corporation, New York, 1982, pp. 297-314.

Some Theoretical Problems of Shot Peening

P. Kuba and V. Sedláček
Technical University
Prague
Czechoslovakia

ABSTRACT

Shot peening has long been used both for cleaning component surfaces and for improving the properties of surface layers. This paper reviews various ways of calculating the sizes of dimples formed by shot impacts under various conditions, dimple depths, the generation and magnitude of stresses and their effect on the deflection of a thin strip shot peened from one side only. The calculation results have been tested experimentally on an AlCu4Mgl alloy; this work has allowed detailed analyses of the coverage, dimple sizes and distribution patterns, and the amount of deflection. Stress calculations were based on the plastic deformation of the surface layer. In the discussion, the newly applied methods are compared with previously published approaches; the relationships between results obtained on flat strips and on cylindrical bars are examined; and procedures are described for determining the peening process parameters by calculations.

INTRODUCTION

Shot peening as a way of enhancing the fatigue properties has lately been adopted on an ever wider scale and not only for steel, but for titanium and aluminium alloys too. New equipment and test methods have been devised, but this has only emphasized the need for a better understanding of the action of shot peening, and for some way of establishing the optimum process parameters by calculations rather than experimentally.

The principal target parameters of this process are the Almen intensity and the coverage. Peening induces compressive stresses in a surface layer of a certain thickness; together with the increased dislocation density, these stresses hamper the nucleation and propagation of microcracks. Several theoretical studies have been devoted to these matters; the present paper concentrates on some selected conclusions and relationships, and on their experimental verification in the shot peening of aluminium alloys.

Coverage

One of the chief tasks in any theoretical analysis of shot peening is to ascertain, and estimate the importance of, the changes which the shot impacts cause on the surface that is undergoing treatment. An equation can be derived for the force which the impinging shot exerts on the surface, and can be used for estimating the dimple sizes. The ratio between the area covered by dimples and the total surface area is known as the coverage, S. Theoretical analyses of shot impingement have been published by several authors |1-5|, who all adopt either an elastic or an ideal plastic approach. For ideally plastic materials, an expression has been formulated which reveals the dimple diameter d resulting from the impact of a spherical particle of diameter D and density ρ at velocity v:

$$d = 2\left(\frac{\rho v^2}{6\bar{p}}\right)^{1/4} D \tag{1}$$

where $\bar{p}$ is the mean deformation resistance at the point of impact, and enters a dimensionless parameter $\rho v^2/\bar{p}$ which is called the damage factor. According to |5|, $\bar{p}$ is given by:

$$\bar{p} = 0.6 + \frac{2}{3} \ln C\, R_p0.2 \tag{2}$$

where C = E.d/Rp0.2.D is a term representing the properties of the treated material, E:Rp0.2, as well as the amount of plastic deformation, d:D. The value of $\bar{p}$ varies from $R_p0.2$ to 3 $R_p0.2$, most commonly being 2.55 $R_p0.2$. The largest of these values is observed only when deformation is fully plastic and the loading rate is very low, as in hardness testing, but is also applicable when hard shot hits relatively soft material. Various alternative expressions have been evolved |10, 11|.

A formula quoted in the literature |8| for diameter d in hardening materials reads:

$$d = K_y\, D\left(\frac{2}{3}\rho v^2 \frac{n_d + 2}{H_d}\right)^{\frac{1}{n_d + 2}} \tag{3}$$

where n_d is the dynamic coefficient of strengthening,

K_y - a correction factor for repetitive impacts, and

H_d - the dynamic hardness number.

The magnitude of K_y is estimated by an empirical expression,

$$K_y = 1 + B_y\,(n_y - 1)^{m_y} \tag{4}$$

where n_y is the number of impacts on an area described by $\pi d^2/4$, and can be ascertained from the coverage; B_y, m_y and n_d are constants characterizing the material's resistance against repetitive impact stressing. The values established in an AlCu4Mgl alloy are H_d = 2350 MPa, n_d = 2.12, B_y = 0.18 and m_y = 0.21.

When several balls in succession impinge on the same spot, the impressions or indentations they create overlap; the probability of at least one shot impacting on area ΔS is defined as

$$P_1 = 1 - e^{-n_y} \qquad (5)$$

Clearly, P_1 varies with coverage S; at S = 98%, n_y comes to 4, at S → 100%, n_y works out as 6.

The dimple depth h is governed by the relationship

$$h = D\left(\frac{\rho v^2}{6\bar{p}}\right)^{1/2} \qquad (6)$$

Stresses in surface layer

The stresses induced by shot peening can arise either by the action of Hertz pressures, as shot impacts on the surface, or else by plastic deformation of the surface layers, caused by tangential forces. The former mode predominates when hard materials are peened with soft shot, the latter when plastic materials are peened with hard shot. The fundamental problem here is to ascertain or predict the stress and strain field perpendicular to the peened surface. Fig. 1 is a schematic illustration of the stress patterns obtained by these two alternative approaches, under the simplifying assumptions commonly employed in such calculations. The literature contains other models too |5, 8, 12, 14|, but in all of them many of the constants that enter the calculations have to be determined experimentally.

The decisive factor is the depth or thickness h_p of the layer affected by the peening. For an AlCu4Mgl alloy, this is estimated by means of an empirical expression |8| :

$$h_p = \frac{5.8\ hD}{d} \qquad (7)$$

When a thin strip is shot peened from one side only, it deflects. The arc height f and radius of curvature R can be exploited for estimating the stress intensity, the estimates being based on the relationship implied by the relevant force equilibrium and moment equilibrium conditions. Several different calculation procedures exist, employing various simplified relations or computer routines. A simple expression for the stress magnitude associated with a stepwise transition at depth h_p has been formulated by Saverin |21| :

$$\sigma = -\frac{4}{3}\ \frac{f}{a^2}\ \frac{E}{1-\nu}\ \frac{H^2}{h_p} \qquad (8)$$

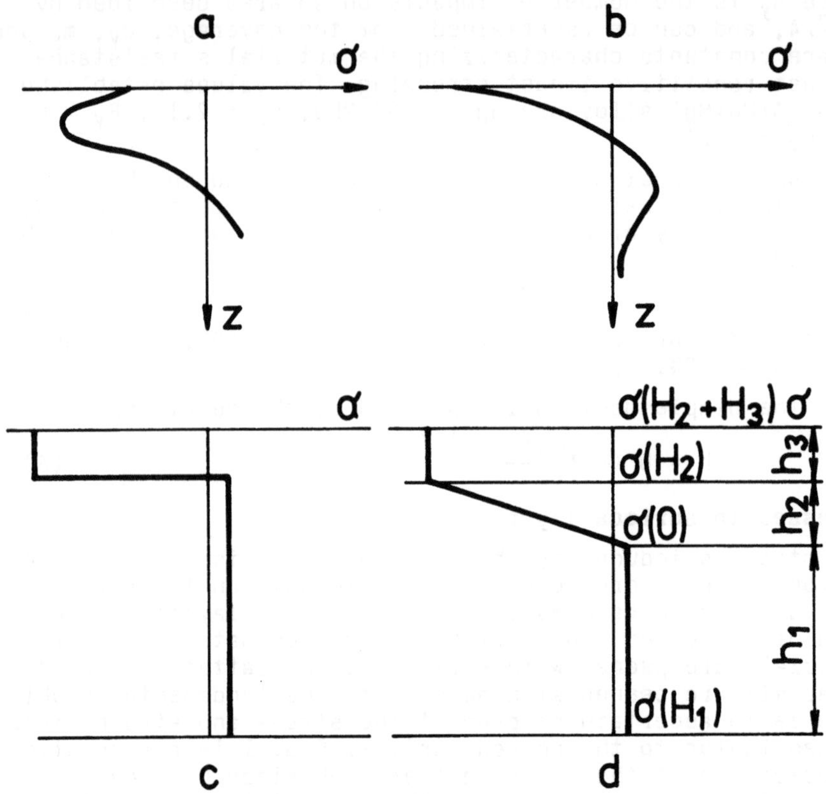

Figure 1. Stresses induced at shot impact sites:
a - by the action of Hertz pressures,
b - by plastic deformation of the surface layer,
c - the simplified scheme underlying the calculations.

It can be demonstrated that this relation remains valid even for the more general cases of stress distribution which require the adoption of an effective depth of the affected zone, h_{ef}. For the stress distribution represented in Fig. 1d, h_{ef} equals $h_3 + 1/2\ h_2$.

Most of the calculation techniques rest on the implicit assumption that before the shot peening, the treated body was free of any internal stresses. Calculations which took into account surface stresses, as well as experimental evidence obtained to corroborate the results of these calculations, all confirmed that their influence was less than 2 to 3% |8|. On the other hand, factors which we must not neglect include stress relaxation at elevated temperatures |14|, or during protracted exposure times, or under the action of external applied forces |23|.

Deflection of thin strips

The dimples caused by shot impacts arise by plastic deformation of the surface layer. The elongation of this layer, i.e.

the change in its volume, induces compressive stresses in the layer. If this treatment is applied to only one side of a strip, that strip will necessarily deflect. The arc height f is proportional to the coverage, and can be established with the aid of a simple equation |5|:

$$f = K a^2 \frac{D}{H^2} \left(\frac{\varrho v^2}{\bar{p}}\right)^{3/4} S^{3/2} \qquad (9)$$

where a is the measured length of the chord,
H - the strip thickness, and
K - the impact effectiveness, which depends on the treated material and the peening conditions.

The literature contains a good many other expressions compiled by various authors; regardless of the approach we adopt, we must bear in mind that the arc height is governed not only by the strip thickness, but also by the properties of the strip material.

Experimental work

Flat specimens of an AlCu4Mgl alloy with an $R_p0.2$ of 430 MPa and E of 71,320 MPa, having the dimensions of the standard Almen strip and thicknesses of 2.38 or 5 mm, were peened on injector equipment, at a pressure of 0,4 MPa, with glass balls that had a mean diameter of 0.26 mm and a density of 3000kg.m^{-3}. The delivery nozzle was 200 mm off the treated surface.

Coverage was assessed, both by the linear intercept and the point counting methods, on five photographs taken in the lengthwise centre line of strips peened for 0.75, 1.75, 3.3, 6 and 12 seconds. The results are presented in Table 1. The point counting method produced less scatter, but the average coverages S ascertained by both methods are in good mutual agreement. On the photograph of a specimen that had been peened

TABLE 1
Coverage after various peening times and their standard deviations

Time	A		B	
(s)	S(%)	σ	S(%)	σ
0.75	50.9	0.5	52.6	1.8
1.75	76.8	0.3	77.8	1.2
3.3	85.5	0.4	85.3	1.2
6.0	94.7	0.2	94.9	0.8
12	97.9	0.2	98.2	0.4

A - line intercept
B - point counting method

for only 0.75 seconds, the area density of dimples was measured by means of seven probes of various sizes; 45 measurements were taken with each of them. Calculations then indicated an impact rate q of 78.7 balls per mm^2 per second. This figure enables us to compute the process duration needed to produce the desired 98% coverage: at $n_y = 4$, the relation

$$t = \frac{4n_y}{\pi d^2 q}$$ yields t = 8 seconds, a time in close agreement with experimental findings.

The dimple size was investigated along with the frequencies of incidence of various dimple classes. Fig. 2 is a summary of the findings made on 432 dimples. The average dimple size was 0.09 ± 0.02 mm. For the experimentally ascertained mean impact velocity of 56 $m.s^{-1}$, calculations render an average diameter of d = 0.097 mm, which is reasonably close to the experimental finding. The scatter in the dimple sizes was largely due to scatter in the size of the unscreened glass balls, which varied in actual diameter from 0.21 to 0.32 mm, and in their impact velocities, which ranged from 45 to 80 $m.s^{-1}$. For the maximum ball size and velocity, the theoretical d_{max} is 0.145 mm; for the minimum values, d_{min} comes to 0.067 mm, or less than half the maximum figure. Some even smaller dimples were detected, but these are attributable to velocity drops caused by collisions between balls, or by rebounds, or by the impingement of previously crushed balls.

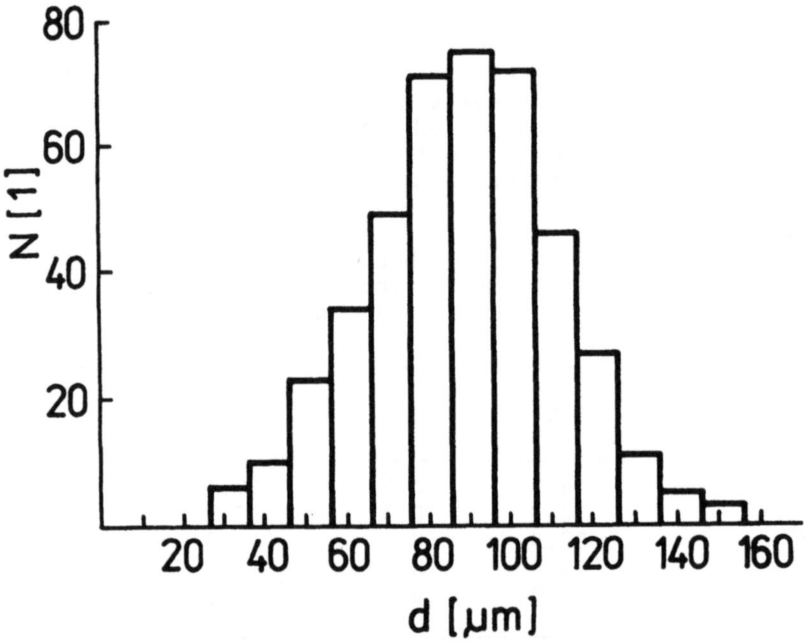

Figure 2. Dimple size distribution after 0.75 seconds of shot peening

Fig. 3 shows how arc height f grows during the peening of strips of two different thicknesses. The curve for strips

2.38 mm thick indicates that, like the 5 mm strips, they reached saturation level at f = 0.075 mm after 3 to 4 seconds, well before the desired 98% coverage was attained. The maximum arc height can also be established by calculations; very good agreement with experimental results was obtained at K = 12.6.

A comparison af arc heights recorded on strips of two different thicknesses revealed that $f_1:f_2=H_2^2:H_1^2$. Thicker specimens, 7 and 10 mm thick, appeared to conform to this law too, but their arc heights, 0.009 and 0.004 mm respectively,were difficult to measure exactly.

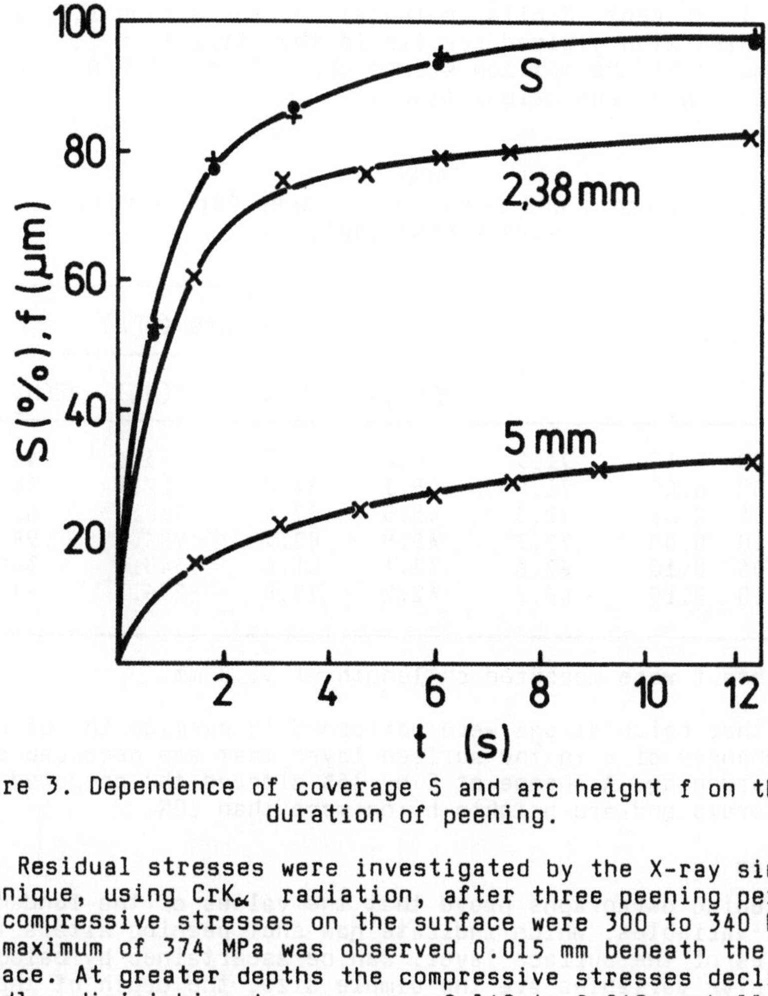

Figure 3. Dependence of coverage S and arc height f on the duration of peening.

Residual stresses were investigated by the X-ray sin ψ^2 technique, using CrK_α radiation, after three peening periods. The compressive stresses on the surface were 300 to 340 MPa, the maximum of 374 MPa was observed 0.015 mm beneath the surface. At greater depths the compressive stresses declined rapidly, diminishing to zero some 0.140 to 0.240 mm bellow the surface, and then giving way to tensile stresses in the strip interior.

Stress calculations were based on a method postulating a constant volume change of the surface layer |13| due to peening, $\Delta V/V = 3\ \varepsilon_3^o$. The calculations were also utilized to investigate various possible stress patterns, which were described by the corresponding changes in thickness h_2 and h_3 as defined in Fig. 1. An example of the resultes thus obtained is shown in Table 2, which lists the figures for a thickness of 2.25 mm and a volume change of ε^0 = 0.005. These figures indicate that tensile stresses within the strip are relatively small; the greatest differences were found in the h_2 layer, a fact that confirms the importance of the actual stress distribution. Results comparable with the experimental findings were gained for instance for a h_2 of 0.05 mm and h_3 of 0.10 mm, i.e. h_{ef} = 0.125 mm. The arc height proved to be practically independent of the stress distribution. Stress calculations founded on the radius of curvature or on the arc height measurements produced results that differed from those quoted above by about 1 per cent. Similar calculations for strips 1.3, 5 or 7 mm thick also yielded results in the literature |14| that in soft material the maximum stress can attain 0.8 $R_p0.2$ were confirmed both by our calculations.

TABLE 2
Arc heights and residual stresses for various stress distributions

Thickness (mm)			f (mm)	Stress (MPa)			
h_1	h_2	h_3		$\sigma(h_1)$	$\sigma(0)$	$\sigma(h_2)$	$\sigma(h_2+h_3)$
2.15	0.00	0.10	71.5	44.8	86.7	453.6	447.5
2.15	0.02	0.08	71.9	45.1	87.0	512.1	507.2
2.15	0.05	0.05	72.3	45.5	87.4	630.0	626.8
2.15	0.10	0.00	72.7	45.9	87.7	986.7	986.7
2.15	0.05	0.10	67.6	43.1	81.4	348.0	342.2
2.10	0.10	0.10	66.7	42.2	77.8	276.7	271.0

Arc heights f were measured on length of 31.8 mm.

Further calculations were performed to examine the effects of the changes of E in the surface layer that are produced by strain hardening. A change of E by 15% altered the computed stress levels and arc heights by no more than 10%.

Discussion

The preceding paragraphs prove that the values of the fundamental physical variables, which indicate how shot peening alters the properties of the surface layer, can be ascertained by calculations. The decisive variables are the dimple size, the depth of the affected zone, and the stress intensity. The dimple size is best estimated by equation (3): this expression incorporates the dynamic hardness value and hence takes into account the strain hardening effect; but it has a drawback in that it necessitates knowledge of the values of empirical constants. The dimple depth or h:d ratio is also a measure of plastic deformation, and a criterion of the geometrical similarity of the peening process and treated material properties. It is better expressed in terms of d:D, for instance by the following relation|8|:

$$\frac{h}{d} = C_o(\frac{d}{D} - B_o) \qquad (11)$$

For an AlCu4Mg1 alloy, C_o is 0.248 and B_o is 0.023. This formulation makes it easier to determine the thickness of the layer affected by plastic strain, as defined by equation (7). As the d:D ratio varies from 0.05 to 0.6, the h:d ratio spans an interval of 0.06 to 0.14; in our case, h:d=0.06 and hence h_p worked out as 0.1 to 0.2 mm, since h_p is governed mainly by the larger dimples. The values obtained by these calculations were in close agreement with those found experimentally.

The compressive stresses are more difficult to determine. The proportion due to Hertz pressures is relatively well defined mathematically and, under some simplifying assumptions, is straightforward to calculate. Calculations based on the surface layer deformations call for knowledge of the properties of thin surface films, and these properties are as yet impossible to establish by experiments; these methods therefore rely on estimates and assumptions not backed by any hard evidence. Procedures utilizing an elastic model can at best yield only very approximate results. We consequently prefer the alternate loading theory |8|, because its approach is closest to the actual sequence of events in shot peening processes. Both calculations and experiments conducted with aluminium alloys have shown that peening entails a cyclic strain hardening process, even though the number of cycles strengthening the surface layer is generally only two or three. Aluminium alloys are known to have anisotropic properties in their surface layers too, but the decisive factor is the residual strain in that layer. Calculations taking this approach have to employ relevant elasticity modules, and must define a boundary between the domains of linear and power-law approximations, which may be located as deep as 0.17 mm. Applied to our experimental conditions, this procedure rendered a surface stress magnitude of 390 MPa; the real surface stressing will naturally be lesser, because the result is distorted by the simplified view taken of the course of plastic deformation.

A reversed procedure is feasible too: we can start out from the assumption that the compressive stress in the surface layer cannot exceed 0.8 $R_p0.2$, and derive our requirements on the depth of the affected zone and on the dimple size from this assumption. After selecting this depth and dimple size accordingly, we must then choose such process parameter values which will enable us to attain these targets, or adapt the parameter values if we require less than the maximum possible stressing.

It would be wrong to regard the compressive stresses as the only significant effect of peening: another important effect is the increased dislocation density caused by plastic deformation. This increment cannot be assessed by calculations, and may vary widely from one spot to the next. Nevertheless the dislocations, their density and their distribution strongly affect above all the fatigue behaviour, i.e. the nucleation and propagation of very short cracks.

All the equations presented above are based on the implicit assumption that the spherical shot impinges on the planar treated surface perpendicularly. In practice, however, this is clearly an exceptional case. Even given an entirely flat surface the shot will still impinge at various angles, because (especially when the nozzle is relatively far from the treated surface) the stream of shot impelled at the surface is conical rather than cylindrical. The different impact angles produce different effects on the treated surface. Even worse complications ensue when this surface is not flat, but shaped, as component surfaces generally are. In that case our calculations must take into account not only different impact angles, but also the way they alter the actual process duration. For example, calculations show that if we are peening a cylindrical surface that rotates during the treatment, we must double the processing time to achieve the same coverage, and increase it by a factor of 2.5 to attain the same Almen intensity, as when we are peening a flat strip.

Relatively little has been published about the way various impact angles influence that outcome of shot peening |22|, and even that little deals chiefly with steel; no such analyses are available for aluminium alloys, nor for complex component shapes. The only measure normally taken in practice in such cases is to check the actual Almen intensity, by fastening Almen strips. The aim is to ensure an equal impact energy and measurements performed in practice confirm that the outcome depends substantially on differences in the shot size and on the actual impact velocity.

To ensure satisfactory results in shot peening operations, the shot must be screened by size; must be as regular and uniform as possible in its particle shape; and care must be taken to maintain a uniform impact velocity by precise adjustment of the delivery pressure and accurate dosing of the shot. That is why in modern shot peening equipment so much effort is devoted to provisions for monitoring and regulating the key process parameters.

Conclusion

Relationships for calculating or estimating the effects of various process parameters on the resultant properties of shot peened surface layers have been experimentally investigated on a type AlCu4Mgl aluminium alloy. The principal factors examined were the coverage, dimple sizes, and the calculated residual stresses. The computed results were in good agreement with experimental findings, but the calculation procedure calls for knowledge of several material constants which have to be ascertained by experiments for every alloy that is to be treated. That limits the usefulness of the existing calculations methods, and stresses the need for their further development and refinement. As shot peening techniques gain ever wider acceptance in industrial practice, calculation methods have an ever greater role to play in determining the process parameters and ensuring compliance with the specified target values in routine operations.

Acknowledgements

The authors wish to thank V. Marek, of the State Reaearch Institute for the Protection of Materials in Prague, for his aid in the peening trials; the staff at the Department of Machining Theory at the Mechanical Engineering Faculty of the Technical University in Prage for assistance with the roughness measurements; and Z. Knesl of the Institute of Physical Metallurgy in Brno for his stress calculations.

REFERENCES

1. Proc. of the 1st Int. Conf. on Shot Peening, ed. H. Niku - Lari, Oxford Pergamon Press, Oxford, 1981.
2. Proc. of the 2nd Int. Conf. on Shot Peening, ed. H. O. Fuchs, Paramon, N.J., 1984.
3. Shot Peening, eds. H. Wohlfahrt, R. Kopp and O.Vöhringer, DGM, Oberursel, 1987.
4. Al-Hassani, S.T.S. In /1/ pp. 583-602.
5. Al-Hassani, S.T.S. In /2/ pp. 275-82.
6. Iida J., In /2/ pp. 283-92.
7. Guechichi, H., et al. In Impact Surface Treatment, ed. S. A. Meguid, Elsevier, London, 1986, pp. 11-19.
8. Rykovskij, B. P. , Smirnov, V.A., and Ščetinin, G. M., Mestnoje upročněnije detalej poverchnostnym naklepom, Mašinostrojenije, Moskva, 1985.
9. Lieurade, H.P., Bignonnet, A., In /3/, pp. 343-59.
10. Kopp., R., Hornaner, K.P., In /1/ pp. 541-54.
11. Kopp., R., In /2/ pp. 6-14.
12. Niku-Lari, A., In /1/ pp. 237-47.
13. Kroupa, F., and Knesl, Z., Strojírenství, 1987, 37, pp. 651-59.
14. Hoffman, J., Scholtes, B., Vöhringer, O. and Macherauch, E., In /3/, pp. 239-46.
15. Müller, P., P. and Urffer, D., In /3/, pp. 49-54.
16. Knotek, O. and Elsing, R., In /3/, pp. 361-68.
17. Sedláček, V., In Basic Mechanisms in Fatigue of Metals, eds. P. Lukáš and J. Polák, Academia, Praha, 1988, pp. 73-81.
18. Lu, J., Flavenot, JF. In /3/ pp. 391-98.
19. Barret, L. and Cailliau, J., In /3/, pp. 399-406.
20. Desvignes, M., Gentil, B. and Khabou, T., In /3/, pp. 369-76.
21. Saverin, M., M., Drobestrujnyj naklep, GNTIML, Moskva, 1955.
22. Ebenau, A., Vöhringer, O., and Macherauch, E., In /3/, pp. 253-260.
23. Kirk, D., In /3/ pp. 213-20.

Fatigue Damage Evaluation of Shot Peened High Strength Aluminum Alloy

Y. Oshida

Syracuse University
Syracuse New York USA

and J. Daly

Metal Improvement Company
Paramus New Jersey USA

ABSTRACT

Shot-peening was applied to high strength aluminum alloy (Al-7050). At various stages during fatigue processes, x-ray diffraction studies including residual stress measurements (by Cr) and dislocation density measurements based on diffraction line analyses (by Cu and Mo) were perfomed. It was found that (1) both surface and interior residual stresses were relaxed as a function of fatigue cycle ratio, (2) relaxation rate of interior residual stress showed a negative slope, (3) the position where the maximum interior dislocation density was developed was close to that where the maximum peak interior residual stresses were generated, (4) surface dislocation density decreases and interior dislocation density increases by increasing fatigue cycles, (5) when ratio of interior to surface dislocation density exceeds 1000, the specimen is close to its fatigue life, and (6) this dislocation density ratio serves a promising fatigue damage indicator.

INTRODUCTION

Shot peening is employed as one of the most effective surface improvements, particularly in fatigue behavior and/or corrosion resistance of metallic materials. This improvement is recognized in enhancing fatigue strength or reducing fatigue crack growth rate, and retarding susceptibility to stress corrosion cracking [1~5]. Since only the surface layer is plastically deformed by shot peening and the bulk material remains undeformed, this inhomogeneous deformation induces compressive residual macrostress in a relatively thin surface layer [6]. It was reported that the maximum residual stress was developed at the surface for soft materials; while the maximum residual stress was observed at subsurface (usually 125~250 μm below the surface) for hard material or soft material peened at high intensities [5,6]. Since residual stresses act as if they are mean stresses in fatigue cycling process, compressive residual stress reduces the mean stress in surface, resulting in that the initiation and early fatigue crack growth are retarded [7,8].

Plastic deformation of the shot-peened surface layer also causes the strain hardening. Increase in fatigue strength is, therefore, a consequence of both compressive residual stresses and increased hardness due to workhardening. It was found that the residual stress is a dominant factor during the fatigue process in hardened steels, but that in softer alloys the improvement results principally from the strain hardening generated during the shot peening [9]. Nevertheless, it is commonly believed that when a compressive residual stress approaches 60% of material's ultimate tensile strength, fatigue strength is improved in the region of 25~30% and 30~55% in the reversed bending fatigue cyclings [10~12]. The localized plastic deforming surface layers, furthermore, increases dislocation density which results in a remarkably reduced ductility [13]. Although positive effect of the surface compressive residual stress and negative effect of the high dislocation density are competing each other on enhancing the fatigue strength, it was found that the latter effect on microcrack propagation is compensated by the former beneficial effect [13].

Since fatigue process is associated with cyclic plastic deformation, it is a commom understanding that residual stresses are not stable and redistributed when they are subjected to repeated loads and that its stability depends on the material's hardness. Consequently, the total fatigue life is highly dependent on a stability of developed residual stresses. There are experimental evidences showing that the residual stress relaxation depends also upon applied stress ratio (i.e, a ratio of applied stress amplitude for fatigue cycling, σ_{app}, to test material's yield strength, σ_Y). Repeated cycles of applied stresses below the yield strength acted much more slowly, and no serious reduction of residual compressive stress was found for applied stresses lower than about one third of the yield strength, even a million cycles of such stress were applied to the shot-peened specimen. It was, however, shown that residual stress relaxation occurs during fatigue for peak cyclic stresses below and near the endurance limit, σ_ω [14~16]. Components designed for high performances are normally operated in a cyclic stress regime ranging from σ_Y to σ_ω. Considerable data have been reported on the relaxation phenomenon taken place within this regime. In general, relaxation begins rapidly within the first few cycles and its rate decreases with increasing fatigue cycle ratio, N/N_F (ratio of fatigue cycles N to number of cycles-to-failure, N_F) [8].

Studies have shown that compressive residual stress induced by shot peening decreases both in surface and subsurface layers as a function of stress and surface depth [5,17]. It was found, by studying changes in residual stresses in shot-peened 4130 steel by x-ray diffraction method, that the peak residual (compressive) stress, σ^p_{rs}, was at 190 μm below the surface and that relaxation of initial residual stresses was taken place both at the surface and the peak position, but the relaxation of surface residual stress, σ^s_{rs}, showed higher rate than that of σ^p_{rs} [17]. It was also reported that both σ^s_{rs} and σ^p_{rs} relaxed linearly with respect to the cycle ratio up to N/N_F = 25%, followed by a slightly reduced rate during remaining fatigue lives [17]. Ni- and Co-based alloys showed similar trends in relaxation of σ^s_{rs} [18]. It was also shown that σ^s_{rs} relaxed linearly from the early stage of fatigue cycling and relaxed rapidly towards the fatigue failures in steels, and that relaxation rate of surface residual stresses increased as a function of σ_{app}/σ_Y, ranging from 66% to 80% [19].

Nonuniform microstrains are induced during surface plastic deformation and are associated with microstructural features such as precipitates, inclusions, grain, and dislocation interactions [20]. These microstresses are also not stable during the fatigue damaging process. Broadened x-ray line diffracted from a cold worked material was sharpened with subsequent fatgiue cycling of the material [13,18,20]. It was found, by studying the fatigue softening in cold worked copper by x-ray back reflection camera method, that the rms (root-mean-square) nonuniform microstrain - observed at 150A in the <100> direction - decreases monotonically as a function of cycle ratio, N/N_F [21]. Data reported by Baggerly et al. [21] indicated that initial dislocation density, ρ^0, of $2.25 \times 10^{10}/cm^2$ reduced to $1.2 \times 10^9/cm^2$ at N_F (= 4.5×10^6 cycles) conducted at ε_p = 0.12%, after microstrains were converted to dislocation densities by procedures proposed by Williamson and Smallman [22]. Noyan et al. [23], studying softening of residual stress-strain distributions in shot-peened α and β dual phase brass, observed that the line breadth decreases monotonically as a function of distance from the surface. Bonnafe et al. [24], studying fatigue damage in Al-2024 and -7075 alloys by x-ray diffraction, found that initial surface dislocation density decreases during the fatigue process by recognizing the decreasing the x-ray diffracted line broadening. It was reported that the initial microstrain decreases linearly from surface layer to 300 μm depth, followed by a constant value in bulk material and that the highly deformed surface layer showed high dislocation density, particularly in Al-8090 alloy [25].

In the above, it is clearly stated that shot peened materials possess compressive residual macrostress in the surface layer which is also strain hardened to some extent. During the fatigue stressing governed by cyclic plastic deformation, it is also recoginzed that progressive changes in macrostress and microstress are taken place, depending on a stress state as well as strain (or stress) amplitudes. These progressive changes in elastic macrostresess and plastic microstresses in shot-peened high strength aluminum alloy are major considerations in the present investigations. It is more important and practical to study the progressive changes in the internal redistribution of residual stresses as well as dislocation densities, by means of nondestructive techniques. These studies will provide us useful information for fatigue damage assessment and for prediction of remaining fatigue lives of important structural alloys, which are shot-peened.

EXPERIMENTAL PROCEDURES

Fatigue test specimens were prepared from a high strength aluminum alloy, Al-7050 T7651 sheet (Cu 2.0~2.6, Mn <0.10, Mg 1.9~2.6, Ti <0.06, Si <0.12, Zr 0.08~0.15, Cr <0.04, Zn 5.7~6.7, Fe <0.15, Al bal.). Figure 1 shows the configuration of test specimens. All four sides of the center test portion were shot peened (done at Metal Improvement Co.,) under following conditions; shot size: MI 230, intensity: 0.005 A, coverage: 100%. Fatigue tests were conducted under a four-point reversed bending mode (R = -1), by using Sonnatg Universal Fatigue Tester operating with the frequency of 30 Hz in air. Selected stress amplitudes were 30, 40 and 50 ksi. Since the yield strength of unpeened Al-7050 is 70.5 ksi (or 50 kg/mm^2), the applied stress ratio, σ_{app}/σ_Y, was calculated to be 42.6% for σ_{app} = 30 ksi, 56.7% for 40 ksi and 70.9% for 50 ksi, respectively.

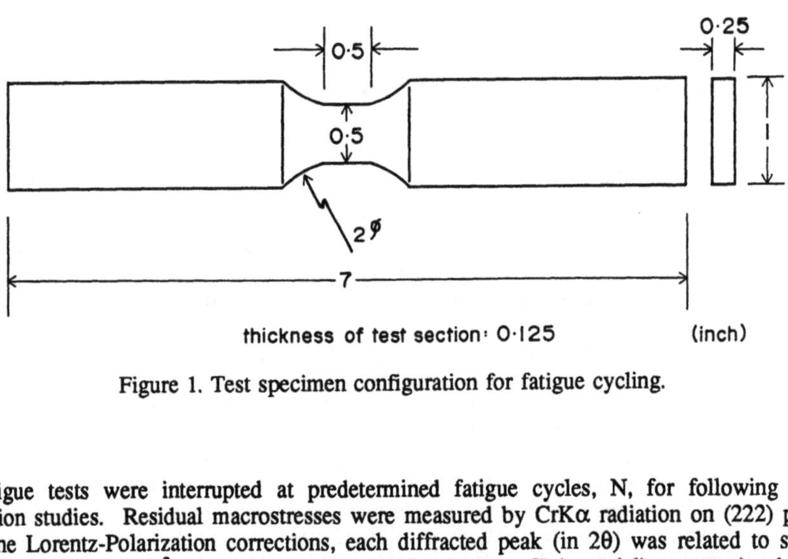

Figure 1. Test specimen configuration for fatigue cycling.

Fatigue tests were interrupted at predetermined fatigue cycles, N, for following x-ray diffraction studies. Residual macrostresses were measured by CrKα radiation on (222) plane. After the Lorentz-Polarization corrections, each diffracted peak (in 2θ) was related to $\sin^2\psi$. Theoretically, 2θ and $\sin^2\psi$ are linearly related with a constant K (material's constant), which is -9.40 kg/mm^2 (or -13.37 ksi). Surface microstress in terms of nonuniform microstrain can also be determined by x-ray diffraction line analysis. For this study, CuKα was radiated on (200)-(400) planes and (111)-(222) planes. In order to investigate progressive changes in residual stresses and dislocation densities, two methods were employed. The first non-destructive method was based on different penetration depth of used x-rays. The effective penetration depth x is defined as; $x = \sin\theta/2\mu$, where μ is the linear and absorption coefficient of the specimen and θ is the Bragg's diffraction angle. Calculated x-ray penetration depth of used x-ray sources in aluminum is listed below. Therefore, by using different x-rays, one will be able to obtain x-ray informations from near surface layer by Cu radiation and subsurface layer (ca. 100μm below the surface) by Mo radiation. It should be noted here that, for example by using Mo radiation on (222) plane, the x-ray information averaged over from the surface to 104 μm beneath the surface can be obtained and not from only the sublayer at 104 μm below the surface.

TABLE 1. X-Ray Penetration Depth [μm]

(hkl)	CuKα	MoKα	CrKα
(111)	12.55	55.33	
(222)	23.89	103.97	11.89*
(200)	14.40	62.72	
(400)	41.20	125.84	

* The affected depth was reported to be about 30 μm [24].

The second destructive method was a consequent removal of the surface layer. Since one side layer removal might cause nonuniform redistribition of residual stresses, both surfaces of fatigued specimen were firstly mechanically polished by using metallographic paper, followed by chemical polishing to remove further the slightly deformed layer by the previous mechanical removal. The chemical polishing was performed with a mixed solution (called R-5 bright dip), composed of H_3PO_4 78%, HNO_3 3.2% and H_2O 18.8%, at 60^0 C for 1 min. Thickness of removed layer was monitored by change in size of indentation indented onto the fatigued specimen surface by using the Vicker's hardness cone.

After non-destructive measurements of residual stress and x-ray diffraction line analyses by Cu and Mo radiations were completed, the fatigued specimen was mounted again on the fatigue tester for further fatigue cycling until the next predetermined cycle. This procedure was repeated many times until specimen was fatigue failed. By knowing the final number of cycles-to-failure of specimen, N_F, predetermined cycles were recalculated to determine the actual cycle ratios, N/N_F.

RESULTS AND DISCUSSIONS

S-N Curve

S-N curve of shot-peened Al-7050 T-7651 is shown in Figure 2, being compared with that obtained from un-peened Al-7050 T-7651 [26]. It was found that 100% coverage shot peening improved the fatigue endurance limit, $\sigma_{(\omega)}$, by approximately 33% and that even in a regime where the stress ratio is between σ_Y and $\sigma_{(\omega)}$ the fatigue strength increased by factor of 2.5~4.

Relaxation of Surface and Peak Interior Residual Stress

Figure 3 shows depth profiles of x-ray residual stresses in shot-peened aluminum alloys. Mark 1 is data for the present study. Marks 2 [27], 3 [28] and 5 [24] are residual stress depth profiles of shot-peened Al-7075 alloys with different heat treatments and different peening energy; while mark 4 is for Al-2024 alloy [28]. Microstructure in Fig. 3 shows the indentation by a hardness indent after surface layer was removed about 100 μm from the original surface. It was found that, as shot-peened condition, Al-7050 shows the residual stress, σ^s_{rs} of approximately -40 ksi and that the peak residual stress, σ^p_{rs}, of -58 ksi was situated at about 70 μm below the surface.

Figure 4 shows progressive changes of σ^s_{rs} as a function of fatigue cycle ratio. Being independent of the stress ratio, it was observed that σ^s_{rs} relaxed monotonically from the beginning of the fatigue cycling up to about 80% of the cycle ratio, followed by rapid relaxation. The onset of remarkable increases in relaxation rate might correspond to a propagation of macrocracks toward to the final very sharp relaxation observed at or near N_F. Therefore, it can be said that the fatigue cyclings between the onset of the residual stress relaxation and N_F are related to the crack propagation stage, N_P. Fatigue cycles consumed from the beginning to the onset points are related to crack initiation period, N_I. As it is generally believed that N_P/N_F is larger at lower applied stress ratio, σ_{app}/σ_Y, the present observations confirm this general trend.

Figure 5 shows progressive changes in residual stress distribution as a function of fatigue cycle ratio. In general, it was found that both the peak residual stresses, σ^p_{rs}, and surface residual stress, σ^s_{rs}, relax by increasing the fatigue cycle ratios for all stress amplitudes.

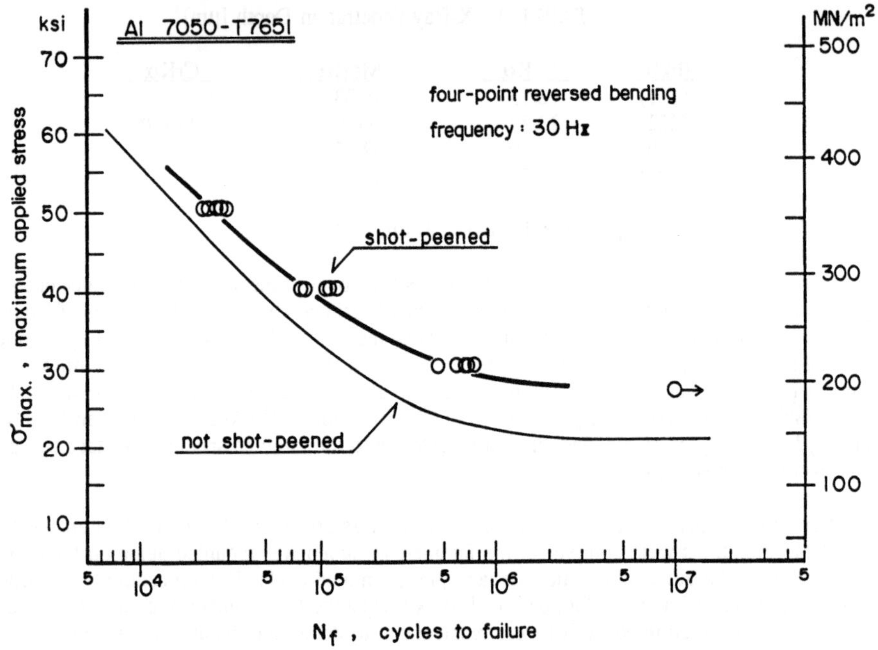

Figure 2. S-N curves of shot-peened aluminum alloy.

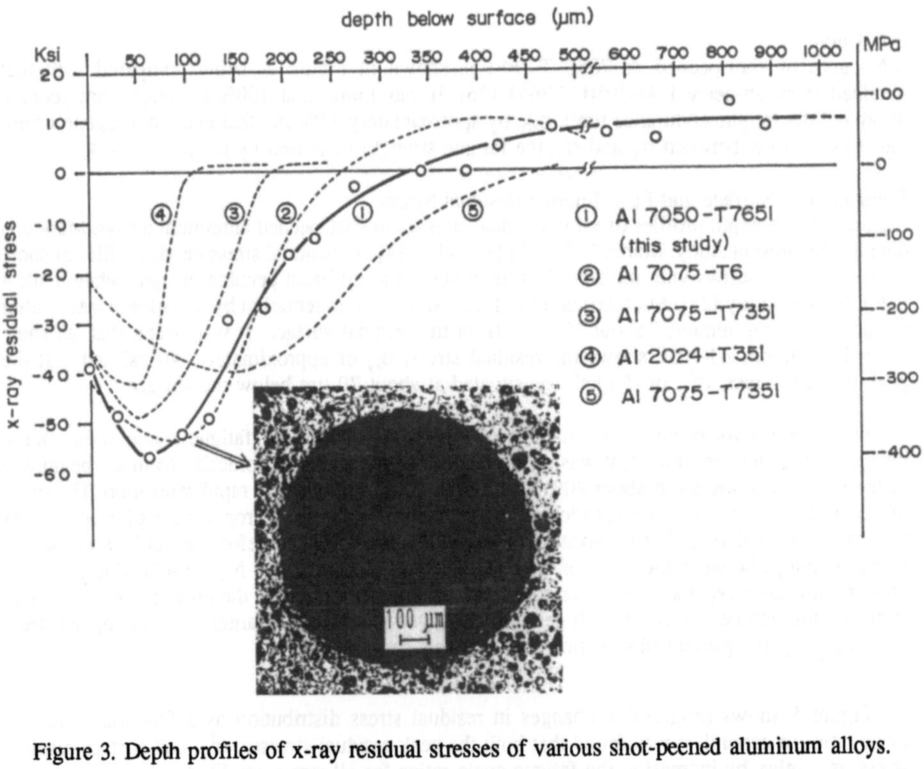

Figure 3. Depth profiles of x-ray residual stresses of various shot-peened aluminum alloys.

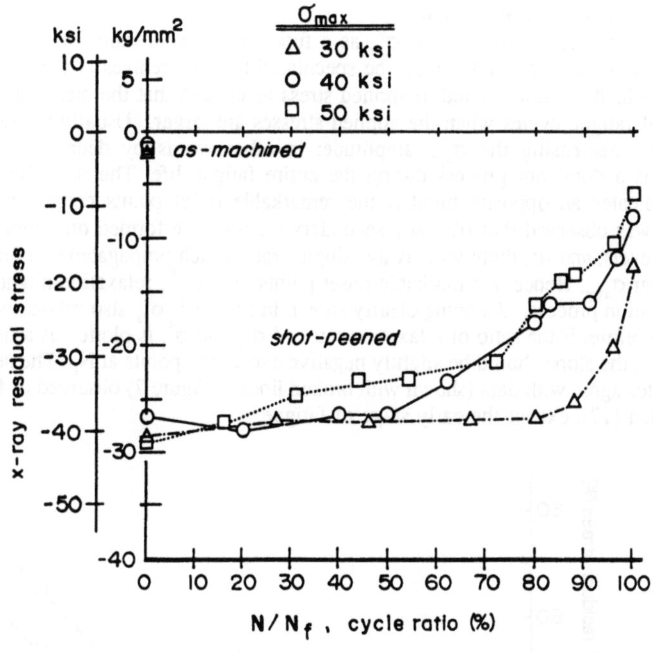

Figure 4. Progressive changes in surface residual stresses of shot-peened Al-7050 alloy, under three applied stress amplitudes.

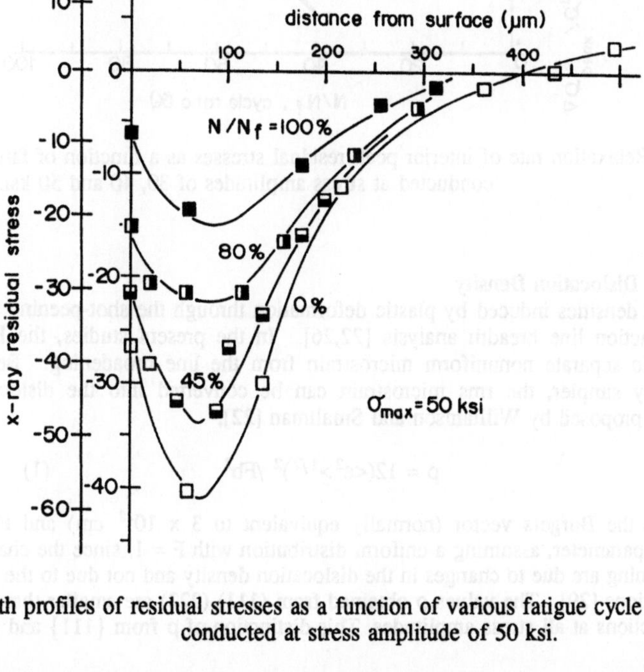

Figure 5. Depth profiles of residual stresses as a function of various fatigue cycle ratios, conducted at stress amplitude of 50 ksi.

Relative Relaxation Rates of σ^s_{rs} and σ^p_{rs}

Figure 6 shows the relaxation rate of σ^p_{rs} as a function of cycle ratio for three different stress levels. With limited data points, it can be speculated that σ^p_{rs} relaxes at the same rate after the onset points which are also related to applied stress levels and that the onset will be appeared at early stage of fatigue cycles when the applied stresses are larger. Usually the ratio of N_I to N_F increases with decreasing the σ_{app} amplitude; in other words by decreasing σ_{app} the crack propagation is a dominant process during the entire fatigue life. The data shown in Figure 6, however, indicates an opposite trend if the remarkable onset points corresponds to the crack initiation. It was observed that (i) many secondary cracks were formed on samples conducted at higher σ_{app} levels, and (ii) there was always single crack which propagated to a final failure when tested at lower σ_{app}. Hence, a remarkable onset points of the σ^p_{rs} relaxation might be responsible to crack initiation process. As being clearly shown in Figure 5, σ^s_{rs} also relaxes with larger rates than σ^p_{rs}. Therefore, if the ratio of relaxation rates of σ^p_{rs} and σ^s_{rs} is plotted as a function of N/N_F (see Figure 7), the slope should be slightly negative except the points at N_F. The negative slope in relaxation rates agree with data (shown with broken lines in Figure 7) observed in 4130 steel when being replotted [17], except the early stage of fatigue.

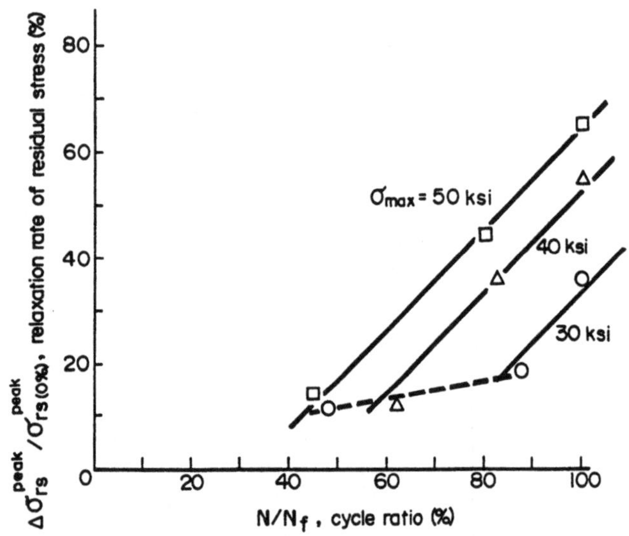

Figure 6. Relaxation rate of interior peak residual stresses as a function of fatigue cycle ratios, conducted at stress amplitudes of 30, 40 and 50 ksi.

Changes in Dislocation Density

Dislocation densities induced by plastic deformation through the shot-peening are calculated by x-ray diffraction line breadth analysis [22,26]. In the present studies, the Hall method was empolyed to separate nonuniform microstrain from the line broadening. Because it appears conceptually simpler, the rms microstrain can be converted into the dislocation density by procedures proposed by Williamson and Smallman [22];

$$\rho = 12(\langle\varepsilon^2\rangle^{1/2})^2 / Fb^2 \qquad (1)$$

where b is the Burgers vector (normally equivalent to 3 x 10^{-8} cm) and F is a dislocation disribution parameter, assuming a uniform distribution with F = 1, since the changes in the x-ray line broadening are due to changes in the dislocation density and not due to the rearrangement of the dislocations [29]. The values ρ obtained from (111)-(222) are smaller than that from (200)-(400) reflections at all stress amplitudes. This distinction of ρ from {111} and ρ from {200} reflections suggests a difference in the damage processes of grains with {100} planes parallel to the surface from those having {111} planes parallel to the surface. From an analysis of effective resolved shear stress, τ_{eff}, for pure bending case, in grains having {111} planes and {100} planes parallel to the surface, it was suggested that more dislocation sources are activated inside grains with {100} planes parallel to the specimen surface [26]. Therefore, in the following data presentations, ρ values from {100} planes are used.

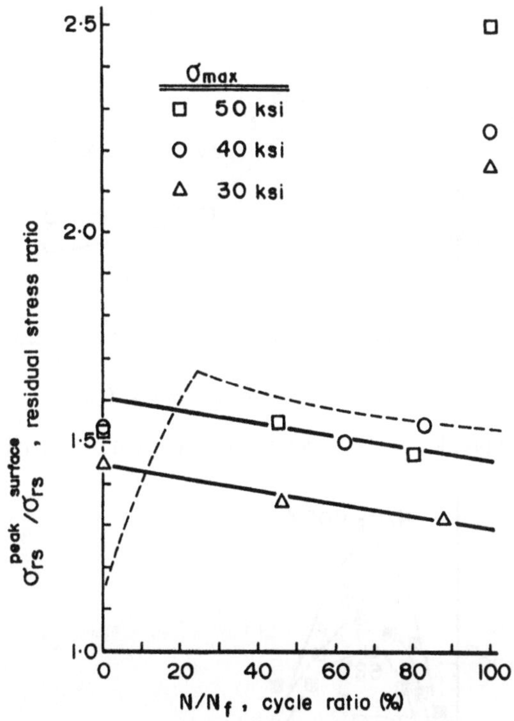

Figure 7. Ratio between the interior peak and surface residual stresses

As machined condition, it was found that the original dislocation density was 2 x 10^{10}/cm^2 [26]; while as shot-peened condition, as shown in Figure 8, original dislocation density, ρ^0, was approximaely 1.5 x 10^{11}/cm^2, averaged over three specimens. Figure 8 shows also depth profiles of dislocation densities, which were obtained from CuKα radiation and by surface layer removal technique. As discussed previously, if MoKα radiation is used, it penetrates deeper than CuKα does (i.e., approximately 100 μm from the surface). Closed marks in Figure 8 show results of dislocation densities obtained by MoKα radiation. Although there might be some doubts about the plotting these data points in terms of depth-scale, in this study, the effective penetration depth was chosen to be 100 μm. Although ρ values by Mo radiation are higher than ρ by Cu radiation, their profiles can be generally considered to be similar. Fatigue damage is entirely governed by a plastic deformation which, in turn, generates dislocations. During a fatigue cycling of work-hardening type of materials, it was found that (i) ρ_s is always higher than ρ_i, (ii) the surface layer behaved as a barrier-like layer [26], and (iii) the barrier became sufficiently strong the local stress fields associated with the accumulation of dislocations in the surface layer pileups exceeded the fracture strength, and cracks were initiated [29]. Of interest is that the peak position where the interior dislocation density shows its maximum value, ρ_i^p is close to the position for σ_{rs}^p. Therefore, the positive effect of compressive residual stress and negative effect of dislocation density seem to be competing factors governing subsequent fatigue cyclings.

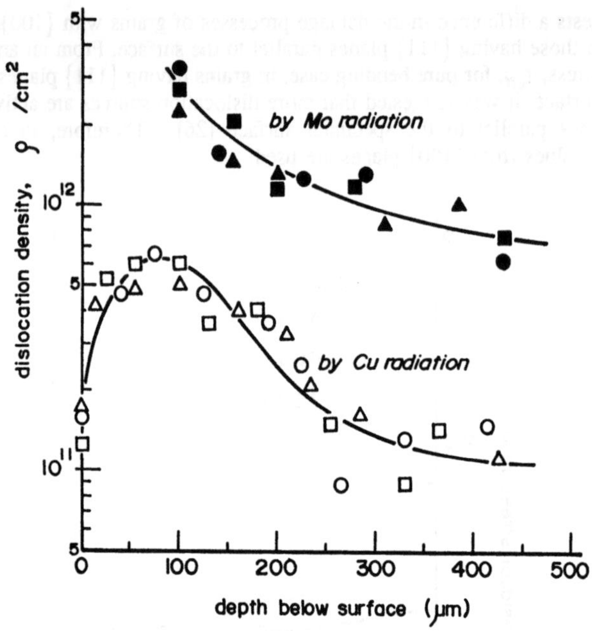

Figure 8. Depth profiles of dislocation density, observed by Cu and Mo radiations.

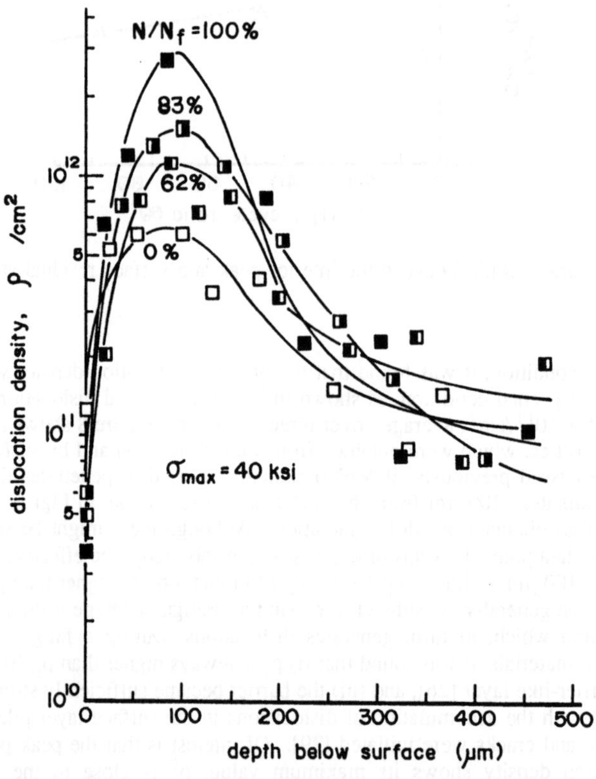

Figure 9. Progressive changes in dislocation density as a function of fatigue cycle ratio, conducted at stress amplitude of 40 ksi.

Figure 9 shows the depth profiles of dislocation densities for different fatigue cycle ratios conducted at σ_{app} of 40 ksi. It is observed that, by fatigue damage processing, surface dislocation density, ρ_s, decreases; while the interior dislocation density at peak, ρ_i, increases. These profiles are obtained by destructive method of a continuous removal of surface layer. As discussed in Figure 8, the peak interior dislocation density can also be measured by nondestructive means of MoKα radiation with deeper penetration. As discussed previously, the surface layer and interior layer behave differently. Fatigue crack(s) will be initiated when the local stress fields associated with accumulated dislocation density exceeds the fracture strength. Figure 10 summaries observed dislocation densities of ρ_s (with open marks) and ρ_i (with closed marks) for three applied stresses. It was found that (i) changes in both dislocation densities seem to be independent of stress amplitudes, (ii) surface dislocation density, ρ_s, shows a decreasing trend; while ρ_i shows an increasing trend, and (iii) dislocation density at the interior peak has a similar increment as unpeened Al-7050 alloy, as shown with broken zones in Figure 10 [26]. Since the dislocation density, ρ_i, increases continuously while the specimen is cyclically worksoftened and the σ^P_{rs} is also continuously relaxed, it appears that the density of obstacles or barriers to the dislocation movement must decrease. It is, therefore, feasible that precipitates or secondary phases of this alloy might be systematically cut by these dislocations and redissolve [29,30].

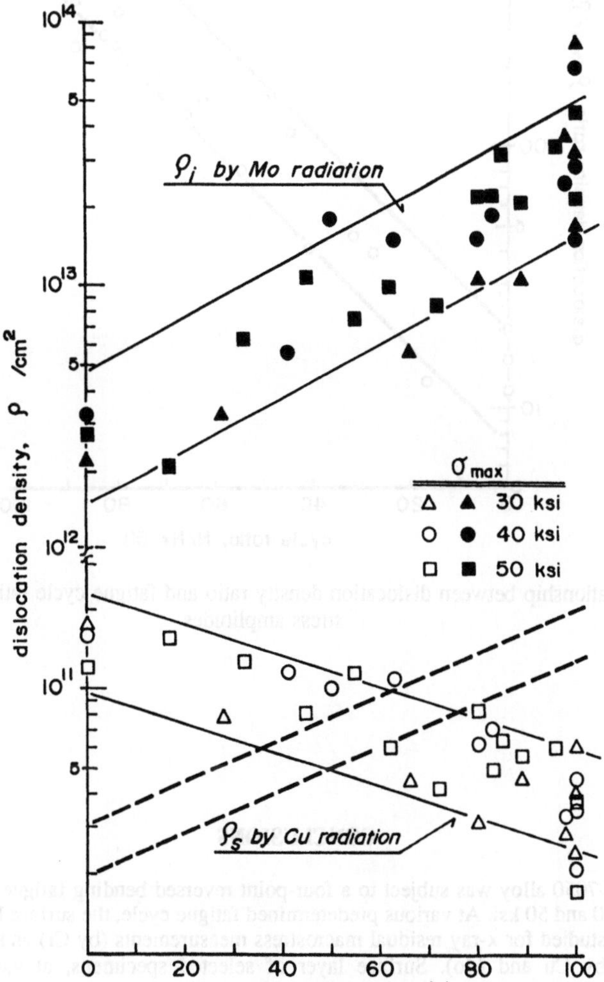

Figure 10. Changes in surface and interior dislocation densities.

Hence the increasing ρ_i seems to be responsible to the major fatigue damage process. Figure 11 is another representation of the dislocation density in the interior peak, in reference to that at the surface layer. The ratio of dislocation density in the interior peak to that at the surface, $\rho_i(Mo)/\rho_s(Cu)$, is plotted as a function of cycle ratio for all applied stress amplitudes. Although the data show scattering, this representation is proposed because it obviates the need for calibration curves, which may be difficult to obtain in practical applications and can be sensitive to manufacturing and processing histories. From Figure 11, it is concluded that if the dislocation density ratio, $\rho_i(Mo)/\rho_s(Cu)$ exceeds 1,000, the fatigue failure is occured, or 98% of fatigue life is consumed, although the observed ratio does not reflect to the true value of ρ_i/ρ_s.

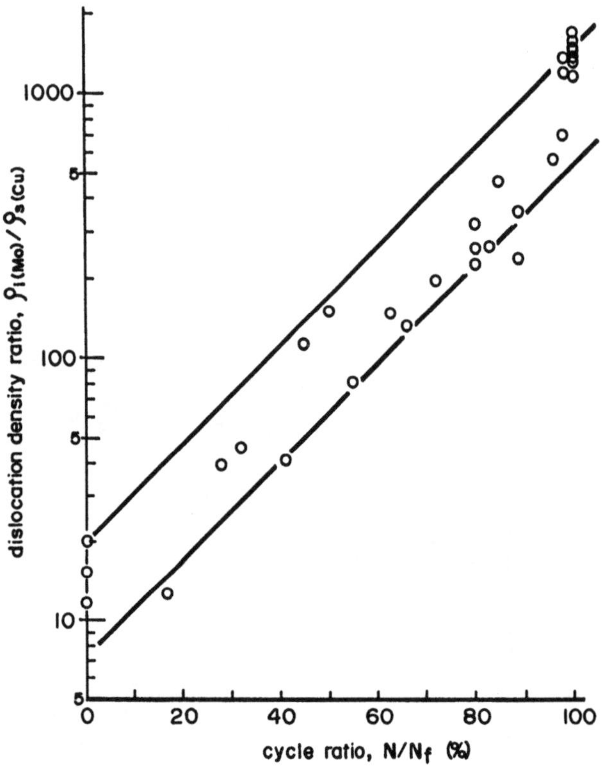

Figure 11. Relationship between dislocation density ratio and fatigue cycle ratio for all tested stress amplitudes

CONCLUSIONS

Shot-peened Al-7050 alloy was subject to a four-point reversed bending fatigue tested at applied stresses of 30, 40 and 50 ksi. At various predetermined fatigue cycle, the surface layers of fatigued specimen were studied for x-ray residual macrostress measurements (by Cr) and x-ray diffraction line analyses (by Cu and Mo). Surface layer of selected specimens, at various N/N_F, was mechanically/chemically removed to obtain the depth profiles of both macrostresses and microstresses. Within limited numbers of data, the followings are concluded.

(1) As shot-peened conditions, surface residual stress, σ^s_{rs} was -40 ksi and the peak internal residual stress, σ^p_{rs} was -58 ksi. During fatigue cycling, both σ^s_{rs} and σ^p_{rs} are relaxed. The relaxation of σ^s_{rs} was faster than that of σ^p_{rs}.
(2) Depth profile of dislocation densities, ρ obtained by the surface layer removal method (by Cu) shows the peak value of ρ at the interior, ρ_i. By using Mo (having deeper penetration depth than Cu), the ρ depth profile shows a similar to that obtained by Cu, although the absolute value of ρ_i by Mo is about 10 times higher than that by Cu.
(3) The interior position which shows maximum residual stress, σ^p_{rs}, was found to be situated close to the region where the interior dislocation density, ρ_i, showes its maximum value.
(4) During fatigue cycling, the surface dislocation density, ρ_s, decreases and the interior peak dislocation density, ρ_i, increases with different rates.
(5) The dislocation density ratio, ρ_i/ρ_s, is independent of applied stress amplitudes. If the ratio $\rho_i(Mo)/\rho_s(Cu)$ is over 1000, the data indicates that 98% or more (~ N_F) of the fatigue life is already consumed or the specimen is fatigue failed. Therefore, the dislocation density ratio can serve a promising fatigue damage indicator for shot-peened high strength aluminum alloy.

Many structural components were installed as un-peened conditions. If the failure probability model [26] and the presently proposed dislocation density ratio indicator are applicable, one can predict the prolonged fatigue life of the component (which was not originally shot-peened and has been fatigue damaged to some extent) after the rejuvenating shot-peening is applied. Studies on effects of shot-peening on rejuvenating the fatigue damaged materials are under progress.

REFERENCES

[1] Feld,P.G. and Gillespie.R.D., Controlled Shot Peening Reduces or Eliminates Stress Corrosion Cracking, Proc. ICSP-2, 1984, pp. 121-125.

[2] Koehler,W., Influence of Shot Peening with Different Peening Materials on the Stress Corrosion and Corrosion Fatigue Behavior of a Welded AlZnMg-Alloy, Proc. ICSP-2, 1984, pp. 126-132.

[3] Kirk,D. and Jarrett,M., Influence of Shot-Peening on Fatigue and Corrosion-Fatigue Behavior of Structural Steel, Proc. ICSP-2, 1984, pp. 133-142.

[4] Harrison,J., Controlled Shot Peening: Cold Working to Improve Fatigue Strength, Heat Treat., 19,(8), 1987, pp. 16-18.

[5] Wohlfahrt,H., Shot Peening and Residual Stresses, in Residual Stress and Stress Relaxation ed. by E.Kula and V.Weiss, Plenum Press, N.Y., 1982, pp. 71-92.

[6] Guechichi,H., Castex,L., Frelat,J. and Inglebert,G., Predicting Residual Stress due to Shot Peening, in Residual Stresses in Science and Technology", 1, 1987, pp. 449-456.

[7] Radhakrishwan,V.M. and Prasad,C.R., Relaxation of Residual Stress with Fatigue Loading, Eng. Fract. Mech., 8, 1976, pp. 593-597.

[8] Jhansale,H.R. and Topper,T.H., Engineering Analysis of the Inelastic Stress Response of a Structural Metal Under Variable Cyclic Strains, ASTM STP 519, 1973, pp. 246-270.

[9] Syren,B., Wohlfahrt,H. and Macherauch,E., The Influence of Residual Stresses and Surface Topography on Bending Fatigue Strength of Machined CK 45 in Different Heat Treatment Conditions, Proc. 2nd Int'l Conf. on Mech. Behvaior of Material, Boston, 1976, pp. 212-235.

[10] Schreiber,R., Wohlfahrt,H. and Macherauch,E., Effect of Shot-Peening Fatigue Strength of the Normalized Steel 16 MnCr 5, Arch. Eisenhuettenwes., 48, 1977, pp. 649-652.

[11] Leutjering,C.G.G., Influence of Shot Peening on Notched Fatigue Strength of Ti-6Al-4V, Proc. ICSP- 2, 1984, pp. 175-180.

[12] Woelfel,M.M., Experimental Determination of Fatigue Strength as Related to Peening Intensity and Specimen Thickness, Proc. ICSP-2, 1984, pp. 365-368.

[13] Gray,H., Wagner,L. and Luetjering,G., Influence of Residual Stresses on Fatigue Crack Propagation of Small Surface Cracks, in Residual Stresses in Science and Technology", 2, 1987, pp. 815-821.

[14] Pattionson,E.J. and Dugdale,D.S., Fading of Residual Stresses due to Repeated Loading, Metallurgia, 66, 1962, pp. 228-230.

[15] Gould,R.W. and Pittella,C.F., An X-Ray Invesitgation of Fatigue Behavior of Cold Worked Aluminum, in Advances in X-Ray Analysis", 16, 1973, pp. 354-366.
[16] Weiss,V., Oshida,Y. and Wu,A., A Note on Fatigue Damage Assessment by X-Ray Diffraction Techniques for a 304 Stainless Steel Specimen, Jour. of Nondestructive Eval., 1(3), 1980, pp. 207-213.
[17] Esquivel,A.L. and Evans,K.R., X-ray Diffraction Study of Residual Macrostresses in Shot-peened and Fatigued 4130 Steel, Exp. Mech., 8, 1980, pp. 496-503.
[18] Boggs,B.D. and Byrne,J.G., Fatigue Stability of Residual Stress in Shot Peened Alloys, Met. Trans., 4, 1973, pp. 2153-2157.
[19] Desvignes,M., Gentil,B. and Castex.L., Fatigue Progressing of Shot Peened Steel, in Residual Stresses in Science and Technology", 1, 1987, pp. 441-448.
[20] James,M.R., The Relaxation of Residual Stresses During Fatigue, in Residual Stresses and Stress Relaxation", 1982, pp. 297-314.
[21] Baggerly,R.G. and Pelloux,R.M.N., X-Ray Analysis of Fatigue Softening in Cold Worked Copper, in Advances in X-Ray Analysis", 12, 1969, pp. 236-248.
[22] Williamson,G.K. and Smallman,R.E., III. Dislocation Densities in Some Annealed and Cold-Worked Metals from Measurements on the X-Ray Debye-Scherrer Spectrum, Philos. Mag., 1, 1956, pp. 34-46.
[23] Noyan,I.C. and Cohen,J.B., An X-ray Diffraction Study of the Residual Stress-Strain Distributions in Shot-peened Two-phase Brass, Materials Sci. and Engrg., 75, 1985, pp. 179-193.
[24] Bonnafe,J.P., Lebrun,J.L. and Maeder,G., X ray diffraction and acoustic emission study of fatigue damage in aluminium alloys, in Residual Stresses in Science and Technology", 2, 1987, pp. 777-784.
[25] Mutoh,Y., Fair,G.H., Noble,B. and Waterhouse,R.B., The Effect of Residual Stresses Induced by Shot-Peening on Fatigue Crack Propagation in Two High Strength Aluminium Alloys, Fatigue Fract. Engng. Mater. Struct., 10(4), 1987, pp. 261-272.
[26] Weiss,V. and Oshida,Y., Fatigue Damage Characterization by X-Ray Diffraction and Failure Probability, EPRI AP-4477, 1986, pp. 16/1-16/18.
[27] "Shot Peeninfg Stress Profiles", ed. by H.O.Fuchs, Metal Improvement Company, NJ, 1986.
[28] Niku-Lari,A., Contraintes Resiuelles et Fatigue des Alliages D'aluminium Grenailles, Proc. ICSP-2, 1984, pp. 102-114.
[29] Kramer,I.R., Feng,C.F. and Wu,B., Dislocation-Depth Distribution in Fatugued Metals, Materials Sci. and Engrg., 80, 1986, pp. 37-48.
[30] Stubbington,C.A. and Foryth,P.T.E., Some Observations on Microstructural Damage Produced by Fatigue of an Aluminium - 7.5% Zinc - 2.5% Magnesium Alloy at Temperatures between Room Temperature and 250^{0}C, Act. Met., 14, 1966, pp. 5-12.

Effect of Shot Peening Process Controls on Consistent Treatment

J. S. Eckersley

Metal Improvement Company
Belleville
Michigan
USA

INTRODUCTION

Shot peening is a cold working process applied to the surface of metal parts to increase their resistance to fatigue, fretting fatigue, stress corrosion cracking, corrosion fatigue and various other tensile stress related metal failures. Shot peening has also been used effectively to form aerodynamic curvatures into large sculptured panels such as aircraft wing skins, some over 90 feet (30 m) long. In all cases, the effectiveness of the process is due to the introduction of residual compressive stresses into the surface of the parts. When a metal surface is indented, a residual compressive stress is induced as the subsurface material attempts to push out the indentation and restore the surface to its former condition. Shot peening, by bombarding the surface with millions of tiny spheres of steel, glass or ceramic, totally covers the surface with indentations, creating an even layer of compressive stress that is at its maximum just below the surface and is at least equal to 50% of the ultimate tensile strength of the metal.

It becomes immediately apparent that, for shot peening to be consistently effective and reproducible, a number of parameters must be controlled. Among the most important:

1. The peening media.
2. The energy of the shot stream.
3. The percentage of indentation of the part surface.
4. The repeatability of the peening machine.

SPECIFICATIONS THAT CONTROL SHOT PEENING

There are innumerable "company" specifications for shot peening particularly, as you might expect, in the aircraft and automotive industries. Most are based on three "industry" specifications:

1. MIL-S-13165, Shot Peening of Metal Parts. (1)
2. AMS 2430 Shot Peening. (2)

3. AMS 2431 Peening Media. (2)

A new specification is almost ready for publication by the SAE. Entitled "Computer Monitored Shot Peening", it will in all likelihood have the designation AMS 2432.

Over the years, these specifications have become increasingly stringent. For instance, in the early seventies, the Revision B of the Military Specification MIL-S-13165 essentially "outlawed" peening with hand held nozzles in favor of the use of automated machines. Now, Revision C, just published, includes the use of computer monitored equipment.

The United States Naval Air Command recently set up a review team to look at shot peening practices. The team reported that "There is no evidence of substantial problems resulting from the use of peening, but the potential for problems where large enhancement is provided, and the wide variation in industry practices, led the team to the conclusion that a policy which imposes controls is needed".(3) They recommended the "immediate" use of computer monitored shot peening for any new design that is dependant on shot peening for life enhancement.

The importance of tight controls cannot be overemphasized because to date there is no non-destructive, production method of determining the quality of the peening on a part after the peening has been performed.

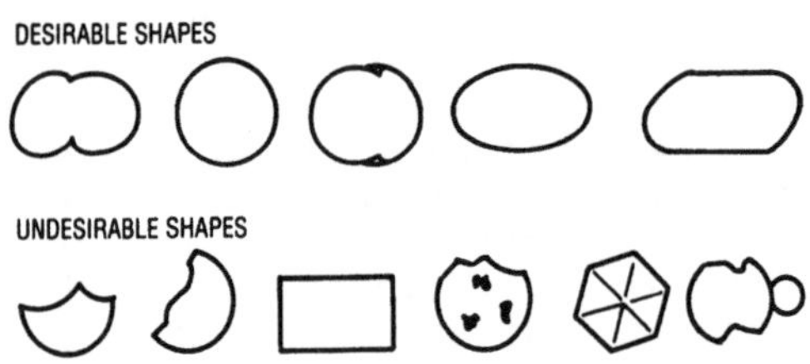

FIG. 1 **PEENING MEDIA**

THE PEENING MEDIA

The Aerospace Materials Specification #2431 covers seven types of peening media: Cast Steel Shot, Regular and Hard; Conditioned Carbon and Stainless Steel Cut Wire; Peening Balls; Glass Shot and Ceramic Shot. The specification imposes limitations on chemistry, size, sphericity, permissible numbers of misshapen or broken particles, soundness (i.e., the absence of hollows, cracks, etc.) and hardness (fig. 1). The limits are tighter than for non-aerospace specifications as might be expected and it is also the first time that just about all the readily available types of peening media are addressed in a single document. AMS 2431 does not address the uses and characteristics of each of the shot types so a review here might be useful. AMS 2431 assigns a separate /# to each type so the same numbering system will be used here:

AMS 2431/1 - CAST STEEL SHOT, REGULAR (45-52 HRC)
A general purpose medium for parts of a hardness less than 52 HRC. (fig. 2)

AMS 2431/2 - CAST STEEL SHOT, HARD (55-62 HRC)
Similar to regular, it is recommended for parts harder than 55 HRC. Breakdown rate is high, particularly when peening hard parts, but the magnitude of compressive stress generated on hard parts is significantly higher than when regular shot is used. Can be used on softer parts with no inconvenience, provided broken particles are kept below specification requirements.

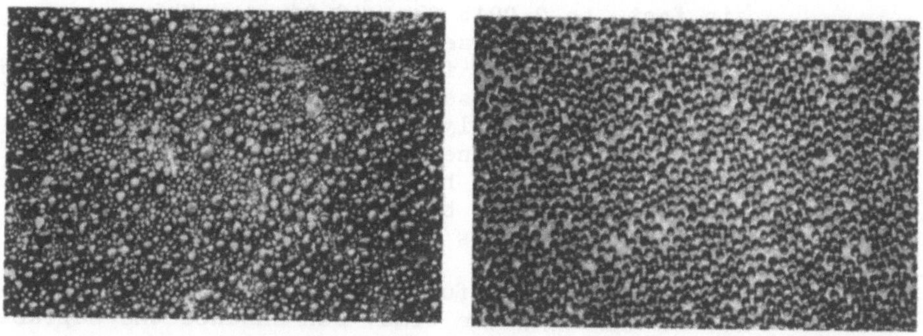

(A) (B)

FIG. 2 (A) UNACCEPTABLE SHOT PEENING MEDIA (B) ACCEPTABLE SHOT PEENING MEDIA

AMS 2431/3 - CONDITIONED CARBON STEEL CUT WIRE SHOT
Cut wire shot is produced by chopping wire into cylinders so that the length equals the diameter and then rounding the sharp corners by repeated blasting against a hardened target. As might be expected, the purchase price of conditioned cut wire shot is much higher than cast steel shot but because it is wrought rather than cast, cut wire shot is very durable. Size uniformity is also a significant feature but small sizes are difficult to produce.

AMS 2431/4 - CONDITIONED STAINLESS STEEL CUT WIRE SHOT
Stainless cut wire shot has all the advantages of the carbon steel variety with the additional one that it does not leave iron deposits on the peened surface. This property is particularly useful when peening nonferrous parts which otherwise must be decontaminated. Comparatively, hardness is low and purchase price high.

AMS 2431/5 - PEENING BALLS
Used almost exclusively for Peen Forming, peening balls can be as large as 1/2 inch (13mm) in diameter, though 1/8 (3mm) and 1/4 inch (6mm) are more common. Toughness to resist breaking is the most important parameter.

AMS 2431/6 - GLASS SHOT
Although the breakdown rate is high, glass shot has some interesting advantages:

1. Chemically inert glass shot will not contaminate the part being peened. It is often used in a secondary peening to reduce the residual iron left by cast shot.
2. On very hard parts, it has been shown that a secondary peening with small glass shot at a low intensity will increase the magnitude of the surface compressive stress, with a corresponding even greater increase in cycles to crack initiation.
3. Glass shot, unlike other media, is available in sizes below a nominal 0.007 inch (0.2mm); down, in fact, to 0.0015 inch (0.04mm). This very small medium is useful for peening into tight radii such as are found in thread roots. A word of caution: the surface in the roots must be smooth because only a very shallow depth of compression is generated by such small shot.
4. Glass, of course, has much less mass than steel and can therefore be used where a low level of energy transfer is required; for instance, on very thin parts.
5. Glass shot is useful as a blast cleaning medium which at, the same time, can enhance the surface properties.

AMS 2431/7 - CERAMIC SHOT
Made by fusing zirconium oxide powder in a glass binder, ceramic shot is very hard and inert. It is more durable than glass but it is also expensive to buy. A lower breakdown rate makes the condition of the ceramic shot easier to maintain than glass in the peening machine.

The above specifications are essentially for purchasing of peening media. However, the soon to be published AMS XXXX "Computer Monitored Shot Peening" specification, which is a user rather than purchasing document, references AMS 2431 by requiring that the shot in the peening machine be maintained to the same standards. Other specifications have their own "condition of shot" tables that specify the maximum amount, by particle count or by weight, of undesirable shapes that can be allowed in the machine charge. In either case, the amount is minimal for good peening practice and dictates that the user either changes the charge often or has machines equipped with devises that automatically maintain the shot condition by removing both sub-size and broken particles. These devises consist of vibrating screens that control the size content and classifying spirals that control the shape content. It is essential that round, correctly sized media be used to provide an even depth of compressive stress in a surface covered with round shallow dimples free from sharp, angular indentations.

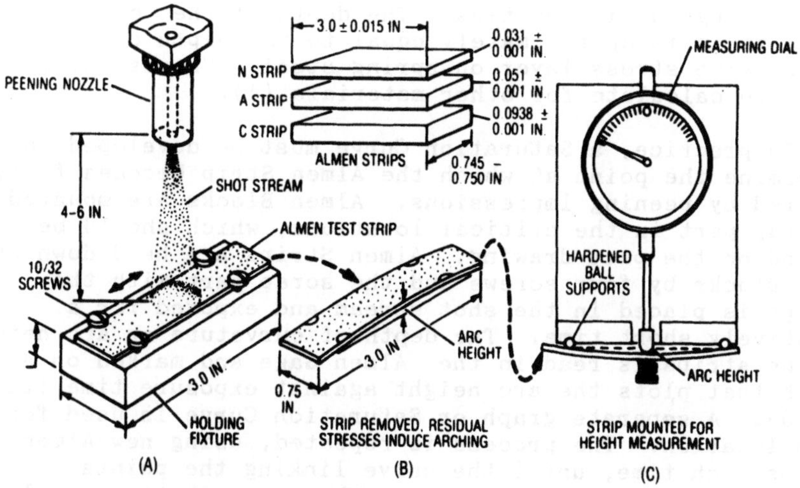

FIG. 3 THE ALMEN STRIP SYSTEM

TRANSFERRED KINETIC ENERGY

Soon after "shot blasting" was determined to be beneficial on valve springs for the Buick and Cadillac engines in 1928, it became necessary to find a means to ensure some degree of repeatability. Those involved knew, of course, that kinetic energy was being transferred by the shot to the springs but to measure the mass and velocity of millions of pieces of shot inside a machine was next to impossible and far from practical. Even if it were possible, three other critical factors would have been left out of the equation since what was of real interest was not really how much energy was in the peening media but how much of this energy would be transferred to the steel of the springs. This transferred energy would also be affected by the hardness of the spring steel, the hardness of the shot and the angle of impingement between the shot and the surface.

John Almen finally developed a system that bears his name and incorporates all the factors plus being sufficiently practical for shop use. The Almen system is still in use and, despite some inherent inaccuracies, has not been improved upon, though efforts are being made to do so. Almen used standardized, flat strips of spring steel (fig. 3) which take on a curvature when exposed to a shot stream on one side. The degree of curvature is proportional to the intensity of the kinetic energy in the shot stream. Once the strip is fully covered with peening impressions, the degree of curvature is relatively unaffected by exposure time. The curvature of the strip is measured on a special Almen depth gage in thousands of an inch or hundredths of a millimeter and that number becomes the Almen Intensity. Three thicknesses of Almen strips (N, A & C) cover the normal range of intensities. The depth of the curvature on the A strip is approximately equal to the depth of the compressive stress layer on spring steel. Charts exist from which to calculate for other materials (4).

In practice, a Saturation Curve must be developed to determine the point at which the Almen Strip becomes fully covered by peening impressions. Almen Blocks are mounted on a scrap part in the critical locations, which should be marked on the part drawing. Almen Strips are held down on the Blocks by four screws and the scrap part with the Strips is placed in the shot stream and exposed for a relatively short time. The depth of curvature or arc height of the strips is read in the Almen Gage and marked on a graph that plots the arc height against exposure time (fig. 4 & 5). A separate graph or Saturation Curve is used for each location. The process is repeated, using new Almen Strips each time, until the curve linking the points flattens out so that by doubling the exposure time results in no more than 10% increase in the arc height. Almen Intensity has then been reached and must fall within the range indicated on the drawing or specification. If it does not, the shot velocity and/or the shot size must be adjusted and a new Saturation Curve prepared before the parts can be peened.

Almen Intensity, then, becomes an elegant method of measuring, relatively, the energy that is transferred from the peening media to the surface of the part. From the Almen Intensity can be calculated the approximate depth of the compressed layer on the part in relation to the part material. However, it must be noted that the exposure time required to reach saturation on the Almen Strip is usually different from the exposure time required to correctly shot peen the part.

FIG. 4 **ALMEN FIXTURE**

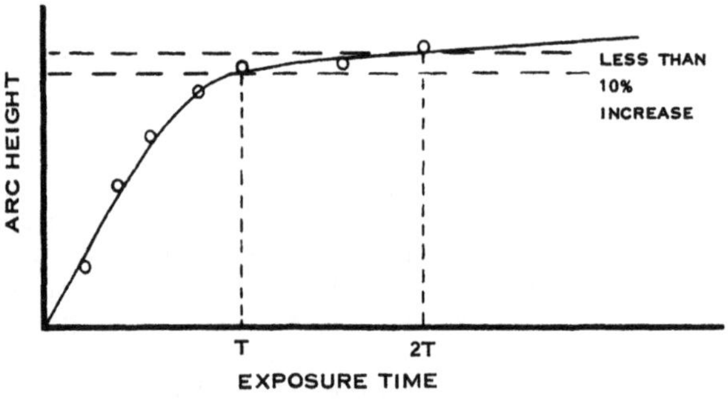

FIG. 5 **SATURATION CURVE**

PEENING COVERAGE

The correct exposure time is governed by a number of factors mostly unrelated to the Almen Strip: (fig. 6)

1. Part hardness. It quickly becomes obvious that a part substantially harder than an Almen Strip will require longer exposure. The peening dimples will be smaller on the harder material and more dimples will be required indicating a longer exposure time to produce full coverage or total obliteration of the surface. Conversely, on softer parts, less exposure is usually required.
2. Shot hardness also affects coverage rate for essentially the same reasons given above. Softer shot will produce smaller dimples on harder parts, requiring more exposure.
3. Shot size. Small shot covers the surface much faster than large. The reverse is true for intensity: higher intensities will result in faster coverage, for the same shot size.
4. Surface condition. If a surface, particularly a hard surface, has visible machine marks, greater exposure will be needed to peen down the high spots so that the low spots can be peened. Sometimes, if the machine grooves are significant, two sizes of shot may be indicated: First, small shot to get into the grooves, followed by large to produce sufficient depth of compression.
5. If the material work hardens substantially, additional exposure time may be indicated.
6. Part geometry also must be taken into account. A fatigue-critical fillet may be "shaded" so that the overall part may get more than full coverage while sufficient exposure time is allowed to get full coverage in the fillet. The additional coverage is not a problem except in the case of nickel-base materials used in jet engines. Extended coverage (over 500%) on these materials has been shown to cause strain hardening and a small loss of fatigue strength when compared to normal coverage. Because jet

(A) (B)

FIG. 6 **SHOT PEENING COVERAGE (A) PARTIAL (B) FULL**

engine components are usually repeened at overhaul to restore compressive stresses and also the complex geometry requires long exposure times to reach, for instance, the pressure faces in the dovetail slots, a method known as "coverage mapping" has been devised. A fluorescent tracer (see below) is used to determine the time required to reach full coverage on the various areas of the part. In production, the part is masked, leaving exposed only the areas requiring maximum peening time. As the part is being peened, the masks are removed to expose, in turn, areas that require less peening time so that all critical areas receive essentially the same coverage.

7. Most specifications allow holes to be peened externally at 45° that is as long as the depth of the hole does not exceed the diameter. If the hole will be peened from both ends, the depth may be equal to twice the diameter. Deeper holes must be peened internally. Usually, holes smaller than 3/8 inch (10mm) are not required to be peened unless they are specifically so designated on the part drawing. Holes down to 0.080 inch (2mm) can be peened successfully but coverage determination inside small holes is very difficult unless a fluorescent tracer is used.

Determination of coverage in many circumstances is not easy even when a decision is made as to how much coverage is appropriate for a given part. 100% coverage is the point at which the surface is totally covered in peening dimples which then translates to exposure time in the peening machine for a given set-up. On relatively soft parts, this visual determination can be achieved with a 10x magnifier which works well unless there are many parts, large surface areas or restricted line of sight (as in holes). On very hard parts (carburized gears, for instance) visual determination is difficult and requires a very practiced eye. Using greater magnification does not help.

A fluorescent tracer lacquer system, such as PEENSCAN (5.), called out in MIL-S-13165, is highly effective (fig. 7). Basically, the part is coated with the lacquer and, after drying, peened at increasing time increments until no more lacquer is visible on the surface under black (UV) light. Because of the high visibility of the fluorescent dies, 100% coverage determination can be quickly achieved. The 100% coverage time can then be factored to suit those circumstances as outlined in 1-7 above. There has been some concern as to the linearity of PEENSCAN on very hard parts, where it is most useful. A simple test using Almen Strips hardened to 58-60 HRC determined that PEENSCAN removal does indeed coincide with saturation and 100% coverage at this hardness level.

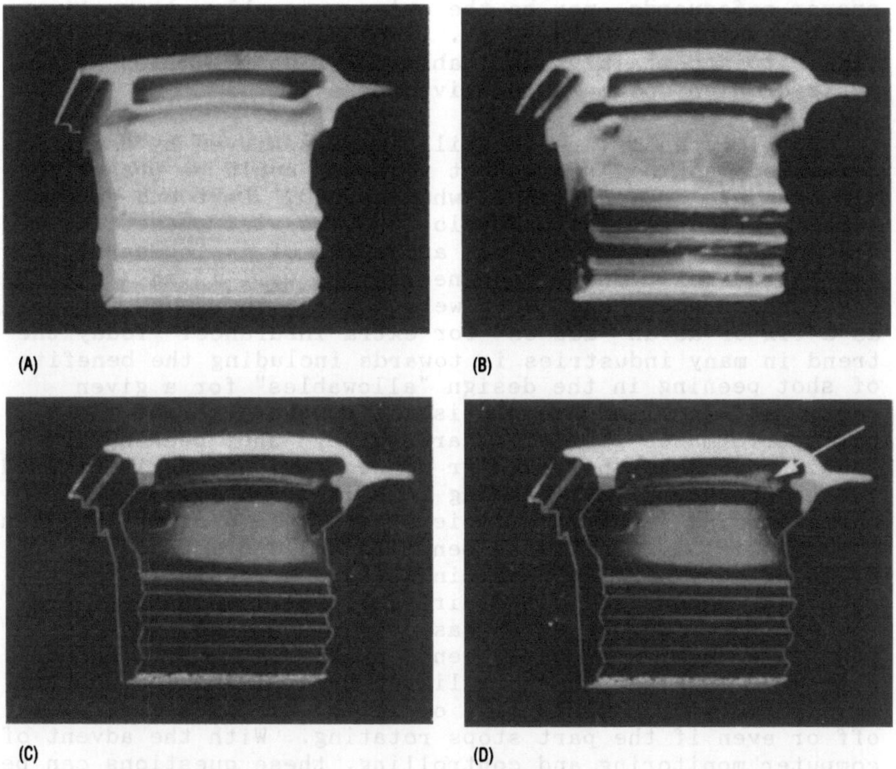

FIG. 7 THE PEENSCAN SYSTEM. (A) COATED UNPEENED. (B) PEENED 15 SECONDS, PARTIAL COVERAGE. (C) PEENED 60 SECONDS, FULL COVERAGE. (D) PEENED 60 SECONDS, IMPROPER NOZZLE ANGLE IN CAVITY (AS SHOWN)

Coverage is the most critical parameter in shot peening: even more so than intensity. Yet deciding how much coverage is correct and then determining that the part has received the correct coverage require considerable experience and know-how. If there is to be an error, it should be on the side of too much coverage rather than too little. Insufficient coverage can create surface tensile stresses which, particularly in the case of stress corrosion cracking, can be disastrous. On the other hand, for high production parts, unnecessarily long exposure times will be cost prohibitive.

REPEATABILITY

Once the intensity and the coverage have been determined, the next concern has to be repeatability. The surface condition, intensity and the coverage must be consistent over the surface of the part and consistent from part to part in a production lot and from one lot to the next; even on different machines. The quality of the shot peening is impossible to measure non-destructively on the part after it is peened. Obviously, peening using a hand held nozzle will not produce consistent results. While manual peening, with proper safeguards, may be the only reasonable alternative for the weldments of a 60 ft. (20m) diameter ammonia storage tank, it is certainly unsuitable for high precision aircraft parts or high volume automotive parts.

Until recently, repeatability was achieved by using automated machines where shot velocity could be preset by air pressure or centrifugal wheel speed. Part and nozzle location, translation and velocity were also preset, as were distance from part to nozzle and angle of impingement. Automated shot peening machines served well in an environment where engineers were content to use shot peening as a fix or as an "add on" for extra insurance. Today the trend in many industries is towards including the benefits of shot peening in the design "allowables" for a given part. An excellent example is the connecting rods for a high performance passenger car engine. Shot peening was "designed in" so that lighter connecting rods could be used, thereby reducing the rotating mass in the engine with corresponding gains in efficiency and fuel consumption. In a commercial aircraft, the benefits are enormous for allowing just a 2% increase in fatigue strength in the machined components of the airframe, the landing gear and the engines. The question has always remained, though: "Was this particular part peened correctly?" Even in an automated machine, there is little hope of knowing whether a nozzle plugs up momentarily, or if the air pressure falls off or even if the part stops rotating. With the advent of computer monitoring and controlling, these questions can be addressed for each individual part.

Computer monitored shot peening machines "read" all these parameters and more, such as shot flow, every second. If any of the parameters goes out of limit, the machine will shut down and warn the operator of the problem. It can only be restarted when the problem is corrected. In addition, the computer can control or vary nozzles on part translation or rotation. Now, hard data can accompany the part to satisfy the designer's need to document the repeatability: on large surfaces, inside small holes, on hard parts or soft, simple or complex.

CONCLUSION

Computers have brought to shot peening the essential factor that has been missing: the positive assurance that the parts are peened correctly. With this assurance, the user industries are moving to take greater advantage of shot peening as they balance lighter designs against longer service lives.

Note: Metric sizes are for reference only and are approximate.

REFERENCES:
1. Complimentary copy of MIL-S-13165 Revision C available from: Metal Improvement Company, 10 Forest Avenue, Paramus, NJ 07652, Tele.: 201/843-7800, Fax number: 201/843-3460.

2. Available from: SAE, 400 Commonwealth Drive, Warrendale, PA 15096, Tele.: 412/776-4841.

3. Navair Review Team Report Naval Air Systems Command.

4. "Shot Peening Stress Profiles"; H. O. Fuchs, Editor. Complimentary copy available from Metal Improvement Company (see 1. above).

5. PEENSCAN is a process licensed by Metal Improvement Company (see 1. above)

Effect of Pre- and Post-Processing upon Shot-Peening Residual Stresses

T. Rasul

Dr A. Q. Khan Research Laboratories
Rawalpindi
Pakistan

ABSTRACT

The process of shot-peening finds extended use in industry to combat fatigue fracture in engineering components. The introduction of compressive residual stresses increases the fatigue life considerably and forms an integral part of the design process. Although the benefits of the shot-peening process are well established, all investigations overlook the history of the component being treated. This increases the scatter in the results and makes them unreliable. Also, although researchers are aware that post-processes performed on treated components reduce and redistribute the peening compressive residual stresses, no experimental confirmation exists.

The following work is the result of an extended investigation into the effect of pre- and post-processing upon shot-peening residual stresses.

Specimens were subjected to controlled turning to produce four different surfaces roughnesses, and the residual stress fields for each were established by a modified centre hole drilling technique. The different specimens were then exposed to the same shot-peening parameters and the resulting residual stress distribution again investigated. A comparison of the two revealed that rough turned surfaces had lower compressive residual stresses after shot-peening and the reason was the initially higher tensile residual stress associated with the rougher surfaces. It was also shown that the maximum amount of residual stress was induced when the surface roughness before and after shot-peening were similar.

For the post-processing effect controlled grinding was studied and the results indicated a straight line reduction in the maximum compressive residual stress as well as a redistribution of the stress field. It was also shown that even after a layer removal equivalent to 50% of the arc-height, the compressive stress still remained compressive. The material investigated was an aircraft alloy 817M40.

INTRODUCTION

The presence of residual stresses is difficult if not impossible to avoid, no matter what the method of manufacture may have been. There are many studies which show that surface tensile residual stresses are detrimental to the fatigue life of engineering components while surface compressive residual stresses are beneficial, e.g., see (1-2) and the references mentioned therein. The shot-peening process in which small spherical steel shots / glass beads are bombarded under air pressure onto the surface of the component, is the most widely used industrial method of imparting beneficial compressive residual stresses in the surface and near surface layers of a prospective component. A typical residual stress distribution of a component before and after shot-peening is given in Figure 1.

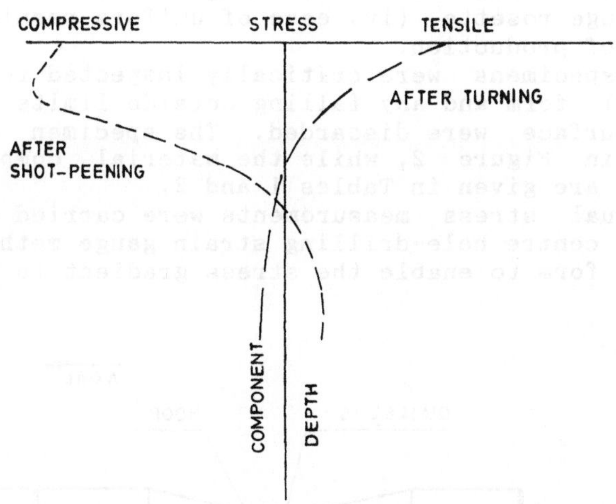

Figure 1. Effect of Shot-Peening on Typical Turning Residual Stresses

Although many studies have shown both theoretically and experimentally that the shot-peening process helps to convert the detrimental surface tensile residual stresses into beneficial compressive residual stresses, all such work has ignored the history of the component that is to be treated, e.g.,see reference (3) and the references mentioned therein. Since the peening residual stresses are superimposed on whatever residual stress field the component has, the stress distribution before shot-peening plays as important a role as the shot-peening residual stresses. Thus to obtain meaningful and reliable results attention must be focused on the effect of different residual stress fields prior to peening, being subjected to the same shot-peening parameters. The results presented in this study were obtained during a detailed analysis carried out by the author (4).

MATERIALS AND METHODS

The material investigated steel 817M40 is a typical aircraft alloy and finds extensive use in industry. Components made from this alloy are also frequently shot-peened to enhance the fatigue fracture life. For steel, both the pre- and post-processing aspects were looked into. The material was selected on the basis of (i) its being an aerospace application material, (ii) its behaviour not having been previously studied, and (iii) its availability.

A close watch was kept on the specimens which for pre- and post-processing effects were produced on a centre-lathe. The factors considered in the selection of the specimen geometry were (i) ease of residual stress measurements, (ii) freedom from distortion on loading, (iii) to permit use of strain gauge rosette, (iv) ease of uniform peening, and (v) economics of production.

All specimens were critically inspected regarding the geometrical form and any falling outside limits or having a damaged surface were discarded. The specimen details are provided in Figure 2, while the material composition and properties are given in Tables 1 and 2.

Residual stress measurements were carried out by the well known centre hole-drilling strain gauge method albeit in a modified form to enable the stress gradient to be measured.

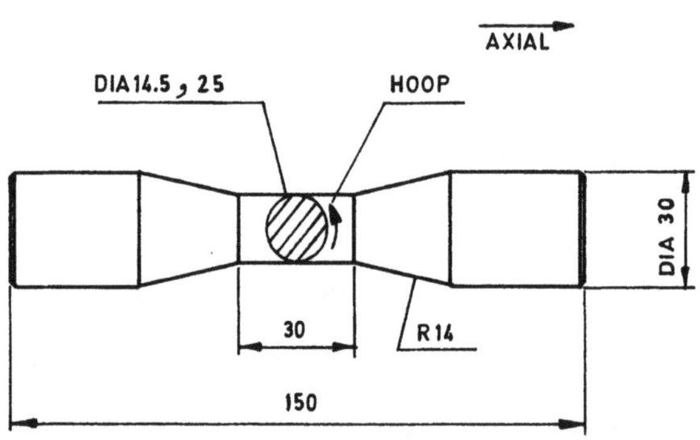

Figure 2. Specimen Dimensions

TABLE 1
Chemical Composition of Test Material

Element	Maximum/Minium
Carbon	0.45/0.35
Silicon	0.35/0.10
Manganese	0.70/0.45
Nickle	1.80/1/30
Chromium	1.40/0.90
Molybdenum	0.35/0.20
Sulphur	0.05/----
Phosphorus	0.05/----

TABLE 2
Mechanical Properties of Test Material

Property	Steel 817M40
Young's Modulus	207 GPa
Ultimate Tensile Strength	1 GPa
Proof Strength 0.1 %	770 MPa
Poissons Ratio	0.28
Hardness	300 Hv

The hole drilling method is well documented e.g. see references (5-7), and the details of the modification are given in reference (4).

RESULTS

Pre-Processing

Steel specimens were produced to four different surface roughnesses varying from a fine 1.2 µm Ra to a rough 6.6 µm Ra. Residual stress measurements were carried out on these and the results for the two extreme roughnesses are presented in Figure 3.

The results have been normalized in both X and the Y directions by dividing the residual stresses with the material yield strength and the hole depth with the specimen thickness. Each point on the graph represents a residual stress measurement point some distance from the surface. Due to time and cost limitations only three specimens for each type were investigated for residual stresses, and only if the three readings were more than 10% apart a further confirming measurement was performed. Besides a series of experiments were also carried out to throw light on the scatter in residual stresses within the same specimen, as well as between two specimens subjected to the same stress producing process e.g., two specimens similary produced and subjected to the same shot-peening parameters,(see chapter 6 of (4)).

As shown in Figure 3 the residual stress distribution for the fine specimen is significantly different to that for the rough specimen, with the rougher specimen having greater

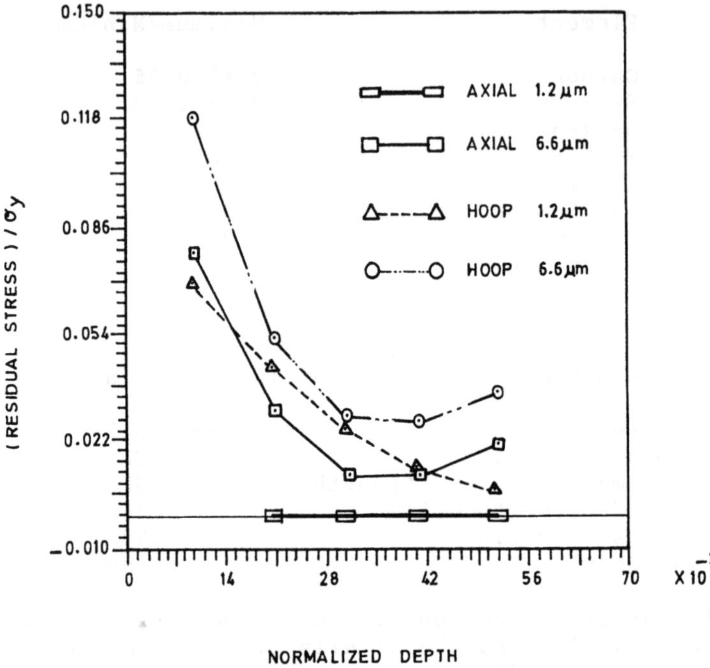

Figure 3. Turning Residual Stresses for Extreme Roughnesses for Steel

tensile residual stresses in general. Figure 4 Shows these stresses together with those for the two specimens of 3.2 and 4.5 µm Ra.

The stresses are the surface stresses normalized by the yield stress and are plotted with the X-axis representing the surface roughness of the specimen. A trend of increasing tensile surface residual stress with increasing surface roughness can be seen in the plot.

The residual stress distribution of these specimens after shot-peening is given in Figure 5. Only the axial stress distribution is given here, the hoop residual stress distribution being omitted; due to shot-peening the residual stresses in both directions become more or less similar in distribution with only slight variations in quantity. It is obvious from the graph that shot-peening has induced compressive residual stresses in the surface and the near surface layers of all four surface roughnesses investigated. The magnitude of the residual stresses is however lower in the very rough surfaces as compared to that for the fine surface.

Post-Peening

For the effect of post-processing controlled grinding

was selected and the residual stresses were measured in a

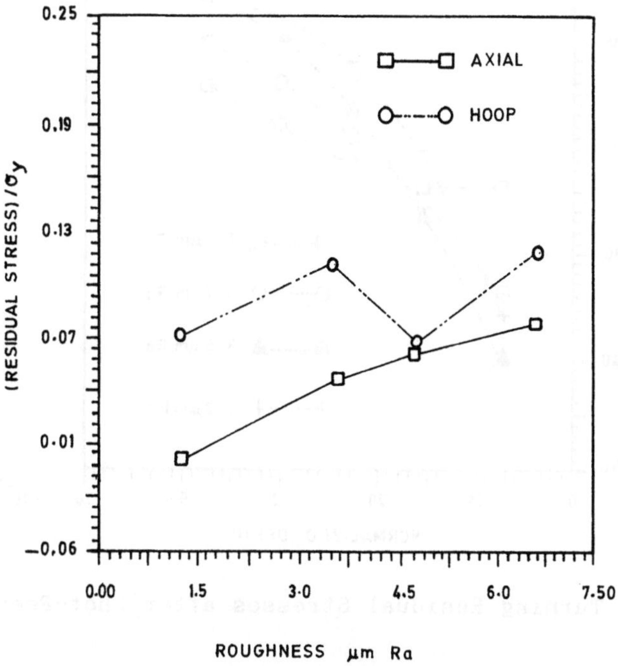

Figure 4. Turning Residual Stresses for Steel

shot-peened component. Residual stresses were then again measured in a component shot-peened and then post-processed to different surface finish by grinding. These results are given in Figure 6.

The stress distribution for lapping is considerably different to that for grinding and could be an experimental error as the result for lapping was available for a single specimen. However the process of lapping as performed could have possibly caused excessive heat to be trapped and hence caused the distribution to vary significantly from that of grinding. This aspect needs further investigation before meaningful conclusions can be drawn. Although it was established that the scatter in residual stress results from specimen to specimen was of the order of 5%, residual stresses were measured on the same one specimen to establish the variation of residual stresses with arc-height. This specimen was shot-peened and the residual stresses measured, it was then ground finished and the stresses measured again. Greater and greater depth was successively removed from the specimen and the stresses measured at each stage. The results are represented in Figure 7.

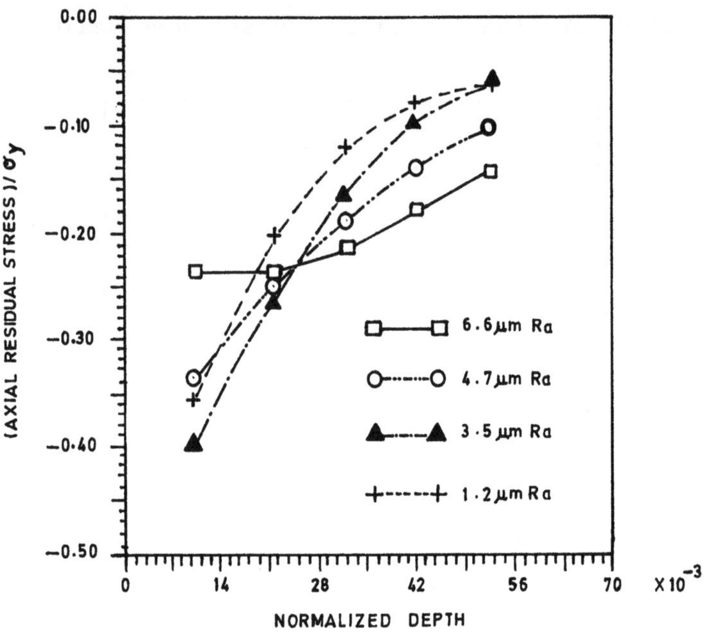

Figure 5. Turning Residual Stresses after Shot-Peening

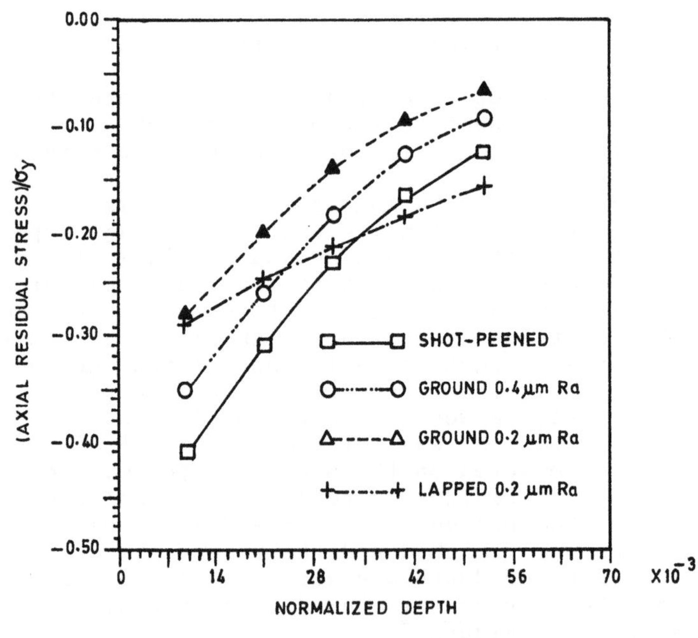

Figure 6. Axial Residual Stress Distribution for Post-Processing (Comparison of different specimen)

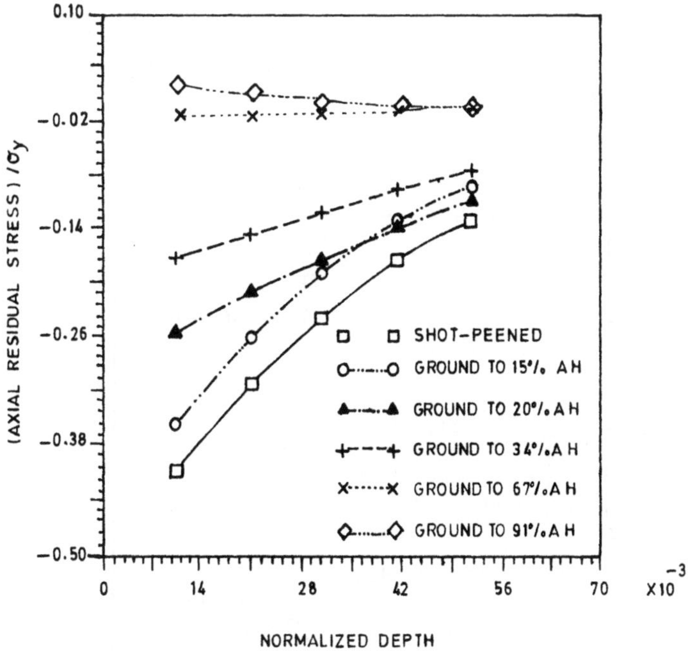

Figure 7. Residual Stress Distribution Versus Arc-Height Removed (Same specimen)

DISCUSSION

Pre-Processing

From Figure 5 it can be clearly seen that despite the initial rather large variation in surface roughness i.e., from 1.2 µm Ra to 6.6 µm Ra, shot-peening of the same parameters has induced compressive residual stresses in all specimens. This, on the one hand shows that it is possible to induce compressive residual stresses in surfaces which are in excess of 6 µm Ra rough and on the other that considerable latitude with regard to surface finish is available in producing surfaces that need to be shot-peened. The near surface compressive residual stresses in the 6.6 µm Ra surface is also not significantly lower (0.27 times σ_y as opposed to 0.43 times σ_y) to that of the 1.2 µm Ra surface. As a matter of fact a surface as rough as 6.6 µm Ra would never be shot-peened with an intensity of 12 A, and the surface stresses of 0.27 σ_y can be increased by shot-peening a rough surface with a higher intensity. Since the cost of producing rough surfaces increases exponentially with a reduction in Ra surface roughness, it is always advantageous to produce rougher surfaces rather than fine if the component is to be shot-peened before use. When the turning residual stresses are subtracted from the shot-peening residual stresses and plotted against the surface roughness, the plot of Figure 8 results.

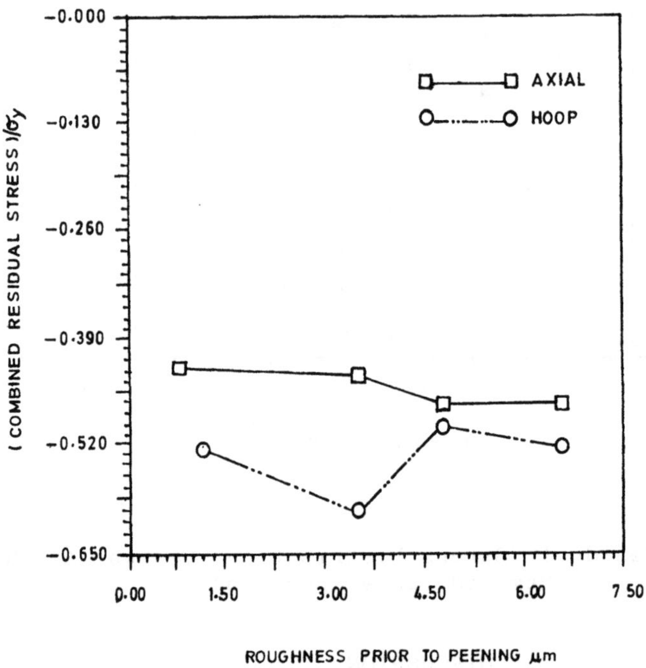

Figure 8. Combined Surface Residual Stresses

The more or less horizontal nature of this plot (it must be emphasized here that no smoothening action has been applied to the data points on the plot) signifies that the lower amount of compressive residual stresses in the rougher surfaces is a consequence of the higher initial tensile residual stresses. Thus, it is important to specify the limit of residual stresses in a component that is to be shot-peened, if repeatable results after shot-peening are essential.

The Ra roughness values before and after shot-peening for a number of different specimens are plotted in Figure 9.

It shows that although the initial roughness of the turned surfaces varied considerably, it was not so with the shot-peened surfaces. It also shows that while fine surfaces are made rougher, rough surfaces are made smoother, and that the effect of the initial turned surface, is in general, carried through the process of peening. Since the turned

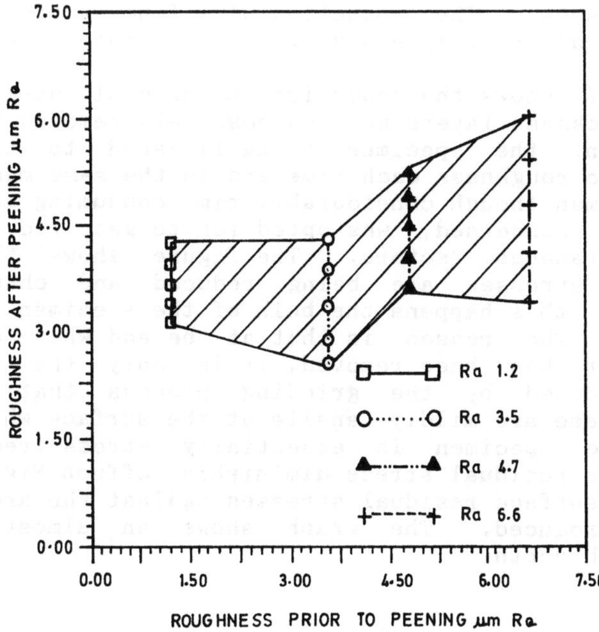

Figure 9. Scatter Band for Roughness after Shot-peening

surfaces of 1.2 µm Ra to 6.6 µm Ra are within 2.5 µm to 6 µm Ra after peening, components for general shot-peening practice need not be produced outside this scatter band.

Post-Processing

As long as the post-processing method is able to produce an acceptable surface finish, and does not induce tensile residual stresses due to the finishing operation, any available technique may be used. Fine turning has been shown to produce near zero stresses and could be used as an economical post-processing method, but the increase in surface hardness due to shot-peening must be kept in mind. Grinding boring and lapping are all finishing processes, and by the very nature of their being finishing processes, the material removal rate is small and hence the influence on the residual stress field is also small.

The reduction in residual stresses due to grinding is because of two reasons; the induction of tensile residual stresses due to grinding and the removal of the shot-peened compressed layer. However it is the effect of the latter which is dominant.

Figure 6 shows the effect of grinding after shot-peening; a reduction in the compressive stress magnitude. This is straightforward as the plastically deformed layer which is responsible for the compressive stresses is being removed. Thus the compressive stresses are reduced to less compressive or in extreme cases even to tensile stresses. The production of a fine surface means greater loss of the compressive residual stresses induced by shot-peening.

Figure 7 shows the reduction in residual stresses as different thickness layers were successively removed from the same specimen, the specimen being finished to the same characteristic roughness each time and in the same manner. A single specimen though considerably time consuming as far as handling was concerned, was opted for to get rid of all possible extraneous factors. The figure shows that the compressive stresses are being reduced and changed to tensile. As this happens the bulk of the specimen becomes stress-free. The reason is that at the end when all the plastic layer has been removed, it is only the residual stresses induced by the grinding process that become dominant. These are mildly tensile at the surface whilst the rest of the specimen is essentially stress-free. To visualize the residual stress diminishing effect Figure 10, showing the surface residual stresses against the arc-height has been produced. The graph shows an almost linear variation with depth.

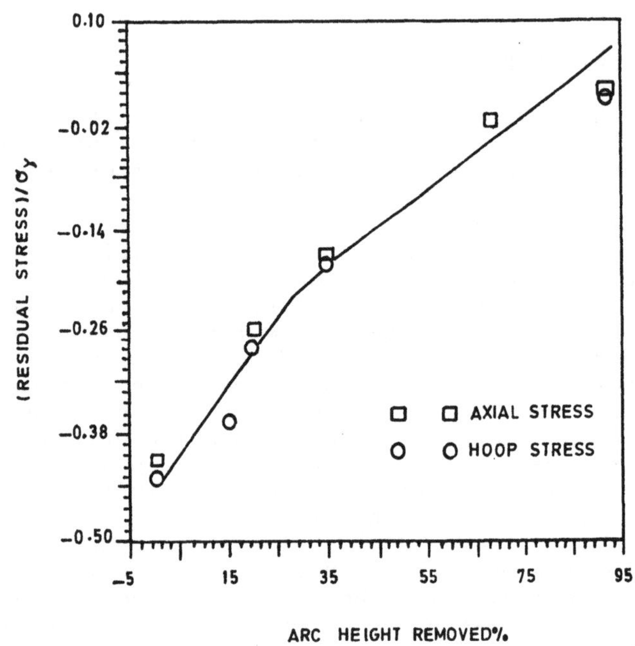

Figure 10. Surface Residual Stresses Versus Arc-Height Removed (Same Specimen)

This variation can however be better represented by two straight lines; the first being of a greater gradient than the second. This is due to the fact that the initial layers have higher stresses, and the removal of these affects the overall distribution more than the removal of layers some distance from the surface.

It is worth noting that even after 50% of the arc-height has been removed, the stresses have not changed to tensile, although they are almost zero. Thus considerable latitude is available to ensure that the rough surface is converted to a smooth enough surface, proper for the use in mind.

CONCLUSIONS

The conclusions presented here relate to controlled turning which was used as a pre-process and controlled grinding which was used as a post-process for this study.

Pre-Processing

By keeping all factors constant and varying the feed rate only, surfaces having different roughness were produced for the examination. These surfaces were subsequently shot-peened and the compressive residual stresses thus induced, measured. All components had compressive residual stresses; however those that had higher initial tensile residual stresses to start with, had lower compressive stresses after shot-peening. It is indicated that the cause of lower compressive residual stresses in rough specimens is the initial higher tensile stress.

Although one would neither use critical surfaces as rough as 6 μm Ra nor shot-peen such surfaces with an intensity of 12 A, it was indicated that shot-peening induced considerable stresses in the roughest surface examined. This result is of considerable importance as the cost of producing components increases dramatically with lower Ra values. Thus, general use components could be produced towards the higher roughness limits studied here, and shot-peened with intensities in excess of 12 A, quite economically.

Also since the shot-peening process reduces the roughness of surfaces in excess of 3.5 μm Ra, and vice versa, there is no advantage in producing surfaces better than 3.5 μm Ra unless it is a consequence of the production method used. This, again, is important as the results have indicated that the maximum lift towards compression in residual stresses, is obtained when the surfaces before and after shot-peening are equally rough.

Post-Processing

The process of controlled grinding was used as a post-processing operation and shot-peened specimens had various thickness layers removed. The resulting residual stress field was measured, and it was clearly indicated that any form of post-surface treatment resulted in a reduction in the compressive stresses as well as a redistribution of the residual stress field. This reduction in stresses was rapid during the initial layers and reduced with deeper layers. The reduction of residual stresses with the arc-height removed can be represented by two straight lines of different gradients; the first having a sharper gradient than the second. It was also indicated that even after 50% of the arc-height had been removed the stresses were still compressive.

Crossover Point In Shot-Peened Specimens

When measurements were carried out on shot-peened specimens, the residual stresses were found to be compressive up to a depth of 0.75 mm. Since such stresses are expected to change from compressive to tensile in the first 1 mm of depth, the possibility of a measurement error was suspected and considered.

An experiment was performed in which residual stresses in a shot-peened Almen strip (6 C) were measured by the same method. The results clearly indicated, that the measurement technique, as being used, was capable of picking up the crossover point, as accurately as any other technique, if not better. These results were also compared with those measured by electro-etching, and the depth of the crossover point agreed to within 0.15 mm while the maximum residual stress distribution for the strip is given in Figure 11.

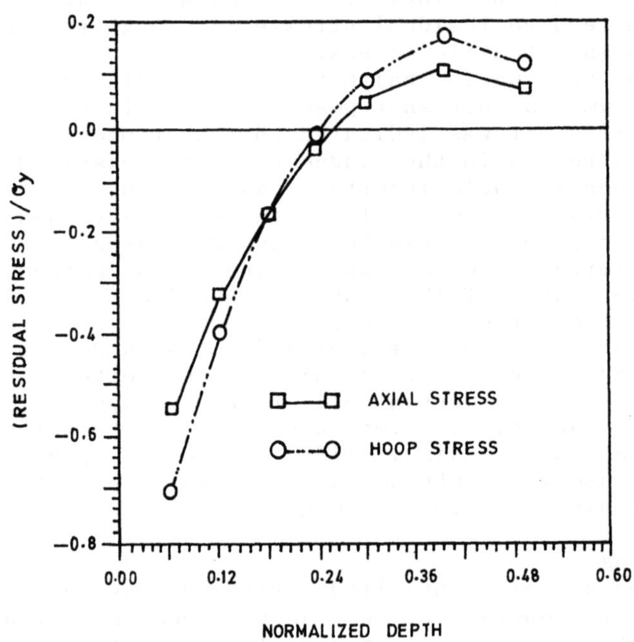

Figure 11. Shot-Peening Residual Stresses in an Almen Strip

It must, however be noted, that the presence of a crossover point in the Almen strip, does not necessarily mean, that there is not one within the range of the hole drilled in the actual specimen. Whilst the depth of the compressed layer is of the same order as compared to the depth of tensile stresses in the 2.4 mm thick Almen strip, it is not so in the 14.5 mm specimens. For the test specimens the compressive stress layer is a much smaller percentage of the tensile stress layer. The hole-drilling technique is primarily an averaging technique, and despite the modification discussed herein, remains an averaging technique. Thus, in very thick specimens it is not possible to accurately obtain the crossover point.

Accuracy Repeatability and Overall Error.

Any set of experiments however well planned and executed, shall not produce results that can be relied upon unless a few words are mentioned about the accuracy and repeatability of the setup. A number of different experiments were performed to check the validity of the results obtained. These pertained to four distinct parts of the study, i.e., mounting of strain gauges, location of specimens in the experimental jig, measurement of hole diameter and hole depth, and the measurement of residual stresses, (for details see chapter 4 of (4)).

During the extensive residual stress measurements using the modified hole-drilling method a careful account of the various errors (as above) involved was made. It is logical that these errors are not directly additive and the overall error cannot be a simple sum of these. Besides, these errors are the maximum observed, and the average values are considerably below those stated. When these maximum errors are added a total overall error of 8.6 percent maximum is obtained. The comparison of the modified hole-drilling technique with other popular methods has also established that the results agree to within 10 percent.

REFERENCES

1. Waterhouse, R.B., Noble, B. and Leadbeater, G., The Effect of Shot-Peening on the Fretting Fatigue Strength of an Age Hardened Aluminium Alloy (2014 A) and an Austentic Stainless Steel (En 58 A), Jnl of Mech Working Tech, Volume 8, Number 2+3, August 1983, pp 147-153

2. Nikulari, A., Influence of Residual Stresses Introduced by Shot-Peening upon the Fatigue Life of Materials, Exptl Tech, March 1983, pp 21-25.

3. HAMMOND,D.W., Fatigue Fracture in the Presence of Shot-Peening Residual Stresses: Initiation, Propagation and Relaxation Phenomena, PhD Thesis, Cranfield Institute of Technology, Bedford, England, 1988.

4. Rasul, T., Effect of Pre- and Post-Processing Upon Shot-Peening Residual Stresses, PhD Thesis, Cranfield Institute of Technology, Bedford, England, 1989.

5. Mathar, J., Determination of Initial Stresses by Measuring the Deformation Around Drilled Holes, Trans ASME, Volume 56, Number 4, 1934, pp 249-254.

6. Rendler, I.J. and Vigness, I., Hole-Drilling Strain Gauge Method of Measuring Residual Stresses, Exptl Mech, Volume 6, Number 12, December 1966, pp 577-586.

7. Flaman, M.T., Brief Investigation of Induced Drilling Stresses in the Centre Hole Method of Residual Stress Measurement, Proc. 4th Int. Congress on Exptl Mech, Boston, May 1980, pp 304-308.

Notation

µm	Micrometer
Ra	Average Roughness Parameter
Oy	Yield Stress
Hv	Vickers Hardness Number
MPa	Mega Pascals
GPa	Giga Pascals
Arc-Height	Deflection in thousandths of an inch of a standard metallic strip (Almen strip) when subjected to shot-peening. Used to compare shot-peening intensities (4).
12 A	Deflection of 0.012 inch in an 'A' strip.

Acknowledgement

The author wishes to thank Professor SA Meguid who is indirectly responsible for the work which appears here. The results for the paper have all been taken from the author's PhD thesis "Effect of Pre- and Post-Processing Upon Shot-Peening Residual Stresses", and that work was supervised by Professor Meguid. The author also expresses his thanks to Dr. A.Q. Khan of Dr. A.Q. Khan Research Laboratories, Pakistan, who very kindly allowed him to take part in this conference.

Surface Kinetics of Hydrogen Entry into Zirconium Films

C. A. Ward, M. B. Elmoselhi*, L. Pataki and B. D. Warr*

University of Toronto
Toronto
Ontario
Canada

**Ontario Hydro*
Toronto
Ontario
Canada

ABSTRACT

A study of D_2 interaction with two vacuum deposited *Zr* films has been conducted. The films were deposited on a glass substrate and were of different thicknesses. An examination of the kinetic uptake of D_2 by the thicker film indicates that surface kinetics are rate-limiting. An equilibrium adsorption isotherm for D_2 adsorbing on each film at 300°C was determined using a technique that neglected the absorption of D_2 during the measurements. The isotherms were found to be consistent, supporting the validity of adsorption isotherm established and the conclusion that surface kinetics are rate-limiting in the D_2 uptake by these thin *Zr* films.

INTRODUCTION

Zirconium alloys are often used in the nuclear industry in circumstances where they are exposed to an environment containing hydrogen. Ideally they would be protected in these circumstances from hydrogen absorption and embrittlement by an oxide layer [1]. However, the oxide can not always be relied on to satisfactorily perform this function because it can break or become porous and allow fresh *Zr* to be exposed to hydrogen. Often when this occurs, there are several rate processes that are set in motion and the final result depends on the relative rates at which each process proceeds.

The objective of the study that we report was to examine the rates of the processes involved in hydrogen absorption by *Zr* and to identify the one that limits the rate. When *Zr* is exposed to a gas phase of hydrogen, we assume there to be four sequential rate processes involved in the transfer of hydrogen to the bulk *Zr* phase: *1)* the transfer of hydrogen molecules from the gas phase to the interphase, *2)* their dissociation there to ad-atoms, *3)* the transfer ad-atoms from the interphase to the bulk *Zr* phase, and *4)* the diffusion of the atomically absorbed hydrogen further into the bulk *Zr*. To examine these rate processes, we experimentally examine the rate of uptake when a vacuum deposited *Zr* film is exposed to D_2 under controlled conditions [2] and use the statistical rate theory approach [3, 4] to analyze the results.

MATERIALS AND METHODS

The experimental apparatus used in this study is shown schematically in Fig. 1. This apparatus was used to deposit *Zr* films on the inside surface of a spherical glass (Pyrex®) vessel under ultrahigh vacuum conditions. The results described below were obtained from two separate films, each of different thicknesses. The basic procedure used for the depositions were basically the same. A sample of purified, thin (~0.1 *mm*) *Zr* was wrapped around a *W* electrode. Each end of the *W* was welded to *Ni* rods that were connected to an ultrahigh vacuum feed-through. The *Ni* rods and *Zr* wrapped *W* were enclosed in the spherical glass vessel as indicated. This glass was also connected to a gas dosing system.

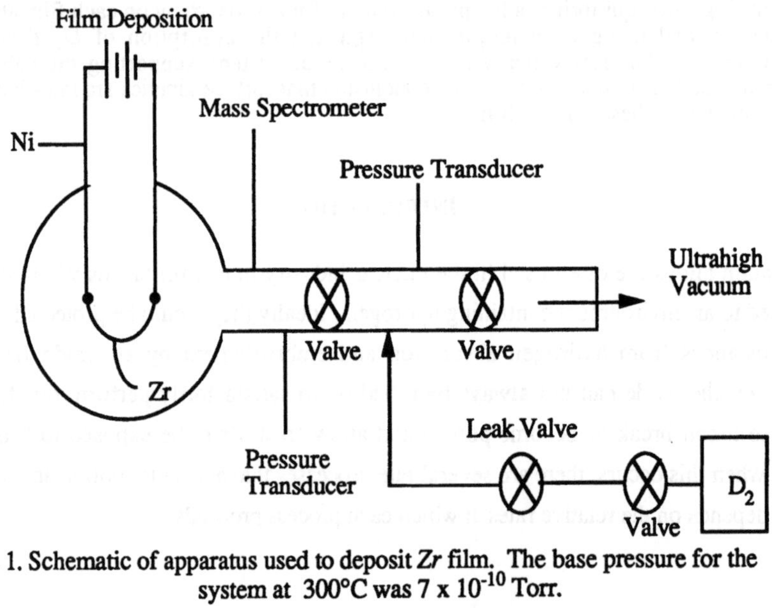

Figure 1. Schematic of apparatus used to deposit *Zr* film. The base pressure for the system at 300°C was 7 x 10^{-10} Torr.

With the spherical vessel immersed in liquid N_2 and evacuated to a pressure of $\sim 5 \times 10^{-10}$ Torr, an electrical current was passed through the *W* electrode, to bring the *Zr* to its melting point (1852°C). After a small portion of the *Zr* had melted, the current was stopped. In one case, a metallic grey film was deposited. It was visibly continuous and completely opaque. A quartz crystal microbalance that was present in the glass vessel (not shown in the schematic) indicated that the film thickness was approximately 33Å. A second film that was deposited in a separate glass vessel was thinner. After the deposition of the latter film, the glass vessel appeared smoky. Following the study with this latter film, the spherical glass vessel was broken and samples of the surface were examined with energy dispersive *x*-ray analysis at a total of 21 positions. The film was found to be continuous, its thickness uniform to within ± 10% and no impurities could be detected.

Different dosing volumes were used with the two films, but in both cases the amount of D_2 present was determined by measuring the pressure and temperature of the dosing volume when it was filled. When the system had been brought to an equilibrium condition with the dosing volume filled with D_2 and the *Zr* film deposited, D_2 was rapidly admitted to the spherical vessel. The pressure in the glass vessel was monitored and was observed to rise suddenly and then slowly diminish as the deuterium was taken up (adsorbed and absorbed) by the *Zr* film. The measured pressure during an experimental run is shown in Fig. 2. The amount of D_2 in and on the *Zr* film, $N^F(t)$, as a function of time was inferred from the amount of D_2 in the dosing system initially, N_o, and the measured value of the pressure in the glass vessel, P_G, using the relation:

$$N^F(t) = N_o - (V_G P_G) / \overline{R}T \tag{1}$$

where $\overline{R}$ is the gas constant, T the temperature and V_G is the volume of the dosing system and the glass vessel. These volumes were measured before the start of the experiments.

RESULTS

The inferred value of N^F(t) for the thicker *Zr* film (~33Å) that was obtained from the measured pressures shown in Fig. 2 is shown in Fig. 3. A series of experiments was performed in which the pressure was successively increased. The plateau values of N^F that were obtained at each pressure are shown in Fig. 4.

Also a second series of experiments of this type was performed with the thinner *Zr* film. The plateau values of N^F that were obtained from this latter series of experiments are also shown in Figure 4 as the darkened rectangles. As may be seen there, the experiments conducted with the thinner film were performed at much lower pressures than were the ones with the thicker film.

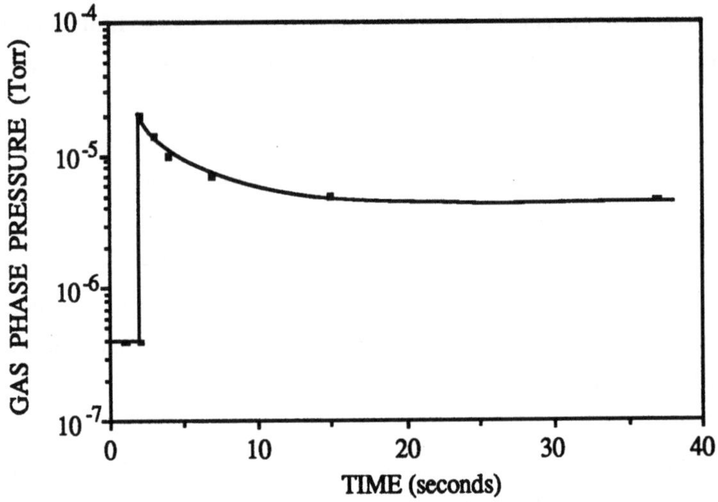

Figure 2. Measured pressure in the spherical vessel before and after (~ 33Å) *Zr* film was exposed to D_2.

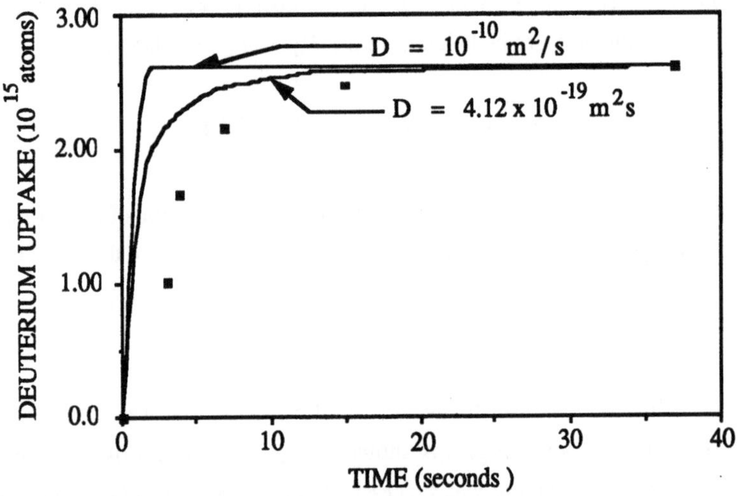

Figure 3. The measured uptake of deuterium by a (~33Å) *Zr* film (solid squares) at 300°C. The labelled solid lines are the predicted uptake, assuming the rate of uptake is controlled by the diffusion into the bulk *Zr*. Note the disagreement, no matter the value of the diffusion coefficient assumed.

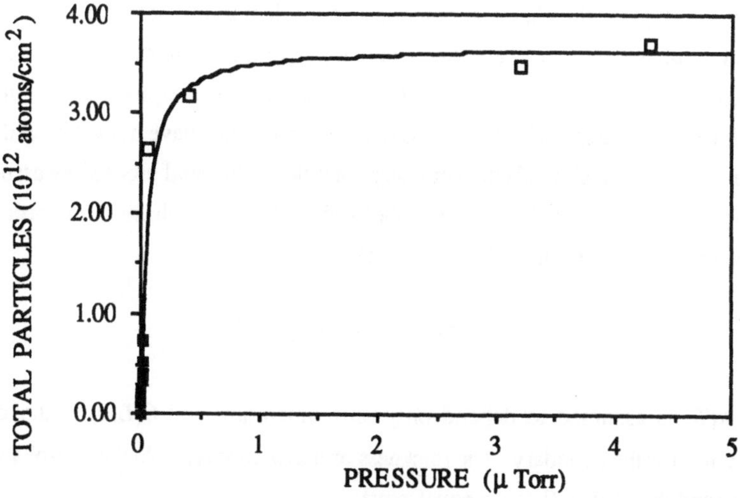

Figure 4. Equilibrium adsorption isotherm for D_2 adsorbing on *Zr* films at 300° C. The data were determined from measurements made with films of different thickness. The open squares are for the thicker film.

IDENTIFICATION OF THE RATE-LIMITING PROCESS IN THE UPTAKE OF D_2

When the plateau value of N^F is measured by the technique described above, one can not be sure how much of the D_2 is present in the bulk *Zr* (absorbed) and how much is adsorbed on the surface. To try to distinguish between these two amounts, we shall consider each of two hypotheses.

A. Infinitely Fast Surface Kinetics

As the first case to be considered, we suppose that the kinetic processes are sufficiently fast so that we may assume that at the boundary of the bulk zirconium substrate, the concentration of atomic deuterium is at all times equal to the equilibrium value corresponding to the pressure in the gas phase. From the theoretical point of view, this requires that throughout the uptake process, the following relations must exist between the chemical potentials.

$$\mu_M^G = \mu_M^I$$

$$\mu_M^I = 2\,\mu_A^I \qquad (2)$$

$$\mu_A^I = \mu_A^S$$

where μ_M^G, μ_M^I are the chemical potentials of the deuterium molecules in the gas phase and the interphase; μ_A^I, μ_A^S are the chemical potentials of the atomically adsorbed and absorbed deuterium. The latter would be evaluated at the interphase but on the bulk *Zr* side. However, the expressions for these chemical potentials have not yet been developed. It is thus necessary to simplify the boundary condition. We shall proceed by assuming that the concentration of deuterium on the bulk *Zr* side of the interphase is proportional to the pressure in the gas phase throughout the uptake:

$$C(L,t) = \kappa(T) P^G \qquad (3)$$

where $\kappa(T)$ is assumed to depend only on temperature and $C(L, t)$ is the deuterium concentration at the boundary. The thickness of the *Zr* film is *L*. At the interphase between the glass and the *Zr* film (i.e. at *x* equal zero)

$$(\partial C(x,t)/\partial x)_o = 0 \qquad (4)$$

and in the bulk *Zr*

$$(\partial C/\partial t) = D(\partial^2 C/\partial x^2 \qquad (5)$$

where *D* is the diffusion coefficient. Conservation of mass in the closed glass vessel requires

$$(V^G/\bar{R}T)\, dP^G/dt + AD\,(\partial C/\partial x)_L = 0 \qquad (6)$$

where *A* is the surface area of the *Zr* film.

$$C(x,0) = 0 \qquad o<x<L \qquad (7)$$

This problem is now of a form that has been solved in [5]. If the following definitions are introduced

$$\tau \equiv (Dt/L^2) \qquad (8)$$

$$R \equiv (x/L) \qquad (9)$$

and if M_∞ is the total amount of deuterium present in the film when the system finally comes to equilibrium, then a parameter α may be defined

$$\alpha = \left[1 - \left(M_\infty \overline{R}T / V^G P_o\right)\right] \left(M_\infty \overline{R}T / V^G P_o\right)^{-1} \tag{10}$$

This equation can be used to evaluate α, since all of the other parameters are known. The solution of the problem given in [5] is in terms of this parameter:

$$(M_\tau / M_\infty) = 1 - \sum_{i=1}^{\infty} \frac{2\alpha(1+\alpha)}{1 + \alpha + (\beta_i \alpha)^2} \exp(-\beta_i^2 \tau) \tag{11}$$

where $M(\tau)$ is the amount in the film at τ and β_i are the roots of

$$\tan \beta_i = -\beta_i \alpha \tag{12}$$

In [5], the first six roots of the above equation are given for different values of α.

For the experimental results shown in Fig. 2, the value of α was 0.24. Using this value of α, and different values of the diffusion coefficient, the calculated uptakes of deuterium by the film as a function of time are also shown in Fig. 3. We would emphasize that in making these calculations, it has been assumed that the surface kinetic processes are infinitely fast so that the rate-limiting process must be diffusion in bulk zirconium. The value of the diffusion coefficient that has been previously reported [6] at 300°C is approximately 1.0×10^{-10} m^2/s. Note that assuming this value of the diffusion coefficient results in a strong over prediction of the rate of uptake.

If the surface kinetic processes were not rate-limiting, equilibrium would have been achieved in less than 5 seconds, according to this prediction. However, the data shows a much slower rate of uptake. Thus one must conclude that the rate-limiting process does not lie in the bulk as assumed in the analysis. Rather, it must lie with the alternative possibility, i.e. with one of the surface kinetic processes.

To emphasize that this conclusion could not have been reached because an incorrect value of the diffusion coefficient was assumed, a second calculation has been performed with a value of the diffusion coefficient that was assumed to be nine orders of magnitude smaller than what is thought to be the correct value. Notice that even for this impossibly small value of the diffusion coefficient, there is still an over prediction of the rate of D_2 uptake by the Zr film.

B. Rate-Limiting Surface Kinetics

We now propose to examine the same uptake data, but we shall suppose the other limiting hypothesis to be valid. Namely, we suppose that during the time of the experiment (~ 40 s)

there is no significant transfer of deuterium into the bulk *Zr*, i.e. all of the deuterium taken up by *Zr* film goes onto the surface. There are then two sequential rate processes with which we must be concerned: *1)* the rate of transfer of molecules from the gas phase to the interphase, and *2)* the rate of dissociation of the adsorbed molecules.

To obtain the expression for the rate of adsorption, we shall adopt the statistical rate theory approach. This approach uses the concept of quantum mechanical transition probabilities to develop the expression for interfacial molecular transport [2] and for the rate of chemical reactions [3]. For the rate of molecular adsorption, J_{GI}, it gives [7]:

$$J_{GI} = K_{GI} \sinh\left[\left(\mu_M^G - \mu_M^I\right) / kT\right] \tag{13}$$

where K_{GI} is an equilibrium property of the system, namely the rate of molecular exchange between the gas phase and the interphase under equilibrium conditions [2, 3]. The same theoretical approach gives as the expression for the rate of dissociation in the interphase, J_D, the following expression [8, 9].

$$J_D = K_D \sinh\left[\left(\mu_M^I - 2\mu_A^I\right) / kT\right] \tag{14}$$

where K_D is another equilibrium property of the system, i.e. the equilibrium exchange rate between adsorbed atoms and molecules

$$D_2 \leftrightarrows 2D \tag{15}$$

Conservation of mass now allows us to write

$$\left(dN_M^I / dt\right) = J_{GI} - J_D \tag{16}$$

and

$$\left(dN_A^I / dt\right) = 2J_D \tag{17}$$

From these latter two equations, one finds

$$d\left(2N_M^I + N_A^I\right) / dt = 2J_{GI} \tag{18}$$

If the total number of particles (i.e. twice the number of molecules, plus the number of atoms) in the interphase is denoted as N^I then, after inserting Eq. (13) into this latter relation, separating variables and integrating, one finds:

$$\int_{N^I(o)}^{N^I(\tau)} \frac{dN^I}{\sinh\left[(\mu_M^G - \mu_M^I)/kT\right]} = 2t K_{GI} \tag{19}$$

where $N^I(o)$ is the total number of particles at the initial instant and $N^I(t)$ is the number present at any time t.

We shall approximate the gas phase as an ideal gas. The expression for the chemical potential of this phase is then

$$\mu_M^G = kT \ln\left(P^G \phi\right) \tag{20}$$

where ϕ is a function only of temperature. The pressure in the gas phase can be expressed in terms of N^I. From Eq. (1), we may now write

$$P^G(t) = P^G(o) - (N^I(t) - N_o^I)(kT / V^G) \tag{21}$$

When this latter expression is substituted into Eq. (20), the expression is obtained for the chemical potential of the molecules in the gas phase in terms of $N^I(t)$.

In order to perform the integral indicated in Eq. (19), it is also necessary to obtain the expression for the chemical potential of the molecules in the interphase. Little fundamental information is available about the nature of this surface [10]. Thus we shall proceed by the method introduced in [11]. The first step in this method is to recognize that the isotherm, such as that shown in Fig. 4, defines a relation between N^I and the pressure in the gas phase, P^G. Secondly, this relation is an equilibrium relation and it arises from the fact that the chemical potentials are equal at each point on the isotherm.

$$\mu_M^G(P_e^G) = \mu_M^I(N_e^I) = 2\mu_A^I(N_e^I) \tag{22}$$

where the subscript e indicates that a property is defined from this isotherm relation. If we now consider the general nonequilibrium circumstance, we may use this relation to define the chemical potential of the adsorbed molecules in the nonequilibrium circumstance, $\mu_M^I(N^I)$, in terms of a fictitious pressure P_e^G that would exist in the gas phase if there were N^I particles on the surface and the gas phase were in equilibrium with the interphase with this number present; thus from Eq. (20), one finds

$$\mu_M^I (N^I) = kT \ln \left[P_e \, \phi \, (T)\right] \tag{23}$$

and after making use of Eq. (20), we may write Eq. (19) in the form

$$2t\, K_{GI} = \int_{N^I(o)}^{N^I} \frac{dN^I}{\left[P^G / P_e^G) - (P_e^G / P^G)\right]} \tag{24}$$

Equation (24) must be solved numerically and for this purpose, it aids to have an empirical expression for P_e^G in terms of N_e^I. We find that the following relation gives a reasonable fit to the measured isotherm data shown in Fig. 4.

$$P_e^G = A\, N_e^L / \, (B - N_e^I) \tag{25}$$

if A and B are chosen to be 5.41×10^{-8} Torr and $1.84 \times 10^{12} / cm^2$ respectively. The calculated isotherm is also shown in Figure 4.

The equations necessary to predict the $N^I(t)$ numerically are now available. We proceed by predicting the time t at which N^I reaches a particular value. A value of N^I is chosen and the integral in Eq. (24) is numerically integrated with the Simpson rule, using Eqs. (21) and (25). The value K_{GI} which gives the least square error in the fit of the data shown in Fig. 3 is found to be 6.9×10^8 molecules / $cm^2\, s$. The fit obtained with this value of K_{GI} is shown in Fig. 5.

In view of the agreement seen there, it is apparent that the second hypothesis leads to an adequate description of the data. Thus assuming the surface kinetics to be rate-limiting leads to a consistent description of the observed uptake of D_2 by the Zr film. This conclusion is supported by the consistency of the isotherm data obtained from the measured plateau uptake on the films of different thicknesses. This consistency may be seen from an examination of the results in Figure 4.

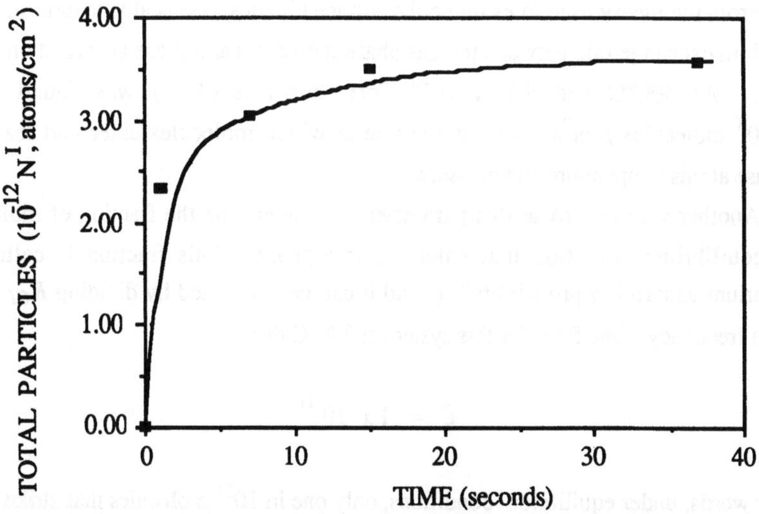

Figure 5. Calculated value of the total number of particles in the interphase using statistical rate theory and neglecting atomic transfer to the bulk *Zr* side of the interphase. The data are the same as that of Figure 3.

DISCUSSION

If one proceeds to examine the observed uptake of D_2 by a 33Å thick *Zr* film (Fig. 3) by assuming that the surface kinetics are infinitely fast so that equilibrium is instantaneously established at the boundary and therefore the rate of uptake is governed by diffusion into bulk *Zr*, then one is led to an over prediction of the rate of uptake. This over prediction is so great that the data can not be fitted by any reasonable value of the diffusion coefficient. By contrast, if one neglects the transfer of atoms from the interphase to the bulk *Zr* during the period of the uptake (~ 40 *s*) and uses a theory of kinetics that has been found to adequately describe the adsorption in other systems [7, 8, 11], then one finds that the theory gives a reasonable prediction of the rate of uptake of D_2 by the *Zr* film. Thus it would appear from the results obtained with this particular film that it is the surface kinetics that controls the uptake.

To examine this further, isotherm measurements were carried out on a much thinner *Zr* film. The isotherm results obtained with each film are shown in Fig. 4. Note that the results fall on one smooth curve even though the thickness of each film were approximately an order of magnitude different. The consistency of the isotherms also suggests that there was no significant transfer into the bulk of the thicker *Zr* film during the period of the measurements; thus supporting the conclusion that the surface kinetics are rate-controlling in the uptake of D_2.

From the theory used to examine the surface kinetics, one is able to determine the equilibrium exchange rate between the gas phase molecules and those adsorbed on the Zr surface. At 300°C and 4.7 x 10^{-6} Torr, this rate (K_{GI}) was found to be 6.9 x 10^8 molecules $/cm^2s$. This is the rate at which molecules enter and leave the interphase at this temperature and pressure.

Another way to look at this parameter is to determine the fraction of molecules under equilibrium condition that enter the interphase. This fraction is called the "equilibrium adsorption probability", ζ, and it can be calculated by dividing K_{GI} by the collision frequency. One finds for this system at 300°C that

$$\zeta = 1 x\ 10^{-11} \tag{26}$$

In other words, under equilibrium conditions, only one in 10^{11} molecules that strike the Zr surface become adsorbed there. This vålue may be contrasted with the "sticking probability" which would be exactly zero in the equilibrium circumstance, but the sticking probability also shows a strong dependence on surface coverage [12]. One of the advantages of the statistical rate theory approach [2, 3] is that the theory correctly predicts the dependence of the adsorption rate on surface coverage, and it is therefore unnecessary to introduce an empirical, coverage dependent function such as the sticking probability.

CONCLUSIONS

In treating the surface kinetics, the assumption was made that there was no transfer of deuterium from the interphase to the bulk zirconium. As seen above, this led to a consistent description of the observed kinetic uptake of D_2 by the 33Å Zr film. If deuterium atoms were transferred from the interphase to the bulk Zr side of the interphase in any significant number, the diffusion study indicates they would have rapidly diffused into the bulk. The fact that consistent isotherm data were obtained from the two films of different thicknesses using the experimental technique described above suggest that the bulk Zr of the thicker film played little role in the uptake during these exposures. Thus there could only have been an insignificant transfer of deuterium to the bulk Zr during these exposure periods. The concept that emerges from this analysis is that while the deuterium readily adsorbs into the interphase of a clean Zr surface as molecules or atoms, the rate of deuterium transfer from the interphase to the bulk Zr side of the interphase is the rate-limiting process in the absorption of D_2 by Zr. This concept is consistent with all of the data available at present for this system.

ACKNOWLEDGEMENT

This work was supported by Ontario Hydro and University Research Incentive Fund of Ontario.

REFERENCES

1. Smith, T., Kinetics and mechanism of hydrogen permeation of oxide films on zirconium. J. Nucl. Materials, 1966, **18**, 323–336.

2. Wedler, G., Papp, H. and Schroll, G. Adsorption of carbon monoxide on polycrystalline nickel films. Surface Science, 1974, **44**, 463–479.

3. Ward, C.A., Findlay, R.D. and Rizk, M., Statistical rate theory of interfacial transport. I. Theoretical development. J. Chem. Phys., 1982, **76**, 5599-5605.

4. Ward, C.A., Effect of concentration on the rate of chemical reactions. J. Chem. Phys., 1983, **79**, 5605-5615.

5. M. D. Mikhailov and M. N. Özisik, Unified Analysis and Solutions of Heat and Mass Diffusion, Wiley, 1984, p. 474.

6. Kearns, J.J. Diffusion coefficient of hydrogen in alpha zirconium, zircaloy–2 and zircaloy–4. J. Nucl. Materials, 1972, **43**, 330–338.

7. Ward, C.A. and Elmoselhi, M.B., Molecular adsorption at a well defined gas-solid interphase statistical rate theory approach. Surface Science, 1986, **176**, 457-475.

8. Findlay, R.D. and Ward, C.A., Statistical rate theory of interfacial transport. IV. Predicted rate of dissociative adsorption. J. Chem. Phys., 1982, **76**, 5624-5631.

9. Ward, C.A., Farahbakhsh, B. and Venter, R.D. Absorption Rate at the Hydrogen-Metal Interphase and its Relation to the Inferred Value of the Diffusion Coefficient. Z. Physikalische Chemie, 1986, **147**, 89-101.

10. Ward, C.A. and Elmoselhi, M.B., The coverage dependence of the heat of adsorption and the predicted wavelength of surface luminescence. Surface Science, 1988, **203**, 463-488.

11. Findlay, R.D. and Ward, C.A. Rate of gas adsorption on a nonideal solid surface: statistical rate theory approach. Submitted for publication.

12. Vacuum Physics and Technology, eds. G.L. Weissler and R.W. Carlson (Academic Press, Toronto, 1979, p. 29.

Abrasive Flow Machining for Automatic Surface Finishing and Capacitance Technology for In-Process Surface and Dimensional Metrology

L. J. Rhoades

Extrude Hone Corporation
Irwin
Pennsylvania
USA

ABSTRACT

Abrasive Flow Machining (AFM) deburrs, polishes and radiuses surfaces and edges by flowing an abrasive media over these areas. The process is particularly used on internal shapes which are difficult to process by other machining techniques. The machine controls the extrusion pressure and the flow volume and speed. Since abrasive action occurs only in areas where the flow is most restricted, tooling is often used to direct the media to the appropriate areas. A variety of finishing results can be achieved with different abrasive types, sizes, concentrations and media viscosities. Capacitance surface texture measurement uses the parallel plate theory to evaluate a three-dimensional area surface. The voids of air spaces on the surface become the dielectric of the formed capacitor. Capacitance surface texture measurement has found a wide range of applications in production, as well as in the laboratory.

INTRODUCTION

Abrasive Flow Machining (AFM) is used to deburr, polish or radius surfaces and edges by flowing a semisolid abrasive media over these areas. The process embraces a wide range of feasible applications: from critical aerospace and medical components to high production volumes of parts. AFM can reach even the most inaccessible areas, processing multiple holes, slots or edges in one operation. Advances in media formulation and tool design coupled with new capabilities in processing and automation have established the abrasive flow process as a way of satisfying tough manufacturing requirements economically and productively.

The AFM process can be used in a wide range of finishing operations. It can process many areas on a single workpiece or multiple parts simultaneously. Inaccessible areas and complex internal passages can be finished economically and productively. Automatic AFM systems can handle thousands of parts per day, greatly reducing labor costs by eliminating tedious hand work. By understanding and controlling the process parameters, AFM can be applied to an impressive range of finishing operations, providing uniform, repeatable, predictable results.

PROCESS DESCRIPTION

In abrasive flow machining two opposed cylinders extrude a semisolid abrasive media back and forth through passages formed by the workpiece and the tooling (Figure 1). The process is abrasive only where the flow is restricted: the extrusion area. The machining action compares to a grinding or lapping operation as the abrasive media gently and uniformly hones the surface or edges. Materials from soft aluminum to tough nickel alloys, ceramics and carbides can be successfully "micro-machined" with this process.

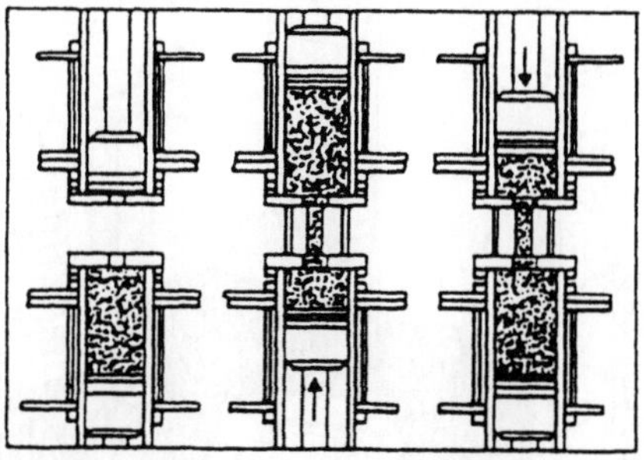

Figure 1. Abrasive Flow Machining Process

Abrasive flow machining can be applied to achieve a wide range of predictable, repeatable results. If the objective of AFM involves uniformly polishing the walls of the restricting passages, in die polishing for example, the media chosen should maintain a uniform flow rate as it travels through the passage (Figure 2). For deburring or radiusing the edges of a passage, a less viscous media that increases flow rate once it enters the passage causes the edges to be abraded more than the passage walls (Figure 3). The type of flow pattern that occurs depends on the machine settings, the media formulation, and the workpiece and tooling configuration.

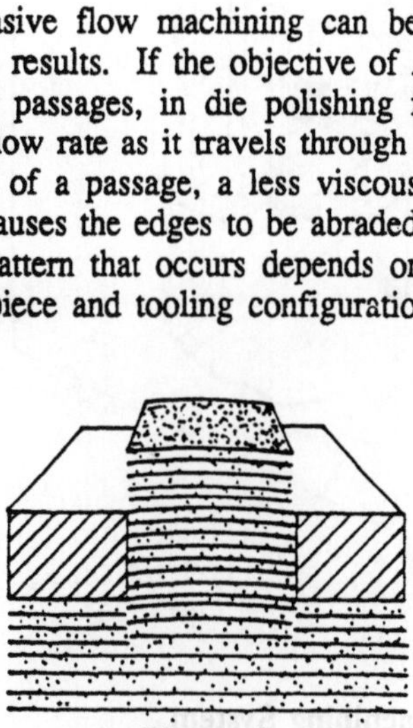

Figure 2.

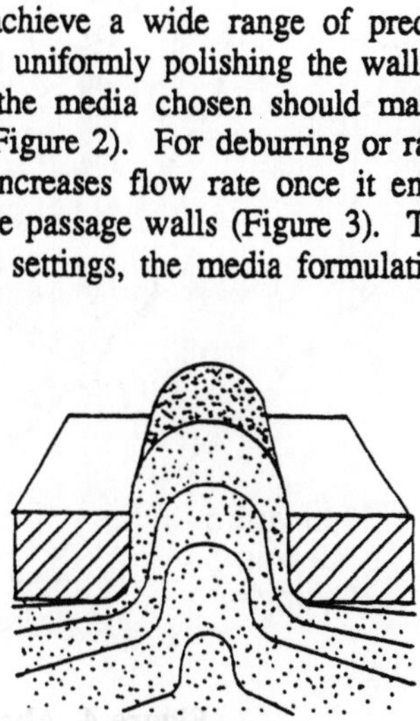

Figure 3.

The media flow pattern can be varied to achieve different finishing results.

PROCESS PARAMETERS

Machine

The abrasive flow machine, available in a variety of sizes, contains two vertically opposed media cylinders which hydraulically close to hold a part or fixture between them. By repeatedly extruding media from one cylinder to the other, an abrasive action is produced wherever the media enters a restrictive passage as it travels through or across the workpiece (Figure 4).

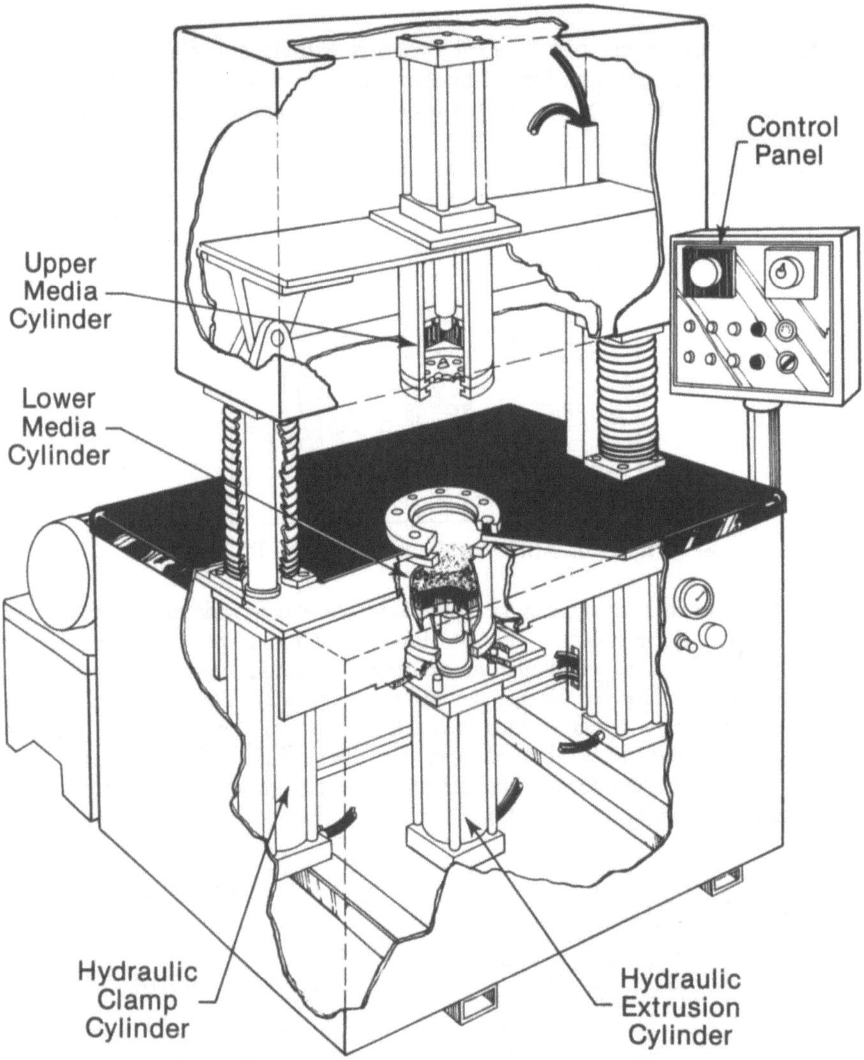

Figure 4. Abrasive Flow Machining System

The machine also controls the extrusion pressure. Machines are available with extrusion pressures that range from 7 to 220 bar (100 to 3200 psi) with flow rates exceeding 380 liters (100 gallons) per minute. The volume of flow depends on the displacement of each media cylinder stroke and the total number of stroke cycles used to complete the job. These variables are both preset at the machine.

Control systems may be added to monitor and control additional process parameters such as media temperature, viscosity, wear and flow speed. AFM systems designed for production applications often include part cleaning and unload/reload stations, media maintenance devices and cooling units. These automated systems can process thousands of parts per day with processing times typically ranging between one and three minutes for each pallet loaded with workpieces (Figure 5).

Figure 5. This automated AFM cell deburrs and polishes auotmotive armature shafts. The system processes 49 parts per fixture in less than 2 minutes and is equipped with automatic burr qualification, part loading, unloading and cleaning.

Tooling

The tooling holds the workpiece in position and directs the abrasive media to the appropriate areas. Many AFM applications require only simple fixturing: dies typically need no special tooling; the die passage itself provides the restriction for the flow path. For external edges or surfaces, tooling is used to restrict the flow between the outside of the part and the inside of the fixture. The tooling may also serve to restrict flow at areas where abrasion is desired or to block the flow through areas to remain unaffected. High production fixtures are designed to facilitate part loading, unloading and cleaning. Often mounted to indexing tables, these fixtures may hold multiple parts for processing in one operation (Figure 6).

Figure 6. Twelve diesel fuel injector nozzles in one fixture are processed simultaneously.

Media

The media is composed of a pliable semisolid carrier and a concentration of abrasive grains (Figure 7). The viscosity of the carrier and the abrasive grain size, type and concentration can be varied to achieve specific finishing results. Higher viscosity, nearly solid media is used for uniformly abrading the walls of large passages. Lower viscosity media is generally appropriate for radiusing edges and for processing small passages. When forced into a restrictive passage, the viscosity of the media temporarily rises, holding the abrasive grains rigidly in place. The media abrades the passages through which it flows only when in this viscous state. The viscosity returns to normal when the thickened portion of media exits the restrictive passage, producing little or no abrasion.

Media Type/ Viscosity	LV Low	LMV Low/Medium	MV Medium	HMV High/Medium	HV High
Passage size*					
Minimum	1/64" (0.4 mm)	1/32" (0.8 mm)	1/16" (1.6 mm)	1/8" (3.2 mm)	1/4" (6.4 mm)
Maximum	1/8" (3.2 mm)	1/4" (6.4 mm)	1/2" (12.8 mm)	1" (25.4 mm)	2" (50.8)

*Indicated passage sizes are widths or diameters, and assume that passage length is two times the width.

Figure 7. An important element of the AFM process is the media, a polymer carrier mixed with abrasives.

The abrasive grains are most commonly made of silicon carbide, although boron carbide, aluminum oxide and diamond may also be used. Particle sizes range from 0.005 mm to 1.5 mm (0.0002 to 0.060 in.). The better the starting finish, the smaller the grit size used for processing. The larger abrasives cut at a faster rate, while the smaller sizes provide finer finishes and accessibility to small holes. The depth of cut made by the abrasive grains at the surface depends on the extrusion pressure applied and on the stiffness of the media as well as on the size of the abrasive grains. Air or vacuum may be used to remove the media from accessed areas. Final traces can be extracted in a solvent wash.

In the AFM process, the abrasive cutting particles break and become dull, and the abraded material becomes part of the abrasive media. The effective life of the media depends on a number of factors including the initial batch quantity, the abrasive size and type, the flow speed and the part configuration. Typically a machine load of media can be used for weeks, processing thousands of parts, before replacement.

PROCESS APPLICATIONS

AFM offers precision, consistency and flexibility to a wide range of applications in aerospace, automotive, production and die finishing. Other applications have been developed in areas as diverse as surgical implants and centrifugal pumps. The process was initially developed to perform critical deburring of aircraft valve bodies and spools, providing burr free internal edges, routinely passing 20X microscopic inspection while producing precisely controlled edge radii. The turbine fuel spray nozzles pictured below are shown before (Figure 8) and after (Figure 9) processing.

Figure 8. **Figure 9.**
Turbine engine fuel spray nozzles use AFM to satisfy critical deburring requirements.

Some additional AFM applications in aerospace include removal of the thermal recast layer in the lasered or EDM'd cooling holes of blades and disks, deburring fuel spray nozzles, and polishing cast surfaces of blades, compressor wheels and impellers (Figure 10).

Figure 10. Milled surfaces of an impeller use AFM to improve the surface finish while retaining the shape within the profile envelope.

The AFM process is used to polish a variety of dies for extruding, drawing, forging, cold heading and compacting, eliminating the expense and inconsistency of handwork. Uniform, controlled stock removal permits accurate sizing of die passages with minimal dimensional change. Typically, original EDM'd finishes of 2.5 μM R_a (100 μinch) are polished to below 0.25 μM R_a (10 μinch) in a ten minute cycle without special tooling requirements. Surfaces finishes are generally improved to one tenth the original finish.

The multi-port aluminum extrusion die shown in Figure 11 used AFM for improving the EDM finish of 1.9 μM R_a (75 μinch) to a 0.18 μM R_a (7 μinch) in a five minute cycle. A close-up of the teeth is shown before (Figure 12) and after (Figure 13) processing.

Figure 11. Multiple passages in one workpiece can be processed simultaneously.

Figure 12.

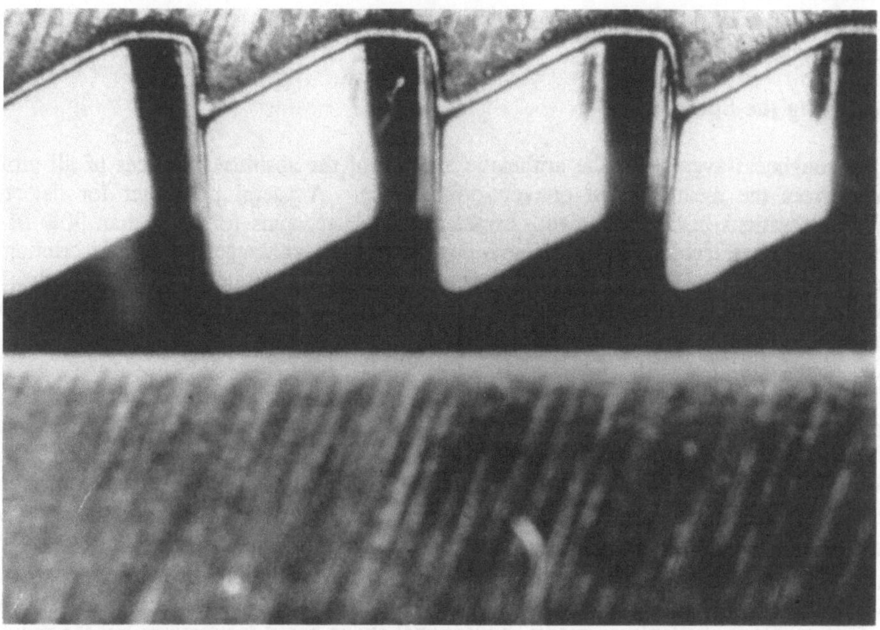

Figure 13.

Close-up of the teeth on the extrusion die in Figure 11 shown before and after processing.

One of the primary advantages of the AFM process lies in the uniformity of the polished surface, especially when compared to tedious, manual finishing methods. This advantage leads to other benefits directly associated with lower labor costs with improved part performance, longer life, less scrap and rework and reduced inspection times. Whereas hand polishing smears surface peaks causing the edges to roll over and producing an inconsistent surface, the AFM process grinds away the high spots of the area, yielding a more uniform, unidirectional surface and consequently a more reliable, longer lasting part. Figure 14 shows the progressive polishing action of the bearing surface on the extrusion die shown above; the processing time between each successive stage is 80 seconds.

Figure 14. The AFM process grinds away high spots resulting in a uniform surface.

CAPACITANCE METROLOGY

One of the most potentially far-reaching of the new developments in state-of-the-art surface metrology utilizes capacitance-based technology. One of the systems has been on the market for nine years and uses a stationary capacitive sensing probe and area-based technology for true three-dimensional surface characterizations. The second system uses a fringe field capacitance sensor for measuring the profile of a surface and is just now being released to the market.

Identifying the Surface

R_a, or roughness average, is the arithmetic average of the absolute distances of all profile points from the mean line in one cut-off distance. A useful parameter for detecting general vibrations in overall height characteristics, R_a accounts for more than 90% of all surface roughness specifications. It provides, however, a rather generalized description of the actual surface. The **ideal** surface parameter would accurately characterize the full contour to its functional performance. This parameter would be expressed as one number, for quick reference and for easy comparison with other surfaces.

Like any complex three-dimensional object, a microsurface is not easily described by a single term. The reasonable alternative involves the use of supplemental parameters such as R_t and R_p in addition to R_a. (R_t defines the maximum peak-to-valley dimension of the profile within the tracing length. R_p, or "smoothing depth," is the distance from the graphical centerline to the highest peak of the profile.) Since each machining process leaves its own characteristic microsurface contour, no predictable relationship between R_a, R_t, and R_p exists. As the conditions of any given machining operation change, the interrelation of these parameters will also change. Derivation of these parameters from a surface profile is available on many conventional stylus-based instruments but the required sensitivity and mechanical complexity of contact stylus instruments limit their application on near or in process applications. The capacitive systems were developed in response to application shortcomings of the existing stylus methodology.

Area-Based Capacitance Assessment

The classic capacitance parallel plate definition states that the electrical capacitance between two conductive plates is proportional to the area of the plates (A) and inversely proportional to the distance separating the plates (d), or:

C = K * p * A/d where C = capacitance
K = mathematical constant
p = dielectric permitivity
A = conducting plate area
d = plate separation

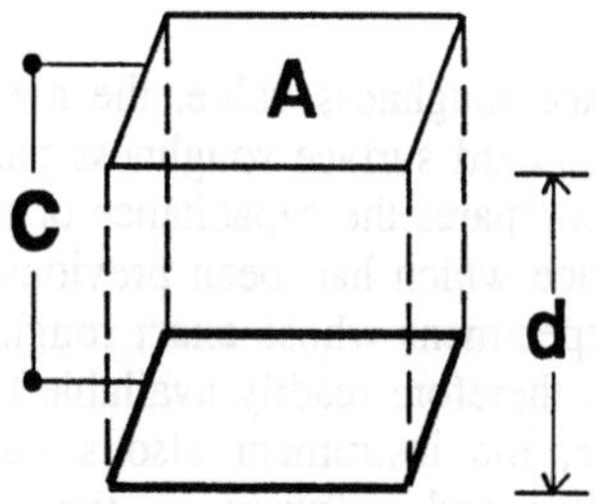

Figure 15. Capacitance parallel plate definition.

By taking a **flexible** capacitor plate, conforming it to a conductive workpiece surface, and measuring the resultant capacitance between the two, an indication of the workpiece texture can be obtained. The voids and spaces between the peaks and valleys of the workpiece surface are a function of surface texture and the air in these voids and spaces becomes the dielectric of the capacitance circuit. The compliant sensor bridges the surface peaks conforming to geometry variations and waviness, measuring only the roughness.

In practice, the compliant sensor is manufactured from thin mylar film with metal vapor-deposited on the back side. The sensor is then produced by a photographic/etch process, similar to printed circuit board fabrication. This simple and versatile procedure enables the production of numerous sensor configurations and inexpensive design and build for custom applications. The sensor film is then attached to open-cell urethane or molded silicon elastomer "compressors" that keep the sensor film in intimate contact with the workpiece surface. The soft cushion/mylar film combination will not mar or scratch the surface, and since the static procedure uses no moving parts, the sensor is very robust. Remaining circuitry measures the capacitance, temperature compensates (capacitance can also be a great temperature sensor), and converts the reading into texture values (Figure 16).

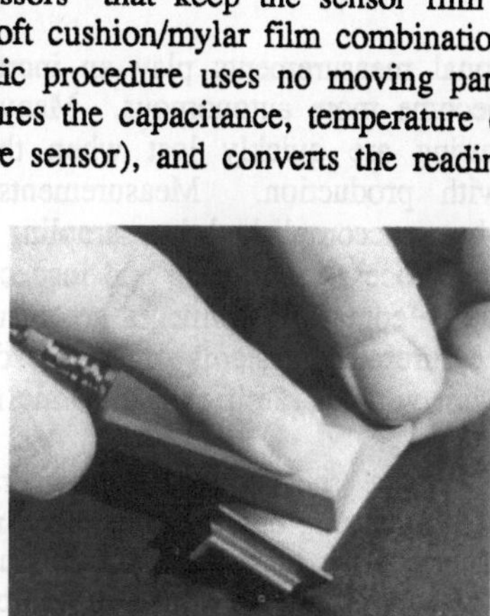

Figure 16. Measuring an airfoil surface with a capacitance instrument.

The capacitance developed between the sensor and a target surface will reflect the total three-dimensional microsurface rather than just the two-dimensional stylus profiles currently used as the standard. Capacitance, as an electrical property, is volume-based, and so accounts for R_p, R_t, as well as R_a in all three dimensions of the microsurface. Yet because of its sensitivity and accuracy, the capacitance of a measured surface can be directly related to the popular R_a parameter by merely comparing its value to the capacitance value of a reference surface (with a known R_a value) that has a similar microsurface signature as the surface being measured. That is, a specimen-based methodology.

In order to obtain a surface roughness value, the average thickness of the airspace dielectric is converted into a standard surface roughness parameter, preferably R_a. To do this, the measuring instrument compares the capacitance of the surface in question with the capacitance of a reference surface which had been previously entered into the instrument. (Conductive reference surface specimens whose exact roughness values have been verified are already widely used and are therefore readily available.) In addition to the capacitance value of this reference surface, the instrument also stores the surface's corresponding roughness value. The circuitry and software in the instrument then compares the capacitance of the of the reference surface (and its corresponding roughness value) to the capacitance of the surface being measured, thus deriving the surface roughness value.

Any reference surface used as the calibration specimen for the instrument—whether a laboratory-verified standard or merely a spare part with a generally "acceptable" roughness—must be machined by the same process as the surfaces to be measured. This is because, as mentioned previously, the precise contour of the microsurface shape or signature of a surface produced by a machine tool will significantly affect that surface's capacitance and, therefore, its roughness as measured by the capacitive instrument.

As a result of motionless assessment, non-marring sensors, portability, durability, and reduced measurement times, area-based capacitance texture measurement has found application into numerous areas including near and in process, yet there are still times that the surface under question needs to be analyzed by its total profile.

Capacitance Profiling

High speed, on-line dimensional measurements play an increasingly important role in manufacturing as processes become more autonomous. Many of the advantages gained through automated manufacturing are quickly lost when the speed of inspection is inadequate to keep pace with production. Measurements which may have been unnecessary, or could have been accomplished by sampling techniques, have become essential elements of automated process monitoring and inspection. Often the adaptation of manual methods or laboratory contact instruments are insufficient solutions to these measurement problems. Measurement systems must be designed to survive in a comparatively harsh environment without sacrificing dynamic range, accuracy or speed.

A newly developed capacitance sensing system answers these problems and has particular advantages in the manufacturing environment. The measurement is sufficiently localized to be applied to surface contour mapping and even surface profilometry (not just average surface roughness). In this technique a translating probe uses a thin metallic strip, laminated between layers of insulating material. The probe is oriented so that the strip is perpendicular to both the surface to be gaged (which forms the second electrode) and the direction of motion. A "fringe" electric field is produced between the perpendicular electrode and the target surface (Figure 17). The perpendicular arrangement of the thin electrode results in a highly localized electric field, sensitive only to the local surface distance. As the probe is moved along the surface, variations in capacitance of the field are generated by virtue of changing distance of the surface profile. The instantaneous capacitance is determined by the local distance of the electrode from the surface. The changes in capacitance modulate an electronic detector providing a signal related to this distance. The advantages of this capacitive sensor include: high speed, durable construction (leading to high reliability), and small sensor size.

The sensitivity of the sensing element may be derived from an analysis of the electric field. For the infinitely thin electrode shown in Figure 17 of width **w**, height **H**, and displaced a distance **h**, above an infinite plane, the capacitance generated is:

$$C = 4*e*w/pi * \ln(2H/h)$$

where it is assumed that the surrounding material is homogenous with dielectric constant **e**. Note that the capacitance is directly proportional to the width **w**, and inversely proportional to the logarithm of the separation distance **h**. As the distance between the electrode and the surface becomes small, the capacitance increases rapidly.

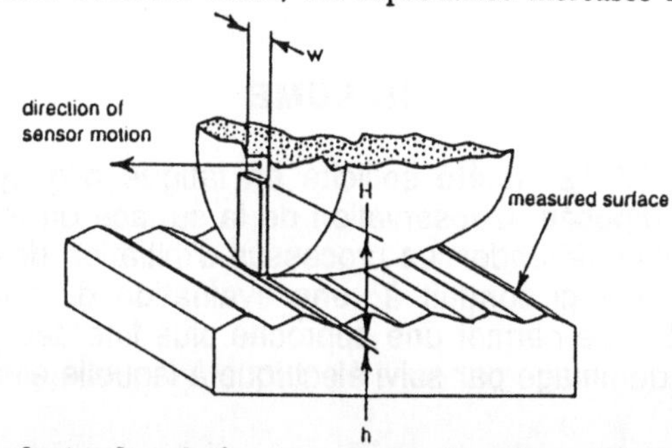

Figure 17. Skid and electrode relative to measured surface. Electrode dimensions are also illustrated. Forward half of skid is not shown.

The electrode is very thin, typically 0.1 micron, and is bonded between thick layers of ceramic base material. The ceramic completely encases the thin electrode, protecting it from contamination and mechanical damage. The ceramic case rests directly on the surface to be measured, broadly contacting the surface—forming a skid that maintains a constant average distance between the tip of the electrode and the surface. The probe is dragged across the surface in the direction perpendicular to the plane of the electrode. The capacitance varies as a function of the distance between the bottom of the electrode and that portion of the surface immediately beneath it. The profile of the surface is inferred from the record of the capacitance and probe position which is determined by an encoder on the motion control stage.

Many of the usual disadvantages of contact stylus systems do not apply to the fringe-field capacitive profiler. Since the skid provides a broad area of contact as it rests against the surface, the required force is minimal and need not be carefully controlled. The probe motion does not cause surface damage. The sensor itself is very robust and since, unlike a contact stylus, the capacitive profiler need not mechanically track the fine surface variations, scanning speed is not limited by inertial effects. Scanning speeds several orders of magnitude higher than those commonly used in contact stylus devices are easily achievable.

Evolution du Dommage Surfacique au Cours d'un Essai de Fatigue Oligocyclique: Comparaison avec des Mesures de Dommage Volumique

J. Charrier, R. Roux et C. Gasc

ENSMA
Poitiers
France

RESUME

Un acier Z2 CND 17 12 a été sollicité en fatigue oligocyclique à chaud en déformation totale imposée. L'observation de la surface du matériau en cours de cyclage a permis d'appréhender les processus d'initiation des microfissures et la quantification de celles-ci conduit à une évaluation du dommage subi par le matériau. Cette méthode permet une approche plus fine des phénomènes qu'une mesure globale du dommage par suivi électrique à laquelle elle a été comparée.

INTRODUCTION

Les sollicitations imposées aux matériaux dans les installations thermiques sont généralement sévères du fait des températures élevées et de la longue durée de vie exigée. Ces sollicitations sont de plus, en général, complexes. Pour les simuler, on les décompose en sollicitations simples fluage, relaxation, fatigue oligocyclique... que l'on associe ensuite de façon à modéliser le comportement réel du matériau (1). Dans ce travail les phases de fluage et de fatigue ont été séparées afin d'évaluer l'influence de l'endommagement créé par un fluage initial sur le comportement ultérieur en fatigue. Le fluage entraîne également un vieillissement par précipitation ; ce vieillissement seul a donc été effectué avant fatigue.

Définition de l'endommagement en fatigue oligocyclique

D'une façon générale, on peut dire qu'un matériau est endommagé lorsque les défauts créés par la sollicitation ne sont plus restaurables. On quantifie cet endommagement à l'aide d'un scalaire D variant de 0 à 1. Le problème consiste à relier ce scalaire à un paramètre physique facilement mesurable traduisant l'endommagement. D dépend donc à la fois du concept de dommage retenu et de la méthode de mesure utilisée. Miner (2) en a donné une première expression en supposant que le dommage est proportionnel au nombre de cycles N effectués. La variation de D, en fonction de N, est donc linéaire. Plus récemment, Lemaitre et Chaboche (3), à partir de la notion de contrainte effective de Kachanov (4) et Rabotnov (5) ont donné une expression du dommage plus complèxe sous la forme d'une double loi puissance (6) :

$$D = 1 - \left(1 - \left(\frac{N}{N_R}\right)^{\lambda}\right)^{\mu}.$$

où λ et μ sont des cœfficients dépendant du matériau mais aussi du paramètre de mesure utilisé (7), et N_R désigne le nombre de cycles à la rupture.

Le nombre de cycles Na à partir du quel D est mesurable dépend de la sensibilité de la méthode de mesure. Il peut varier de 0,8 N_R lorsqu'on utilise comme paramètre les variations de module d'élasticité où la chute de charge maximum, à 0,5 N_R lorsqu'on utilise la méthode de variation du potentiel électrique (8) qui est une des plus précises.

L'amorçage des fissures s'effectue généralement à la surface du matériau bien que l'on ait observé dans certains cas des initiations aux interfaces inclusion-matrice (9) ou sur des défauts internes tels que les micropores (10-11-12). En dehors de ces situations particulieres, l'observation des fissures en surface donne de précieux renseignements sur les mécanismes de déformation et sur l'état d'endommagement du matériau.

METHODE EXPERIMENTALE

Des essais de fatigue oligocyclique ont été réalisés,à 600°C, sur un acier inoxydable Z2 CND 17 12 (316L SPH) pour différents états structuraux :

- hypertrempé : état de référence
- vieilli 600 H à 600°C, avant fatigue
- flué 600 H à 600°C sous 250 MPa, avant fatigue.

Les traitements de vieillissement et de fluage ont été réalisés sous vide secondaire afin d'éviter toute oxydation susceptible de perturbé les résultats (13-14). Les éprouvettes lisses de forme cylindriques de diamètre 8 mm et de longueur utile 16 mm ont été polies électrolytiquement (poli miroir). Les essais ont été réalisés en traction compression à déformation cyclique totale imposée ($\Delta\varepsilon_T/2$ = 0,7 %). La contrainte en haut de cycle, au régime stabilisé est de 250 MPa, la même qu'en fluage non interrompu. Une première série d'essais a permis de suivre l'endommagement par la méthode du suivi électrique. Au cours d'essais interrompus, le fût des éprouvettes a été observé au microscope électronique à balayage (figure 1) et la longueur des fissures a été mesurée. Afin de ne pas perturber les essais, le cyclage a été interrompu à F et Δl = 0. La surface analysée était de 400 mm^2, soit toute la partie utile de l'éprouvette. Le diamètre des grains étant de 60 µm, seules les fissures de longueur supérieures ou égales à 100 µm (soit 1,5 fois le diamètre de grain) ont été répertoriées de façon à ne pas prendre en compte d'éventuels déchaussements de grains.

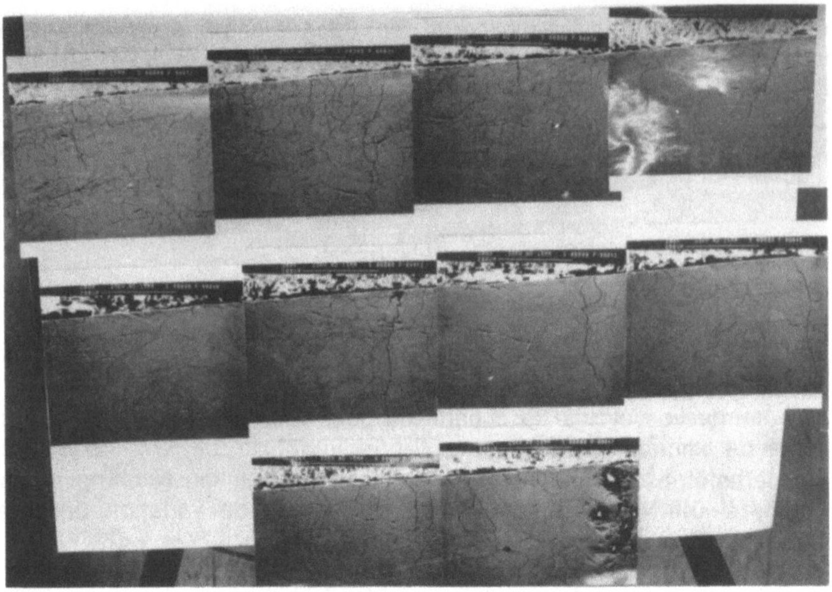

Figure 1. Exemple de fissures superficielles apparaissant sur le fût des éprouvettes après N cycles.

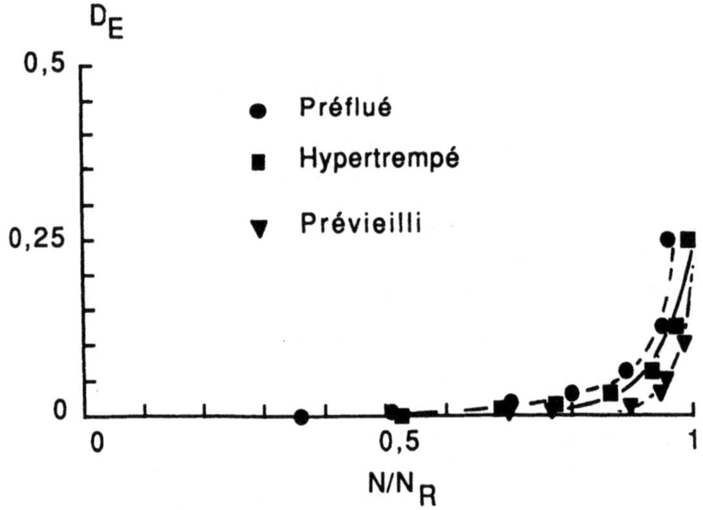

Figure 2. Courbes de dommage obtenues par la méthode du suivi électrique.

RESULTATS ET INTERPRETATION

Mesures électriques

Les courbes d'endommagement du matériau, pour les trois états, mesuré par le suivi électrique sont données à la figure 2. On peut considérer l'endommagement évalué par cette méthode comme un endommagement volumique. En effet, les mesures sont réalisées en courant continu haché de fréquence 10 Hz, pour éviter tout échauffement parasite ; l'effet de peau est donc faible. Ces courbes montrent que l'état prévieilli s'endommage le moins rapidement, puis vient l'état hypertrempé et l'état préflué, pour lequel l'amorçage est détecté à mi-durée de vie.

Comptage de fissures

Les longueurs des fissures superficielles ont été groupées par classe de 100 μm, ce qui a permis de tracer les histogrammes représentés figure 4 :

Nombre de fissures / classe = F(classe,nombre de cycles)

Pour les différents états, chacun des histogrammes correspond à la valeur moyenne de trois essais ; la dispersion des mesures est peu importante sur les premières classes. Dans la mesure du possible, on s'est efforcé de suivre l'évolution de chaque fissure mais cela n'a pas toujours été possible. D'une manière générale, on constate que l'évolution des fissures des premières classes n'est pas continue, et que certaines fissures ne se propagent pas apparemment en surface. Il n'a pas été possible de repérer la fissure principale : en effet, certaines "grandes fissures" se bloquent alors qu'une "petite" peut croître très rapidement au cours du cyclage. Quelques coupes métallographiques nous ont permis de voir que la pénétration des fissures était peu importante, de l'ordre de la grosseur de grain au maximum. Cette cinétique de croissance a été schématisée sur la figure 3.

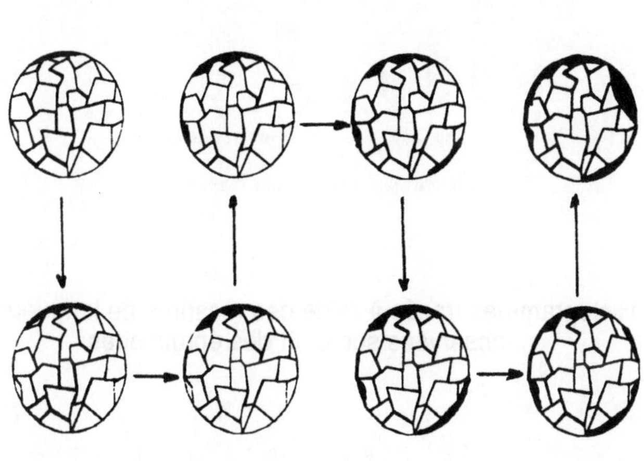

Figure 3. Schéma de croissance des fissures superficielles dans une section d'éprouvette d'après des observations sur des coupes.

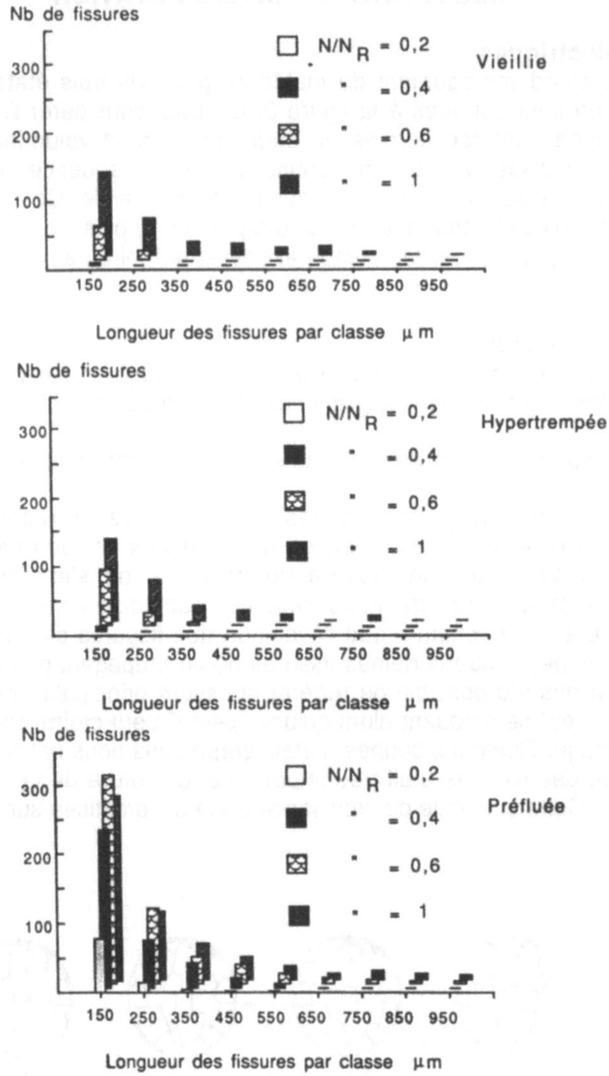

Figure 4. Histogrammes tracés à l'aide des mesures de longueurs des fissures observées sur le fût des éprouvettes.

Dans ce cas, la densité de fissures n'est pas un bon paramètre puisqu'elle peut décroître en fin d'essai soit par coalescence de microfissures, soit par baisse du taux d'initiation de nouvelles fissures (figure 4). De la même façon si la longueur cumulée Lc augmente toujours en cours d'essai, cette augmentation peut être occasionnée soit par l'augmentation de la longueur des fissures, soit par l'augmentation du nombre de fissures. Pour ces deux raisons, on a préféré utiliser,comme paramètre de dommage surfacique, la longueur moyenne :

$$\bar{L} = \frac{\Sigma L}{\text{nombre de cycles}}$$

que nous avons relié au dommage par la relation :

$$D_{FS} = \frac{\bar{L} - 150}{\bar{L}_{rupture} - 150}$$

La valeur à l'amorçage de 150 µm a été adoptée pour tenir compte de la première classe de longueur (100 - 200 µm) que si L < 150 µm, $D_{FS} = 0$. Les courbes d'endommagement ainsi déterminées sont données figure 5. La figure 6 représente la fraction de durée de vie en fonction de $\bar{L}$, tout état confondu.

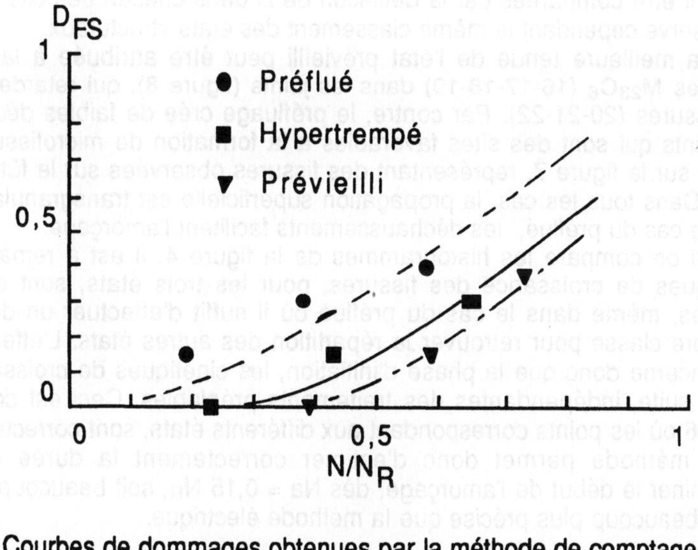

Figure 5. Courbes de dommages obtenues par la méthode de comptage des fissures superficielles.

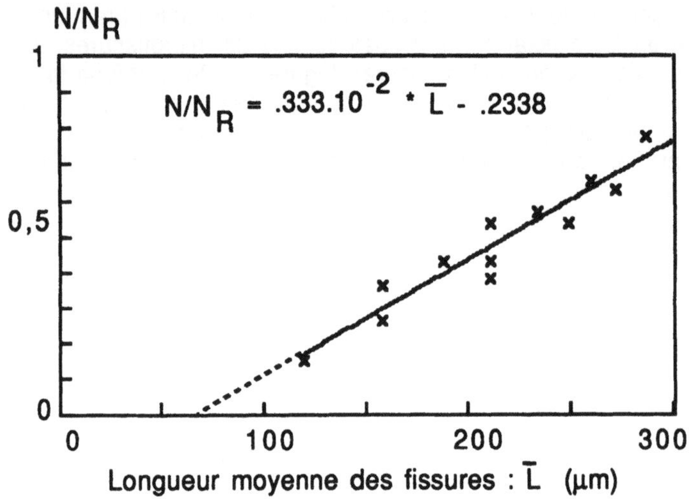

Figure 6. Espérance de vie en fonction de la longueur moyenne des fissures superficielles.

DISCUSSION

Pour un nombre de cycles donné, les valeurs des dommages des figures 2 et 5 ne peuvent être comparées car la définition de D dans chacun des cas est différente. On observe cependant le même classement des états structuraux.

La meilleure tenue de l'état prévieilli peut être attribuée à la présence de carbures $M_{23}C_6$ (16-17-18-19) dans les joints (figure 8), qui retarde la nucléation des fissures (20-21-22). Par contre, le préfluage crée de faibles déchaussements aux joints qui sont des sites favorables à la formation de microfissures. Ceci est illustré sur la figure 7, représentant des fissures observées sur le fût pour les trois états. Dans tous les cas, la propagation superficielle est transgranulaire, même si, dans le cas du préflué, les déchaussements facilitent l'amorçage.

Si on compare les histogrammes de la figure 4, il est à remarquer que les cinétiques de croissance des fissures, pour les trois états, sont en réalité très voisines, même dans le cas du préflué où il suffit d'effectuer un décalage de la première classe pour retrouver la répartition des autres états. L'effet du préfluage ne concerne donc que la phase d'initiation, les cinétiques de croissance seraient, par la suite, indépendantes des traitements préalables. Ceci est confirmé par la figure 6 où les points correspondant aux différents états, sont correctement alignés. Cette méthode permet donc d'estimer correctement la durée de vie et de déterminer le début de l'amorçage, dès $N_a = 0{,}15\ N_R$, soit beaucoup plus tôt et de façon beaucoup plus précise que la méthode électrique.

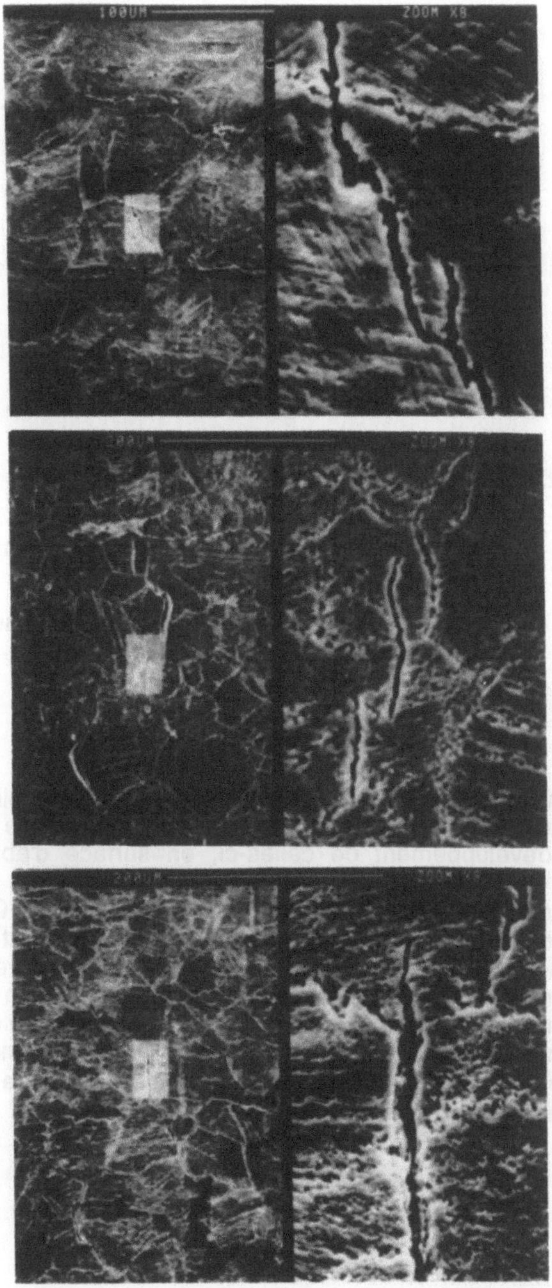

Hypertrempé

Vieilli 600 H 600°C

Flué 600 H 600°C
250 MPa

Figure 7. Fissuration de surface dans les trois états structuraux étudiés.

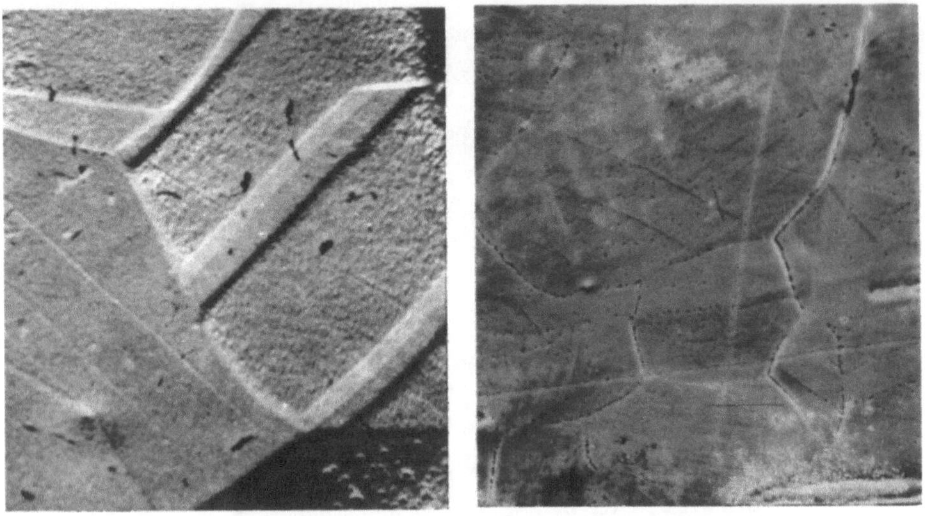

a) Hypertrempé b) Vieilli

Figure 8. Carbures $M_{23}C_6$ dans les joints de grains, après vieillissement de 600 H à 600°C sous vide (G = 3500).

CONCLUSIONS

L'acier inoxydable austénitique Z2 CND 17 12, sollicité en fatigue oligocyclique, à 600°C, en déformation totale imposée, présente en surface de nombreuses microfissures. Le développement de celles-ci, en surface d'abord, puis en profondeur, conduit à la formation d'une fissure principale que l'on peut détecter dès 50 % de la durée de vie par la méthode du suivi électrique. Cependant, la mesure des fissures superficielles et leur comptage permettent d'évaluer le dommage subi par le matériau et de prévoir plus précisément la durée de vie, dès 15 % du nombre de cycles à la rupture. Cette méthode d'évaluation du dommage est, de plus, très sensible à l'état structural du matériau et permet de déterminer en même temps la nature de l'amorçage. Bien que plus lourde, cette méthode donne des résultats beaucoup plus complets et plus significatif qu'une méthode de détection volumique.

REFERENCES

1. Min, K.B. and Raj, Hold time effects in high temperature fatigue. Acta Met., 1978, Vol. 26, 1007-1022.

2. Miner, M.A., Journal of Applied Mechanics, 1945, 67, 159-164.

3. Lemaitre, J., Chaboche, J.L., Journal de Mécanique Appliquée, 1978, Vol. 2,3.

4. Kachanov, L.M., Izr Akad Nauk SSE Otd Tech Nauk, 1958, 26-31.

5. Rabotnov, Y.N., Proc. Int. Cong. Appl. Mech. Stanford, 1968, Springer 1969.

6. Chaboche, J.L., Revue Française de Mécanique, 1964, 50-51.

7. Roux, R., Thèse Docteur-Ingénieur Poitiers France, 1986, 99-116.

8. Cailleteau, G., Policella, H., Baudin, G., C.R. Acad. Sciences Paris, 292, série 2.

9. Lankford, J., Kusengerger, F.N., Metal. Trans., 1973, Vol. 4, 553-559.

10. Gell, M., Leverant, G.R., Fatigue at elevated temperature. A.S.T.M.-S.T.P., 1973, 520, 37-67.

11. Leverant, G.R., Gell, M., Trans. Met. Soc. AIME, 1969, 245, 1167.

12. Duquette, D.J., Gell, M., Met. Trans., 1972, 3, 1899-1905.

13. Chalant, G., Remy, L., Proc. Vth ICSMA, Aachen, 1979, Vol. 2, 1195-1200.

14. Gayda, J., Miner, R.V., Met. Trans., 1983, 14A, 2301-2308.

15. Levaillant, C., Thèse d'Etat U.T.C. France, 1984, Vol. 1, Chap. 1.

16. Deleury, M.J., Thèse Docteur-Ingénieur ENSM Paris France, 1980.

17. Aubert, M.D., Tech. RMA, 1981, 81, 703.

18. Weis, B., Strickler, R., Met. Trans., 1972, Vol. 3, 851-866.

19. Lai, J.K., Mat. Sci. and Ingi., 1983, 61, 101-109.

20. Driver, J.H., Met. Sci. Journal, 1971, 5, 47-50.

21. Runkle, J.C., Pelloux, R.M., A.S.T.M. - A.T.P., 1979, 675, 153-173.

22. Davidson, J., Balladon, P., Honnarat, Y., Wache, X., Mem. Sci. Rev. Met., 1973, LXX, 7-8, 543-558.

The Influence of Fine Structure, Morphology and Composition of Alloy and Oxide on the Growth of Cr_2O_3 Scales

M. Hirbod

ABB Stal AB
Finspong
Sweden

ABSTRACT

The kinetics of morphological and structural development of Cr_2O_3 scales on Fe-Cr alloys, with and without yttrium additions, under isothermal conditions at 1000°C have been studied in an attempt to elucidate the growth processes. The steady state scale on the yttrium-free alloy is continuous and highly convoluted, with a porous, equiaxed grain structure. It is suggested that such a configuration results from the transport processes in the scale and from development of high compressive stresses, leading to lateral growth of the oxide. In particular, both inward movement of oxygen and outward movement of chromium occur, enabling new oxide to form within the bulk scale with accompanying oxide dilation. The presence of active elements, such as yttrium, in the alloy has a marked effect on the oxidation process.

In particular, there are considerable changes in the microstructure and the adherence of the scale resulting in a considerable decrease in the amount of scale spalling upon cooling or thermal cycling. The influence of the active element is accounted for in terms of its effect on the transport processes in thescale, i.e. the supression of chromium diffusion.

This enables oxygen to penetrate to the scale/metal interface where new oxide is formed without significant stress development.

INTRODUCTION

Low oxidation rates are generally achieved by the formation of a scale which acts as a barrier to the reactants, and which should have the following properties. The concentration of ionic defects which enable transport should be low and the oxides should remain stoichiometric. The vaporization rate of the oxides forming the scale should be insignificant. The scale should be virtually free of pores, cracks or other paths for short-circuit diffusion. Furthermore, the scale should adhere to the underlying alloy even when rapid temperature changes occur. In practice, Cr_2O_3 and Al_2O_3 such as developing on most high-temperature alloys represent such protective scales.

There is a considerable amount of information in the literature on the beneficial effect of certain alloying additions (e.g. Y, Hf and other rare earth elements) on the oxidation behaviour of Al_2O_3- and Cr_2O_3-forming alloys. Rare earth elements have been reported to affect the direction of predominant diffusion through the scale, the reaction rates and the scale adhesion, but there is still no fundamental understanding of the processes involved. It is particularly evident that the morphologies and microstructures of the oxides formed, and the nature of stress in these oxides, are still poorly understood (1,2,11). This has stimulated the present study using transmission electron microscopy in conjunction with scanning electron microscopy, to investigate the oxide grain structures and substructures of thick oxide scales. An attempt has been made to explain and interpret the development of the observed oxide morphologies and substructures and to correlate these with the observed oxidation kinetics and oxide adhesion.

RESULTS

Development of the Cr_2O_3 Scale Morphology on Fe-Cr and Fe-Cr-Y Alloys at 1000°C

<u>Fe-28%Cr-Alloys</u>: After the initial stages of the transient oxidation, in those regions where oxide had spalled on many specimens, oxidation of some of the smooth surfaces of the voids had taken place to a greater or lesser extent. Some of these areas were oxide-free, while others contained only a few oxide nuclei, or even the formation of a complete fine-grained, inner oxide layer had taken place, Fig. 1a.

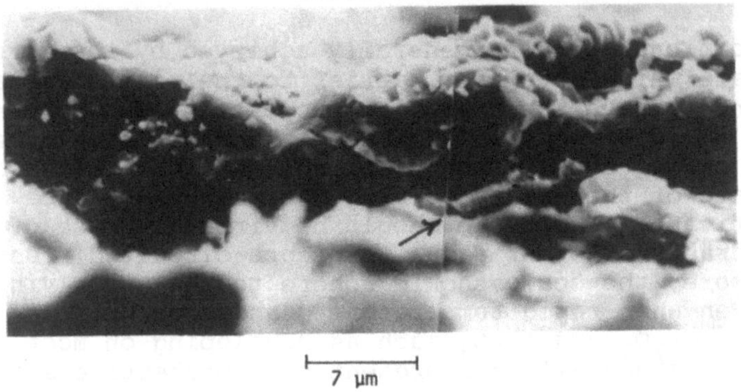

Figure 1a Fracture section of the scale formed on Fe-28%Cr after 24 h oxidation at 1000°C

Examination of SEM fractograph cross-sections shows the outer oxide layer has a highly convoluted oxide morphology, with an equiaxed structure. The inner oxide layer at the alloy/oxide interface appeared fine-grained with a compact structure.

Figure 1b shows the fracture section of a scale after 48h oxidation. The innermost part of the oxide scale consisted mainly of fine grains, and indicated that the scale remained in contact with the alloy surface over the particular area examinated. There are several regions where contact between the outer and inner parts of the oxide scales has been lost with typical "balloon" configurations. In areas where the scale apparently exhibited a single layer, the compact equiaxed oxide grains, with transgranular fracture sections, made demarcation of the layers impossible.

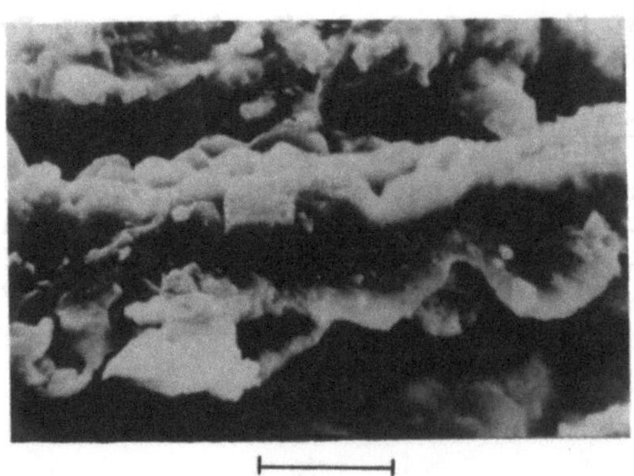

Figure 1b Fracture section of the scale formed on Fe-28%Cr after 48 h oxidation at 1000°C

Large cavities were present at the alloy/oxide interface, Figs. 1a-c. Also small intergranular pores were observed, randomly distributed from the inner/outer layer interface to partway through the outer layer. These pores were situated mainly at the grain boundaries of the outer equiaxed oxide layer.

Figure 1c Fracture section of the scale formed on Fe-28%Cr after 140 h oxidation at 1000°C.

The TEM micrographs shown in Fig. 2a-c show the microstructure of the oxide grains extracted from the gas/scale and alloy/scale areas shown in Figs. 1a-c respectively. The boundaries of the polycrystalline oxide grains consist of individual lines; Figs. 2a-c. Figs. 2a and b show details of resolved grain-boundaries and sub-grain boundaries, containing pores and line defects. The boundary planes in this region are all nearly normal to the plane of oxide. The dark spots in Fig. 2a are considered to be due to the intersection of dislocations steeply normal to the surface plane of oxide. The white spots appearing at the surface of the Cr_2O_3 oxide grains in Fig. 2b were possibly due to impurities, or to some oxides with different structures.

The microstructural features of the oxide scale cross-section were revealed by electron transmission microscopy of thin sections, prepared by ultramicrotomy, and of extracted replicas. The polycrystalline morphology of the oxide scale is evident in Fig. 2c. The minor portion of the observed porosity in the oxide scale was intragranular, while most pores occur at triple points. In addition, there were possible indications of planar defects in the oxide scale with characteristic dark and light fringes as observed in the centre of Fig. 2c. Dislocations, both singular and networks, were typical of the oxide scale cross-sections.

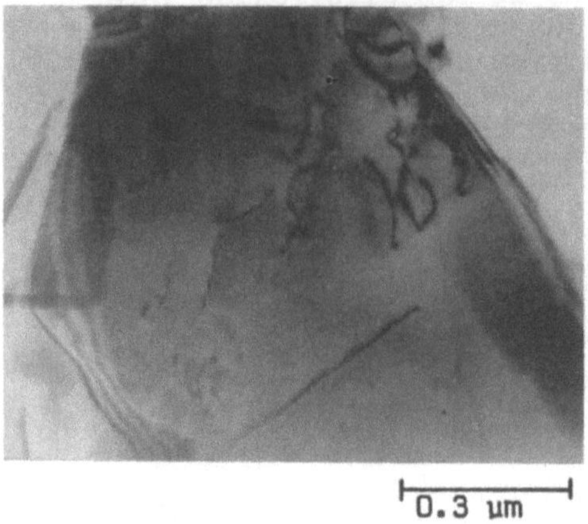

a) Bright field

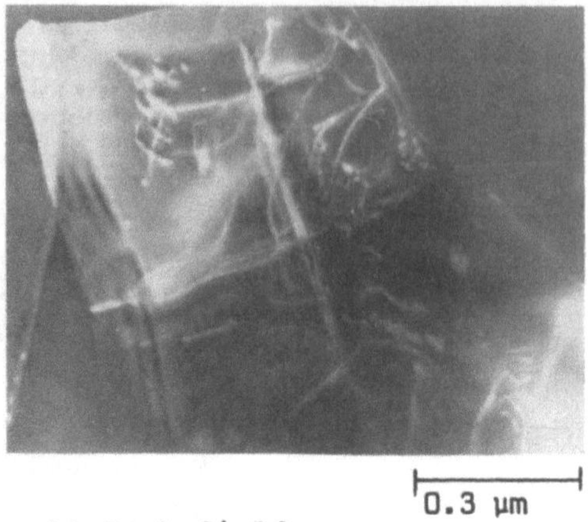

b) Dark field

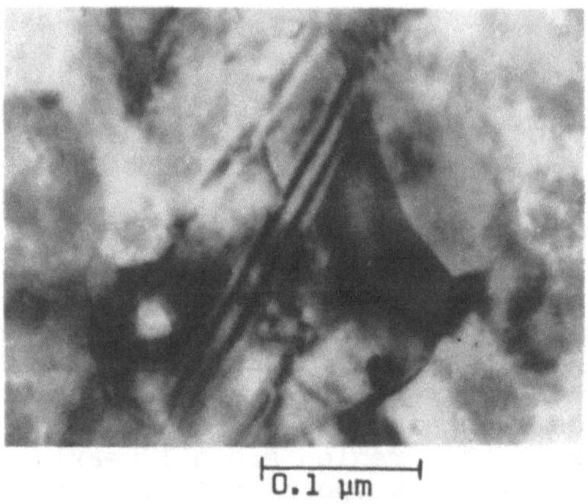

c) Bright field

Figures 2a-c Extraction carbon replicas of the gas/oxide metal/scale interfaces and ultramicrotomed cross section, respectively from Fe-28%Cr after oxidation for 48 h at 1000°C.

With increasing oxidation time, the structure of the inner oxide layer coarsened. After 24h the inner oxide layer covered the alloy surface completely. This was revealed by the "loops and troughs" configuration of the inner layer, dictated by the voids and ridges of the alloy surface, c.a. Figs. 1a-c. After longer oxidation times, such "loops and troughs" were not observed: instead a less convoluted inner layer had developed. At this stage of oxidation there were also indications that this inner layer was attached to the inner surface of the oxide convolutions, Fig. 1c. The inner layer was approximately two to three grains in depth and remained nearly constant throughout the scale growth.

Fe-28% Cr-1% Y Alloys: The oxide scale of Fe-28Cr-1Y after 1h at 1000°C was composed of fine crystallites with some larger crystals protruding from the scale, Fig. 3a. Two types of oxide morphologies were observed: large, slab-like grains which were mainly located along the abrasion markings, and fine grains with a granular morphology which filled the space between the former type.

The thin oxide layer was generally well adherent to the alloy, showing only occasionally signs of a spalling due to mechanical deformation during cooling. Fig. 3b illustrates the underlying alloy after detachment of the oxide, showing the imprints of the oxide grains.

Upon increasing the oxidation time, the scale was composed of a thin layer of compact, adherent, fine-grained oxide, with isolated agglomerates of large-grained, thicker oxide.

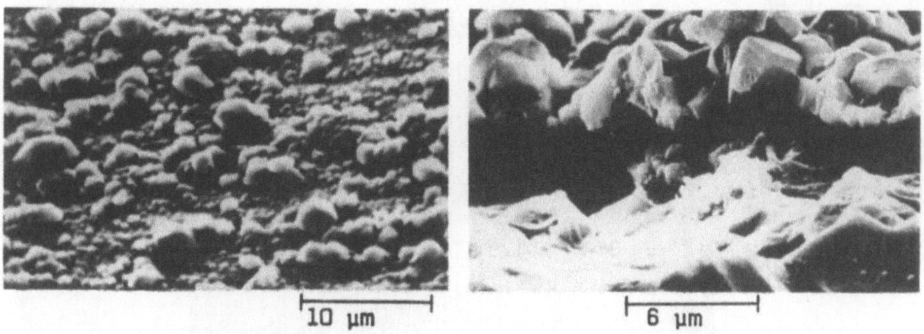

a) 24 h b) 140 h

Figures 3a,b Surface and fracture section of the oxide scale formed on Fe-28%Cr-1%Y alloy after oxidation at 1000°C.

It was further noticed that there was a considerable increase in the number and size of the larger grains, while the increase in size of the finer grains was less apparent, as can be seen in Figs. 3a and b. Dispersive X-ray analysis of these two oxides indicated no apparent compositional differences, both being high in chromium, with no detectable traces of yttrium.

Examination of the extracted replicas from several specimens revealed evidence for the formation of yttrium-enriched oxides in the integranular oxide zone, intersecting the oxide scale at the alloy surface. Yttrium-enriched oxide was not observed in the scale near the oxide/gas interface.

Dark and bright field TEM examination of the oxide crystallites indicated the absence of porosity in the scale. Whereas only a limited region of subgrains appear to be in a reflecting position, Fig. 4a, most of the relatively few dislocations in this same area show reversed contrast. Figures 4b,c demonstrate the existance of small crystallite structures, mainly in oxide grain boundaries. There are also indications of such microstructural features within the plane of the oxide grains, appearing as bright spots in the centre of the electron micrograph shown in Fig. 4c. An extremely low dislocation concentration was usually observed in the oxide grains. Not only were they relatively sparse in number, but also relatively short in length, as can be seen in Fig. 4c.

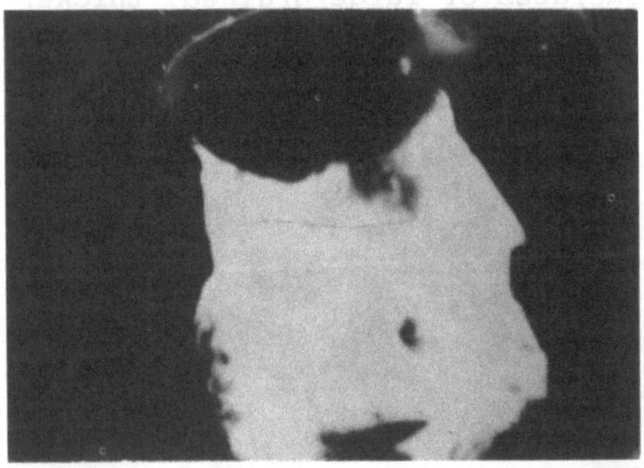

a) Dark field 0.45 um

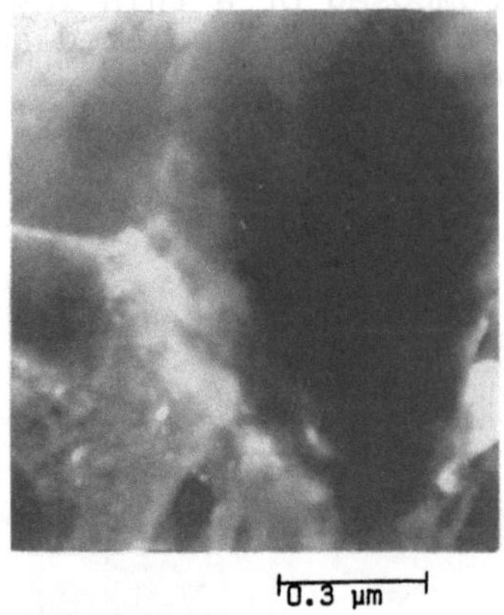

b) Dark field

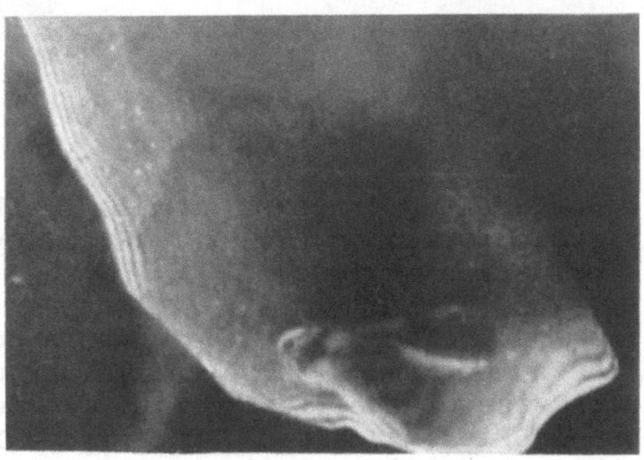

c) Dark field 0.4 um

Figures 4a-c Extraction carbon replicas of scale/gas, scale/alloy interface and fractured scale on Fe-28%Cr after oxidation for 48h, respectively

DISCUSSION

Fe-28Cr

The present general observations are consistent with the postulated growth mechanism for Cr_2O_3 scales on undoped Fe-Cr alloys via simultaneous oxygen diffusion inwards along grain boundaries, and chromium outward migration through bulk and grain boundaries.

It has been proposed (3-5) that in the initial stages of oxidation, nuclei of Cr_2O_3, and possibly iron oxides and mixed oxides, grow and spread until they cover the alloy surface as a continuous, thin film. Thereafter beneath this film, individual grains of Cr_2O_3 grow mainly at their outer interface, with the generation of newly formed grains both at this interface, and within the existing oxide layer. Continued oxidation results in growth of the scale in directions normal to the specimen surface, and also in lateral directions. Consequently, the oxide scale develops a ridged or convoluted morphology from the very early stages of oxidation, which progressively coarsens with continued isothermal oxidation. There are several reported studies (6-10) in which the development of a highly convoluted Cr_2O_3 has been observed. The formation of oxide within the existing oxide layer results in the development of highly compressive stresses. It is possible that these stresses are responsible for the dislocation structures in Cr_2O_3 observed in this study. These examinations (1) also indicated a greater density of dislocations in the zone adjacent to the central equiaxed layer and the inner region of the oxide scale, compared with the lower dislocation density observed in the outer regions. These observations are consistent with the existence of higher compressive stresses within the oxide scale (11-15), in the regions where possibly new oxide is being formed. However, the oxide also continues to grow at the scale surface, but the relatively high density of dislocations in the oxide at the gas/scale interface was surpirsing, since the oxide which is formed at this outer surface is usually expected to be formed in a stress-free condition.

The observations presented here on the formation and growth of the fine-grained inner layer can possibly be explained if it is assumed that the inward diffusion of oxygen through the outer oxide layer, takes place via easy diffusion paths.

The microstructure of the cross-section of the outer layer contained a large number of grain and subgrain boundaries, with a high degree of intergranular porosity, which may act as such diffusion paths. As suggested previously, some inward diffusion of O^{2-} ions would seem to be required to produce the growth stresses. Caplan and Sproule (4) indicate the importance of anion grain boundary diffusion in the growth of Cr_2O_3. The fact that the coarse grains of the outer layer are composed of small crystallites, and thereby involving a high number of sub-grain boundaries, further indicates a significant contribution of oxygen through grain boundaries, to provide a source of growth stresses. As oxidation proceeds, the occasionally observed intergranular pores, Figs. 1a-c, 2c, possibly aided by the growth stresses, can establish short-circuit paths connecting the alloy/oxide interface with the outer surface, thereby allowing direct access of oxygen to the alloy surface. However, the slow growth of the finer grains in the later stages of oxidation (after 24h), possibly indicates a limited supply of oxygen through the outer layer, which is the rate-controlling step for the growth of the inner layer. For the formation of the fine-grained inner layer, other factors may also be important: e.g. the hot and oxide-free condition of the exposed alloy which promotes nucleation of a fine-grained oxide.

Fe-28%Cr-1%Y

The incorporation of rare earth elements, from the alloy, into the oxide scale had been previously reported by several workers (11-15).

In the present study, the observation of the two types of oxide grain morphology, and the effects of this morphology on the oxidation rate and the scale adhesion, can be explained if it is assumed that yttrium incor-porated into the Cr_2O_3 scale suppresses outward chromium diffusion in the oxide grains.

After the impingement of the oxide nuclei in the initial transient stage, very early in the development of the oxide scale, yttrium is incorporated into the "fine-grained" oxices, and so the outward diffusion of chromium occurs to a lesser extent than in the "large-grained" oxides. As oxidation proceeds, the amount of yttrium incorporated into the oxide scale progressively increases, and consequently, outward chromium diffusion in the oxide grains progressively decreases. This already explains the slower rate of thickening of the "fine-grained" oxide on Fe-28%Cr-1%Y during oxidation. Furthermore, at the alloy/ oxide interface the bases of the oxide grains are replicated in the alloy surface (see Fig. 2b). This is attributed to the inward growth of the "fine-grained" oxide, which proceeds, largely by O^{2-} ion diffusion along the oxide grain boundaries.

The "large-grained" oxides, with a discontinuous, agglomerate morphology, grow in certain locations, mainly at or near the gas/oxide interface, from the early beginning of the oxidation process. It is implied that the growth and development of the "large-grained" oxide morphology is a direct result of the more rapid outward chromium diffusion. It is proposed that this occurs because insufficient yttrium is incorporated into the oxide during the early oxidation stages to suppress completely the outward diffusion of chromium.

Recent studies of the solute segregation in MgO and $-Al_2O_3$ (9-12, 13-15) have shown cases where the solute segregated to the grain boundaries. A reduction in the sintering rate was observed, and it is proposed that this was caused by a reduction in the rate of grain boundary transport. In the present case, exposure of the yttrium-containing alloy and the yttrium-free alloy to similar oxidising conditions resulted in the development of a smaller Cr_2O_3 grain size on the former alloy over a considerable proportion of the surface, and thus to an increase in grain-boundary area.

Hence, the oxidation rate of the yttrium-containing alloy should have been higher since further short-circuit paths were available for the inward diffusing O^{2-} ions. In fact, the opposite occurred, and the yttrium containing alloy oxidised at a much lower rate. This suggests that the rate of the predominant outward diffusion of Cr^{3+} was strongly reduced. It is possible that the segregation of yttrium on dislocations, or other short circuit diffusion paths, as suggested by Golightly et al. (2, 14), effectively blocks the transport of Cr^{3+} ions in the oxide grains. Also, the lack of observations of pores or microcracks in the transmission electron microscopical examination of these scales, might be similarly due to the formation of complexes with segregated yttrium into the oxide.

SUMMARY AND CONCLUDING REMARKS

Fe-28% Cr

The simultaneous diffusion of oxygen ions in the oxide grain boundaries and of chromium ions, probably in the oxide grains and grain boundaries, leads to the formation of new oxide at, or adjacent to, oxide grains within the scale as well as at the scale/gas interface.

This causes a lateral growth of the oxide accompanies by the development of highly compressive stresses. The existance of relatively high dislocation densities in these regions are considered to be microstructural manifestations of this lateral growth. Lateral growth further results in the detachment of the oxide scale from the alloy by the creation of cavities between the alloy and the scale, and the development of a convoluted oxide morphology which coarsens with increasing oxidation time, as represented schematically in Fig. 5a.

Fe-28%Cr-1%Y

The convoluted scale morphology formed on the yttrium-free alloys, was not observed on the Y-containing alloy. The addition of yttrium to the alloys lead to the development of an irregular scale, as represented schematically in Fig. 5b. This oxide scale configuration is considered to prevent development of compressive growth stresses and formation of a convoluted oxide morphology, as observed for the almost uniform and continuous oxide scale formed on the yttrium-free alloys. Any lateral growth of the isolated regions of "large-grained" oxide would not be suppressed by surrounding scale and the oxide would merely develop over the much thinner, adjacent "fine-grained" oxide.

Detailed transmission electron microscopy of the cross-sections of the oxide scale grown on Y-free alloy did not show any considerable dislocation density. This is in agreement with the above suggestion of the absence of growth stresses within the oxide scale during the oxidation process.

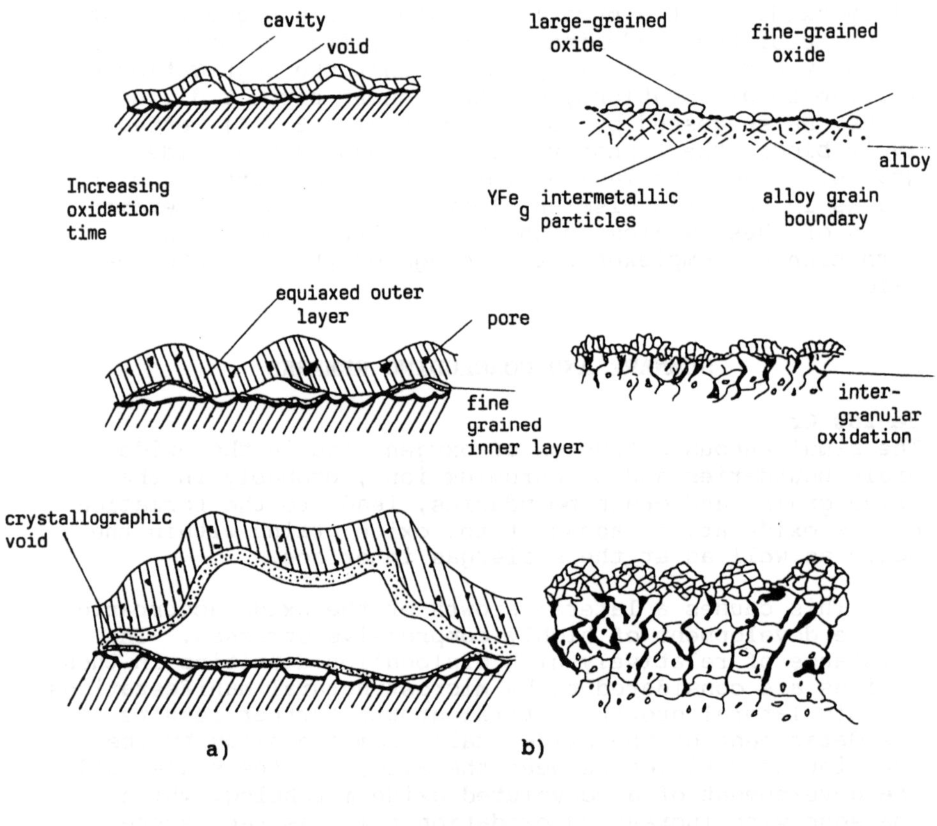

Figures 5a & b Schematic diagram showing progressive development of the main features of the oxide scale and of the alloy/oxide interface during isothermal oxidation of a Fe-28%Cr alloy, a) and Fe-28%Cr-1%Y at $1000^{o}C$, (b) respectively

ACKNOWLEDGEMENTS

The author would like to express his most sincere gratitude to:
Professor G.C. Wood and Dr F.H. Scott for their valuable advice and encouragement during the supervision of this work, and for their patience during the preparation of this study.
Dr F.A. Golightly for helpful discussions and constant help.
Mrs. R. Nygren for typing the manuscript.

REFERENCES

1. HIRBOD, M., Ph.D. Thesis University of Manchester, England, 1980.
2. STOTT, F.H., WOOD, G.C., HIRBOD, M., GOLIGHTLY, F.A., "Environmental Degradation of High Temperature Materials", Proc. of the Institution of Metallurgists Conference, Douglas, Isle of Man, 1980, No. 15.1
3. WOOD, G.C., STOTT, F.H., J.Mat.Sci.Tech., 1987, 3, 519.
4. CAPLAN, D., SPROULE, G.I., Oxid.Met. 1975, 9, 459
5. LILLERUD, K.P., KOFSTAD, P., J.Electrochem. Soc. 1980, 127, 2397
6. SKELDON, P., CALVERT, J.M., LEES, D.G., Oxid.Met. 1987, 28 (1/2), 109
7. GOLIGHTLY, F.A., STOTT F.H., and WOOD, G.C., J.Electrochem, 1979, 126, 1037
8. STOTT, F.H., Mater.Sci.Techn., 1988, 4, 437
9. FOX, P., TATLOCK, G.J., Mater.Sci.Tech., 1988, 4, 439
10. STOUT, J.H., GERBERICH, W.W., S. LIN and LII, Proc.Symp. on Fundamental Aspects of High Temperature Corrosion, II, The Electrochem. Soc., Boston, 1986, 173
11. ROMANARAYANAN, T.A., and PETKOVIC-LUTON, R., ibid. 75
12. SMIALEK, J.L., and GIBALA, R., NASA Technical Memorandum 79259, Cleveland, Ohio, 1979.
13. BENNET, M.J., Proc. 10th Int. Cong.Met.Corr., Oxford and IBH Publishing Co, P.V.T. Ltd., New Delhi, 1987, 4, 3761
14. KIM, Y.K., PRYZBYLSKI, K., and YUREK, G.J., 259, ibid Ref. 20.
15. KING, W.E., PARK, J.H., ROUTHORT, J.L., and GORETTA, K.C., Oxid.Met. 1988, 29, 314, 217

Effect of Surface Texture Geometry on the Dynamic Performance of Hydrostatic Bearings

Z. S. Safar, M. O. A. Mokhtar and T. A. Osman

Cairo University
Giza
Egypt

ABSTRACT

This paper presents a numerical study of dynamically loaded thrust bearing taking into consideration the effect of surface geometry. Non-parallelism between the bearing and the runner surface is also considered since this is a common problem in hydrostatic thrust bearing, which is caused by machining and assembly errors, elastic and thermal deformations and vibrations. The effect of axisymmetric surface texture in macro-scale is represented by assuming a preassigned sinusoidal undulations with different amplitude and frequencies. The modified Reynolds equation for misaligned thrust bearing with squeeze being included and the film geometry representing a rough surface, is solved numerically using Gauss Siedel iterative technique to obtain the pressure distribution, hence the bearing performance characteristics, namely; load capacity, bearing stiffness and damping coefficient in the form of load, stiffness and damping factors. The results are obtained for both multi-recess and mid-circular recess bearings taking into consideration different values of macro frequencies, micro amplitude ratios and macro amplitude ratios superimposed on smooth parallel or misaligned reference bearing.

INTRODUCTION

Some investigators have paid attention to the surface topography, it has been reported from time to time that the influence of surface topography on the lubricated bearing is considerable. The clarification of the squeezing mechanism of lubricant, which flows between rough surfaces, is quite important for the improvement of the dynamic behavior. Roughness effects in porous circular squeeze-plates with arbitrary wall thickness was studied by Prakash and Tiwari [1]. They compared between their exact solution and an earlier approximate analysis to determine the range of influencing parameters for which the approximate solution was acceptable. In the study of damping characteristics of squeeze films under dynamic (cyclic) load, Younes and Mokhtar [2] presented a paper, fora circular hydrostatic thrust bearing with different geometrical configurations and surface texture irregularities on both the macro-and micro scales and subjected to sinusoidal harmonic excitations .Compared to the performance of a flat and smooth disc, results indicated that the micro-and macro surface undulations play a major role in dictating the squeeze action damping characteristics. In a later work by Mokhtar and Younes [3], they observed that the concave and convex surfaces provide respectively lower and higher damping loads when compared to a flat smooth circular disk.

The surface undulation effect on the squeezing mechanism of lubricant was examined by Marui and et al [4]. They introduced an approximate equation of squeeze flow for rough surface. They concluded that the concrete means for improving the dynamic characteristics of the slider in hydrodynamic or boundary lubrication may be clarified through the design parameters and factors.

There are only few works published on the study of multi-recesses bearing . The effect of various design parameters e.g concentric recess to supply pressure ratio, L/D ratio, recess to bearing area ratio and number of recesses are studied for capillary and orifice compensated bearings, by Ghosh and Majumdar [5] The dynamic performance of multi-recess hybrid-operation oil journal bearings was made by Chen and et al [6], and they computed the stiffness and damping coefficients of capillary compensated bearing with four recesses

The purpuse of this paper is to study the dynamic performance for both multi-recess and mid-circular recess bearings taking into consideration the effect of micro and/or macro undulations superimposed on smooth parallel or misaligned reference bearing .

ANALYSIS

Film thickness variation

Assuming the macro-and micro-texture of the bearing surface to be represented by superimposed sinusoidal undulations with different amplitudes and frequencies, the variation of the oil film thickness of the tilting bearing surfaces could be written in the form:

$$H=1 + H_v \sin \omega t + C R (1-\cos\theta) + H_1 \sin (R \pi/2 \pm \pi/2) + H_2 \sin (2\pi R S_x + \Gamma) \quad (1)$$

The pressure equation for incompressible isoviscous fluid with the flow being assumed laminar is written as:

$$\frac{\partial}{\partial R}\left(R H^3 \frac{\partial P}{\partial R}\right) + \frac{\partial}{R\partial \theta}\left(H^3 \frac{\partial P}{\partial \theta}\right) = B_n R \frac{\partial H}{\partial \theta} + S_n \frac{\partial H}{\partial T} \quad (2)$$

By using the substitution $P=S\ H^{-1.5}$ together with the finite difference expressions, equation (2) can be rewritten as follows :

$$S(i,j)=\{C_2\ S(i+1,j) + (2-C_2)\ S(i-1,j) + C_3(S(i,j+1)-S(i,j-1)) - C_1\ G\}/C_4 \quad (3)$$

where:

$$C_1=(\Delta R/R)^2$$

$$C_2=1 + \Delta R/(2R)$$

$$C_3=C_1/(\Delta\theta)^2$$

$$C_4=2(1+C_3-C_1 F/2)$$

$$F = \left(-\frac{3}{4 H^2}\right)\left[\left(\frac{\partial H}{\partial \theta}\right)^2 + 2 H \left(\frac{\partial^2 H}{\partial \theta^2}\right) + R^2 \left(\frac{\partial H}{\partial R}\right)^2 + 2 H R^2 \left(\frac{\partial^2 H}{\partial R^2}\right) + 2 H R \left(\frac{\partial H}{\partial R}\right)\right] \quad (4)$$

$$G = \left(\frac{R^2}{H^{1.5}}\right)\left[B_n \frac{\partial H}{\partial \theta} + S_n \cos \omega t\right]$$

For the simple cases where a perfectly flat and smooth surfaces is assumed the tilting terms are reduced Equation (2) with the film thickness variation equation

(1) for tilting bearing surfaces could be solved using Gauss Seidel iterative technique to obtain the bearing The pressure equation is subjected to the boundray conditions shown in fig. (1) for the two cases considered .

Bearing performance Characteristics

Load Capacity

By integrating the pressure distribution over the bearing area, the load capacity in nondimensional form could be computed as follows:

$$W = \int_{R=0}^{R=1} \int_{\theta=0}^{\theta=2\pi} P(i,j)\, R\, d\theta\, dR \qquad (5)$$

Therefore, by using Simpson rule the load capacity equation (6) is calculated. The solution is repeated at three different time domain to calculate bearing stiffness and damping coefficient after Szeri [7] .

Bearing Stiffness And Damping Coefficient

In order to achieve a better understanding of bearing stiffness and damping coefficient, a linearized-load coefficients have been adopted by Szeri [7] as follows :

$$dW = W - W_o = \left(\frac{\partial W}{\partial H}\right) H + \left(\frac{\partial W}{\partial H^{\cdot}}\right) H^{\cdot}$$

The partial derivatives of the load are the system stiffness

$$\text{Bearing Stiffness} = S_t = -\left(\frac{\partial W}{\partial H}\right) \qquad (6)$$

$$= \frac{-[W(h_o + 2\,dh,\ 0) - W(h_o,\ 0)]}{2\,\Delta h}$$

and the system damping

$$\text{Damping coefficient} = C_D = \left(\frac{\partial W}{\partial H^{\cdot}}\right) \qquad (7)$$

$$= \frac{[W(h_o,\ 2\,dh^{\cdot}) - W(ho,\ 0)]}{2\,\Delta h^{\cdot}}$$

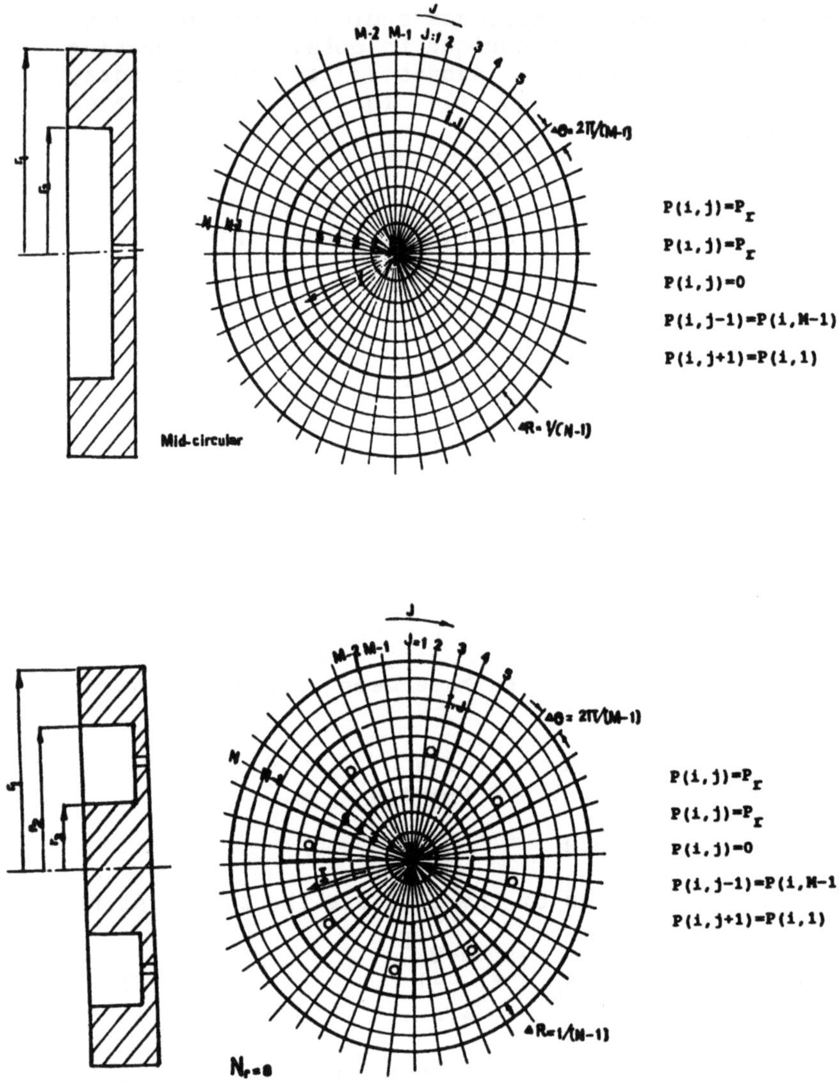

Fig.(1) Configuration of Hydrostatic Thrust Bearing

The design parameters are fixed as such: squeeze number S_n=10, radii ratio R_r=0.7, recess radii ratio R_{r1}=0.1, angular frequency $\omega t=200^o$ with film thickness amplitude H_v=0.25 and capillary tube with coefficient Z=10. In case of tilting bearing the tilt parameter and the bearing number are taken as C=0.5 and B_n=60.

RESULTS AND DISCUSSION

In case the bearing surface has been altered by macro or (and) micro undulations described by equation (1) , the resulting load capacity bearing stiffness damping coefficient will differ than that of the flat and smooth reference bearing for each specific geometry by the load factor stiffness factor and damping factor defined as :

$$W_R = W/W_o$$

$$S_{tR} = S_t/S_{to}$$

$$C_{DR} = C_D/C_{Do}$$

As result of the stiffness and damping definitions, both stiffness and damping factors give the same values, however, the proceeding discussions will be confined to the load and stiffness factors .

Micro surface undulation

The micro surface undulation contains a series of very small concave and convex areas distributed over the circumferential of the sill bearing surface. Consequently the concave shape leads to increase the oil film thickness which in turn results in relatively lower values in the load and stiffness factors .

Figures (2) through (5) show the variation of the load and stiffness factors with micro undulation frequencies (S_x) and different micro amplitude ratios H_2 superimposed on smooth parallel and tilted bearing surfaces. These plots are developed for the case of multi-recess bearing (Nr=8) and that with mid-circular recess .

The analysis has shown a general decrease in load factor, stiffness factor and consequently damping factor irrespective of the micro undulation amplitude and frequency. This is existent for either load and stiffness where the factors have values less than one .

With referance to parallel bearing surfaces, it is

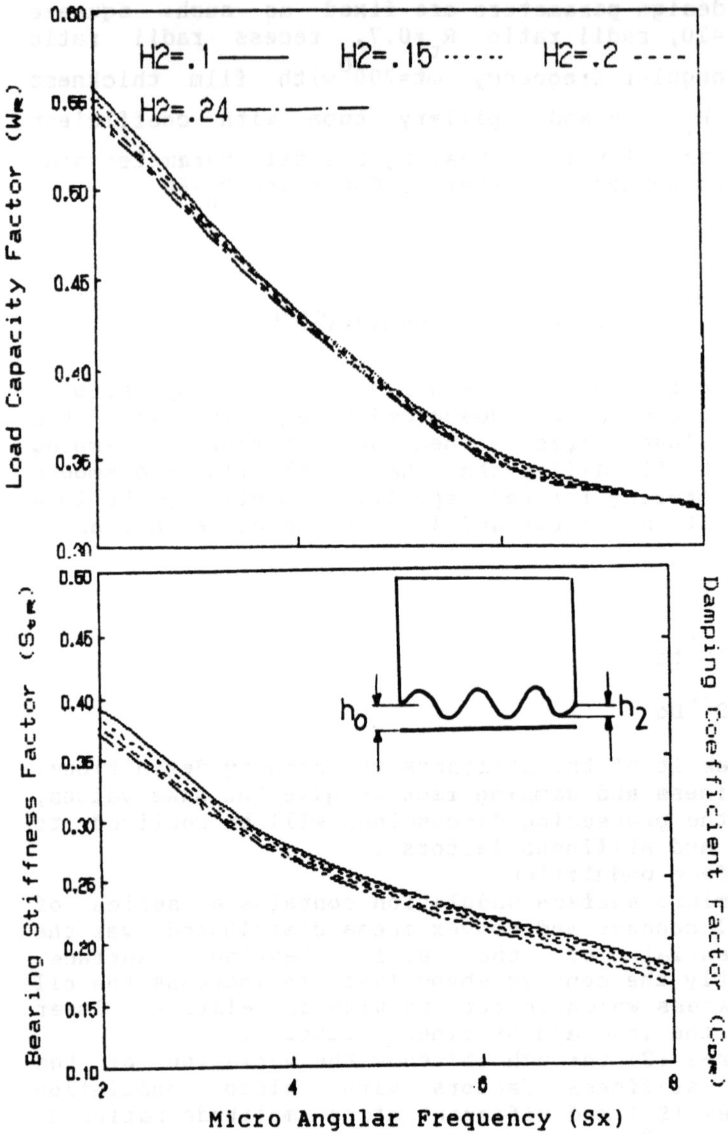

Fig.(2) Variation of load, stiffness and damping factors as function of micro angular frequency with different film thickness micro amplitude ratio and with 8 recesses parallel bearing

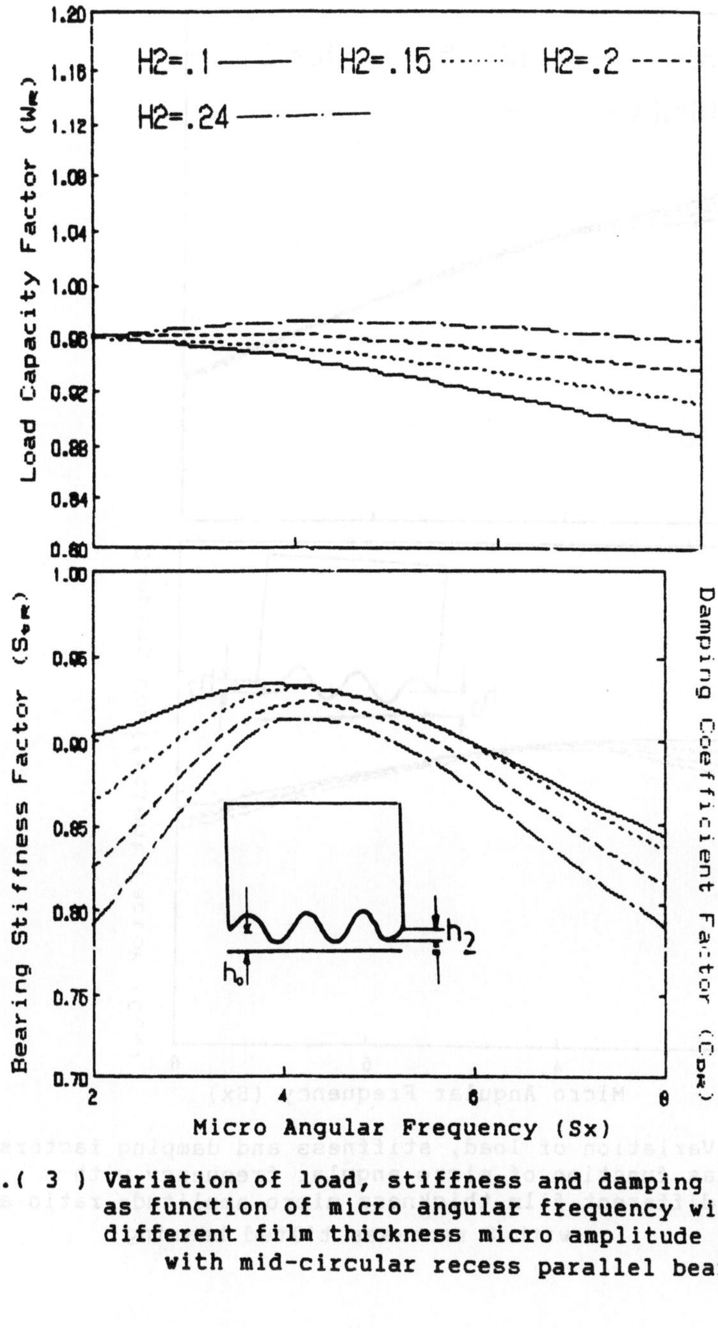

Fig.(3) Variation of load, stiffness and damping factors as function of micro angular frequency with different film thickness micro amplitude ratio and with mid-circular recess parallel bearing

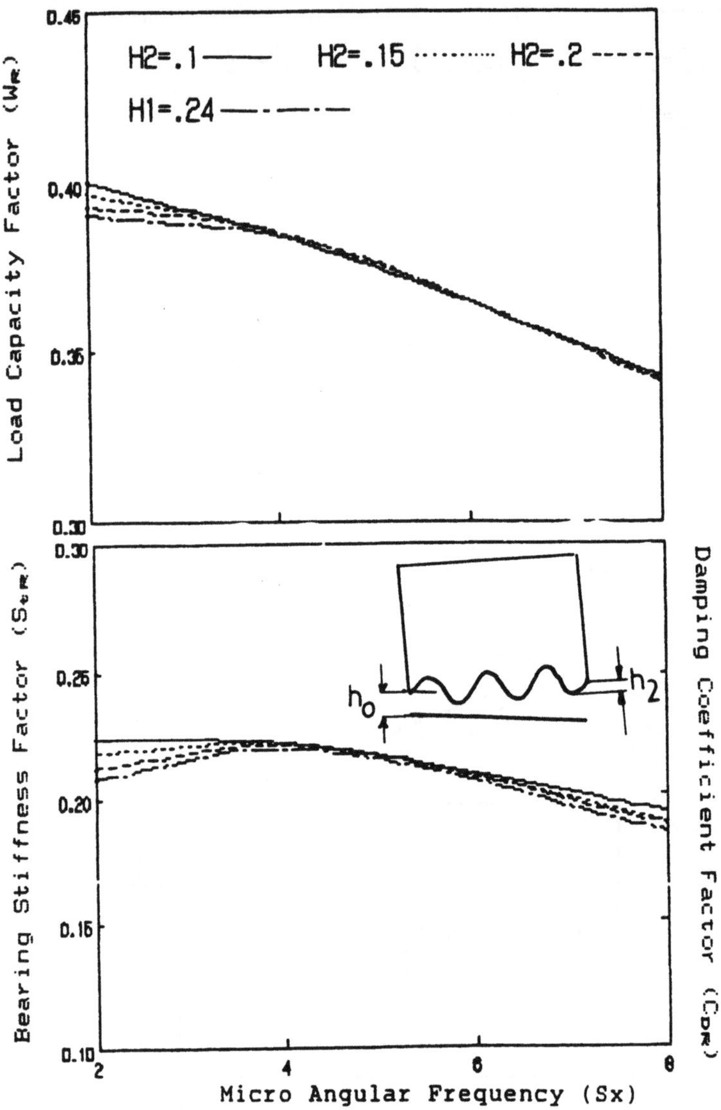

Fig.(4) Variation of load, stiffness and damping factors as function of micro angular frequency with different film thickness micro amplitude ratio and with 8 recesses tilted bearing

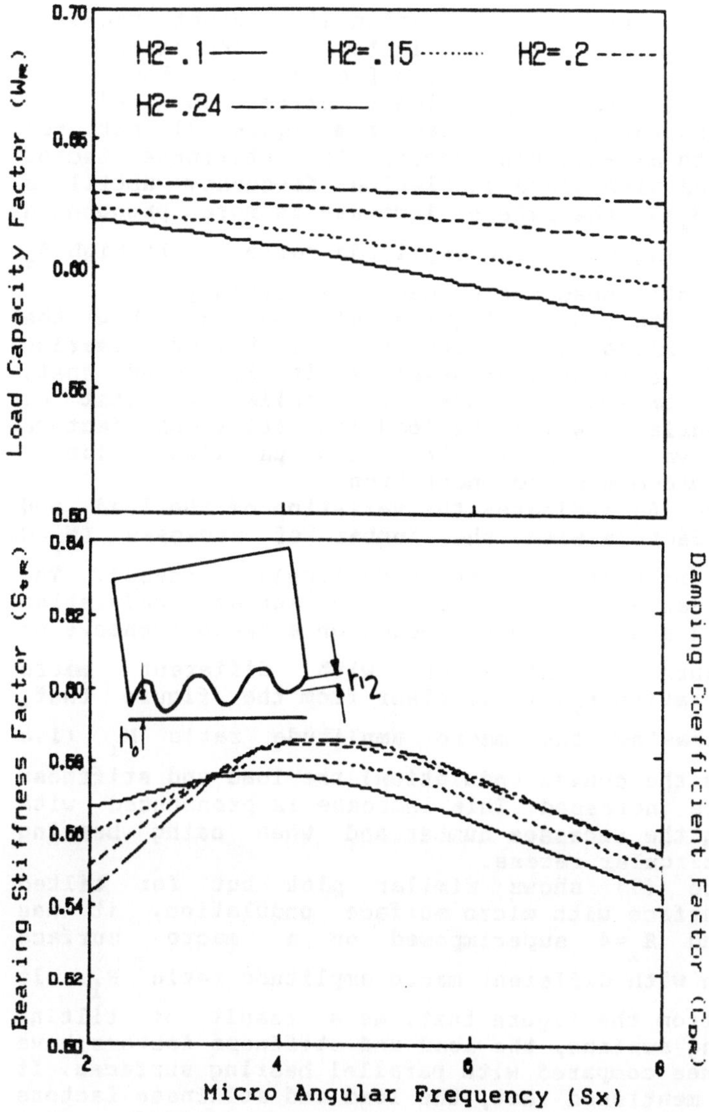

Fig.(5) Variation of load, stiffness and damping factors as function of micro angular frequency with different film thickness micro amplitude ratio and with mid-circular recess tilted bearing found from figure (2), for bearing with eight recesses that, the load and stiffness factors are decreased with increasing the micro undulation frequencies. The micro amplitude ratio has insignificant effect on load and stiffness factors. It is clear from figure (3) that, for bearing with mid-circular recess, the stiffness factor is increased with micro undulation frequency uptill a frequency S_x=4, the rate of increase is more pronounced with increasing the micro amplitude ratio H_1. At high S_x values the stiffness factor decreases rapidly .

On the other hand, figures (4) and (5) show the effect of micro undulation on the tilted bearing surface, for both type of bearings, it is found that, the curve behavior trends are similar to that of parallel surface, while the load and stiffness factors have lower values compared to that of parallel surface.

Macro and micro surface undulation

Figure (6) indicates the variation of the load and stiffness factors with the number of recesses (N_r=0 refers to the bearing with mid-circular recess). The parallel bearing surface has micro surface undulation with H_2=0.2 and S_x=4 superimposed on a macro (concave or convex) surface undulation with different macro amplitude ratios H_1. It is clear from the figure that, with decreasing the macro amplitude ratio H_1 (i.e increasing the convex undulation) the load and stiffness factors are increased. This increase is pronounced with decreasing the recesses number and when using bearing with mid-circular recess.

Figure (7) shows similar plot but for tilted bearing surface with micro surface undulation, it has H_2=0.2 and S_x=4 superimposed on a macro surface undulation with different macro amplitude ratio H_1. It is found from the figure that, as a result of tilting the bearing surface, the load and stiffness factors have lower values compared with parallel bearing surfaces. It should be mentioned that, the load and stiffness factors are increased with decreasing the macro amplitude ratio H_2=-0.5. This increase is noted at bearing with mid-circular resess. On the other hand, the stiffness factor has higher values with H_1=-0.5 and bearing with four recesses .

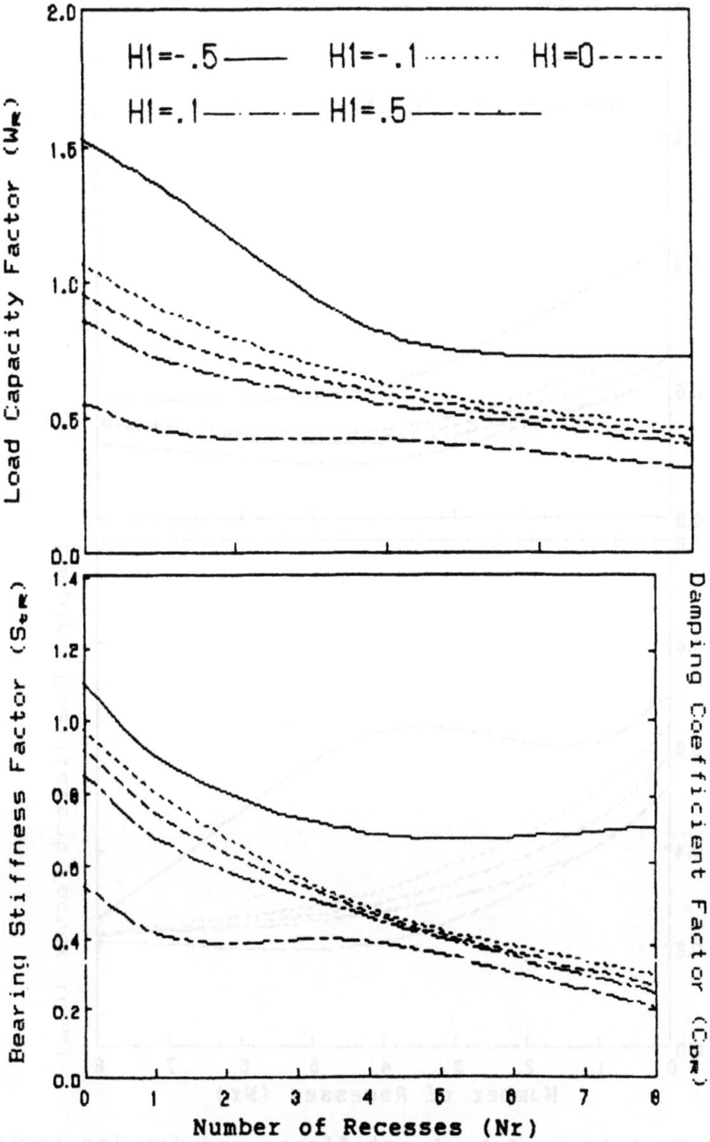

Fig.(6) Variation of load , stiffness and damping factors as function of number of recesses with different film thickness macro amplitude superimposed on micro parallel bearing surfaces

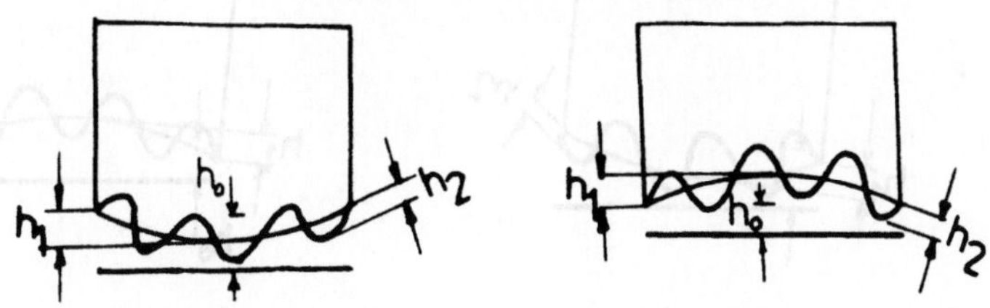

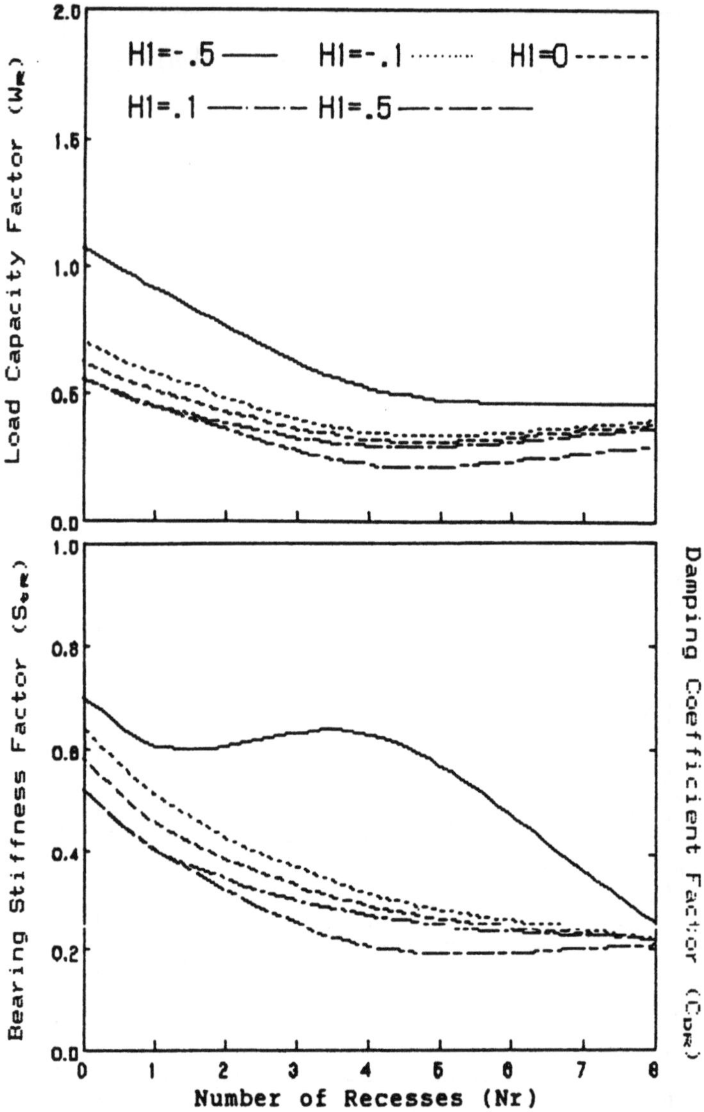

Fig.(7) Variation of load, stiffness and damping factors as function of number of recesses with different film thickness macro amplitude superimposed on micro tilted bearing surfaces

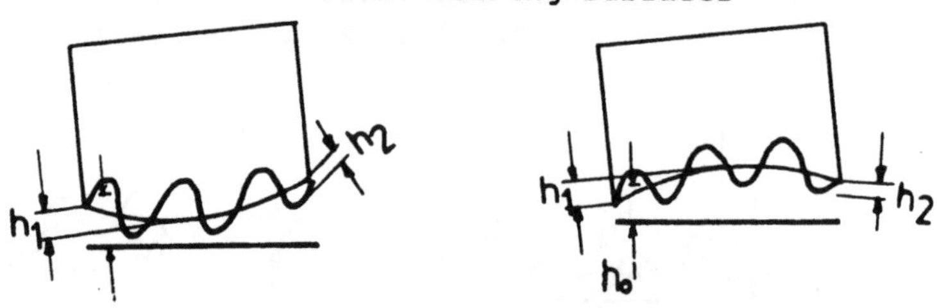

CONCLUSIONS

It should be mentioned from the previous results and discussion that,

-Micro surface undulation silghtly reduces the load, stiffness and damping factors in parallel or tilted for bearings. However, the micro amplitude ratio has insignificant effect on load, stiffness and damping factors.

- Smooth convex macro surface undulation or even convex macro surface undulation superimposed on micro undulation results in an increase in the load, stiffness and damping factors, in the two cases of parallel or tilted surfaces reference. This result is pronounced when using mid-circular recess bearings .

REFERENCES

1.Prakash, J. and Tiwari, K."Roughness Effect in Pourous circular Squeezed-Plates with Arbitrary Wall Thikness" ASME Journal of Tribo., Januray 1983, V. 105, P. 90-95 .

2.Younes, Y.K. and Mokhtar, M.O.A."Analysis of Damping Characteristics of Squeeze Films Under Dynamic (Cyclic) Loading" Tribology International, June 1984, V. 17, No. 3, P. 139-145 .

3.Mokhtar, M.O.A. and Younes, Y.K."Surface Waveness Effect on Damping of Oil Squeeze Circular Disks" 4th Cairo University MDP Conf., Dec. 1988, P. 287-293 .

4.Marui, E. and et al "On the Squeeze Effect of Lubricant Between Two Rough Surfaces" ASME Journal of Tribo., Oct. 1987, V. 109, P. 696-703 .

5.Ghosh, M.K. and Majumdar B.C."Design of Multirecess Hydrostatic Oil Journal Bearings" Tribo. International, April 1980, P. 73-78 .

6.Chen, Y.S., Wu, H.Y. and Xie, P.L."Stability of Multirecess Hybrid-Operating Oil Journal Bearings" ASME Journal of Tribo., January 1985, V. 107, P. 116-121 .

7.Jeng, M.C. and Szeri, A.Z."A Thermohydrodynamic Solution of Pivoted Thrust Pads: Part III Linearized Force Cofficients" ASME Journal of Tribo. V. 108, April 1986, P. 214-218 .

Study of Tools for Hot Rolling of Steel Alloy Tubes

S. Uchida, I. Masuda and K. Watanabe

Nippon Steel Corporation
Kitakyushu
Japan

ABSTRACT

Main problems in rolling of steel alloy tubes are not only heavy mill load and poor ductility of the tube materials but also short tool life. It is, therefore, important to properly choose tool materials and methods of surface treatment in order to obtain longer tool life.

Lives of tool materials such as refractory metal and high alloy steel were investigated on experimental models of a press roll piercer and an elongater. Methods of surface treatment such as electrodeposited composite coating and aluminizing were also tested.

The following results were obtained : (1)high alloy steel with electrodeposited composite coating or aluminizing stood hot wear but deteriorated with sticking in rolling of 18%chromium-8%nickel stainless steel. (2)Molybdenum was the most excellent against hot wear and sticking among the tested materials.

Since molybdenum was expensive, some methods were developed to repair used molybdenum tools by hot isostatic pressing(HIP) and low pressure prasma spray(LPPS). Posibilities of those methods were confirmed on the experimental mills.

1. INTRODUCTION

As the seamless pipes are used in severe environment of oil digging, demands for alloy tube and high alloy tube are incresed. Alloy tubes are manufactured by hot extruding process or hot rolling process. Hot extruding process is suitable to manufacture tubes of materials of poor workability, but it is difficult to make long size produts and its cost is high. Therefore, hot rolling process is preferable. In this case, we have to solove such problems as not only heavy mill load caused by high deformation resistance, poor ductility of the tube material but also short tool life. Among them, tool life is the most difficult and important problem. Today, plugs for a piercer and an elongator have average service life of about 300 tubes in making low carbon steel, but they are damaged by high temperature and stress after only 2 or 3 tubes in making 13% chromium steel. In the case of higher alloy tube, a plug is supposed to be damaged after rolling half length of a tube. As a result, products are not made properly because of poor inner quality. It is, therfore, important to properly choose tool materials and methods of surface treatment in order to obtain longer tool life. They were investigated mainly on experimental models of a press roll piercer and an elongator.

2. DURABILITY OF TEST PLUGS

A plug is exposed in a hard thermal condition in experimental model mill rolling, since its size is smaller than that of a practical mill. We can, therefore, estimate relative durability through rolling alloy steel tubes of a smaller quantity. Tool materials such as refractory metal and heat resisting metal whose surface is treated against heat and sticking were investigated.

We discuss here of the test results of the tool materials as follows. 1) TZM molybdenum; it consists of a dispersion strengthening matrix with a very small amount of titanium, zirconium and carbon, and its strength is about 500MPa at such high temperature as 1000 degrees centigrade(℃). Molybdenum oxide is expected to have lubrication effect. 2) Heat resisting steel and hot tool steel treated by aluminizing and electrodeposited composite coating. They were compared with plugs practically used in rolling low carbon tubes. Treatment by vapour deposite process is excluded because film provided by that method is too thin, 1-2μm thick, to endure heat attack and stress through piercing and rolling.

2.1 Durability of the test plugs in press roll piercing

2.1.1 Press roll piercing conditions

18%Chromium(Cr)-8%Nickel(Ni) stainless steel heated at 1200℃ was used as pierced material. Inlet billet size was 80mm square and 500mm long. It was pierced to 93mm in outside diameter, 23.25mm in thickness. Fig.1 shows dimensions of a plug.

2.1.2 Materials and surface treatment of test plugs

The following plugs were tested. (1)TZM molybdenum composed of 99.4%Mo, 0.5%Ti, 0.08%Zr, 0.02%C, which was preheated at about 400℃ before using, because its impact value is low at lower temperature, as shown in Table 1. (2)Aluminized and diffusion annealed alloy composed of 53.5%Fe-17%Cr-24%Ni-5%Al-0.5%Si. (3)Hot tool steel composed of 5%Cr-1%Mo coated with boron nitride by electrodeposition method. (4)Steel alloy which contains 3%Cr-1%Ni, and which was treated in the air with steam at temperature about 900℃ to provide scaled surface.

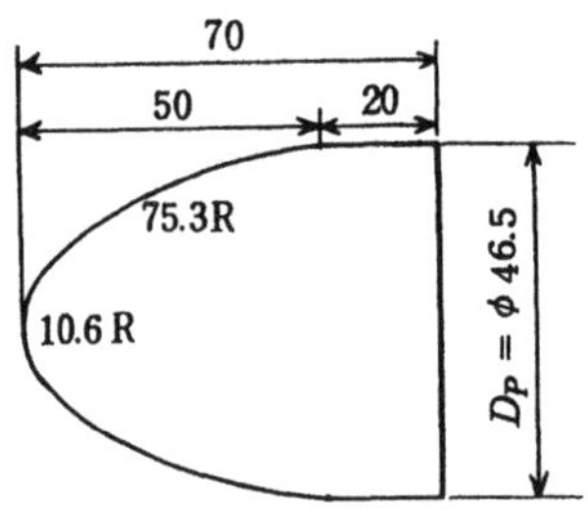

Fig. 1. Dimensions of press roll piercer plug(unit:mm)

TABLE 1
Impact value of TZM

Temperature	Impact value
room temp.	4.2N·m/cm^2
400℃	42.6N·m/cm^2

2.1.3 Durability of the test plugs in piercing on an experimental mill

(1)Surface conditions of the test plugs after piercing.

Test plugs after piercing one 18%Cr-8%Ni stanless steel are compared with the initial ones as shown in Fig.2. A TZM plug showed no sticking and damage after piercing. It was, then, tried 10 more times without any damage. Aluminized plug was not peeled off or not damaged, but it was covered partially with pierced material. Plug coated with boron nitride was partially sticked and damaged. Its damage is considered to be caused by the weakness of substratum as 5%Cr-1%Mo steel. Scale treated plug was partially sticked and damaged too.

TZM was the best among the test materials. Higher alloy than pierced material was needed for plug against damage. It was hard to prevent a plug from sticking by alminizing or coating with boron nitride.

(2)Piercing loads

Piercing loads, such as mandrel thrust force to support plug, roll torque, roll separating force and pusher force, are shown in Fig.3. Mandrel thrust force was about 784,000N for the plugs of TZM, aluminized alloy and scale treated steel alloy, and about 882,000N for the one coated with boron nitride. The latter is seemed to have higher friction coefficient than others. Though we expected that molybdenum oxide would act as a lubricant, it was not confirmed for TZM plug. The reason for that was considered that plug was exposed in too little oxygeon during piercing for molybdenum oxide to be formed sufficiently.

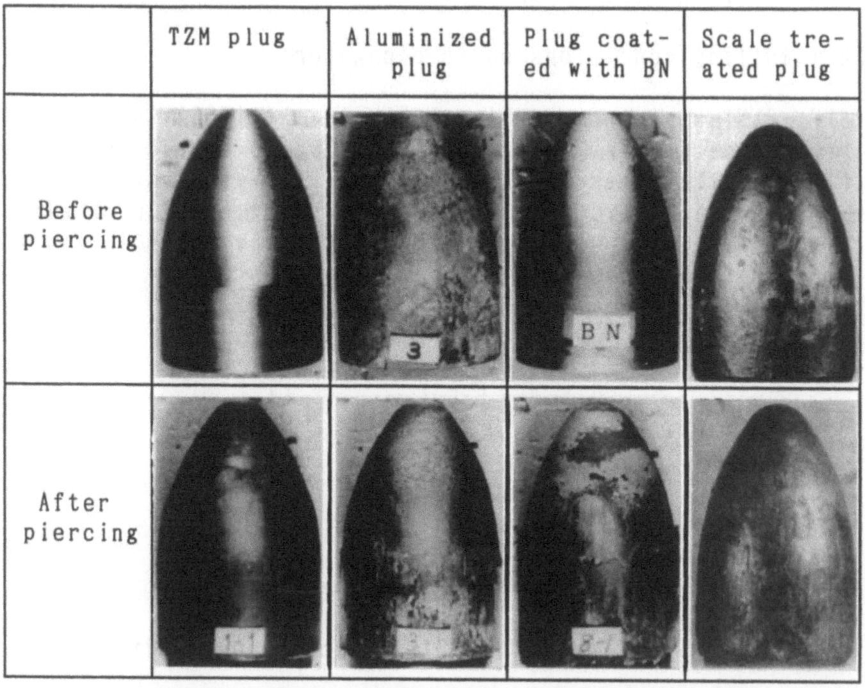

Fig. 2. Surface conditions of test plugs after piercing one stainless steel tube.

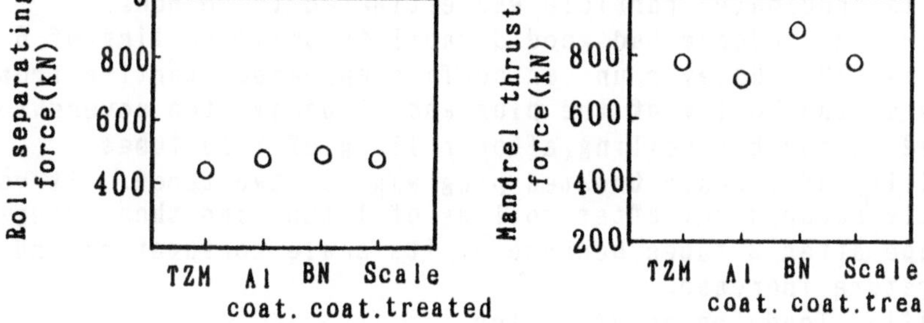

Fig. 3 Piercing loads on a press roll piercer.

2.2 Durability of test plugs on an experimental elongator

We investigated mainly plugs of molybdenum such as TZM and sintered pure molybdenum on an experimental elongator, because it showed long life on an experimental press roll piercer.

2.2.1 Rolling conditions on an elongator

18%Cr-8%Ni stainless steel tube heated at 1200℃ was used for elongated material. Inlet size was 75mm in outside diameter, 14.25mm in thickness, 500mm in length. It was elongated to 75mm in outside diameter, 5mm in thickness. Plug 60mm in outside diameter was used. Dimensions of a plug are shown in Fig.4.

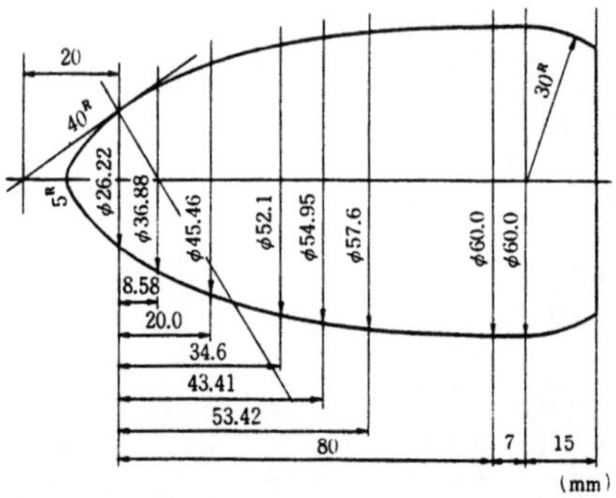

Fig. 4 Dimensions of elongator plug.(unit:mm)

2.2.2 Durability of test plugs in rolling on an experimental elongator

(1) Surface condition of test plugs after rolling.

Test plugs after rolling 18%Cr-8%Ni stainless steel tubes are compared with the initial ones as shown in Fig.5.

TZM plug had no sticking and no damage after rolling, judging from its smooth surface. Its durability was certained more than 13 tubes. Its surface temperature reached at 900-1000℃ in rolling. Volume of smoke occurred during cooling. The plug surface was gradually covered with precipated particles like salt, when it was exposed in the air after rolling. From X-ray analysis, precipated particle was estimated to be MoO_3.

Sintered molybdenum had good durability until rolling of 6 tubes as TZM. However, uneven surface appeared, starting from cracks at the bottom of the plug and it became too severe to be used in further rolling, after rolling of 7-10 tubes.

Durability of a scale treated plug was for two tubes. It was slightly scraped out after rolling of 1 tube and then changed in shape after 2 tubes because of its scale consumption and temperature increase.

(2) Rolling loads on an elongator

Rolling loads, such as mandrel thrust force, roll torque, roll separating force, are shown in Fig.6. A TZM plug and sintered molybdenum plug had 10% less in load than a scale treated plug. There was no difference between a TZM plug and a sintered molybdenum plug. Molybdenum oxide formed during rolling was seemed to have acted as a lubricant in elongator rolling.

(3) Inner surface conditions of the tubes

Inner surface conditions of the tubes rolled by three test plugs were shown in Fig.7. A TZM plug showed no trouble and smooth condition after rolling of 13 tubes. A sintered molybdenum plug had hollows after rolling of 7 tubes on account of deformation of the plug itself. And a scale treated plug had a severe uneven surface caused by scraping after rolling of 2 tubes.

(4) Consumption of a molybdenum plug.

A molybdenum plug was heated up to high temperature during rolling because of its high thermal conductivity and was covered with its oxide. Molybdenum oxide was of so high vapor pressure that plug was consumed by its sublimation in the air. Its consumption after rolling was measured and was shown in Fig.8. Its consumption was about 2.5 gram per one tube rolling (equals to 0.1 weight percentage), and there was little change in outer diameter of the plug after rolling of 13 tubes.

(5) Comparison TZM with sintered molybdenum.

There were differences in property at high temperature between TZM and sintered molybdenum. Change in micro structure at high temperature of the both are shown in Fig.9. TZM was consisted of a dispersion strengthening matrix with so very small elements that there was no change in micro structure at 1200℃. Sintered molybdenum was brittle at the same temperature, because its grain size became large. It is difficult to expect durability of sintered molybdenum, as temperature of a molybdenum plug reached more than 1200℃ in continuous rolling.

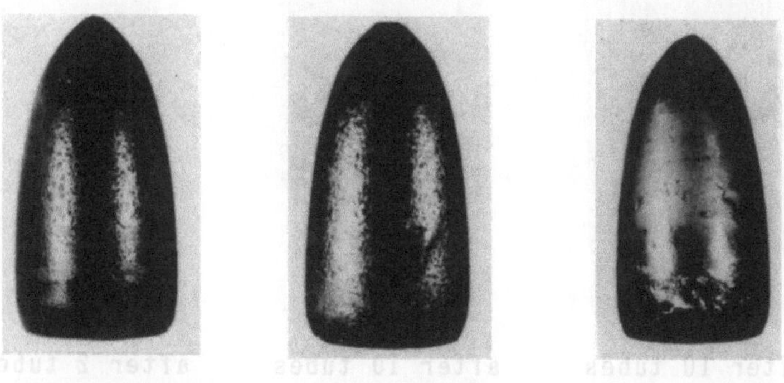

after 10 tubes TZM plug — after 10 tubes Sintered Mo plug — after 2 tubes Scale treated plug

Fig. 5. Surface conditions of test plugs after elongator rolling.

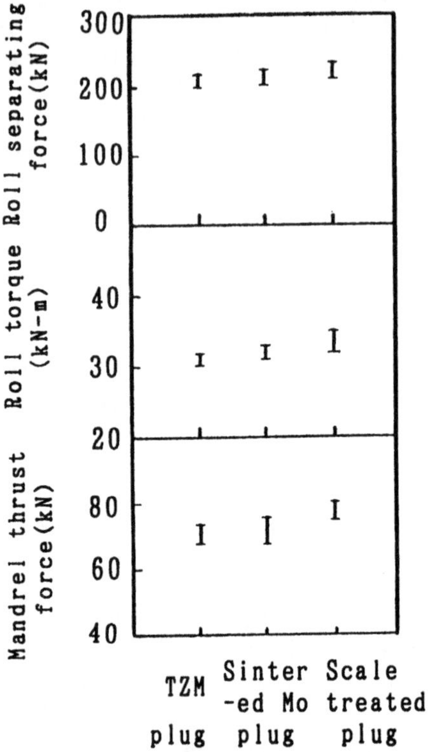

Fig. 6. Rolling loads on an elongator.

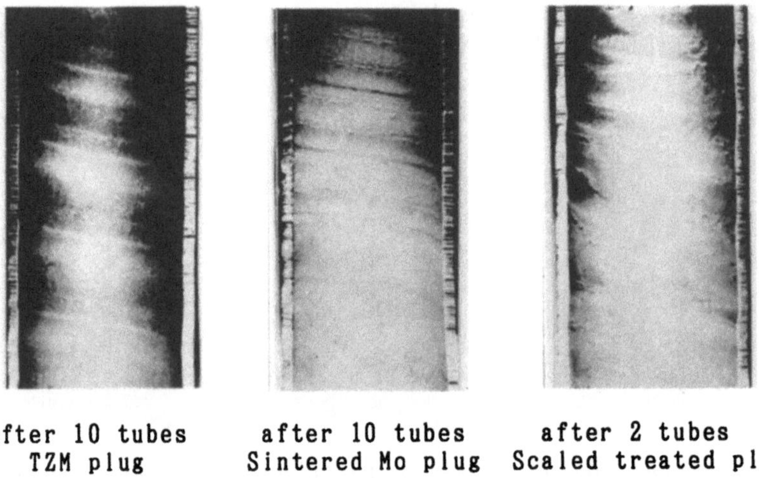

Fig. 7. Inner surface conditions of the tubes rolled by an elongator.

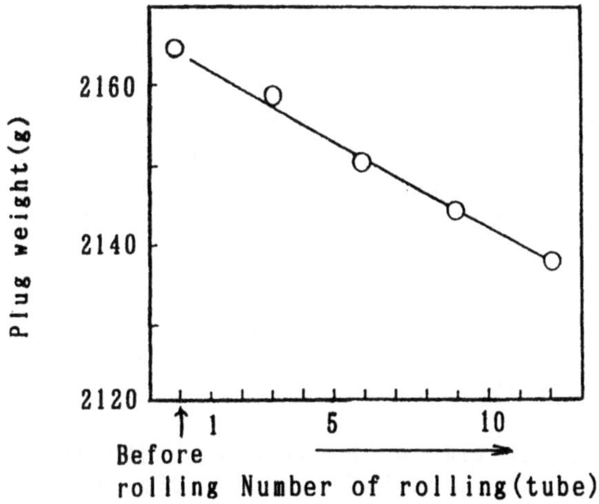

Fig. 8. Change in weight of TZM plug.

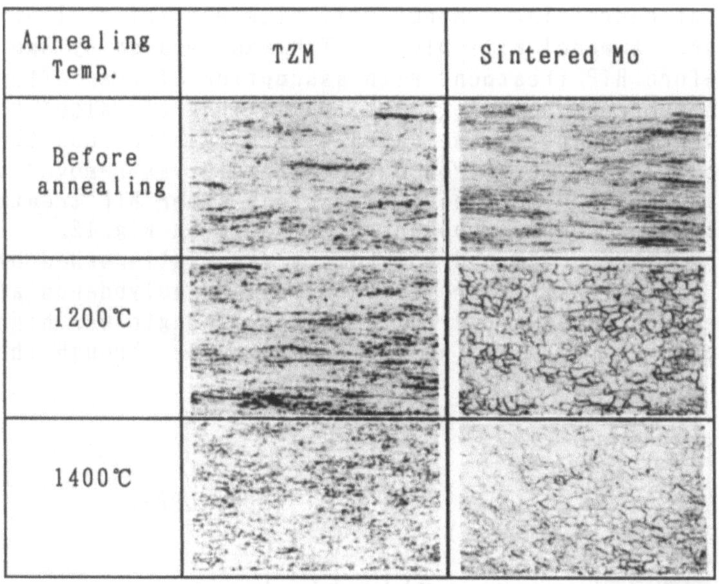

Fig. 9. Changes in micro structure of TZM and sintered Mo at high temperature.

3. METHODS OF REDRESSING A MOLYBDENUM PLUG AND REUTILIZING TECHNICS

High durability of a molybdenum plug was confirmed by the results mentioned above. It was desirable to redress the used plug, since molybdenum was expensive. Some methods were developed to redress by hot isostatic pressing (HIP) and low pressure plasma spray (LPPS).

3.1 Redressing method of molybdenum plug by HIP and LPPS.

3.1.1 Redressing conditions by HIP

Effect of HIP treatment time on bonding strength of small test pieces of TZM was investigated beforehand. Dimensions of the test pieces are shown in Fig.10. Temperature of HIP treatment was set at 1350℃ with pressure 120MPa. Bonding strength was evaluated by tensile test. In comparison of treatment time of 3 hours and 5 hours, the former was better as shown in Fig.11. It was because longer treatment time made grain size larger at high temperature as 1350℃. As a result, HIP treatment conditions of redressing a model mill plug was set at 1350℃ X 120MPaX3hour. A model mill plug of TZM was resaved by 3mm in diameter before HIP treatment with assumption of wear after rolling. Molybdenum powder (about 100μm in grain size) was filled around the plug in the capsule of mild steel, and the capsule was welded by electron beam. Capsule was removed and the plug was machined to proper dimensions after HIP treatment. Micro structure of bonding boundary is shown in Fig.12.
The boundary between TZM and molybdenum was distingushed by difference in micro structure between TZM and molybdenum after HIP treatment. It was found that bonding strength was high, because there were crystalline grains expanding through the boundary.

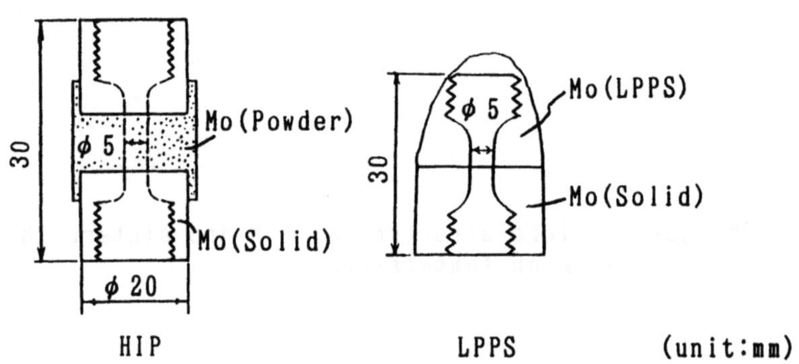

Fig.10. Dimensions tensile test pieces for HIP and LPPS.

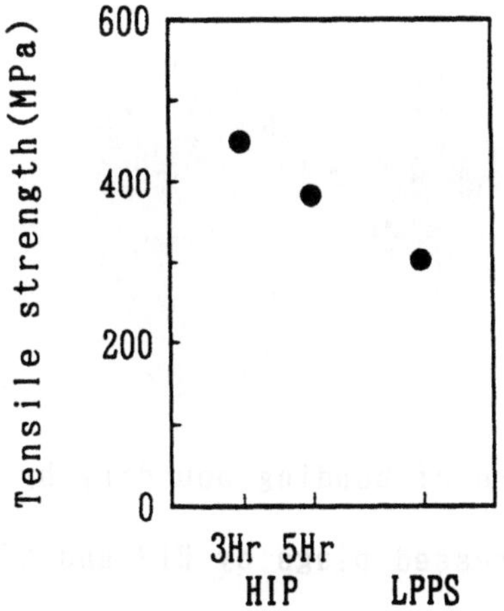

Fig.11. Tensile strength of bonded parts by HIP and LPPS.

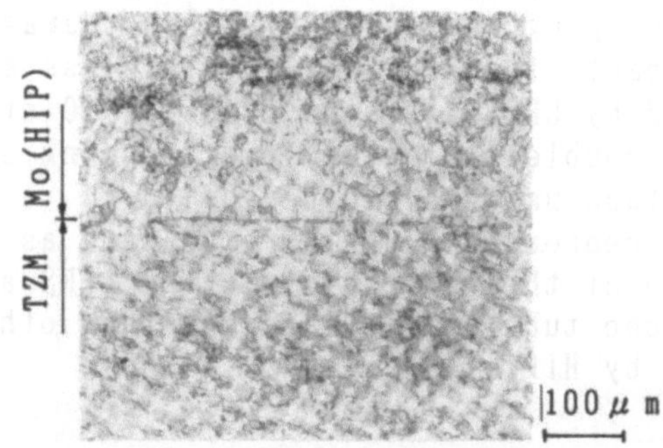

Fig.12. Micro structure of bonding boundary by HIP.

3.1.2 Redressing conditions by LPPS

A model mill plug of TZM was resaved by 3mm in diameter before LPPS treatment, as in HIP case. It was sprayed with pure molybdenum powder of 50 μm in grain size in the LPPS chamber controlled at pressure, 20-30 torr, after arc cleaned. The treated plug was used for rolling without machining. Boundary bonding strength was about 300MPa. This value was about 60 percent of that by HIP. It was, however, much higher than that by atomospheric pressure plasma spray(APPS) which was usually dozens MPa. Boundary micro structure of LPPS is shown in Fig.13. There was no diffusion on its boundary, differently from micro structure by HIP. Layer of LPPS was quenched so that its hardness was higher by 60 in micro vickers hardness unit than that of the substratum. It was, therefore, expected that the layer by LPPS surpassed that by HIP against wear because of its high hardness.

Fig.13. Micro structure of bonding boundary by LPPS.

3.2 Durability of redressed plugs by HIP and LPPS

Molybdenum plugs redressed by HIP and LPPS were tested on a model elomgator in aforesaid rolling conditions. Surface conditions of the test plugs after rolling are shown in Fig.14. As for the plug redressed by HIP, no damage was found in its redressed part after rolling 10 stainless steel tubes. The plug redressed by LPPS was also used for 10 stainless steel tubes without no trouble. However, lower bonding strength by LPPS might limit tube material to be rolled.

Rolling loads on redressed plugs were as same as on the new ones. Consumption of the redressed plug by LPPS after rolling was 1.5 gram per one tube, and it was less than others including the one redressed by HIP.

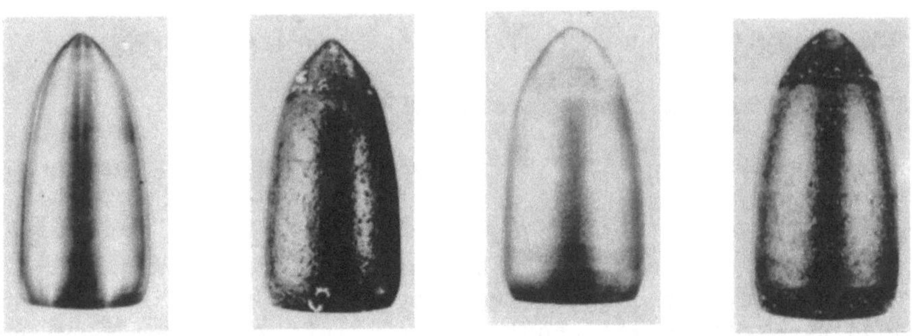

Before rolling After 10 tubes Before rolling After 10 tubes
Redressed by HIP Redressed by LPPS

Fig.14. Surface conditions of redressed plugs by HIP and LPPS after rolling.

4. CONCLUSIONS

Lives of tool materials such as molybdenum and redressed molybdenum,high alloy steel whose surface was treated were investigated on experimental models of a press roll piercer and an elongator. The following results were obtained: (1)higher alloy steel than piercing material with aluminizing and electrodeposited composite coating stood hot wear but deteriorated with sticking in rolling of 18%chromium -8%nickel stainless steel. (2)TZM,molybdenum with a dispersion strengthening matix,was the most excellent against hot wear and sticking among the tested materials. Mill load on an elongator was lower than in case of other plug materials. The reason for that is considered that molybdenum maintains high strength at high temperature over a thousand degrees centigrade and that it is covered with oxidized gas of its own. (3)Optimum conditions of redressing used molybdenum plugs by HIP and LPPS were obtained. The lives of the redressed plugs were comparable with that of the new ones. (4)Redressing by LPPS gave slightly poorer boundary bonding than by HIP,but better surface against wear.

Quality of Surface Coating by Submerged Arc Welding

R. Kejžar

ZRMK Institute for Materials
Ljubljana
Yugoslavia

ABSTRACT

Along with development of agglomerated powders, also submerged arc surfacing has become a very interesting cladding technique for surfaces subject to wear and corrosion. Agglomerated powders, which contain alloying elements, make possible single-pass surfacing of structural steel with different surfacing compositions. They are selected with regard to stresses causing wear. In surfacing, good quality of surfacings, without pores, inclusions and cracks in the surfacing iteself and in the heat affected zone, should be assured.

INTRODUCTION

The quality of surfaces subject to wear grows in importance. The lifetime depends on steels and metal alloys applied. Manufacture of tools out of high-alloyed steels or special nickel-based and cobalt-based alloys, which would assure high wear resistance, would be very expensive. Surfacing, however, makes possible to use the above mentioned expensive alloys,which often have to be imported, only for cladding of surfaces and working edges subject to wear, the remaining part of a machine or of a tool being made of a cheap, common and easily procurable structural steel or tool steel.

Surfacing processes have been developed out of various welding processes. They are selected with regard to workpieces which are to be renewed or whose surfaces subject to wear are to be refined as well as with regard to equipment and filler materials available. Gas surfacing, manual arc surfacing and TIG surfacing are still predominantly used in renewal of worn-out and damaged machine parts and tools. The main cause for a mass application of these processes is uncomplicated and cheap equipment which is used by maintenance personnel also for welding. Surfacing of thicker layers and longer worging edges as well as larger surfaces, however, is with the above-mentioned processes quite time-consuming. In these cases, semi-automatic and automatic techniques of gasshielded surfacing and of submerged arc surfacing seem to be more prospective (1-7).

In surfacing, it is not intended to melt the parent metal. The main purpose is to clad a working surface with a wear-resistant layer whose joint with the parent metal has to be of good quality. The tool surface (parent metal) should be melted only so far as to achieve faultless joining of the surfacing with the parent metal.

Strong melting of the parent metal is a disadvantage of arc surfacing processes. With the introduction of filler materials in the molten pool andin the arc, melting of the parent metal is reduced. Thus the direct arc influence on workpieces is prevented and diminished respectively (see Fig. 1.)

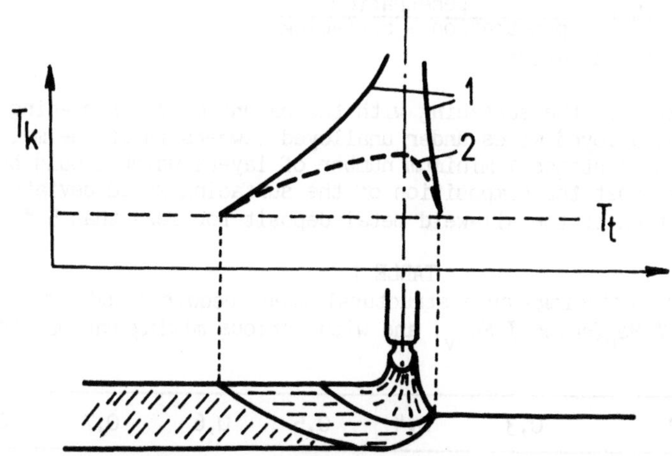

Figure 1. Diagram of temperature distribution in the molten pool

T_k - molten pool temperature

T_t - melting temperature

1 - direct arc influence on melting of the parent metal

2 - indirect arc influence on melting of the parent metal

Deep penetration is a characteristic of submerged arc welding. In surfacing under alloyed agglomerated powders, melting of the parent metal is partly reduced by the introduction of the metals which are in the welding powder, but it is still very intensive. Mixing of the surfacing with the parent metal is very important as far as the surfacing composition is concerned. The influence of the surfacing dilution with the parent metal is very obvious from the following equations (8-10):

- Calculation of a multi-layer surfacing:

$$\% Me_n = \% Me_{\check{c}v} - K^n (\% Me_{\check{c}v} - \% Me_{OM}) \qquad /1/$$

- Calculation of the number of surfacings required under the condition that: $\% Me_n \geq 0{,}95 \ \% Me_{\check{c}v}$

  (chemical composition of the above surfaced layer deviates from the composition of an all-weld metal deposit for 5 % at the most)

$$n = \frac{\log 0{,}05 \ . \ \% Me_{\check{c}v} - \log (\% Me_{\check{c}v} - \% Me_{OM})}{\log K} \qquad /2/$$

In surfacing on a low-alloy structural steel, the equation 2 gets a lot simpler:

$$n = \frac{\log 0,05}{\log K} \qquad /3/$$

Symbol explanation:

% Me_n, % $Me_{\check{c}v}$ and % Me_{OM} - element contents in the surfacing "n", in the all-weld metal deposit and in the parent metal

K - mixing ratio ($K = \frac{\text{penetration}}{\text{penetration + surfacing}}$)

n - number of surfaced layers

In intensive mixing of the surfacing with the parent metal, surfacing of high-alloyed layers with alloyed wires under unalloyed powders is of the multi-layer type . Table 1 states a minimum number of layers which should be surfaces in order that the composition of the surfacing would deviate from the composition of the all-weld metal deposit for less than 5 %.

TABLE 1
Number of surfacings on a structural steel required under the condition: % $Me_n \geqslant 0,95$ % $Me_{\check{c}v}$, and with various mixing ratios "K"

Mixing ratio "K"	0,3	0,4	0,5	0,6	0,7	0,8
Number of layers "n"	2,49	3,27	4,32	5,86	8,40	13,43
(according to equation 3)	3	4	5	6	9	14

Surfacing under alloyed agglomerated powders makes possible to obtain the composition of a surfaced layer required already in a single-layer cladding, even in surfacing on a structural steel. The dilution of the surfacing, caused by melting of the parent metal, and burn-off of alloying elements should be replaced by alloying of the surfacing by means of an alloyed agglomerated powder.

RULES OF ALLOYING OF THE SURFACING IN SURFACING UNDER ALLOYED AGGLOMERATED POWDERS

Already the first trials of single-layer surfacing on structural steel with unalloyed wires under alloyed agglomerated powders showed that the surfacing alloying, by means of powder, with carbon, silicon, manganese, chrome, molybdeum, vanadium and tungsten is relatively simple and very effective.

Table 2 gives theoretical compositions and results of analyses of one-layer surfacings in surfacing with unalloyed wire "EPP 2" (0,1 % C ;0,1 % Si; 1,0 % Mn) on a structural steel (0,17 % C;0,24 % Si; 0,55 % Mn) under agglomerated powders chosen (11).

TABLE 2
Theoretical and actual composition of one-layer surfacings on a structural steel

Wire/Powder	Theoretical composition EPP 2/U-Mo1	EPP 2/BM-V	Actual composition EPP 2/U-Mo1	EPP 2/BM-V
% C	0,92	1,75	0,44	1,16
% Si	0,19	0,18	0,84	0,93
% Mn	1,38	0,50	0,70	0,44
% Cr	8,02	12,80	6,13	9,31
% Mo	1,06	7,30	0,71	5,25
% V	0,58	4,34	0,41	2,05
% W	-	9,14	-	6,03
Remarks	K = 0,60	K = 0,35	UtopMo 1	BRM 2

The difference between the actual and theoretical element contents in the surfacing is a result of burn-off, i.e. oxidation, of alloying elements during surfacing. There is only silicon pick-up. For surfacing of high-speed steels, the SiO_2 content will have to be reduced even in case of a basic powder in order to obtain surfacings with the silicon content under 0,5 %.

In alloying of the surfacing by means of agglomerated welding powder, there are certain limitations. It is not recommendable to manufacture agglomerated powders by adding difficult to-melt metals in powder form (tungsten with melting point at 3410°C and molybden with melting point at 2625°C). Neither larger quantities of ferrotungsten are recommedable because in welding under high-agglomerated powders prepared in this way, unhomogeneous surfacings with metal inclusions are obtained (9, 10, 12, 13).

In surfacing of high-alloyed layers on a structural steel, quality surfacings will be obtained only if a part of alloying elements (above all tungsten) are alloyed to the surfacing also by means of the wire.

In surfacing under alloyed agglomerated powders, the surfacing composition depends on the rate of melting of the welding wire, of the alloyed agglomerated powder and of the parent metal.

Melting of the wire depends on welding parameters and on the composition and electric conductivity of the wire (see Fig.2 and Table 3).

Welding efficiency "η" depends primarily on the metal content in the alloyed agglomerated powder. Also the welding wire (in inverse proportion - see Table 3) and welding voltage have an important influence on the welding efficiency. The welding efficiency increases along with the welding voltage rise.

Composition of the all-weld metal is determined by the ratio between the welding wire melted-off and the metals which come into the surfacing from the welding powder (account should be taken also of the burn-off-oxidation of alloying elements during surfacing). In Table 4, some compositions

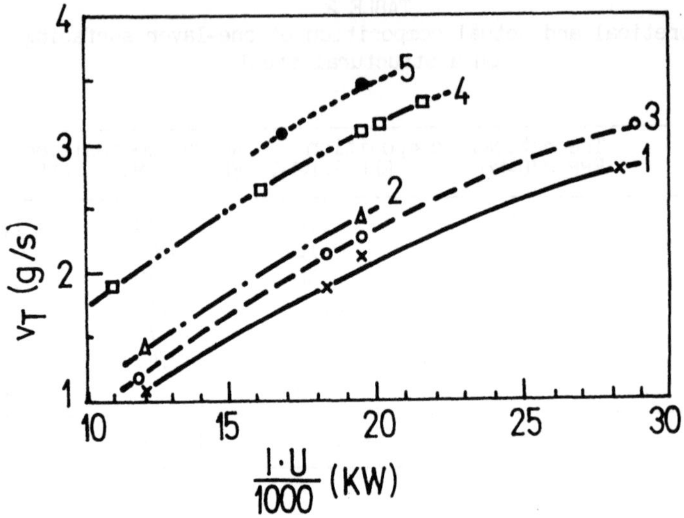

Figure 2: Melting "v_T" of some welding wires as it depends on welding parameters

1 - EPP 2 (0,1 %; 0,1 % Si; 1,0 % Mn; Fe basis)

2 - BRM 2 (0,9 % C; 4,0 % Cr; 5,0 % Mo; 6,5 % W; 1,9 % V; Fe basis)

3 - Nickeline (pure Ni)

4 - NIMONIC (20 % Cr; 20 % Co; 6 % Mo; 2 % Ti; 0,5 % Al; Ni basis)

5 - Ravnin 4 (0,2 % C; 19 % Cr; 1 % Mn; Ni basis)

TABLE 3

Melting "v_T" and welding efficiency "η" in the"wire/powder" combination chosen and with the follwing parameters 520 A and 38 V (19,8 kW)

Welding wire	v_T (g/s)	"η" in combination with powder	
EPP 2	2,11	135	U-Mo 6
BRM 2	2,48	115	U-Mo 6
Nickeline	2,30	185	NiMo 1
NIMONIC	3,10	170	NiMo 1
Ravnin 4	3,48	160	NiMo 1

of the all-weld metal surfacings with selected welding parameters and combination "wire/powder" are given.

The welding parameters being chosen, the welding efficiency "η" determines the ratio between melting of the welding wire and metal components in the welding powder.

TABLE 4
Compositions of the all-weld metals

Combination wire/powder	I (A)	U (V)	η (%)	C (%)	Cr (%)	Mo (%)	Co (%)	W (%)	V (%)	Al (%)	Ti (%)	Basis
EPP2,Ø3/U-Mo4	620	43	160	0,81	7,9	1,8	-	-	1,7	-	-	Fe
Nickelin,Ø3/NiMo1	520	38	185	-	22,3	19,1	-	4,2	-	-	-	Ni
NIMONIC, Ø3/NiCo1	520	38	170	-	19,0	15,0	28,0	-	-	0,2	0,8	Ni

$$\eta = \frac{G_n}{G_{\check{z}}} \qquad /4/$$

$$G_n = G_k + G_{\check{z}} \qquad /5/$$

$$G_k = G_{1.d} - 0{,}15_{\check{z}l} \frac{xM_{Me}}{M_{Me_xO_y}} \qquad /6/$$

$$G_{\check{z}l} = \frac{G_m + G_{v.s}}{0{,}85}$$; In the welding slag, there is about 15 % of metal oxides which result from burn-off of the alloying elements during welding. /7/

The welding parameters chosen (I, U) determine also melting of the parent metal and in this way indirectly also mixing of the surfacing with the parent metal.

$$G_n + G_{OM} = f\ (I, U) \qquad /8/$$

$$K = \frac{G_{OM}}{G_{OM} + G_n} \qquad /9/$$

Symbol explanation:

G_m, $G_{1.d}$ and $G_{v.s}$ - parts by weight of mineral compounds, of metal alloding admixtures and of dry matter of water glass in the welding powder.

$G_{\check{z}l}$, G_k, G_n, $G_{\check{z}}$ and G_{OM} - calculated parts by weight of the forming welding slag, of the metals which alloy the surfacing from the powder, of the surfacing and of the melted wire and parent metal.

M_{Me}, $M_{Me_xO_y}$ - atomic and molar weights of the metals and of the metal oxides which oxidize during welding.

η - welding efficiency

K - mixing ratio

Remark: Melting of the parent metal and consequently also the mixing ratio "K" increase along with the preheating temperature "T_p".

ONE-LAYER SURFACING OF A WORKING EDGE

In the manufacture of knives, working edges are surfaced. High-efficiency sumberged arc surfacing is problematic due to formation of a large weld pool. With the application of a water-cooled Cu ingot mold, however,high-efficiency one-layer surfacing of working edges is very easy to carry out (see Fig. 3).

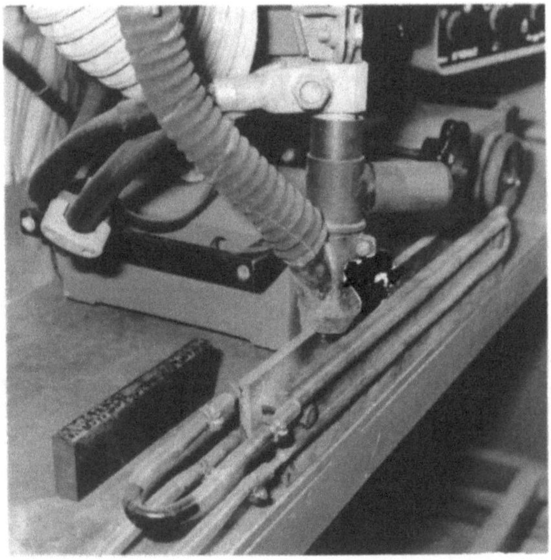

Figure 3. Surfacing of working edges by applying Cu-ingot mold; magn. 3 x

Surfacing was carried out on steel Č. 5742(56 NiCrMoV7; 0,55 %C; 1,1 % Cr; 1,7 % Ni; 0,5 % Mo and 0,1 % V)with the unalloyed welding wire EPP 2 under the alloyed agglomerated powder U-Mo 4. Dimensions of workpieces were as follows: 300 x 50 x 20 mm. In surfacing of a 20 mm large edge with the following parameters: 620 A, 43 V, 3 m/min, the welding efficiency was about 160 % and the mixing ratio of the parent metal K = 0,35.

The composition foreseen of the one-layer surfacing on steel Č.5742, which was calculated on the basis of the composition of all-weld metal for surfacing with the unalloyed wire EPP 2 under powder U-Mo 4, agrees well with the actual composition of the one-layer surfacing (see Table 5).

In surfacing with the parameters 620 A, 43 V and welding speed of 3 m/min, nicely shaped surfacings are obtained. The surfaced specimen can be seen in Fig. 3 leftward of the Cu-ingot mold. The cross-section of the surfacing is shown in Fig. 4.

TABLE 5
Compositions of all-weld metal and the calculated and actual compositions of the one-layer submerged arc surfacing on steel Č. 5742 with wire EPP 2 and powder U-Mo 4 with the parameters: 620 A, 43 V (η = 160 %, K = 0,35)

	C (%)	Si (%)	Mn (%)	Cr (%)	Mo (%)	V (%)	Ni (%)
All-weld metal	0,81	0,95	1,12	7,87	1,80	1,67	0,11
Calculated composition	0,72	0,62	0,73	5,50	1,34	1,12	0,67
Actual composition	0,76	0,68	0,53	5,53	1,52	1,29	0,79

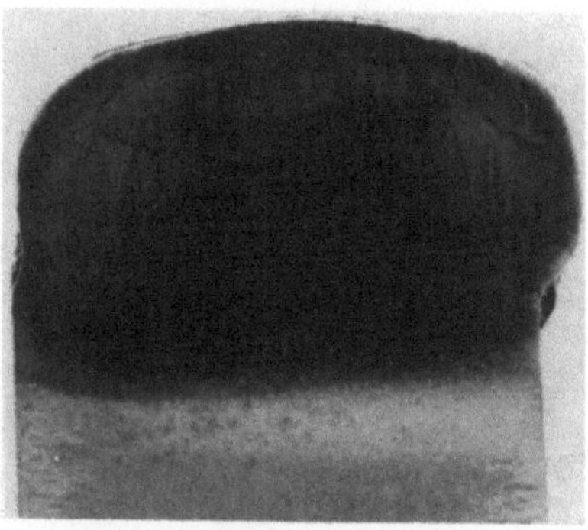

Figure 4. Appearance of the surfacing
magn. 3x

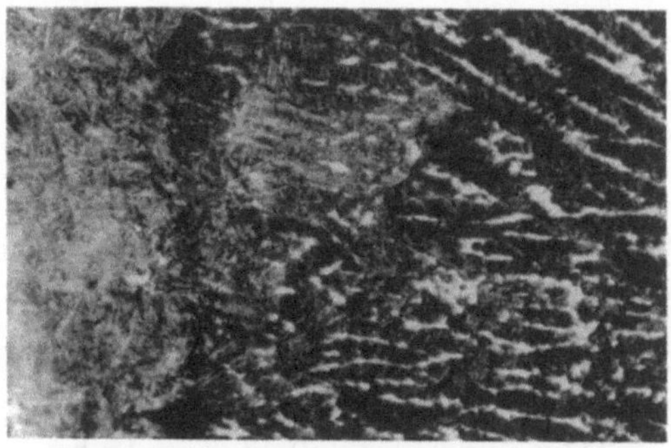

Figure 5. Structure of the transition zone parent metal/surfacing
magn. 100 x

The surfacings are hard and wear-resistant (see Table 6) and faultless (without inclusions, porosity and micro craks) in the surfacing itself and in the HAZ (see Fig. 5) already in the surfaced state.

TABLE 6
Hardness of the surfacing

Location of the surfacing	Weld face	Middle	Root	HAZ	Parent metal
Hardnes (HV)	630/770	660/770	615/720	230/715	226/234

SURFACING OF WORKING PLANES

In surfacing of planes and in assessing the quality of one-layer and multi-layer surfacings, overlapping of surfacings, which usually reaches 30 %, should be taken into account. In one-layer surfacing of a plane, only the first surfacing is welded on the parent metal alone, while the second and the following surfacings partly (by 30 %) overlap the previous surfacing (see Fig. 6).

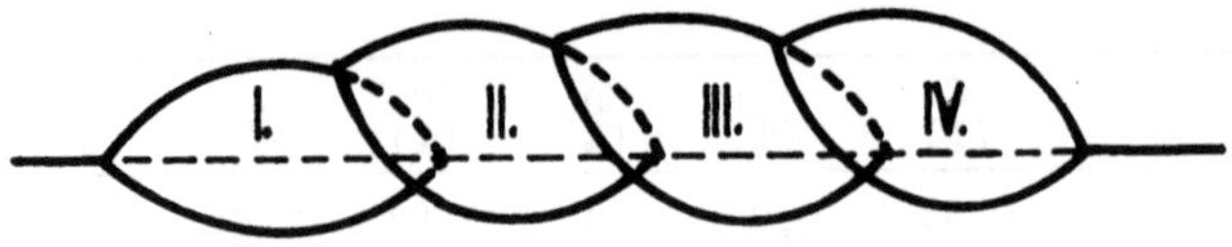

Figure 6.Scheme of a one-layer cladding of a plate

A one-layer surfaced plate is therefore more richly alloyed than the first surfacing (see Table 7).

TABLE 7
Composition of a one-layer surfacing on steel Č.5742 with wire "Nickeline,Ø 3" under powder "NiMo1" (preheating temperature = 400°C, I = 520 A U = 38 V)

Surfacing	C (%)	Cr (%)	Mo (%)	W (%)	Fe (%)	Ni (%)
I.	0,22	13,8	11,7	2,5	38,9	33,3
II.	0,18	15,4	13,0	2,8	31,6	37,1
III.	0,17	15,5	13,2	2,8	31,1	37,5
IV.	0,17	15,6	13,2	2,8	31,1	37,6

Table 7 shows one-layer surfacing of a Ni-alloy on steel. The iron share in the one-layer surfacing is very high. Special Ni-alloys, which may contain only limited quantities of iron (e.g. max. 5 % Fe), should be surfaced on steel in several layers. It is advisable to surface first an intermediate layer of nickel or Ni-alloy by shallow penetration process (e.g. TIG process) and only on this layer a wear-resistant alloy under alloyed agglomerated powders (10).

Surfacing of an intermediate layer by shallow-penetration TIG process (K = around 0,15) and 30 % overlap of the surfacings on steel Č. 5742 and one-layer surfacing on the intermediate layer with wire "Nickeline" under powder "NiMo1" are shown in Table 8.

Table 8 shows that when combining TIG surfacing and surfacing under alloyed agglomerated powders, i.e. in two-layer surfacing, quality Ni-claddings with low iron content are obtained in steel cladding.

TIG surfacing of hot-forging tools is very difficult and timeconsuming. In practical usage, it is more convenient to apply automatic submerged arc surfacing to the tools which are to be surfaced in a preheated state (see Fig.7 and Table 9).

Figure 7. Surfacing - cladding of a working plane of a hot forging tool

TABLE 8

Compositions of the intermediate layer surfaced by TIG process (wire: "Nickeline" and of the one-layer surfacing on the intermediate layer with wire"Nickeline,Ø 3" under powder "NiMo1" with 30 % overlapping of the surfacings (preneating temperature = 400° C, I = 520 A, U = 38 V)

Process		C (%)	Cr (%)	Mo (%)	W (%)	Fe (%)	Ni (%)
TIG	K = 0,15	0,07	0,1	0,1	Ø	10,8	89,0
submerged arc	K = 0,40	0,02	14,9	12,8	2,8	4,0	65,5

TABLE 9

Compositions of the first and the second intermediate layers surfaced with wire"Nickeline,Ø 3" under powder "NiMo1" with 30 % overlapping of the surfacings and composition of the upper wear-resistant layer surfaced with wire "NIMONIC, Ø 3 ", under powder "NiCo1" with 30 % overlapping of the surfacings (preheating temperature = 400°C, I = 520 A, U = 38 V)

Layer no.	Wire/powder	C (%)	Cr (%)	Mo (%)	Co (%)	W (%)	Al (%)	Ti (%)	Nb (%)	Fe (%)	Ni (%)
1	Nickeline /NiMo1	0,17	15,6	13,2	-	2,8	-	-	-	31,1	37,6
2	Nickeline /NiMo1	0,05	20,5	17,5	-	3,8	-	-	-	8,9	49,8
3	NIMONIC /NiCo1	0,02	19,5	15,7	20,5	1,1	0,2	0,6	0,8	2,4	39,7

Table 9 shows that also in three-layer surfacing - cladding of steels under alloyed agglomerated powders - quality claddings of the special Ni-alloy (NIMONIC) with low iron content are obtained.

CONCLUSION

Surfacing under alloyed agglomerated powders is a very interesting and prospective process of refinement of working edges and of cladding surfaces subject to wear. Various wear-resistant and high-alloyed iron-base alloys can be one-layer surfaced on low alloy structural steels. Dissolution of the surfacing caused by melting of the parent metal and burn-off of alloying elements is substituted by alloying the surfacing through alloyed agglomerated powder.

The surfacing composition selected can be assured if melt-off rate of the welding wire, the melting rates of the welding wire and of the parent metal are known. Investigations showed that by the selection of the welding wire and of the welding parameters, the ratio between the melting-off of the welding wire and the melting of the welding powder,which is decisive for the all-weld metal composition, is determined. The welding parameters and the preheating temperature selected determine the mixing of the surfacing with the parent metal (mixing ratio "K"), which is, together with the all-weld metal composition, decisive for the surfacing composition.

The special Ni-alloys which may contain only limited quantities of iron (e.g.max. 5 % Fe) should be surfaced on a steel in several layers. It is advisable to surface first an intermediate layer of nickel or Ni-alloy by shallow-penetration process and only on this layer a wear-resistant alloy under alloyed agglomerated powders.

The surfacing technology selected should assure that the surfacings are of good quality, without defects, undesired structures and cracks in the surfacing itself and in the heat affected zone.

REFERENCES

1. E. Pertender, E. Noninger (Prevod L. Knez): "Borba proti obrabi z varjenjem" Varilna tehnika, Ljubljana, 23 (1974) 3/4, 67/71.

3. "Schweisszusatzwerkstoffe für Werkzeugstähle" Böhler.

3. "Weld Surfacing and Hardfacing" The Welding Institute, Abington, 1980.

4. R. Kejžar:"Održavanje navarivanjem" Održavanje mašina i opreme - OMO 9 (1980) 9,499/503: "Building-up by Welding Maintenance "Proceedings of 5th European Maintenance Congress, Opatija, 1980, 345/367.

5. R. Kejžar: "Izdelava in vzdrževanje orodij z navarjanjem" Varilna tehnika, Ljubljana, 32 (1983), 4, 81/83.

6. R. Kejžar: "Obnavljanje rezilnih robov in delovnih površin orodij" Varilna tehnika, Ljubljana, 30 (1981) 2, 42/45.

7. R. Kejžar:"Razvoj varilnih postopkov obnavljanja in oplemenitenja Strojniški vestnik - Tribologija, Ljubljana, 31 (1985) 7/8, 179/184.

8. R. Kejžar, D. Brezovar: "Platiranje z navarjanjem" Zbornik referatov s posvetovanja "Tribologija v teoriji in praksi", Ljubljana, 1986,131/145.

9. R. Kejžar: "Applicability of Building-up Processes to Manufacture and Restoration of Tools". Proceedings of the International Conference on the Joining of Materials, JOM-4, Helsingør, 1989, 26/36.

10. R. Kejžar: "Hardfacing by Submerged Arc Welding" Proceedings of the 2 nd International Conference on Tooling "Neue Werkstoffe und Verfahren für Werkzeuge", Bochum, 1989, 301/314.

11. R. Kejžar, D. Brezovar: "Built-up Wear-Resistant Claddings" Proceedings of 4th Tribology Conference, Budimpešta, 1987, 100/103.

12. R. Kejžar, L. Kosec: "The Weld Alloying Transfer of Alloyed Elements from the Electrode to the Weld at the Manual Arc Welding" IIW Doc.SG 212-463-79, Ljubljana, 1979.

13. R.Kejžar: "New Prospectives of Cladding by Submerged Arc Building-up" Proceedings of International Conference on Welding Technology in Developing Countries-Present Status and Future Needs, Roorkee, 1988.

Surface Degradation and Stress Corrosion Cracking of Bronze in Moist Oil

E. Lunarska

Institute of Physical Chemistry
Warsaw Poland

S. H. Shim and Z. Szklarska-Smialowska

Ohio State University
Columbus Ohio USA

ABSTRACT

The possibility of stress corrosion cracking of two kind of high strength bronzes in a hot moist engine oil has been investigated using slow strain rate tensile test. Effect of chemical composition and microstructure of materials as well as residual stresses on surface degradation and stress corrosion cracking has been studied.

INTRODUCTION

Bronzes undergo surface degradation in natural and synthetic oil, however, scarce information can be found on susceptibility to SCC of bronzes in those environments. The main corrosive agent in oil is water present in it which contains dissolved products of oil oxidation, such as carboxylic and specially sulfonic acids(1). In pure water (2-4) and water containing $SO_3^=$, $SO_4^=$ ions (2,3,5) and SO_2 (1,6) SCC of Cu and Cu alloys has been observed and studied. No definite mechanism of SCC under those circumstances has been established, and selective dissolution (2,7) oxide film cleavage (4,7) as well as adsorption induced cracking (3) have been proposed.
The aim of present work was to check the response of two high strength bronzes to the modified engine oil environment and to distinguish factors affecting their susceptibility to SCC.

EXPERIMENTALS

The chemical composition of used material is shown in Table 1. Shape and dimensions of plain, notched and grooved tensile specimens are shown in Fig.1. Notches were machined on one flat wall of plain specimens and grooves were made by pressing the plain specimens against a proper die. Some notched specimens were stress relieved by heating at 190°C for 1 hr.
Tensile tests were carried out with a strain rate of $8.3\times10^{-7}s^{-1}$ in air at RT, in oil at 107°C, in moist oil at 107°C and in distilled water at 96°C.

TABLE 1
Chemical composition of bronze specimens, wt %

Material	Cu	Sn	Pb	Zn	Si	Cr
A	89	4	4	3	-	-
B	95.4	1.5	-	-	3	0.1

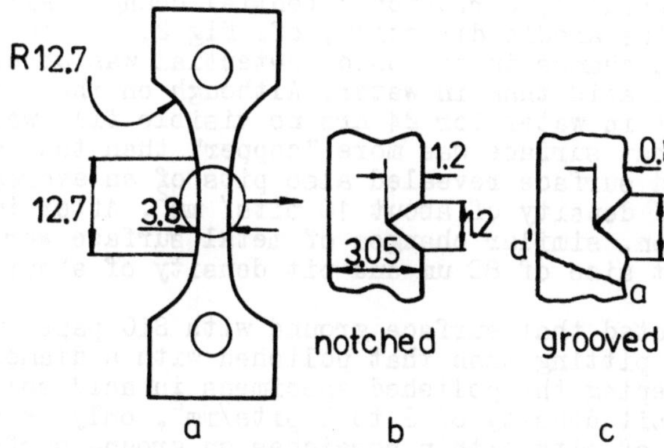

Figure 1. Sketches of specimens used for slow strain rate tests. a - plain specimen, b - notched, c - grooved

The moisture in the oil was maintained by bubbling into corrosion cell a nitrogen gas passing through the washer filled with distilled water. Bubbling started 0.5 hr before straining and was continued during all tensile test. The water content in the oil thus attained was estimated at about 0.5 to 1 wt%.
After tensile test, each specimen was washed with acetone in ultrasonic cleaner, and fracture and side surfaces were examined under a scanning electron microscope. In some cases, the corrosion products were removed by etching in dilute HCl with inhibitors in an ultrasonic cleaner. The corrosion products were examined by using EDS.
The change in corrosion potential of used materials was recorded during immersion of specimens in distilled water and in 0.1 wt% solution of p-toluenesulfonic acid ($CH_3C_6H_4SO_3H$) at 90° for up to 24 hrs. That solution simulated the water fraction in oil (1). After test the specimen surface was examined by optical microscope.
Microstructure of studied alloys was observed after etching in 2g $K_2Cr_2O_7$ + 4 ml H_2SO_4 + 1 ml HCl + 50 ml H_2O solution.

RESULTS

Microstructure

Material A exhibits the average grain size of about 10 um. Pb rich grains of similar size are homogeneously distributed as a second phase of a volume fraction of 0.078, not forming the continuous network. Material B contains some inclusions in a uniform postdendritic in origin structure.

Corrosion Test

Figure 2 shows the change in corrosion potential during immersion in water and sulfonic acid solution of studied materials.

In case of material B, corrosion potential attained the stable value after 20 to 30 min of exposition in both environments. Surface of material B after immersion for 24 hrs in water was covered with a blue-reddish film. After keeping in solution for 24 hrs no change of metal surface has been visible.
In case of material A, corrosion potential changed all the time shifting into the anodic direction, cf. Fig 2, curves A. The more pronounced change in corrosion potential was observed in toluenesulfonic acid than in water. Although on the surface of material A kept in water for 24 hrs no visible film was observed, the color of that surface was more "copper" than that of base metal. Corroded surface revealed also pits of an average size about 50 um and density of about 10 pits/ mm^2. After immersion in acid solution, similar changes of metal surface were observed with pit size of 80 um and pit density of about 15 pits/ mm^2.
It should be noted that surface ground with SiC paper was more susceptible to pitting than that polished with a diamond paste. For example, keeping the polished specimens in acid solution for 24 hrs causes pit density of 2 to 5 pits/mm^2, only. Preferrential formation of pits within scratches on ground surface has been often observed.

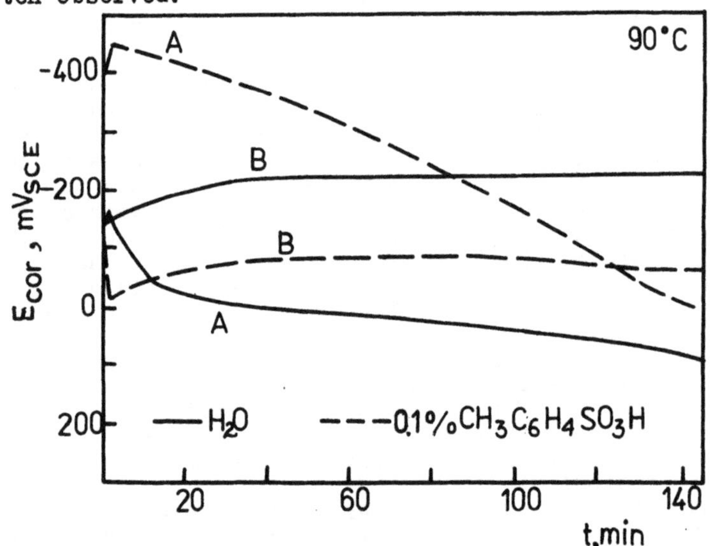

Figure 2. Change of corrosion potential of materials A and B during immersion in water and p-toluenesulfonic acid solution at 90°C

Slow Strain Rate Test

The typical results of the slow strain rate tests for all experimental conditions are presented in Tables 2 and 3, for materials A and B, respectively.
In case of notched and grooved specimens, the occurrence of SCC

TABLE 2
Results of slow strain rate tensile tests for material A

specimen	testing conditions	σ_{uts} MPa	fracture strain	fracture time	fracture mode
plain	air, RT	401	0.179	59.8	D
plain	oil, 107°C	445	0.24	81.2	D
plain	moist oil, 107°	421	0.22	72.7	D+B
notched	moist oil, 107°	396	0.028	9.4	D+B
notched and stress relieved	moist oil, 107°	160	0.082	27.5	D
grooved	moist oil, 107°	380	0.16	52.8	D+B
grooved	water, 96°C	406	0.19	64.3	D+B

D - ductile fracture
B - brittle fracture

TABLE 3

Results of slow strain rate tensile tests for material B

specimen	testing conditions	σ_{uts} MPa	fracture strain	fracture time	fracture mode
plain	air, RT	500	0.34	113.5	D
plain	oil, 107°C	541	0.362	120.6	D
plain	moist oil, 107°	534	0.367	122.3	D
notched	moist oil, 107°	556	0.082	27.3	D
notched and stress relieved	moist oil, 107°	486	0.062	20.8	D
grooved	moist oil, 107°	538	0.298	99.3	D+B
grooved	water, 96°C	548	0.326	108.5	D

D - ductile fracture
B - brittle fracture

was mostly established on the base of fractographic examination, since time-to-failure values of slow strain rate test did not indicate clearly the SCC effect for those specimens.
In both materials, slight increase in strength and plasticity was observed at testing in oil at 107°C as compared with air at RT.

In case of material A, SCC (estimated from a change of time-to-failure and from fractographic observations) took place for plain, notched and grooved specimens in moist oil and for grooved specimens in water, cf. Table 2. Stress relieving of notched specimens eliminated the SCC effect (Table 2).
For material B, stress corrosion cracking was observed only in case of grooved specimens tested in moist oil, cf. Table 3.

Fractography

Fracture surface of both material tested in air at RT and in oil at 107°C was fully ductile.
Typical examples of fracture surface and side surface of material A specimens underwent SCC in moist oil are shown in Figures 3 and 4, respectively.
The fracture surface of plain, notched and grooved specimens has mainly ductile appearence with the outer parts covered with a thick layer of corrosion products, cf. Fig. 3a . After removing of that layer, brittle intergranular fracture surface becomes visible, cf. Fig. 3 b. Characteristic features of that fracture surface are: 1) regularly spaced at about 1 um striations seen on grain boundaries and 2) spots leaving by disappeared Pb rich grains.

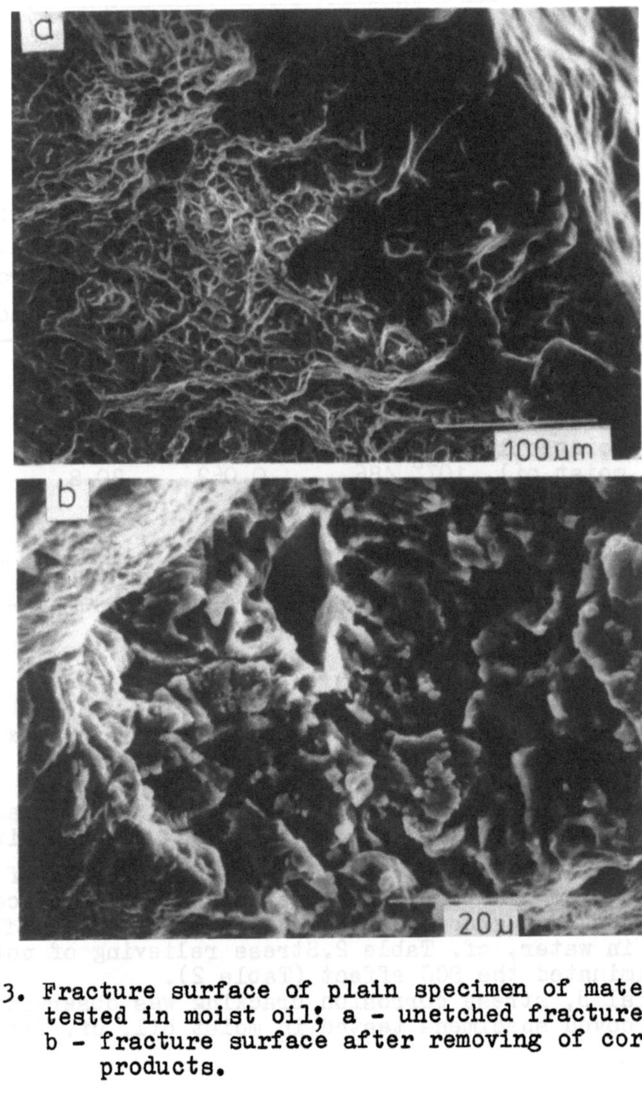

Figure 3. Fracture surface of plain specimen of material A tested in moist oil; a - unetched fracture surface, b - fracture surface after removing of corrosion products.

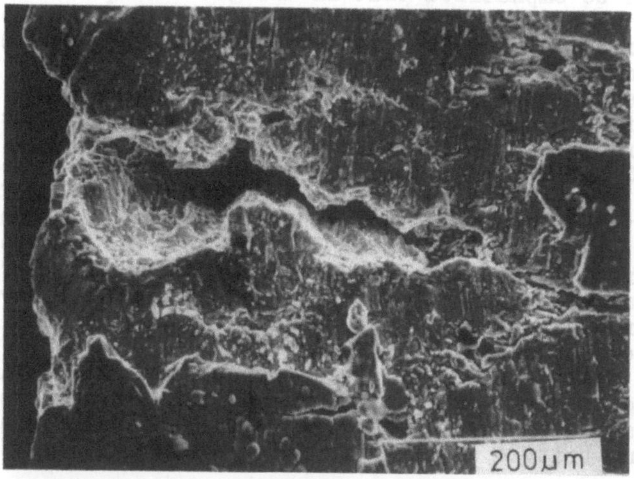

Figure 4. Side surface of plain specimen of material A tested in moist oil

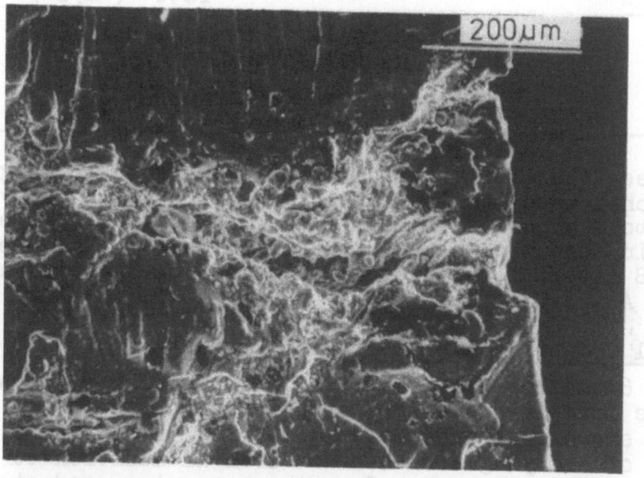

Figure 5. Side surface of grooved specimen of material B tested in moist oil

On the side surface of specimens of material A tested in moist oil, the cracks initiating on the outer part of specimen are seen, cf. Fig. 4. Corrosion products are visible inside the cracks and on the side surface. In some places, the pits are also seen on the side surface with the small cracks originated from them, cf. Fig. 4 .

It should be emphasized that in all grooved specimens tested, fracture plane did not correspond to the smallest cross-section, but occurred outside the groove following the plane schematically shown in Fig. 1c as plane a - a′.
Grooved specimens of material A tested in hot water showed ductile fracture with some sites of brittle appearence adjacent to the grooved wall of a specimen. The mode of those parts was mixed, IG and TG spots could be observed.
Material B exhibited fully ductile fracture surface for all environments and specimen type exept the grooved specimen tested in moist oil. As seen in Fig. 5, similarly as in material A, cracks are initiated on the outer part of specimen. The cracks are filled with the corrosion products, but the layer of corrosion products seen on side surface is not so thick as in case of A material, cf. Figures 4 and 5.
In Table 4, the values of crack propagation rate estimated taking into account time-to-failure and maximum observed depth of crack visible on fracture surface are presented for specimens exhibiting brittle cracking in moist oil.

TABLE 4

Estimated crack propagation rate in moist oil

Material	specimen	crack propagation rate (m/s)
A	plain as rolled	0.78×10^{-9}
	notched as rolled	1.9×10^{-9}
	grooved	1.8×10^{-8}
B	grooved	2×10^{-9}

It is seen that for material A estimated crack propagation rate for notched specimen is twofold greater and for grooved specimen about an order of magnitude greater than that estimated for plain specimen. Crack propagation rate for grooved specimen of B material is an order of magnitude lower than that for similar A specimen.

Chemical Composition of Corrosion Products

In Fig. 6 the low energy part of X-ray energy dispersion spectra for base materials and for corrosion products formed in moist oil are shown. For base materials (Fig. 6, curves a, c) small signals of alloying elements are seen in addition to the copper signal. EDS spectra taken from the corrosion products reveale strong S signal in case of both materials (cf. Fig.6, curves b, d). Analysis of chemical composition of corrosion product formed on material A in moist oil was done taking into account intensity of SK_{α} ,SnL_{α}, CuK_{α} ,ZnK_{α} and PbL_{α} signals and assuming that only those elements are present in corrosion products. The average composition of corrosion products thus estimated was in wt% : 9.36 S, 1.28 Sn, 64.3 Cu, 2.98 Zn, 27.5 Pb. The enrichment in Pb and depletion in Sn in comparison with base metal is obvious.
In case of corrosion products formed on A material in hot water the Cu signal was seen; no clear signals of other elements were observed.

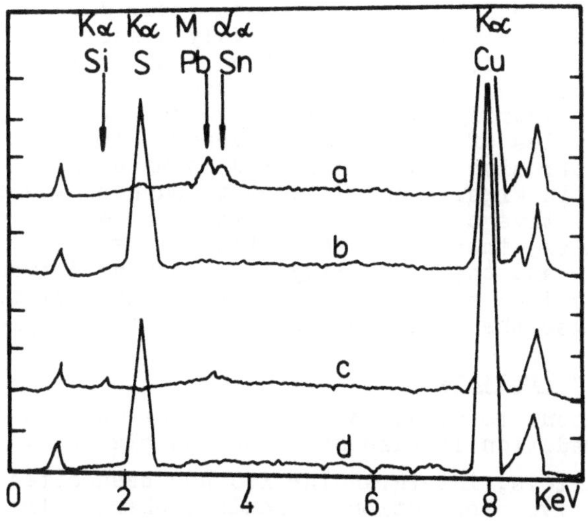

Figure 6. EDS spectra for studied base materails A (curve a) and B (curve c) and for corrosion products formed in moist oil on material A (curve b) and on material B (curve d)

DISCUSSION

Surface Degradation

As follows from the present results, material A undergoes surface degradation in hot water, p-toluenesulfonic acid and in moist oil. Shift of corrosion potential into anodic direction during the immersion in electrolytes (cf. Fig. 2) reflects the change in chemical composition of metal surface. Similar observations have been reported for Pb bronze in benzesulfonic acid (1) as accounted for the selective dissolution of Pb and enrichment of surface in Cu. The observed change in color of material A after immersion in electrolytes seems to support the concept of Cu enrichment but not due to dissolution of Pb only , since Pb forms a separate phase. The effect of dissolution of that phase on potential and composition of metal surface could be expected only in case of dissolution of Pb rich grains emerging the surface. However, microscopic examination revealed the majority of Pb rich grains unetched. Therefore, the selective dissolution of other elements from the solid solution (Zn, Sn) could be also involved.That possibility is supported by the comparison of Pourbaix diagrams for alloyung elements (8).
In case of metal B no drastic change in surface due to immersion in hot water and sulfonic acid have been observed.
Presence of high S content in corrosion products formed on both materials during testing in moist oil indicates that the reaction of metal with sulfonic acids plays an essential role in corrosion process. However, the corrosion in moist oil is more complex, and other oil components, specifically the carboxylic acids should be also involved.

According to (1) the following processes occur during the corrosion of metals in moist oil: selectively dissolved metal ions react with the products of oil oxidation, and as a result, carboxylates, carbonates, hydroxides, sulfonates and sulfates form in the bulk of electrolyte. Corrosion products remain in environment as particles of 1 to 10 um or deposit on metal. The morphology of corrosion products formed on materials A and B in moist oil shown in Figs. 4 and 5 seems to support the above concept. The relative content of alloying elements in corrosion products indicates, therefore, the ability to selective dissolution of elements and the solubility of compounds formed. Sn pressumably dissolves sèlectively from the alloy but forms soluble compounds, so the depletion of Sn in corrosion products occurs.

Stress Corrosion Cracking

As shown in present results, material A being more susceptible to surface degradation is also more susceptible to SCC than B one.

In case of Cu alloys, intergranular SCC has been often associated with selective dissolution mechanism, whereas transgranular SCC with film-induced-cleavage mechanism (7).
The dissolution mechanism seems likely to operate in case of SCC of material A in moist oil. IG fracture surface with regular striations marking the crack arrests as well as complete dissappearing of Pb rich grain seen in Fig. 3b indicate the preferrential dissolution of second phase and the material adjacent to grain boundaries in course of crack propagation. More intensive dissolution of interfacial material than that of grain body is due to the difference in the chemical composition of those parts. Although no network of Pb-rich phase has been observed, the segregation of Pb, Sn and Zn at grain boundaries in very probable (2). The preferrential dissolution along grain boundaries is more pronounced at stress corrosion since stresses emphasize selective dissolution (9) and environment within the crack is more aggressive (10).

In case of material B, dissolution processes within the crack formed in moist oil are very extensive as may be concluded from the amount of corrosion products filled the crack, cf. Fig. 5. However, the appearence of fracture surface beneath the corrosion products is undefinite, and exact conclusion on SCC mechanism cannot be drawn. As shows postdendritic structure of B material, segregation of Sn and thus its selective intergranular dissolution is probable. However, presence of Si inhibits SCC similarly as has been found in case of brasses (11).

Cold Work Effect

Cold work increases susceptibility of Cu alloys to the SCC, the effect is accounted for the creating of residual stresses (6) or for the formation of highly defected structure (10) due to cold deformation.
From the behaviour of differently treated specimens of material A, some conclusions on the effect of cold work on SCC may be drawn.
Results obtained for notched and notched and stress relieved specimens (cf. Table 2) show that the notch effect on SCC is mainly due to work hardening of metal subsurface layer during notch mashining and not due to stress concentration.

In case of grooved specimen, the highest level of cold work reveals the material in the bottom of a groove. However, the cracks form and propagate outside the groove, as marked in Fig. 1 c (plane a - a′). This means that not the cold work level itself but the created residual stresses affect the susceptibility to SCC, as has been also shown for Inconel in (12). At the bottom of groove, compressive residual stresses operate, whereas maximum tensile stresses should exist on the boundary between plastically deformed and undeformed material. And crack propagate along that boundary, similarly as found in (6) for brass.
As follows from Table 4, increasing amount of cold work in subsurface material layer increases the readiness of SCC. Specially susceptible to SCC are grooved specimens (Tables 2,3 and 4) so this modification of tensile specimens could be used to examine the behaviour of relatively resistant materials in not very aggressive environments.

CONCLUSIONS

1. Hot moist engine oil causes SCC of Sn-Pb-Zn and Si-Sn high strength bronzes due to the action of sulfonic and carboxylic acids.
2. Sn-Pb-Zn bronze is more susceptible to surface degradation and to SCC than Si-Sn one probably due to higher probability of selective dissolution of alloying elements.
3. Susceptibility to SCC increases by cold work of bronzes due to creating of tensile residual stress.

ACKNOWLEDGMENTS

The work was supported by Rockwell International Corporation Troy, Michigan.

REFERENCES

1. Bratkov,A.,Churshukov,E., Teoreticheskie osnovy zashchity metallov ot korrozii v topliva ch. In Teoreticheskie osnovy khimmotologii,ed.A.Bratkov, Khimia, Moscow, 1985,pp278-309
2. Parkins,R.N.,Rangell,C.M.,Yu,J.,Metall.Trans.,1985,16A,1671-1681
3. Hinz,M.B.,Nettleton,L.J.,Heldt,L.A.,Metall. Trans.,1985, 16A, 971-978
4. Hardie,D.,Ebtehay,K.,Br.Corr.J.,1987, 22, 202- 206
5. Torchio,S., Mazza,F.,Corrosion Sci.,1986, 26, 813-826
6. Han,M.K.,Beavers,J.A.,Goins,W.,Mater.Perf. 1987,26, 20-24
7. Gassa,L.M.,Vilche,J.R.,Corrosion Sci.,1985, 25, 145-157
8. Pourbaix,M.,Atlas of Electrochemical Equilibria in Aqueous Solutions, Pergamon Press, NY, 1966
9. Pickering,H.W.,Byrne P.J.,Corrosion, 1973, 29, 325-331
10. Torchio,S.,Corrosion Sci., 1986, 26, 133-151
11. Bobylev,A.V.,Korrozionnoe rastreskivanie latuni,Metalurgizdat, Moscow, 1956
12. Totsuka,N.,Lunarska,E.,Cragnolino, G.,Szklarska-Smialowska, Z.,Scripta Metall., 1986, 20, 1035-1040

On Magnetron Sputtered Gradient and Multilayer Coatings

O. Knotek, J. Jungblut and F. Löffler

RWTH Aachen
Aachen
Federal Republic of Germany

Abstract

In the study, multilayer and gradient coatings were used to exploit the positive characteristics of ceramic coatings by increasing adhesion. Titanium hard material coatings such as TiN, TiC and TiAlVN proved suitable as a base layer on the substrates. Ceramic materials such as SiC, Si(C,N), Ti(B,N) and Al_2O_3 were then used for the outer coatings. Different methods were used for the transitions between the base and outer layers, both ionic etching and transition via a gradient layer playing an important role.

Introduction

New applications are constantly being found for coated tools. Following the successful use of Ti-based coatings for the machining of steel, certain wear-protective ceramic coatings are already being used to machine cast iron. Owing to the differences in machining behaviour, this imposes new demands on the coating materials. Ceramic hard material films possess high wear resistance to the stresses involved, owing to their good material properties. Their adhesion in the substrate-coating composite is, however, too low unless an intermediate layer is used. It is therefore necessary to develop multilayer coatings which ensure favourable composite adhesion.

Magnetron sputtering was used to produce the multilayer coatings. This process enables extremely complex coating sequences to be realized successfully. Coating compositions can be varied over a wide range by using differing target materials and admitting reactive atmospheres.

Structure and Deposition of the Gradient and Multilayer Coatings

The multilayer coatings consist of at least a base layer and an outer layer. Between these layers, there may be other layers with certain characteristics. The aim of combining layers is to achieve a suitable composite system, whose characteristic profile could not be attained by a single coating layer.

Multilayer coatings with as many as 1000 individual layers can be manufactured in order to exploit particular characteristics of phase boundaries. This technology has been applied to TiC-TiB_2 /1/ and TiN-Al_2O_3 coatings /2/. These coatings are intended to achieve an optimum hardness-toughness ratio /3/.

The present paper describes deposition of three different outer layers. In one case, the material used was silicon carbide, with extremely high hardness in the range of 6000 HV 0.05 and good tribological characteristics. In the other case, aluminium oxide and coatings in the Ti-B-N system were used as the outer layer. Apart from good tribological characteristics, Al_2O_3 possesses good chemical resistance, even at high temperatures.

The varying requirements on the constituents of the multilayer coatings are shown in *Fig. 1*. It will be apparent that both outer layer materials meet the basic requirements. The interface must essentially provide high adhesion for the two adjacent layers. Reactions between the two sides at higher temperatures, leading to destruction of the composite, also have to be avoided.

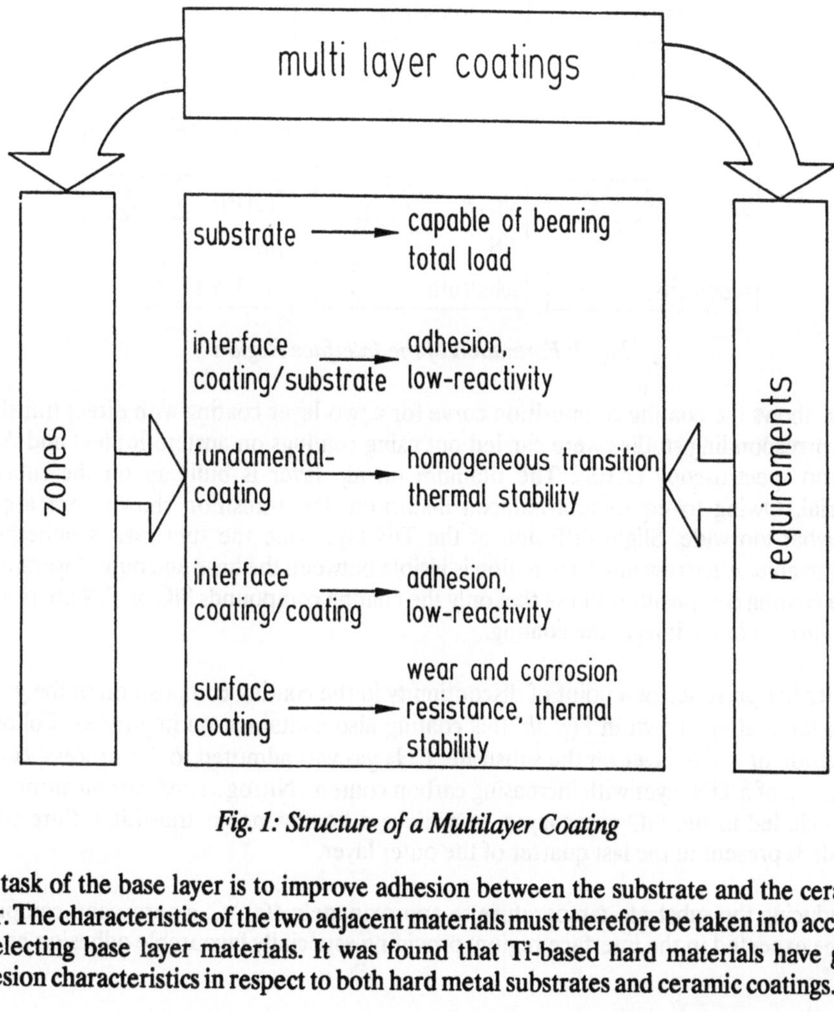

Fig. 1: Structure of a Multilayer Coating

The task of the base layer is to improve adhesion between the substrate and the ceramic layer. The characteristics of the two adjacent materials must therefore be taken into account in selecting base layer materials. It was found that Ti-based hard materials have good adhesion characteristics in respect to both hard metal substrates and ceramic coatings. The form of interface structure is a further significant influencing variable for the final result; this can be varied over a wide range using the magnetron sputtering process. The three significant differences in formation of the interface region are shown in *Fig. 2*. The simplest option for multilayer coating manufacture is to deposit different individual layers without further influencing the interface region. Distinct improvements are, however, found with coating systems with a graded transition. This enables discontinuities in characteristics to be avoided. Intermediate etching also leads to good improvement of adhesion, by pre-activating the previous layer. A combination of graded transition and intermediate etching can be achieved by the magnetron sputtering plant, and yields extremely interesting results.

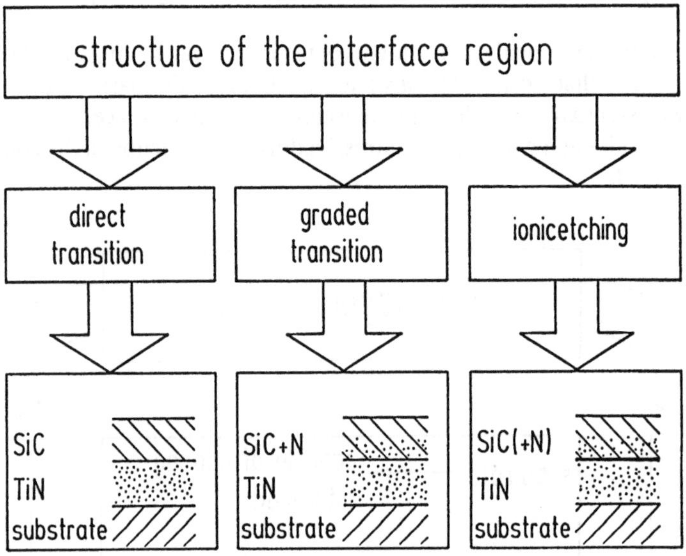

Fig. 2: Formation of the Interface Region

Fig. 3a shows the coating composition curve for a two-layer coating with direct transition. The corresponding studies were carried out using coatings on austenitic steel and Auger electron spectroscopy (AES). The titanium nitride layer is built up on the substrate material. Owing to equipment-induced distortion, the transition shown here appears somewhat too wide. Slight diffusion of the TiN layer into the substrate is nonetheless recognizable. A narrow interface region is visible between the base and outer layers. Study of the coating composition shows that only the coating compounds SiC or TiN are present. There are no impurities in the coating.

Despite the presence of a point of discontinuity in the coating composition of the graded two-layer coating shown in *Fig. 3b*, this coating also contained no impurities. Following deposition of a TiN layer on the substrate, CH_4 gas was admitted to the process, causing formation of a TiN layer with increasing carbon content. Nitrogen and carbon atoms were also included in the SiC outer layer, to achieve a homogeneous transition. Pure silicon carbide is present in the last quarter of the outer layer.

By reducing the substrate temperature to approximately 400 °C, a paracyano compound may be expected in the interface region, providing especially favourable adhesion characteristics. Since this graded coating was also subjected to a series of other tests, its characteristic profile will be described in greater detail in the next section.

Characteristics of Gradient and Multilayer Coatings with an SiC or Al_2O_3 Outer Layer

A shallow concavity was ground in the coated specimens, enabling us to measure the thickness of the individual layers and determine hardness distribution throughout the coating depth. The reliability of the procedure was validated by surface grinding of the coatings.

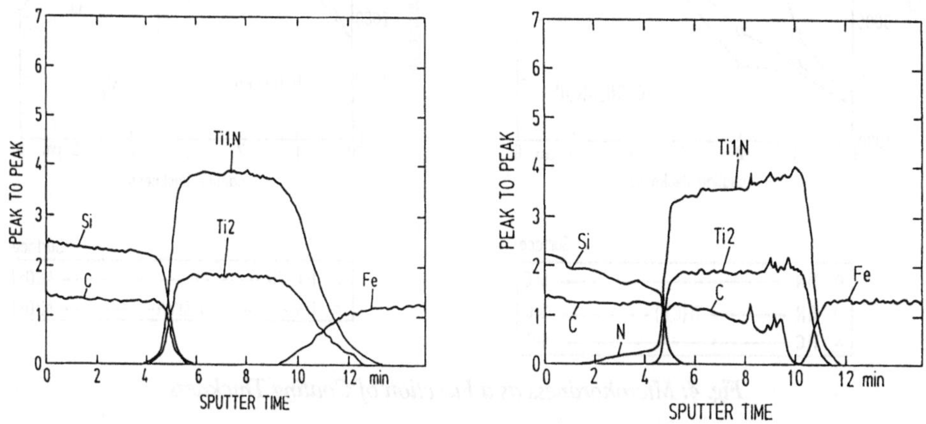

Fig. 3: AES Depth Profile
a) SiC-TiN coating with direct transition
b) Si(C,N)-Ti(C,N) coating with graded transition

The total coating thickness for all specimens was roughly 6 μm. A comparison of hardness throughout the coating thickness yielded similar curves for direct-transition layers, as shown in *Fig. 4a*. The base layer possesses the usual hardness for titanium hard material coatings /4/. The SiC layer also exhibits its known, extremely high hardness. This hardness was also measured in SiC layers deposited at temperatures of roughly 1100 °C, using the CVD process /5/. The temperature for magnetron sputtering was approximately 600 °C. In the interface region, where measurement is extremely difficult, there was a steep increase in hardness. This transition is also observable with the Ti(C,N) layer. The somewhat earlier rise in hardness results from a different thickness distribution of the individual layers. The adhesion of these two-layer coatings is extremely good, at 50 N. The ceramic coatings show amorphous deposition. The TiC, TiN and Ti(C,N) phases were demonstrated by X-ray microstructural analysis.

The aluminium oxide outer layer shows lower microhardness, as indicated in *Fig. 4b*. This results, however, in an extremely homogeneous hardness curve throughout the coating depth, with positive effects on other mechanical coating properties. Apart from the titanium nitride base layer, a TiAlVN layer was deposited on the substrate. In this case, improved addition of the Al_2O_3 layer is expected, especially from the aluminium constituents.

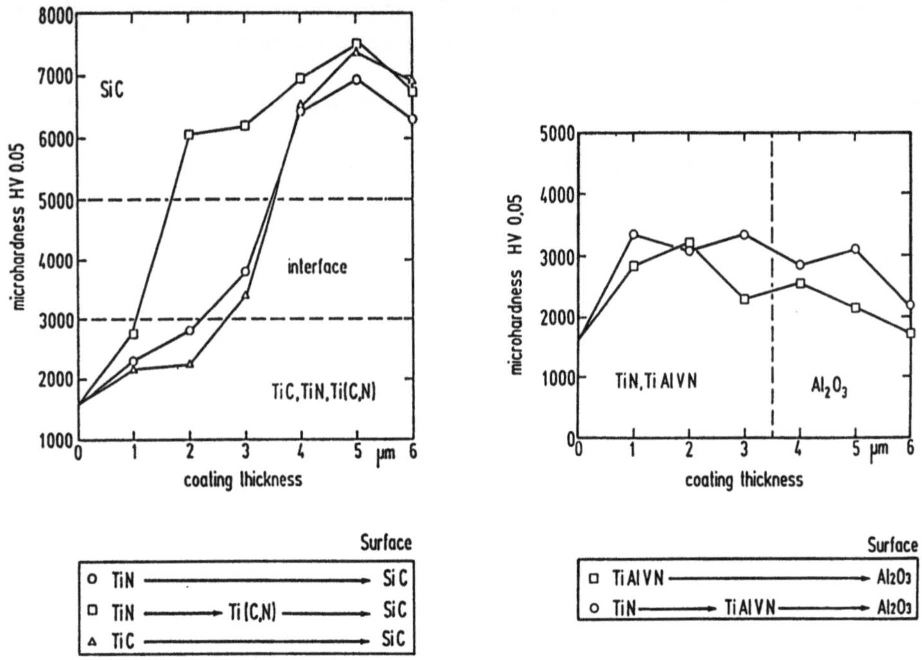

Fig. 4: Microhardness as a Function of Coating Thickness

The three-layer TiN-TiAlVN-Al_2O_3 coating has a critical load of 30 N in the scratch test, whereas the two-layer TiAlVN-Al_2O_3 shows only extremely slight spalling at the scratch edges even at 90 N.

This coating exhibited a hardness increase in the outer layer region following three hours of heat treatment at 1000 °C. An Al_2O_3 and an AlO phase. No reaction with the substrate was detected, even after heat treatment, confirming the good thermal stability indicated by other results.

The good tribological characteristics of the SiC coatings are recognizable in *Fig. 5*. Different multilayer coatings were examined for wear in the stylus-disc tribometer. P 1000 grain SiC abrasive paper was used as the counterpart. The SiC layer hardness contributes to the high abrasive resistance, since it is much greater than that of compact SiC materials /6/.

The wear/tool-life turning test provides a further source of application-related results. Wear mechanisms include mechanical abrasion, erosion of the cutting edge through cold pressure welding and chemical stresses such as oxidation, diffusion and plastic deformation. Owing to this combination of wear processes, this test yields considerable information on the resistance of composite systems. Both Cm60V steel and GGG 60 cast iron were used as machining counterparts. As shown in *Fig. 6*, machining tests were made with CVD-coated hard metals in addition to the various PVD coatings. CVD coatings are the best proven currently available. The two-layer TiAlVN-Al_2O_3 showed itself well able to compete. In the graded layer with the structure Ti(C,N)-Si(C,N), crater depth was reduced to one quarter of the unetched value following intermediate ionic etching. This is attributable partly to the additional cleaning and activation of the surface of the first layer, and partly to possible paracyano formation, since the substrate temperature was lower in this case.

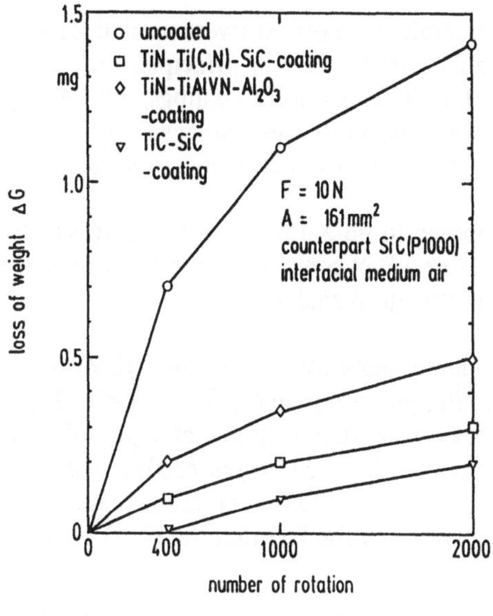

Fig. 5: Wear Behaviour in the Stylus-Disc Tribometer

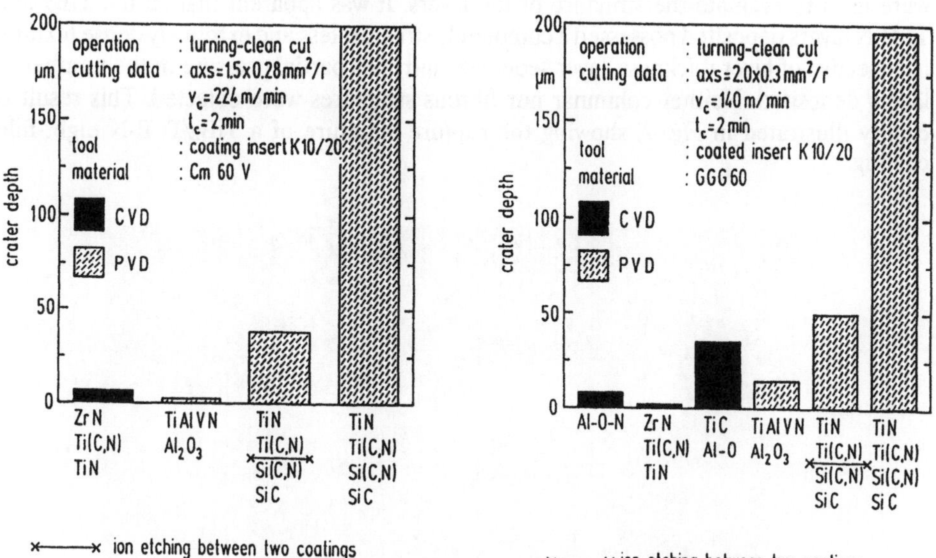

Fig. 6: Crater Depth after Wear/Tool-Life Turning Test

Characteristics of Gradient and Multilayer Coatings in the Ti-B-N System

Gradient and multilayer coatings in the Ti-B-N system are produced in order to achieve good adhesion while retaining great hardness. Double gradient layers and two- to eight-fold multilayer coatings were deposited. The objective of the investigations was to determine the optimum layer sequence. In some multilayer coatings, TiN was used as the bottom and

TiB_2 as the top layer (designation TiN/TiB_2), in others TiB_2 formed the bottom layer and TiN the top layer (designation TiB_2/TiN).

The Bragg angle and the strength of the reflexes were used to identify the phases occurring in the layers. Phase analysis showed that, in principle, only the TiB_2 phase could be demonstrated in both the TiB_2/Ti-B-N double coatings, the TiB_2/Ti-B-N eight-fold coatings and the TiB_2/TiN or TiN/TiB_2 multilayer coatings. Neither other Ti-B-N compounds such as TiB or Ti_2B_5 nor pure boron or titanium phases were identified.

All the Ti-B-N coatings proved to be X-ray amorphous, irrespective of nitrogen partial pressure. Neither TiB_2, TiN or a B-N compound, nor a putative ternary Ti-B-N structure was demonstrated by X-ray structural analysis.

The TiN phase was also clearly demonstrated in the TiB_2/TiN or TiN/TiB_2 multilayer coatings. Other titanium-nitrogen compounds, for example Ti_2N, did not occur. The characteristic TiN phase peaks were demonstrated at angles of reflection of $2\Theta = 43.04°$ and $2\Theta = 49.95°$. These characterize the (111) or (200) plane, the intensity of the (111) orientation growing as compared to the (200) orientation as the number of layers increased. This behaviour may be explained by the fact that the (111) plane and the (001) plane of TiB_2 are coherent.

Scanning electron microscope images of rupture surfaces produced under tensile stress were used to evaluate the structure of the layers. It was apparent that all the TiB_2 and Ti-B-N layers deposited possessed a completely structureless and extremely dense texture, irrespective of layer thickness, layer sequence, nitrogen partial pressure or the number of layers deposited. Neither columnar nor fibrous structures were detected. This result is clearly illustrated in *Fig. 7*, showing the rupture structure of a TiB_2/Ti-B-N eight-fold coating.

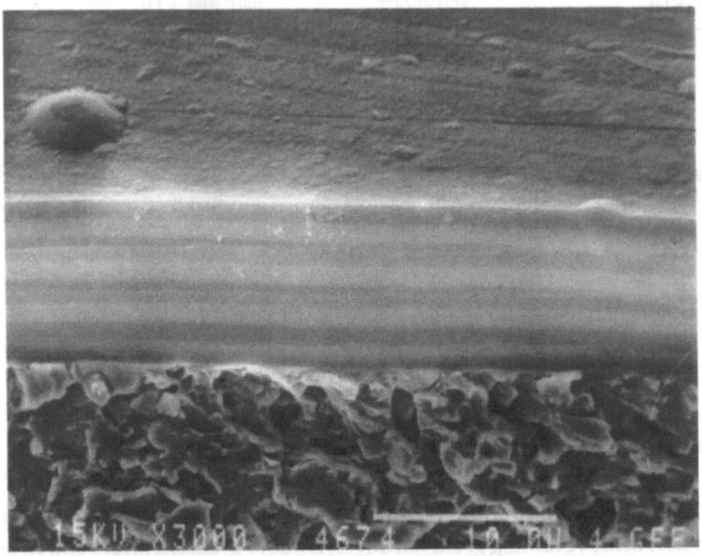

Fig. 7: Rupture Structure of a TiB_2/TiN Eight-Fold Coating

Electron beam microanalysis (ESMA) of some specimens revealed the quantitative composition of the coatings and enabled further conclusions to be drawn on the phases present in the coatings, especially their stoichiometry. It was also used to determine the type and concentration of other elements not demonstrated by X-ray analysis. The results of ESM analysis are shown in *Fig. 8.*

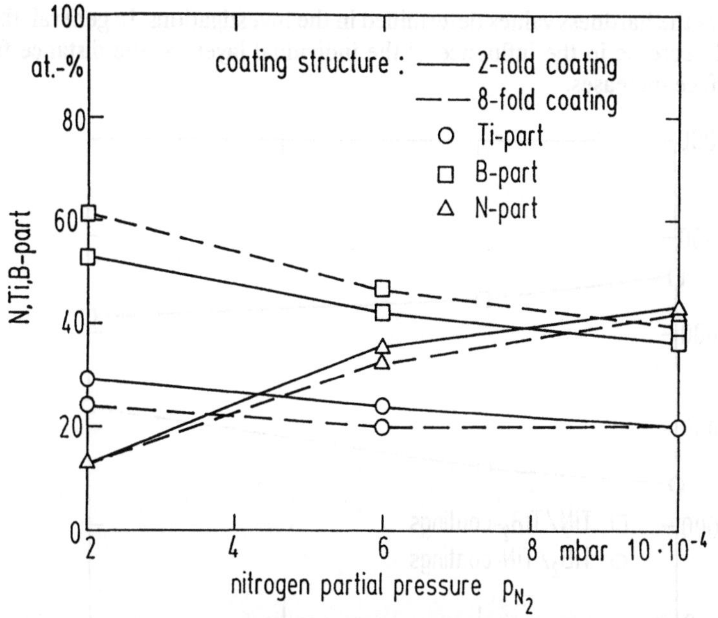

Fig. 8: Electron Beam Microanalysis of Ti-B-N Intermediate Layers as a Function of Nitrogen Partial Pressure and Layer Structure (TiB_2/Ti-B-N Coating System)

Of particular interest is the fact that in both the TiB_2/Ti-B-N double layer coatings and the corresponding eight-fold coatings there was not only a continuous rise in the argon component with increasing nitrogen partial pressure, but this was substantially higher in the eight-fold than in the comparable double layer coatings.

Whereas the increase in argon concentration with rising nitrogen partial pressure may be attributed to strengthening of the "dig-in" effect, the increased argon content of the Ti-B-N layers of the eight-fold as opposed to the double coatings is explained by the lower thickness of these layers.

The coating rate for all coating variants in the Ti-B-N system was approximately 6 μm/h.

Hardness measurements of these hard coatings should take into account the influence of the substrate material or previous layer. In order to exclude this influence, a minimum coating thickness dependent on the coating and substrate hardness is required. In general, the indentation diagonals should not exceed 70 % of the coating thickness, while new results for Vickers hardness measurements of TiN coatings indicate that a coating thickness greater than or equal to the expected indentation diagonals should be ensured.

If this problem with single layer coatings is transposed to the multilayer coatings deposited in this instance, it may be concluded that not only should a minimum coating thickness be maintained in order to exclude the influence of the substrate material, but that the different influences of the individual layers in the multilayer coatings on microhardness values need to be taken into account in order to ensure correct evaluation.

Fig. 9 indicates the hardness values determined in the investigations. In general, they show the expected decrease in the influence of the individual layers as the distance from the substrate surface increases.

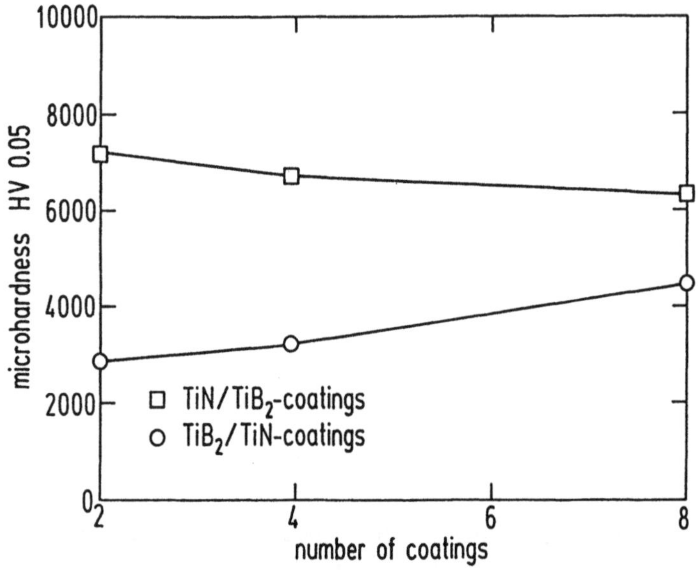

Fig. 9: Microhardness of Sputtered TiB_2/TiN or TiN/TiB_2 Multilayer Coatings as a Function of the Number of Layers

For example, TiB_2/Ti-B-N double and gradient coatings with a low TiB_2 sputtering time (7.5 min) attain exclusively lower hardness values than coatings produced with a TiB_2 sputtering time of 30 min. This behaviour is explained by the greater microhardness of TiB_2 single layer coatings as compared to Ti-B-N coatings. A similar explanation may be found for the greater hardness values of the TiN/TiB_2 as opposed to the TiB_2/TiN coatings (*Fig. 9*) if the considerably greater microhardness of TiB_2 as compared to TiN is taken into account. With the same TiB_2 component, the TiB_2 intermediate layers in the TiN/TiB_2 layers are nearer to the overall surface of the layer and thus exert a considerably greater influence on microhardness than the TiB_2 intermediate layers in the TiB_2/TiN coatings.

Summary

Magnetron sputtering permits troublefree deposition of multilayer and gradient coatings. The process allows such coatings to be influenced over a wide range. Following appropriate optimization processes, high-purity ceramic layers can be deposited on titanium carbide layers, with good adhesion and tribological properties. Multilayer coatings of this kind exhibit good resistance in machining tests.

Literature

/1/ Schulz, H.: Thesis, University of Karlsruhe, 1987
/2/ Leyendecker, T.: Thesis, RWTH Aachen, 1985
/3/ Holleck, H.: Möglichkeiten und Grenzen einer gezielten Stoffauswahl für verschleißfeste Hartstoffschichten, Zeitschrift für Werkstofftechnik 17, 1986, 334-349
/4/ Knotek, O.; W. Bosch: TiN(C,N) Coatings by Reactive Sputtering, Metal Powder Report, Vol. 39, No. 7, July 1984
/5/ Brütsch, R.: Chemical Vapour Deposition of Silicon Carbide and its Applications, Thin Solid Films, 126, 1985, 313-318
/6/ Shaffer, P.T.B.: Materials Index No. 1, Plenum Press, New York, 1964, 107-111

Hydrophobic Reorientation of Glow-Discharge Treated Polyethylene as Measured by Contact Angles

H. C. van der Mei, I. Stokroos, J. M. Schakenraad and H. J. Busscher

University of Groningen
Groningen
The Netherlands

ABSTRACT

In order to test the possibility of hydrophobic reorientation of oxygen glow-discharge treated polyethylene (5 min, 50 W), we have repeatedly oxidized polyethylene with time intervals of 4 days. During these time intervals water contact angles were measured every day to monitor the hydrophobic recovery of the surface. After the first oxidation the contact angles dropped from 105 to 74 degrees and hydrophobic recovery was complete already after 4 days. However, after 3 treatments the minimum contact angle observed decreased to 25 degrees and hydrophobic recovery became incomplete. After extremely long recovery times the maximum water contact angle observed was still 50 degrees. These observations give evidence in support of the dynamic behaviour of polymer surfaces and indicate that repeated glow-discharge treatment of surfaces after hydrophobic recovery can enhance the effect of the treatment.

INTRODUCTION

Radio Frequency Glow Discharge (RFGD) treatments are often used to improve the adhesion of polymer surfaces [1-5]. Oxygen plasma treatments of polymers greatly decrease the hydrophobicity of the surface which is advantageous for many technological, industrial and medical applications [6, 7]. However, the effects of such a treatment are often transient and the hydrophilicity obtained is lost in time.

Hypotheses to account for this hydrophobic recovery are either based on subtle phenomena as the dynamic behaviour of polymers at surface [8] or ordinary mechanisms as surface contamination and molecular dissociation. Owen *et al.* [9] summarized several possible mechanisms for the hydrophobic recovery of polymers, being:

1. Reorientation of surface hydrophilic groups away from the surface
2. Migration of untreated polymer chains from the bulk

3. External contamination of the polymer surface
4. Changes in surface roughness
5. Loss of volatile oxygen-rich species to the atmosphere

To us, the possibility of external contamination seems inadequate in general, since a completely hydrophilic glass surface with an initially zero water contact angle only becomes as hydrophobic as a 18 degrees water contact angle after exposure to the atmosphere, as illustrated in Figure 1.

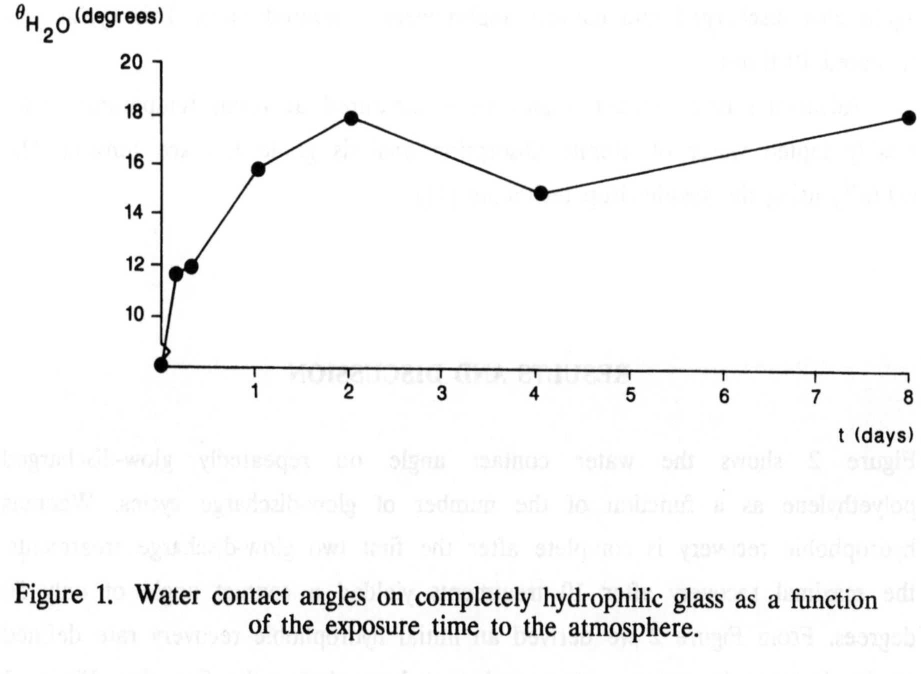

Figure 1. Water contact angles on completely hydrophilic glass as a function of the exposure time to the atmosphere.

It is the aim of this paper to investigate whether polymer surface dynamics (possibilities 1 and 2) can account for the hydrophobic recovery of oxygen glow-discharge treated polyethylene. To this end, water contact angles are measured on repeatedly glow-discharged treated surfaces.

MATERIALS AND METHODS

Polyethylene sheets (thickness 1 mm) were obtained from TKG Kunststoffe, Darmstadt, West-Germany and cut to pieces of 1 by 4 cm. Samples were cleaned by rinsing with ethanol (96%). Thus obtained sample surfaces were essentially free of oxygen. (O/C = 0.002, as measured by X-ray photoelectron spectroscopy) and possessed a stylus surface roughness R_A [10] of 1.0 μm.

These samples were glow-discharged at day zero in a PLASMOD at 50 Watt and 10-15 mbar oxygen pressure during 5 min. Subsequently, water contact angles were measured every day during a 4 days period. After 4 days a sample was again glow-discharged and contact angles were measured daily. This cyclus was repeated 10 times.

Advancing type contact angles were measured at room temperature with freshly tapted water of atomic absorption analysis grade (surface tension 72.8 $mJ.m^{-2}$) using the sessile drop technique [11].

RESULTS AND DISCUSSION

Figure 2 shows the water contact angle on repeatedly glow-discharged polyethylene as a function of the number of glow-discharge cycles. Whereas hydrophobic recovery is complete after the first two glow-discharge treatments, the maximal recovery after 10 treatments yielded a contact angle of only 50 degrees. From Figure 2 we derived an initial hydrophobic recovery rate defined as the increase in water contact angle per hour during the first day. Figure 3 demonstrates the initial hydrophobic recovery rate as a function of the number of RFGD treatments.

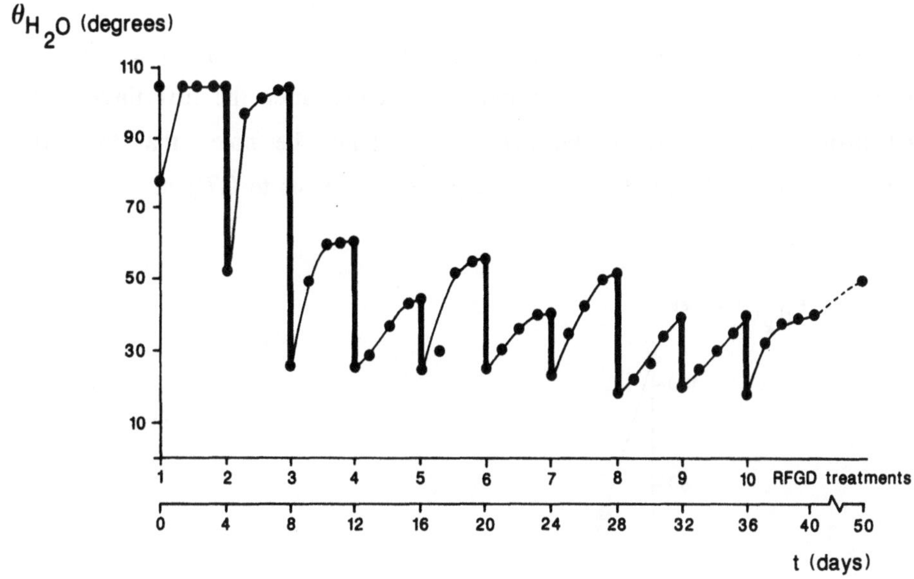

Figure 2. Water contact angles Θ_{H_2O} on repeatedly glow-discharged polyethylene as a function of the number of RFGD treatments.

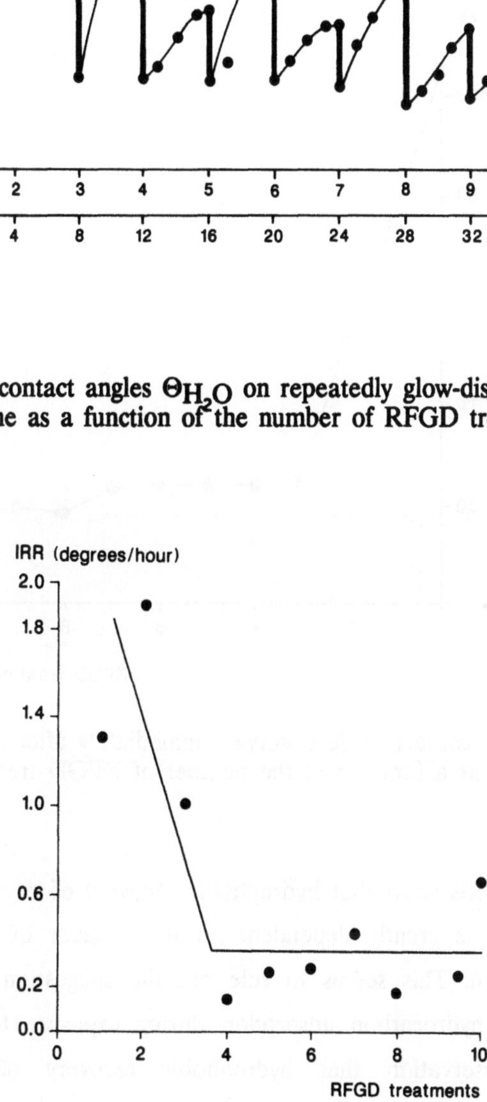

Figure 3. Initial hydrophobic recovery rate (IRR) (degrees/hour) as a function of the number of RFGD treatments.

Figure 4, in which the contact angle observed immediately after glow-discharge treatment is plotted as a function of the number of cycles, shows that the minimum contact angle after a treatment decreases after the first three cycles and then becomes constant. Furthermore, it should be noted that the stylus surface roughness slightly decreased during the treatments to 0.7 μm.

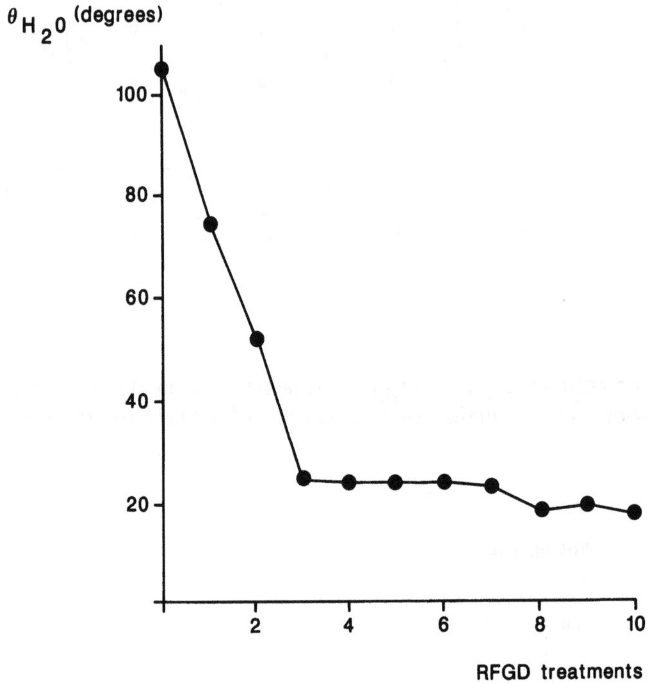

Figure 4. The water contact angle observed immediately after a glow-discharge treatment as a function of the number of RFGD treatments.

These observations show that hydrophobic recovery of oxygen glow-discharge treated polyethylene is greatly dependent on the number of times the RFGD treatment is repeated. This seems to rule out the suggestion that hydrophobic recovery is due to hydrocarbon adsorption during exposure to the atmosphere [12]. Also the observation that hydrophobic recovery of RFGD treated polyethylene encompasses contact angles from 40 to 60 degrees, whereas atmospheric contamination of glass yields a water contact angle of 18 degrees (see Figure 1), supports this conclusions.

The recovery rate is highest after the first RFGD treatments (see Figure 3), hypothetically because of the relatively high number of hydrophobic groups in comparison with the number of hydrophilized groups in the surface region, facilitating fast reorientation. Since eventually the hydrophilized region becomes thicker and thicker as the number of RFGD treatments increases, migration of hydrophobic groups to the surface will become slower (see also Figure 3) and incomplete (see Figure 2), as indicated by the results in this paper. More detailed investigation however, with other surface sensitive techniques (e.g. X-ray photoelectron spectroscopy or reflective infrared spectroscopy) is needed to verify this hypothesis.

ACKNOWLEDGEMENTS

The authors are greatly indebted to M.H. Uyen for technical assistance and to M.P. Schakenraad-Dolfing for the preparation of this manuscript.

REFERENCES

1. Chow, B.C. and Whittle, D.J., Wettability and adhesion studies of plasma-treated plastic substrates for gelatin holograms. J. Adh. Sci. Technol., 1988, 2, 363-8.

2. Yasuda, H., Bumgarner, M.O., Marsh, H.C. and Morosoff, N., Plasma polymerization of some organic compounds and properties of the polymers. J. Pol. Sci., 1976, 14, 195-224.

3. Vargo, T.G., Gardella, J.A., Meyer, A.E. and Baier, R.E., Radio Frequency Glow Discharge Oxidation/Hydrilysis of Polytetrafluorethylene and Polyvinylidenefluoride Membrane Surfaces. Macromolecules, 1989, submitted.

4. Ruckenstein, E. and Gourisankar, S.V., Preparation and characterization of thin film surface coatings for biological environments. Biomaterials, 1986, 7, 403-22.

5. Chawla, A.S. and Sipeha, R., Characterization of plasma polymerized silicone coatings useful as biomaterials. J. Biomed. Mat. Res., 1984, **18**, 537-45.

6. Fakes, D.W., Davies, M.C., Brown, A. and Newton, J.M., The surface analysis of a plasma modified contact lens surface by SSIMS. Surf. Interf. Anal., 1988, **13**, 233-6.

7. Van Wachem, P.B., Interactions of cultured human endothelial cells with polymeric surfaces. In Interactions of cultured human endothelial cells with polymeric surfaces, Thesis, Groningen, The Netherlands, 1987, pp. 31-46.

8. Andrade, J.D., Gregonis, D.E. and Smith, L.M., Polymer surface dynamics. In Surface and Interfacial Aspects of Biomedical Polymers, ed. J.D. Andrade, Plenum Press, New York, 1985, pp. 15-39.

9. Owen, M.J., Gentle, T.M., Orbeck, T. and Williams, D.E., Dynamic wettability of hydrophobic polymers. In Polymer Surface Dynamics, ed. J.D. Andrade, Plenum Press, New York and London, 1988, pp. 101-10.

10. Busscher, H.J., Van Pelt, A.W.J., De Boer, P., De Jong, H.P. and Arends, J., The effect of surface roughening of polymers on measured contact angles of liquids. Coll. Surf., 1984, **9**, 319-31.

11. Neumann, A.W. and Good, R.J., Techniques of measuring contact angles. In Surface and Colloid Science, eds. R.J. Good and R.R. Stromberg, Plenum Press, New York and London, 1979, pp. 31-91.

12. Gerenser, L.J., X-ray photoemmision study of plasma modified polyethylene surfaces. J. Adhesion Sci. Tech., 1987, **1**, 303-18.

Hall Effect and Resistance Measurements on Implanted and Annealed Silicon Layers

C. Christofides

University of Toronto
Toronto
Ontario
Canada

ABSTRACT

A study of the effects of the annealing temperature on arsenic implanted silicon layer (at high dose and energy) is reported. Hall effect and resistance measurements, at room temperature (300 K) have been employed to characterize arsenic-implanted non-annealed and annealed (200 - 1100 °C) silicon films. This technique allows one to probe the annihilation processes of damage layer defects as a function of annealing conditions. A significant modification of the carrier scattering mechanism occurs for annealing temperatures around 550 °C, which is manifested as a drastic change in the behavior of the temperature dependence of the transport parameters. The activation energy of the recovery process of the ionic implantation damage, found to be about 0.6 *eV*, is attributed to a local reconstruction of the implanted layer.

INTRODUCTION

One disadvantage of ion implantation is its generation of defects in the lattice, due to collisions between impinging ions and crystal atoms [1]. The concentration of these defects greatly influences the electronic, optical, and thermal properties of semiconductors. In general it is very difficult to identify the various kinds of defects; however, for the characterization of materials, it is essential to know the defect concentration, energy of formation, energy of dissolution (of complex defects), and mechanism of interaction with carriers.

High implantation doses lead not only to the creation of many localized defects, but also to the formation of an amorphous layer. It is generally known that most of the damage induced by ion implantation can be removed by annealing the crystal after implantation. In fact, the annealing is mainly expected to repair the damage caused by ion implantation [1].

According to Gibbons [1], if the implantation dose is much greater than the critical dose Φ_c, [2] thirty minutes of annealing between 550 and 600 °C will help diffuse more than 90% of the implanted ions from interstitial positions to substitutional ones, where they will be electronically active. If the implantation dose Φ is approximately equal to the critical dose Φ_c, 50 to 90% of the implanted ions become electronically active for an annealing period at 550 °C. In this case, in order to approach the ideal crystal state, we must anneal the layer between 800 and 1100 °C. If Φ is smaller than Φ_c, after an annealing around 550 °C only 10% of the implanted ions take substitutional positions [3]. Due to the small population of implanted ions and vacancy sites, the probability of vacancy-interstitial ion recombination is small. Finally, in the case of very small implantation dose ($\Phi << \Phi_c$ annealing for 30 minutes at 350 °C is sufficient to activate the majority of impurities (80 to 90%).

The aim of this paper is to analyze the main reconstruction mechanisms of silicon layers damaged by arsenic ion implantation at high dose and energy. Towards this end, resistivity and Hall effect measurements at room temperature have been conducted on unannealed and annealed samples. Finally, in order to find the activation energy of the recovery process of the ionic implantation damages, we used the measured transport coefficients.

Experimental Details

Two-inch diameter silicon wafers were implanted through a thick oxide layer, at room temperature, with arsenic (dose $\Phi = 8\times10^{14}\ cm^{-2}$; energy $E = 120\ keV$). For the n-type Si substrates (8 $\Omega\cdot cm$) used in this work, the implantation depth of the arsenic was about 0.3 μm. The samples were then thermally annealed (t=1 hour, $200\leq T_a \leq 1000$ °C) in an inert atmosphere (N_2). The impurity profile, as obtained by a SUPREM simulation and the resulting annealing profiles obtained both by a SUPREM simulation and spreading resistance (see Fig.1) measurements [4] were shown in Fig. 2.

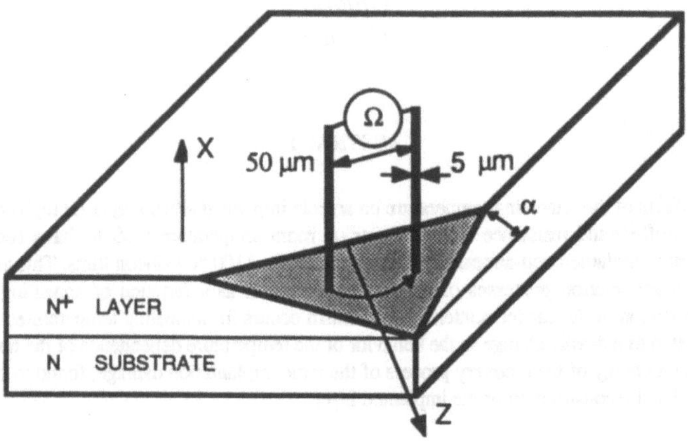

Figure 1 Spreading resistance measurements for angle-lapped semiconductor sample.

In this section, the method of measurement which were used to study the transport properties of the implanted silicon samples will be described. This method has been used to measure the dc sheet resistance R_s Eq. (1) gives the sheet resistance R_s as a function of two successive Van der Pauw transresistances R_1 and R_2 [5]:

$$R_s = \frac{\pi \phi}{2\ ln 2}(R_1 + R_2) \tag{1}$$

where ϕ is the *Van der Pauw* function.

The Hall mobility μ_H can now be obtained from R_s and the Hall voltage V_H appearing between two non-successive contacts; V_H arises from the application of a magnetic field B perpendicular to sample surface [5]:

$$V_H = \frac{V(+B) - V(-B)}{2} \tag{2}$$

where $V(+B)$ and $V(-B)$ are the magnetic field induced voltages perpendicular to the sample in two different directions. In our study B was 0.6 *Tesla*. Now we can write:

$$\mu_H = \frac{V_H}{I\,B\,R_s} \tag{3}$$

where I is the injected current through the contacts

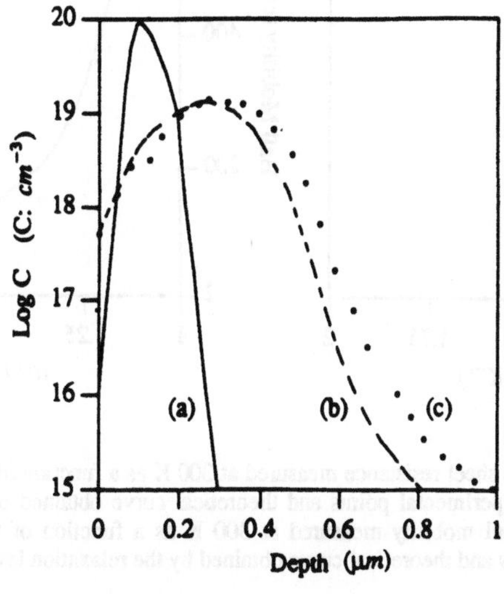

Figure 2 Impurity profiles of the arsenic implanted silicon layers obtained by SUPREM simulations: (a) before annealing, and; (b) after thermal annealing at 1100 °C for 1 hour; and (c) by spreading resistance measurements after the same annealing as in (b) [dots].

Discussion and Interpretation

The transport coefficients (ρ, R_s and μ_H) measured at 300 K have been used in order to study the annealing kinetics of the layer as a function of the annealing temperature, T_a, and the annealing time, t.

Figures 3(a) and (b) show the variation of the sheet resistance R_s, and the variation of the Hall mobility μ_H, respectively, measured at 300 K, as a function of the inverse of the annealing temperature T_a. A strong modification of the material appears to occur at about T_a= 420 °C; near this temperature the post-annealing resistance varies from 10^2 to 5.5×10^5 Ω. The resistance for the unannealed sample was approximately 5.5×10^5 Ω. In our case, the annealing kinetics are consistent with a local annealing process in which the sheet resistance and Hall mobility are described by the relaxation law [6-8]:

$$R_s = (R_{si} - R_{sf})\ exp(-t/\tau) + R_{sf} \tag{4a}$$

and

$$\mu_H = (\mu_{Hi} - \mu_{Hf})\ exp(-t/\tau) + \mu_{Hf} \tag{4b}$$

where R_{si} and μ_{Hi} are the initial sheet resistance and initial Hall mobility (prior to annealing) respectively; R_{sf} and μ_{Hf} are the final sheet resistance and final Hall mobility (after annealing at 1100 °C) respectively,

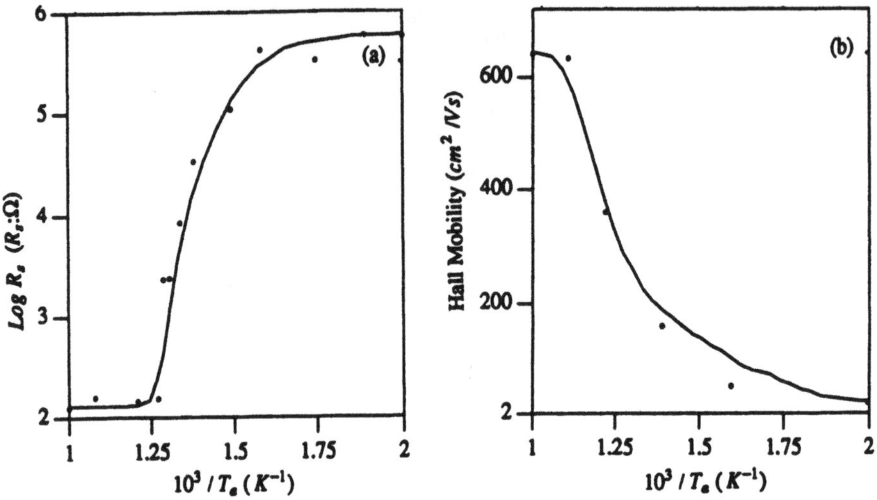

Figure 3 (a) Variation of the sheet resistance measured at 300 K as a function of the inverse of the annealing temperature (experimental points and theoretical curve obtained by the relaxation law). (b) Variation of the Hall mobility measured at 300 K as a function of the annealing temperature (experimental points and theoretical curve obtained by the relaxation law).

t is the annealing duration, and τ the relaxation time of the local annealing process. τ is energy activated:

$$\tau = \tau_o \; exp(E_a/kT_a) \tag{5}$$

where E_a is the activation energy. The solid curves of Figures 3(a) and (b) have been obtained from Eqs. (4a) and (4b), using E_a=0.65 eV and τ_o = 3.6×10^{-6} hr. The small value of E_a is most likely attributable to a local migration of point defects [1] and differs from the usual value (2-3 eV) which is characteristic of a mechanism involving propagation of the recrystallization front from the substrate towards the surface [8]. Rearranging Eqs. (4a) and (4b), the reduced sheet resistance and reduced Hall mobility could be defined as:

$$Q_R = \frac{R_{si} - R_{sf}}{R_s(T_a) - R_{sf}} = exp(t/\tau) \tag{6a}$$

and

$$Q_\mu = \frac{\mu_{Hi} - \mu_{Hf}}{\mu_H(T_a) - \mu_{Hf}} = exp(t/\tau) \tag{6b}$$

From Eqs. (5) and (6a) the natural logarithm of the reduced sheet resistance can be written as:

$$\ln(Q_R) = t/\tau_o \; \exp(-E_a / k\, T_a) \tag{7a}$$

from Eqs. (5) and (6b) the natural logarithm of the reduced Hall mobility can be written as:

$$\ln(Q_\mu) = t/\tau_o \; \exp(-E_a / k\, T_a) \tag{7b}$$

Figures 4(a) and (b) show the plots of ln(Q_R) and ln(Q_μ) in the semilogarithmic spaces as a function of $1/T_a$. The characteristic slopes of ln[ln(Q_R)] and ln[ln(Q_μ)] are proportional to the activation energy E_a. According to the experimental slopes, the activation energy is approximately 0.6 eV and, as expected, is in agreement with the E_a value which to obtain the solid curves of Figs. 3(a) and (b).

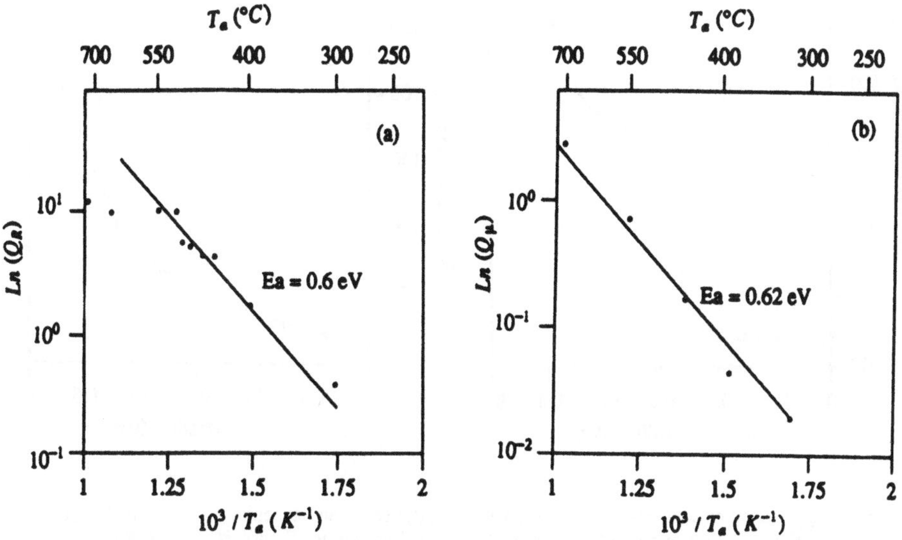

Figure 4 (a) $Ln(Q_R)$ variation with inverse of annealing (1 hour) temperature T_a: determination of the activation energy of the relaxation time. (b) $Ln(Q_\mu)$ variation with inverse of annealing (1 hour) temperature T_a: determination of the activation energy of the relaxation time.

From the spreading resistance measurements [7], $Q_R(T_a)$ was calculated as a function of depth X, and then the activation energy E_a was found for each X using Eq.(7a). E_a was found to remain approximately independent of X ($E_a = 0.3\text{-}0.45\ eV$ for $0 \leq X \leq 0.6\ \mu m$).

In Fig. 5(a) the variation of transition temperature T_t as a function of depth X is depicted, where T_t is the annealing temperature for which the value of the initial resistance of the unannealed sample is reduced by a factor 10 after an isochronal annealing of one hour. Note that this transition temperature T_t increases as a function of depth X in the layer; it is evident that a higher temperature is required in order to annihilate defects far from the layer surface.

In Fig. 5(b), we present the variation of relaxation time τ as a function of depth X, deduced from $Q_R(X)$, using Eq. (5) and some data obtained with spreading resistance measurements [7]. Fuller [6] has given a theoretical relationship for $\tau(X)$, based on local annealing processes:

$$\tau(X) = \frac{k\,T_a}{4\,\pi\,q^2}\,\frac{1}{D\,N(X)} \tag{8}$$

where D is the diffusivity of the neutralizing atoms [$D = D_o exp(-E_a/kT_a)$], k is the Boltzmann constant and $N(X)$ is the density of atoms which aid in the neutralization of the defect or complex during the annealing.

Since $\tau(X)$ experimentally increases with depth [Fig.5(b)], and E_a is almost constant with X, we can conclude from the Eq.(8) that: (i) the diffusing species is probably the same throughout the implanted layer because, as we noted, E_a is independent of depth X (it appears that the defects are interstitial silicon atoms). (ii) The density of diffusing atoms, $N(X)$, decreases with depth X, since D is approximately constant with X. This result confirms that the density of defects decreases with X, as one would expect.

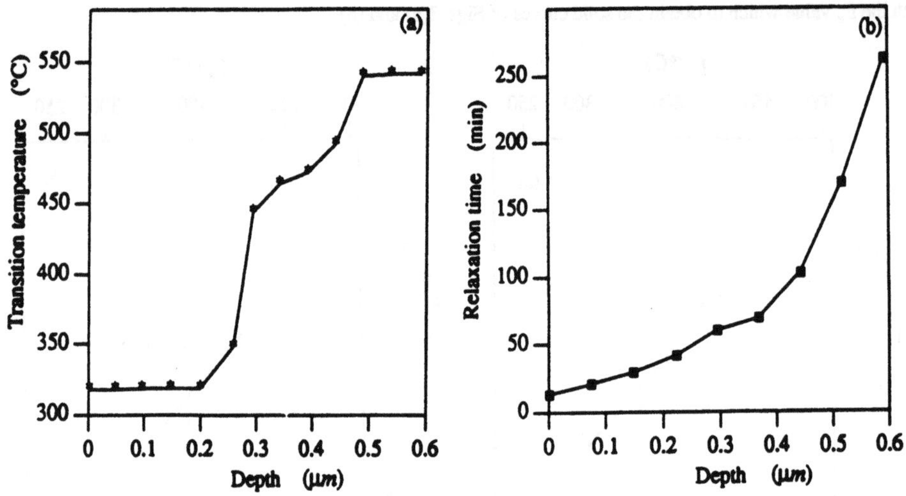

Figure 5 (a) Transition temperature as a function of depth (isochronal annealing: $t = 1$ hour; T=300 K). (b) Relaxation time as a function of depth (isothermal annealing: T_a=400 °C; T=300 K).

Summary and Remarks

A study of the annealing kinetics of ion implantation-induced defects has been made by means of electronic transport investigations. The transport properties of samples annealed under different conditions (isochronal annealing: t=1 hour, $200 \leq T_a \leq 1100$ °C) where studied in order to examine the annihilation kinetics of these defects. Towards this end we have used two electronic methods, thus allowing a more exhaustive analysis of the experimental results: The spreading resistance method provided us with information regarding the depth-dependence of the transport parameters in the implanted layer and the Van der Pauw method allowed us to study the influence of annealing to the transport coefficients.

The main results of our study can be summarized as follows: (1) As for the defect annihilation processes in layers damaged by ion-implantation, we determined a relatively small activation energy (0.4 to 0.6 *eV*), which is probably characteristic of a local lattice rearrangement mechanism involving the migration of interstitial silicon atoms. The small value of the activation energy (<3 *eV*) indicates that the mechanism consisting of the propagation of a recrystallization front from the substrate towards the surface was not dominant in our case. (2) These measurements confirmed that the density of defects decreased with the depth of the implanted layer. This conclusion was reinforced by the independence of activation energy with the depth, which indicated a decreasing density of defects with the depth, through the relaxation time [$\tau(X)$] relationship.

The analysis of the transport properties of arsenic implanted silicon layers provided new information about the annealing kinetics of defects and inhomogeneities induced by ion implantation. We have shown that transport coefficient measurements constitute a very suitable means of investigating the annihilation of residual defects and the occurance of structural changes in such implanted layers after thermal annealing at temperatures both below and above the current threshold temperature of recrystallization.

In crystalline semiconductors transport properties are relatively well interpreted by the classical transport theories. Amorphous semiconductors have also been the object of several theoretical papers which allow one to explain their transport mechanisms in a qualitative way. On the other hand, it is difficult to interpret the transport properties of intermediary materials, namely inhomogeneous materials which are hybrids of completely crystalline and amorphous materials; our plan involved using the classic models of solid crystals, in addition to those suited to amorphous materials. Implanted and unannealed, or insufficiently annealed silicon layers are typical of these inhomogeneous materials; our experiments have indicated the necessity of further developing the transport models which are used to extract useful information from the experimental data.

ACKNOWLEDGMENTS

The experimental work presented in this paper has been done in *LPCS-ENSERG* Laboratory (Grenoble, France). The author would like to thank Drs. H. Jaouen and G. Ghibaudo for their help during the measurements.

REFERENCES

1. J.F. Gibbons, Ion Implantation in Semiconductors (Part I), *Proc. IEEE* **60**, 1062 (1972).
2. S. Prussin, D.Margolese, and R. Tauber, Formation of Amorphous layers by Ion Implantation. *J. Appl. Phys.* **57**, 180 (1985).
3. L.A. Christel, J.F. Gibbons, and T.W. Sigmon, *J. Appl. Phys.*, **52**, 7143 (1981).
4. J.R. Ehrstein, *Nondestructive Evaluation of Semiconductor Materials and Devices*, edited by J.N. Zemel (Philadelphia, Pennsylvania, 1979).
5. L.J. Van der Pauw, A Method of Measuring Specific resistivity and Hall Effect of Discs Arbitrary Shape, *Philips Res. Rep.*, **13**, 1 (1958).
6. C.S. Fuller, in *Semiconductors*, edited by N.B. Hannay (Reinhold, 1959), Ch.5.
7. C. Christofides, Etude des Proprietes de Transport dans des Coushes de Silicium Desordonnes par Implantation Ionique a Forte Dose - Influence du Recuit Thermique. *Ph.D Thesis*, (LPCS - ENSERG, Grenoble, France, Dec. 1986).
8. C. Christofides, H. Jaouen, and G. Ghibaudo, Etude de silicium implante a l'arsenic par effet de transport. Influence du recuit thermique., *Revue Phys. Appl.*, **22**, 407 (1987).

Diffusion of Carbon in Carbon Steels and Graphitic Cast Irons Heat Treated by Concentrated Energy Sources

V. Sedlacek

Technical University
Prague Czechoslovakia

J. Horyna and J. Zamecnik

Technical University
Liberec Czechoslovakia

ABSTRACT

This paper examines the diffusion of carbon under the conditions induced by rapid heating and cooling of carbon steels and graphitic cast irons and attempts a physico-metallurgical and mathematical formulation of the process. The paper presents a model for determining the diffusion path lengths of carbon in the metal matrix and relations describing the redistribution of carbon in space and time. Theoretic examination is used for determining critical heating rates for carbon steels and graphitic cast irons during their heat treatment by concentrated energy sources, particularly by lasers.

INTRODUCTION

Many research workers have lately concluded that rapid heating and cooling alters the mechanism and kinetics of the phase transformations of metals. These changes are presumed to be closely related to the fundamentals of mass transfer by diffusion.

ANALYSIS AND DESCRIPTION OF PROBLEM

Phase transformation $\alpha \rightarrow \gamma$ in carbon steels and graphitic cast irons during rapid heating

Transformation hardenings of carbon steels and graphitic cast irons have in common the fact that the metal matrix austenitizates in a very short time proceeded with a phase transformation as a result of rapid cooling by heat removal within a component. In the metal matrix of steels the carbon presented at initial stage, as a rule as cementite, the matrix of cast irons can be even only ferritic and it is graphite that serves as a source of carbon for austenite saturation. In a ferrite-pearlitic steels the new phase nucleates prior on the ferrite-cementite phase boundary and the growth of new phase is controlled by the diffusion of carbon from cementite into growthing austenite. After nucleation the steel changes into the three phases system of ferrite-austenite-cementite with two phase boundaries, namely austenite-ferrite and austenite-cementite. On these phase boundaries temperature dependent concentrations of carbon exist

c(F-A)...carbon concentration in ferrite on the boundary with austenite,
c(A-F).. carbon concentration in austenite on the boundary with ferrite,
c(A-C).. carbon concentration in austenite on the boundary with cementite.

The composition of cementite on the boundary with austenite c(C-A) and ferrite, as well, equals practically to 6,68 wt% C. Rapid heating of graphitic cast irons is a similar case. The metal matrix of these graphitic cast irons is ferritic, pearlitic or ferrite-pearlitic with graphitic systems uniformly scattered. Then the new phase nucleates prior on the ferrite-graphite or ferrite-cementite phase boundary and the growth is controlled by the diffusion of carbon from graphite or from cementite into growthing austenite. Graphitic cast iron changes into the four phases system of ferrite-austenite-cementite-graphite with three phase boundaries of austenite-ferrite, austenite-cementite and austenite-graphite. On these phase boundaries the above mentioned temperature dependent carbon concentrations exist including carbon concentrations in the area of graphitic systems, i.e.

c(Fg-G) ..carbon concentration in ferrite on the boundary with graphite,
c(A-G) .. carbon concentration in austenite on the boundary with graphite.

The composition of graphite on the boundary with austenite c(G-A) is practically 100 wt %C. Significance of symbols is obvious from Fig.1.
The growth of austenite may be explained as the diffusion of carbon into ferrite, too. The diffusion of carbon into ferrite can exist together with the presence of a concentration gradient. Carbon is thus taken away from the carbide or graphite boundary and the process of diffusion tends to increase the contents of carbon on the ferrite-austenite boundary. That is why the ferrite-austenite phase boundary moves in the direction of diffusion, i.e. towards ferrite and austenite-cementite phase boundary towards cementite or austenite-graphite towards graphite.

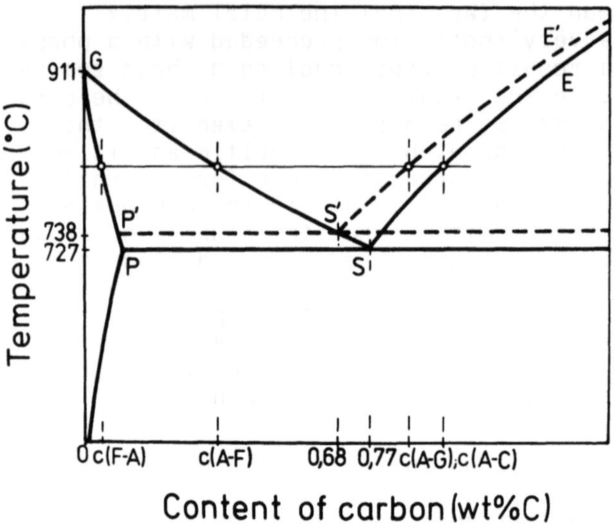

Figure 1. Significance of symbols is obvious.

From results of our theoretical and experimental research works it follows that certain heating rate may alter the mechanism of austenitization of carbon steels and graphitic cast irons because temperatures above Ac1 may induce independent changes in ferritic sections of the matrix into carbon-free austenite without diffusion interactions with cementite or graphite. From concentration conditions point of view this mechanism may be considered diffusionless. This may be accepted under the condition that during super rapid heating, in Ac1-Ac3 temperature interval, a certain amount of ferrite changes into austenite by the interaction with cementite or graphite.

The general idea of the austenitization mechanism during rapid heating of carbon steels and graphitic cast irons is then as follows. The existence of first austenite nuclei on the ferrite-cementite or ferrite-graphite phase boundaries starts the diffusional movement of the austenite-ferrite phase boundary towards ferrite. The prior existence of austenite in the intersection areas of moving diffusional interphases with ferrite subboundaries is only natural. At the same time austenite can also exists outside areas of moving diffusional interphases, namely on the boundaries of ferrite grains and subgrains, in case of favourable thermodynamic conditions. If the amount of heat supplied is sufficient, the process developes at the rate the size of which depends on the diffusional velocity of carbon. The existence of new austenitic systems is also possible inside ferrite, namely without any supply of carbon from cementite or graphite, i.e. by diffusionless transformations. If the distance between two graphitic particle is short the intersection of diffusionally moving shells terminates the

process of austenitization in this place. If the distace is long the temperature remarkably rises above Ac3 before austenitic shells intersect and ferritic grains will transform into low carbon austenite by the diffusionless mechanism. The higher heat quantity is supplied to the transforming volume the higher part of ferrite will transform into low carbon austenite by the diffusionless mechanism.
This analysis also concerns the austenitization between pearlitic grains and graphitic particle and pearlite.

Mathematical analysis of diffusion of carbon in carbon steels and graphitic cast irons during rapid heating.
Suppose an elementary particle of carbon moveable in the process of diffusion. The movement occurs by jumping the particle from the intersticial position to the neighbourhood one. As illustrated in Fig.2., the particle will be found in the distance of the vektor $\vec{r}(t)$ from initial position after some time. After the jump of particle from the point B to the neighbourhood point C by $\vec{dr}$ distance this vektor changes into $\vec{r}(t+dt)$, where dt is the time required for jump from B to C. The distance between A and C may be described by equation for the particular direction α of jump

$$r^2(t+dt) = r^2(t) + (dr)^2 - 2.r(t).dr.\cos\alpha \qquad (1)$$

The angel α ranges from 0 to π , i.e. the point C is located at the sphere surface of the B center and the dr radius.
If all posible jumps in the surroundings of the point B are taken in account, i.e. $\alpha \in \langle 0;\pi\rangle$, it leads to the equation

$$\overline{r^2(t+dt)} = \overline{r^2(t)} + \overline{(dr)^2} \qquad (2)$$

If the time interval of jump dt is considered to be the same in the average for each of jumps and the distance dr corresponds to the interatomic distance d, we obtain from (2)

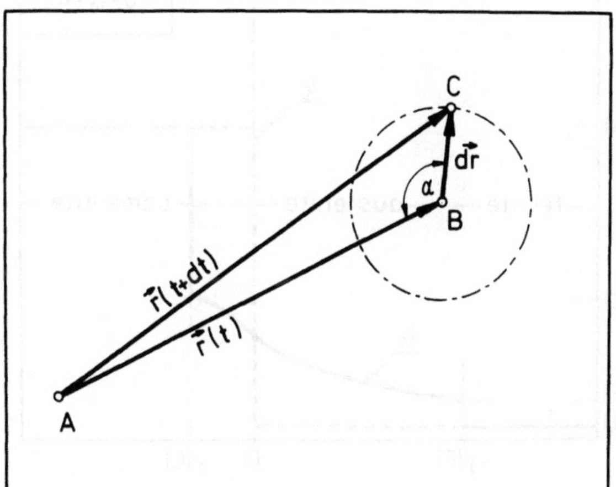

Figure 2. Jump of particle from the point B to the point C.

$$\overline{r^2(k+1)} = \overline{r^2(k)} + d^2 \qquad (3),$$

where k is a number of jumps of the size d. Consequently, after k jumps the mean square of the diffusion path length is

$$\overline{r^2(k)} = k \,.\, d^2 \qquad (4)$$

Deviding the whole sequence of jumps into the certain groups, i.e. $k = k_1 + k_2 + \ldots + k_n$, we finally obtain

$$\sqrt{\overline{r^2(k)}} = \sqrt{\overline{r^2(k_1)} + \overline{r^2(k_2)} + \ldots\ldots + \overline{r^2(k_n)}} \qquad (5),$$

where $\sqrt{\overline{r^2(k)}}$ is the mean diffusion path length of particles after k jumps. To calculate of the mean diffusion path length of particle in accordance with the above mentioned equations the knowledge of the temperature distribution in time of the examined place of the material is necessary.
The description of the ferrite-austenite-cementite system in accordance with Fig.3. can be obtained through solution of Fick´s laws. The equation (6) describes the distribution of the carbon concentration in austenite

$$c = c(A\text{-}C) - [c(A\text{-}F) - c(A\text{-}C)]\,.\,\frac{\operatorname{erf}\left(\frac{x}{2\sqrt{Dt}}\right) - \operatorname{erf}\beta(C)}{\operatorname{erf}\beta(F) + \operatorname{erf}\beta(C)} \qquad (6)$$

Motion equations of the phase boundaries are given by the expressions

$$y(C) = 2.\beta(C)\sqrt{Dt} \quad \text{and} \quad y(F) = 2.\beta(F)\sqrt{Dt} \qquad (7)$$

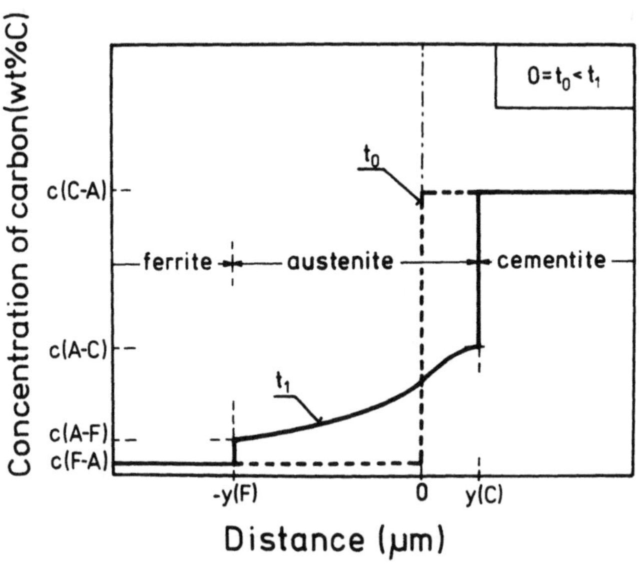

Figure 3. System of ferrite-austenite-cementite

where $\beta(C)$ and $\beta(F)$ are dimensioless koeficients for the description of which the system of the transcendental equations must be solved:

$$\sqrt{\pi}.\beta(F).[\mathrm{erf}\beta(F)+\mathrm{erf}\beta(C)].e^{\beta(F)^2} = \frac{c(A-C)-c(A-F)}{c(A-F)-c(F-A)}$$
$$\sqrt{\pi}.\beta(C).[\mathrm{erf}\beta(F)+\mathrm{erf}\beta(C)].e^{\beta(C)^2} = \frac{c(A-C)-c(A-F)}{c(C-A)-c(A-C)} \quad (8)$$

For the ferrite-austenite-graphite system in accordance with Fig.4. we obtain analogically

$$c = c(A-G)-[c(A-Fg)-c(A-G)].\frac{\mathrm{erf}\left(\frac{x}{2\sqrt{Dt}}\right)-\mathrm{erf}\beta(G)}{\mathrm{erf}\beta(Fg)+\mathrm{erf}\beta(G)} \quad (9)$$

Motion equations of the phase boundaries and the system of the transcendental equations for description of dimensionless koeficients $\beta(Fg)$ and $\beta(G)$ we may obtain analogically from (7) and (8).

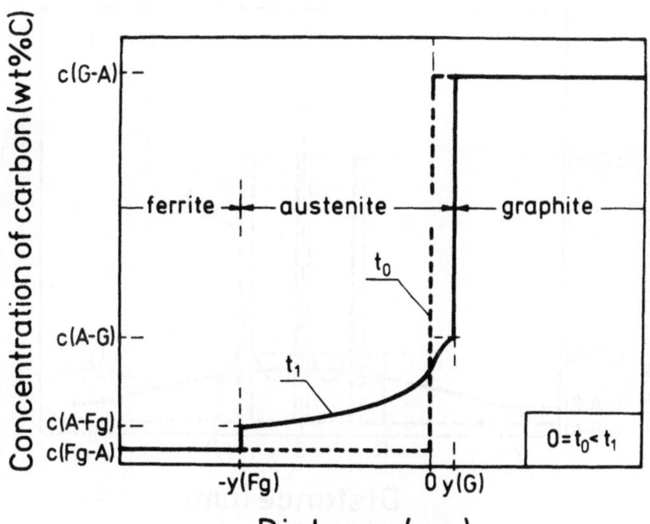

Figure 4. System of ferrite-austenite-graphite.

Carbon concentrations are not constant on the austenite--ferrite, austenite-graphite and austenite-cementite phase boundaries during rapid anizothermal heating. If we presume that concentrations c(C-A), c(F-A), c(Fg-A), c(G-A), c(A-F), c(A-C), c(A-Fg) and c(A-G) change with temperature in accordance with the diagrams Fe-C and $Fe-Fe_3C$ during heating then futher mathematical treatment of

transcendental equations leads to $\beta(F)$, $\beta(C)$, $\beta(Fg)$ and $\beta(G)$ as a function of temperature.
If we devide the temperature distribution of examined area in unique time intervals Δt the mean diffusion distance of the phase boundary from initial position results according to (5)

$$\bar{y} = \sqrt{\sum_{i=1}^{n} \left[2\beta(i)\sqrt{D(i).\Delta t}\right]^2} = 2\sqrt{\Delta t}.\sqrt{\sum_{i=1}^{n} \left[\beta(i)^2 D(i)\right]} \quad (10)$$

where $\beta(i)$ and $D(i)$ correspond to the temperature obtained.
Two phases system of austenite-ferrite models the situation when pearlit has austenited in the ferrite--pearlitic matrix with the average concentration of carbon $c(P\rightarrow A)$ and austenite diffusionally propagates inside ferrite surroudings. The solution of Fick's laws in both two and three dimensional spaces leads to the equations desribing the redistribution of carbon concentration during this process of austenitization. Presuming transformated pearlit to be the globule of the radius R then, in accordance with Fig.5.,

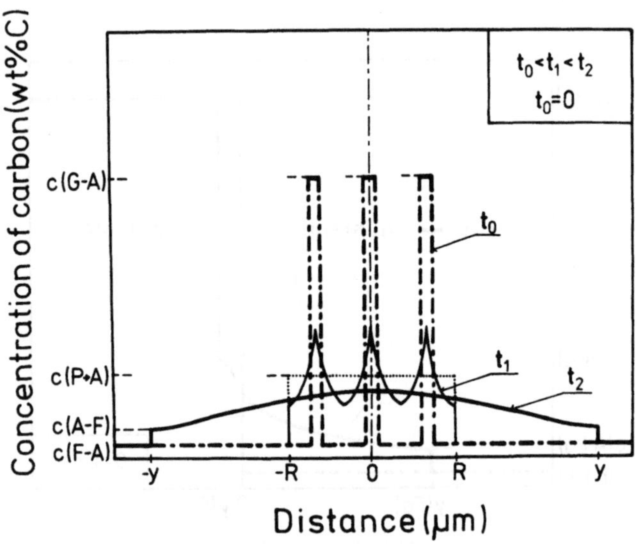

Figure 5. Redistribution of carbon concentration in system of austenite (pearlite)-ferrite.

the redistribution of carbon concentration is modelled by the system of equations (11):

1. For the area $0 \leq x < R$:

$$c = \frac{c(P\rightarrow A)}{2} . \left\{ \text{erf}A + \text{erf}B + \frac{y(F)}{x\sqrt{\pi}} . \left[\exp(-A^2) - \exp(-B^2)\right] \right\}$$

2. For the area $R \leq x < R + y(F)$:

$$c = \frac{\frac{c(P-A)}{2}.\left\{erfC+\frac{y(F)}{x\sqrt{\pi}}.\left[exp(-C^2)-1\right]\right\}+c(F-A)\sqrt{\pi}.B.exp(B^2).(erfB+1)}{1+\sqrt{\pi}.B.exp(B^2).(erfB+1)}$$

3. For the area $x \geq R + y(F)$: (11)

$$c = c(F-A)$$

where $A = \frac{R+x}{y(F)}$, $B = \frac{R-x}{y(F)}$ and $C = \frac{2.R}{y(F)}$

Detection of critical heating rates for carbon steels and graphitic cast irons during rapid heating

From analysis in the second part of this paper it is known that the annihilation of the austenite-ferrite phase boundary can happen in two cases:

1. The heating rate was such a case that the austenitization was solely ensured by the diffusion of carbon into ferrite, i.e. the diffusionally propagating shells of high carbon austenite have intersected and consequently ferrite totally disappeared from the system.
2. The heating rate was such a case that during rapid heating the part of ferritic phase was transformed into low carbon austenite by diffusionless transformation and then the diffusionally propagating shell of high carbon austenite has intersected this low carbon austenite.

In both cases when ferrite disappears in the moment the character of the diffusion of carbon changes and diffusional uniforming of the carbon concentration starts in austenite. From the fact, that the diffusion coefficient of carbon is approximately 1000 times higher in ferrite compared to austenite at the same temperatures, it follows that heating of material is more advantageous at the rate when austenitization happens solely by the diffusion of carbon in ferrite. The heating rate higher than the critical one substantially delays the attainment of a uniform carbon concentration and rapid cooling produces the heterogeneous martensite-ferritic structures.

From above mentioned analysis the critical heating rates were detected for carbon steels. The average distance among nearest outside cementite plates of two neighbourhood pearlitic colonies was taken as the mean statistic distance of pearlitic colonies detected from the structure of carbon steels after normalizing.

The detection of critical heating rates for graphitic cast irons comes from the fact that the futher dissolution of graphitic carbon into austenite undesirable after complete austenitization of the whole volume of the matrix because the higher contents of carbon in austenite the higher suspectibility to cracks during quenching. It is also necessary to mention that from the metallographical point of view cast iron is very heterogeneous material, for its microstructure closely related to the conditions of its existence. Consequently, the detection of critical heating rates for graphitic cast irons was limited to ferritic and pearlitic cast irons of diferent sizes of graphitic systems. Resultant kinetic diagram is shown in Fig. 6. The size of graphitic systems, let us say the size of nodules and lamellas was presumed 20 μm. The kinetic diagram of

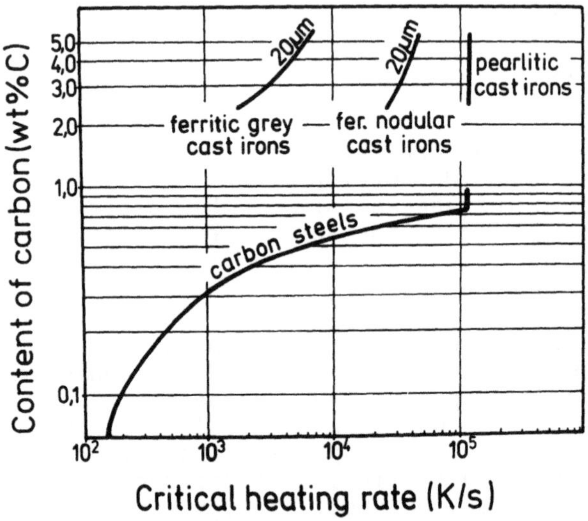

Figure 6. Kinetic diagram of carbon steels and cast irons.

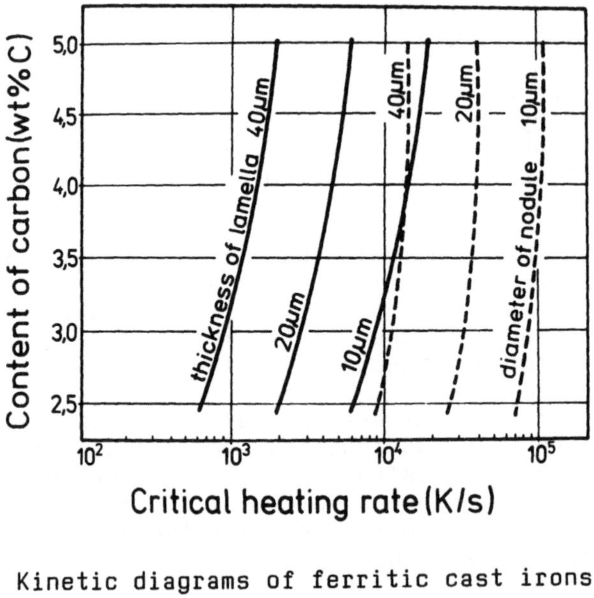

Figure 7. Kinetic diagrams of ferritic cast irons.

ferritic cast irons of different sizes of nodules and lamellas are in Fig. 7.
For detection of the critical heating rates we took into account the effects of different heating rates on the temperature at which austenitization commences.
The results achieved in the above mentioned theoretical analysis have been fully corroborated experimentally.

CONCLUSIONS

This paper presents the physico-mettalurgical and mathematical formulation of the diffusional mass transfer in carbon steels and graphitic cast irons treated by rapid heating.
In correspondence with the results of theoretical analysis of the diffusion of carbon in carbon steels and graphitic cast irons during rapid heating the kinetic diagrams were compiled. These may serve in practice for determining optimum heating rates for trasformational surface hardening of these materials by concentrated energy sources, such as laser. Conclusions corroborate that heating rates higher then those set out in this diagram substantially delay diffusional saturation of the metal matrix by carbon concentration and the attainment of a uniform carbon concentration by diffusion and rapid cooling produces the heterogeneous martensite-ferritic structures. On the contrary lower heating rates reduce the effeciency of exploration of concentrated energy sources.

REFERENCES

1. Horyna, J. and Zamecnik, J., Diffusion of carbon in carbon steels during rapid heating and cooling. Research work, Department of Material & Metallurgy, Technical University, Liberec, October 1988.

2. Horyna, J., Zamecnik, J., Korinek, M. and Kucera, J., Diffusion of carbon in graphitic cast irons during rapid heating and cooling. Research work, Department of Material & Metallurgy, Technical University, Liberec, January 1990.

3. Horyna, J., Study of kinetics of austenitization of chosen steels during rapid resistance heating. Research work, Department of Material & Metallurgy, Technical University, Liberec, February 1978.

APPENDIX

For more informations related to the above mentioned problem, please contact Mr. J. Horyna and Mr. J. Zamecnik at Technical University in Liberec.

Influence of Electrochemical Honing Methods on Load Capacity of Surfaces

A. F. Budzyński

Akademia Techniczno-Rolnicza
Bydgoszcz Poland

and J. Kozak

Politechnika Warszawska
Warsaw Poland

ABSTRACT

Combined electrochemical machining (ECM) with abrasive honing (HS) in a process of electrochemical honing (ECH) provides new opportunities in this kind of surface treatment. This paper deals with the experimental results pertaining to grindability and wear of surfaces treated by electrochemical honing in comparison with the surface treated by conventional honing methods. The microgeometry of electrochemically treated surfaces revealed that surface details differ considerably in comparison with the surface machined by grinding itself.

INTRODUCTION

In the case of machining holes, shaped in the form of rotational cylinders (d, l), the diameter increment δd of hole should has a constant value. This condition for the most popular two-directional honing (TDH) is difficult to fulfill. On the basis of pattern in Figure 1a it is evident that in case if:

$$l_p > \frac{l_h}{\mathrm{tg}\ \frac{1}{2}\alpha} \tag{1}$$

where: l_h - length of apparent (segmental) shaped tool,
l_p - length of tool trace displacement,
α - intersection angle of machining marks.

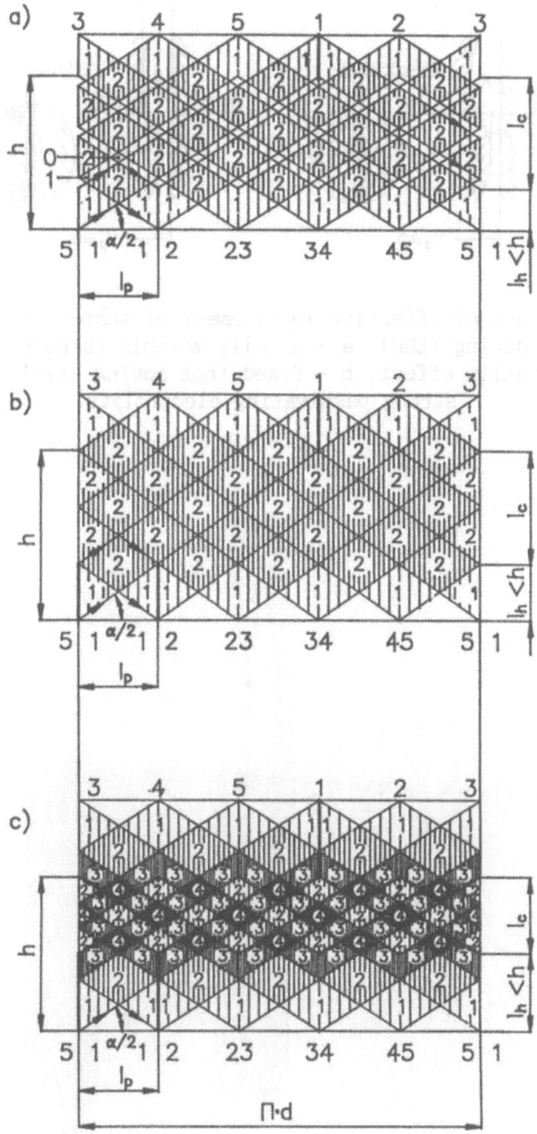

Figure 1. Differentiation of effect's numbers n_t of segmental tool l_h ($l_h < h$) for the case: h = const, α = const, l_p = const.

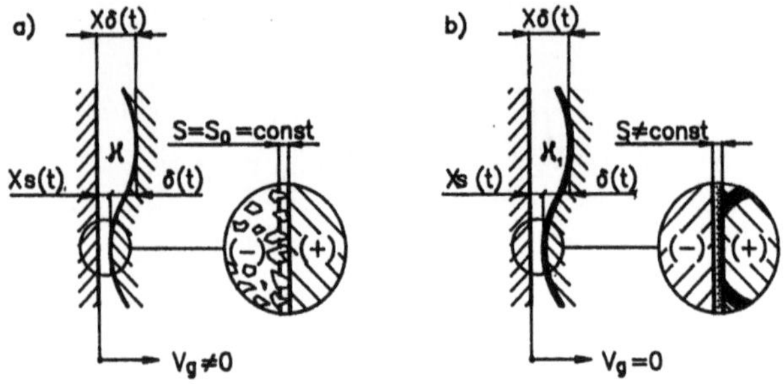

Figure 2. The means of effective improvement of shape deviations δ in electrochemical honing (ECH): a - axially movable cathode and electrolyte with weak passivation effect, b - fixed (not moving axially) cathode and strong passivating electrolyte.

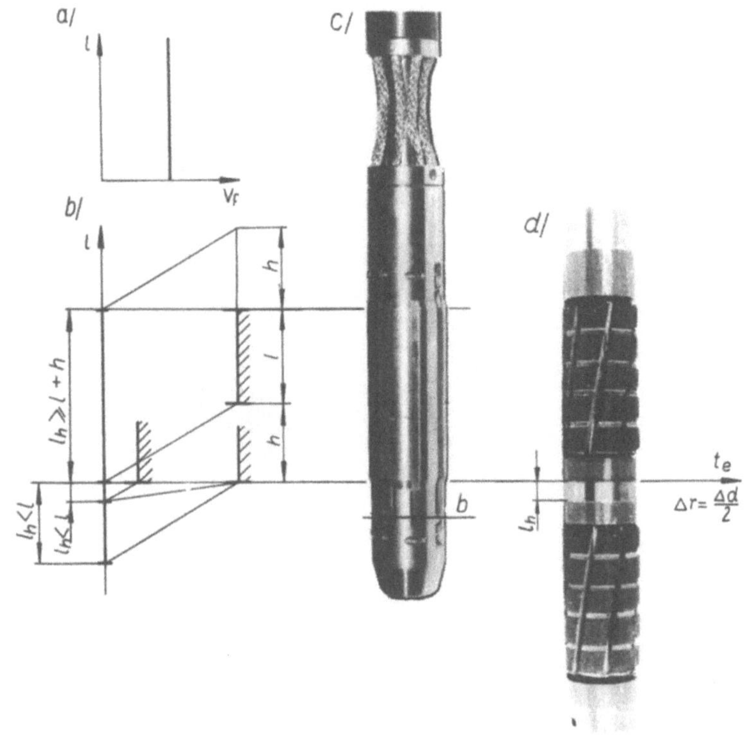

Figure 3. Constant time t_e of cathode effect and possibilities of its practical realization: a - function $v_f = f(l)$, b - time t_e of cathode effect or diameter increment δd of machined hole, (c,d) - tools of author's constructions, enabling realization of constant time t_e of cathode effects.

then there are many places on machined surface where tool does not effect at all and also some places where it effects only once or twice. In such case, during one machining cycle defined as:

$$0 \leq (\pi d - \Sigma l_p) < l_p \tag{2}$$

the number of effects n_t of the tool on the length $l_c = h - l_h$ will be: $n_t = 0, 1, 2$. Analysing in the same way the case shown in Figure 1b it can be noticed that for

$$l_p = \frac{l_h}{tg\ \frac{1}{2}\alpha} \tag{3}$$

on all surface under treatment on the length $h - l_h$, $n_t = 2$. In Figure 1c it is proved that if:

$$\frac{l_h}{2tg\ \frac{1}{2}\alpha} < l_p < \frac{l_h}{tg\ \frac{1}{2}\alpha} \tag{4}$$

than $n_t = 2, 3, 4$. From the next examples it is evident that for

$$l_p = \frac{l_h}{2tg\ \frac{1}{2}\alpha} \tag{5}$$

$n_t = 4$, and for

$$\frac{l_h}{3tg\ \frac{1}{2}\alpha} < l_p < \frac{l_h}{2tg\ \frac{1}{2}\alpha} \tag{6}$$

$n_t = 4, 5, 6$. A sequence of these combinations leads to settle the following regularity:

$$l_p = \frac{l_h}{mtg\ \frac{1}{2}\alpha} \quad \ldots \quad n_t = 2m \tag{7}$$

$$\frac{l_h}{(m+1)tg\ \frac{1}{2}\alpha} < l_p < \frac{l_h}{mtg\ \frac{1}{2}\alpha} \quad \ldots \quad n = 2m,\ 2m+1,\ 2(m+1), \tag{8}$$

where: m - natural number.

This deduction convince that two-directional abrasive honing (TDH) in the natural way can lead to deformation of the machined hole macrogeometry.

After careful analysis of the other element's shaping methods known it can be seen that electrochemical machining (ECM) in natural manner (Figure 2) can be use to improve the macrogeometry of machined hole. This brought strightly to introducing of electrochemical honing (ECH). From Figure 2 it is evident that for effective improvement of shape error δ (deviation), both in the function of time t and in the function of δr, the machining should be performed with slit S_0 as narrow as possible and with maintaining this value constant during all the period of machining. These conditions lead to the successity of using an axial moving cathode which follows the displacement (X_S) of machining surface what, in turn, causes increasing of the hole diameter by δd or using electrolytes, which cause strong passivation of the surface (Figure 2).

The tools enabling the constant number n_t of abrasive sticks effects and constant time t_e of cathode effect are shown in Figure 3. The investigations of these tools have proved their great practical usability. By the means of these tools it is possible to obtain very little shape deviations of machined surfaces (δr, δw, δp), what is not possible to obtain by traditional abrasive honing (HS). Not explained remains the problem of microgeometry of machined surface; in particular an overetching of surfaces, typical for ECM, which appears as the change of load capacity of the surface treated by means of ECH.

INVESTIGATED FACTORS, RANGE OF THEIR VARIATIONS AND METHODS OF MEASUREMENT

The investigations of the effect of surface machining method on the surface load capacity were carried out according to the method and on test stand described in [6].

The following factors were chosen as variable factors:

- electrochemical honing (ECH), abrasive (traditional) honing (HS), mixed honing (ECH + HS) - understandable as ECH with short (30 seconds) abrasive honing, finishing the machining of hole;
- material's feature of cylinders (steel, cast iron);
- constructional feature of piston seals (cast iron, rubber).

Output parameters, which enable estimation of grindability and wear of the honed surface, include:

- diameter increment δd of tested cylinders, which evidence the grindability of tested surfaces;

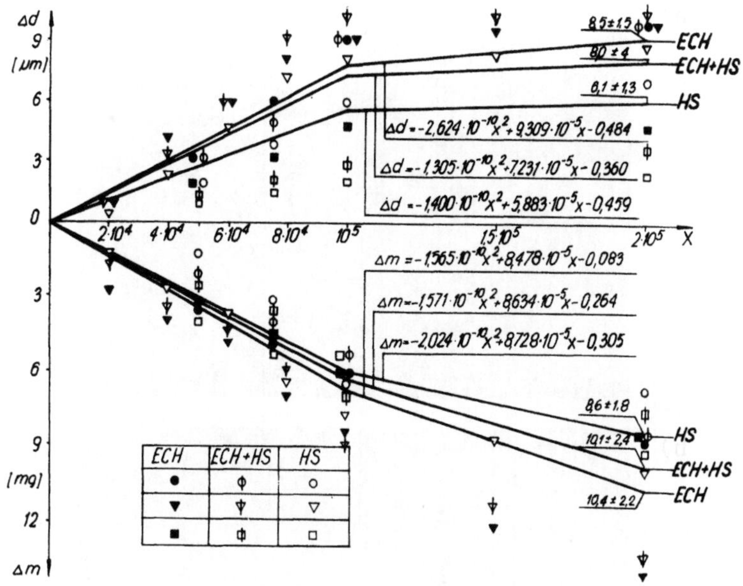

Figure 4. Diameter increment δd of steel cylinder and mass decrement δm of cast iron seal in function of number x of piston strokes depending on methods of machining.

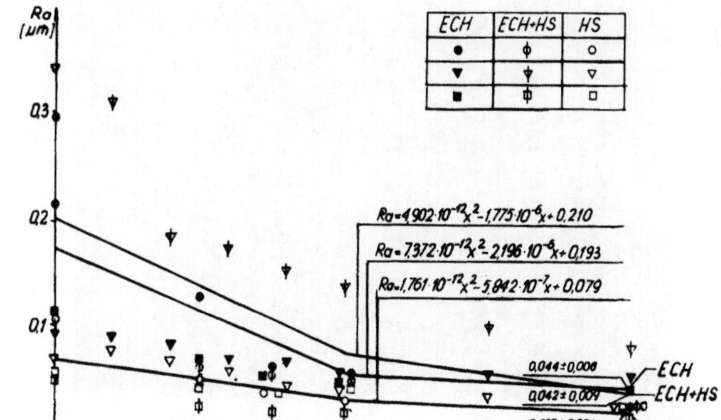

Figure 5. Change of cylinder surface roughness in function of number x of piston strokes for the kinematic pair: steel cylinder - cast iron seal.

a)
b)
c)

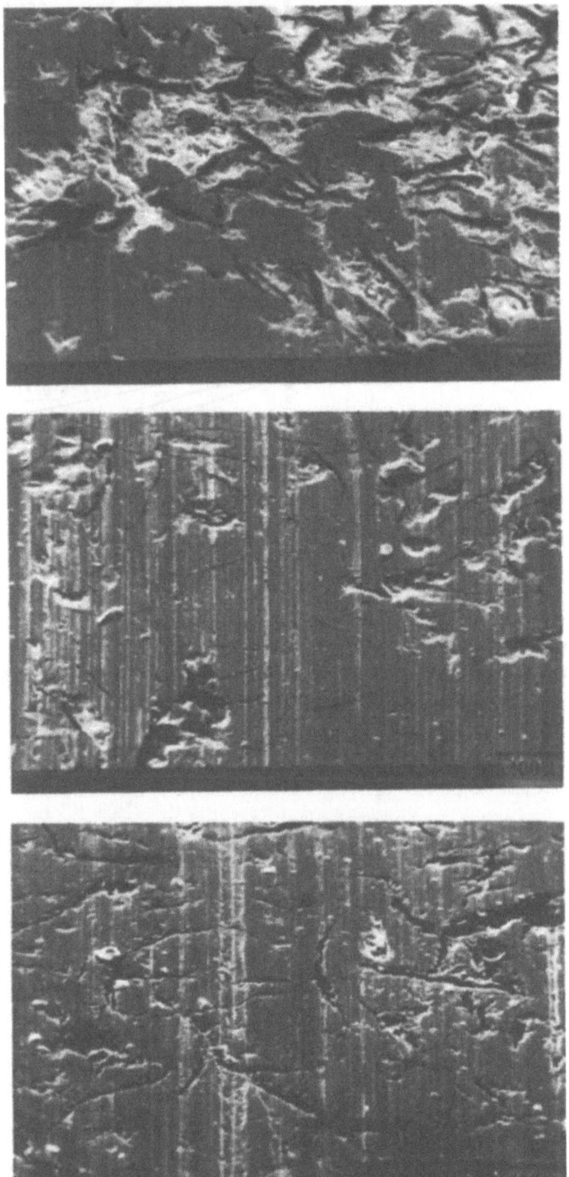

Figure 6. The surface of cast iron cylinder machined by one-direction transverse honing after working with cast iron seal ($2 \cdot 10^5$ piston strokes): a - ECH, b - ECH + HS, c - HS; mag. 160x.

- mass decrement δm of seal rings, which characterize the wear of machined surfaces;
- roughness of tested surfaces (Ra, Rmax, NL20, NL50) and leaks of medium through piston seal in the cylinder, reflecting the grinding-in of the tested kinematic pair and boundary values for its correct performance.

TEST RESULTS AND DISCUSSION

In this paper, considering its limited volume, only some test results are referred. More detailed discussion of this problem can be found in other works [1÷11].

In Figure 4 is shown the diameter increment δd of steel cylinder mating with cast iron seal in a function of piston strokes (X) for the different honing methods (ECH, HS, ECH + HS). On this diagram are also shown the mass decrements δm of the seal in relation to δd.

In Figure 5 are shown the changes of Ra parameter as a function of number X of piston strokes for the kinematic pair consisted of steel cylinder and cast iron seal and in Figure 6 there is shown the surface of cast iron cylinder machined by one-directio transverse honing (TH) with ECH, HS and jointly ECH + HS, investigated after $2 \cdot 10^5$ piston strokes. It is worth to notice the remaining overetching (Figure 6a) after electrochemical honing, which decreases the load capacity of tested surface and lack of such overetching on the test samples machined by abrasive honing (HS).

After careful examination of test results by significance testing of particular factor's influence (using Snedecor's F-test) it is possible to ascertain that:

- in all investigated cases the number X of piston strokes has essential effect on the cylinder diameter increment δd and on the mass decrement δm of seals, what is obvious and confirms only correctness of experiments and calculations;
- the means of machining (ECH, ECH + HS, HS) have no substantial influence on the diameter increment δd of steel cylinders but have great influence on δd of cast iron cylinders. In all evaluated cases the diameter increment δd of cylinder machining by abrasive honing (HS) was the least. In

three out of four cases the δd values of cylinder shaped by joint using of ECH + HS were between δd values for HS and ECH;

- the means of treatment (ECH, ECH + HS, HS) have essential influence on the mass decrement δm of seals only in the case of rubber ones;
- the means of treatment (ECH, ECH + HS, HS) have essential influence on the change of parameter Ra i.e. roughness of the cast iron cylinder surface. The surfaces of steel cylinders are characterized by low values of Ra directly after machning but undergo small changes in relation to number X of piston strokes.

CONCLUSIONS

The feature of electrochemical honing process, appearing in uneven digestion (etching) of the phase constituents of machining material, causing increased grindability and wear of steel and cast iron is now used for finishing treatment of cylinder surfaces made wholly of aluminium with high silicone content (AlSi17Cu7). During the electrochemical honing of such cylinders, the process is continued until very hard cristals of silicone (height 1 ÷ 3 μm and hardness about 1200 HV) will appear on the machined surface. The surface of cylinders machined by ECH can contain sufficient film of lubricant between protruding silicone cristals thus contributing, according to our investigatios,to considerable improvement of durability of this kinematic pair in relation to machning methods used till now.

REFERENCES

1. Budzyński, A.F., Theoretical principles and technical factors of electrochemical honing. In Proceedings of ISEM6, Kraków, 1980.

2. Budzyński, A.F., Research on optimization of the process of electrochemical honing (ECH). In Proceedings of ISEM7, Birmingham, 1983.

3. Budzyński, A.F., Electrochemical honing - studies on the shape precision of holes being processed. In Proceedings of ISEM8, Moscow, 1986.

4. Budzyński, A.F., Theoretical and experimental research on the process of electrochemical honing (ECH). In Proceedings of ISEM9, Nagoya, 1989.

5. Budzyński, A.F., Theoretical and experimental research on the process of electrochemical honing (ECH). Assistant professor's dissertation, Technical University, Warsaw, 1982.

6. Budzyński, A.F., The influence of directivity of machining marks on grindability and wear of the honed surface. Surface Engineering, Toronto, 1990.

7. Karwowski, K., Method and the control system diminishing the rolling errors of holes undergoing honing. Doctor's dissertation, Technical University, Warsaw, 1982.

8. Orobiej, J., Examination of finishing processing parameters of the holes 0.1 ÷ 0.5mm in diameter. Doctor's dissertation, Technical University, Warsaw, 1984.

9. Seroka, St., Honing of holes making possible the obtaining of rolling area parallel to the axis of the geometrical structure. Doctor's dissertation, Technical University, Warsaw, 1982.

10. Wocianiec, R., Influence of certain parameters, especially of the tool shape, on the precision of shape and size of the holes undergoing honing. Doctor's dissertation, IOS, Kraków, 1982.

11. Zakościelny, St., Influence of the coarseness directivity and methods of production on wear and grindability of the structure undergoing honing, based on hydraulic servo-motors. Doctor's dissertation, Technical University, Gliwice, 1981.

Evaluation of Acid Rain Resistant Clear Coatings for Stone Structures

B. S. Sitaramam and V. R. Puttagunta

Lakehead University
Thunder Bay
Ontario
Canada

ABSTRACT

All over the world, monuments and other structures made of marble, limestone and similar materials are being deteriorated by acid rain and other atmospheric pollutants. Although damage caused by corrosion to metallic and non-metallic components of industrial products has received considerable attention from the scientific community, and protective coatings have been developed, no significant effort has been made to protect stone structures which require clear protective coatings.

This paper presents the results of a physical model study to evaluate the performance of some clear coatings for protection of stone structures from damage caused by acid rain. An experimental set up was designed to simulate the acid rain phenomenon in the laboratory. The set up is an accelerated weatherometer so that results can be obtained in a reasonable time frame. After exposing the test samples to the weatherometer, for a few weeks, the coatings were tested for their adhesion, resistance to weathering, hardness and gloss. Comparative results for various coatings are presented. Polysiloxane, copolymer resin and single pack polyurethane are not suitable whereas two pack acrylic polyurethanes promised good results.

INTRODUCTION

Probably the most discussed air pollution problem during the past decade is acid rain and this will continue in the next century. Heavy Industrialization and urbanization, where emissions of sulphur and nitrogen oxides are high and it is with these emissions that the most acid precipitation is thought to originate. The deposition of acidic materials on building structures occurs through acidic snow, rain, fog, dry particles and gases. Reports of acid rain are common throughout Europe, Japan, USA, Canada etc., where pH values of 3.8 -4.5 becoming almost the norm. In the Eastern United States acid rain has a pH less than 5.0 and consists of 65% sulphuric acid, 30% Nitric acid and 5% other acids. In recent times the pH of rain in Quebec province was as low as 3.7. The effects of acid rain on stone structures have been amply documented. Exfoliation of marble and limestone monuments and buildings and pitting of granite stone work are common all over the industrial world (1).

Gauri (2) studied the effect of SO_2 on marbles impregnated with epoxy resins and certain acrylics. Hydron (Polyhydroxy methyl methacrylate) over Epi- Rez 510 impregnated specimens have offered best results in their studies. Polyurethane coating materials are suitable as coatings on marble (3). Styrene ethylacrylate copolymers, acrylic polymers, polysiloxanes are also good protective coatings on stone like materials (4,5). Polyurethane coatings are known for their resistance to acid, acid rain and pollutants such as bird excrements (6). A number of clear coatings of polyurethane are available in the market which can resist the acid rain and other atmospheric pollutants (7). Acrylic polymers with hydroxy functional groups cross-linked with polyisocyanates had an overall performance superior to that of polyurethane and two pack epoxy coatings (8). Accelerated tests conducted with panels coated with vinyl, epoxy polyamide or urethane paints in circulating tap water and cold synthetic sea water showed that a two pack urethane primer and a top coat paint system met all the requirements of resistance to weathering (9). Polyurethane coatings were found to be very suitable for subsonic rain erosion protection (10). Polymers were best and epoxy ester resins were worst in accelerated tests and no differences were observed in natural tests after several months (11). Accelerated tests using irradiation and water sprays are apparently sufficient for determining resistance to natural weathering of coatings (12).

As there is no reference to accelerated weatherometer tests on these coatings under acid rain conditions, we decided to do these tests. Since no inexpensive accelerated weatherometer is available in the market for these tests, a simple accelerated weatherometer was assembled in the laboratory, as described in method No. 26 of Specification DEF/1053, which was briefly described by Morgans (13), with a few modifications to meet our conditions.

ACCELERATED WEATHEROMETER

The unit consists of a vertical steel drum of 24 in. diameter, with an open top. The coated marble panels are placed in a gallery round the periphery of the drum and are exposed continuously to ultraviolet light from a 450 W high pressure mercury vapour lamp of Hanovia type which delivers 15% of its intensity in UV-C region (2200 A-2800 A), 15% in UV-B region (2800 A-3200 A), 15% in UV-A region (3200 A-4000 A), 45% in visible region (4000 A-7000 A) and 10% in infrared region (10000 A-14000 A) with a total radiating energy of 175.8 W. The lamp was protected from acid water spray with a Pyrex glass envelope. Pyrex absorbs the UV radiation below 2800 A; this radiation is not found in the earth's atmosphere since it is absorbed by the atmosphere (14). Air was passed through the envelope to remove the heat from Infrared radiation. A 3 rph motor was mounted on the lid of the drum whose rotating shaft was replaced with a stainless steel tube. The top end of the tube was connected to an overhead tank which carried acid water of 3.5 ± 0.2 pH. To the other end a shower head was fixed to spray acid water on the marble specimens. The acid water (pH 3.5 ± 0.2) was prepared from distilled water of about 5.3 pH and an aqueous solution containing H_2SO_4 and HNO_3 in the ratio 65:30 by volume. The UV lamp and acid water were used for 18 h a day and for the rest of the 6 h air was blown through the unit, as suggested by Morgans (13). The steel drum and the lid were painted with three coats of metal primer and three coats of white anti-rust polyurethane paint. The used acid water was collected in a container, neutralised with sodium hydroxide before draining out. The temperature during the light and spray periods was maintained at $35 \pm 5°C$.

MARBLE, LIMESTONE SAMPLES, COATINGS AND EXPOSURE

The marble and limestone samples were obtained from the Geology Department of Lakehead University. The samples cut to the size of 4x2x0.25 in., were polished, cleaned by ultrasonic cleaning unit and dried. Seven coating samples were obtained from the market. They are 1. Polysiloxane, 10-20% in mineral sprit, 2. Copolymer resin 10-20% in solvent, 3. and 7. are single pack polyurethanes and 4, 5, and 6. are two pack polyurethanes of acrylic polyol and aliphatic polyisocyanates.

The marble and limestone samples were coated using a brush with the above coatings to a thickness of about 1 mil. The impregnation technique described by Gauri (2) was also tried by diluting the polyurethane coatings with reducer. As the coatings have very good adhesion to the substrates and were not easily peelable, a 1 mil thick coatings were applied on teflon sheets and were subjected to the same environment to study the infrared spectra of the films.

The samples were exposed to the ultraviolet light and acid water for 500, 1000, and 1500 h and their gloss, hardness, adhesion, IR and SEM were studied. The gloss was studied as per ASTM D4449-85 ('Standard method for visual evaluation of gloss differences between surfaces of similar appearance'). The hardness was studied as per ASTM D3363-74 ('Standard test method for film hardness by pencil test'). The adhesion was studied as per ASTM D3359-87 ('Standard methods for measuring adhesion by tape test'). Infrared spectra of the free films obtained from the teflon sheets was studied by Beckman 4250 spectrophotometer. SEM micrographs were obtained by Hitachi-S 570 model.

RESULTS AND DISCUSSION

The polysiloxane coating No.1 and copolymer resin coating No. 2., both have failed in 250 h. Brush coatings completely deteriorated in 250 h. Whereas the samples coated by Gauri's technique (2) deteriorated in 500 h. The copolymer resin coating had yellowed, and was brittle after 500 h. The infrared spectra of polysiloxane coating and copolymer resin coating could not be obtained because after 500 h the films were so brittle that they could not be lifted off the teflon sheets. The single pack polyurethane coatings 3 and 7 yellowed to some extent and lost gloss to a good extent after 500 h. After 1000 h the coatings have developed cracks, blisters and the films started delaminating from the substrate. The infrared spectra of the single pack polyurethanes showed the evidence of oxidation, loss of urethane and loss of phthalate functionality. These results were very much comparable to those obtained by Bauer (15). In the case of two pack acrylic polyurethanes after 500 h no visual defects were observed. After 1000 h in the case of coating 4 a slight yellowness was observed, but the rest of the coatings, 5 and 6, are flawless. It was observed that only the unreacted isocyanate peak was missing after the exposure to the UV light and acid rain, but otherwise there was no difference in the spectra even after 1500 h. SEM micrographs of coatings are presented in fig 1 and they show some blistering after 500 h and severe cracks after 1000 h in case of single pack polyurethene. The two pack polyurethene film developed small cracks after 500 h and much bigger cracks after 1000 h.

The pencil hardness, adhesion and gloss are presented in Table 1.

The results obtained by our weatherometer are very much comparable to the results obtained by the other commercial units. It is known that the damage caused by the accelerated weatherometer is many times greater than the damage caused by natural weathering (16). No attempt has been made to predict natural weathering from these studies as one should not try to predict the number of hours of laboratory exposure that might equal a year of natural weathering (17).

The UV-source used in this weatherometer was a high pressure mercury vapour lamp of 450 W. When relative spectral energy distribution of this lamp is compared with that of standard lamps in use, such as Sunshine Carbon arc lamp, Xenon arc lamp of 6500 W, FS-40 fluorescent sun lamp, and Sun light, it was found that, mercury vapour lamp emits a greater radiation in the UV-B region, than Sunshine Carbon arc lamp, Xenon arc lamp and Sun light but less than FS-40 lamp. It is this region

TABLE 1

Gloss, Hardness and Adhesion values of the Coatings

No.	Coating material	Gloss after 0	500	1000	1500	Hardness after 0	500	1000	1500	Adhesion after 0	500	1000	1500
Hours			Hours				Hours						
1	Polysiloxane	1	--	--	--	4H	--	--	--	5B	--	--	--
2	Copolymer Resin	1	--	--	--	4H	--	--	--	5B	--	--	--
3	Single pack Polyurethane	1	0	0	--	3H	5H	2H	--	5B	1B	0B	--
4	Two pack acrylic Polyu-urethane	1	0	0	0	5H	6H	6H	6H	5B	4B	4B	4B
5	Two pack acrylic Poly-uretane	2	1	1	0	5H	6H	6H	6H	5B	4B	4B	3B
6	Two pack acrylic Poly-uretane	2	1	1	0	5H	6H	6H	6H	5B	4B	4B	4B
7	Single pack polyurethane with UV-absorber	3	1	0	--	4H	6H	2H	--	5B	3B	0B	--

Gloss: 0= None (indistinguishable), 1= Very slight (barely distinguishable), 2= Moderate (easily seen), 3= Considerable.

Hardness: Scratch hardness, The hardest pencil that will not rupture or scrath the film. The pencils are made in GT. Britain REXEL CUMBERLAND, Derwent Graphic

Adhesion: 5B= None of the squares of the lattice is ditached, 4B= Less than 5% of the area is affected, 3B= 5-15% of the area is affected, 2B= 15-35% of the area is affected, 1B= 35-65% of the area is affected, 0B= Greater than 65% area is affected.

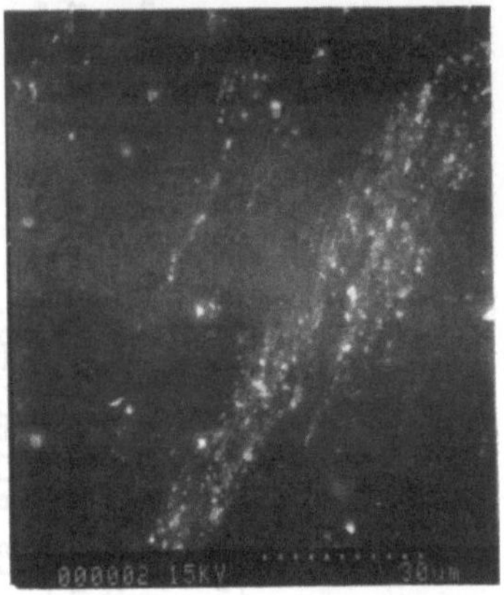

(a)

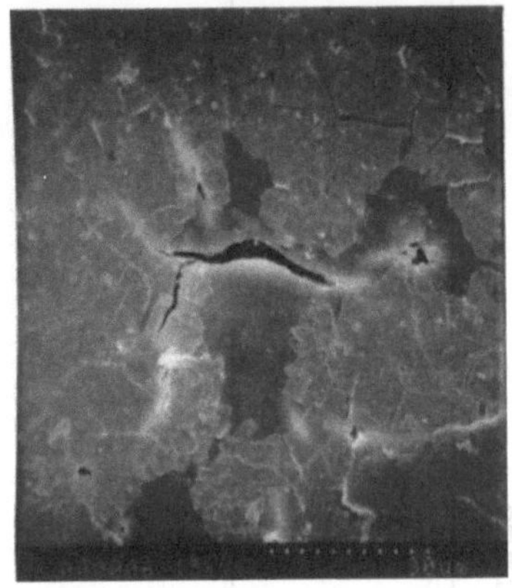

(b)

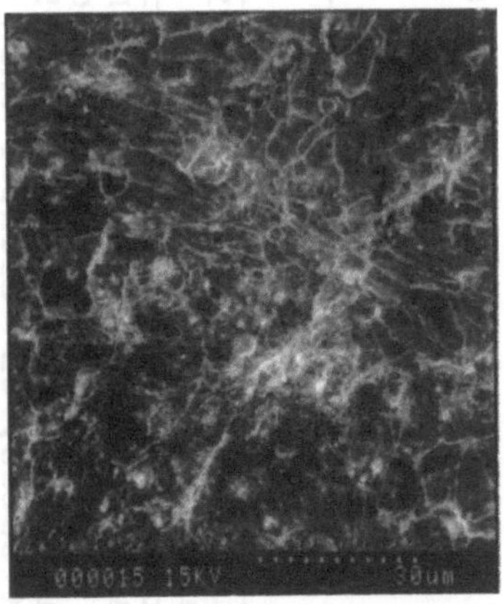

(c)

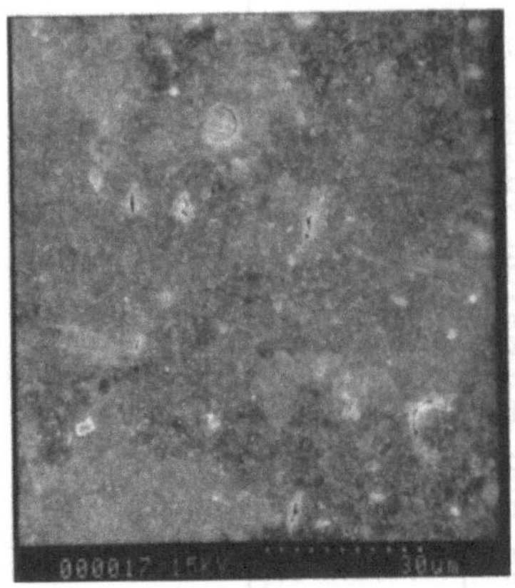

(d)

Fig 1. SEM micrographs of clear polyurethane coatings exposed to UV and acid water (a) Single pack polyurethane after 500 h exposure, (b) single pack polyurethane after 1000 h exposure (c) two pack polyurethane after 500 h exposure and (d) two pack polyurethane after 1000 h exposure.
(magnification in all cases 1000x)

that causes severe polymer damage. Mercury vapour lamp emits radiation more or less the same as that of FS-40 lamp but less than that of Sunshine Carbon lamp, Xenon arc lamp and Sun light in the UV-A region which also causes polymer damage. So it can be said that the polymer degradation caused by mercury vapour lamp is nearly the same as that caused by commercial lamps or Sunlight.

In most of the commercial accelerated weatherometers the water used is mostly distilled water or deionised water whereas in this unit an acid water of 3.5± 0.2 pH was used. When both UV light and acid water are used in combination the polymer degradation expected is more than in any other commercial weatherometers. In actual weathering conditions, the first few spells of rain are more acidic than the later spells. The later spells of rain, wash the stronger acid from the coated surface and thus the coatings are not exposed to strong acidic rain for a long time. In this weatherometer with a 3 rph spray head, the coatings are exposed to 'acid rain' of 3.5± 0.2 pH every twenty minutes. It was observed that droplets of acid water deposited on the coating evaporate to some extent and leave a relatively concentrated acid on the surface for some time before the next spray washes the acid down and leaves the fresh acid on the surface. Because of this reason also more damage is expected to the polymer coating. In the case of single pack polyurethanes the polymer damage is found to be much faster. In the case of two pack acrylic polyurethanes not much damage occurred to the polymer but some small cracks were observed as early as 500 h of exposure and more severe cracks after 1000 h of exposure.

On the contrary most of these two pack acrylic polyurethanes have done very well in the regular accelerated weatherometer studies done by others. The coatings did quite well up to 1000 h of exposure in QUV unit and Weather-O-Meter studies. In the natural weathering conditions they performed well up to 3 years. As these two pack acrylic polyurethanes have good adhesion to marble and limestone, and pocess good scratch hardness after exposing to acid rain and UV light, these may prove to be useful on marble and limestone structures to combat the atmospheric pollutants and acid rain.

CONCLUSIONS

The data show that the polysiloxane and copolymer resin coatings do not have a good resistance to acid rain and weathering. The single pack polyurethanes because of their initial light yellowness and the yellowing nature after exposure and rapid deterioration in UV and acid rain environment are not suitable for coating on marble and limestone structures. The two pack acrylic polyurethanes are best suited for this purpose, coatings 4, 5 and 6 have promised good results. Good transparency and non-yellowing nature of these coatings are encouraging. The reduction in gloss and cracking are the only flaws in these coatings. By adopting the technique described by Gauri (2) we can apply a thin coat of the acrylic polyurethane on marble and limestone structures and protect them from the onslaught of atmospheric pollutants and acid rain.

REFERENCES

1. McGraw Hill Concise Encyclopedia of Science and Technology, McGraw Hill Inc., New York, 1984, pp. 10, 11.

2. Gauri, K.L., Dodener, G.C., Lipscomb, N.T. and Sarma, A.C., Reactivity of Treated And Untreated Marble Specimens in an SO_2 Atmosphere. Studies in Conservation, 1973, **18**, 25-35.

3. Continental de Marcmoles, S.A., Application of Protective Coatings to Marble. Span. Patent 367, 670, 18 Feb. 1970, 4pp.

4. Gaertner, K.H., Preservation agent for all stone like materials (Synthetic and natural) for example concrete, clay, sand stone, marble, plastered walls etc., Ger. Offen. Patent. 2, 546, 879, 18 Oct. 1975, 4pp.

5. Post, M.A., Godette, M., and Campbell, P.G., Anti Graffiti Coatings. Modern Paint and Coatings 1978, **68**(2), 28.

6. Potter, T.A. and Williams, J.L., Coatings Based on Polyurethane Chemistry: An Overview and Recent Developments. Journal of Coatings Technology, 1987, **59** (749), 63-72.

7. Durable paint systems protect against Environmental attack. Modern Paint and Coatings, 1988, **78** (10), 150.

8. North, A. G., Room Temperature Curing Acrylics. Journal of Paint Technology, 1971, **43** (557), 44-50.

9. Field, D.E., and Alexander, A. L., Accelerated Tests for Coatings Exposed to Cold High Humidities. Accel. Tests Mar. Topside. Submerged Coatings Symp. 1966 (Pub. 1969) 1-3.

10. Moraveck, J.F., Polyurethane Coatings for Subsonic Rain Erosion Protection. SAMPLE Tech. Conf. Proc. 2nd. 1970, 425-39.

11. Rothova, B., Kupilikiva, A. and Nemec, M., Characteristics of Coating Materials on aging. Koroze. Ochr Mater., 1970, **14** (2), 27-8.

12. Papernorth, W., Application-Technological prerequisits for A Reproduceable Weathering Test. Farbe und Lack, 1970, **76** (5), 467-71.

13. Morgans, W.M., Outlines of Paint Technology, Charles Griffin and Co., London, 1969, pp. 384.

14. Bernnan, P., and Fedor, C., Sulight, UV and Accelerated Weathering. Presented at the Society of Plastics Engineers Automotive RETEC. Nov. 1987, pp. 1-8.

15. Bauer, D.R., Gerlock, J.L. and Dikie, R.A., Rapid, Reliable Tests of Clear Coat Weatherability, A proposed protocol. Progress in Organic Coatings, 1987, 15, 209-221.

16. McEwen, D.J., Verma, M.H. and Turner, R. O., Accelerated Weathering of Automotive Paints Measured by Gloss and Infrared Spectroscopy. Journal of Coatings Techology, 1987, **59** (755), 123-29.

17. American Society for Testing and Materials. Specification - Designation G53-84, pp. 962-66.

Computer Modelling of the Abrasive Flow Machining Process

A. J. Fletcher, J. B. Hull, J. Mackie* and S. A. Trengove

Sheffield City Polytechnic
Sheffield
UK

**Extrude Hone Ltd*
Milton Keynes
UK

ABSTRACT

Abrasive flow machining (AFM) is a process for finish machining of edges and surfaces by the controlled extrusion of an abrasive laden viscoelastic polymer across these edges or surfaces.

Since the initial development of the process in the mid-1960's, to deburr hydraulic control blocks used in the radar-guided target acquisition systems for Phantom Jets, the process has gained widespread acceptance across the broad spectrum of modern manufacturing industry. Applications now range from high technology aerospace components to high volume automotive components [1].

The Authors are currently involved in an extensive research and development programme into the abrasive flow machining (AFM) process. The aim of the research programme is to achieve a better understanding of the the thermal and fluid flow properties of polymers used in AFM, in order to establish a basis for the control and optimisation of the process.

The results reported here include the rheological characteristics of the Extrude Hone MV polyborosiloxane material, a description of the AFM process and consideration of the relationship between these rheological properties and the process.

INTRODUCTION

Abrasive flow machining is a technique used for the final stages of the machining of engineering components. The process may be used to deburr or modify the intersections of machined surfaces, or polish machined areas, resulting in a significant improvement to the surface finish. Additional benefits of the process include surface texture modifications and the removal of surface layers which have been subject to mechanical or thermal damage.

The process utilises the viscoelastic polymer polyborosiloxane, which is mixed with fine abrasive particles and forced through or around the component using an abrasive flow machine, thus achieving the machining action.

The process has been developed over the last 20 years and is used extensively for a variety of engineering components from high volume operations such as fuel injection nozzles or sewing machine eyes as well as the machining of specialised components of very complex geometry such as turbine blades and hydraulic control blocks.

The aim of the research project is to achieve a better understanding of the thermal and fluid flow properties of polyborosiloxanes, in order to establish a basis for the control and optimisation of abrasive flow machining, which is the main industrial use for these materials.

The research programme will include the determination of the rheological properties of the polyborosiloxane both on its own and when mixed with grit particles as well as consideration of heat generation and flow of the materials while passing through orifices of different diameters. This will be supported by the measurement of the relevant physical parameters. Subsequently, it is intended to use the information obtained to calculate velocity and stress profiles during abrasive flow machining.

This paper describes the abrasive flow machining process and considers the characteristics of Extrude Hone MV grade polyborosiloxane in relation to the machining process.

DESCRIPTION OF THE PROCESS

Process development

The Extrude Hone abrasive flow machining process was initially developed in the mid 1960's in order to find a more effective method of deburring hydraulic control blocks , which at the time were being deburred by hand. The blocks were used at the centre of the hydraulic control system for the radar guided target acquisition systems of Phantom jets, and contained 83 intersecting holes in a block no bigger than 150 cm3.[1]

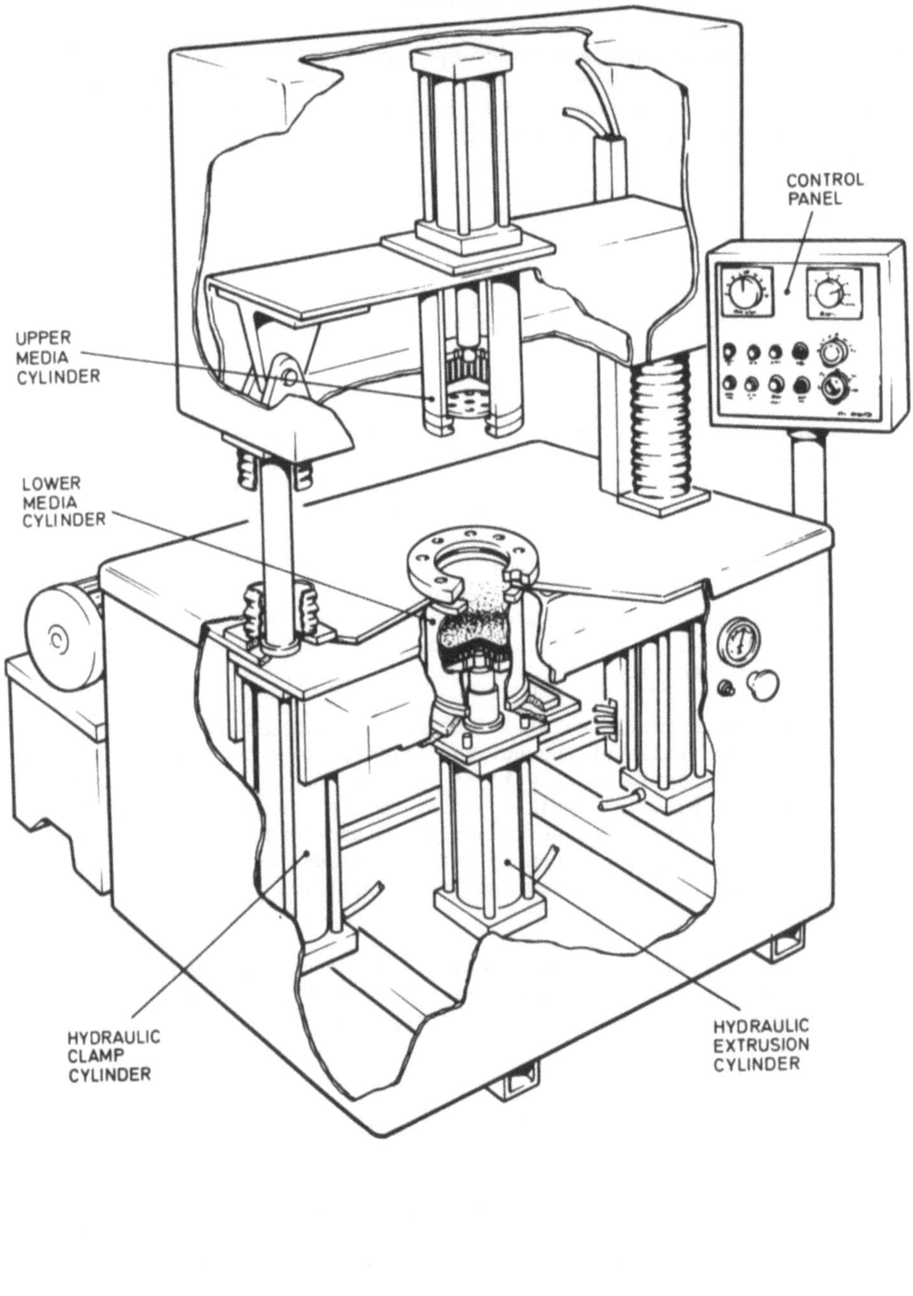

Figure 1. Schematic of a typical abrasive flow machine.

By 1968 the process had been successfully developed for application to the hydraulic control blocks by the Extrude Hone Corporation. The process was rapidly diversified into a number of additional applications such as aerospace components and die and mould polishing operations.

The Extrude Hone Corporation currently have over a thousand system applications throughout the world, and support an ambitious research and development effort aimed at meeting escalating requirements for precision machining in aerospace, automotive, die-making, and other major markets.

The abrasive flow machine[2]
All abrasive flow machines, such as the one shown in figure 1[3], are based on a common configuration. The equipment must clamp and seal the component in such a way that the polymer abrasive mixture is contained within the system. In addition it must pump a pre-selected quantity of abrasive media through the fixture at a specified pressure . To this end cylinders containing the abrasive mixture are mounted above and below the component.

The components, or fixture holding one or more components, are placed on top of the lower cylinder. A moveable rail is then lowered until it clamps the part or fixture between the two cylinders.

Abrasive media is forced hydraulically from the lower cylinder through, or across, the part and into the upper cylinder. Flow is then reversed to complete one cycle. When the appropriate number of cycles has been completed the moveable rail is raised to release the part or fixture.

Extrude Hone abrasive flow machines are available in a range of sizes depending on the specific application. These machines are available with adjustable controls for media flow pressure, media flow rate, and cycle times.

Machines may be operated manually or automatically from a control panel. Additional control systems may be added to measure parameters such as media temperature, viscosity, wear and flow speeds. Auxiliary systems for such operations as fixture handling, additional and refeed media supply to the process area, and media cooling systems are also available.

The fixtures or tooling
The function of the tooling is simply to hold the workpiece in position and to contain and direct the flow of media. In designing tooling, the areas where abrasion is desired are first identified. The tooling is constructed in such a way that the flow of media through or across these areas is restricted. For a workpiece with straight-through passages, such as a tube, an internal spline, a valve plate, or an extrusion die, the tooling merely holds the parts in place between the cylinders and allows the through-passages to restrict the flow. In many die polishing operations, no tooling is required.

To process external edges or surfaces, the part is contained within a flow passage ,so that flow is restricted between the outside of the part and the interior of the flow passage.

Counterbores, recessed areas, or even blind cavities can be processed using a restrictor or mandrel to fit inside the components and restrict the flow at the desired areas.

Any number of parallel restrictions can be processed at once. In many cases, two (occasionally more than two) successive restrictions in the same flow stream can be abraded, so long as the cross-sectional areas of the successive restricting passages are equal. However, if there are two successive restrictions in the same flow stream with significantly different cross-sectional areas, only the more restricted passages will be abraded.

Very complex parts, such as valve bodies, may require processing in two or more operations, processing some areas in one operation and others, with different flow paths in the next. Very large parts can be processed in sections [4].

The abrasive media
It is the abrasive laden polyborosiloxane material which transfers the machine extrusion force to the workpiece edges and surfaces to achieve the finishing action.

The polyborosiloxane material is available in a range of viscosities depending on the specific application.

The abrasive, which is dispersed evenly throughout the polyborosiloxane will also be selected depending on the application requirements. Typical abrasives used include silicon carbide, aluminium oxide and boron carbide. The machining characteristics of the polyborosiloxane/abrasive mixture can be modified further by alteration of the size range of abrasive used and the ratio in which it is mixed with the polymer.

EXPERIMENTAL PROCEEDURE

All results reported in this paper refer to Extrude Hone MV grade polyborosiloxane without abrasive addition.

The rheological evaluation of this material was completed on a Davenport capillary extrusion rheometer [5].

THE RHEOLOGICAL PROPERTIES OF THE MV POLYBOROSILOXANE AND THEIR RELATIONSHIP TO THE AFM PROCESS

The polyborosiloxane material must flow sufficiently readily at room temperature for it to follow the contours of complex surfaces. At the same time, it must be sufficiently rigid to allow the transmission of the pressure from its point of application to the grit particles. Unless both these conditions are met the honing operation will not occur.

Furthermore, the applications of the process tend to involve selective machining at points of restriction. It is often undesirable that universal honing should occur.

The amount of metal removal achieved during abrasive flow machining has been investigated by consideration of the wall shear stress, τ. The wall shear stress for a circular restriction may be calculated using equation 1, below [6].

$$\tau = \frac{pr}{2l} \qquad (1)$$

Where p is the pressure drop along the restriction,
and r and l refer to the radius and length of the restriction respectively.

Thus for a given restriction geometry the shear stress is a linear function of the pressure across the restriction. The significant relationship between extrusion pressure and metal removal has been shown previously [7].

As the shear rate of the polymer increases in the vicinity of restrictions the relationship between shear rate and shear stress is very important.

The shear rate in the vicinity of a restriction during abrasive flow machining is determined by the piston speed, the cross-sectional area of the piston and the cross-sectional area of the restriction.

This becomes clear when the shear rate equation is considered for a circular restriction [6].

$$\dot{\gamma} = \frac{4Q}{\pi r^3} \qquad (2)$$

Where $\dot{\gamma}$ is the shear rate,
Q is the volumetric flowrate of polymer/abrasive mixture, and hence is determined from the piston velocity and cross-sectional area, and,
r represents the radius of the circular restriction.

Thus the shear rate increases if either the radius of the restriction decreases, the piston velocity increases, or a larger diameter piston(AFM) is used.

The relationship between wall shear stress and shear rate for MV polyborosiloxane has been determined by capillary rheometery and can be seen below, in figure 2.

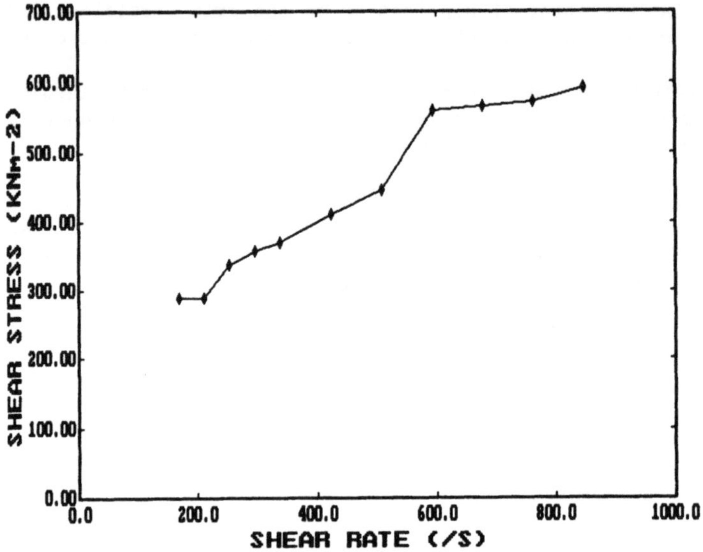

Figure 2. The variation in wall shear stress with shear rate for MV polyborosiloxane without abrasive.

The slope of this curve, which is a measure of the coefficient of viscosity of the material, reduces as the shear rate increases i.e. the material is pseudoplastic. The shear viscosity shear rate relationship for the MV polyborosiloxane can be seen in figure 3.

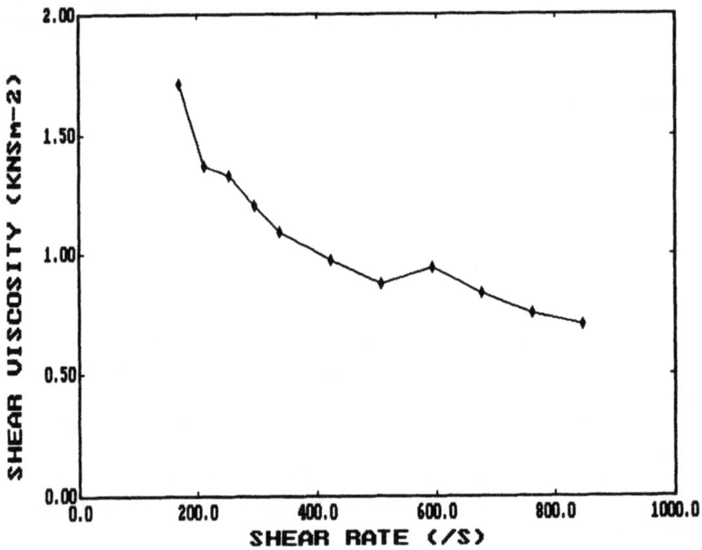

Figure 3. The variation in shear viscosity with shear rate for MV polyborosiloxane without abrasive.

Thus, the material flows more readily in restrictions as the coefficient of viscosity will be lower on account of the increased shear rate. At the same time, the wall shear stress will be higher. The success of the Extrude Hone abrasive flow machining operation may be associated with these properties.

Another important variable to consider with respect to the abrasive media property requirements is the variation in wall shear stress with time.

The variation in wall shear stress with time for various shear rates has been investigated on the Davenport capilliary rheometer.

The variation in wall shear stress with time at a high and low shear rate respectively can be seen in figures 4a and 4b.

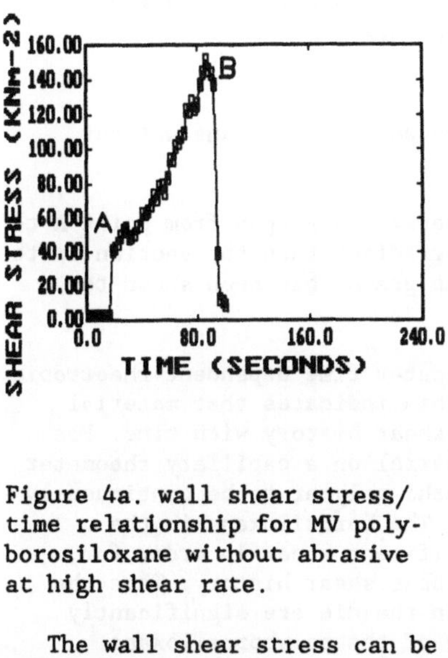

Figure 4a. wall shear stress/ time relationship for MV polyborosiloxane without abrasive at high shear rate.

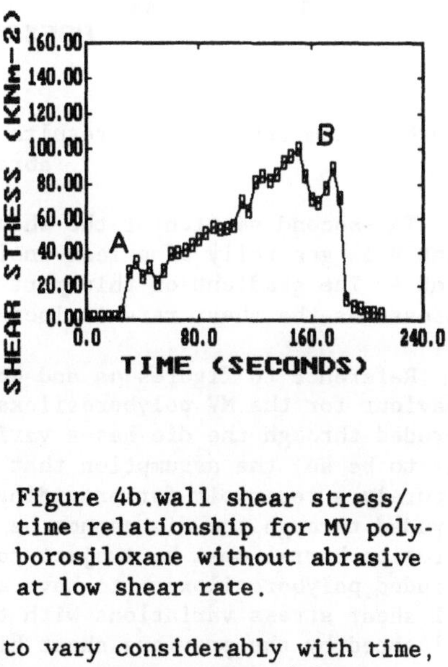

Figure 4b.wall shear stress/ time relationship for MV polyborosiloxane without abrasive at low shear rate.

The wall shear stress can be seen to vary considerably with time, despite the shear rate being maintained at a constant rate (piston speed, piston cross-section and die cross-section all constant).

Results from the capillary rheometer show a two stage increase in shear stress. The first, a very sharp increase may be associated with elastic compression of the polyborosiloxane in the barrel up to point A. The shear stress value of point A has been shown to increase as the shear rate increases. Some simple barrel compression tests for MV borosiloxane have been completed and confirm that the material has a degree of compressibility (see figure 5).

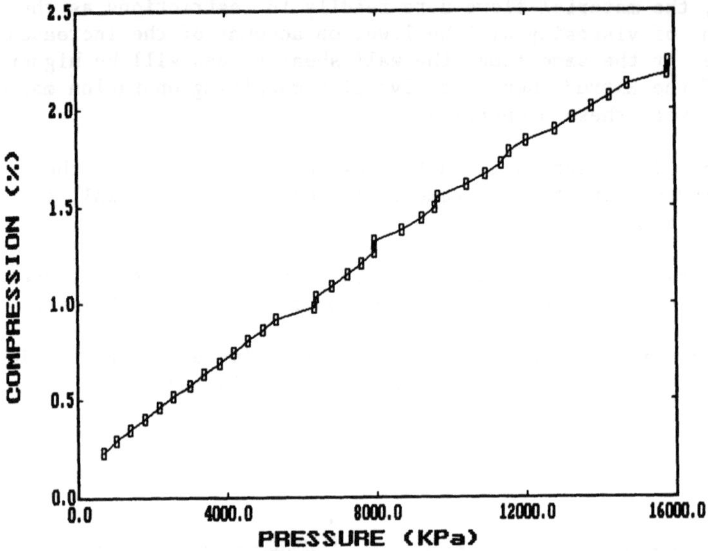

Figure 5. Compression test results for MV polyborosiloxane without abrasive.

The second section of the shear stress/time graphs from point A to point B is generally of a less inclined gradient than the section up to point A. The gradient of this part of the graph has been shown to increase as the shear rate is increased.

Reference to figures 4a and 4b indicates time dependent rheotropic behaviour for the MV polyborosiloxane. This indicates that material extruded through the die has a variable shear history with time. For this to be so, the assumption that the barrel on a capillary rheometer is purely a reservoir for providing 'fresh' polymer to be continuously extruded through the die cannot be true. The borosiloxane in the extrusion barrel must be subject to significant shear in order for extruded polyborosiloxane to have a variable shear history. Thus the wall shear stress variations with time in the die are significantly influenced by the previous shear history of the polyborosiloxane.

In order to investigate the time dependent properties of the polyborosiloxane material further, a rheological evaluation using a concentric cylinder rheometer is currently under investigation (results to be published). This type of rheometer enables close control of the polymer shear history.

This time-dependent wall shear stress at constant shear rate for MV polyborosiloxane may have considerable effect on the abrasive flow machining operation. The duration of the piston stroke (stroke capacity) will determine the maximum wall shear stress obtained in one stroke.

It would therefore be expected that greater machining action would be achieved as a result of longer piston stroke durations, due to the higher wall shear stresses generated.

CONCLUSIONS

Abrasive flow machining is a highly successful finish machining process used in a wide range of engineering fields. However, as with all finish machining processes, further optimisation and control of the process may be achieved from greater understanding of the mechanisms of the machining action.

To date, research has shown that the rheology of the MV grade polyborosiloxane contributes significantly to the success of the process. The pseudoplastic nature of the polymer enables selective machining in areas of restriction. In addition, the rheotropic nature of the material indicates that the AFM piston stroke capacity may be significant in determining the machining action achieved.

REFERENCES

1. L.J.Rhoades, True grit: The Extrude Hone story, Pittsburgh high technology, November-December 1988.
2. L.J.Rhoades, Abrasive flow machining and its use, Non-traditional machining conference, December 2-3 1985.
3. Extrude Hone sales brochure.
4. Extrude Hone technical booklet.
5. Rheological properties of thermoplastics with a capillary rheometer, ASTM standard D3835-79.
6. F.N.Cogswell, Polymer melt rheology, George Godwin Limited, 1981, pp24-28.
7. Robert.E.Williams, M.S.Thesis, University of Nebrasca, 1989.

ACKNOWLEDGEMENTS

The Authors wish to express their gratitude to The Extrude Hone Corporation for their continued financial, material and technical support of this project. One of the Authors (S.A.T.) also wishes to acknowledge the financial support of the Science and Engineering Research Council during the course of this research programme.

Theoretical and Practical Study of On-Line Identification for the Grinding Burn

Huang Ren, Li Zhiyuan, Shi Xiurong, Ding Ruilian and Mao Yuliang

Southeast University
Nanjing
People's Republic of China

ABSTRACT

The grinding burn is one of most serious problem in the field of surface integrity. However, previous works are based on the experimental methods for investigating residual stress, microhardness and metallograph in the laboratory conditions and visual methods for identifying the burned colour in the pratical situation. So far there has not yet been any studies successfully applied to the on-line conditions. Based on the characteristic analysis for the grinding sparks temperature collected by an infrared measuring instrument, this paper proposes a new recognition techique to identify the grinding burn in the on-line conditions. Experiments show that this recognition technique is available to the mass production of ball bearing.

INTRODUCTION

It is well known that the essence of grinding burn is a heat damage which causes changes of surface integrity of the workpiece. The main defects of the laboratory method is usually destructive test from which one can do experiment with sample test only and can not examine the defects for every workpiece. Therefore, people would run the risk of the machine assembled with reject hidden defects in the ground surface layer and the machine would have an inherent weakness which may lead up the functional faults and catastrophic failure during the machine operation. Since the grinding temperature exerts an important influence on the surface integrity, a number of attempts have been made to measure the temperature generated at the interface between the wheel and workpiece with a thermocouple inserted in a small hole which ends just below the workpiece surface. Although this method is a classical and typical one, in general, the machine parts can not be allowed to make such a hole for inserting thermocouple and the interface is always covered by the workpiece and wheel, so this method is difficult to be used for on-line conditions. In view of this facts above, it is necessary to change the traditional idea of measuring interface temperature to measuring grinding sparks temperature for identifying the grinding states(burned and unburned) during the operation. In order to realize this purpose, our attention will be focused on the following problems:

1. the measuring methods and signal processing;
2. the regularities of the temperature between the grinding sparks and the grinding zone,
3. the identification methods of grinding states.

MEASURING METHODS FOR GRINDING SPARKS TEMPERATURE

Description of Measuring Method

In the grinding process, the sparks radiate from the interface of the wheel and workpiece. As shown in Fig.1(a), the grinding sparks rays are intensive near the tangent AA of the interface but scatter away from AA. The distribution of the sparks temperature is shown in Fig.1(b). An innumerable spark rays forms a sparks flow. If the measured area of sparks flow is large, the average temperature (but not the temperature of a single spark) of the measured area perpendicular to the tangent AA (for short, it is called grinding sparks temperature) may be varied seriously, and as a result, one can not get a logical data series that can represent the regularity of grinding process. Thus, the measured area should be small and as near AA as possible. For this reason, a special infrared measuring instrument HWX-2 for small area was developed. Considering the sparks temperature is usually about 1000℃, the measuring temperature range was designed in 600℃—1500℃, the measured area is $0.4\times0.6mm^2$ and

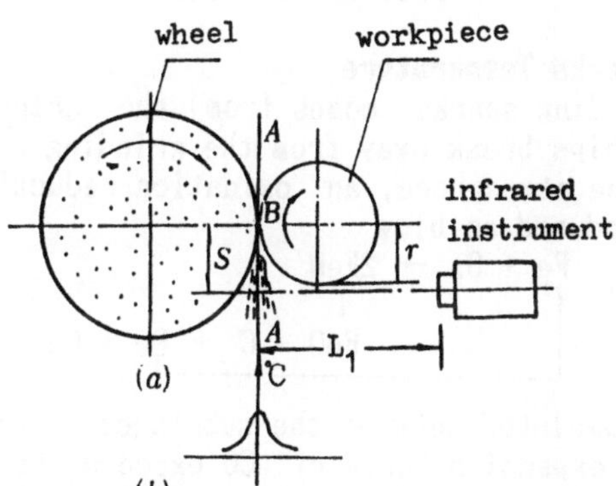

Figure 1. Grinding sparks flow and its temperature distribution in the cross section

the response time is less than 500 μs. At the lower range of the measured temperature, the resolution of this instrument is about ±1°C. The grinding sparks heat may be also transmitted by an optical fibre from the measured area to the infrared detector if necessary.

Experimental Equipments

In order to investigate the temperature measured by infrared instrument, a thermocouple (1) was set at the measured point of the sparks flow. In addition, another thermocouple (2) was inserted in the workpiece to compare the regularity of the temperatures between the workpiece and sparks flow (Fig.2)

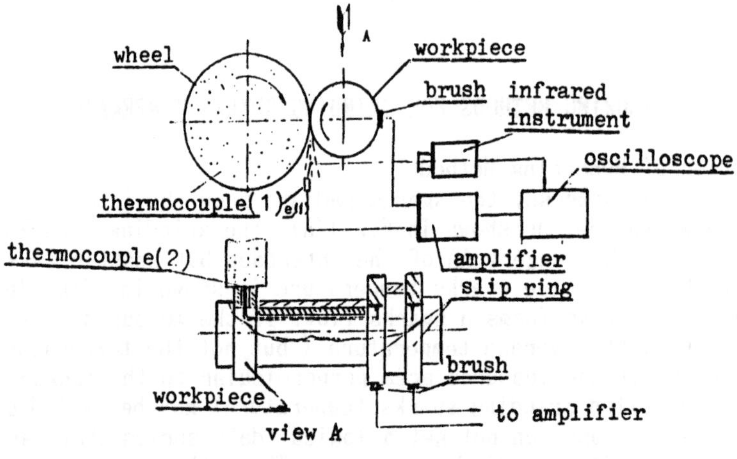

Fig.2 Experimental equipments

RELATIONSHIP OF TEMPERATURES BETWEEN THE GRINDING SPARKS FLOW AND GRINDING ZONE

Formation of Sparks Temperature

The heat of grinding sparks comes from the grinding zone. When the scorching hot chips break away from the grinding zone and contact with the oxygen in the atmosphere, an oxidation-reduction process will be continuously produced as blow

$$Fe + O_2 \rightarrow 2FeO$$

$$FeO + C \rightarrow Fe + CO$$

where CO is accumulated between the substance of chip and the film FeO. As soon as the expansion force of CO exceeds the strength of the film FeO, the CO will be released and burst into sparks. The measured area of the sparks flow is selected at the position bc (Fig.3) where the temperature of grinding sparks flow changes with distance S (Fig.3a). In the present work, S = 30mm. Indeed, the temperature accepted by the infrared detector is a statistic average value which includes the heat of all the sparks through the measured area at time t, denoted by $\{x_t\}$. For continuous time t(t=1,2,..., N), a stochastic data series will be got (see Fig.7).

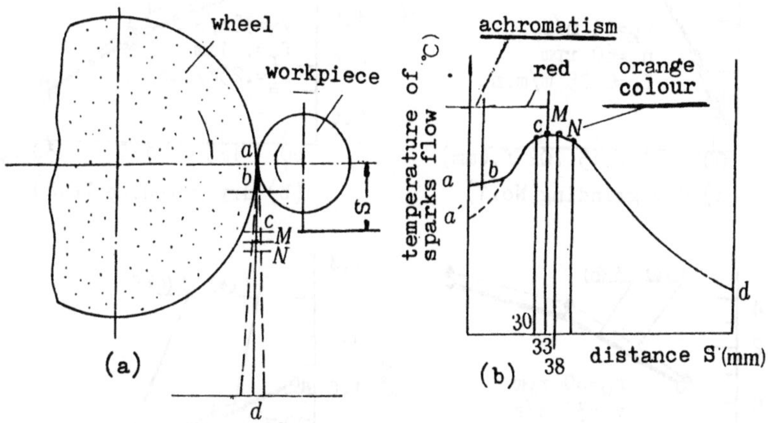

Fig.3 Grinding sparks temperature distribution in the longitudinal section

Similarity and Relational Grade of the Temperature of Grinding Sparks Flow and Grinding Zone

Similarity analysis: The temperature of grindign sparks flow is collected by an infrared and a fibre/infrared instrument, the temperature of grinding zone is clooected by a thermocouple. Since the units of these methods are different, the original signals should be normalized as

$$x_k = x_t/\bar{x}_t \tag{1}$$

where $\bar{x}_t=\frac{1}{N}\sum_{t=1}^{N} x_t$ is the average of the original signals. Fig.4 shows that the regularities of grinding sparks temperature have good similarity as the temperature as grinding zone.

Relational grade analysis: By using Grey System theory [6], the similarity can be expressed numerically with relational grade. Suppose there are m subseries of original signals $\{x_k^{(o)}(t)\}$ where k = 1,2,... N_k; N_k is a real number. In the present work, m = 2 represents two methods(infrared and fibre/infrared). The upper mark(o) expresses original data series. The data series measured by thermocouple is used as a reference series $\{x_o^{(o)}(t)\}$ for comparision. As mentioned above, the original data series should be normalized by Eq.(1), then, the corresponding normalized series $\{x_k(t)\}$ and $\{x_o(t)\}$ can be got as shown in Fig.4. The absolute difference at t = l may be expressed by

$$D_{ok}(l) = \{ \mid x_o(l) - x_k(l) \mid \},\ l=1,2,\ldots N_k$$

Thus, it follows that the minimum and maximum differences in the series $D_{ok}(l)$

$$D_{ok}(min) = \min_l \{ \mid x_o(l) - x_k(l) \mid \}$$

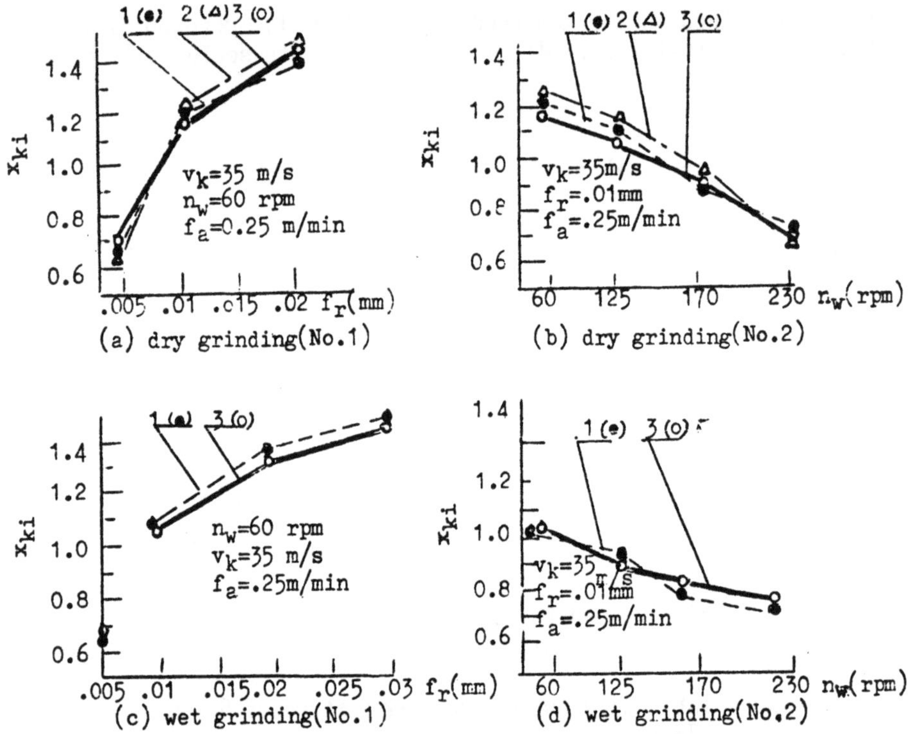

Figure 4. Comparision of similarity of normalized data by using three methods 1—fibre/infrared, 2—infrared, 3—thermocouple

$$D_{ok}(max) = \max_{l} \{ | x_o(l) - x_k(l) | \}$$

and the minimum and maximum differences of the m subseries

$$D(min) = \max_{k} \{ D_{ok}(min) \}$$

$$D(max) = \max_{k} \{ D_{ok}(max) \}$$

then, the relational grade coeffecient of series $\{xo(t)\}$ and $\{x_k(t)\}$ is defined as

$$r_{ok}(l) = \frac{D(min) + V}{D_{ok}(l) + V}$$

where

$$V = \begin{cases} \text{a given real number, if } \forall k \in \{1,2,\ldots m\}, \{x_k(t)\} \text{ is parallel to the reference } \{x_o(t)\}; \\ c\,D(max), \quad \text{if } \exists x_k \in \{x_t\}, \ \{x_k(t)\} \text{ does not be parallel to the reference } \{x_o(t)\}; c \in (0,1) \end{cases}$$

relational grade is defined as

$$R_{ok} = \frac{1}{N_k} \sum_{l=1}^{N_k} r_{ok}(l)$$

The results of experiments (Tab.1) show that the relational grade R_{ok} of the fibre/infrared instrument is larger slightly than that of the infrared instrument since the fibre ends at the position in the sparks flow. However, if the distance L_1(Fig.1) is decreased within 100mm and the detector is protected from the cooling water by a tube, the relational grade will be increased to 0.870—0.885.

Table 1
Relational grade of grinding sparks temperature to grinding temperature

grinding conditions	No.	fibre/infrared instrument	infrared instrument	
			$L_1 = 307$ mm	$L_1 = 100$ mm
dry grinding	1	0.857	0.590	-
(without cooling)	2	0.845	0.546	-
wet grinding	1	0.791	-	0.885
(with cooling)	2	0.669	-	0.870

STATE IDENTIFICATION OF THE GRINDING BURN

Procedure of the Identification

The procedure of the identification is based on the Pattern Recognition. The idea of this paper can be generalized as following block diagram.

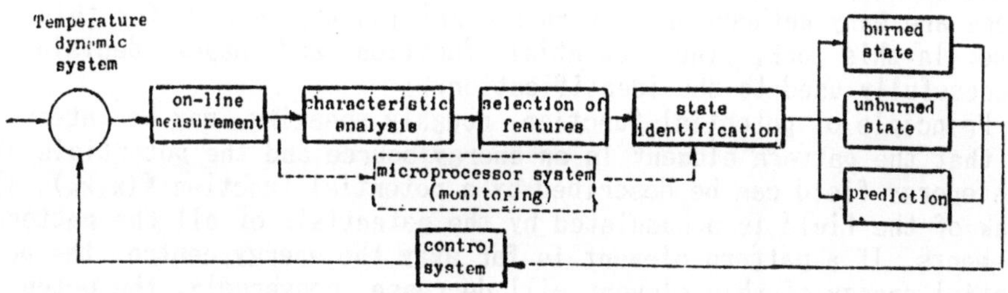

Figure 5. Block diagram of the identification for grinding burn

Signal Processing

Because the $\{x_t\}$ is a stochastic process, it is difficult to identify the burned and unburned states by direct viewing. The purpose of signal processing is to transform the measured signal into some characteristics related to the burned or unburned states by using following methods such as:

1. Time Series Analysis is a new scientific approach to transform the original signal into mathematic models. The results of experiments show that the grinding sparks temperature may be described by AR(n) model

$$x_t - a_1 x_{t-1} - a_2 x_{t-2} - \dots - a_n x_{t-n} = e_t \qquad (3)$$

where n is the model oerder. For grinding sparks temperature, n = 6-8, the parameters a_i and residual e_t include useful informations which can express the burned and unburned states.

2. Spectrum analysis is a traditional method to transform the original signal into characteristics in frequency domain. This method can reveal the informations of main harmonic components in the stochastic process. The result of experiments shows that the main energy of the sparks temperature is in the range of lower frequency ($f \leq 200Hz$). When the ground surface is burned, the higher frequency components ($f = 600$-$800Hz$) sometimes may occur.

3. Statistic analysis is also a traditional method to transform the original signal into statistic characteristics such as mean, autocovariance, kurtosis coefficient etc.

Selection of Features

The signal processing above, indeed, is a process of characteristic analysis, the purpose of characteristic analysis is to discard useless informations (including noise of the system) and retain the useful informations related to the grinding states. In order to simplify the computation of identification, one should select some characteristics (in general, 2-3.) as features to constitute a pattern space for describing the grinding states. The result of experiments shows that the mean, variance, on step autocorrelation, kurtosis coefficient, model parameter a_1 and residual e_t may be prior to be selected as the features.

State Identification

There are many methods of Pattern Recognition may be used for this purpose. In this work, the potential function and Bayes decision are successfully used to the identification.

1. Principle of potential function, roughly speaking, may be interpreted as that the pattern element is an energy source and the potentials of the energy field can be described by a potential function $K(x, x_k)$. The peak of the field is accumulated by the potentials of all the pattern elements. If a pattern element is far away the energy centre, the potential energy of this element will decrease, conversely, the potential energy will increase. The potential function is constituted in the process of continuous iterations and learning. According to the distribution of pattern samples, the potential energy centre and field strength in the pattern space of two clusters w_1 and w_2 will be gradually formed in this process, i.e. every state of w_1 and w_2 has its energy peak, therefore, a valley between peaks of two clusters is consequently formed. This valley separates two states of clusters, i.e. the decision function. The general expression of iteration can be written in the form

$$K_{k+1}(x) = K_k(x) + C_{k+1}K(x, x_{k+1}) \tag{4}$$

As mentioned above, the decision function follows

$$d_{k+1}(x) = d_k(x) + C_{k+1}K(x, x_{k+1}) \tag{5}$$

where $K_k(x)$ is a potential function at k-th step iteration, and

$$K(x, x_{k+1}) = \sum_{i=1}^{\infty} h_i^2 f_i(x) f_i(x_{k+1}) \tag{6}$$

$f_i(x)$ is an orthogonal function, $0 \leq h_i \leq 1$

$$C_{k+1} = \begin{cases} 0, & \text{if } x_{k+1} \in w_1 \text{ and } K_k(x_{k+1}) > 0, \\ 0, & \text{if } x_{k+1} \in w_2 \text{ and } K_k(x_{k+1}) < 0, \\ +1, & \text{if } x_{k+1} \in w_1 \text{ and } K_k(x_{k+1}) < 0 \\ -1, & \text{if } x_{k+1} \in w_2 \text{ and } K_k(x_{k+1}) > 0 \end{cases} \quad (7)$$

2. Bayes decision method is based on the conditional probability. If the priori probability density function $p(w_i)$ can be computed from statistic data and the conditional probability density $p(x/w_i)$ of the given state w_i can be computed from pattern samples investigated, then, the posteriori probability density function $p(wi/x)$ may be expressed by Bayes theorem

$$p(w_i/x) = \frac{p(x/w_i)\, p(w_i)}{p(x)} = \frac{p(x/w_i)\, p(w_i)}{\sum_{j=1}^{2} p(x/w_j)\, p(w_j)}$$

$i, j = 1, 2,$

Therefore, the decision rule may be described as

$$\begin{aligned} p(w_1/x) &> p(w_2/x), & x \in w_1 \\ p(w_1/x) &< p(w_2/x), & x \in w_2 \end{aligned} \quad (8)$$

Obviously, Eq.(8) can be also written in this form

$$\begin{aligned} p(x/w_1)p(w_1) &> p(x/w_2)p(w_2) & x \in w_1 \\ p(x/w_1)p(w_1) &< p(x/w_2)p(w_2) & x \in w_2 \end{aligned} \quad (9)$$

APPLICATIONS OF ENGINEERING PRACTICE

Grinding Conditions

This system is applied to a mass production line of ball bearing on the machine for grinding the concave surface of the internal ring shown as Fig.6.

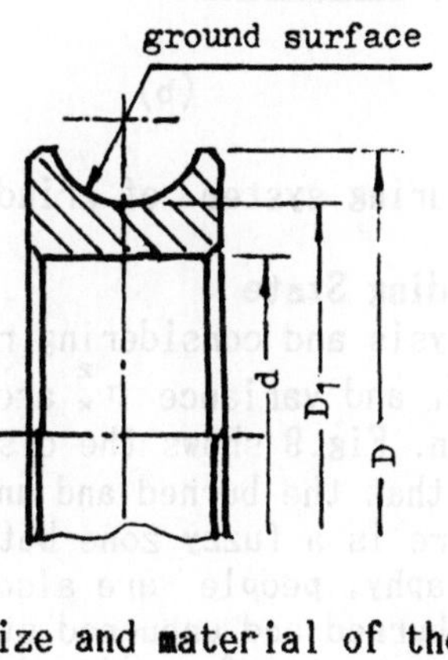

Figure 6. Size and material of the workpiece

The grinding cycle consists of following steps: mounting workpiece in the jig on the spindle → rapid feed to the wheel → ough grinding→ pause→precision grinding→non-spark grinding→back-off→ unloading,..., the mounting and unloading operations are performed by a robot and the monitoring signals are collected from the precision grinding operation.

Measuring and Monitoring Systems

The measuring system is shown as Fig.7 from which the signals are recorded by a taps recorder and transmitted to the computer IBM PC/AT by which the pattern samples will be trained for determining parameters of the decision function, then, stored in a hardware EPROM 2732. The main functions of identification are realized by a single chip microprocessor system (MCS-51) as shown in Fig.8 which consists of CPU 8031,EPROM 2732, and an external memory 6116 for storing sample data and the results of computation.

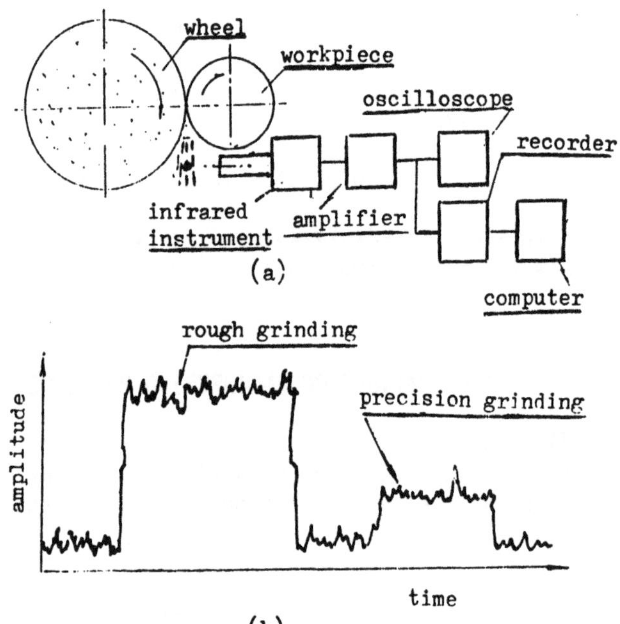

Figure 7. Measuring system of grinding sparks temperature

Identification of grinding State

By characteristic analysis and considering requiements of real time monitoring, the mean $\bar{x}_t$ and variance σ_x^2 are selected as the features for pattern recognition. Fig.9 shows the distribution of the pattern samples.It is obvious that the burned and unburned states can be clearly separated. However,there is a fuzzy zone between two states. In fact, in the field of metallography, people are also difficult to make sure a distinct boundary for burned and unburned states. Practical experiments prove that the relative errors of on-line identification by using two methods above are within 5%.

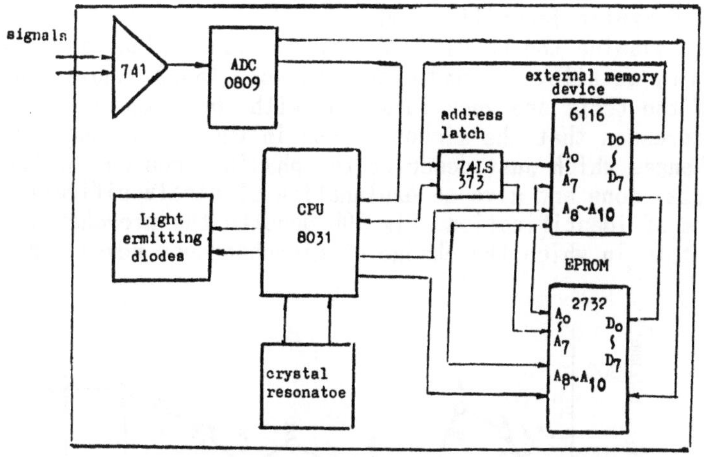

Figure 8. Monitoring system for identifying grinding states

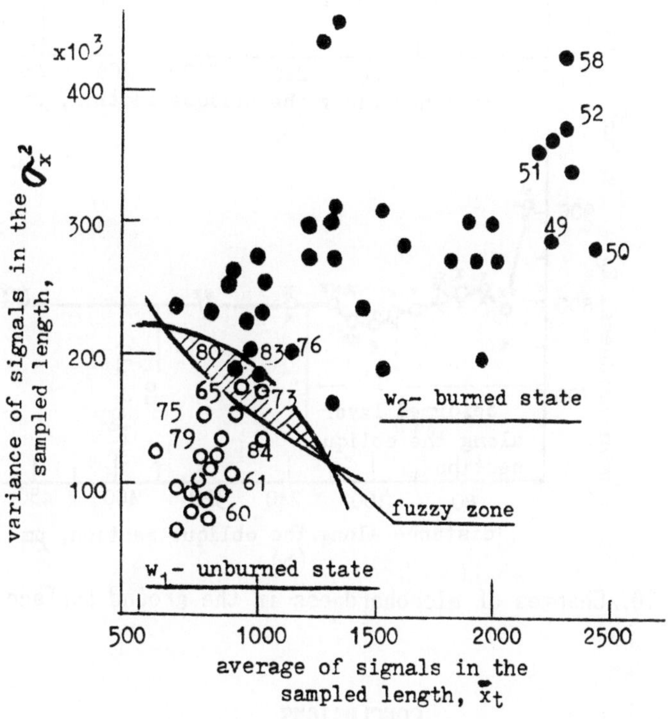

Figure 9. Distribution of pattern samples of two states

Checks for Computer Identification
In order to examine the results of computer identification some samples are tested by metallograph and microhardness. Tab.2 shows that the results of the tests are very identical with the computer identification. Fig.10a expresses that the microhardness in the fuzzy zone reveals non-regular changes which just describe the physical reasons of the formation of fuzzy zone and give an explanation of misclassification during the process of identification. Fig.10b denotes the microhardness of unburned state in which the change of microhardness reveals more regularity.

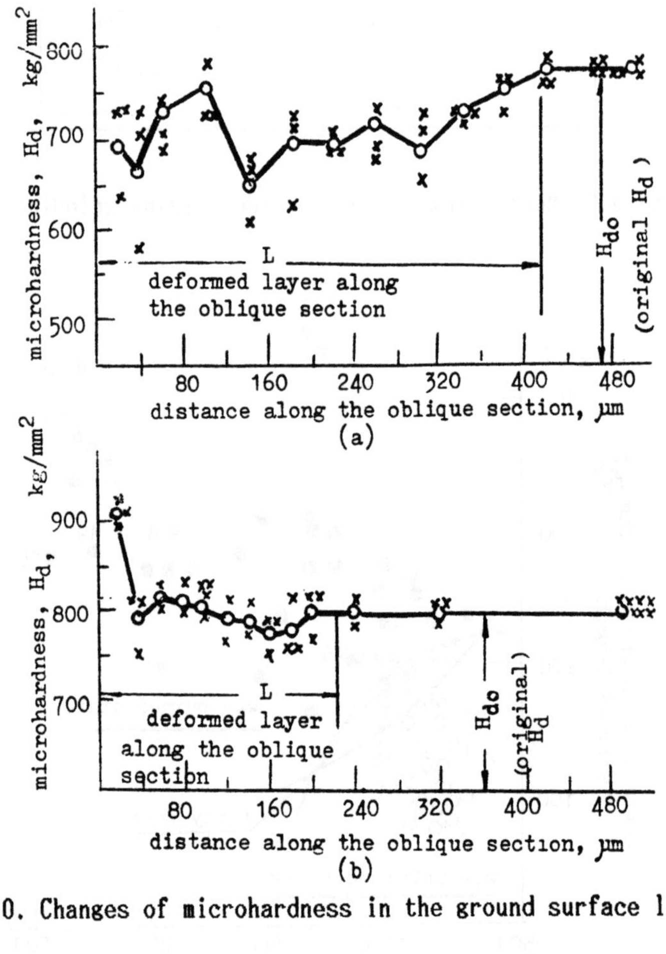

Figure 10. Changes of microhardness in the ground surface layer

CONCLUSIONS

The conclusions we can draw from this work are

1. The grinding sparks temperature is a stationary stochastic process. On this basis, a serious of researches on the instrument developments, similarity of sparks temperature and grinding temperature, signal

processing, recognition technique and monitoring system development were carried out by authors [1-5].

2. By using Grey System theory, it proves that the sparks temperature has good similarity to that of the grinding temperature. This is an important basis for identifying the grinding states via sparks temperature.
3. Since the grinding sparks temperature is a stochastic process, there are many methods can be used for analyzing the sparks temperature process and some of features can be selected for representing changes of grinding states.
4. By using the theory of Pattern Recognition, it proves that the potential function and Bayes decision method are successfully applied to the identification of grinding burn.
5. Engineering practice shows that this recognition technique is available to the mass production of ball bearing.

Table 2. Results of on-line identification and test by metallograph

sample No.	on-line identification	test by metallograph	test by microhardness
49	burned	x x x	d d d, h h
50	burned	x x x	d d d, h h
51	burned	x x x	d d d, h h
52	burned	x x x	d d d, h h
58	burned	x x x	d d d, h h
60	burned	0	+ , h h
61	unburned	0	+ , h h
75	unburned	0	+ , h
76	unburned	x x	d , h h h
* 73	burned	0	+ , h
* 65	burned	x x	d d , h h
* 80	burned	x x	d d d, h
* 83	burned	x x	d , h h h
84	unburned	0	+ , h h

x——burned (slight, medium, serious); 0——unburned
d——microhardness decreased (slight, medium, serious)
+——microhardness increased; h——distribution of microhardness inhomogenous (slight, medium, serious), *——in fuzzy zone.

6. Since the grinding burn has important influence on the surface integrity, experiments also prove that this technique of identification may be used to determine the behaviour of deformed layer, such as the residual stress performance and hardening states in the ground surface layer.

REFERENCES

1. Huang Ren, Shi Xiurong, Research on the grinding sparks temperature

via infrared measuring system. Journal of Nanjing Institute of Technology, No.3 1983.

2. Huang Ren, Surface integrity of precision grinding and on-line monitoring, Machine Fault Diagnosis and Operation Condition Monitoring, Southeast Unversity Press, 1988, pp.135-220.

3. Huang Ren,Shi Xiurong, A feasibility study of on-line monitoring for the burn of ground surface layer via grinding sparks temperature, Proceedings of the Second International Metal Cutting Conference, 1985, pp. 484-502.

4. Huang Ren, Theoretical study of on-line identification for the grinding burn, Journal of Nanjing Institute of Technology, No.2, 1988, pp. 1-12.

5. Huang Ren, Research on the methods of on-line identification for the grinding burn, Journal of Nanjing Institute of Technology, No.2, 1988, pp. 13-20.

6. Deng Julong, Relational space of grey system, Fuzzy Mathematics, No.3 1985, pp.

7. Zhou Guanxiong, Computer pattern recognition, Hua Zhong University of Science and Technology Press, 1986.

Sand Erosion and Cavitation Behaviour of Hydraulic Turbine Runner Materials

A. S. Rao and D. Kung

Powertech Labs Inc.
(formerly B.C. Hydro Research & Development)
Surrey
British Columbia
Canada

ABSTRACT

Cavitation and sand erosion behaviour of hydraulic turbine runner materials were studied using a vibratory cavitation test apparatus. Unlike the cavitation damage, the mass loss of specimens exposed to sand erosion show a linear relation with respect to duration of the test. The sand eroded specimens did not show any cavitation related damage.

INTRODUCTION

Cavitation erosion is a form of surface wear, commonly found on hydraulic turbine runners in hydro electric generating stations. This type of erosion is caused by the pressure fluctuation of water in the hydraulic turbine system. One of the approaches to mitigate cavitation erosion on hydraulic turbines is to fabricate the runner out of cavitation resistant materials. This subject has been extensively researched over many years and significant progress has been made to identify cavitation resistant materials. Currently ASTM G-32(1) standard procedure is used to rank the cavitation resistance of the material.

When the water flowing through the turbine contains silt or sand particles, a dramatic increase in surface wear has been observed. There are two schools of thoughts on this phenomenon. Matsumura et al [2] concluded that this type of surface wear is not caused by cavitation but by repeated impacts of sand or solid particles. On the other hand Hengyun et al [3] and Liu [4] found that the combination of abrasive wear due to sand particles and cavitation erosion is responsible for the dramatic increase in surface wear of hydraulic turbines. Hengyun et al [3] hypothesized that the sand particles in the water impinge upon the surface of the runner at high velocities under the action of cavitation.

Our investigation was initiated to improve the understanding of the cavitation testing in sandy water and to study the behaviour of turbine runner materials under the combination action of cavitation and abrasive wear.

EXPERIMENTAL METHODS

Materials

A hot rolled carbon steel (ASTM A-36), a high strength alloy steel, (Astralloy), a high strength low alloy steel (T-1) and an age hardened stainless steel (17-4PH) bars were used in this study. The chemical composition of these materials is shown in Table 1. Test specimens as recommended in the ASTM G-32 standard were machined from the bar stock materials. The hardness of these specimens was measured using a brinell hardness tester and these values are listed in Table 2.

TABLE 1
Material Specifications

Specimen	C	Mn	Si	Ni	Cr	Mo	P	S	Cu
Carbon Steel (ASTM A-36)	.24	.9	.4	-	-	-	.04	.05	-
High Strength Alloy Steel (Astralloy)	.20/ 0.29	.8/ 1.2	.20/ .4	3.25/ 4.0	1.25/ 2.0	.25/ .5	.015	.010	-
HSLA (T-1)	.10/ .20	.60/ 1.0	.15/ .35	.7/ 1.05	.4/ .65	.4/ .6	.035	.04	.15/ .50
Age Hardened Stainless Steel (17-4 PH)	.07	1.0	1.0	3.05/ 5.0	15.5 17.5	-	.04	.03	3.0/ 5.0

Cavitation Tests

The cavitation tests of A-36 and 17-4 PH materials were conducted in accordance with the ASTM G-32 standard method. A peak-to-peak amplitude of 50 μm (0.002 in.) was used in these tests. The test specimen was vibrated at 20 kHz in a beaker of distilled water maintained at room temperature (22 ± 1°C). The mass of the specimen due to cavitation was determined by carefully weighing the specimen at regular intervals.

Sand Erosion Tests

A modification of the vibratory cavitation test apparatus of the ASTM G-32 method was used in these tests. Figure 1 shows the schematic representation of the experimental set up. A standard cavitation test specimen as shown in the ASTM G-

32 standard was prepared from the bar stock of the above mentioned materials. A slurry containing 200 grams of 40 to 100 mesh size of Ottawa sand in 600 cc of distilled water was used as the fluid medium. The slurry was placed in a funnel

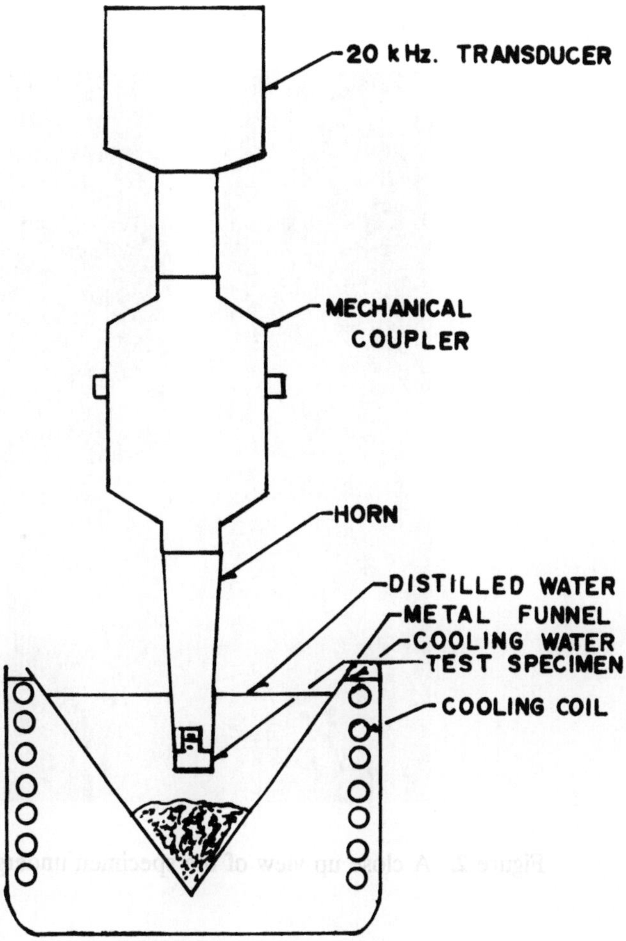

Figure 1 Sand Erosion Test Appatatus

shaped galvanized steel container. A stand off distance of 19 mm (0.75 in) between the surface of the test specimen and the top level of the sand was used during the test. The sand slurry was replaced at regular intervals. The specimen was vibrated axially at 20 kHz and the temperature of the slurry was maintained at room temperature (22 ±1°C) using cooling coils. A close up view of the specimen undergoing sand erosion is shown in Figure 2. (A conical glass container has been used for taking the photograph only). As seen from this figure, the sand particles are drawn to the surface of the specimen during the test. The test specimen is therefore exposed to the condition of vibratory cavitation as well as the accelerated sand erosion. At regular intervals the test specimen was removed from the test apparatus, thoroughly cleaned, dried and weighed. The mass loss of the specimen was plotted against test duration. The eroded surface of specimens were examined using a Scanning Electron Microscope. The early stages of the damage of the specimen were examined under an optical microscope.

Figure 2. A close up view of the specimen undergoing sand erosion

RESULTS

The mass loss curves generated through the sand erosion test are shown in Figure 3. These curves show a linear relation with respect to time and this relationship can be expressed as:

$$Y = mX \tag{1}$$

where: Y = mass loss (mg)
X = test duration (hrs)
m = slope which represents the erosion rate (mg/hr)

The mass loss curves generated through cavitation tests for A-36 and 17-PH materials are shown in Figure 4. The erosion damage on these specimens was confined only to the testing surface. No other parts of the specimen or the testing

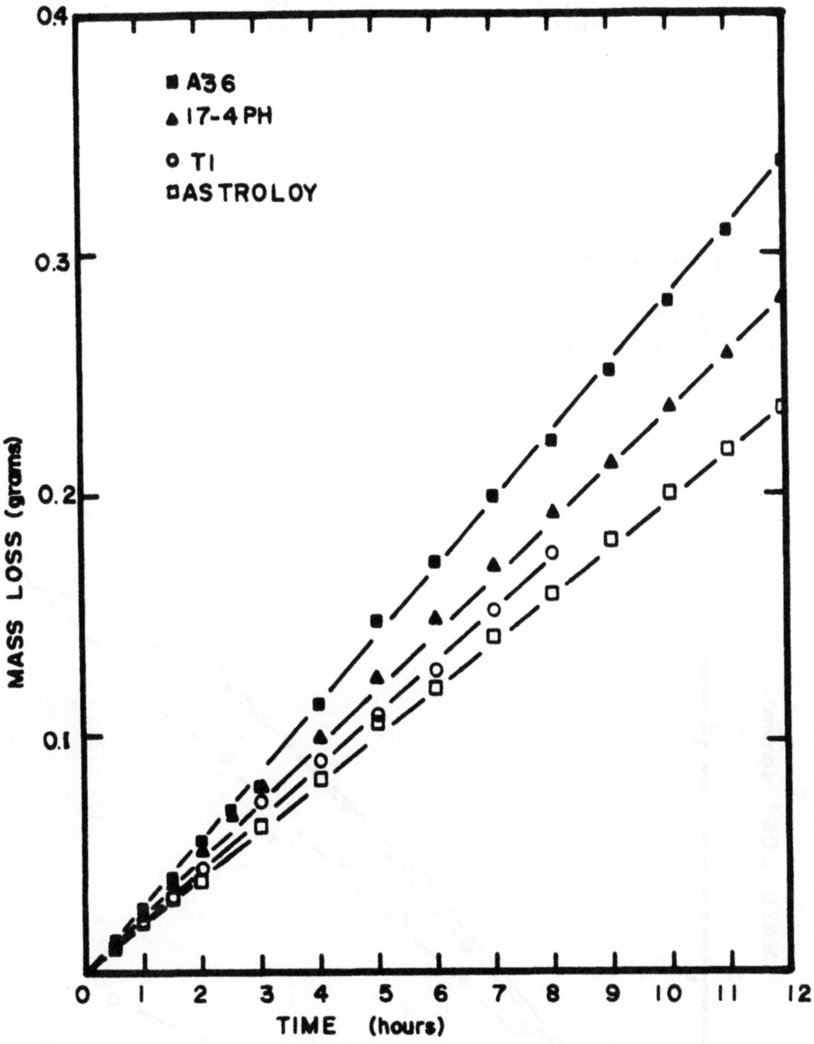

Figure 3. Mass loss curves of specimens exposed to sand erosion

apparatus experienced erosion damage. The mass loss curves due to cavitation are characterized by an incubation period where little or no mass loss occurs. This is followed by a steady state erosion region where the mass loss remains linear with respect to time and finally the mass loss curves show a decreasing erosion rate zone. The steady state erosion rate of specimens due to cavitation and sand erosion is shown in Table 2.

TABLE 2
Erosion Rates of Different Materials

Specimen	Hardness (BHN)	Cavitation Erosion Rate (mg/hr)	Sand Erosion Rate (mg/hr)
A-36	121	28	28
17-4PH	370	11	24
T-1	400	-	23
Astralloy	430	-	20

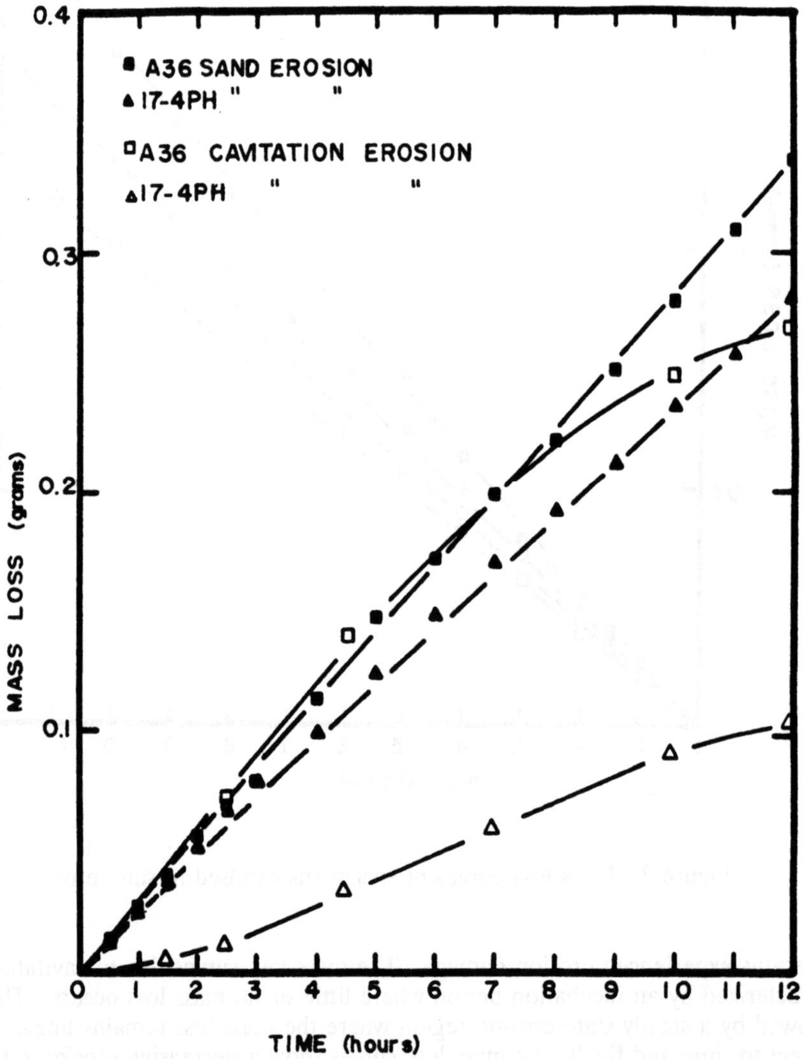

Figure 4. Mass curves of two specimens exposed to cavitation and sand erosion

Figure 5 shows typical damage to the surface of the specimens at the early stages (10 and 20 seconds) of the sand erosion test. Indentation marks of the sand particles are clearly seen on these photomicrographs. Figure 6 is a macrograph of the eroded surface of two A-36 carbon steel specimens; one was exposed to vibratory cavitation for 12 hours while the other specimen was exposed to sand erosion for the same duration. The cavitated specimen shows a combination of holes (pits) and flat areas on the surface whereas the specimen exposed to sand erosion shows a surface similar to that of a sand blasted carbon steel plate. Figure 7 shows the scanning electron micrographs of specimens exposed to sand erosion.

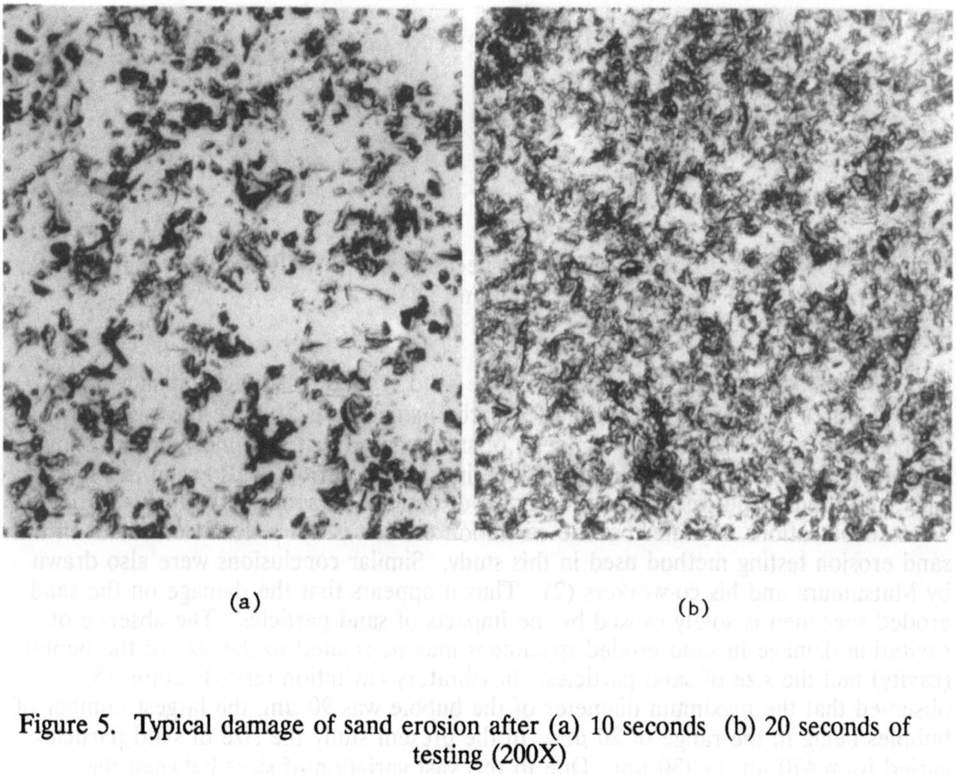

(a) (b)

Figure 5. Typical damage of sand erosion after (a) 10 seconds (b) 20 seconds of testing (200X)

DISCUSSION

The mass loss curves (Figure 3) of sand eroded specimens show a linear relationship between the cumulative mass loss and the duration of the test. Matsumura and his coworkers (2) observed similar behaviour on their sand erosion tests. The surface appearance of the sand eroded carbon steel (A-36) specimen matches with that of the specimen tested by Matsumura et al (2). These observations clearly indicate that the vibratory sand erosion testing apparatus used in this study is a simple and inexpensive apparatus for conducting accelerated sand erosion tests on materials. Unlike the method used by Matsumura, this apparatus does not require a pump to suspend the solid particle in the liquid.

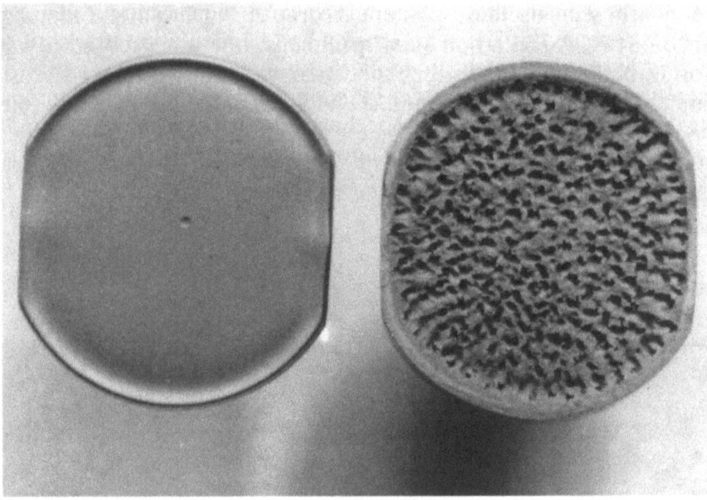

Figure 6. Damaged surface of A-36 test specimen due to (left) sand erosion (right) cavitation (3X)

The damaged surface (Figure 6) of the sand eroded test specimen shows the same features as those of a sand blasted carbon steel plate. Unlike the cavitation damage, the mass loss curves (Figure 3) of sand eroded test specimens do not show any incubation period. The absence of the incubation period is also seen in Figure 5. The sand eroded specimens show the onset of damage as soon as the test is initiated. These observations indicate that the cavitation erosion is not a dominant factor in the sand erosion testing method used in this study. Similar conclusions were also drawn by Matsumura and his co-workers (2). Thus it appears that the damage on the sand eroded specimen is solely caused by the impacts of sand particles. The absence of cavitation damage in sand eroded specimens may be related to the size of the bubble (cavity) and the size of sand particles. In vibratory cavitation tests, Brunton (5) observed that the maximum diameter of the bubble was 90 μm, the largest number of bubbles being in the range of 20 μm. In the present study the size of sand particles varied from 420 μm to 130 μm. Due to this vast variation of sizes between the bubble and the sand particle, the bubbles may not get a chance to implode on the surface of the test specimen. It is also possible that the bubbles implode on the surface but the impact of the larger particles, being far more severe, overrides the impact of the cavitating bubbles.

From Table 2 it is seen that the hardness of the materials has an influence on the sand erosion. The harder material shows better sand erosion resistance than the softer materials. However, a direct correlation between these two factors could not be established with the limited data. A close observation of scanning electron micrographs (Figure 7) of the sand eroded specimen indicates that the mechanism of erosion due to the sand slurry is similar to the mechanism proposed by Levy (6) on the erosion of ductile materials due to solid particle impact. Levy (6) has shown that ductile metals erode by the formation and subsequent removal of small platelets of extruded and forged materials. Figure 7 shows the evidence of the formation, flattening and fracture of extruded platelets on the damaged surface.

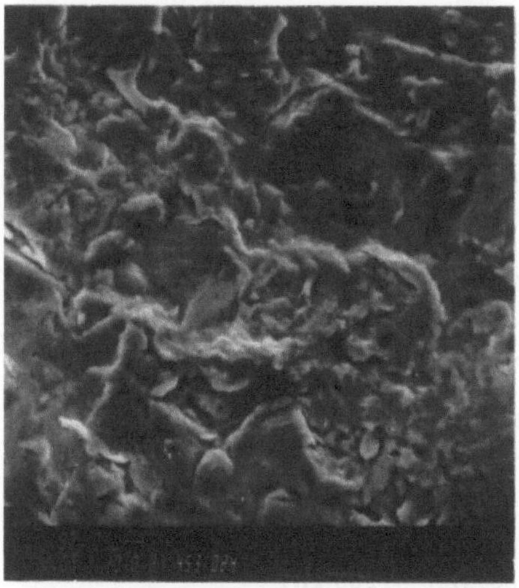

(a)

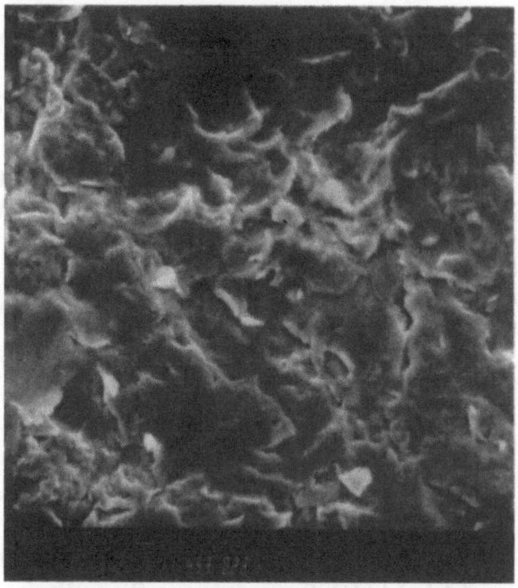

(b)

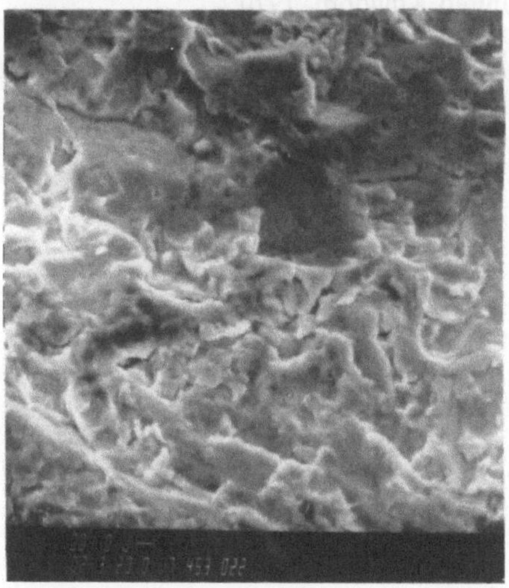

(c)

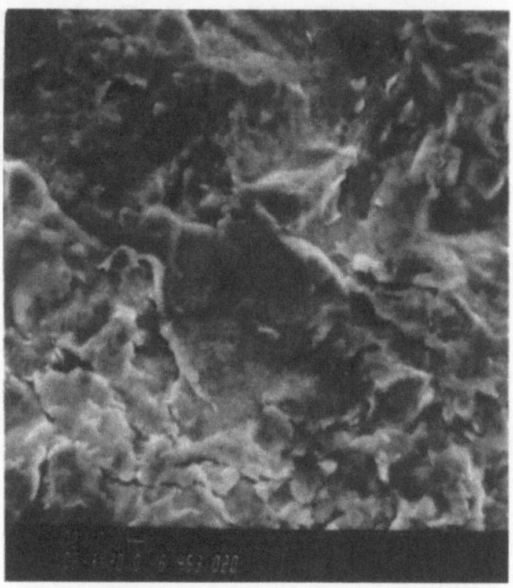

(d)

Figure 7. Scanning Electron Micrographs of specimens exposed to sand erosion
(a) A-36 (b) T-1 (c) 17-4PH (d) Astralloy

CONCLUSIONS

1. The vibratory cavitation apparatus can be modified to conduct accelerated sand erosion tests in liquid.

2. The sand eroded specimens of this study did not show cavitation related damage. It is concluded that the major damage was caused by the impacts of sand particles.

3. The hardness of the materials has an influence on the sand erosion.

REFERENCES

1. ASTM G-32-85, "The Standard Method of Vibratory Cavitation Erosion Test," Annual Books of ASTM Standards, Vol. 03.02, 1989.

2. Matsumura, M., Oka, Y., Hatanaka, H. and Yamawaki, M. "Vibratory Sand-Erosion Testing Method, Proc. 5th Conf. on Erosion by Solid and Liquid Impact, Sept. 1979, Cambridge, England, Paper 53.

3. Hengyun, J., Fengzhen, Z., Shiyun, L., and Chenzhao, H., "The Role of Sand Particles on the Rapid Destruction of the Cavitation Zone of Hydraulic Turbines", Wear, 112 (1986) 199-205.

4. Liu, Y. "Damage of Hydraulic Turbines Operating in Silt-Laden River", Proc. 14th IAHR Symposium - Trondheim, 1988, Vol. 2, 683-693.

5. Brunton, J.H. Proc. Int. Conf. Rain Erosion., 3rd (A.A. Fyall, ed.), p. 821, Royal Aircraft Establishment, Farnborough, United Kingdom, 1970.

6. Levy, A.V. "Erosion Mechanisms in Ductile and Brittle Materials", Proc. 6th Int. Conf. on Erosion by Liquid and Solid Impact, 1983, (eds. Field, J.E. and Corney, S.) Cavendish Lab., Cambridge, England, Paper 39.

Improvement of Fatigue Resistance of Steel Parts after Various Surface Treatments

J. Barralis and L. Castex

ENSAM
Aix-en-Provence
France

ABSTRACT

This article will describe four surface modification treatments including shot peening, laser shock processing, nitriding and laser transformation hardening. Depth profiles of induced residual stresses and peak diffraction widths will be comprehensively discussed.It will be shown that an accurate prediction of the fatigue properties of steel specimens is obtained when Sines's multiaxial fatigue criterion analysis is applied to experimental data.

INTRODUCTION

Surface-modification technology using various surface treatments has gained full acceptance and found widespread application for enhancing the service lifetime of industrial components subjected to high level cyclic loadings. Actual approaches differ from each other in the basic principles and techniques they rely on to induce compressive prestressing and hardening in a definite superficial layer of the processed part, such a situation being claimed to improve fatigue resistance.

The purpose of the present paper is to develop and compare surface modifications related to four different types of treatments involving either i) **mechanical effects** resulting from particle interaction in shot peening and from pressure wave interaction in laser shock processing or ii) **thermometallurgical effects** resulting from phase transformation without composition change in laser transformation hardening and from phase formation due to composition change in nitriding.

The major processing variables are briefly surveyed for each treatment and induced effects are described, especially those concerning:
i) the gradient of the generated residual stress field *vs*. position in the component as obtained by an X-Ray diffraction technique,
ii) the density of crystalline defects in relation to the depth profiles of peak diffraction widths or FWHM obtained by using the same technique.

The advantages and the drawbacks of the various treatments with respect to fatigue behaviour are thoroughly discussed introducing Sines's multiaxial fatigue criterion. Two fully worked illustrative examples concerning nitriding and shot peening are presented.

TREATMENTS

Investigated treatments are listed in table 1.

TABLE 1
Investigated treatments

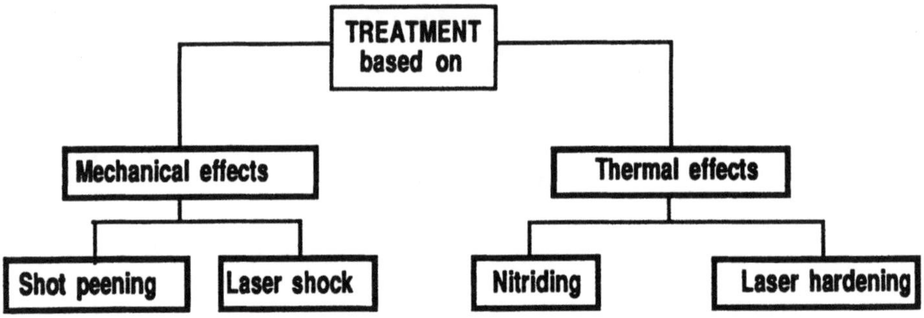

Processing variables

Treatments based on mechanical effects

These treatments are applied to work-hardenable materials.

Shot peening

Effectiveness of prestressing by shot peening is governed by two predominant controlling factors:

- The average kinetic energy of shots whose influence is illustrated in figure 1 for 35CD4 steel specimens (French standardization).
- The rate of coverage which concerns almost exclusively the materials with a high rate of work-hardening (1).

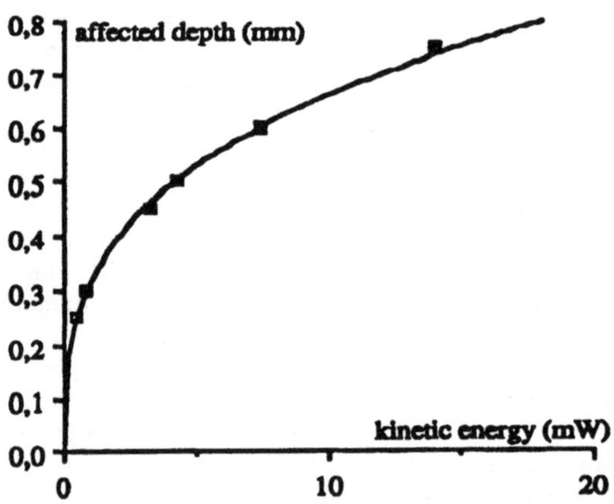

Figure 1.Affected depth as a function of kinetic energy of shots.

Laser shock

The effectiveness of this process is based on the possibility of obtaining a confined plasma region during some tens of nanoseconds and meanwhile avoiding thermal effects. The active process is the plastic deformation generated along the path of an elastic-pressure wave propagating from the surface inwards. The effective depth mainly depends on the transferred momentum per unit area usually expressed in Mbar.ns as illustrated in figure 2 (2).

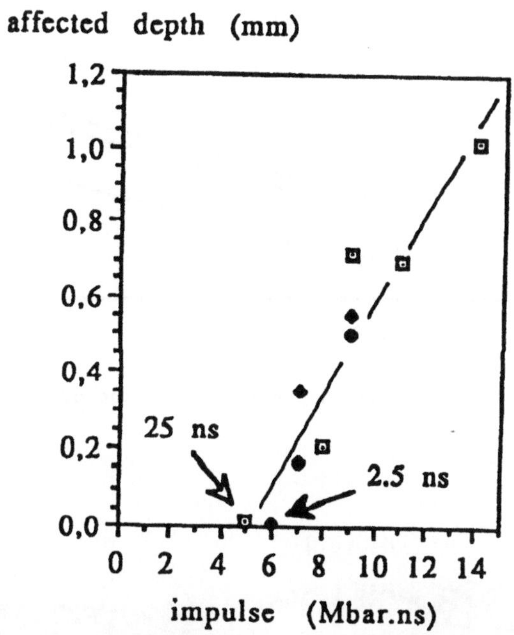

Figure 2. Change in affected depth *vs.* transferred momentum.

Treatments based on thermometallurgical effects

Nitriding

Nitriding is a surface treatment that introduces nitrogen into the surface of steel in the ferritic condition. Steels that are nitrided generally are medium-carbon steels containing sufficient amounts of nitride-formers alloying elements such as chromium, vanadium and molybdenum. Like all diffusion-controlled processes nitriding depends on temperature, time and nitrogen content at the interface atmosphere-material (3).

Laser transformation hardening

In laser transformation hardening the beam energy is deposited in a thin absorbing pre-applied coating. The highest temperature is reached at the surface and workpiece is heated by conduction inwards from the surface. A thin layer is austenitized and usually self-quenching cooling rates are sufficient to ensure austenite to martensite transformation. It must be emphasized that very high heating and cooling rates are involved in this process.

The predominant factor influencing the extension of the affected depth seems to be related to the factor (absorbed power density x interaction time)(4). However some investigators have reported a (power density x sqare root of interaction time) dependency (figure 3). This discrepancy is likely to result from difficulties arising from dwell-time measurements.

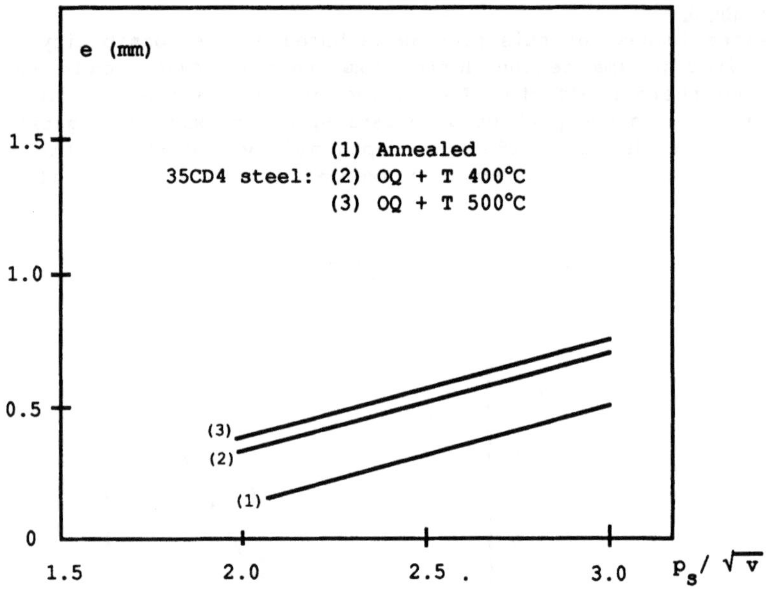

Figure 3.Case depth as a function of ($P_s/\sqrt{v}$)

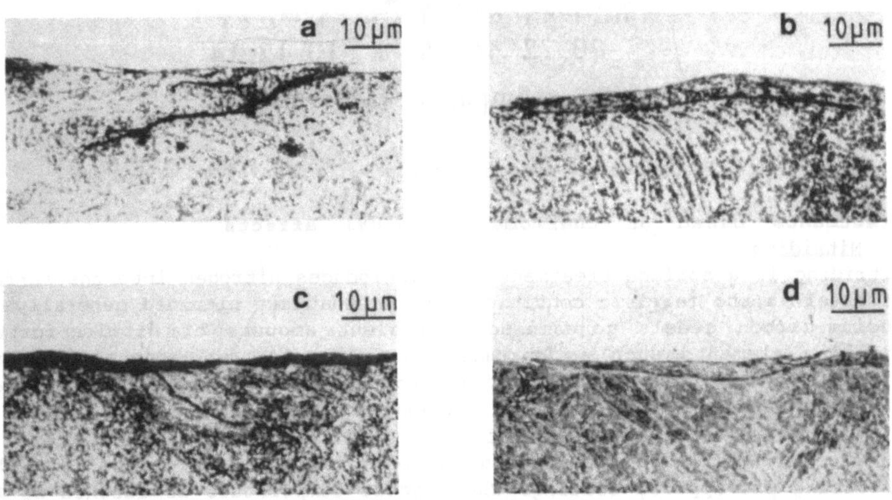

Figure 4.Sections showing microcracking in shot-peened 35CD4 samples.
a) OQ+T,S110 - b) OQ + T, S230 - c) NA, S230 - d) OQ, S110 shots

Induced effects

It is important to separate "superficial effects" involving thicknesses in the range of a few tens of micrometers and "in-depth effects" which generally are responsible for the improvement in fatigue life.

Superficial effects
They concern exclusively shot peening and nitriding.

Shot peening

Impacting shots on the surface cause a skin 10 to 30 μm thick to stretch considerably. This phenomenon is sometimes denoted "hammering"(5) and has two main consequences:

i) increased roughness (orange peel feature),

ii) extreme strain hardening of the affected layer.

The first resulting effect is not very detrimental and perhaps beneficial in some specific applications: reduction of noise in gear settings, enhanced lubrication. However it is important not to exceed the local elongation to fracture of the material. In that case microcracking is prone to occur adversely affecting fatigue properties. It is possible to avoid the onset of microcracking (figure 4) by thoroughly controlling impact velocity to be lower than a limit value characterizing the material (figure 5).

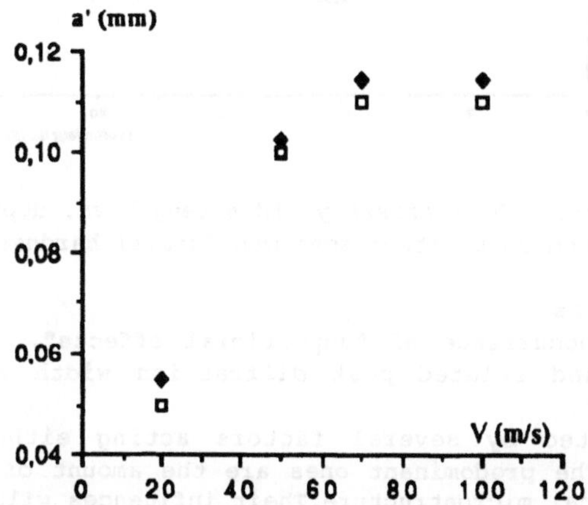

Figure 5.Calculated impact diameters a' as a function of shot velocity
S230 shots - 35CD4 steel - Hardness 30 HRC.
Estimated limit value of shot velocity: 58 $m.s^{-1}$

The second resulting effect induces either hardening or softening of the affected layer depending on the material having high or low hardness in its initial state. The boundary is situated at about 50HRC level. This phenomenon modifies the local characteristics of the steel as illustrated in figure 6 in the case of hardening.

Nitriding

During gas nitriding a compound layer from about 5 to 50 μm thick is formed as an outermost layer that consists of a mixture of Fe_4N (γ) and $Fe_{2-3}N$ (ε). In polyphase compound layers high stress fields develop in the transitional region between the two lattice structures which make the layer friable in addition to the inherent brittleness of ε-nitride. Such a layer acting as a source of microcracks should be removed usually by grinding, sand blasting or chemical dissolution. It must be noticed that when applying plasma nitriding careful monitoring of process variables enables to obtain monophase γ' compound layer or no layer at all.

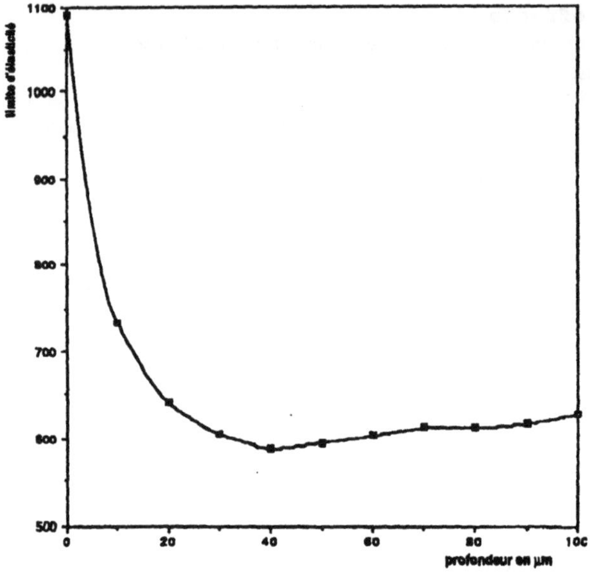

Figure 6.0.2% offset yield strength vs. depth.
Shot peened 35CD4 steel specimen Initial hardness 50HRC

"In-depth" effects

Disregarding the occurrence of "superficial effects", depth profiles of residual stress and related peak diffraction width are schematically plotted in figure 7.

FWHM is affected by several factors acting either separately or simultaneously . The predominant ones are the amount of strain hardening and the metallurgical microstructure.Their influences will be now discussed in detail for each of the four types of treatments.

<u>Treatments based on "mechanical effects"</u>

As previously stated mechanisms involved in establishing residual stress fields originate from hertzian pressure for shot peening and from elasto-plastic pressure wave for laser shock processing.Each phenomenon results in localized plastic deformation of the component.

<u>Residual stress fields</u>

Shot peening induces compressive residual stresses in layers .1 to.6 mm thick, the maximum compressive stress being located at about .02 to .1 mm from the surface.

Laser shock generates a triangle-shaped stress distribution whose maximum in compression is located at the very surface. Affected depths are in the range from .2 to 2 mm.

<u>FWHM</u>

Changes in FWHM due to processing depends on the initial hardness of the materials.The limit value is situated at about 50HRC.

FWHM is found to increase when the material is in low hardness initial condition and is found to decrease when the reverse happens.

Increase in FWHM is due to plastic deformation which in turn results in increasing the dislocation density and leads to a refinement of the substructure.

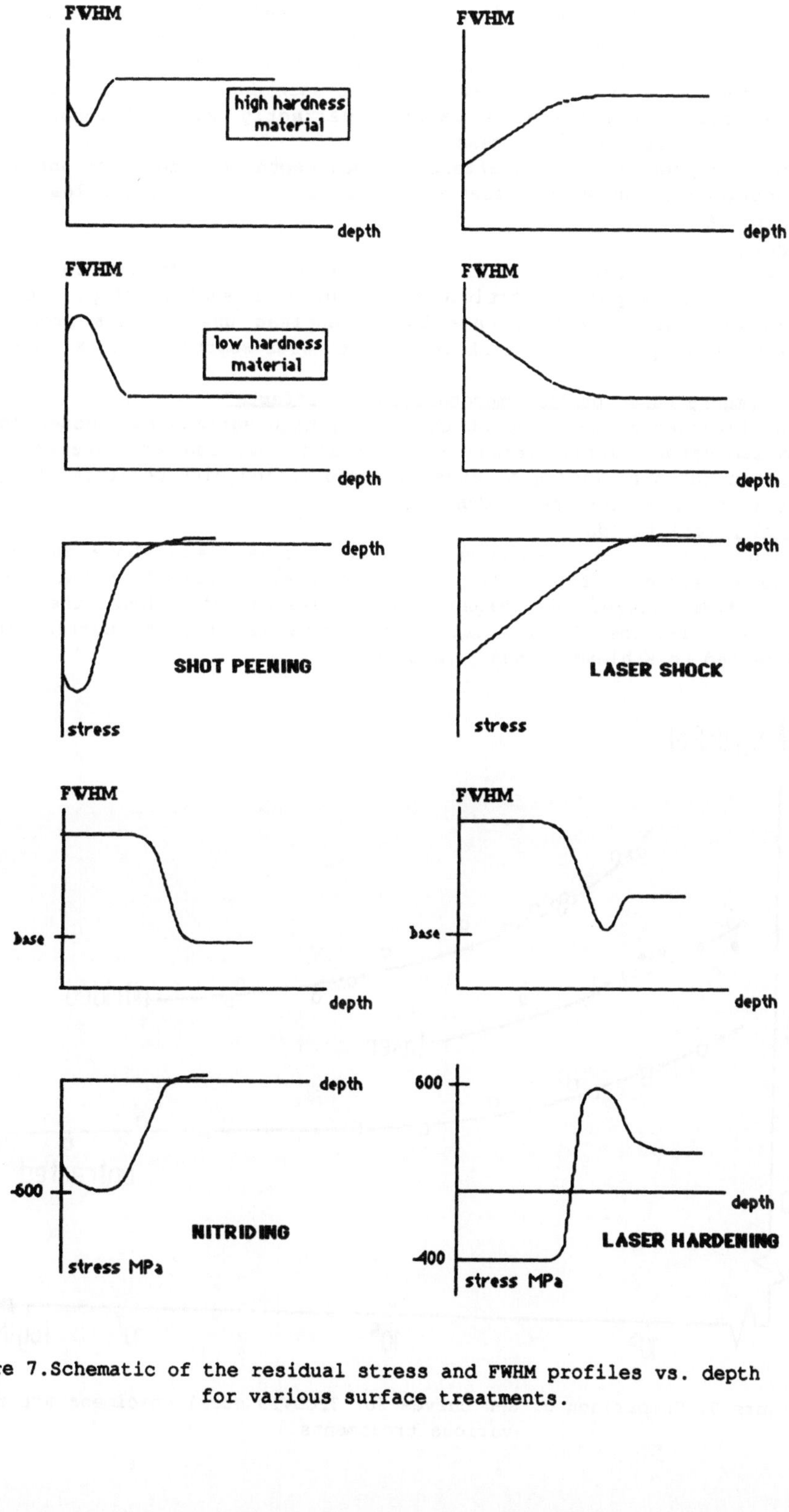

Figure 7.Schematic of the residual stress and FWHM profiles vs. depth for various surface treatments.

Decrease in FWHM also is linked to plastic deformation but in this case leads to rearrangement of the dislocations in excess as-created during the heat treatment used to give the material a high hardness value. If shot peening intensity is high the FWHM can increase due to extreme superficial work-hardening.FWHM profiles matches perfectly with residual stress profiles for laser shock processing.

For shot peening changes affect a lower depth and are important in the accomodation region where plastic deformation occurs during loading and unloading (6).

Discussion

It may be concluded that laser shock processing appears to be more effective with respect to fatigue behaviour than shot peening. The first results are very incentive ones but widespread use of this process is impaired by the prohibitive cost of adapted laser sources at this time.

Treatments based on "thermometallurgical effects"

Thermal treatments are aimed at conferring high surface hardnesses to low or medium-carbon steel grades using solid solution and precipitation phenomenon as in nitriding or phase transformation with positive change in volume as in laser surface hardening.

Residual stress fields

Nitriding induces a compressive residual stress field whose maximum is located at sublayer sites. It depends on steel composition but also on treating temperature. The higher the temperature the lower the average stress level is. The affected depth varies from .2 to .6 mm. Current trends are expected to achieve cases 1mm deep.

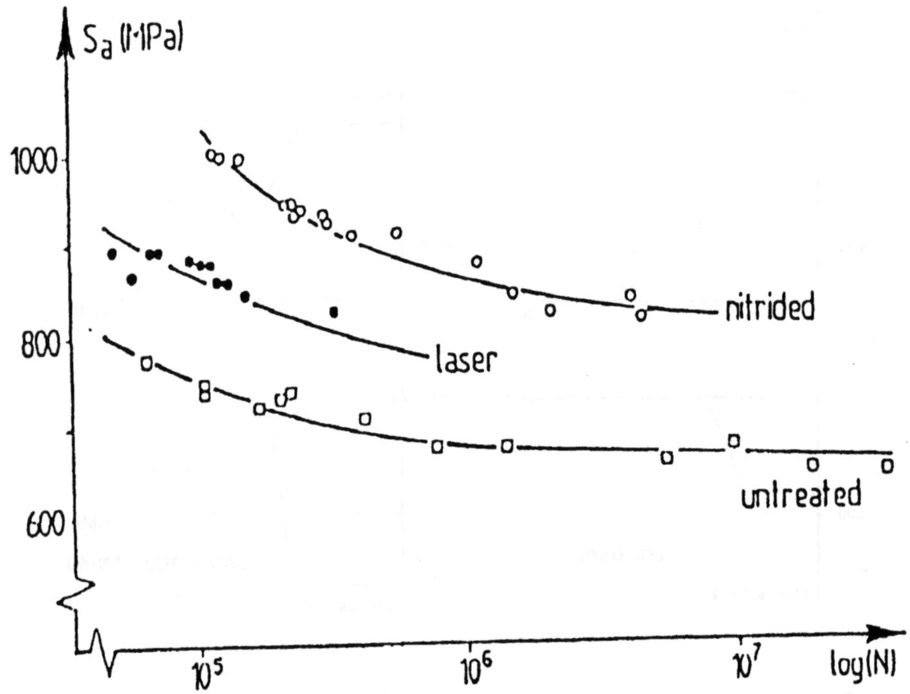

Figure 8. Comparison of S-N curves for 32CDV13 steel specimens after various treatments.

Laser transformation hardening induces profiles showing a compressive plateau region extending from .5mm to 2mm from the surface followed inwards by a steep gradient up to a tensile peak 1.5 times the compressive value and then the stress downs slowly to zero value in the base metal.

FWHM

In either case FWHM depth profiles perfectly match with residual stress depth profiles.

Concerning laser hardening a region softer than the base metal which is coincident with the tensile peak is typical of this treatment. This phenomenon could be attributed to overtempering of the substrate.

Discussion

It can be concluded that nitriding leads to stress profiles that perfectly meet the requirements with respect to fatigue resistance. On the other hand laser hardening generates deeper case depths but the presence of a tensile peak in the depleted hardness zone appears to adversely affect endurance limit as shown in figure 8.

UTILIZATION OF MULTIAXIAL FATIGUE CRITERIA

Usually fatigue properties are discussed in terms of a load in a single direction, however in surface prestressed parts the stress states of interest are generally multiaxial. The representation of the behavior of the material can be extended to these cases by using a multiaxial criterion e.g Sines's criterion (7).

Fundamentals

At any point of the material subjected to a cyclic load the components of the stress tensor can be calculated using the basic equations of solid mechanics. In the interest of brevity only proportional in-phase loads will be considered for which it can be stated that the actual stress tensor is merely the addition of a mean stress tensor σ^m and of an alternating stress tensor σ^a:

$$\sigma(t) = \sigma^m + \sigma^a \sin \omega t$$

Sines's criterion can be plotted in the T_a - P_m plane.

P_m is the mean value of the hydrostatic pressure during cycling. This scalar quantity is equal to the first invariant of σ^m :

$$P_m = 1/3\ \sigma_u{}^m = I_1{}^m$$

T_a is a shear stress equal to the square root of the second invariant of the deviatoric part of σ^a:

$$T_a = (J_2{}^a)^{1/2}$$

$$6\ J_2{}^a = (\sigma_{11}{}^a - \sigma_{22}{}^a)^2 + (\sigma_{11}{}^a - \sigma_{33}{}^a)^2 + (\sigma_{22}{}^a - \sigma_{33}{}^a)^2 + 6\ \sigma_{12}{}^{a\,2} + 6\ \sigma_{23}{}^{a\,2} + 6\ \sigma_{13}{}^{a\,2}$$

It is to be noted that T_a is related to the alternating octahedral shear stress but is a more convenient quantity to deal with.

Sines's criterion is represented by a limit curve, actually a straight line whose equation is:

$$T_a + \alpha P_m = \beta$$

The coefficients α and β are determined for a given material from two fatigue tests, usually a fully reversed torsion test (R = -1) from which a fatigue resistance t at 10^7 cycles is obtained and a fluctuating bending test (R=0) from which a fatigue resistance b at 10^7 cycles is obtained.

It can be shown that:

$$\alpha = (\ t\ -\ b/\sqrt{3}\ )\ (\ 3/\ b) \quad \text{and} \quad \beta = t$$

It has been observed that α is almost independent of the material and has an average value of about 0.25, on the contrary β is very dependent on the metallurgical condition of the material.

Figure 9 gives a graphical representation of the limit curve which divides the T_a - P_m plane in two domains marked (I) and (II). The domain marked (I) is the safety region and the domain marked (II) is the failure region.

In addition the alternating load path can be superimposed in the T_a - P_m plane. If any point of the path falls outside the limit curve *i.e* in region (II) fatigue failure will occur. Conversely for safety to be ascertained the whole path must lie below the limit curve *i.e* in region (I).

Effect of residual stresses

Residual stresses act as an extra hydrostatic pressure and will only modify the tensor of the mean stresses σ_m. The alternating stress tensor depends only on the applied loading and is not affected by residual stresses.

As an example, consider a biaxial isotropic residual stress field as obtained after nitriding or shot peening. The corresponding stress tensor is:

$$\underline{\sigma}^R = \begin{vmatrix} \sigma & 0 & 0 \\ 0 & \sigma & 0 \\ 0 & 0 & 0 \end{vmatrix}$$

For a fully reversed bending test the mean hydrostatic pressure is no longer zero but becomes equal to $2\,\sigma/3$. As stated before, the alternating component remains unchanged. The presence of residual stresses merely results in shifting the load path by an amount equal to $2\,\sigma/3$, rightwards for tensile residual stresses and leftwards for compressive residual stresses as illustrated in figure 9b. It is clear that compressive residual stresses are highly beneficial to the fatigue behavior.

The same basic ideas can be applied to more complex loading conditions.

Effect of hardening

The increase of the hardness of a steel corresponds to an increase of its tensile stress. It is well established that there exist direct relationships between t and TS, for the steels of interest it can be written :

$$t = (1\sqrt{3})\ TS^{0.857}$$

On the other hand, changes in hardness are correctly featured by the changes in FWHM. The following relationship between t and FWHM has been derived from experimental data (8) :

$$t = t_0\ (FWHM/FWHM_0)^n$$

where the subscript zero refers to the properties of the base material and n is a coefficient depending on the material and on the selected treatment. A very good prediction of t is obtained for nitrided steel assuming n=0.5.

An increase of the hardness is taken into account by displacing the boundary upwards or downwards in the Sines's diagram, since β = t. It must be emphasized that the limit curve is a local criterion attached to each location in the affected case.

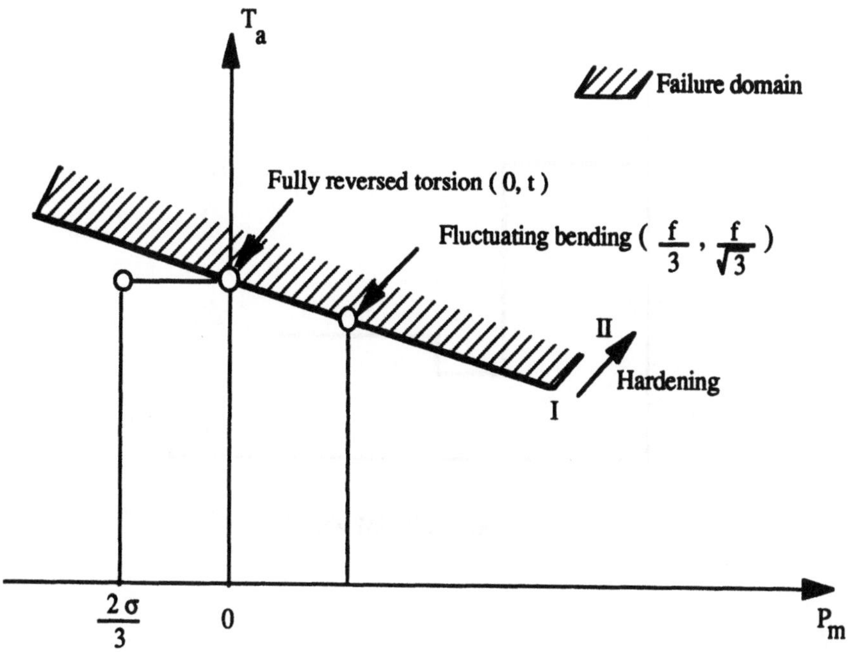

Figure 9. Schematic Sines's criterion diagram.

DISCUSSION ON THE INFLUENCE OF THE VARIOUS TREATMENTS ON FATIGUE BEHAVIOUR

Sines's diagram analysis properly accounts for the combined effects of residual stresses and hardness on fatigue behavior after various surface treatments are applied.

It seems interesting to develop a simplified example based on the following assumptions.

Residual stresses and FWHM profiles *vs*. depth are given in figure 10. Schematically they correspond to those obtained after laser transformation hardening. The residual stress tensor can be written as follows:

$$\sigma^R = \begin{vmatrix} \sigma_{11}^R & 0 & 0 \\ 0 & \sigma_{22}^R & 0 \\ 0 & 0 & 0 \end{vmatrix}$$

where $\sigma_{11} = \sigma_{22} = \sigma$ for a plane geometry.

External sollicitation corresponds to a bending test whose stress tensor is:

$$\underline{\sigma}^a = \begin{vmatrix} \sigma_{11} & 0 & 0 \\ 0 & 0 & 0 \\ 0 & 0 & 0 \end{vmatrix}$$

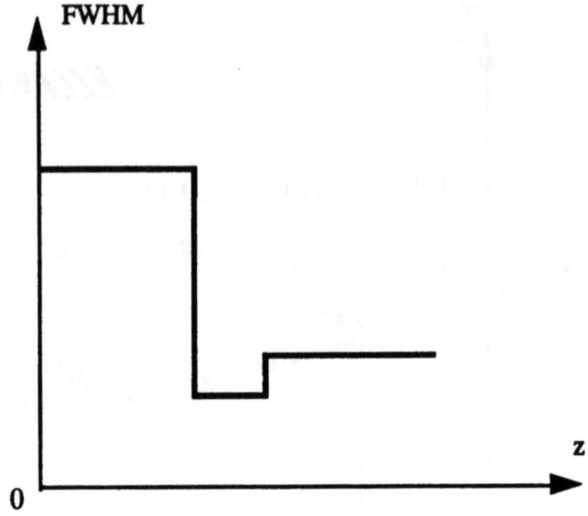

a) FWHM = f(z)

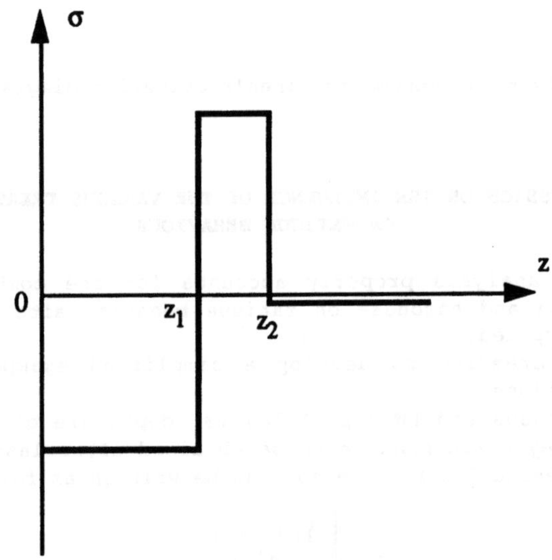

b) σ = f(z)

Figure 10. Illustrative example of residual stress and FWHM depth profiles.

Figure 11 illustrates the successive features of the Sines's diagram.

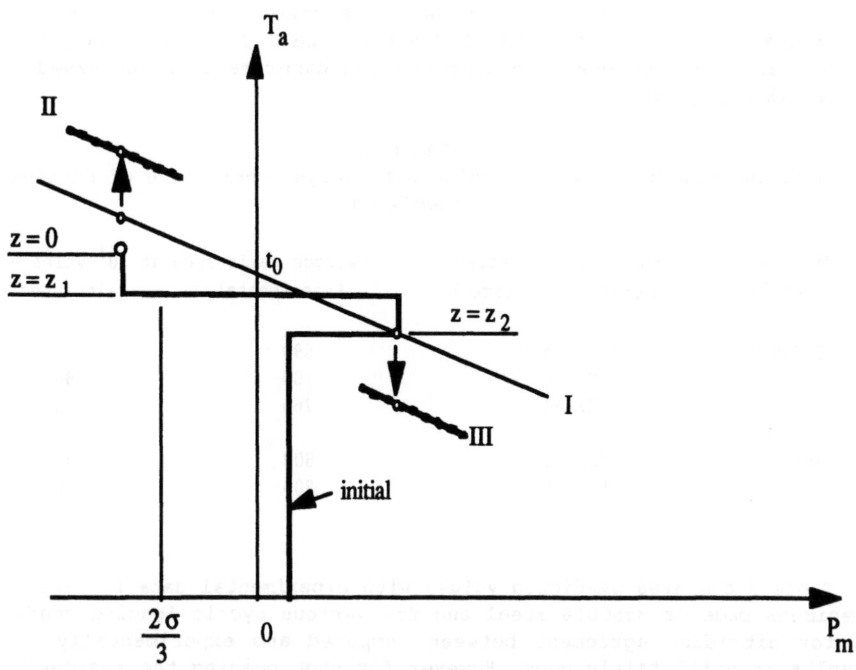

Figure 11. Sines's criterion diagram in relation with the situation illustrated in figure 10.

For the base material, limit curve is in position marked (I). For the treated material, between z=0 and $z = z_1$ hardening occurs, limit curve moves upwards to position marked (II), so enlarging the safety domain, between $z= z_1$ and $z = z_2$ softening occurs, limit curve moves downwards to position marked (II), so restricting the safety domain.

Considering the related load paths, it is clear from the diagram that:

i) in the initial condition, failure occurs at the surface,

ii) in the processed condition failure occurs at a sublayer location,

iii) in order not to penetrate the failure domain the tensile peak should be eliminated or homogeneous hardening should be achieved.

Shot peening and laser shock processing both improve fatigue resistance when applied to steels with initial hardnesses lower than about 50HRC. The same conclusion is true for nitriding.

Improvement is less important for shot peening and laser hardening when applied to steels with initial hardnesses exceeding about 50HRC. In this case the increase in fatigue lifetime due to prestressing is counteracted by the decrease in hardness.

It is to be noticed that for medium values of hardness ranging from 30 to about 50 HRC prohibitive shot peening times could result in a final decrease in hardness. It is frequently observed that these steels are prone to superficial microcracking.

On the other hand laser transformation hardening does not allow so encouraging conclusions as it is impaired by the occurrence of a tensile stress peak and dramatic softening.

A model based on Sines's criterion has been developed and applied to predict the fatigue resistance after various surface treatments.

Table 2 compares predicted values with experimental data for nitrided specimens made of 32CDV13 and of 35NCD16 steels (French standardization).

The agreement between the computed and experimentally observed results is remarkably good (8).

TABLE 2.
Experimental and predicted values of fatigue resistance of nitrided specimens

Material (Steel)	Processing variables Temperature °C/ Time h	Fatigue resistance at 10^7 cycles - MPa Experimental	Calculated
32CDV13	520/ 25	690	670
	550/ 25	700	680
	550/ 50	765	760
35NCD16	520/ 25	800	790
	550/ 25	800	790

Table 3 compares predicted values with experimental data for shot-peened specimens made of 35NCD16 steel and for various cyclic loading conditions. As for nitriding agreement between computed and experimentally observed results is still fairly good. However for shot peening the residual stress state to be taken into account is the stabilized one reached after the adaptation stage. In fact adaptation occurs during the some first tens of cycles under compressive conditions which cause plastic deformation to take place. Conversely when there exist only tensile stresses no change in the initial residual stresses field observed.

TABLE 3
Experimental and predicted values of fatigue resistance of shot peened specimens

Load condition	Fatigue resistance at 10^7 cycles - MPa Experimental	Calculated
Torsion	370	310
Bending	570	550
Bending torsion	380	380

CONCLUSION

To date it is feasible to use Sines's multiaxial fatigue criterion to predict the improvement in fatigue lifetime for parts submitted to various surface treatments. The results of the simulation correspond very well to observation in practice provided there exist no detrimental side phenomena such as microcracking. For the current model to be applied the initial stress state and hardness profiles are to be properly accounted for and experimentally determined.

It turns out that the achievement of a high level of compressive prestressing and hardening without any damage of the outermost layer is the major prerequisite for a surface modification treatment to be effective in upgrading the fatigue properties of steel parts.

REFERENCES

1. Guechichi, H., Castex, L., Frelat, J. and Inglebert, G., Predicting residual stresses due to shot-peening, Second Int. Conf.on Impact Treat. Processes, Editor S.A. Meguid, Elsevier, 1986, pp. 11-22.

2. Fournier, J.,Génération d'ondes de choc par laser pulsé de forte énergie. Applications mécaniques et métallurgiques. Thèse de Doctorat, Ecole Polytechnique, Palaiseau, France, 1989.

3. Barralis, J., Castex, L. and Chaize, J.C., Influence des conditions de traitement sur la distribution des phases et des contraintes résiduelles dans les couches nitrurées, Revue Francaise de Métallurgie, 1986, pp. 629-642.

4. Chabrol, C., Etude des contraintes internes résultant de l'interaction laser-matière dans le cas du durcissement superficiel d'alliages ferreux, Thèse de Docteur-Ingénieur, INSA, n°IDI 1.8514, Lyon, 1985.

5. Wohlfahrt, H., The influence of peening conditions on the resulting distribution of residual stress, Second Int. Conf. on Shot Peening, Editor H.O. Fuchs, Chicago, 1984, pp.316-331.

6. Cao, W. and Castex, L., Mechanical properties of shot peened material and evolution of the residual stress profile in fatigue, Metal behaviour and surface engineering, Editors S. Curioni, B. Waterhouse and D. Kirk, i.i.t.t., 1989, pp.291-296.

7. Sines, G., Behaviour of metals under complex static and alternating stresses, Metal Fatigue, Editors G. Sines and J.L. Waisman, Mc Graw Hill, 1959.

8. Deperrois, A., Castex, L., Dang Van, K., Merrien, P., and Bignonnet, A., High cycle fatigue life of surface hardened steels under multiaxial loading, Inter. Conf. Fatigue 90, Honolulu, 1990.

Relationship between Ductility and Wear Resistance of High Mn-Steels

N. A. A. Rahim Aly and M. R. El-Koussy

Cairo University
Giza
Egypt

ABSTRACT

The aim of the present work is to study the relationship between ductility, yield stress and abrasive wear resistance of high manganese steels containing from 0.53 to 0.95% C and from 12.5 to 18.3% Mn and about 1% Mo. The specimens were studied in the as cast state. Factorial analysis was applied in this work. The two investigated factors were carbon and manganese content. Abrasive wear resistance was studied in wet sand using a laboratory tumbling mill. The relative weight loss was calculated after 240 hours of testing. Some specimens were subjected to compression test. The obtained stress strain curves were used to calculate the yield stress, $\sigma_{0.2}$, the strain hardening coefficient, n, and the strength coefficient, K. The strain hardening coefficient was used as a measure of ductility.

As a result of this experiment four regression equations were obtained. Using these equations the relative weight loss, the strain hardening exponent, the yield stress and strength coefficient were predicted for composition which were not experimentally investigated. The predicted and experimental values were used to construct four charts showing the changes of the investigated parameters with changing carbon content from 0.3 to 1.0% and manganese content from 6 to 24%.

INTRODUCTION

Manganese steels with carbon content lower than 1.0% alloyed with about 1% Mo are widely used as wear resisting steels [1]. The manganese content of these steels ranges from 6 to 20% and they usually do not need heat treatment since they do not develope large grain boundary carbides. The wear resistance of steels depends on their chemical composition, mechanical properties, microstructure, surface to volume dimensions as well as on the wear conditions [2-9]. The rating of steels may change if the wear conditions change from low stress abrasion to gauging abrasion with heavy impact. The high wear resistance of austenitic manganese steels is usually explained by exceptional work hardening behaviour which is the result of low stacking fault energy, dynamic strain aging and stress-induced martensitic transformation [3]. This depends on the amount of compressive stress and shear stress on the surface. Stress-induced martensite creates wear resistant renewable surface layers, however these layers should not be too thick in order to prevent brittle fracture on the surface [3]. The chemical composition of austenitic manganese steels controls the stability of austenite, so it should control also the wear behaviour. Even small change of manganese content from 5 to 6% may affect significantly the wear resistance [3]. It was also shown that austenitic steels containing from 5 to 7% Mn had greater wear resistance than steels with greater manganese content [1-6].

Stress-induced martensite was also observed in the as-cast white irons containing 15-30% Cr [12]. These irons have matrix structures which are essentially austenitic. The hardness of the austenitic matrix of these types of white irons may increase from 400 to 900-1000 HV at wearing surfaces. Such marked increase in hardness was attributed to the formation of stress-induced martensite in regions of eutectic austenite close to eutectic M_7C_3 carbides [10]. Stress-induced martensite formation is also known for austenitic stainless steels.

However in cases where austenite is very stable, as in the case of some high manganese steels the formation of stress-induced martensite was not observed [10,11].

Many attempts have been conducted to find correlation between hardness and wear resistance. Some workers [4] concluded that the hardness appeared to be the most important metallurgical factor relative to the wear resistance. However, in many cases, it has been observed that harder material had lower wear resistance [4,12]. Usually the plotting of the hardness values against the wear resistance results in a very wide band indicating that the hardness is not the only important factor for wear resistance. Thus, it is important to find correlation between wear resistance and other mechanical properties.

The aim of the present work was to find the relation between wear resistance and mechanical properties other than hardness. Compression test was chosen to determine yield stress, strain hardening exponent and strength coefficient. The material used for these study was austenitic manganese steels having various amounts of carbon and manganese, i.e. having various ability for forming hard stress-induced martensite layer. These steels have relatively well known wear behaviour from industrial and experimental data. These data were used to explain the experimental and predicted values of the wear resistance obtained in the present work.

MATERIALS AND EXPERIMENTAL TECHNIQUES

The materials used in the present work were alloys of high manganese steel. The chemical composition of the investigated alloys are given in Table (1).

These alloys were prepared in an induction furnace and then cast in sand in the form of 20x20x250 mm bars. The specimens were studied in the as cast state since this type of steel does not need heat treatment.

TABLE 1
Chemical composition of the investigated alloys

Alloy No.	Elements, Weight %							
	C	Mn	S	P	Si	Ni	Cr	Mo
1	0.65	14.3	0.015	0.055	0.94	-	-	1.3
2	0.92	18.3	0.025	0.04	0.37	-	-	1.1
3	0.53	18	0.015	0.054	0.9	-	-	1.5
4	0.95	12.5	0.022	0.05	0.87	-	-	1

Abrasive wear resistance was studied in wet sand using a laboratory tumbling mill. The relative weight loss was calculated after 240 hours of testing by the relation $\Delta w/w_0$, where Δw-weight loss, w_0-initial weight. The sand was changed every 8 hours. The specimens used were cut into 20x20x30 mm bars, using three test pieces for each composition.

In order to determine the yield stress and ductility, 20 mm cubic specimens were prepared and carefully polished. Strain gages were attached to four sides of these specimens to measure longtudinal and lateral strains. These specimens were subjected to compressive test. To reduce the coefficient of friction special papers were applied between the machine platens and the specimens. The obtained stress strain curves were used to calculate the yield stress, $\sigma_{0.2}$, the strain hardening coefficient, n, and the strength coefficient, K. The strain hardening coefficient was used as a measure of ductility.

Statistical factorial analysis was applied in this work. It should be mentioned that the designed chemical composition was 0.5% and 0.95% for carbon content and 12.0 and 18.0% for Mn content. However, the actual composition deviated from the designed values, that is why the level of carbon content (X_1) used in the statistical analysis was taken as -1 for 0.59% and +1 for 0.95%. For the same reason the level of the manganese content (X_2) was taken as -1 for 13.4% Mn and +1 for 18.15%.

The statistical analysis of the experimental data resulted in four regression equations showing the effect of carbon and manganese separately as well as their combined effect on the wear resistance, yield stress, strain hardening exponent and the strength coefficients. The relationship between wear resistance and the investigated mechanical properties was examined.

In order to check the validity of the obtained regression equations and charts three more alloys were prepared with chemical composition given in Table (2).

It should be noted that alloy No. 1 of Table (2) has almost the same composition of alloy No. 4 of Table (1). This alloy was used as a control alloy and its wear resistance was taken as unity.

TABLE 2
Chemical composition of additional check alloys.

Alloy No.	Elements, Weight %					
	C	Mn	Si	Mo	S	P
1	0.98	12.9	1.3	1.08	0.015	0.085
2	0.99	7.44	0.84	1.39	0.015	0.065
3	0.86	6.18	0.6	1.12	0.018	0.06

RESULTS AND DISCUSSION

The statistical analysis of the obtained experimental results (Table 3) led to regression equations of the form :

$$Y = \overline{Y} + b_1X_1 + b_2X_2 + b_{12}X_1X_2 \tag{1}$$

The following equations were obtained :

1. **For relative weight loss (WL)**

$$Y_{WL} = (73.14 - 4.12\ Y_1 - 1.42\ X_1 + 4.31\ X_1X_2)\ 10^{-3} \tag{2}$$

2. **For yield stress (YS)**

$$Y_{YS} = 271.25 + 8.75\ X_1 + 3.75\ X_2 - 33.75\ X_1X_2 \tag{3}$$

3. **For strain hardening exponent (n)**

$$Y_n = 0.257 - 0.007\ X_1 + 0.017\ X_2 + 0.042\ X_1X_2 \tag{4}$$

4. **For strength coefficient (K)**

$$Y_K = 1187.5 + 87.5\ X_1 + 87.5\ Y_2 + 187.5\ X_1X_2 \tag{5}$$

TABLE 3
The obtained experimental data.

Alloy Composition		Yield Stress MPa	Strain Hardening Exponent, n	Strength Coefficient, K MPa	Relative Weight Loss $\Delta w/w_o$
C%	Mn%				
0.65	14.3	225	0.29	1200	0.083
0.53	18.0	300	0.24	1000	0.0715
0.95	12.5	310	0.19	1000	0.0661
0.92	18.3	250	0.31	1550	0.0729

From the data given in Table (3) and in Fig. (1), it is clear that the lowest weight loss during the wear test coincides with the greatest value of the yield stress and the lowest values of strain hardening coefficient, n, and the strength coefficient, K. On the other hand, the greatest weight loss was obtained in the alloy having the lowest yield stress (Fig. (1-b)). The greatest wear resistance was obtained in alloy having lowest Mn content (12.5%) and highest carbon content (0.95%).

The relation between the experimental values of the strength coefficient, K, and the relative weight loss, $\Delta w/w_0$, given in Fig. (1-c) is less clear than that obtained in Fig. (1-a) and b. These results show that their exists a tendency for the wear resistance to increase with decreasing the strain hardening exponent and the strength coefficient and with increasing the yield stress.

The obtained regression equations show that for all the investigated properties combined effect of Mn and C, given by b_{12} in the regression equation (Eq. 1) is greater than the separate effect of carbon or Mn (b_1 or b_2). This means that the effect of carbon or manganese cannot be discussed separately from each other so it can be concluded that the tendency of the effect of carbon content on the investigated properties depends on the Mn content and vice versa. The regression equations were used to calculate predicted values of relative weight loss, yield stress, strain hardening exponent and the strength coefficient for carbon contents ranging from 0.3 to 1.00 and manganese contents from 6 to 24%. The obtained calculated values, together with the experimental values were used to construct charts showing the change of the mentioned properties with changing manganese and carbon contents.

From Fig. (2), it is clear that the greatest relative wear resistance is obtained in two cases :

1. When Mn content is in the range of 6 to 8% and carbon content is about 1%.

2. When Mn content equals 24% and carbon content amounts to about 0.4%.

It is known [3] that former compositions tends to give stress induced α-martensite when subjected to severe wear. It is also known that the latter composition tends to give ε-martensite induced by deformation [3,11]. This may explain the high wear resistance of these two compositions. At high carbon and manganese content (1% C and 24% Mn) the austenite becomes more stable and less able to develop stress induced martensite, so the wear resistance is low for this composition. High carbon content (about 1%) decreases the stacking fault energy of high manganese steels containing more than 20% Mn [11]. At the same time, it is known that the amount of ε-phase decreases with stacking fault energy. It was reported [11] that steel containing above 0.6% C and 20% Mn did not develop ε-phase because ε-phase is suppressed strongly by carbon content. That is why lowering carbon content increases the wear resistance of Mn-steels containing more than 20% Mn. Alloys containing 0.3-0.4% C and about 6% Mn may develop martensite easily upon cooling from the austenitic range. Even cooling rate of 0.03°C/sec. was reported to develop structure consisting of 30% martensite [11]. According to Ingo Schmidt

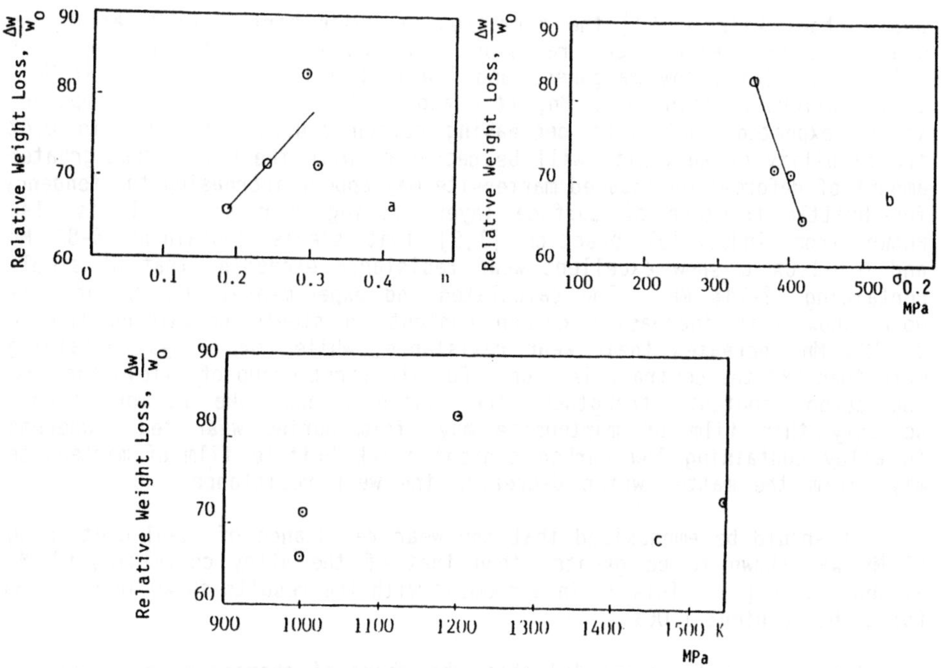

Figure 1 : Relationship between relative weight loss $\Delta w/w_o$ and strain hardening coefficient, n, yield stress, $\sigma_{0.2}$, and strength coefficient, K. (experimental results)

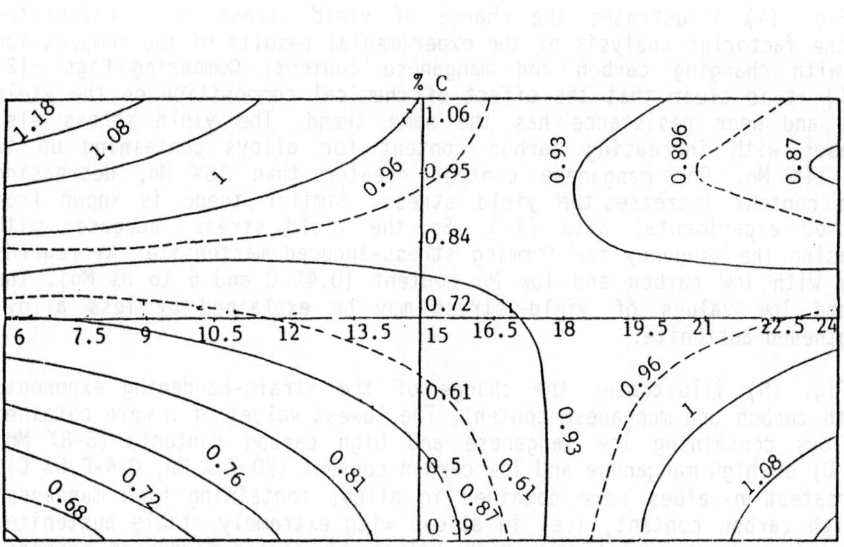

Figure 2 : Effect of carbon and manganese content on the relative wear resistance [relative weight loss of specimens No. 4 of Table 1 divided by relative weight loss of each specimen]

and Michael Wilhelm [3] the formation of thick martensite layer during wear test may reduce wear resistance. Probably, this is the case for alloys containing low manganese and low carbon content (0.4% C, 6% Mn). If the alloys containing 6% Mn, 1% C and 6% Mn, 0.4% C will be compared, it is expected that with decreasing carbon content from 1.0 to 0.4% the stability of austenite will be decreased with the result that greater amount of deformation induced martensite may appear increasing the tendency for brittle fracture of surface layers during wear test. It is also known from industrial practice [1,5] that steels containing 6-8% Mn and 0.8-1.0% C show excellent wear resistance exceeding that of steels containing 12-14% Mn. The calculated and experimental results of this work show that increasing carbon content in steels containing from 6 to 12% Mn increases their wear resistance, while for alloys containing more than 18% the contrary is true. For the first group of alloys increasing carbon content strengthens the austenite and make it more stable so only thin film of martensite may form during wear test, whereas in alloy containing low carbon content thick brittle film of martensite may form the matter which decreases the wear resistance.

It should be emphasised that the wear resistance of steel containing 5% Mn was shown to be greater than that of the alloy containing 6% Mn at about 1% C [3]. This is in agreement with the results shown in Fig. (3) for carbon content 1.06%.

Thus it can be concluded that the chart of chaning wear resistance with chemical composition, given in Fig. (2), is in agreement with known industrial practice as well as with published experimental results concerning the stability of austenite [3,11].

Fig. (4) illustrates the change of yield stress $\sigma_{0.2}$, calculated from the factorial analysis of the experimental results of the compression test with changing carbon and manganese content. Comparing Figs. (2) and (1) it is clear that the effect of chemical composition on the yield stress and wear resistance has the same trend. The yield stress also increases with increasing carbon content for alloys containing up to about 14% Mn. For manganese content greater than 18% Mn, decreasing carbon content increases the yield stress. Similar trend is known from published experimental data [13]. So the yield stress increases with increasing the tendency for forming stress-induced martensite. As regards alloys with low carbon and low Mn content (0.4% C and 6 to 8% Mn), the obtained low values of yield stress may be explained by less alloy-strengthened austenite.

Fig. (5) illustrates the change of the strain-hardening exponent, n, with carbon and manganese content. The lowest values of n were obtained in alloys containing low manganese and high carbon contents (6-8% Mn, 0.9-1 C) or high manganese and low carbon content (20-24% Mn, 0.4-0.5% C). The greatest n-values were obtained in alloys containing high manganese and high carbon content, i.e. in alloys with extremely stable austenite. On the other hand, relatively high values of strain-hardening exponent were calculated for alloys containing ~0.4% C and 6% Mn. Thus, it is clear that low n-values correspond to high wear resistance.

From Fig. (6), it can be concluded that low strength coefficient values, K, correspond to high wear resistance. These low K values are obtained as a result of low strain hardening coefficients of the corres-

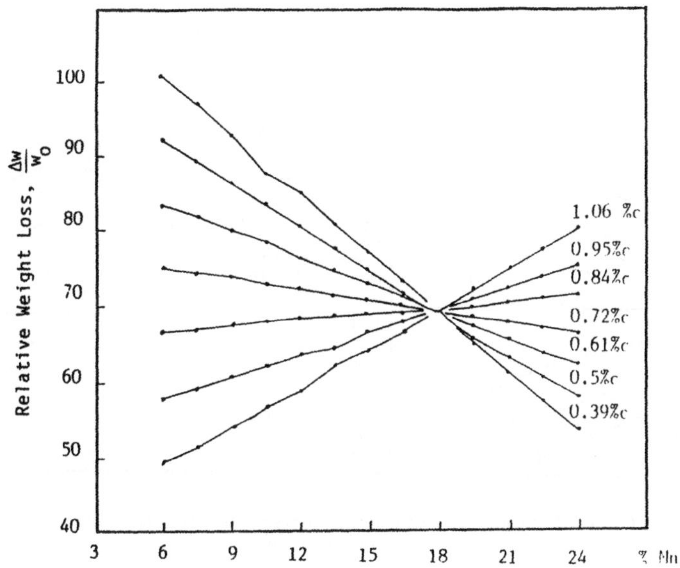

Figure 3 : Effect of manganese content on the relative weight loss at different levels of carbon values

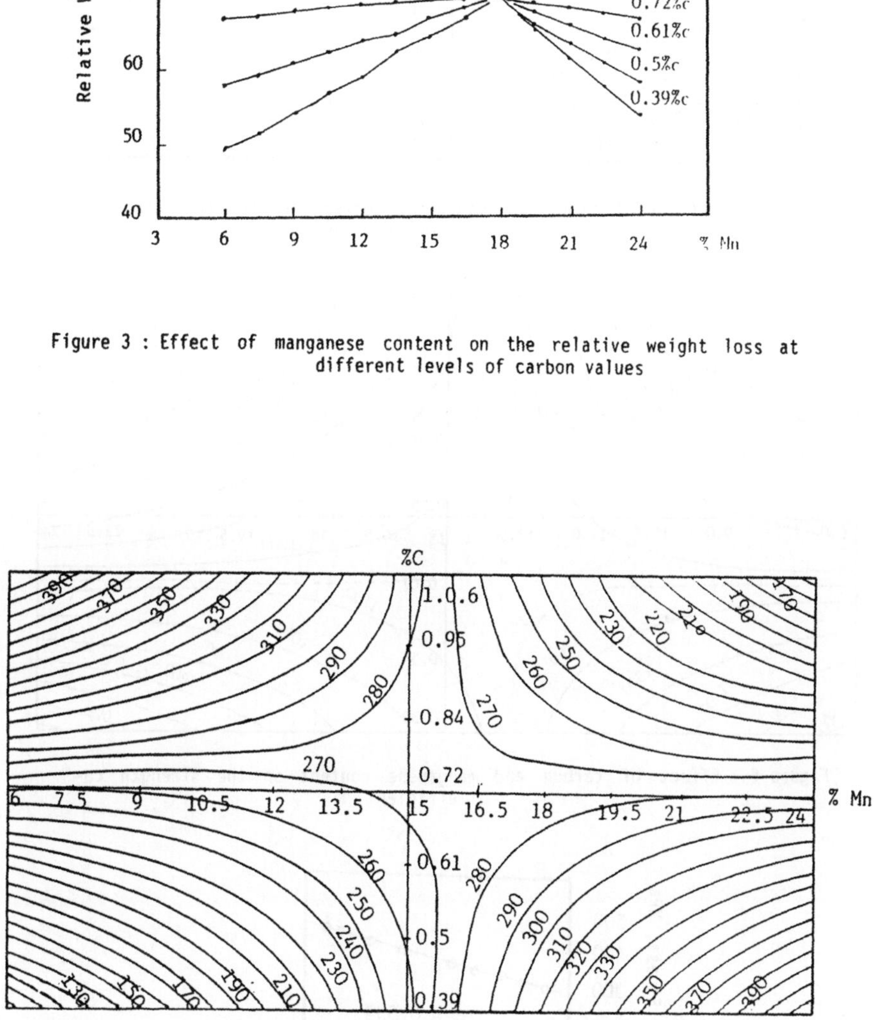

Figure 4 : Effect of carbon and manganese content on the yield stress, $\sigma_{0.2}$

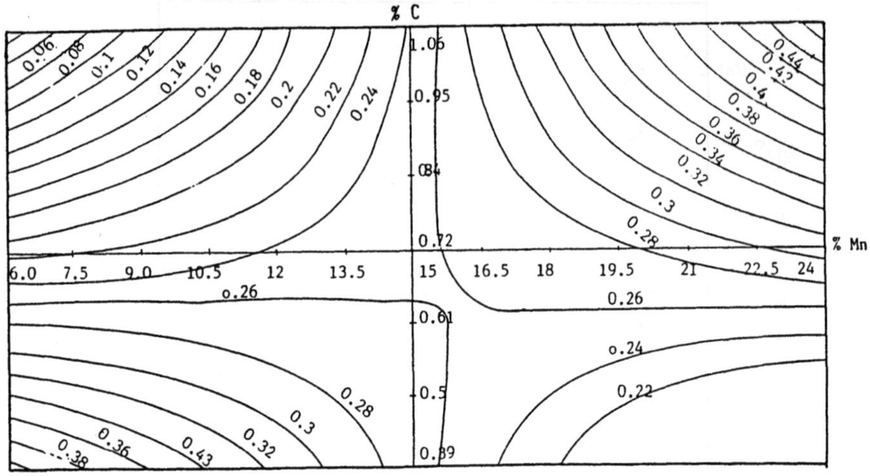

Figure 5 : Effect of carbon and manganese content on the strain hardening exponent, n

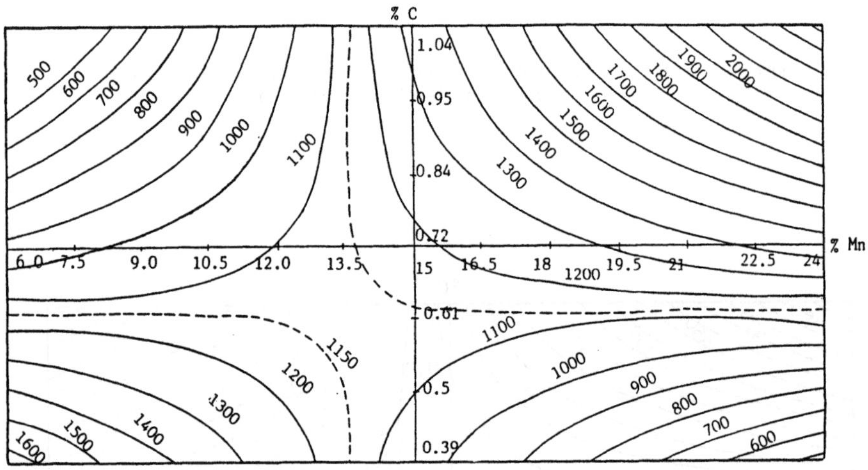

Figure 6 : Effect of carbon and manganese content on the strength co-efficient, K

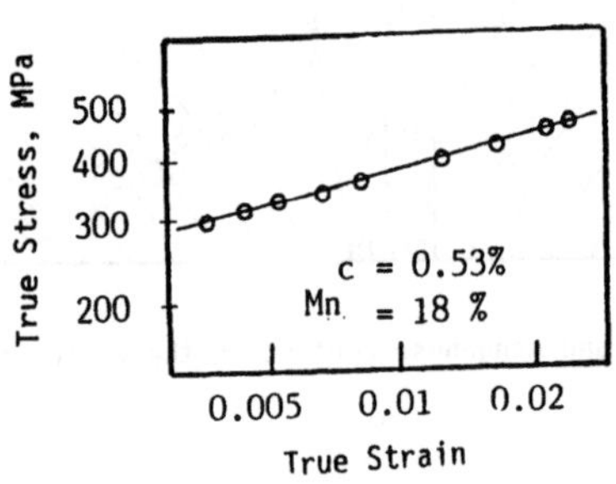

Figure 7 : Straight lines obtained on plotting true stress vs. true strain (log-log scale)

ponding alloys.

Fig. (7) shows the log-log scale plot of true stress and true strain for some investigated alloys. Relatively good straight lines were obtained.

Thus, the results of the present work show that there is marked correlation between the results of wear test and compression test. Greater wear resistance is obtained in alloys having higher yield strength, lower strain hardening coefficient, n, i.e. lower ductility.

Probably one of the mechanisms leading to wear in the conditions of the present work is surface fatigue. For the same stresses acting during wear process greater plastic strains will be produced in steels with lower yield stress and greater ductility which may lead to more marked low-cycle high strain fatigue and fracture of surface layers.

When values of the yield stress,strain hardening exponent and strength coefficient were plotted against weight loss. Bands were obtained which were always narrower at greater values of yield stress and at lower values of n and K. The bands were also narrower when they were plotted for steels with the same range of carbon content and different manganese content.

In order to check the accuracy of using factorial analysis and wide extrapolation used in this work three additional alloys were prepared (Table 2) and their wear resistance was measured by the same procedure as for other specimens. Wear resistance alloys No. 1 in Table (2) and No. 4 in Table (1) was taken as unity and the weight loss of these specimens was divided by the weight loss of other specimens and the ratio was taken as relative wear resistance. The relative wear resistance of the check alloys is given in Table (4).

TABLE 4

Relative wear resistance of alloys given in Table 2.

Alloy No.	Composition		Relative Wear Resistance		Error %
	C %	Mn %	Experimental	Calculated	
1	0.98	12.9	1	1.03	+3
2	0.99	7.44	1.08	1.17	+8
3	0.86	6.18	1.18	0.97	-18

The maximum error of -18% was obtained for alloy No. 3 with predicted relative wear resistance lower than the experimental one. Relatively good prediction was observed for alloy No. 23 with an error of only 8%. Since the composition of alloy No. 2 and especially No. 3 (Table 2) lie far from the experimental alloys (Table 1) used in the calculation of the regression equation, it is expected that more accurate results can be obtained for alloys lying closer to the experimental composition given in Table (1). It also seems that the effect of lowering manganese content from 12 to 6% at carbon level of 0.9-1.0% on the wear resistance is stronger than predicted by the obtained regression equations.

CONCLUSIONS

1. The obtained four charts showing the predicted effect of carbon and manganese contents on the wear resistance, yield stress, $\sigma_{0.2}$, strain hardening exponent, n, and strength coefficient, K, are in general agreement with known industrial practice, published experimetanl results and theories of austenite stability in austenitic manganese steels. These charts can be used for comparing the properties of many steels containing from 6 to 24 Mn% and from 0.39 to about 1% C.

2. The lowest weight loss (highest wear resistance) coincides with the greatest value of the yield stress and the lowest values of strain hardening coefficient and the strength coefficient (K) obtained from compression test.

3. The tendency of the effect of carbon on the investigated properties depends on the Mn content and vice-versa. Wear resistance and yield stress tend to increase, and strain hardening exponent tends to decrease, in alloys with known strong tendency for forming thin layers of stress induced α or ε martensite.

4. Alloys having high wear resistance and yield stress as well as low strain hardening exponent are :

   a) Alloy with 6 to 8% Mn and about 0.95% C known for their tendency of forming stress-induced α-martensite.

   b) Alloys with 24% Mn and about 0.4% C known for their tendency for forming stress-induced ε-martensite.

REFERENCES

1. Gouin, C.E., Manganese Steels that Don't Require Heat Treatment, Climax Molybdenum Company, Colorado, U.S.A., 1976.

2. Cleemann, J.O., Grinding Media in Tumbling Mills, Zement-Kalk-Gips-Nr. 7, 1987.

3. Schmidt, I. and Wilhelm, M., Friction-Induced Martensitic Transformation of Hot-Rolled Austenitic Mn-Steels, Zmetakle Bd. 78, 1987, H. 2, pp. 129-130.

4. Albright, D.L. and Dunn, D.J., Wear Behavior of Iron and Steel Castings for the Mining Industry, Journal of Metals, November 1983.

5. Miller, A.E., Wear in Steels, Journal of Metals, October 1983.

6. Herman, A., Fabert, Jr., Materials for Wear Resistance, Metals Engineering Quarterly, November 1975.

7. Lekhachein and Shewman, P.G., Effect of Hardness on the Solid Particle Erosion Mechanisms In AISI 1060 Steel, Wear, 89, 1983, pp. 291-302.

8. Kashcheev, V.N., Some Aspects of Improving Abrasive Resistance, Wear 89, 1983, pp. 265-272.

9. Vermeulen, L.A., Howat, D.D. and Gough, C.L.M., Theories of Ball Wear and the Results of a Marked Ball Test in Ball Milling, Journal of the South Africa Institute of Mining and Metallurgy, August 1983.

10. Pearce, J.T.H., The Use of Transmission on Electron Microscopy to Study the Effect of Abrasive Wear on the Matrix Structures of a High Chromium Cast Iron, Wear 89, 1983, pp. 333-344.

11. Votenova, T.F., High Manganese Steels and Alloys, Moscow, Metallurgia, 1988.

12. Aly, N.A. and El-Koussy, M.R., Effect of Microstructure on Abrasive Wear of Some Cast Irons and Steels, Second International Conference on Mining and Metallurgical Engineering, Suez Canal University, March 20-22, 1989.

13. Handbook der Souderstalkunde Eduard Houdremont, Springes-Verlag, 1956, pp. 425 (Russian Ed.).

The Use of Shot Peening as a Surface Preparation Treatment for the Electrodeposited Chromium Plating

J. Lu, J. Reby, J. F. Flavenot, G. Vasseur and B. Sutter

CETIM·
Senlis
France

ABSTRACT

The present paper is intended to show the residual stresses obtained with different chromium plating processes . A new measurement method is used to evaluate the residual stresses distribution . The effect of a shot peening pretreatment and of a degassing post treatment were studied . It can be seen that using a combination of the different processes , it is possible to optimize the residual stresses profiles which are very important for the fatigue and corrosion resistance of the coating layer and the chromium electroplating material .

INTRODUCTION

The use of protective coatings in industrial applications is rapidly increasing in many new areas. The adhesive properties of the coating are of prime importance since they limit the coatings useful life. One important factor which contributes to coating failure is the residual stress produced during its manufacture.

The sources of stress-induced failure can be divided into two factors. Firstly, residual stresses present in the coating, and secondly, mechanical stresses generated during the service of the component. Although residual stress in coatings and substrates is often considered, very little written material exists concerning its actual measurement.

The use of hard chromium plating as an eletrodeposited surface layer on other metals is now a well established industrial practrise. The chromium layers are deposited in order to take advantage of the high hardness, corrosion resistance and the wear resistance or the low coefficient of friction of such deposits. But in most cases, the electrodeposits of chromium are often extensively cracked. Their formation is due to the residual stress in the eletrodeposit as it is formed [1]-[5].

The present study shows the modification of the the residual stresses distribution and microstructure in hard chromium plating by shot peening. The shot peening is used as a finishing treatment before the chromium plating. The residual stress distribution obtained for different processes will be shown. Two techniques will be studied: conventional electrodeposition and eletrodeposition by pulsed current. The results of residual stresses concerning the plating layer with and without shot peening will be analysed.

PROCESS OF THE ELETROPLATED CHROMIUM

In our study, the chromium was deposited from aqueous solution of chromic acid. The following bath composition was used : 250 g/l of Cr_2O_3 , 2.5 g/l of H_2SO_4. The bath temperature was about 50 °C ± 2 °C.

One characteristic of the deposit made from this bath, by direct current plating (30 to 40 A/dm^3) is its tendency to spontaneous cracking. Using a pulse reversed current process developed by CETIM [6-8], crack free chromium deposits can be produced . The currrent shape shown in fig.1 is described by the following parameters :

T cathodic = Duration of the cathodic impulsion
T anodic = Duration of the anodic impulsion
I an = Anodic current density
I catho = Cathodic current density

In our work, the following plating procedure was used : Surface preparation (grinding or shot peening), Precleaning, Electrolytic cleaning, Sulfuric acid etching, Chromium plating, With or without degassing.

The deposits produced were grey, mat in appearance, but readly polished. Surface of DC and pulsed reverse current deposits are shown in fig.2.

The resistance of corrosion of such crack free deposit shows high protective value as compared with conventional coating according to table 1 (chromium plating on steel 835 M30).

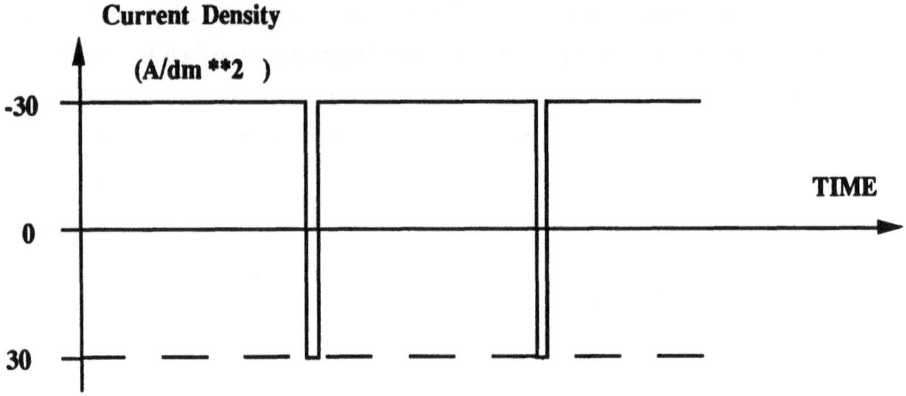

Figure 1. Chromium plating by pulsed reverse current. Electrolysis current

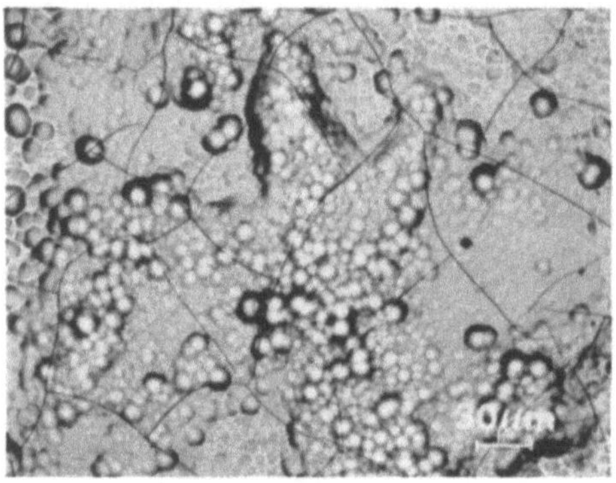

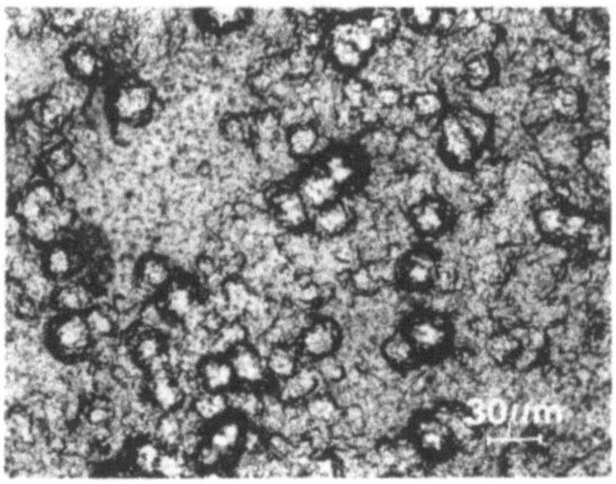

(a) (b)

Figure 2 : Microstructure of Cr coating surfaces, (a) By conventional method (DC) : $I=30A/dm^2$, (b) By pulsed reverse current

Electrolysis Current	Thickness of deposit	Rugosity CLA	Salt spray * resistance
DC	15 µm	0.14	12 h
Pulsed reverse current	15 µm	0.14	300 h

Table 1 * Below the ASTM's standard ASTM B117-85

SAMPLE PREPARATION CONDITIONS

In the previous paragraph, we have described the process of the electrodeposited chromium coatings. The advantage of the pulsed current method is the possibility to minimize the microcrack formation. But in [9], it is shown that the fatigue strength is not higher for the case of the coating obtained by pulsed current method. The [10] is concerned with the development and understanding of shot peening as controlled process, applied prior to hard chromium and nickel deposition. The main effect of shot peening is to introduce the compressive residual stresses to stop the propagation of the crack initiated in the coating layer. Up now, very few litterature have shown a systematic study of residual stresses in different combination of coating system . Table 2 shows the different case studied in this work. The substrate is a 30CND 8 grade steel and the chemical composition is shown in table 3.

Surface Preparation condition \ Chromium process	Conventional chomium plating	Pulsed current chromium plating
Without shot peening	CASE 1	CASE 2
with a pretreatment of shot peening	CASE 3	CASE 4
with a pretreatment of shot peening and a post treatment of degassing		CASE 5

Table 2 Sample preparation conditions

C	Mn	Si	Ni,max	Cr	Mo	P, max	S, max
0.3	0.34	0.35	2.4	2.15	0.7	0.03	0.025

Table 3 Chemical composition of 30CND8 grade steel

Tables 4 and 5 show respectively the shot peening conditions used and the eletrodeposited conditions.

Shot diameter (mm)	Almen Intensity	Coverage
0.6	0.2-0.25mm A	125 %

Table 4 Shot peening pretreatment condition

T cathodic	T anodic	Degassing treatment
2s	10 ms	200 °C during 3H

Table 5 Electrodeposited condition

RESIDUAL STRESSES MEASUREMENT METHOD

Concerning the measurement techniques, different methods can be used. The reference [2] shows a large review of these techniques applied to chromium platings layers. One of these methods cited is the use of a thin metal strip plated on one side only, in order to determine the residual stresses in that plate by the measurement of deflection it caused to the strip. He has also mentionned the possibility to use X-ray diffraction method. In a study carried out in CETIM [9], it was shown that it is possible to evaluate the residual stresses by X-ray diffraction method. But the microcracked structure of the chromium plating induce some errors of calculation of stress. Another difficult is the time-consuming task for the layer removal In this study, the measurement technique used is the incremental hole drilling method. The principle of this method is largely developed in following references [11] -[13].

RESULTS AND DISCUSSIONS

Fig.3 shows the residual stresses obtained in a layer manufactured by a chemical chromium plating. In the coating layer, the tensile residual stresses are observed up to a depth of 200 µm , compressive residual stresses are obtained from this depth. It can be seen that the level of the residual stresses measured are very important. The fluctuation of the level of residual stresses can be explained as following : in fact, the level of residual stresses depends on the number of cracks in the chromium coating. The coating manufactured by conventional chromium plating is microcracking. When the hole depth corresponds to a high cracks density zone, the level of the stresses is smaller (fig.4).

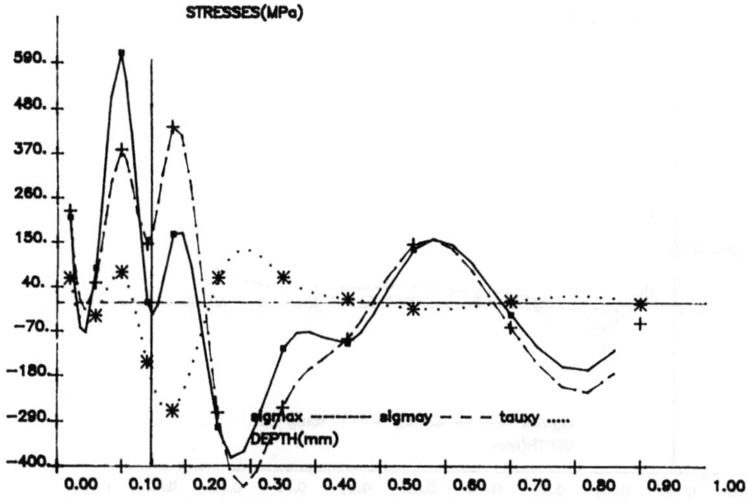

Figure 3 Residual stresses measured in a plating obtained by conventional method.

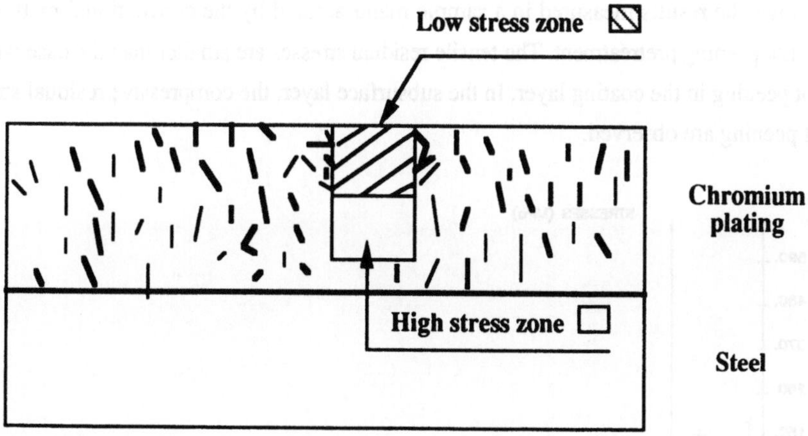

Figure 4 Mechanism of the oscillation of the residual stresses through a hard chromium plating.

The [2] has shown a very instructive illustration of the relationship between stress produced and the cracking that occurs in chromium coating, as its thickness increases .

Fig.5 shows the residual stresses measured in a sample manufactured by the pulsed current process. It can be seen that the level of residual stresses is smaller than a conventional plating. As described previously, this process induce a cracking free coating. So the corrosion resistance is more important for this kind of coating.

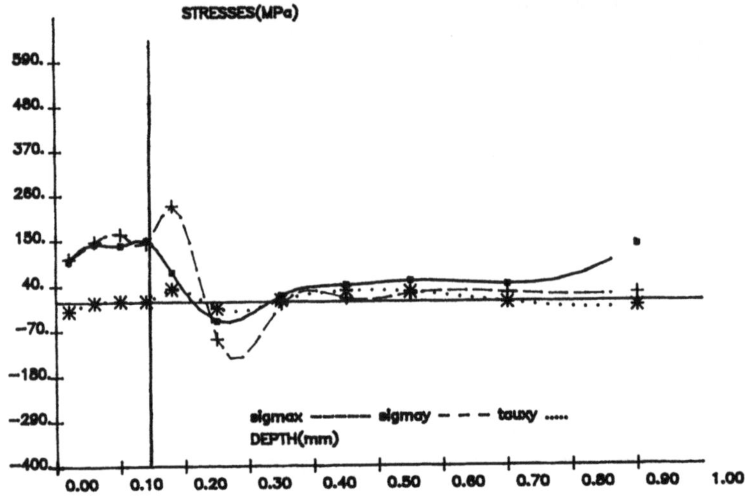

Figure 5 Residual stresses obtained on an sample manufactured by pulsed current method.

Fig.6 shows the results measured in a sample manufactured by the conventional method but with a shot peening pretreatment. The tensile residual stresses are smaller than the case without the shot peening in the coating layer. In the subsurface layer, the compressive residual stresses of shot peening are observed.

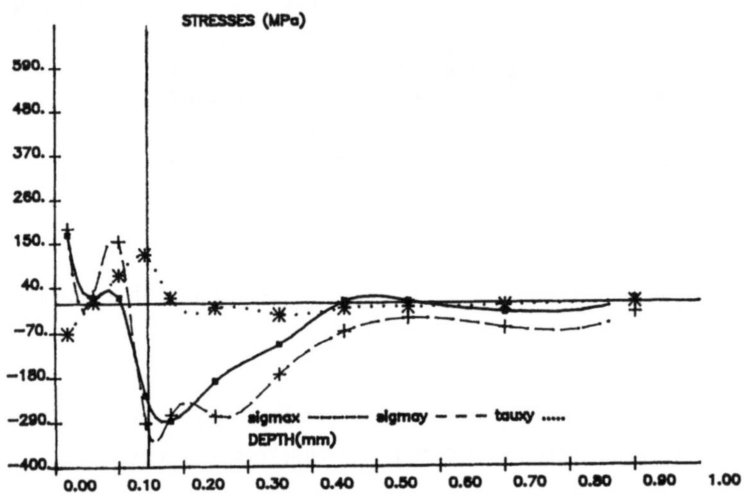

Figure 6 Residual stresses distribution measured on a sample obtained by conventional process with a shot peening pretreatment.

The fig.7 shows the residual stresses measured on a sample manufactured by the pulsed current process with a shot peening pretreatment. On the surface, the tensile residual stresses have the same level as compared with the case without the shot peening. On the subsurface, a compressive residual stresses is observed. So the residual stresses of shot peening are not relaxed during the chromium electroplating.

Fig.8 shows the residual stresses obtained on a sample manufactured with the pulsed current process using a shot peening pretreatment and a degassing post treatment. It can be seen that this treatment decreases the surface tensile treatment .

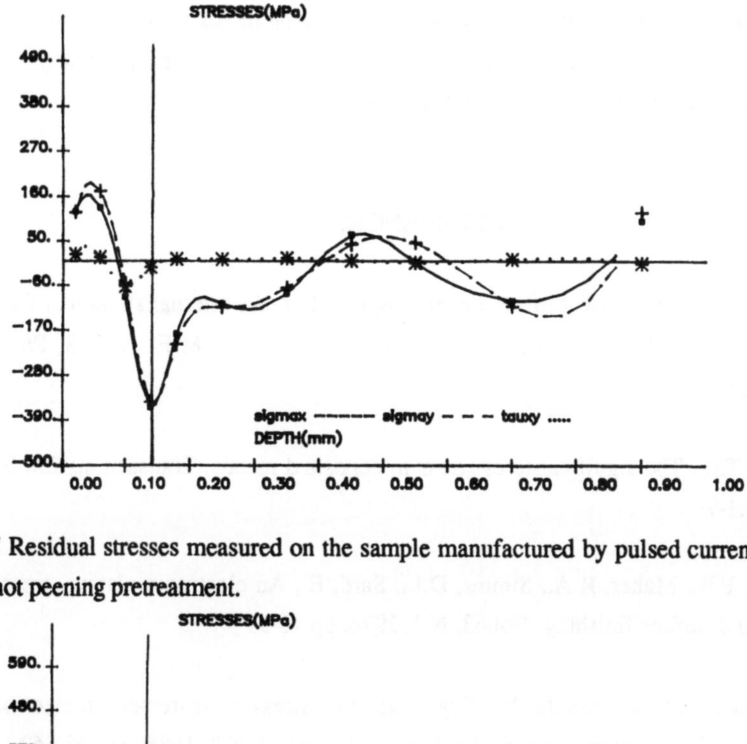

Figure 7 Residual stresses measured on the sample manufactured by pulsed current process with a shot peening pretreatment.

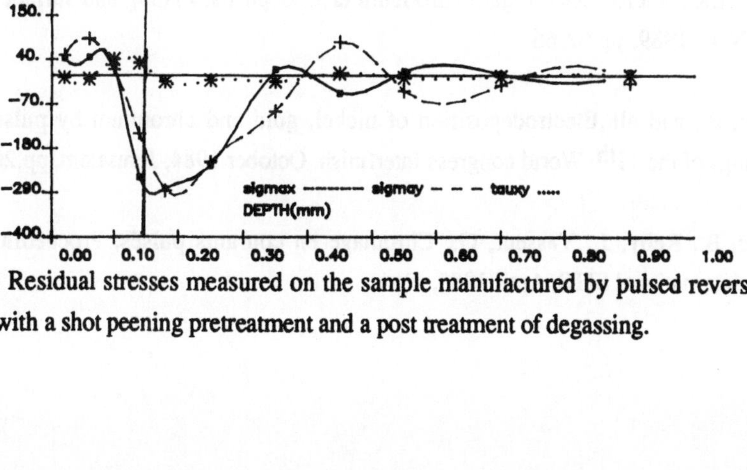

Figure 8 Residual stresses measured on the sample manufactured by pulsed reverse current method with a shot peening pretreatment and a post treatment of degassing.

CONCLUSION

This study shows that it is possible to optimize the residual stresses in hard chromium electroplating. The shot peening process is an efficient pretreatment to introduce the compressive residual stresses in substrate. The chromium plating does not modify these stresses. The pulsed current process introduce a smaller and more regular tensile residual stresses in the plating layer. The degassing is also a post treatment which is benefical for the residual stresses distribution in the plating layer. Moreover further, the fatigue test must be carried out to evalute the effect of shot peening pretreatment in the case of pulsed current process. The previous study [9] shows that the corrosion resistance is better for this case but the fatigue test shows the fatigue resistance were not increased using a simple pulsed current process in comparison with the conventional method.

REFERENCES

1. Dennis, J.K., The effect of the underlying metal on the internal stress and formation of microcracks in electrodeposited chromium, Transactions I.M.F. ,Vol.43, Part 2, 1965, pp.84-96.

2. Such, T.E., The significance of stress in deposited metals, Transactions I.M.F., Vol.66, 1988, pp.1-7

3. Koch, F.B., Maher, R.A., Simms, D.L., Sard, R., An electronic plating stress monitor, Plating and surface finishing, Vol.63, N.1, 1976, pp.46-51

4. Kuramoto, M., Kobayashi, Y., Nagasawa, J.I., Stress measurement in electrodeposited chromium, Nondestructive control (In Japenese) , Vol.37, N.8, 1988, pp.636-642

5. Jones, A.R., Microcracks in hard chromium eletrodeposits, Plating and surfacefinishing, Vol.76, N.4, 1989, pp.62-66

6. Sutter, B., and all.,Electrodeposition of nickel, gold and chromium by pulse plating, Proceedings of the 11th World congress Interfinish, October 1984, Jerusalem, pp.201-206

7. Sutter, B., Reby, J., Vasseur, G., Chromage en courants pulsés, Proceedings of the Congres International SITS, April 1985, Paris

8. Sutter, B., Reby, J., Vasseur, G., Chromage dur en courant periodiquement inversé, Proceedings of the 12th World Congress Interfinish, October 1988, Paris, pp.369-377

9. Sutter, B., Reby, J., Vasseur, G., Essais industriels chromage campagne 1987 , Internal Report CETIM , 1990

10. Crocker, G.S., Shot peening for the metal finishing.

11. Niku-Lari, A., Lu, J., Flavenot, J.F., Measurement of the residual stresses distribution by the incremental hole-drilling method, Experimental Mechanics, June 1985, pp.175-185

12. Lu, J., Flavenot, J.F., Measurement of the residual stress distribution in the multilayer materials by the incremental hole drilling method, Proceedings of 2nd international conference of residual stress, Edited by Back and All., Elsevier Applied Science, Nov. 1988, pp.103-109

13. Lu, J., Flavenot, J.F., The effect of the finishing treatments on the residual stresses distribution in advanced coating materials, Proceedings of the 1989 SEM Conference on Experimental Mechanics, June 1989, pp.927-934

Analytical Solutions of Crack Developing at Rail Joint Due to Impacting

Fan Jashen

Yunnan Institute of Technology
Kunming People's Republic of China
(visiting scholar at the US National Center for Earthquake Engineering Research, State University of New York at Buffalo)

and Chong-Shien Tsai

State University of New York at Buffalo
Buffalo New York USA

Abstract

Applying the Hertz formula for impacting of two elastic spheres and considering the rail to be circular with infinite radius, we obtain the formula to calculate the impact force on the joint of the rail exerted by the wheels of the locomotive. Considering the joint portion of the rail to be cantilever with holes, we obtain the formula for calculating the stress conentration around the bolt holes at the joint.

1. Introduction

The joint of rail is the weak portion of railway. While the train passes over it, the wheels of the locomative yield huge impact force on it and cause serious damages especially on the severe winter, according to a statistic figure of China, in the coldest region at the northeast area of China, in December, January and February, the damaged rails in these three month reach 67% of the total amount of a year, especially such damaged rails as shown in Fig. 1 are about 30 – 50% of the damages occurred at the severe winters. This kind of damage always occur at the rail end and abreast the coming locomotive.

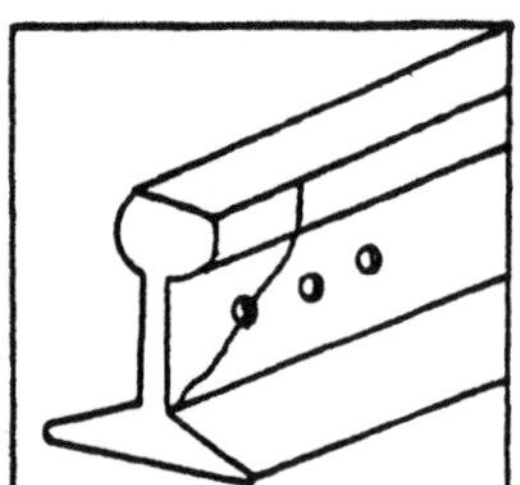

Fig. 1

The thermal gap between two rail end has its maximum value 21.6mm in Chinese regulation which is the main source to cause huge impact force by the wheels of locomotive at the joints of the rails.

2. The calculation of the impact force

According to the Hertz impact formula between two elastic sphere, the maximum impact force is [1] :

$$P_{max} = n^{\frac{2}{5}}\left(\frac{5}{4}\frac{V_0^2}{\mu}\right)^{\frac{3}{5}} \tag{1}$$

Here in Eq. (1), we have

$$\left\{\begin{array}{ll} \mu = \frac{m_1+m_2}{m_1 m_2}, & n = \frac{4}{3\pi(\theta_1+\theta_2)}\sqrt{\frac{R_1R_2}{R_1+R_2}} \\ \theta_i = \frac{\lambda_i+2\xi_i}{4\pi\xi_i(\lambda_i+\xi_i)}, & \lambda_i = \frac{2\xi_i\nu_i}{1-2\nu_i},\ (i=1,2) \end{array}\right\} \tag{2}$$

Here in Eqs. (1) and (2), m_1 and m_2 are the masses of these two spheres respectively, V_0 is the relative impact velocity at the initial instant of impacting, R_1 and R_2 are the radii of these two spheres respectively , ξ_i and ν_i are the shear modulus and Poisson's rario of the sphere respectively. Take the rail to be the second sphere with infinite radius, now only the radius of the wheel R_1 appears in the following formulae, we take R_1 to be R without ambiguity, thus we have:

$$n = \lim_{R_2\to\infty}\frac{4}{3\pi(\theta_1+\theta_2)}\sqrt{\frac{R_1R_2/R_2}{(R_1+R_2)/R_2}} = \frac{4}{3\pi(\theta_1+\theta_2)}\sqrt{R_1} = \frac{4}{3\pi(\theta_1+\theta_2)}\sqrt{R} \tag{3}$$

The wheel and rail are made of steel, we have $\nu_1 = \nu_2 = \frac{1}{3}$, $\xi_1 = \xi_2 = 7.938 \times 10^{10}(N/m^2)$, after calculation we obtain $\theta_1 = \theta_2 = (3\pi \times 7.938 \times 10^{10})^{-1}(m^2/N)$.

As seen from Fig. 2, if the velocity of the train is V, then the impacting veloeity V_0 of the wheel to the rails joint would be:

$$V_0 = \frac{V}{R}\sqrt{\delta^2 + (\Delta y)^2}$$

Since the ralative vertical displacment Δy between two adjecent rails ends is smaller than the gap $\delta(\Delta y \ll \delta)$, so we have

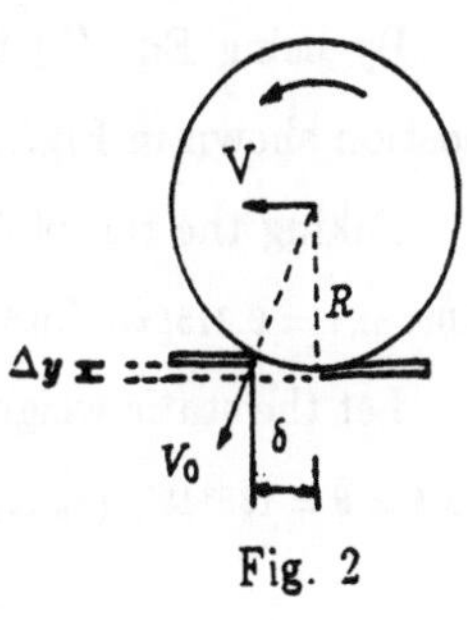

Fig. 2

$$V_0 = \frac{V}{R}\delta \tag{4}$$

Substituting Eqs. (2),(3) and (4) into Eq.(1), we obtain:

$$P_{max} = \frac{3340}{R}(\frac{V\delta}{\sqrt{\mu}})^{\frac{4}{5}} \quad (5)$$

Eq.(5) is the formula to calculate the maximum impact force, the units must be, time-sec. force-N and length-m.

The following example gives numerical results. V= 22.22 m/sec, R=0.875m, $m_1 = 2635kg$ -mass of the locomotive under the damping spring, $m_2 = 99.27kg$ - mass of the rail joining the impact, $\delta = 0.006m$. Using these values, we obtain $P_{max} = 159936N$. We will give an explanation for the portions joining impact of the locomotive and the rail in the last section.

3. Stress Concentration around the Screw Hole at the Joint

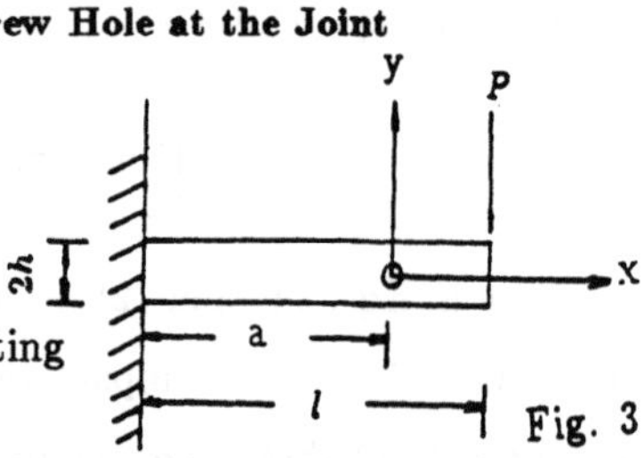

Fig. 3

In many countries, the joint of two rails are connection of two cantilever beams, thus we consider it as shown in Fig. 3 to be the calculating diagram. Where l is the length from the rail end to the nearest tie center, a is the length of the hole nearest the rail end to the center of the nearest tie, 2h is the height of the rail. If σ_θ is the hoop tensile stress of the hole, r is the radius of the hole and J is the moment inertia of the rail cross section, we have[2] :

$$\sigma_\theta = \frac{P_{max}(l-a)r}{J}(\sin\theta - \sin 3\theta) - \frac{P_{max}}{J}\left[(2h^2 - r^2)\sin 2\theta + r^2 \sin 4\theta\right] \quad (6)$$

From $\frac{\partial\sigma_\theta}{\partial\theta} = 0$, we obtain:

$$(l-a)r(\cos\theta - 3\cos 3\theta) + (2h^2 - r^2)cos2\theta + 4r^2\cos 4\theta = 0 \quad (7)$$

By using Eq. (7) to find the loction of $(\sigma_\theta)_{max}$, we see that it is just the crack postion shown in Fig. 1

Taking the rail of 50kg/m as an example: $J = 1.489 \times 10^{-5}(m^4)$, $2h = 0.14m, L - a = 0.056m, r = 0.0155m$. Substituting these values into Eq.(7), we get $\bar{\theta} = 139°15'$

Let the static weight of a wheel be $102949N$, then $P_0 = 102949 + 159936 = 2.63 \times 10^5 N$, as $\theta = \bar{\theta} = 139°15'$, $(\sigma_\theta)_{max} = 1.7032 \times 10^8 N/m^2 > [\sigma] = 1.372 \times 10^8 N/m^2$ - allowable stress.

3. Conclusions

The Hertz formula for two elastic spheres impact is derived from the second law of Newton $F = m\frac{d^2s}{dt^2} = m\frac{dv}{dt}$, so it is not the static contact formula. Also from this Hertz's theory, the required time t to complete the impact process is

$$t = 2.94(\frac{5}{4}\frac{1}{n\mu})^{\frac{2}{5}}V_0^{-\frac{1}{5}} = 9.79(\frac{m_1 m_2}{m_1 + m_2})^{\frac{2}{5}}(V\delta)^{-\frac{1}{5}} \tag{8}$$

The required time t^* for the wheel to go over the gap between two rails ends is:

$$t^* = \frac{\delta}{V} \tag{9}$$

Substituting the above datas into Eqs. (8) and (9), we obtain $t = 1.16\times10^{-4}(sec)$, $t^* = 7.2 \times 10^{-4}(sec)$, $t > t^*$,

The maximum impact force occurs before the wheel has passed the gap, therefore Eq. (5) is correct.

(2) The propagation velocity of the stress wave is $5.8 \times 10^3(m/sec)$ in the rail. The value $t^* = 7.2 \times 10^{-4}$ at this time, the stress wave has propagated a distance of 4.176 (m) in the rail, which is approximately the effective length of rail in impact 3.5(m).

(3) The more effective method to decrease the damages of rails is to decrease the effective length of the rail during the impact. In order to prvent crawling of rail, decreasing the length of the gap between two rails ends is essential. It is noted that $P_{max} \propto m^{\frac{3}{5}}$ and $P_{max} \propto \delta^{\frac{6}{5}}$. If we have $\delta = 1.6cm$ together with crawling distance 0.5cm, then rail damage as shown in Fig. 1 would occur.

(4) According to Ref.[5], taking the rail a beam of uniform cross section with infinite length either supported by many elastic points or on continuous elastic foundation of Winkler model. Calculated results show that the relative error between these two assumptions is $5 - 7\%$ and the later assumption is used much more. Under this more used assumption, when a vertical concentrated load P exterts at the end of a rail in the static case, the governing equation of rail is :

$$\frac{d^4y}{dx^4} + 4k^4y = 0 \quad (k = (\frac{u}{EJ})^{\frac{1}{4}}) \tag{a}$$

Where x is the abscissa measured from the end of the rail, u is the coefficient of the foundation, EJ is the flexural rigidity of the rail and k is the rigidity ratio between the rail and the foundation, k takes the value $(9 \sim 18) \times 10^{-5}(m^{-1})$.

The solution of Eq. (a) is

$$y = \frac{pk}{2u} e^{-kx}(coskx + sinkx)$$

Thus we have the following conclusion: When x is equal to $\frac{3\pi}{2k}$ or $\frac{2\pi}{k}$, the values of bending moment, shear force and deflection are so small which can be neglected for the part of the rail far from this distance ($3 \sim 3.5m$).

In the dynamic case, the main stress wave in the rail is bending wave which has the same velocity as longitudinal wave[3]. When the wheel just pass the gap, this wave propagates a distance of $4.176m$ which is near the length of rail to be affected by a concentrated vertical load P.

The portion of locomotive joining this impact can be considered to be the mass under the damping springs, because the dynamic effect above the springs has been absorbed by damping springs, the vertical acceleration of this portion is very small compared with the portion under the springs. This assumption was adopted by Ref.[4] and we also adopt it in this work.

Finally, the Hertz Formulae for impact of two spheres are really plane problem of impact, thus whether these two impacting bodies are two spheres, two cylinders or one sphere and one cylinder, the impact takes place at the plane through the connecting line of the centers of these two bodies and perpendicular to the ground, the projection of a sphere or a cylinder on this plane is a disc without difference, attention should be paid to correct calculation of their volumes and masses, there is no reason to blame the application of Hertz formulae in this paper.

Acknowledgement

This work has be encouraged and supported by Professor Robert L. Ketter, Director of the U.S. National Center for Earthquake Engineering Research.

References

1. Chien Wei-Zang, et al, Theory of Elasticity (in Chinese). Scientific Publishing House, 327-329, (1956)
2. G. N Saven, Stress Concentration around Holes (Chinese Translation from Rus-

sian) Scientific Publishing House, 93,(1958)

3. J. Engelbrecht, Nonlinear Wave Processes of Deformation in Solids, Pitman Advanced Publishing Program, 97, (1983).
4. Zhou Hong Nie, et al., Study on Calculation of Strength of Rail Joint, Proceedings of the 1963 Railway Scientific Symposia(in Chinese), Vol. 15, the People's Railway Publishing House, 15-20, (1964).
5. Compiled by the Structure & Buiness of Railway (in Chinese), 259-261, the People's Railway Publishing House, (1962).

An Investigation into Surface Generation during Machining High Strength Low Alloy Steel

A. Basu, T. Chandra and R. Rudziejewski

University of Wollongong
Wollongong
New South Wales
Australia

ABSTRACT

This paper deals with the study of surface generation of a Ti-bearing HSLA steel used extensively in off-shore platforms and transportation industries. Experiments were carried out to study the variation of surface roughness with different cutting conditions and a statistical computer package program known as SAS was used to determine mathematical empirical relations between surface roughness, speed, feed and depth of cut. The chips obtained during the cutting operations were collected and the variations of chip forms with cutting conditions were also observed. During tool wear studies, variation of surface roughness with tool wear and cutting time was investigated and the machined surfaces consisted of varieties of surface damages which degrade with tool cutting time.

INTRODUCTION

High Strength Low Alloy (HSLA) Steels are relatively new types of modern engineering materials in which strength and toughness are greatly increased by addition of microalloying elements such as chromium, vanadium, niobium, titanium or molybdenum. The use of these microalloying elements and sophisticated controlled rolling techniques have provided the opportunity to obtain the desired properties.

The development of these steels is very extensive and they are widely used in different fields such as pipe-line application, heavy industrial machinery, ship building, off-shore steel structures, etc. Four common characteristics of structural steels are versatility, economy, strength and endurance. The HSLA steels have higher values of yield strength, ultimate tensile strength, greater resistance to atmospheric corrosion and lower impact transition temperatures. Because of its lighter weight structure HSLA steels are classified among the most important and metallurgically interesting modern engineering materials. From this group of HSLA steels, one of which has been strengthened with titanium will be tested for its machinability which depends on material properties and machining conditions. Microstructures of steels containing a soft matrix with hard second phases, such as ferritic-pearlitic steels, shall have better machinability [1-4] than a single phase steels with soft matrix. The present HSLA steel consists of single phase fine ferrite structure with very high UTS (720 MPa) and 480 MPa yield strength. The steel was air-colled below Ar_1 temperature after it had finished rolling in order to achieve these mechanical properties.

This paper highlights the influence of cutting speed, feed, depth and machining tool time on surface generation (mainly surface roughness) in conjunction with chip form study. Predicting machining parameters which will satisfy roughness and chip breakability is a main goal of this work. To achieve this aim, this work is carried out in three parts:

1) Experimental turning operations on a CNC lathe machine in which two different types of indexable cutting inserts are used in two categories of operations viz. light roughing, semi-finishing and finishing operations.

2) Experimental Data obtained are collected and used in a computer program called Statistical Analysis System (SAS) to find mathematical relations between different variables: surface roughness, speed, feed and depth of cut.

3) Machined surfaces are examined under a Scanning Electron Microscope and photomicrographs of rough surfaces are taken to find surface damage which degrades with tool cutting time.

EXPERIMENTAL PROCEDURES

To conduct experimental works the following conditions are used:

Cutting conditions

Semi-finishing
and light roughing: speed V (m/min) = 75, 100, 140, 180, 220, 260, 300, 400, 500

	feed f (mm/rev)	= 0.05, 0.02, 0.4
	depth of cut d (mm)	= 0.5, 1, 2, 4, 6
Finishing:	speed V (m/min)	= 75, 150, 200, 300, 400, 500, 700
	feed f (mm/rev)	= 0.02, 0.05, 0.08, 0.11, 0.14
	depth of cut d (mm)	= 0.1, 0.3, 0.5

Work materials

HSLA steel containing 0.01% Ti, 0.08% C, 1.5% Mn and 0.009% N.

Indexable inserts used

Coromant's carbide inserts are used:

1) CNMG 120408 QM GC 415 (for semi-finishing and light roughing).
2) TNMG 160408 QF GC 415 (for finishing)

with a triple coating of Ti (C_1 N), $AL_2 O_3$ and TiN, the total thickness is approximately 9 mm.

Experiment

Experiments are repeated three times with measurements of surface roughness for every experimental point at three places along the circumference.

Equipment

1) HITEC-TURN 20S11 CNC Lathe machine
2) SURFCOM 550 AD Surface Roughness Measuring Instrument
3) Scanning Electron Microscope was employed to study the surface topography of machined surfaces generated at different speeds and cuts.

RESULTS AND DISCUSSION

Experimentally determined Ra values for combinations of cutting speed, feed and depth in semi-finishing and light roughing are shown in Figure 1. Figure 2 shows some results of chip forms and chip breakability.

The surface roughness values for lower depth vary significantly with speed. However the effect of feed is more clear. The best results have been obtained with feed 0.05 mm/rev but inspection of the chip form (Figure 2) shows that it does not give broken chips in the range d ≈ 0.5mm (not shown in Figure 2) to 6mm. Acceptable chip breakability appears with d ≈ 1 - 4 mm and f ≈ 0.2 - 0.4 mm/rev in wide range of speeds including semi-finishing as a final operation. Breakability of chips with higher depth should be solved by further investigation using different chip breakers.

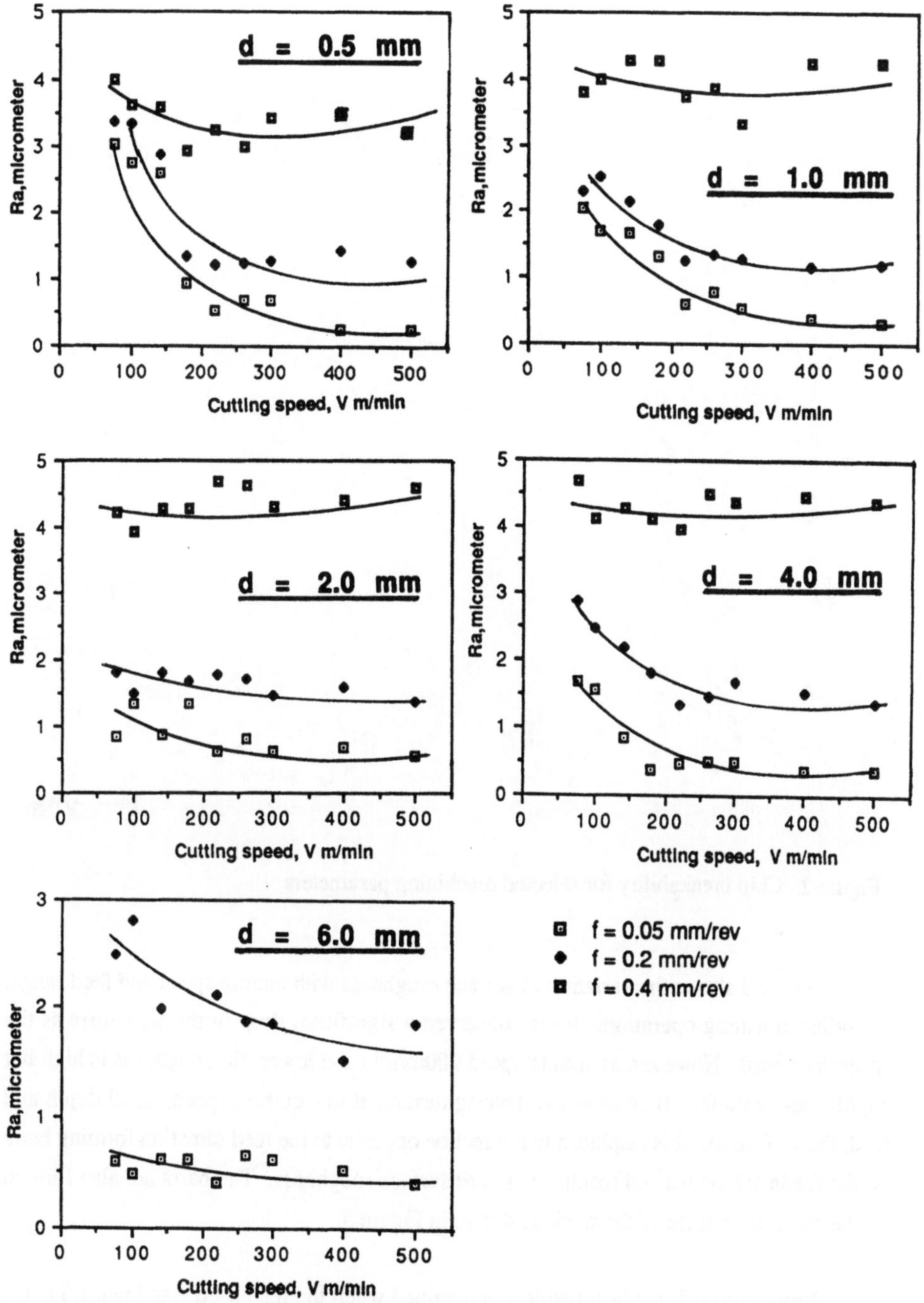

Figure 1. Influence of cutting speed on machined surface roughness in semi-finishing and light roughing operation.

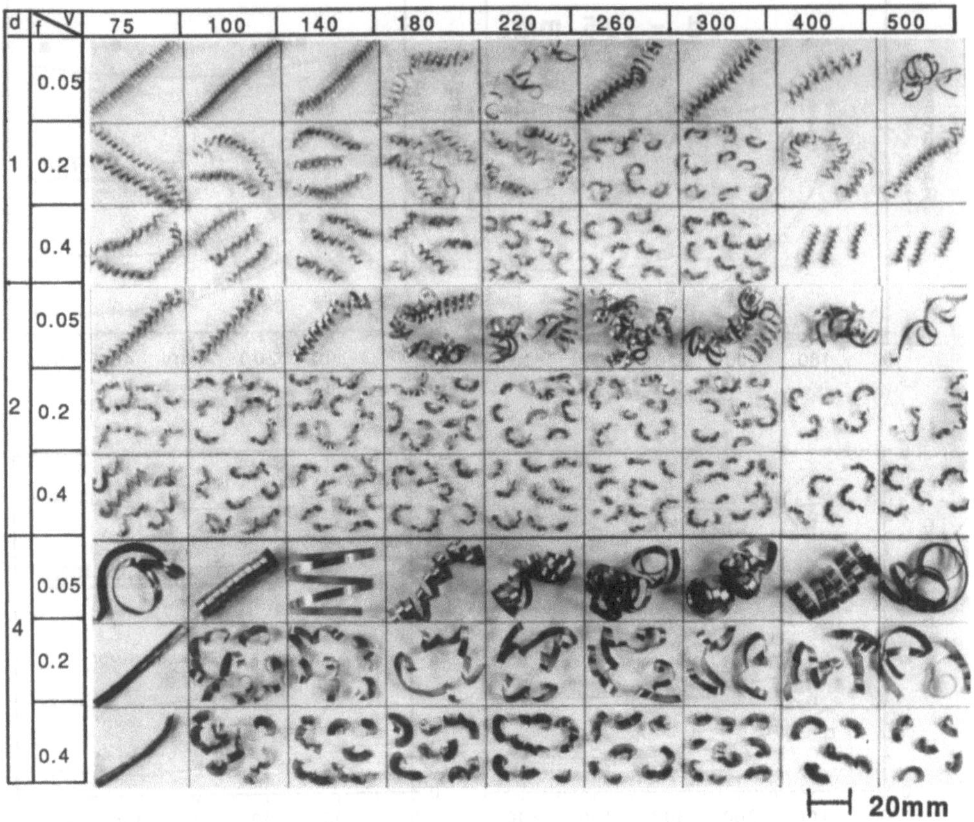

Figure 2. Chip breakability for selected machining parameters

Figure 3 shows the variation of surface roughness with cutting speed and feed range, including finishing operation. It was observed a significant drop in the R_a values as the speed increased. However, at cutting speed 200m/min and lower, the roughness is high but slightly less with feed 0.14 mm/rev. During turning at low cutting speed, small depth and feed, the surface metal is displaced in a direction opposite to the feed direction forming barrs on the feedmark ridges and resulting in poor surface roughness. The barrs are also formed on the back flat surface of the work as shown in Figure 5.

From Figure 3, the best result was obtained when the feed used was lowest, i.e. $f \approx 0.02$ mm/rev, where a R_a value of 0.18 µm was recorded at a cutting speed of 500 m/min, when $d \approx 0.1$ mm. The use of higher speed does not give better results (for V = 700 m/min, f = 0.02 m/min, d= 0.1 mm, R_a = 0.28 µm) as shown in Figure 3. The higher values of d and f are also acceptable in finishing operation if chips do not procure practical problems in machining.

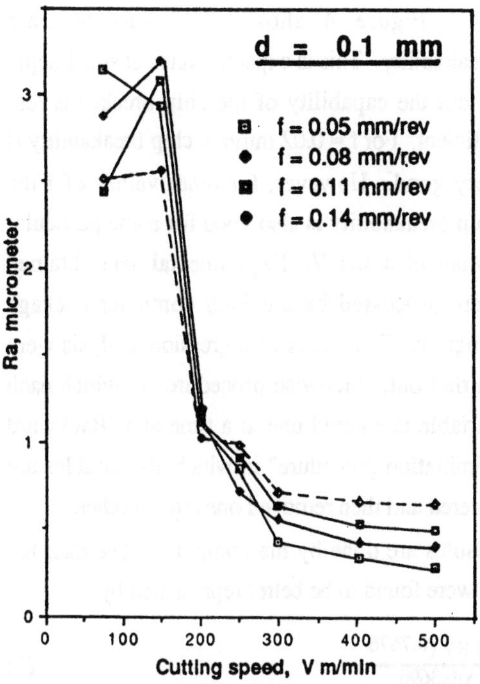

Figure 3. Influence of cutting speed and feed on surface roughness in finishing operation.

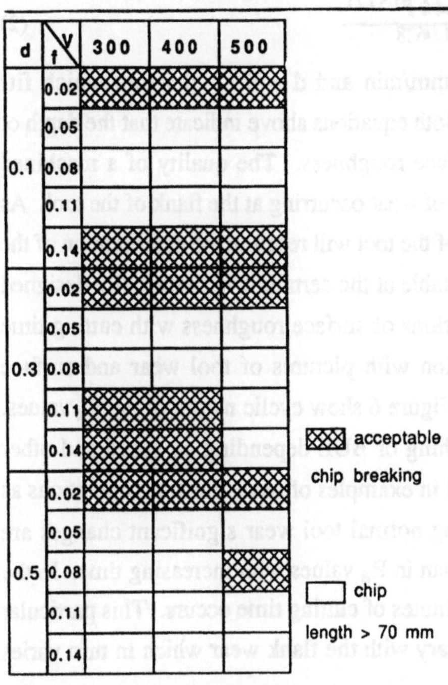

Figure 4. Chip breakability in finishing operation.

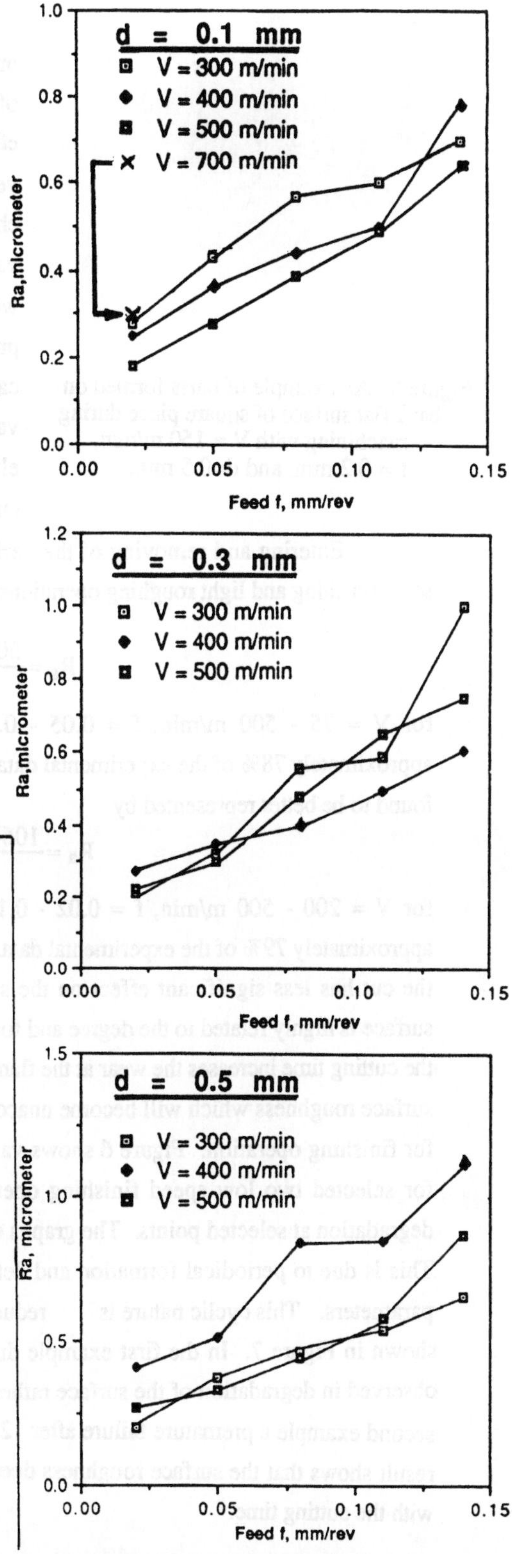

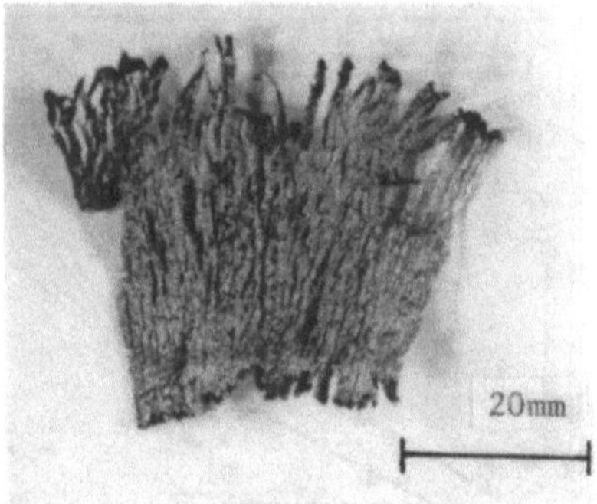

Figure 5. An example of barrs formed on back flat surface of square piece during machining with V = 150 m/min, f = 0.2 mm, and d=0.5 mm.

Figure 4 shows the chart of chip breakability. This is expected since at small depth of cut the capability of the chip breaker is less efficient. For f ≈ 0.02 mm/rev chip breakability is very good. However, for other values of f the chip breakability is also good for some particular values of d and V. Experimental data obtained were processed by the SAS computer package program. Two types of regression analysis were carried out, "Stepwise procedure" in which each variable is entered one at a time and "Backward elimination procedure" in which all variables are entered and then removed one after another.

Entering and removing of the variables are done by the computer. The data for semi-finishing and light roughing operations were found to be better represented by

$$R_a = \frac{50.89\, f^{0.7676}}{V^{0.3696}} \tag{1}$$

for V = 75 - 500 m/min, f = 0.05 - 0.4 mm/min and d = 0.5 - 6 mm which fits approximately 78% of the experimental data. Similarly, the data for finishing operation was found to be better represented by

$$R_a = \frac{1061.644\, f^{0.5121}}{V^{1.0638}} \tag{2}$$

for V = 200 - 500 m/min, f = 0.02 - 0.14 mm/min and d = 0.1 - 0.5 mm which fits approximately 79% of the experimental data. Both equations above indicate that the depth of the cut has less significant effect on the surface roughness. The quality of a machined surface is highly related to the degree and form of wear occurring at the flank of the tool. As the cutting time increases the wear at the flank of the tool will result in the deterioration of the surface roughness which will become unacceptable at the certain level, especially designed for finishing operation. Figure 6 shows variations of surface roughness with cutting time for selected two low speed finishing operation with pictures of tool wear and surface degradation at selected points. The graphs of Figure 6 show cyclic nature of the R_a values. This is due to periodical formation and detaching of BUE depending on speed and other parameters. This cyclic nature is reduced in examples of semi-finishing operations as shown in Figure 7. In the first example during normal tool wear significant changes are observed in degradation of the surface rather than in R_a values with increasing time. In the second example a premature failure after 12 minutes of cutting time occurs. This particular result shows that the surface roughness does vary with the flank wear which in turn varies with the cutting time.

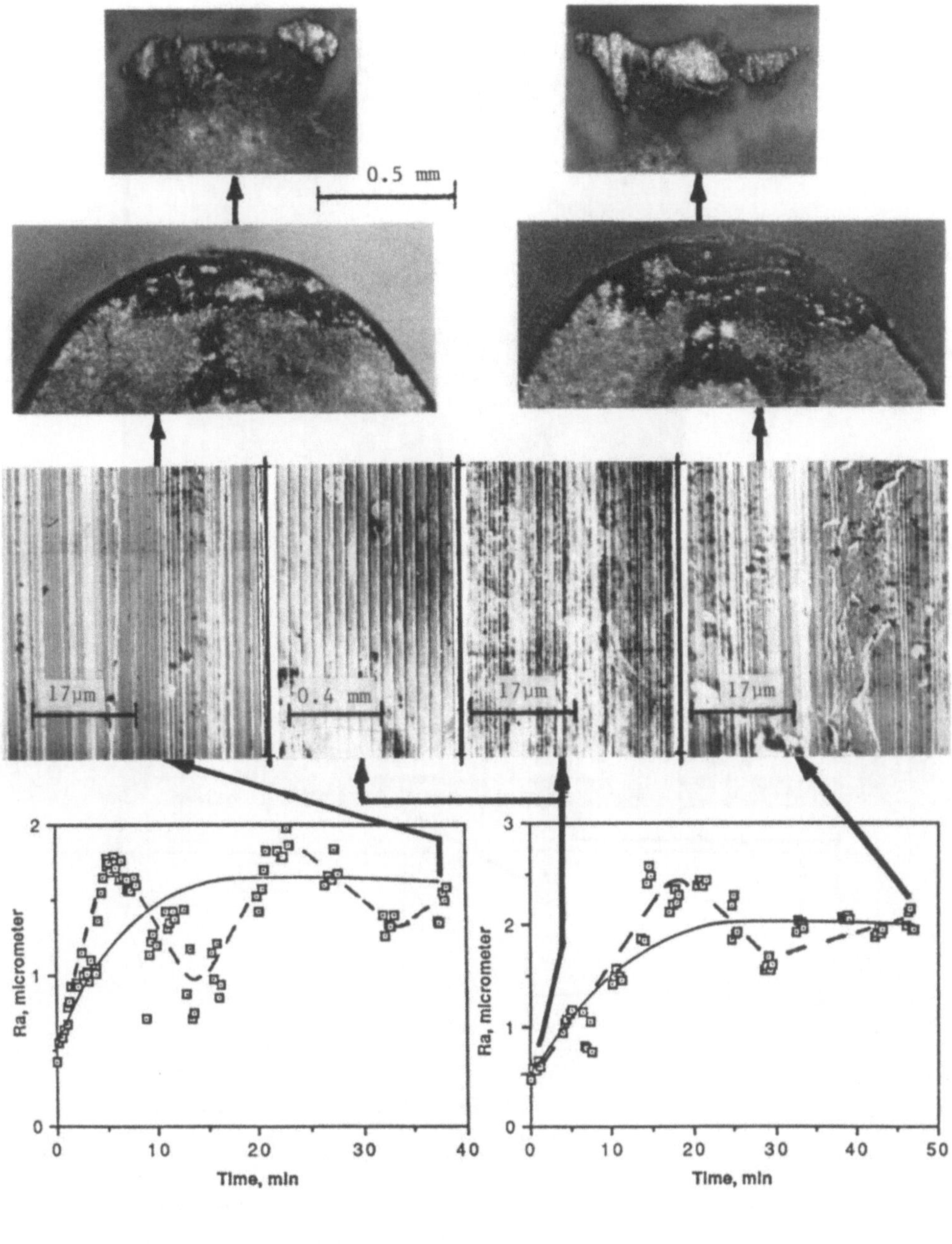

Figure 6. Influence of cutting time on surface roughness, surface degradation, flank wear and crater wear for selected points in two examples of finishing operation with V = 300 m/min, d = 0.3 mm and f = 0.1 mm/rev.

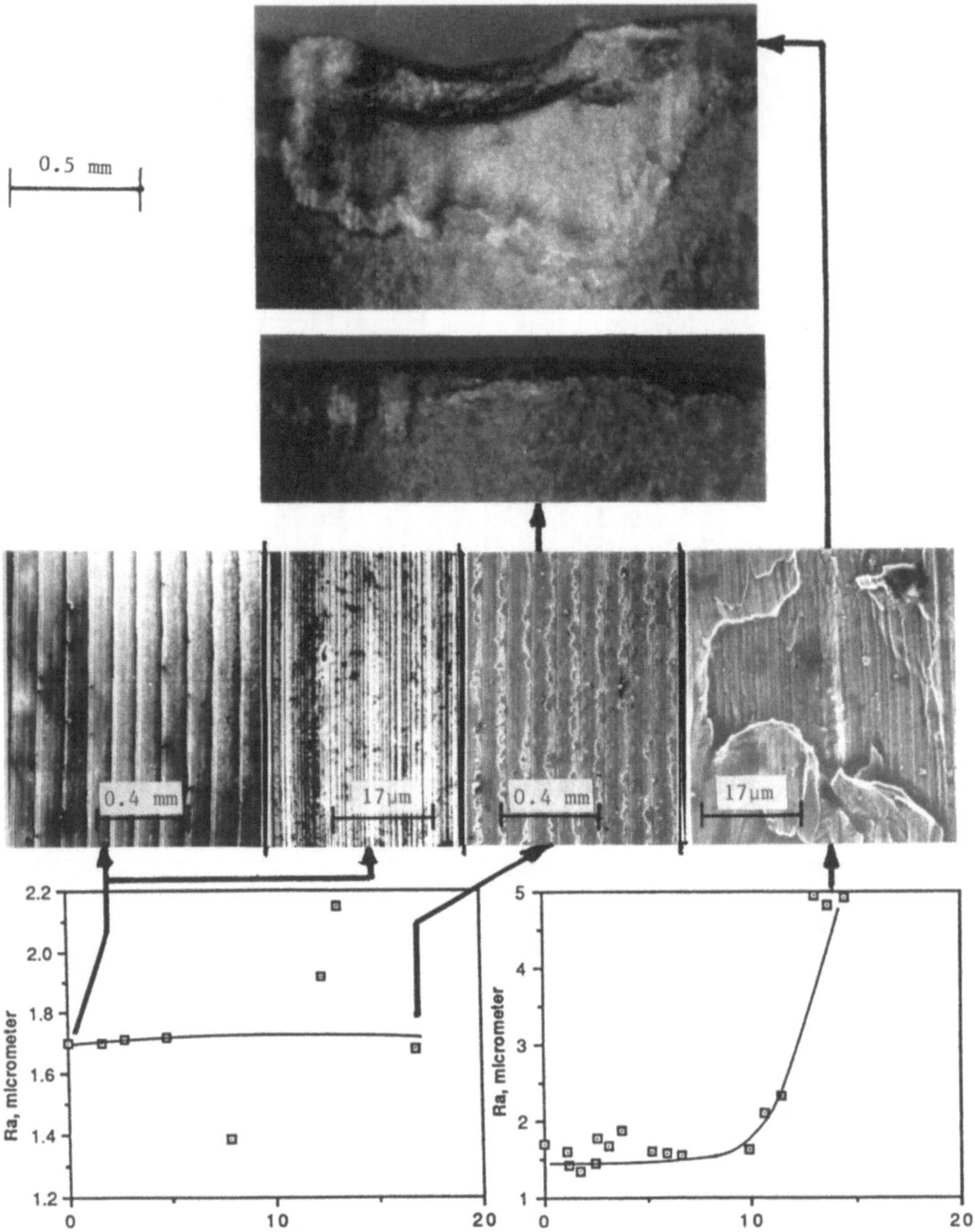

Figure 7. Influence of cutting time on surface roughness, surface degradation and flank wear for selected points in two examples of semi-finishing operations with V = 150 m/min, d = 2 mm and f = 0.2 mm/rev.

Roughness is not the only parameter to estimate machined surface quality. In planning semi-finishing (as a final operation) and finishing operations should also be considered in surface degradation for quality of machined surface.

CONCLUSION

From the analysis of the results the following conclusions can be written:

1) HSLA steel containing 0.01% Ti, 0.08% C, 1.5% Mn and 0.009% N can be machined with good results in finishing ($R_a \geq 0.18\ \mu m$) and semi-finishing operations.
2) Finishing operation should be realised with V = 300 - 500 m/min, d < 0.5 mm and f < 0.15 mm/rev depending on chip breakability, surface roughness and productivity. The prediction model of roughness is the following: $R_a = \frac{1061.644\, f^{0.5121}}{V^{1.0638}}$.
3) Semi-finishing as a final operation should be realised with d = 1 - 4 mm, f = 0.2 - 0.4 mm/rev, and wide range of speeds 150 - 500 m/min depending on productivity, roughness and tool life time. The prediction model of roughness is the following $R_a = \frac{50.89\, f^{0.7676}}{V^{0.3696}}$.
4) Depth of the cut has less significant effect on the surface roughness, also the initial breakdown of the coating on the tool does not contribute serious increase of roughness.
5) Machined surfaces carry a variety of geometrical features which further degrade with increasing cutting time.
6) Some further works are necessary to justify our present investigation and gather more machinability data which would be helpful in automated operations.

REFERENCES

1. Araki, T., Yamamoto, S. and Nakajima, H., Improvement of chip disposability of ferritic steels by containing martensitic microstructures. Proc. Int. Conf. High Productivity Machining, Materials and Processing, New Orleans, Louisiana, May, 1985.
2. Backer, L. and Charlier, P. High strength drawn bars of microalloyed steel with improved machinability for mechanical engineering. Proc. Int. Conf. High Productivity Machining, Materials and Processing, New Orleans, Louisiana, May, 1985.
3. Araki, T. and Yamamoto, S. Journal of Japan Society of Precision Engg. (SEIMITSU KIKAI) 1976, 42, p 68-73.
4. Araki, T., Yamamoto, S. and Kanao, M. Influence of metallurgy on machinability, Proc. Int. Symp. Tokyo, 1977, [SI], pp 483-484.

Development of Finish Turning Inserts: Present Trends and Future Prospects

I. S. Jawahir, N. Uddin and C. K. Jun

Carboloy Inc.
Warren
Michigan
USA

ABSTRACT

This paper outlines a brief review of the recent world-wide trends in cutting tool inserts development and presents a summary of results highlighting the ever increasing role of cutting tool development in coping with the requirements of surface engineering, at large. Basic mechanisms governing the chip flow and chip curl in finish turning operations are illustrated with supporting experimental results on chip control and surface finish. An outline on tool coating trends and on advances in new cutting materials technology is also provided, along with future prospects for the use of new chip groove geometries and advanced materials processing techniques in finish turning operations.

INTRODUCTION

Present trends in manufacturing, which now leads towards "near net shape" manufacture, have placed a strong emphasis on finish machining operations. The success of a finish turning operation (which is normally performed under high speed, low feed and low depth of cut conditions), among other factors, very heavily depends on the correct choice of the cutting tool. Increasing role of cutting tools in advanced manufacturing systems, incorporating finish turning operations, has been clearly seen from the complexity of machining operations and the need for maintaining a close dimensional control and surface quality in component manufacture. A combination of advanced tool materials and chip groove geometries, together with various coating techniques, determine the application range of a cutting tool.

Shaw and Crowell [1] have defined finish turning operations as operations, typically performed under a depth of cut of 1.5mm and a feed of less than 0.15mm/rev. Having shown that the surface finish, tool life and dimensional accuracy are major areas of interest in finish machining, they have investigated the problem of surface finish production on machined surfaces resulting from tool nose radius effects. Nakayama et al [2] have followed up this work and provided some useful findings on the temperature effects and built-up edge (BUE) formation in finish turning. Japanese researchers have recently shown that innovative tool designs can improve the surface finish significantly [3]. Optimal cutting edge geometry and methods of producing cutting edges for increased tool life have been shown in the works by Agnew [4], Mayer and Stauffer [5], and Lenz et al [6]. Several researchers have presented their findings on improving the dimensional accuracy in finish turning operations, most notably: Tashlitskii and Kushner [7] and Tashlitskii et al [8]. The CIRP (International Institution for Production Engineering Research), through its various Scientific Technical Committees (STC), has produced a number of valuable reports and papers on finish turning operations.

A special report, produced in 1987 by the STC on cutting [9], highlights the need for chip control, in addition to the previously shown three other aspects of finish machining, in advanced machining systems. Nakayama et al [10], through an extensive work on chip groove geometry, have shown the design feature of a cutting tool with a curved rake face as a means for breaking "thin" chips. Jawahir [11] has studied the effect of tool restricted contact length in chip control suggesting various means of achieving improved chip control and surface finish through the effective use of this tool restricted contact length in appropriate ratios with the machine feed. Nedes [12] has investigated the finish turning problem when machining with a conventional grooved chip breaker and shown the effective use of tool geometry. The use of ceramic cutting tools in finish turning operations has been shown by Gruss [13] with different tool entry modes and the resulting chip control features. Continuous research on advanced tool materials and coating techniques has now given us several innovative designs of chip grooves on cermet tool material and multi-layer coated carbides with specific enrichments. There is also a continuing trend in high speed machining requiring advanced tooling.

In what follows, the authors have attempted to highlight the role of finish turning insert production in the overall scene of advanced manufacture and then to describe the state-of-the art developments in this area. Results of recent experimental findings on chip control and surface finish produced with the representative samples of currently commercially available innovative tool inserts are also presented with an outline of future directions.

BASIC REQUIREMENTS IN ADVANCED MACHINING

The basic objective of an advanced machining system is producing high quality products at minimum cost and maximum productivity. This objective is greatly influenced by a number of functional requirements of the machining process (as shown in Figure 1). The interactive role of the various elements of the functional requirements can be applied to a typical finish turning operation as follows.

Surface Finish

Improved surface finish will be directly reflected in the improved quality while being contributed by the chip control and the dimensional accuracy in finish turning operations.

Dimensional Accuracy

Greater dimensional accuracy is again a direct measure of the product quality which is affected by the tool life, in addition to the surface finish (in finish turning operations) with the prolonged use of a cutting tool.

Tool Life

Increased tool life will reduce the cost of manufacture while positively contributing to the quality of machining. Cutting power requirements affect the tool life due to the generation of elevated temperatures in machining. This, i.e. reduced tool life, in turn affects the dimensional accuracy (again in continuous mode operations) despite the cutting conditions being light in finish turning.

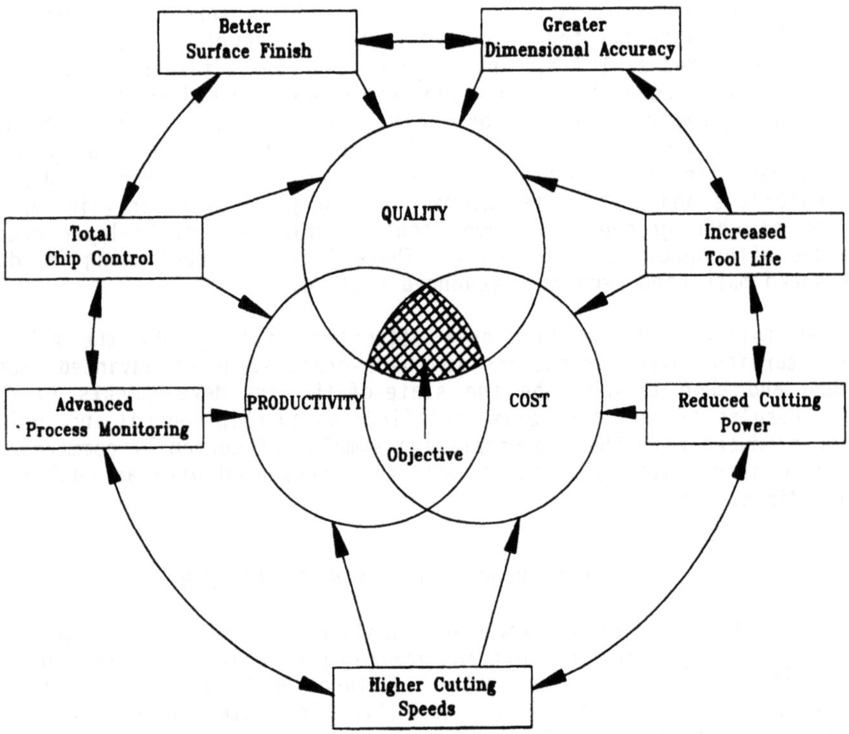

Figure 1. Functional requirements of an advanced machining system.

Cutting Power

Reduced cutting power reduces the cost of machining. Increased cutting speeds increase the power requirements. A compromise is usually made in machining. Although cutting power is not a major concern in finish turning, its interrelated aspects, such as higher cutting speed, directly contribute to this.

Cutting Speed

Increased cutting speed is a key to higher productivity and lower cost of machining. However, the provision of advanced tooling and an appropriate process monitoring system are essential in achieving the above goals. Finish turning operations are generally high speed operations.

Process Monitoring

Advanced techniques are required in monitoring the tool life, chip control, dimensional accuracy and surface finish. Implementing an advanced monitoring system will directly contribute to the productivity enhancement, in general.

Chip Control

The use of correct tools and right cutting conditions are mandatory requirements for breaking the chips into small and handleable sizes and shapes. Efficient chip control results in machining efficiency which is a productivity factor. Chip control also contributes to quality improvements. The central role of chip control in machinability assessments has been shown previously [11].

It is well known that chip control in finish turning operations is a major problem. Maintaining consistent and predictable surface finish in finish turning operations has also been a basic problem. Various means adopted by different cutting tool manufacturers in overcoming these problems are described in the following sections.

FINISH TURNING INSERTS: PRESENT STATUS

Conventional theories of metal cutting seem to be ineffective in predicting the chip flow and chip breaking mechanisms in finish turning operations. Therefore, the "try and see" method adopted in designing cutting tool inserts for finish turning operations has now given to the manufacturing world several totally different chip groove geometries (see Figure 2). Whether or not the various "customer appealing" features included in these designs are necessary, from the functional point of view, remains unanswered.

A cursory look at the features incorporated in typical tool inserts shows:

- Narrow tool face land (fixed or variable).
- Rounded cutting edge.
- Wavy patterns on the cutting edge, tool flank and rake faces.
- Rake face groove (cavity and/or lump) of arbitrary shapes.
- Various means of obstructions (to chip flow) on the tool face.

These various chip forming geometries are recommended by different tool manufacturers for the same set of cutting conditions and work materials. This only shows that there is no "uniquely" defined geometry as yet that could be designed and developed from the fundamental understanding of the metal cutting theory and practice.

Figure 2. Samples of some commercially available chip forming tool inserts for finish turning.

Predictability of chip breaking at consistent surface finish has now become an essential requirement for unattended machining systems. It is possible to fully predict the effects of chip forming geometries and varying cutting conditions only when the basic chip flow mechanics is well understood in finish turning; for example, the effect of cutting edge (hone radius etc.), frictional properties at the tool/chip interface, built-up edge formation etc. The use of more than one tool geometry for a given set of cutting conditions/work material combination now opens up a new research area for fundamental research in finish turning. This need is supported by the following experimental findings on specific aspects.

Surface Finish Variations

Three different tool geometries were used in a finish turning operation of steel 1045 (Bhn=185) at 160m/min cutting speed and 0.2mm depth of cut with varying feeds from 0.06 to 0.15 mm/rev. Two sets of tool inserts having two different tool nose radius values (but the same groove geometry) were used in the experiments. Measured values of the surface roughness (R_a) were plotted against the feed (see Figure 3).

As seen in Figure 3, experimental R_a values at low feeds are much higher than the theoretical values. A close observation of the machined surfaces indicate some signs of built-up-edge deposits randomly. Tool inserts with higher nose radius values produce lower R_a values (i.e. better surface finish) than the ones with lower nose radius, only at high feeds. Also a negative trend is observed with increase in feed when higher nose radiused tool inserts are used. This trend is very difficult to explain unless a very careful metallographic study is carried out to see the actual mechanisms of chip formation. The role of chip grooves (size and shape) and the tool cutting edge sharpness (i.e. hone radius) are believed to be the most likely causes for this definite trend.

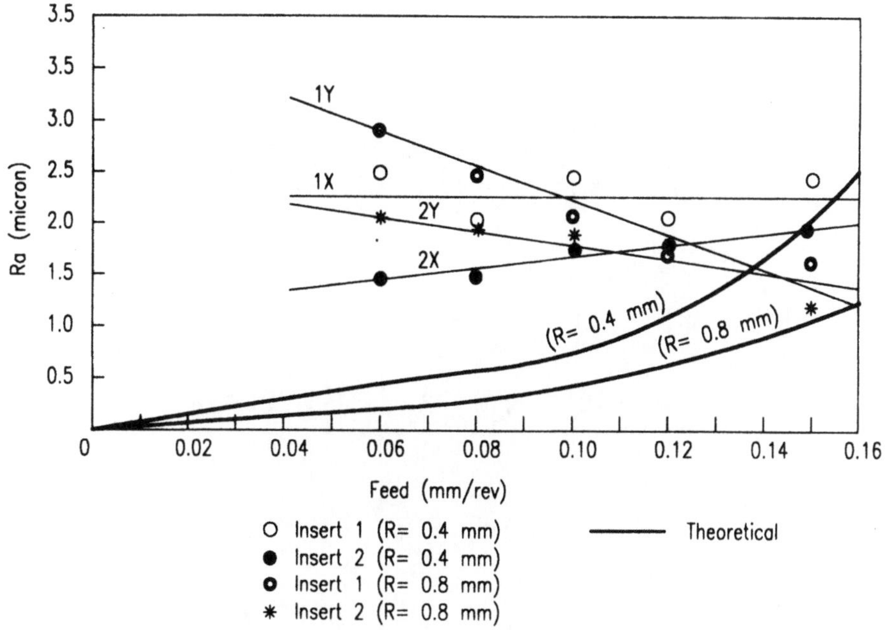

Figure 3. Surface finish variations in finish turning with inserts having two different chip groove geometries and their comparison with theoretical values.

Chip Control Comparison

Machining experiments were conducted with selected representative tool inserts from both carbide and cermet groups under the following cutting conditions:

Cutting speed	= 229 m/min (750 sfpm)
Feed	= 0.056 - 0.313 mm/rev. (0.0022 - 0.0123 ipr)
Depth of cut	= 0.25 - 2.54 mm (0.01-0.1 inch)

The work material used was AISI 1045 steel with 187 Brinell hardness. Chip charts were prepared and the collected chips were mounted on these charts in two sets (see Figures 4 and 5).

Figures 4 and 5 show basically a similar pattern of chip formation in terms of chip breakability. In both cases low feed and low depth of cut conditions are identified as the "problem area" where chip breaking is not possible with the tools considered. In comparison, cermet tools appear to produce more acceptable chip forms/shapes, particularly with tools having lower nose radius (see Figure 5 (c)) values. Further tests conducted with this low nose radius tool (TNMG 331), with a number of randomly selected tool inserts manufactured by different tool manufacturers, confirm the improved chip breakability with both cermet and carbide tools although the "thin" chip breaking (i.e. chip breaking at low feeds and low depths of cut) was not fully achievable.

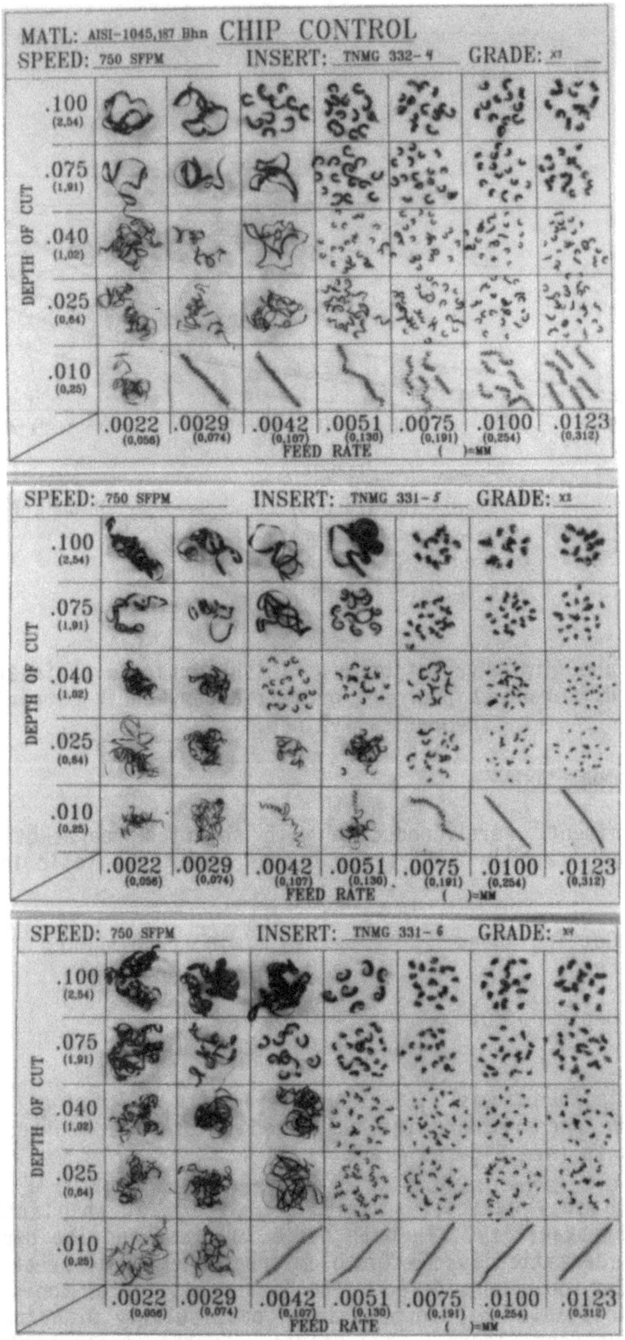

Figure 4. Three typical chip control charts produced in finish turning with carbide inserts.

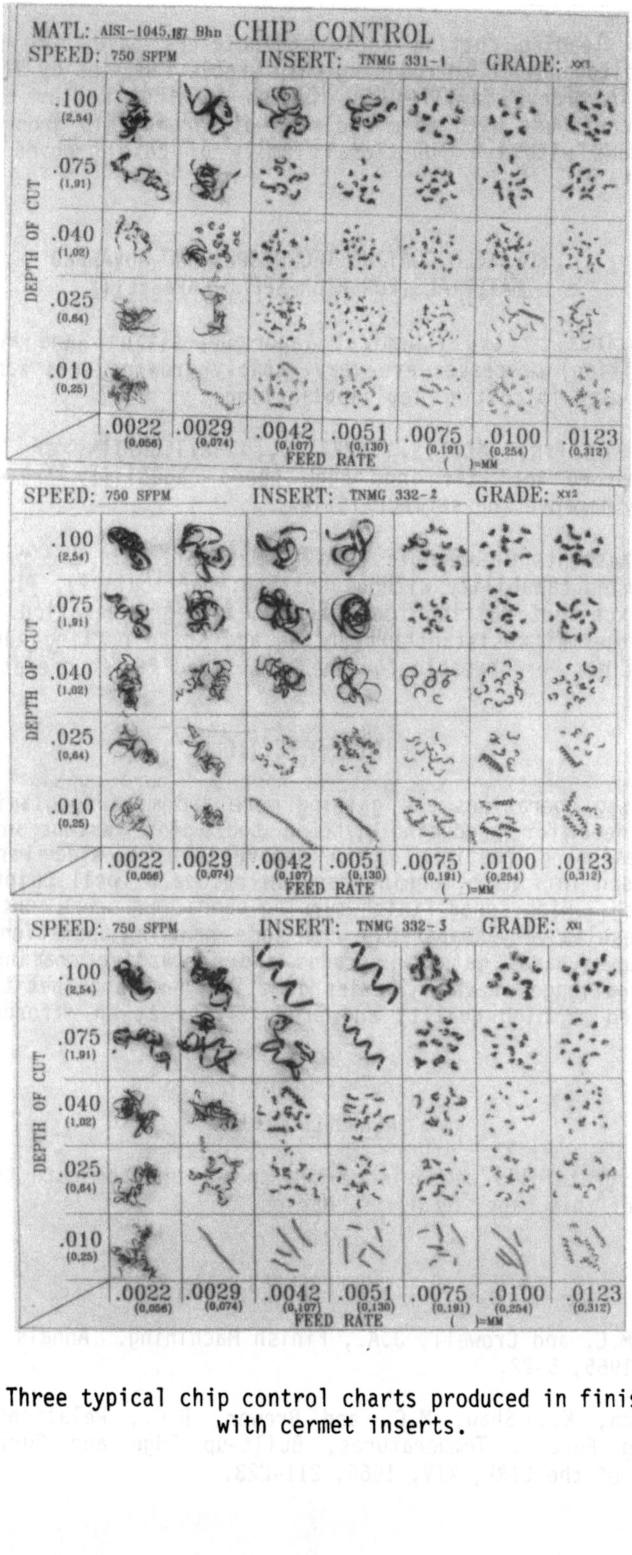

Figure 5. Three typical chip control charts produced in finish turning with cermet inserts.

Chip Groove Families

A number of leading cutting tool manufacturers have, in recent times, adopted a philosophy in introducing chip groove families by function. For example, chip groove families for finish turning, medium turning, etc. This trend continues with more and more different chip groove geometries being produced within a functional family of chip grooves for various applications.

ADVANCED COATING TECHNIQUES AND ADVANCED MATERIALS FOR HIGH SPEED CAPABILITY

Surface coatings by CVD (Chemical Vapor Deposition) and PVD (Physical Vapor Deposition) processes are very rapidly growing areas and have been extensively used in cutting tool applications.

Thin layers of TiN, TiC/TiN, TiC/Al_2O_3, TiC/TiCN/TiN, or TiC/Al_2O_3/TiN coatings improve tool life and high speed capability tremendously and consequently improve the surface finish.

Recent advancements in ceramics and cermet materials are providing high speed finishing capability without coating; particularly, TiC/TiN cermets which are excellent cutting tools for the finish operation. Because of their high thermal resistant properties and sharp cutting edges, cermets without coatings, are capable for the high speed finish operations.

FUTURE DIRECTIONS

Finish turning operations are gaining more and more popularity with the continuing need for producing cutting tools for various work material groups. Designing "optimum" chip grooves for a wide range of work materials used in finish turning has now become a focal point in cutting tool research. High speed finish turning operations will continue to grow producing results very comparably with some grinding operations. Development of "super hard" material grades and innovative coating techniques will be a challenge of the nineties in a world-wide competitive environment targeted at high quality and productivity at an affordable minimum cost.

ACKNOWLEDGEMENT

The authors gratefully acknowledge the speedy and accurate typing of the manuscript of this paper by Ms. K. Marson.

REFERENCES

1. Shaw, M.C. and Crowell, J.A., Finish Machining. Annals of the CIRP, XIII, 1965, 5-22.

2. Nakayama, K., Shaw, M.C. and Brewer, R.C., Relationship Between Cutting Forces, Temperatures, Built-up Edge and Surface Finish. Annals of the CIRP, XIV, 1966, 211-223.

3. Shikata, H. and Nagashima, K., New Cutting Tools With Curved Edges and Faces Improve Productivity of Machining. SME Technical Paper, MR 86-452, 1986, 13 pages.

4. Agnew, J., The Importance and Methods of Carbide Edge Preparation, SME Technical Paper, MR 73-905, 1973, 31 pages.

5. Mayer, J.E., Jr. and Stauffer, D.J., Effects of Tool Edge Hone and Chamfer on Wear Life. SME Technical Paper, MR 73-907, 1973, 20 pages.

6. Lenz, E., Moskowitz, D., Mayer, J.E., Jr. and Stauffer, D.J., Optimal Edge Geometry for Maximum Tool Life. ASME Paper, 77-WA/Prod-43, 1977, 9 pages.

7. Tashlitskii, N.I. and Kushner, V.S., Cemented Carbide Tools for Finish-turning Steels. Vesnik Mashinostroeniya, 54 (5), 1974, 60-63.

8. Tashlitskii et al, Finish Turning of Interactable Alloys Using Tools With a Finishing Edge. Vestnik Mashinostroeniya, 58 (8), 1978, 63-65.

9. Kahles, J.F., Machinability Data Requirements for Advanced Machining Systems, CIRP Technical Report, Annals of the CIRP, 36 (2), 1987, 523-529.

10. Nakayama, K., Arai, M. and Kondo, T., Cutting Tools With Curved Rake Face - A Means for Breaking Thin Chips. Annals of the CIRP, 30 (1), 1981, 5-8.

11. Jawahir, I.S., The Chip Control Factor in Machinability Assessments: Recent Trends. J. Mechanical Working Technology 17, 1988, 213-224.

12. Nedes, C., Characteristic Parameters of Chip Control in Turning Operations With Indexable Inserts and Three-dimensionally Shaped Chip Forms. Annals of the CIRP, 38 (1), 1989, 75-79.

13. Gruss, W.W., Turning of Steel With Ceramic Cutting Tools. Proc. Conf. on Advances in Tool Materials for Use in High Speed Machining, Scottsdale, Arizona, USA, February 1987, pp 105-115.

Machining Residual Stresses

T. Rasul

Dr A. Q. Khan Research Laboratories
Rawalpindi
Pakistan

ABSTRACT

As part of an extended investigation into the effects of pre- and post-processing upon shot-peening residual stresses, the residual stresses due to a controlled turning process were investigated in an aircraft alloy 817M40.

Cylindrical specimen were produced on a centre lathe with all machine parameters fixed except for the feed rate. A variation in the feed rate was used to produce different surface roughnesses varying from 1.2 µm to 6.6 µm Ra. The residual stress field was established by the well known centre hole drilling method albiet with a modification which helped to establish the variation of residual stresses with depth.

The results indicated a definite trend in axial residual stress distribution with respect to the surface roughness, with a rough surface having higher tensile residual stresses. This trend was however not repeated for the hoop or circumferential stresses. The maximum tensile residual stress varied from 90 MPa tensile in the 6.6 µm Ra surface to 50 MPa tensile in the 1.2 µm Ra surface.

Although residual stresses are generally ignored in design, either due to lack of experimental results or for ease of application, the results indicated that some attention must be given to machining residual stresses as these can enhance or reduce the fatigue life of the component. In view of the wide variety of metals, and types of application which necessitate the use of specific machining operations, it is unrealistic to perform an extensive programme covering these applications. Accordingly, at this stage of the work it is difficult to draw a general conclusion. The study also indicates that turning operations produce important changes in the crystalline structure, associated with the large plastic deformation encountered during the machining process.

INTRODUCTION

The process of shot-peening is widely used to induce compressive residual stresses in engineering components subjected to fatigue loading. Since the compressive residual stress distribution after shot-peening is a superposition of the residual stress state prior to peening, and the peening residual stresses, the residual stresses before peening play an important role in determining the overall usefulness of the product. This role has been conveniently ignored for ease of application as well as due to lack of experimental data.

Nevertheless as has been shown in a recent study (1), these stresses play an important role and need to be specified and controlled if repeatable peening is to be achieved. During the investigations carried out in this regard, machining residual stresses due to controlled turning were measured, and are reported here.

MATERIALS AND METHODS

The material investigated 817M40 steel was a typical aircraft alloy and finds extensive use in the aerospace industry. Some investigations were also carried out on Aluminium 7075 alloy, and since the number of specimens tested were too few to allow meaningful conclusions to be drawn, have not been reported here.

Using a centre lathe, controlled turning was carried out on steel to produce specimens as shown in Figure 1.

All machine paramters shown in Table 1 were kept constant and the feed rate only was varied to produce specimens having four different surface roughnesses. These ranged from a fine finish of 1.2 µm Ra to a rough finish of 6.6 µm Ra. The material composition and mechanical properties appear in Tables 2 and 3.

The main decision regarding the geometry of the specimen was whether to use a flat or a cylindrical surface. Although a flat surface was easier to generate, easier in strain gauge application and economical in use, it was not selected. A flat surface causes problems in context of uniform peening over a large enough area to glue strain rosettes. Besides a large number of engineering components are cylindrical, and as such a lot of research work is aimed at and around cylindrical specimens. Results can thus conveniently be compared with other experimental work. Also round specimens allow easier variation and control of surface finish, which was a parameter being investigated in the detailed study mentioned in reference (1). The specimens were critically inspected for geometric variation to ensure that all had essentially the same form.

Residual stress measurements were carried out by the

well known centre hole-drilling method (2-4), albiet with a modification explained below. The normal hole-drilling procedure is an averaging method and gives the average residual stress from the surface upto the depth of the drilled hole. Also as opposed to a twist drill or an end mill, the air abrasive method produces a hole which increases in diameter as it deepens. Thus to obtain the residual stress distribution with respect to depth some form of modification has to be introduced. Also instead of drilling the hole to its full depth, once, better results are obtained when the

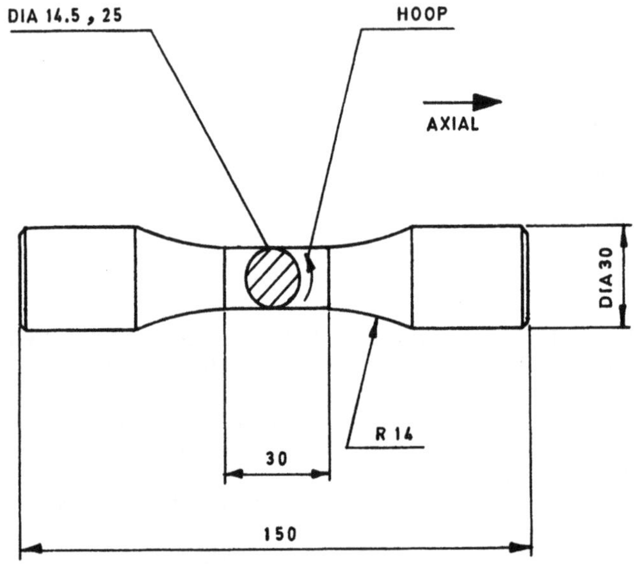

Figure 1. Specimen Dimensions

TABLE 1
Machine Parameters for Turning

Turning Speed	1000 rpm
Lubricant	Soluble Oil
Tool	VBMM 1604 04-53 (0.4 mm tip radius)
Depth of Penultimate cut	0.5 - 0.63 mm
Depth of Final cut	0.25 mm
Feed Rate (6.6 µm Ra)	0.48 mm per revolution
Feed Rate (5.7 µm Ra)	0.28 mm per revolution
Feed Rate (3.7 µm Ra)	0.18 mm per revolution
Feed Rate (1.2 µm Ra)	0.04 mm per revolution

TABLE 2
Chemical Composition of Test Material

Element	Maximum/Minimum
Carbon	0.45/0.35
Silicon	0.35/0.10
Manganese	0.70/0.45
Nickel	1.80/1.30
Chromium	1.40/0.90
Molybdenum	0.35/0.20
Sulphur	0.05/----
Phosphorus	0.05/----

TABLE 3
Mechanical Properties of Test Material

Property	Steel 817M40
Young's Modulus	207 GPa
Ultimate Tensile Strength	1 GPa
Proof Strength 0.1 %	770 MPa
Poissons Ratio	0.28
Hardness	300 Hv

hole is drilled in increments. The incremental method of hole-drilling has been in use for some time, and although it gives more meaningful results as compared to the one-step drilling method, it lacks in the sense that the transition of stresses from compressive to tensile cannot be picked up. To get an idea of the error involved with incremental drilling the following explanation is provided.

Assuming that four different diameter holes are drilled, each in one increment only, the strain relaxation shall be as that given in Figure 2.

The larger hole response is given by curve 1, as the larger the hole the greater is the strain relaxation. If on the other hand the hole is drilled in four increments, such that the hole diameter increases from that associated with curve 4 through to that with curve 1, the relaxed strain curve shall lie within these extremes. The second point is obtained by the intersection of the curve 4 and the end of the first increment, the third by that of curve 3 and the second increment and so on. The strain response is thus given by the red curve in Figure 2. Since the true response is shown by the curve 1, the area enclosed respresents the error introduced if hole widening is ignored.

One method of reducing the error, is to have a large number of intervals so that the red curve is shifted up towards the curve 1, but this solution is limited by practicality. The second, is to perform a large number of experiments to statistically arrive at a polynomial expression relating the hole widening effect to the depth, and use this relation to arrive at the hole diameters, given

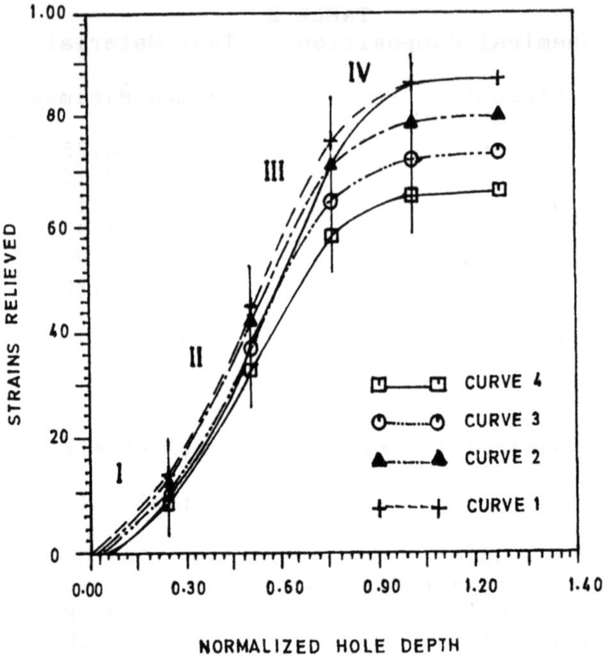

Figure 2. Incremental Curves for Four-Step Drilling

the final diameter. Some work has been done and both manual and computer based techniques exist as in references (5) and (6).

Another method of taking care of this difficulty is detailed as follows, and is based on the relationship provided by Soete and Vancrombrugge (7). It was suggested that the original relationships as given in equations (1) to (3) below for calculating the residual stresses are also valid in the differential form.

$$\sigma 1 = (-E/2K1)\ (P+R/n) \quad (1)$$

$$\sigma 2 = (-E/2K2)\ (P-R/n) \quad (2)$$

Where $P = \epsilon/m$

$n = 1 + v\ K2/K1$

$R = (\epsilon 1 + \epsilon 3) + (2\epsilon 2 - \epsilon)$

$\epsilon = \epsilon 1 + \epsilon 3$

$m = 1 - v\ K2/K1$

$$\phi = -0.5 \tan ((\epsilon - 2\epsilon 2)/(\epsilon 3 - \epsilon 1)) \quad (3)$$

Thus the constants K1, K2, and the strains $\epsilon 1$, $\epsilon 2$ and $\epsilon 3$ can be replaced by their differentials with respect to the hole depth. Instead of using the absolute strain values at various depths, the instantaneous changes in strain, as represented by the slopes of the strain versus depth plots, are used to calculate the stress.

As an example, the strain values are first used to create a plot against the hole depth by fitting a curve through the points, and keeping in mind that (i) the fit should be smooth so that experimental errors are smoothed out, and (ii) it should fit more closely the initial points. The reason for (ii) is obvious that a fit generally cannot be established to include all data points with a reasonable smoothening action. Also we are more interested in the top surface and initial layer which are relevant to this study. The slopes are then calculated at predetermined depth intervals and the process is repeated for the three strains and the constants K1 and K2.

The hole widening effect is thus taken care of as it is not the absolute values but the instantaneous values that are used in the basic equation. Although the technique remains an averaging technique, the averaging effect is reduced, and the modified method is capable of picking up the residual stress variation with depth.

RESULTS

As a result of the experiments performed on the four specimens, the residual stress fields for the two extreme roughnesses studied are given in Figures 3 and 4.

These figures show that the surface tensile residual stresses are greater than those found at some depth from the surface.

The combined effect of all the four roughnesses is given in Figure 5, and shows that the initial surface roughness, produced as a result of turning, has a considerable effect on the subsequent residual stress distribution. It is evident from this figure that generally the finer the finish to which a component has been produced, the less the severity of plastic deformation that it is subjected to, and the less intense the residual tensile stresses induced within it.

From Figure 5 it can also be seen that the axial residual stresses tend to follow a definite pattern with respect to the surface roughness, whereas the hoop residual stresses do not. This is primarily due to the tool drag forces associated with this direction, especially if the tool blunts, and has been corroborated experimentally (8). These residual stresses have no definite pattern and the effect becomes pronounced as the diameter of the component is decreased; the surface curvature becoming more pronounced. On the contrary as the diameter of the component is increased, the hoop residual stresses follow a trend similar to the one shown by the axial residual stresses.

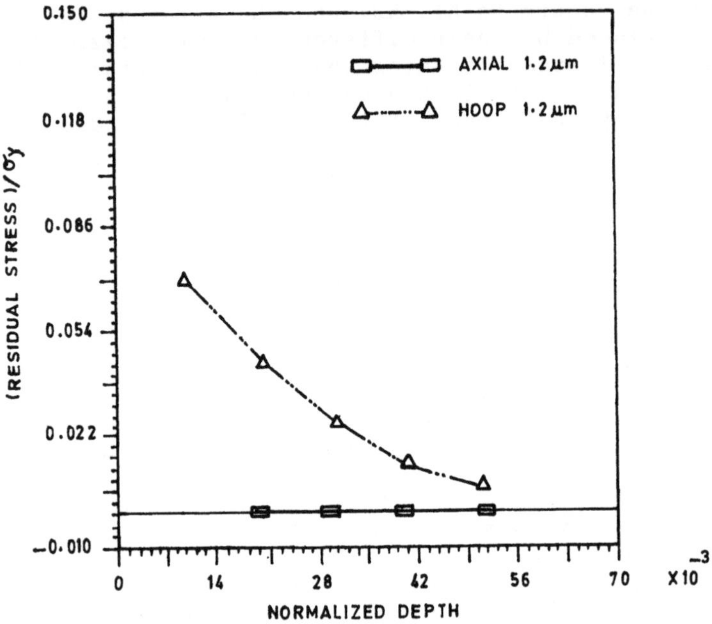

Figure 3. Turning Residual Stresses for Steel 1.2 μm Ra

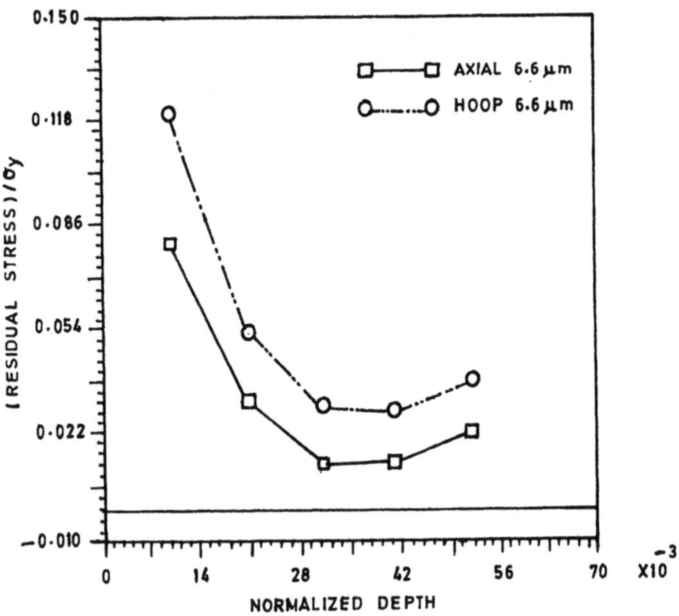

Figure 4. Turning Residual Stresses for Steel 6.6 μm Ra

How the surface integrity of engineering components is influenced by various factors can best be explained by the diagram in Figure 6.

The three main factors are the material, the geometry and the state of stress. Each in itself is a multiple effect of a number of subfactors. Thus, when a rough or fine surface is mentioned, it is not only the surface roughness as given by the average roughness values that should be considered, but also all the other factors given in Figure 6.

As an example, two surfaces of the same material and having the same roughness value, may have different states of work hardening, depending on how fast or slow the rate of metal removal was, and hence a different surface integrity. It is for this reason that the specimen preparation procedure be known precisely as given in Table 1.

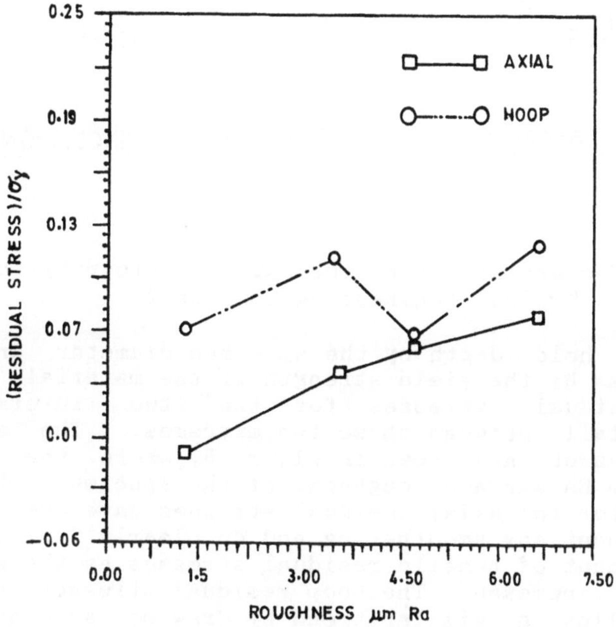

Figure 5. Turning Residual Stresses for Steel

DISCUSSION AND CONCLUSIONS

Looking at Figures 3 and 4 it can be seen that as the surface roughness is increased, the axial residual stress increases from a maximum of 55 MPa to a maximum of 92 MPa. The hoop residual stress for the fine surface is almost zero while for the 6.6 μm Ra surface it falls from a maximum of 62 at the surface to a minimum of 9 MPa at a depth of .45 mm. The scale in both the X and Y directions has been normalized by

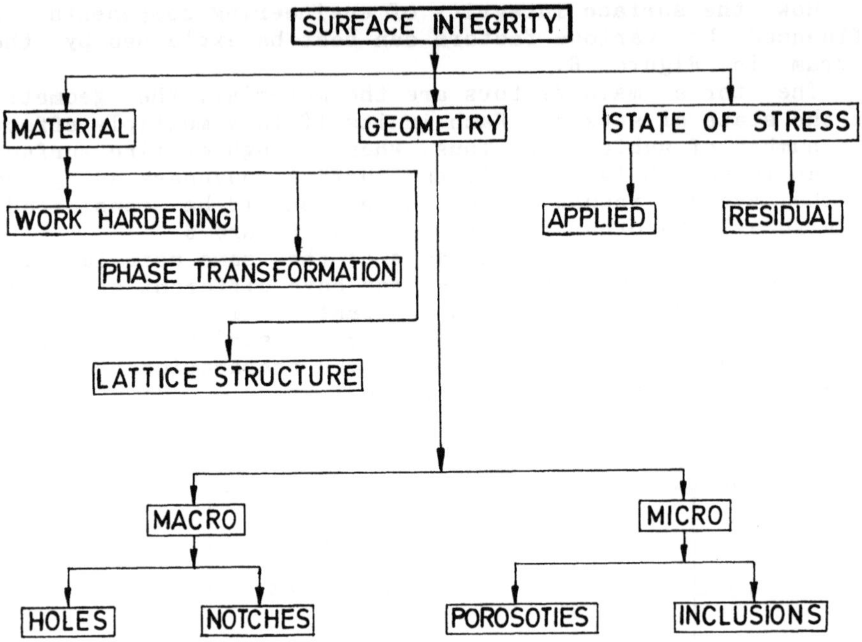

Figure 6. Factors Affecting the Surface Integrity of a Typical Engineering Component

dividing the hole depth by the specimen diameter and the residual stress by the yield strength of the material.

The residual stresses for the two intermediate roughnesses fall between these two extremes. The surface residual stresses are shown in Figure 5, where the X-axis represents the Ra surface roughness of the specimen. The four data points for the axial residual stresses have been joined together without any smoothening and do clearly indicate the increasing amount of tensile residual stresses as the surface roughness is increased. The hoop residual stresses can be shown to follow a similar trend by drawing a mean line through the four data points. However for reasons already mentioned, these stresses generally do not follow the trend shown by the axial residual stresses. The residual stresses with depth for the 1.2 and 6.6 µm Ra roughness specimen are given in Table 4 and 5. These are typical values for one set of specimens, and the results of the other two sets investigated lie within 5 percent of these values. The depth is in milimeters and the stresses in Mega Pascals.

Extrapolation of the results presented here to apply to other materials or even other type of steels should be done with care. Even application of these results to steel 817M40 machined in a manner different to that given in Table 1 would not be logical. However, for any material machined consistently in any conventional or non-conventional manner, trends similar to the ones shown here can be expected, i.e., as the surface roughness is increased, the damage imparted to the component in terms of tensile residual stresses is increased.

TABLE 4

Residual Stresses for the 1.2 μm Ra Specimen

Depth	Axial Stress	Hoop Stress
0.15	- 0.1	53.9
0.3	1.5	34.2
0.45	- 1.2	19.9
0.6	0.8	10.8
1.0	1.3	5.9

TABLE 5

Residual Stresses for the 6.6 μm Ra Specimen

Depth	Axial Stress	Hoop Stress
0.15	61.3	91.3
0.3	23.1	44.0
0.45	9.5	21.9
0.6	9.1	19.9
1.0	10.8	26.8

Accuracy and Repeatability and Overall Error.

Any set of experiments however well planned and executed, shall not produce results that can be relied upon unless a few words are mentioned about the accuracy and repeatability of the set up. A number of different experiments were performed to check the validity of the results obtained. These pertained to four distinct parts of the study, i.e., mounting of strain gauges, location of specimens in the experimental jig, measurement of hole diameter and hole depth, and the measurement of residual stresses, (for details see chapter 4 of (1)).

During the extensive residual stress measurements using the modified hole-drilling method a careful account of the various errors (as above) involved was made. It is logical that these errors are not directly additive and the overall error cannot be a simple sum of these. Besides, these errors are the maximum observed, and the average values are considerably below those stated. When these maximum errors are added a total overall error of 8.6 percent maximum is obtained. The comparison of the modified hole-drilling technique with other popular methods has also established that the results agree to within 10 percent.

REFERENCES

1. Rasul, T., Effect of Pre- and Post-Processing upon Shot-Peening Residual Stresses, PhD Thesis, Cranfield Institute of Technology, Bedford, England, March'1989.

2. Beaney, E.M., and Procter,E., A Critical Evaluation of the Centre Hole Technique for the Measurement of Residual Stress, Strain, Volume 10, Number 1, 1974.

3. Beaney, E.M., Accurate Measurement of Residual Stress on any Steel Using the Centre Hole Method, Strain, Volume 12, Number 3, July 1976, PP 99-106.

4. Procter, E., and Beaney, E.M., Recent Developments in Centre-Hole Technique for Residual-Stress Measurement Exptl Tech, Volume 6, Number 6, 1982, PP 10-15.

5. Beaney, E.M., CEGB, Berkeley, U.K., Private Communication, 1988.

6. Derbyshire, J., CERL, Leatherhead, Surrey, U.K., Private Communication, 1988.

7. Soete, W., and Vancrombrugge, R., Determination of Residual Stresses Below the Surface, in Residual Stresses in Metals and Metal Construction, Ed., W.R. Osgood, Reinhold Publishing Corporation NY, 1954, PP 331-334.

8. Mellington,J., Dowty Rotol,UK, Private Communication, 1988.

Notation

μm	Micrometer
MPa	Mega Pascal
Ra	Average roughness parameter
$\sigma 1, \sigma 2$	Maximum and minimum residual stress, MPa.
E	Modulus of elasticity, MPa
ϵn	Strain read by the nth element of the strain gauge $\mu\epsilon$.
$\mu\epsilon$	Microstrains
ϕ	Angle of maximum stress with direction of $\epsilon 1$, radians.
K1,K2	Hole-drilling calibration constants.
Hv	Vickers hardness number.

Acknowledgement

The author wishes to thank Professor SA Meguid who is indirectly responsible for the work which appears here. The results for the paper have all been taken from the author's PhD thesis "Effect of Pre- and Post-Processing Upon

Shot-Peening Residual Stresses", and that work was supervised by Professor Meguid. The author also expresses his thanks to Dr. A.Q. Khan of Dr. A.Q. Khan Research Laboratories, Pakistan, who very kindly allowed him to take part in this conference.

Effect of Cold Deformation on Boronizing Processes and Wear Behavior of Low Carbon Steel 1020

B. L. Jiang, W. Liu and T. C. Lei

Harbin Institute of Technology
Harbin
People's Republic of China

ABSTRACT

The effect of room temperature rolling from zero to 80% reduction in thickness on the kinetics of boronizing in solid medium and the wear behavior of boronized layers was studied on plain carbon steel 1020. It was concluded that prior cold deformation accelerates the kinetics of boronizing and significantly improves the wear resistance of boronized steel. Therefore, the deformation-boronizing can be used successfully to elongate the wear life of machine components with the benifit of decreasing both the boronizing temperature and time for obtaining predicted case depth.

INTRODUCTION

Boronizing as a technological process for improving the resistance of steels against wear, acidic corrosion and high temperature oxidation began to be used since the seventies of the century. Although detailed studies have been made on the formation and structure of the boronized layers [1] and the growth of the boride crystals [2,3], the kinetics of boronizing and the factors controlling the kinetic processes are not quite clear. On the other hand, cold forming of steel components as a class of effective and economic technologies is becoming more and more widely used in machinebuilding and in many cases high resistance against wear and corrosion is required. Therefore, it seems important to study the kinetic process of boronizing and to clarify the effect of cold deformation on the kinetics and the microstructure and properties of deformation-boronized steels. This was just the aim of the present investigation.

MATERIAL AND METHODS

The industrially standard plain carbon steel 1020 was used for investigation. Rods of 20 mm diameter were hot forged into flat bars of different thicknesses and then annealed at 900 C.

After that the bars were cold rolled and machined to 6x15x25 mm specimens with the degrees of cold deformation varying from zero to 80%.

Boronizing of the specimens was performed in boxes with solid medium 2%B_4C +5%$NaBF_4$+0.5%NH_4Cl +92.5%SiC in a batch furnace with a heating rate of 6 degrees per minute to temperatures 600 to 900 C with holding times from zero to 8 h. The XRD techniques were used for examining the structure of the boride phase. In order to observe the very beginning of the formation of borides at the surface of specimens by TEM a single-side thinning technique was used(4) and the observations were performed on Philips-CM12 electron microscope. The Hitachi 570 scanning electron microscope was used for examining the microstructures with lower magnifications.

The wear behavior of boronized and deformation-boronized specimens were studied in a ping-ring wear tester. The rings 40 mm in diameter were made from a hardmetal with HRC 75.5 and the pins are the tested specimens with 10x4 mm of the wearing surface. wear tests were performed under a normal load of 130 N and a sliding speed 1.5 m/s by dry sliding.

RESULTS AND DISCUSSION

1.Nucleation and growth of borides

The XRD analysis of the borides nucleated at temperatures from 600 to 900 C are Fe_2B. The TEM photograph shown in Fig.1 indicates that the Fe_2B crystals nucleate preferentially on grain boundaries of the ferrite matrix of the steel with a mean size of the nuclei of about 200 nm.

The characteristics of the growth of Fe_2B crystals are shown in Fig.2. It can be seen that in the beginning period of time the boride nuclei grow randomly from the surface into the inner part of the specimen in forms of columns with the preference of their own ⟨002⟩direction as indicated in (2,3)(Fig.2a a). With prolonging the boronizing time crystals with 002 direction not perpendicular to the surface of specimen meet with each other and stop to grow (Fig.2b) and the crystals with the ⟨002⟩direction perpendicular to the surface of specimen can grow continuously forming the increasing layer depth (Fig.2c,d). A scheme showing the stages of crystal gowth of borides is given in Fig.3 which needs no further explantation. Here,one can find that the density of effective nuclei (i.e. number of nuclei per unit surface area with the ⟨002⟩direction perpendicular to the surface of specimen) should be the controlling factor of the growth of boride crystals about which will be discussion in the later.

2.Kinetics of the growth of boronized case depth

The kinetics of the growth of boronized layer of steel 1020 is shown in Fig.4. The kinetic curves can be divided into two stages. The first stage (before s) obeys a parabolic law in the form

$$X^2 = Kt \qquad (1)$$

where X is the case depth,t is time and K is a constant.

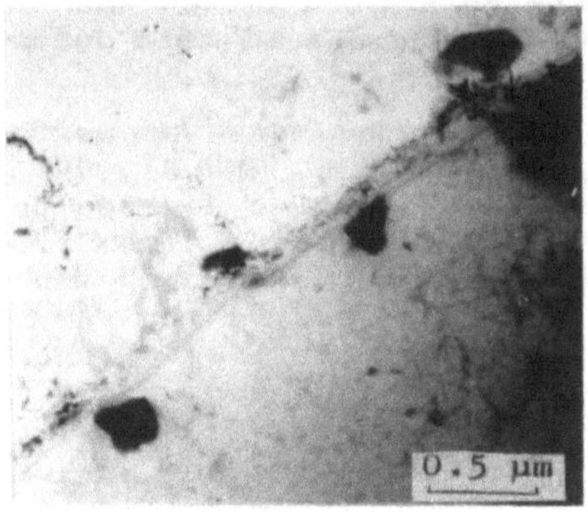

Figure 1. TEM photograph of steel 1020 heated to 600 C in boronizing medium and immediately cooled to room temperature.

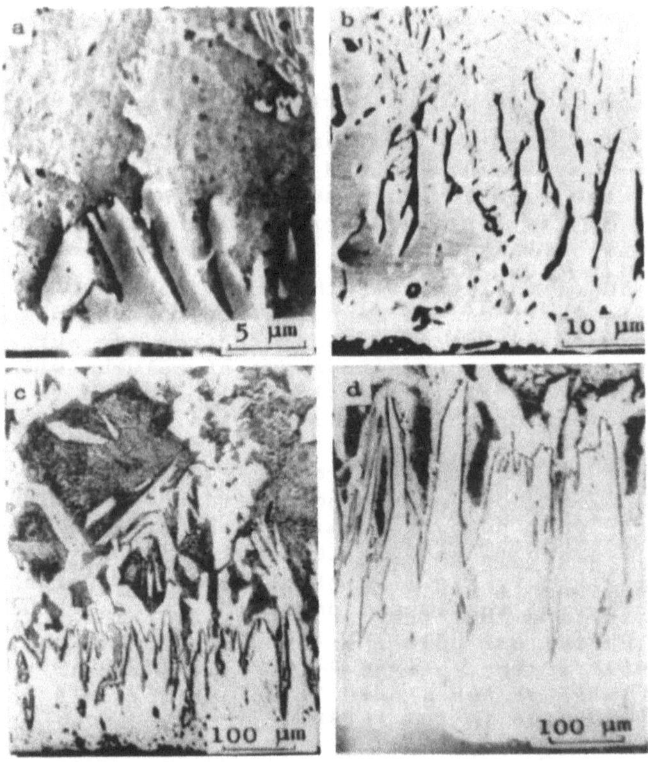

Figure 2. SEM photographs of steel 1020 heated to 900 C in boronizing medium showing the growth of boronized layer:
a- immediately cooled to 20 C; b- held for 10 min;
c- held for 1 h; d- held for 5 h.

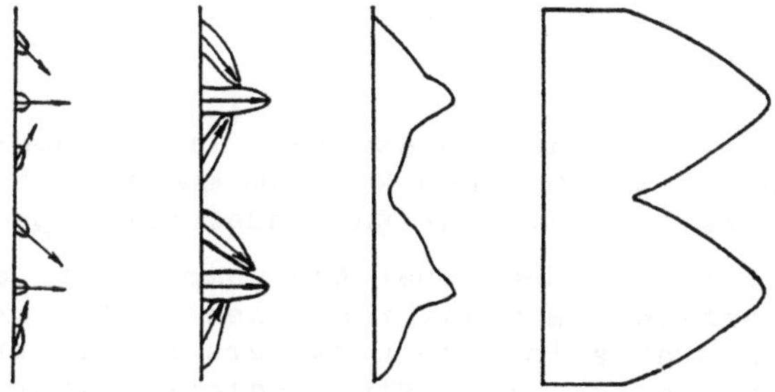

Figure 3. Four steps of the growth of the boronized layer. Arrows indicate the 002 direction of the Fe_2B crystals.

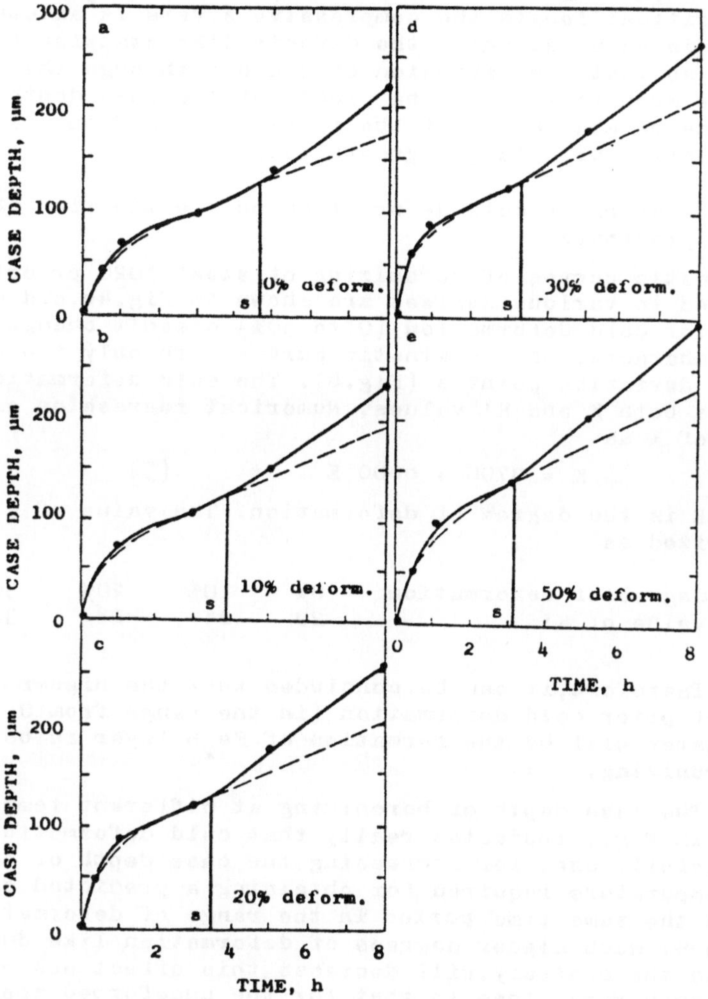

Figure 4. Measures (——) and calculated (---) kinetic curves of boronizing at 900 C for the steel 1020.

The second stage (after s) may be described by a straight line relationahip as

$$X = K't + b \qquad (2)$$

where K' is the slope of the straight line. Numerical regression shows that K = 3700 when X in um and t in hours and K'= 20 (for the case of Fig.4a of the undeformed specimen).

Galibois[5] has found that the compressive stress at the Fe_2B/ -Fe interface gradually increases as the crystal grows due to the difference in both structure and lattice parameters of the two phases. Strunk[6] indicated that B atoms diffuse much faster in Si crystals having screw dislocations. In this investigation defects like stacking faults were found in the Fe_2B crystal in the second stage of its growth (Fig.5). Therefore,it can be concluded that when the Fe_2B crystal grows to a critical length the compressive stress is accumulated to a certain value at which the defects like stacking faults are formed so that the diffusion of B atoms through the crystals becomes much faster and the growth of the case depth deviates from the parabolic law of the first stage and enters into the second stage with larger growth rate.

3.Effect of prior cold deformation on the kinetics of boronizing processes

The kinetic curves of boronizing of steel 1020 priorly cold deformed to various degrees are shown in Fig.4b,c,d and e). The prior cold deformation (0 to 50%) doesn't change the two-stage character of the kinetic curves with only the changes of the deviation point s (Fig.6). The cold deformation increases both K and K' values. Numerical regression gives the value of K as

$$K = 3700 + 6600\,E \qquad (3)$$

where E is the degree of deformation. The value of K' may be summarized as

degree of deformation	0%	10%	20%	30%	50%
value of K'	20	24	28	30	34

Therefore,it can be concluded that the higher the degree of prior cold deformation (in the range from 0 to 50%) the faster will be the formation of Fe_2B layer in both stages of boronizing.

The case depth of boronizing at different temperatures shown in Fig.7 indicates really that cold deformation can be successfully used for increasing the case depth or decreasing the temperature required for obtaining a predicted case depth within the same time period in the range of deformation from 0 to 50%. Much higher degrees of deformation like 80% in this case,on the contrary,will decrease this effect and makes the case depth very close to that for the undeformed specimen.

It has been mentioned formerly that the density of nuclei (more exactly,the density of effective nuclei) is the controlling factor of the kinetics of boronizing process. The effect of cold deformation on the density of nuclei is shown

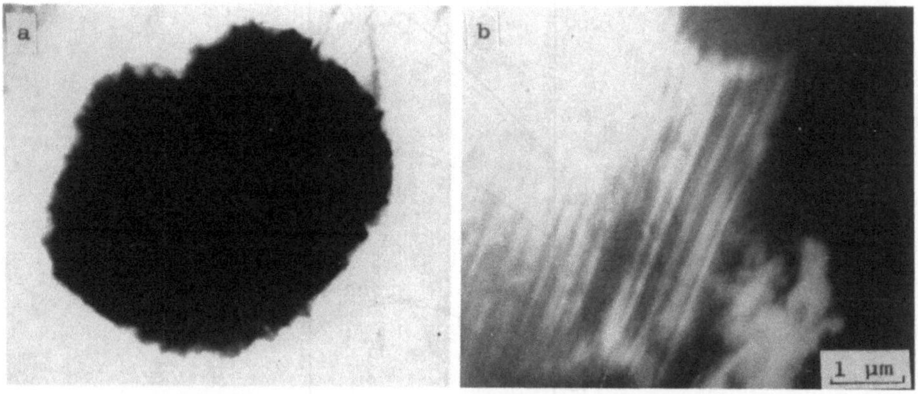

Figure 5. TEM photographs showing the fine structure of Fe_2B crystals: a- 900 C boronized 30 min; b- 900 C 5 h.

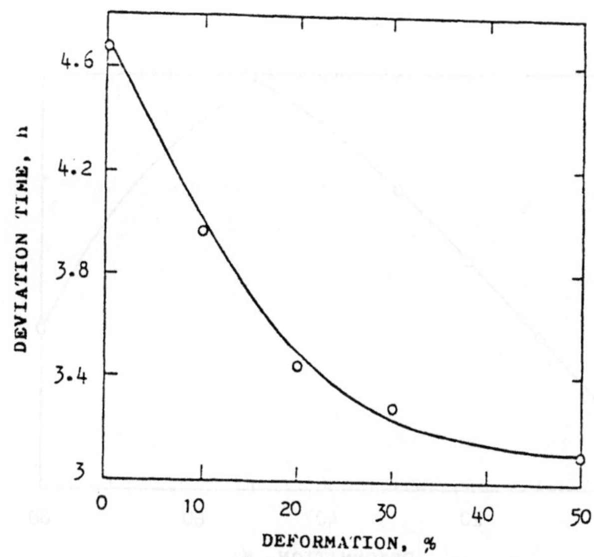

Figure 6. Time of deviation from parabolic law of the first stage vs degree of deformation for 900 C boronizing.

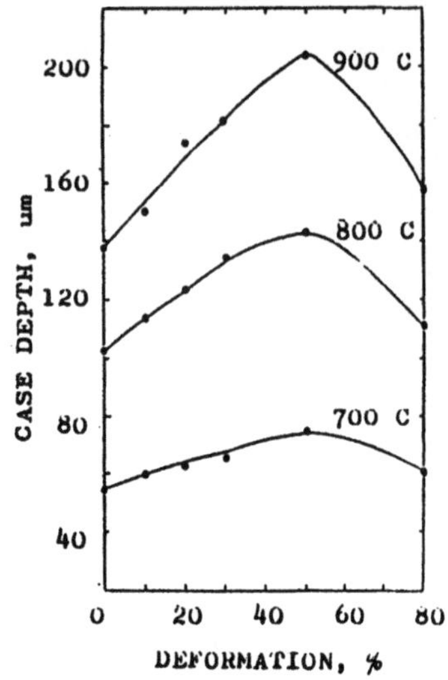

Figure 7. Case depth vs degree of prior cold deformation for steel 1020 boronized at different temperatures for 5 h.

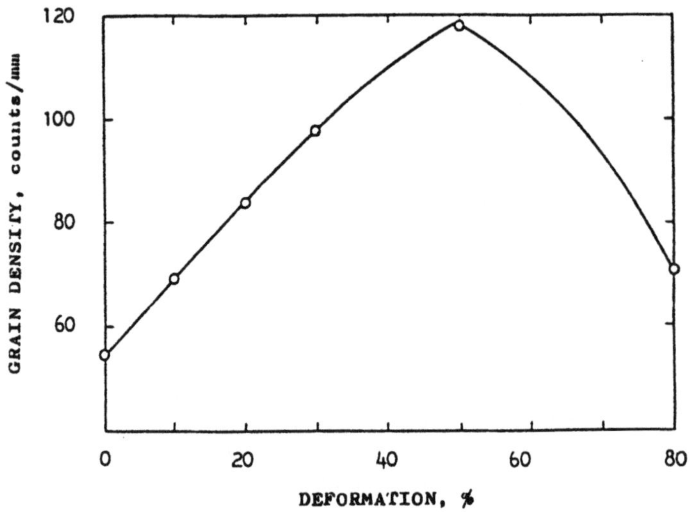

Figure 8. Grain density vs degree of deformation for the borides formed in the surface layers of steel 1020 boronized for 5 h at 900 C.

in Fig.8 indicating that cold deformation increases the density of nuclei when the degree of deformation incteases from 0 to 50% and decreases to a value close to that for the undeformed state when the degree of deformation equals to 80%. It has been shown in Fig.1 that in the undeformed specimen the borides nucleate preferentially on grain boundaries. The nucleation of borides in 20 and 50% deformed specimens occurs preferentially on dislocations within the grains (Fig.9 a,b) but the 80% deformed specimen has the borides nucleated again on grain boundaries (Fig.9c). It was shown that the borides nucleate at relatively lower temperatures (600 C or lower) at which the permeation of B atoms can effectively retard the restoration processes of the deformed steel leaving a higher dislocation density within the deformed ferrite grains. But extremely high degree of deformation like 80% makes the recrystallization processes much easier to occur. Therefore,a 80% prior cold deformation has almost no effect on the density of boride nuclei and case depth.

4.Effect of prior cold deformation on wear behavior of boronized steel

With the increase of case depth prior cold deformation will undoubtedly improve the wear resistance of boronized steel. The weight loss during dry wearing shown in Fig.10 indicates that for the undeformed specimen the transition from slight wear to severe wear takes place at 9 min testing while 20 and 50% deformation delays this transition to 12 min and gives much less weight loss and as a result much higher wear resistance of the steel.

CONCLUSIONS

1.The boronizing process of steels can be divided into two stages, The first stage obeys the parabolic law and the second stage obeys a straght line relationship between the case depth and time.

2.In undeformed specimens the boride nucleates preferentially on grain boundaries of the ferrite matrix. The density of nuclei is the controlling factor of the kinetics of boronizing.

3.Prior cold deformation in the range of 0 to 50% can be successfully used for improving the case depth and thus the wear resistance of the boronized steel. 80% deformation due to the recrystallization of the deformed material has almost no effect on the density of nuclei,the case depth and the wear resistance.

4.Deformation-boronizing with the degree of deformation less than 50% can be effectively used for improving the wear resistance of the steel with the benifit of decreasing both the boronizing temperature and time for obtaining predicted case depth.

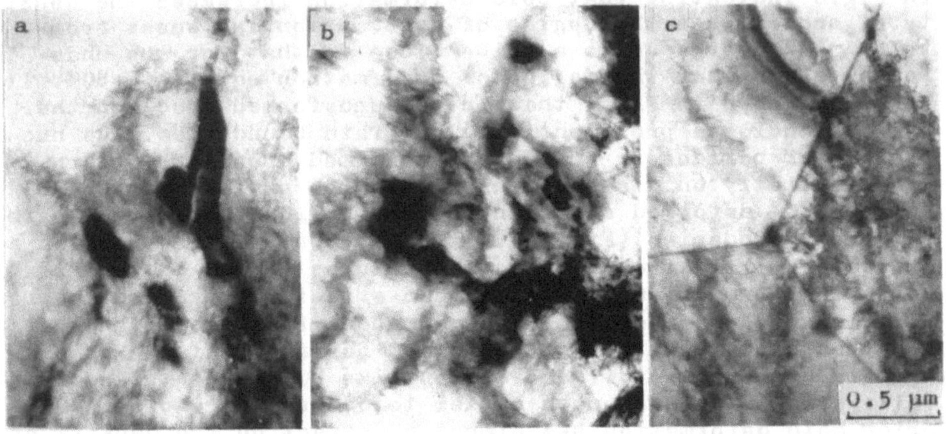

Figure 9. TEM photographs showing the nuclei of borides in the surface layer of steel 1020 boronized at 600 C with zero holding time:
a- 20% cold deformation; b- 50 % cold deformation;
c- 80% cold deformation.

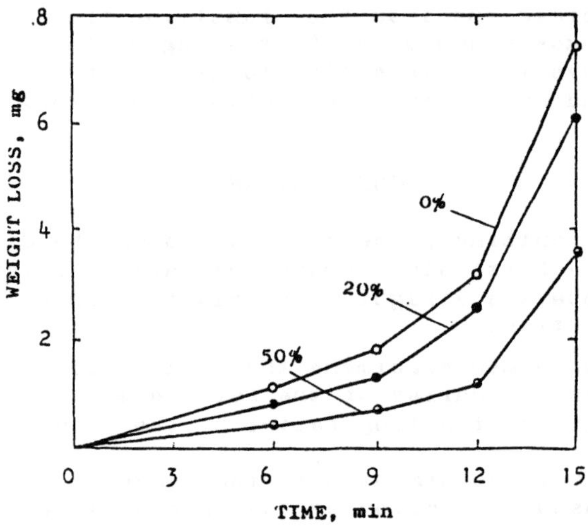

Figure 10. Weight loss by dry wearing of steel 1020 priorly cold deformed to various degrees and then boronized at 900 C for 3 h.

REFERENCES

1.Schaabar,O. Harterei-Techn.Mitt.,1967,22,1.
2.Kunst,H. and Schaaber,O. Harterei-Techn.Mitt.,1967,22, 286.
3.Lin,Z.Z. Metal Science and Technology,1988,7(1),49.
4.Jiang,B.L. Doctor's thesis,Harbin Institute of Technology,Harbin,PR China,1989.
5.Galibois,G. and Boutenko,O. Acta Met.1980,28,1753.
6.Strunk,H. and Gossele,H. Appl.Phys.Lett.,1979,34,530.

Index of Contributors

MIX
Papier aus verantwortungsvollen Quellen
Paper from responsible sources
FSC® C105338
FSC
www.fsc.org

If you have any concerns about our products,
you can contact us on
ProductSafety@springernature.com

In case Publisher is established outside the EU,
the EU authorized representative is:
Springer Nature Customer Service Center GmbH
Europaplatz 3, 69115 Heidelberg, Germany

Printed by Libri Plureos GmbH
in Hamburg, Germany